Lecture Notes in Computer Science 8353

Commenced Publication in 1973
Founding and Former Series Editors:
Gerhard Goos, Juris Hartmanis, and Jan van Leeuwen

For further volumes:
http://www.springer.com/series/7407

Ivan Lirkov · Svetozar Margenov
Jerzy Waśniewski (Eds.)

Large-Scale Scientific Computing

9th International Conference, LSSC 2013 Sozopol, Bulgaria, June 3–7, 2013 Revised Selected Papers

Springer

Editors
Ivan Lirkov
Svetozar Margenov
Institute of Information
and Communication Technologies
Bulgarian Academy of Sciences
Sofia
Bulgaria

Jerzy Waśniewski
Department of Informatics
and Mathematical Modelling
Technical University of Denmark
Kongens Lyngby
Denmark

ISSN 0302-9743
ISSN 1611-3349 (electronic)
ISBN 978-3-662-43879-4
ISBN 978-3-662-43880-0 (eBook)
DOI 10.1007/978-3-662-43880-0
Springer Heidelberg New York Dordrecht London

Library of Congress Control Number: 2014941220

LNCS Sublibrary: SL1 – Theoretical Computer Science and General Issues

Printed on acid-free paper

Springer is part of Springer Science+Business Media (www.springer.com)

Preface

The 9th International Conference on Large-Scale Scientific Computations (LSSC 2013) was held in Sozopol, Bulgaria, during June 3–7, 2013. The conference was organized and sponsored by the Institute of Information and Communication Technologies at the Bulgarian Academy of Sciences.

Plenary Invited Speakers and Lectures:

- P. Bochev, Optimization-Based Modeling — a New Strategy for Predictive Simulations of Multiscale, Multiphysics Problems
- M. Falcone, Recent Advances in the Approximation of Optimal Control Problems via Dynamic Programming
- M. Mascagni, Monte Carlo Methods and Partial Differential Equations: Algorithms and Implications for High-Performance Computing
- G. Haase, Multiple-GPU AMG Solver Environment for Biomedical Applications
- J. Pasciak, Variational Formulations of Problems Involving Fractional Order Differential Operators

The success of the conference and the present volume are the outcome of the joint efforts of many partners from various institutions and organizations. First thanks to all the members of the Scientific Committee for their valuable contribution forming the scientific face of the conference, as well as for their help in reviewing contributed papers. We specially thank the organizers of the special sessions. We are also grateful to the staff involved in the local organization.

Traditionally, the purpose of the conference is to bring together scientists working with large-scale computational models in natural sciences and environmental and industrial applications, and specialists in the field of numerical methods and algorithms for modern high-performance computers. The invited lectures reviewed some of the most advanced achievements in the field of numerical methods and their efficient applications. The conference talks were presented by researchers from academic institutions and practical industry engineers including applied mathematicians, numerical analysts, and computer experts. The general theme for LSSC 2013 was Large-Scale Scientific Computing with a particular focus on the organized special sessions.

Special Sessions and Organizers:

- Numerical Modeling of Fluids and Structures — J. Adler, X. Hu, P. Vassilevski, L. Zikatanov
- Computational Electromagnetics — U. Langer
- Control and Uncertain Systems — M. Krastanov, V. Veliov
- Monte Carlo Methods: Theory, Applications and Distributed Computing — I. Dimov, M. Nedjalkov, J.M. Sellier

- Recent Advances in High-Dimensional Approximation for PDEs with Random Input Data — C. Webster
- Theoretical and Algorithmic Advances in Transport Problems — P. Bochev, D. Ridzal
- Applications of Metaheuristics to Large-Scale Problems — S. Fidanova, G. Luque
- Modeling and Numerical Simulation of Processes in Highly Heterogeneous Media — O. Iliev, R. Lazarov, J. Willems
- Large-Scale Models: Numerical Methods, Parallel Computations and Applications — K. Georgiev, Z. Zlatev
- Numerical Solvers on Many-Core Systems — G. Haase
- Cloud and Grid Computing for Resource-Intensive Scientific Applications — A. Karaivanova, T. Gurov, E. Atanassov

More than 150 participants from all over the world attended the conference representing some of the strongest research groups in the field of advanced large-scale scientific computing. This volume contains 74 papers by authors from more than 25 countries.

The 10th International Conference LSSC 2015 will be organized in June 2015.

January 2014

Ivan Lirkov
Svetozar Margenov
Jerzy Waśniewski

Organization

Scientific Committee

E. Atanassov	Institute of Information and Communication Technologies, BAS, Bulgaria
P. Bochev	Sandia National Labs, USA
I. Dimov	Institute of Information and Communication Technologies, BAS, Bulgaria
S. Dimova	Sofia University, Bulgaria
M. Falcone	Cittá Universitaria, Roma, Italy
S. Fidanova	Institute of Information and Communication Technologies, BAS, Bulgaria
K. Georgiev	Institute of Information and Communication Technologies, BAS, Bulgaria
T. Gurov	Institute of Information and Communication Technologies, BAS, Bulgaria
G. Haase	University of Graz, Austria
O. Iliev	ITWM, Germany
A. Karaivanova	Institute of Information and Communication Technologies, BAS, Bulgaria
M. Krastanov	Sofia University, Bulgaria
J. Kraus	University of Duisburg-Essen, Germany
U. Langer	Johannes Kepler University Linz, Austria
R. Lazarov	Texas A&M University, USA
I. Lirkov	Institute of Information and Communication Technologies, BAS, Bulgaria
S. Margenov	Institute of Information and Communication Technologies, BAS, Bulgaria
M. Mascagni	Florida State University, Tallahassee, USA
M. Neytcheva	Uppsala University, Sweden
M. Paprzycki	Systems Research Institute, PAS, Poland
J. Pasciak	Texas A&M University, USA
D. Ridzal	Sandia National Laboratory, USA
P. Vassilevski	Lawrence Livermore National Laboratory, USA
V. Veliov	TU-Vienna, Austria
J. Waśniewski	Technical University of Denmark, Denmark
Z. Zlatev	Aarhus University, Denmark
L. Zikatanov	Pennsylvania State University, USA

Contents

Plenary and Invited Papers

Numerical Modeling of Fluids and Structures

Control and Uncertain Systems

Applications of Metaheuristics to Large-Scale Problems

Contributed Papers

Plenary and Invited Papers

Stepping into Fully GPU Accelerated Biomedical Applications

Caroline Mendonca Costa[2], Gundolf Haase[1](✉), Manfred Liebmann[1], Aurel Neic[1], and Gernot Plank[2]

[1] Institute for Mathematics and Scientific Computing, University of Graz, Graz, Austria
gundolf.haase@uni-graz.at
http://www.uni-graz.at/~ghaase
[2] Institute of Biophysics, Medical University of Graz, Graz, Austria

Abstract. We present ideas and first results on a GPU acceleration of a non-linear solver embedded into the biomedical application code CARP. The linear system solvers have been transferred already in the past and so we concentrate on how to extend the GPU acceleration to larger portions of the code. The finite element assembling of stiffness and mass matrices takes at least 50 % of the CPU time and therefore we investigate this process for the bidomain equations but with focus on later use in non-linear and/or time-dependent problems. The CUDA code for matrix calculation and assembling is faster by a factor up to 90 compared to a single CPU core. The routines were integrated to CARP's main code and they are already used to assemble the FE matrices of the bidomain model. Further performance studies are still required for the bidomain-mechanics model.

1 Introduction

During the last years GPUs became very attractive to reduce simulation time by porting linear solvers to the accelerator card. Due to the large problem size multigrid methods have been preferred by parts of the community. GPU accelerated geometrical multigrid has been carefully investigated by several authors, see [5,6] for structured grids and its further development for locally structured grids in [3] as examples for success in a monolithic code. Starting from third party demands and large unstructured discretizations, the algebraic multigrid (AMG) has to be applied. Here, the authors proved one order of magnitude acceleration by using GPUs [7,16] also in the multi-GPU context [13]. While the AMG setup in our code still remains on the CPU there is an interesting attempt to move also the AMG setup completely onto the GPU [1] but only with an acceleration of two.

The situation changes when the linear solvers are embedded into a larger framework. When solving the Bidomain equations to simulate cardiac electrophysiology via the Finite Element Method, the stiffness and mass matrices have

I. Lirkov et al. (Eds.): LSSC 2013, LNCS 8353, pp. 3–14, 2014.
DOI: 10.1007/978-3-662-43880-0_1,

to be assembled only once, since the spatial domain is not modified during computation. Thus, in this case, the assembly of the FE matrices it is not a bottleneck of computation. On the other hand, when solving the bidomain-mechanics model, which involves non-linear elasticity, the FE matrices must to be updated at each Newton step, as the spatial domain is deformed, which becomes very expensive as the system increases in size. Therefore, it is the goal of this paper, to implement a highly efficient FE assembly routine using CUDA [14] to increase performance when solving this model. This paper investigates the GPU acceleration for the FE matrix computation in the bidomain model in order to study whether a GPU implementation might pay off for the real challenging bidomain-mechanics equations [15].

We will introduce the bidomain model in Sect. 2 providing the equations for the numerical tests. Section 3 starts with a brief primer on the simulation code CARP and presents the strategy how to reduce critical data transfer between CPU and GPU memory in the non-linear solver when AMG is used as linear solver therein. The improvement of the FE matrix calculations by vectorization and GPU acceleration is described in Sect. 4. The paper finishes with speedups regarding the matrix computation and assembling on GPU and with some conclusions.

2 The Bidomain Model

The bidomain equations in the elliptic-parabolic form are given by

$$\begin{bmatrix} -\nabla\cdot(\boldsymbol{\sigma}_i+\boldsymbol{\sigma}_e)\nabla\phi_e \\ -\nabla\cdot\sigma_b\nabla\phi_e \end{bmatrix} = \begin{bmatrix} \nabla\cdot\boldsymbol{\sigma}_i\nabla V_m + I_i \\ I_e \end{bmatrix} \tag{1}$$

$$I_m = (\nabla\cdot\boldsymbol{\sigma}_i\nabla\phi_i)$$

$$I_m = C_m\frac{\partial V_m}{\partial t} + I_{ion}(V_m,\boldsymbol{\eta}) - I_i \tag{2}$$

$$\frac{d\boldsymbol{\eta}}{dt} = f(t,\boldsymbol{\eta}) \tag{3}$$

$$V_m = \phi_i - \phi_e \tag{4}$$

where ϕ_i and ϕ_e are the intracellular and extracellular potentials, respectively, $V_m = \phi_i - \phi_e$ is the transmembrane voltage, $\boldsymbol{\sigma_i}$ and $\boldsymbol{\sigma_e}$ are the intracellular and extracellular conductivity tensors, respectively, β is the membrane surface to volume ratio, I_m is the transmembrane current density, I_e are extracellular stimuli applied in the extracellular space, I_i is an intracellular current stimulus, C_m is the membrane capacitance per unit area, and I_{ion} is the membrane ionic current density which depends on V_m and a set of state variables, $\boldsymbol{\eta}$ which is defined by f.

At tissue boundaries, no flux boundary conditions are imposed for ϕ_i, with the potential ϕ_e and the normal component of the extracellular current being continuous. At boundaries of the conductive bath surrounding the tissue, no flux boundary conditions for ϕ_e are imposed.

Combining the interstitial and bath spaces into the extracellular space, the bidomain equations can be written as follows

$$
\begin{aligned}
-\nabla\cdot\boldsymbol{\sigma}_e\nabla\phi_e &= \nabla\cdot\boldsymbol{\sigma}_i\nabla\phi_i + I_e && (5)\\
\beta I_m &= \nabla\cdot\boldsymbol{\sigma}_i\nabla\phi_i \\
I_m &= C_m\frac{\partial V_m}{\partial t} + I_{ion}(V_m,\boldsymbol{\eta}) - I_i && (6)\\
\frac{d\boldsymbol{\eta}}{dt} &= f(t,\boldsymbol{\eta}) && (7)\\
V_m &= \phi_i - \phi_e && (8)
\end{aligned}
$$

with no-flux boundary conditions imposed for ϕ_i and ϕ_e.

The matrix representation for the FE discretization for the bidomain equations, written for V_m and ϕ_e only, is given by

$$
\begin{aligned}
K_{ie}\phi_e &= -P(K_iV_m) - M_eI_e && (9)\\
K_iV_m &= -\beta M_iI_m - K_i(P^T\phi_e) && (10)
\end{aligned}
$$

where K_* and M_* are stiffness and mass matrices, respectively, with $* = e|i$ being either the extracellular space, Ω_e, or the intracellular space, Ω_i, P is a prolongation operator from Ω_i to Ω_e and its transpose, P^T, is a restriction operator from Ω_e to Ω_i.

3 GPU Strategy for Non-linear FE Solvers

3.1 CARP Environment

The CARP environment [18,19] (Cardiac Arrhythmia Research Package) is a collection of various contributors for the detailed simulation of cardiovascular phenomena, see Fig. 1 for the software scheme. The gray box in the center contains the kernel for the linear algebra that has to be combined with the non-linear iteration in case of the bidomain-mechanics model. The FE assembly routine is implemented within the CARP environment. The assembly involves the module FEM, which comprises all the finite element computations, particularly the stiffness and mass matrices assembly, and contains the "Matrix Market", which comprises matrix basic operations and is implemented within the Module FMatrix. This module is subject to GPU acceleration in this paper for the matrices resulting from the bidomain equations (5). This is meant as a study whether a GPU implementation might pay off for the real challenging bidomain-mechanics equations, see [15]. We use unstructured tetrahedral FE meshes with linear test functions.

3.2 Draft for a Non-linear Solver on GPUs

In the context of a non-linear setting where we have to solve frequently a linear system as

$$K\left(u^{\text{old}}\right)u^{\text{new}} = f\left(u^{\text{old}}\right). \qquad (11)$$

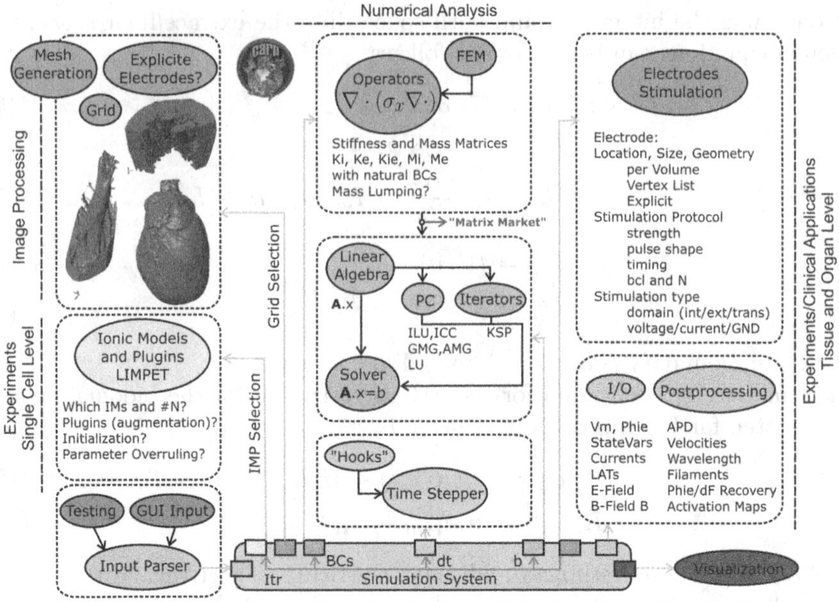

Fig. 1. Structure of the CARP code.

It makes not much sense to accelerate the application of the linear solver (cg with AMG preconditioning) by a factor of 10 on the GPU when the setup of the AMG solver as well as the re-calculation and the re-assembling of the stiffness matrix $K\left(u^{\text{old}}\right)$ are still performed on the CPU. Additionally we have to avoid costly data transfer between CPU and GPU memory. We should also take into account that calculations are very fast on the GPU in contrast to the slow performing search and reorder routines.

The CARP code does not use spatial adaptivity and therefore the topology of the mesh will remain unchanged throughout the non-linear computation, and even during the outer time integration. Therefore, we have to determine the matrix pattern only once in a matrix setup on the CPU, afterwards transfer that pattern once to the GPU and perform (re-)calculation and the (re-)assembling of the stiffness on the GPU repeatedly. Clearly, that requires that the mesh information and the material properties are also available in GPU memory. Section 4 will report on first experiences regarding the matrix calculation on GPU.

The AMG preconditioner setup contains parts which do not accelerate well on a GPU [1] so a closer look at it is necessary. The AMG setup consists of the following parts:

1. Find coarse/fine nodes.
2. Determine interpolation pattern.
3. Calculate interpolation matrix entries.
4. Determine coarse matrix pattern.
5. Calculate coarse matrix entries.

Due to CPU performance issues items 2./3. as well as items 4./5. are handled usually in one routine. In the CARP context we can again assume that material anisotropies will not change dramatically, i.e., that the coarse/fine splitting will remain the same in all (or many) non-linear steps. The same will be assumed for the pattern determinations in items 2. and 4. This indicates a splitting of the setup such that items 1., 2. and 4. are still performed once on the CPU while items 3. and 5. can be handled very efficiently on the GPU. This splitting is still subject to investigation.

4 FE Matrix Calculation on GPU

4.1 Stiffness Matrix

The entries of stiffness matrices K from (9) and (10) are computed as

$$K = \{K_{i,j}\} = \sum_{e=0}^{nElem-1} K_e = \sum_{e=0}^{nElem-1} \left(-C_e\, G_e\, C_e^T\, vol_e\right), \tag{12}$$

where K_e is the stiffness matrix of element e, C_e is the matrix of basis coefficients, G is the matrix of conductivity tensors and vol is the element volume [4,9,10]. Using tetrahedral elements, the basis coefficients of each element are given by the inverse of the matrix

$$C_e^{-1} = \begin{bmatrix} 1 & x_1 & y_1 & z_1 \\ 1 & x_2 & y_2 & z_2 \\ 1 & x_3 & y_3 & z_3 \\ 1 & x_4 & y_4 & z_4 \end{bmatrix} \tag{13}$$

The conductivity tensor of each element is computed as

$$G = g_f\; f \times f + g_s\; s \times s + g_n\; n \times n, \tag{14}$$

for the orthotropic case, where f, s, n are the longitudinal, transverse and normal eigenaxis, g_f, g_s, g_n are the principal eigenvalues. The volume of each tetrahedra is computed as $vol = \frac{|det(C_e)|}{6}$.

The mass matrix $M = \{M_{ij}\}$ in (10) is also computed element wise and simplifies to

$$M_{ij} = \begin{cases} \sum_{e=0}^{nElem-1} M_{e,ij} = 2 \text{ factor } vol_e, & i = j \\ \sum_{e=0}^{nElem-1} M_{e,ij} = \text{factor } vol_e, & i \neq j \end{cases} \tag{15}$$

for linear elements with the factor depending on simple material coefficients.

4.2 The FMatrixArray Structure

The current interface between matrix element calculation and its accumulation into the global matrix consists of a structure FMatrixArray containing a large array of size *number of elements* × *number of nodes per element* with some additional information that stores all the element matrices. This allows to calculate the elements matrices in parallel on many-core chips without fatal data races. Additionally, all the information needed for element matrix calculation as coordinates and material coefficients is also stored as FMatrixArray structures with 1D arrays of appropriate size. Although this approach requires additional temporal memory it outperforms the classical approach without redundant storing of input data and of accumulating the local entries directly into the global matrix by a factor of 5 on a single CPU core.

Specialized explicit expressions were written to compute matrix determinant and matrix inversion, which are the most expensive routines. An example of the vectorized code using an FMatrixArray `ent` to compute the determinant is shown below for a triangular element. The code for a tetrahedra looks the same just much longer. Note that only local variables and explicit expressions are used.

Listing 1.1. Code to compute matrix determinant (vol_e) for a triangular element

```
switch (rows)
  case 3: {
    for(int i = 0; i < nmats; i++) {
      const Real a11 = ent[i*matSize],      a12 = ent[1+i*matSize],
                 a13 = ent[2+i*matSize],    a21 = ent[3+i*matSize],
                 a22 = ent[4+i*matSize],    a23 = ent[5+i*matSize],
                 a31 = ent[6+i*matSize],    a32 = ent[7+i*matSize],
                 a33 = ent[8+i*matSize];
      det[i] = (a12*a23 - a13*a22)*a31
             - (a11*a23 - a13*a21)*a32
             + (a11*a22 - a12*a21)*a33;
  }
```

This calculation of the determinant belongs to the volume computation in Listing 1.2. The code below gets the element list and the node list as input parameters and computes all local stiffness matrices after the appropriate setup of volume, C_e from (13) and G from (14) for each element. The Listing 1.1 is representative for the data handling in all subroutines involved.

Listing 1.2. Code to compute element stiffness matrices K_e

```
int fl_tetFillLocalStiffnessMatrixArray_
   (const ElemList *elst, const NodeList *nlst,
          FMatrixArray *nodes,  FMatrixArray *coeffs,
          FMatrixArray *g,      FMatrixArray *lK,      Real *vols)
{
  fl_fillNodes(elst, nodes, nlst);              // Fill coordinates
  fl_computeVolumes(nodes, vols);               // Compute Volumes
  FMatrix_InvArray(nodes, coeffs);              // Comp. Basis Coefficients C_e
  Real pevs[] = {1.0, 1.0, 1.0};                // simple material parameters
  fl_getCondTensorGPU(elst, g, pevs);           // Comp. Conductivity Tensor G
  fl_integrateStiffness(lK, g, coeffs, vols);   // Comp. local matrices
}
```

The calculation of the mass matrices is handled the same way.

4.3 Implementation of the CUDA Kernels

Non-coalesced Memory Allocation and Access. Matrix entries (stored linearly) are reordered as shown in Fig. 2, so that each thread has access to the first element of its corresponding matrix in cache memory. In the GPU implementation the stride value is 32, which is half the number of maximal threads per block, i.e. 64 threads per block. When allocating memory for an array in non-coalesced format the size of the memory chunk will depend also on the size of this stride. Keeping arrays in both formats is important to obtain maximum performance in both CPU and GPU, as the non-coalesced format is inefficient in the CPU, but the most efficient in the GPU. Therefore, whenever we copy data over to or from the GPU, a conversion routine has to be used.

Data Structure on the GPU. In order to copy data to the GPU, some changes were required in the Element list and Nodes list structures implemented in the standard code. In the standard code, this lists are implemented as general structures holding detailed information about each element and node in the mesh. In the CUDA version, this lists are implemented as one-dimensional arrays, which size varies depending on the element type and number of elements in the mesh. More detailed information is given below and in Fig. 3, where N and L are the global and local indices describing each element, respectively; Lon, Sheet and Sheet normal are the fiber orientation arrays; and Pts and Exp. Pts are the arrays of points in regular and exploded format, respectively.

Element list: array of nodal (local or global) indices describing each element, it is copied in non-coalesced form to the GPU.

Axes lists: arrays of longitudinal, sheet and sheet normal fiber orientations, it is copied to the GPU in non-coalesced form.

Nodes list: an array of point coordinates is copied to the GPU in coalesced form. Can be copied using the regular format, where each point is unique in the list and the element list is used to access the nodes of each element, or in the exploded format, where the points are duplicated and copied in element index order. The latter one uses the nodal indices to describe the elements and the nodes are duplicated in the array. In this case, the element list is not copied to the GPU, as the nodes list can be accessed sequentially.

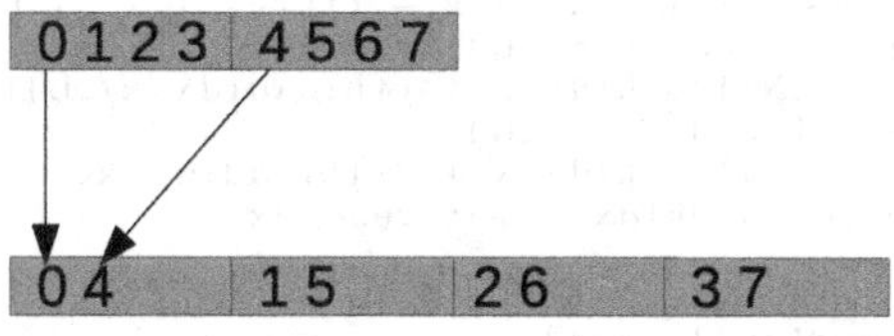

Fig. 2. Coalesced (top) *versus* non-coalesced (bottom) memory access. Matrix entries are reordered using a predefined stride size.

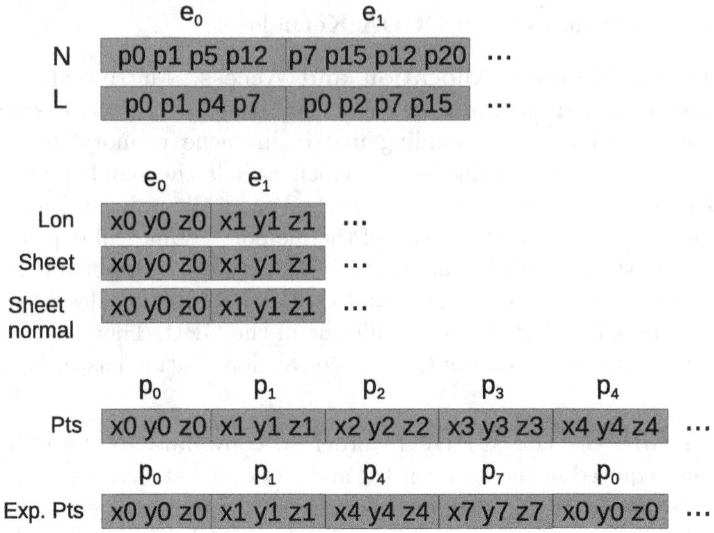

Fig. 3. Data structure organization in the GPU. Elements, axes and nodes lists are structured as linear arrays.

Device Routines. For each kernel, an interface routine was implemented in C, in which a structure with all kernel parameters is assembled before the kernel is called. The kernel is then called with $(p.nElem + \_N - 1)/\_N$ blocks and $\_N$ threads per block. In this case, the total number of threads might be larger than the number of elements, but only $nElem$ threads execute the calculations.

In the device routines, i.e. the kernels, the start indexes are computed by each thread for each array that is accessed in the routine. In the example below (code 1.3), mStart, lStart and pStart stores the initial index of mPtr, dLon, and pevs, respectively, for each thread. Local variables are used to access the register directly. The calculations are done using explicit expressions to avoid loops in the kernel routine. Again the following listing is representative for all kernel routines needed in the element matrix calculations.

Listing 1.3. Kernel to compute conductivity tensor G

```
__global__ void
__device_fl_getCondTensorGPU(_device_fl_getCondTensorGPU_params p)
{
 int mStart = p.msz*_N*blockIdx.x + (threadIdx.x/_L)*p.msz*_L
              + (threadIdx.x%_L);
 int lStart   = 3*_N*blockIdx.x + (threadIdx.x/_L)*3*_L
              + (threadIdx.x%_L);
 int pStart   = 3*_N*blockIdx.x + 3*threadIdx.x;
 int maxsize  = _N*blockIdx.x + threadIdx.x;

 if(maxsize < p.nElem) {
   double f1 = p.dLon[lStart];
   double f2 = p.dLon[lStart+=_L];
   double f3 = p.dLon[lStart+=_L];

   double gf = p.pevs[pStart++];
```

```
      double gs = p.pevs[pStart];

      // (gf-gs)* f x f + gs * I
      p.mPtr[mStart]      = f1*f1*(gf - gs) + gs;
      p.mPtr[mStart+=_L] = f1*f2*(gf - gs);
      p.mPtr[mStart+=_L] = f1*f3*(gf - gs);
      p.mPtr[mStart+=_L] = f1*f2*(gf - gs);
      p.mPtr[mStart+=_L] = f2*f2*(gf - gs) + gs;
      p.mPtr[mStart+=_L] = f2*f3*(gf - gs);
      p.mPtr[mStart+=_L] = f1*f3*(gf - gs);
      p.mPtr[mStart+=_L] = f2*f3*(gf - gs);
      p.mPtr[mStart+=_L] = f3*f3*(gf - gs) + gs;
  }
}
```

Further Optimization. The computation of the matrix inverse for the $4 - by - 4$ (tetrahedra) case exhibited register spilling, which was removed by rearranging computations such that compiler generated temporary expressions have been reused. Additionally we removed the asserting function, used to stop the computations in case the determinant is zero. Memory (de)allocation overhead was identified when computing the stiffness and mass matrix. Thus, the memory (de)allocation calls were moved outside the main loop. Therefore, it is only done once and it is not included in the final assembly times. When using the nodes list with elemental indexing, as described in Sect. 4.3, the array with the points was accessed via texture cache to compensate for the overhead of accessing the nodes in non-sequential order. This approach saves memory, as the points in the nodes list are not duplicated, and computation time is only marginally affected.

4.4 Global Assembly

The accumulation of the element matrices stored in an FMatrixArray structure into one global matrices implemented within the parallel toolbox [12,13]. Therein the element matrix entries will be reordered according to their global row and column indices such that these entries can be accumulated for each global matrix entry in parallel afterwards. Again, the permutation vector for this mapping is determined in an a priori setup. Another approach which would save a lot of temporary memory requires the coloring of the finite elements such that no data races will appear in the accumulation process. This has been applied successfully on vector processors [17] as well as on GPUs [2] general many-core environments [11]. Our own improved version of these parallel matrix accumulations is still ongoing research.

5 Results

We used the following configuration for our experiments. The CPU was an Intel Xeon E5645 with 6 cores (12 threads), clock Speed of 2.4 GHz, 12 MB of L2-cache and 24 GB DDR3 memory with 32 GB/s of memory bandwidth. The GPU

Table 1. Assembly time and speedup for two matrices in double precision.

n elements	Time in seconds				Speedup	
	Stiffness		Mass		Stiffness	Mass
	CPU	GPU	CPU	GPU		
12,500	0.007289	0.000216	0.002725	0.000131	33.74	20.80
50,000	0.034600	0.000517	0.013445	0.000239	66.92	56.25
112,500	0.078769	0.001038	0.030778	0.000439	75.88	70.11
450,000	0.312762	0.003712	0.124095	0.001373	84.26	90.38
1,250,000	0.745111	0.010108	0.300238	0.003620	73.71	82.93

is an NVidia GTX 680 with 1536 CUDA cores, 1006 MHz Base Clock and with 2048 GB GDDR5 memory with 192.2 GB/s of memory bandwidth. The Performance was measured with one vectorized CPU core and compared to the GPU CUDA code for tetrahedral meshes of different sizes. The .cu files, where the GPU kernels are implemented, are compiled with NVCC using architecture sm_20 and -O3 option. The .c files for the CPU are compiled with GCC, using options -g -O3 -std=gnu99, but when compiling for the GPU, the output file of the kernels and the CUDA libraries must be linked to the resulting output file of the .c files.

The numerical tests have been performed for the bidomain equations (5) and several discretization of the unit cube. The assembling process has been performed 10 times in order to get average run times and all the data needed have been initialized and transferred before the timing started. Table 1 presents the assembly times of the CPU and GPU as well as the related speedups achieved which are also depicted in Fig. 4. It can be concluded that the GPU speedup is approximately 80 for larger numbers of tetrahedrals and up even 90 for the assembling of the simpler mass matrices. If we assume a perfect speedup of 8

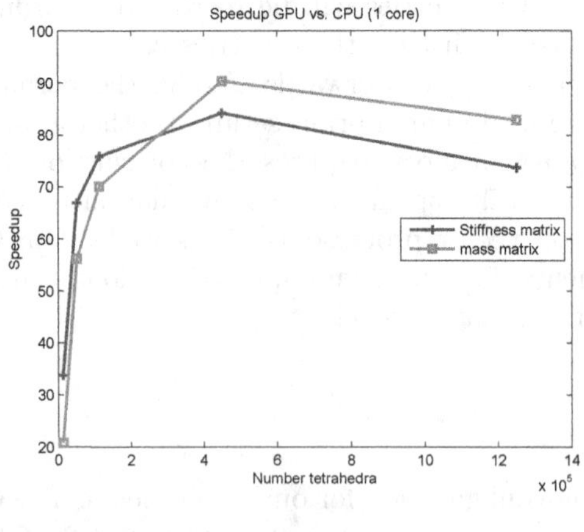

Fig. 4. Speedup of GPU vs. one CPU core for assembling routines in double precision.

when all 8 CPU cores are used then a quite good GPU speedup of 10 still remains. Our algorithms are bandwidth limited and therefore the 8 CPU cores sharing that bandwidth will perform worse. Therefore, even a (good) consumer card as the GTX 680 achieves a significant speedup for matrix calculation and assembling in double precision.

6 Conclusion

The implementation of the FE matrices assembly using vectorized code for the CPU and CUDA for the GPU has proved to be highly efficient, with the CUDA code reaching a maximum speedup of 90. On the other hand, it appears that there is a limit in performance when using CUDA, as the speedup drops when a mesh with more than 450000 elements is used. This might be due to memory bandwidth limitation, as only a limited number of threads can have access to the cache memory at the same time. The work presented is still ongoing.

Nevertheless, we expect that vectorized code as well as the CUDA codes will increase performance when used within the mechanical model, where the matrices have to be re-assembled at each step. Moreover, the modifications required in the main branch to include the new implementation are minimized by modularity, and only a few routines within FEM and FMatrix have to be modified. The next step will consist in applying the described methodology to mechanical problems and test the bidomain-mechanical problem [15] on NVidia's Tesla 20K with eight time more double precision compute units available. Together with the already available a priori calculation of the matrix patterns and the splitting of the AMG preconditioner setup the whole non-linear iteration in the solution process of the bidomain-mechanical problem should run on the GPU.

References

1. Bell, N., Dalton, S., Olson, L.N.: Exposing fine-grained parallelism in algebraic multigrid methods. SIAM J. Sci. Comput. **34**(2), C123–C152 (2012)
2. Cecka, C., Lew, A.J., Darve, E.: Assembly of finite element methods on graphics processors. Int. J. Numer. Methods Eng. **85**(5), 640–669 (2011)
3. Geveler, M., Ribbrock, D., Göddeke, D., Zajac, P., Turek, S.: Towards a complete FEM-based simulation toolkit on GPUs: unstructured grid finite element geometric multigrid solvers with strong smoothers based on sparse approximate inverses. Comput. Fluids **80**, 327–332 (2013)
4. Gockenbach, M.S.: Understanding and Implementing the Finite Element Method. SIAM, Philadelphia (2007)
5. Göddeke, D.: Fast and accurate finite-element multigrid solvers for PDE simulations on GPU clusters. Ph.D. thesis, Technische Universität Dortmund, Fakultät für Mathematik, May 2010. http://hdl.handle.net/2003/27243
6. Göddeke, D., Strzodka, R., Mohd-Yusof, J., McCormick, P.S., Wobker, H., Becker, C., Turek, S.: Using GPUs to improve multigrid solver performance on a cluster. Int. J. Comput. Sci. Eng. **4**(1), 36–55 (2008)

7. Haase, G., Liebmann, M., Douglas, C.C., Plank, G.: A parallel algebraic multigrid solver on graphics processing units. In: Zhang, W., Chen, Z., Douglas, C.C., Tong, W. (eds.) HPCA 2009. LNCS, vol. 5938, pp. 38–47. Springer, Heidelberg (2010)
8. Jónasson, K. (ed.): PARA 2010, Part II. LNCS, vol. 7134. Springer, Heidelberg (2012)
9. Jung, M., Langer, U.: Methode der finiten Elemente für Ingenieure. Lehrbuch, 2nd edn. Springer Vieweg, Wiesbaden (2013)
10. Larson, M.G., Bengzon, F.: The Finite Element Method: Theory, Implementations and Applications. Texts in Computational Science and Engineering, vol. 10, 1st edn. Springer, Heidelberg (2013)
11. Markall, G.R., Slemmer, A., Ham, D.A., Kelly, P.H.J., Cantwell, C.D., Sherwin, S.J.: Finite element assembly strategies on multi-core and many-core architectures. Int. J. Numer. Methods Fluids **71**(1), 80–97 (2013)
12. Neic, A., Liebmann, M., Haase, G., Plank, G.: Algebraic multigrid solvers on clusters of CPUs and GPUs. In: Jónasson [8], pp. 389–398
13. Neic, A., Liebmann, M., Hötzl, E., Mitchell, L., Vigmond, E., Haase, G., Plank, G.: Accelerating cardiac bidomain simulations using graphics processing units. IEEE Trans. Biomed. Eng. **59**(8), 2281–2290 (2012)
14. NVIDIA Corporation. CUDA programming guide 5.0 (2012). http://docs.nvidia.com/cuda/cuda-c-programming-guide/index.html
15. Pathmanathan, P., Whiteley, J.P.: A numerical method for cardiac mechanoelectric simulations. Ann. Biomed. Eng. **37**(5), 860–873 (2009)
16. Rocha, B., Campos, F., Plank, G., Weber dos Santos, R., Liebmann, M., Haase, G.: Simulations of the electrical activity in the heart with graphic processing units. Concur. Comput. Pract. Exp. **23**, 708–720 (2011)
17. Tracy, F.T.: Optimizing finite element programs on the cray X1 using coloring schemes. In: Proceedings of the 2004 Users Group Conference, DOD_UGC '04, pp. 329–333. IEEE Computer Society, Washington, DC, USA (2004)
18. Vigmond, E., Hughes, M., Plank, G., Leon, L.: Computational tools for modeling electrical activity in cardiac tissue. J. Electrocardiol. **36**, 69–74 (2003)
19. Vigmond, E., Plank, G.: Cardiac arrhythmia research package (2009). http://carp.meduni-graz.at

Recent Results in the Approximation of Nonlinear Optimal Control Problems

Maurizio Falcone(✉)

Dipartimento di Matematica, SAPIENZA - Università di Roma,
P.za Aldo Moro, 2, 00185 Roma, Italy
falcone@mat.uniroma1.it

Abstract. This survey paper presents recent advances for the numerical solution of Hamilton-Jacobi-Bellman equations related to optimal control problems. The Dynamic Programming approach suffers for the "curse of dimensionality" and the solution of the nonlinear partial differential equations characterizing the value function of optimal control problems in high dimension is out of reach. However, a combination of various techniques can circumvent this difficulty and find the solution of optimal control problems up to dimension 10, a range of dimensions which could be enough for many applications. We illustrate here some of these techniques: patchy domain decomposition, fast marching and fast sweeping and an acceleration method based on the coupling between value and policy iteration. Numerical examples will illustrate the main features of those methods.

1 Introduction

The numerical solution of partial differential equations obtained by applying the Dynamic Programming Principle (DPP) to nonlinear optimal control problems is a challenging topic that can have a great impact in many areas, e.g. robotics, aeronautics, electrical and aerospace engineering. Indeed, by means of the DPP one can characterize the value function of a fully nonlinear control problem (including also state/control constraints) as the unique viscosity solution of a nonlinear Hamilton-Jacobi equation, and, even more important, from the solution of this equation one can derive the approximation of a feedback control. This result is the main motivation for the PDE approach to control problems and represents the main advantage over other methods, such as those based on the Pontryagin minimum principle. It is worth to mention that the characterization via the Pontryagin principle gives only necessary conditions for the optimal trajectory and optimal open-loop control. Although from the numerical point of view the control system can be solved via shooting methods for the associated two point boundary value problem, in real applications a good initial guess

The author wish to acknowledge the support obtained by the following grants: AFOSR Grant no. FA9550-10-1-0029, ITN - Marie Curie Grant no. 264735-SADCO.

I. Lirkov et al. (Eds.): LSSC 2013, LNCS 8353, pp. 15–32, 2014.
DOI: 10.1007/978-3-662-43880-0_2,

for the co-state is particularly difficult and often requires a long and tedious trial-error procedure to be found.

In this paper we focus our attention on efficient methods to implement the DP approach for nonlinear control problems governed by ordinary differential equations. In particular, our presentation will be centered on the *minimum time problem*, which is associated to the following Hamilton-Jacobi-Bellman equation

$$\begin{cases} \max_{a\in A}\{-f(x,a)\cdot\nabla u(x)\}-1=0\,, & x\in\mathbb{R}^d\backslash\mathcal{T} \\ u(x)=0\,, & x\in\mathcal{T} \end{cases} \tag{1}$$

where d is the dimension of the state, $A\subset\mathbb{R}^m$ is a compact set defining the admissible controls, $\mathcal{T}$ is the target set to be reached in minimal time and $f:\mathbb{R}^d\times A\to\mathbb{R}^d$ is the dynamics of the system. For this classical problem the value function $T:\mathbb{R}^d\to\mathbb{R}$ at the point x is the minimal time to reach the target starting from x (note that $T(x)=+\infty$ if the target is not reachable). For numerical purposes, the equation is solved in a bounded domain $\Omega\supset\mathcal{T}$, so that also boundary conditions on $\partial\Omega$ are needed. A rather standard choice when one does not have additional informations on the solution and deals with target problems is to impose state constraints boundary conditions.

The techniques used to obtain a numerical approximation of the viscosity solution of Eq. (1) have been mainly based on Finite Differences [12,22] and Semi-Lagrangian schemes [15,17]. It is rather important to note that traditional approximation schemes presented for example in [12] and [15] are based on a fixed point iteration scheme, which computes the solution at each node of the grid at every iteration. Denoting by M the number of nodes in each dimension and considering that the number of iterations needed for convergence is of order $O(M)$, the total cost of this full-grid scheme is $O(M^{d+1})$. We easily conclude that this algorithm is very expensive when the state dimension is $d\geq 3$, although it is rather efficient for low dimensional control problems as shown in [15] (see also the book [17]).

The "curse of dimensionality" is a typical drawback of Dynamic Programming and can not be eliminated. However, several techniques have been introduced in order to solve the DP equations in a rather high dimension (see [10] for a first tentative in this direction). Typically $1\leq d\leq 10$ is an interesting range which can allow to solve many problems coming from applications, moreover a model reduction technique can be applied to the original dynamics in order to get a new dynamical system of lower dimension still catching the behavior of the dynamics. This remark is the main motivation which has driven the search for new computational techniques aimed to accelerate convergence and/or to reduce the memory allocation requirements.

Let us give some examples. One possible strategy is based on the decomposition of the domain Ω. The problem is actually solved in subdomains Ω^j, $j=1,\ldots,R$, whose size is chosen in order to reduce the number of grid nodes to a manageable size. Therefore, rather than solving a unique huge problem, one can solve R smaller subproblems working simultaneously on several processors. This produces a simple parallel algorithm. Depending on the choice of the subdomains

Ω^j we can have some overlapping regions or a number of interfaces between the subdomains. The presence of interfaces and/or overlapping regions is a delicate point, since at each iteration of the algorithm it will be necessary to exchange information between processors to properly define the values at the interfaces. Without this communication the result will not be correct. The interested reader can find in the book [26] a comprehensive introduction to domain decomposition techniques, whereas for an application to Hamilton-Jacobi equations we refer to [8,18]. In this approach the choice of the division into subdomains is aimed to choose rather simple boundaries and geometries (typically an hypercube is divided into small hypercubes). A recent improvement has been made in [9] trying to adapt the geometry to the optimal dynamics of the system in order to obtain a subdivision made by "almost" invariant subdomains (the patches), this allows to eliminate the transmission load due to the exchange of informations between different processors. Previous patchy decompositions based on different ideas have been proposed first by Navasca and Krener in [20].

Another proposal to reduce the computational effort is the so-called Fast Marching method introduced in [25,27]. While the full-size grid is always allocated, the computation is restricted to a small portion of the grid, thus saving CPU time. The cost of this method is of order $O(M^d \log M^d)$. In the original version, the Fast Marching method was derived for the Eikonal equation, corresponding (under a suitable change of variable) to Eq. (1) with $f(x,a) = a$ and $A = B_d(0,1)$, the unit ball in $\mathbb{R}^d$ centred in 0 (see [14] for details). Despite the efficiency of the Fast Marching method, at present its application to more general equations of the form $H(x, u(x), \nabla u(x)) = 0$ is not an easy task and it is still under investigation (see [7,11,13,23]) because the causality principle which is behind the ordering of the grid nodes is not easy to detect for general control problems. Other methods have been proposed exploiting the idea that one can accelerate convergence by alternating the order in which the grid nodes are visited giving rise to the so-called "sweeping methods". These methods do not require a special ordering of the grid nodes and are somehow blind, so it could be difficult to prove that they converge after a finite number of sweeps. However, they are easy to implement and they have been shown to be efficient for the Eikonal equation [28] and, more recently, for rather general Hamiltonians [24].

The third method is based on a coupling between two classical methods: value and policy iteration. It is well known that the value iteration is globally convergent but the rate of convergence is rather slow, whereas the policy iteration is locally convergent with a super-linear (or quadratic) rate of convergence. Then, a natural idea is to combine these methods in order to obtain a globally convergent method which starts using the value iteration to switch into the policy iteration when it reaches a "small" neighborhood of the solution.

The survey is organized as follows. Section 2 is devoted to the general presentation of two computational methods: the value iteration and the policy iteration. The semi-Lagrangian scheme associated to these methods will be the first building block for the following improvements. Section 3 is devoted to the patchy domain decomposition method. Section 4 will briefly sketch Fast Marching and Fast sweeping methods. Finally, Sect. 5 is devoted to an acceleration method based on the coupling between value iteration and policy iteration.

2 Two Classical Algorithms for Dynamic Programming

In this section we will summarize the basic results for the two methods as they will constitute the starting point for our new algorithms. The essential features will be briefly sketched, more details can be found in the original papers and in some monographs, e.g. in the classical books by Bellman [6], Howard [19] and for a more recent setting in the framework of viscosity solutions in [3,15]. Let us present the method for the *minimum time problem* where the dynamics is

$$\begin{cases} \dot{y}(t) = f(y(t), \alpha(t)) \\ y(0) = x \end{cases} \tag{2}$$

where $y \in \mathbb{R}^d$, $\alpha \in \mathbb{R}^m$ and $\alpha \in \mathcal{A} \equiv \{a : \mathbb{R}_+ \to A, \text{ measurable}\}$. If f is Lipschitz continuous with respect to the state variable and continuous with respect to (x, α), the classical assumptions for the existence and uniqueness result for the Cauchy problem (2) are satisfied. To be more precise, the Carathéodory theorem implies that for any given control $\alpha(\cdot) \in \mathcal{A}$ there exists a unique trajectory $y(\cdot; \alpha)$ satisfying (2) almost everywhere. Changing the control policy the trajectory will change producing a family of infinitely many solutions of the controlled system (2) parametrized with respect to α.

In the minimum time problem one has to drive the controlled dynamical system (2) from its initial state to a given target $\mathcal{T}$. Let us assume that the target is a compact subset of $\mathbb{R}^d$ with non empty interior and piecewise smooth boundary. The major difficulty dealing with this problem is that the time of arrival to the target starting from the point x and applying the control strategy α, denoted by $t(x, \alpha(\cdot))$, can be infinite at some points (if the strategy does not bring to $\mathcal{T}$), i.e.

$$t(x, \alpha(\cdot)) := \begin{cases} \inf\limits_{\alpha \in \mathcal{A}} \{t \in \mathbb{R}_+ : y(t, \alpha(\cdot)) \in \mathcal{T}\} & \text{if } y(t, \alpha(t)) \in \mathcal{T} \text{ for some } t, \\ +\infty & \text{otherwise}, \end{cases} \tag{3}$$

As a consequence, the minimum time function defined as

$$T(x) = \inf_{\alpha \in \mathcal{A}} t(x, \alpha(\cdot)) \tag{4}$$

is not defined everywhere if some controllability assumptions are not satisfied. In general, this is a free boundary problem where one has to determine at the same time, the couple (T, Ω), i.e. the minimum time function and its domain. Nevertheless, by applying the Dynamic Programming Principle and the so-called Kruzkov transform

$$v(x) \equiv \begin{cases} 1 - \exp(-T(x)) & \text{for } T(x) < +\infty \\ 1 & \text{for } T(x) = +\infty \end{cases} \tag{5}$$

the minimum time problem is characterized in terms of the unique viscosity solution of

$$\begin{cases} v(x) + \sup_{a\in A}\{-f(x,a)\cdot Dv(x)\} = 1 & \text{in } \mathbb{R}\backslash\mathcal{T} \\ v(x) = 0 & \text{on } \partial\mathcal{T}\,, \end{cases} \tag{6}$$

The *semi-Lagrangian scheme* for the approximation of (6) is obtained coupling a discretization in time along the trajectories with a local reconstruction in space via interpolation. Several coupling are possible and the interested reader can find in [17] all the details. Here we just sketch the one dimensional case where the integration along the trajectory is obtained using the Euler method. We introduce a grid G on Ω with nodes x_i, $i = 1,\dots,N$. Without loss of generality, throughout this paper we will assume that the numerical grid G is a regular equidistant array of points with mesh spacing denoted by Δx. We also denote by $\mathring{G}$ the internal nodes of G and by ∂G its boundary, whose nodes act as *ghost nodes*. We map all the values at the nodes onto a N-dimensional vector $U = (U_1,\dots,U_N)$. Let us denote by $h_{i,a} > 0$ a (fictitious) time step, possibly depending on the node x_i and control a, and by $k = \Delta x > 0$ the space step. For every internal node of the grid we follow the dynamics using one step of the Euler scheme [4,5] then we compute the values at the points $x_i + h_{i,a}f(x_i,a)$ via an interpolation operator denoted by $I[U]$ [15]. Finally, we obtain the following scheme in fixed point form for of (6)

$$U = F(U)\,, \tag{7}$$

where $F : [0,1]^N \to [0,1]^N$ (due to the Kruzkov change of variable) is defined componentwise by

$$[F(U)]_i = \begin{cases} \min_{a\in A}\{I\,[U]\,(x_i + h_{i,a}f(x_i,a)) + h_{i,a}\} & x_i \in \mathring{G}\setminus\mathcal{T}\,, \\ 0 & x_i \in \mathcal{T}\,, \\ 1 & x_i \in \partial G\,. \end{cases}$$

The interpolation operator $I[U] : \Omega \to \mathbb{R}$ extends the discrete value function U to the whole space Ω. In order to fix the ideas, one should think to the linear interpolation in $\mathbb{R}^d$ described in [10] but other choices are possible [17]. We choose the time step $h_{i,a}$ such that $|h_{i,a}f(x_i,a)| = k$ for every $i = 1,\dots,N$ and $a \in A$, so that the point $x_i + h_{i,a}f(x_i,a)$ falls in one of the first neighboring cells. In the simplest case, the minimum over A is evaluated by direct comparison, discretizing the set A with N_c points but other (more expensive and accurate) methods are available. Note that defining $F(U) = 1$ on ∂G corresponds to impose state constraint boundary conditions. The final iterative scheme reads

$$U^{(n+1)} = F(U^{(n)})\,, \qquad U^{(0)} = \begin{cases} 0 \text{ on } \mathcal{T} \\ 1 \text{ otherwise} \end{cases}. \tag{8}$$

We refer to [15,17] for details on the building blocks of this construction and for the convergence analysis. With the discrete value function U in hand, we can obtain a feedback map $\Phi_h : \Omega \to A$ just defining

$$\Phi_h(x) := \arg\min_{a\in A}\{I[U](x + h_{x,a}f(x,a)) + h_{x,a}\}\,. \tag{9}$$

Under rather general assumptions (see [16]), it can be shown that this is an approximation of the feedback map constructed for the continuous problem. A detailed discussion on the construction of feedback maps via the value function is contained in [3, p. 140–143]. It is important to note that weak convergence results apply also for Lipschitz continuous value functions.

Then, the *value iteration* based on the semi-Lagrangian method leads to following iterative scheme:

Data: Mesh G, Δt, initial guess V^0, tolerance ϵ.
forall the $x_i \in \mathcal{T}$ **do**
  set $V_i = 0$
end
forall the $x_i \in \partial G$ **do**
  set $V_i = 1$
end
while $||V^{k+1} - V^k|| \geq \epsilon$ **do**
  forall the $x_i \in G$ **do**

$$V_i^{k+1} = \min_{a \in A}\{e^{-\Delta t} I\left[V^k\right](x_i + \Delta t f(x_i, a)) + 1 - e^{-\Delta t}\}$$

  end
  $k = k + 1$
end

Algorithm 1: (VI) Value Iteration method for minimum time problem

Here V_i^k represents the values at a node x_i of the grid at the k-th iteration and I is an interpolation operator acting on the values of the grid.

Algorithm 1 is referred in the literature as the *value iteration method* because, starting from an initial guess V^0, it modifies the values on the grid according to the nonlinear rule in the loop. It is well-known that the convergence of the value iteration can be very slow, since the contraction constant $e^{-\Delta t}$ is close to 1 when Δt is close to 0. This means that a higher accuracy will also require more iterations. Then, there is a need for an acceleration technique in order to cut the link between accuracy and complexity of the value iteration. Note that similar ideas can be applied to other classical control problems with small changes [17].

A classical acceleration technique is the *approximation in the policy space* (or policy iteration), it is based on a linearization of the Bellman equation. This method is due to Howard [19] and dates back to the origin of dynamic programming. First, an initial guess for the control for every point in the state space is chosen. Once the control has been fixed, the Bellman equation becomes linear (no search for the minimum in the control space is performed), and it is solved as an advection equation. Then, an updated policy is computed and a new iteration starts. This leads to the following algorithm.

Note that the solution of the policy evaluation step can be obtained either by a linear system (assuming a linear interpolation operator) or as the limit

$$V^k = \lim_{m \to +\infty} V^{k,m}, \tag{10}$$

Data: Mesh G, Δt, initial guess V^0, tolerance ϵ.
forall the $x_i \in \mathcal{T}$ **do**
  set $V_i = 0$
end
forall the $x_i \in \partial G$ **do**
  set $V_i = 1$
end
while $||V^{k+1} - V^k|| \geq \epsilon$ **do**
  Policy evaluation step:
  forall the $x_i \in G$ **do**

$$V_i^k = \Delta t + e^{-\Delta t} I\left[V^k\right]\left(x_i + \Delta t f\left(x_i, a_i^k\right)\right)$$

  end
  Policy improvement step:
  forall the $x_i \in G$ **do**

$$a_i^{k+1} = \arg\min_a \left\{\Delta t + e^{-\Delta t} I\left[V^k\right](x_i + \Delta t f(x_i, a))\right\}$$

  end
  $k = k + 1$
end

Algorithm 2: (PI) Policy Iteration method for the minimum time problem

of the linear time-marching scheme

$$V_i^{k,m+1} = \Delta t + e^{-\Delta t} I\left[V^{k,m}\right]\left(x_i + \Delta t f\left(x_i, a_i^k\right)\right). \tag{11}$$

Although this scheme is still iterative, the lack of a minimization phase makes it faster than the original value iteration. The sequence $\{V^k\}$ turns out to be monotone decreasing at every node of the grid. At a theoretical level, policy iteration can be shown to be equivalent to a Newton method, and therefore, under appropriate assumptions, it converges with quadratic speed (see [21]). On the other hand, convergence is local and this may represent a drawback with respect to value iterations.

3 The Patchy Domain Decomposition

In this section we introduce our new domain decomposition method for solving equations of Hamilton-Jacobi-Bellman type, in particular (6). The main feature of the new method is the technique we use to construct the subdomains of the decomposition, which are approximate "patches" in a sense inspired by Ancona and Bressan [2] in their study of feedback stabilization. They introduced and investigate the properties of a particular class of *discontinuous feedbacks*, the so-called *patchy feedbacks*.

The following definition gives the fundamental concept of a patch.

Definition 1. *Let $\Omega \subset \mathbb{R}^d$ be an open domain with smooth boundary $\partial\Omega$ and f be a smooth vector field defined on a neighborhood of $\overline{\Omega}$. We say that the pair*

(Ω, f) is a patch *if Ω is a positive-invariant region for f, i.e. at every boundary point $y \in \partial\Omega$ the inner product of f with the outer normal n satisfies*

$$\langle f(y), n(y) \rangle < 0.$$

By means of a superposition of patches, we get the notion of a patchy vector field on a domain $\Omega \subset \mathbb{R}^d$ and they have shown that these can be used to define discontinuous feedbacks stabilizing the system. However, their method is not constructive so an effort has been made to transform this approach into an algorithm. Clearly, from the numerical point of view, the approximation will produce patches which will be "almost invariant" with respect to the optimal dynamics driving the system. Their boundaries can be rather complicated, but this has the advantage that we do not need to apply any transmission condition between them.

Following [9], let us introduce two rectangular (structured) grids. The first grid should be rather *coarse* because it is used for preliminary (and fast) computations only. It will be denoted by $\widetilde{G}$ and its nodes by $\widetilde{x}_1, \ldots, \widetilde{x}_{\widetilde{N}}$, where $\widetilde{N}$ is the total number of nodes. We will denote the space step for this grid by $\widetilde{k} := \Delta x_{\text{coarse}}$ and the approximate solution of the Eq. (6) on this grid by $\widetilde{U}_P$.

The second grid is instead *fine*, being the grid where we actually want to compute the numerical solution of the equation. It will be denoted by G and its nodes by $x_1, \ldots, x_N$, where N is the total number of nodes ($N >> \widetilde{N}$). We will denote the space step for this grid by $k := \Delta x_{\text{fine}}$ and the solution of the Eq. (6) on this grid by U_P. We also choose the number R of subdomains (patches) to be used in the patchy decomposition and we divide the target Ω_0 in R parts denoted by Ω_0^j, with $j = 1, \ldots, R$.

The patchy method can be described as follows.

Patchy Algorithm

Step 1. (Computation on $\widetilde{G}$). We solve the equation on $\widetilde{G}$ by means of the classical domain decomposition algorithm (e.g. where the subdomains are rectangles). For coherence we choose the (static) decomposition made by R subdomains (as the number of patches). This leads to $\widetilde{U}_P$.

Step 2. (Interpolation on G). We define the function $U_P^{(0)}$ on the fine grid G by interpolation of the values $\widetilde{U}_P$. Then, we compute the approximate optimal control

$$\widetilde{a}^*(x_i) = \arg\min_{a \in A} \{ I[U_p^{(0)}](x_i + h_{i,a} f(x_i, a)) + h_{i,a} \}, \qquad x_i \in G. \tag{12}$$

Even if $\widetilde{a}^*$ is defined on G, we still use the symbol "tilde" to stress that optimal controls are computed using only coarse information. We delete $\widetilde{G}$ and $\widetilde{U}_P$.

Step 3. (Main cycle) For every $j = 1, \ldots, R$,

Step 3.1. (Creation of j-th patch). Using the (coarse) optimal control $\widetilde{a}^*$, we find the nodes of the grid G that have the part Ω_0^j of the target in their numerical domain of dependence. This procedure defines the j-th patch, naturally following the (approximate) optimal dynamics. This step will be detailed later in this section.

Step 3.2. (Computation in j-th patch). As initial guess we initialize the j-th solution equal to $+\infty$ on the j-th patch and equal to 0 on the part Ω_0^j of the target. Then, we apply iteratively the scheme (8) in the j-th patch until convergence is reached. Finally, the j-th solution is copied in the matrix that will contain the global solution U_P.

Details on Step 3.1. The basic idea we adopt here is to divide the whole domain starting from a partition of the target only, and let the dynamics make a partition of the rest of the domain. To this end we use the approximation of the optimal control given by $\widetilde{a}^*$ to obtain a domain decomposition fully compliant to the dynamics. More precisely, we divide the target Ω_0 in R parts, each associated to a colour indexed by a number $j = 1, \ldots, R$. Assume for instance that Ω_0 is a ball at the center of the domain and focus on the subset of the target with a generic colour j, denoted by Ω_0^j, see Fig. 1(a). The goal is to find the subset of the domain Ω which has Ω_0^j as numerical domain of dependence. To do that, we initialize the grid nodes with the values ϕ_i as follows:

$$\phi_i = \begin{cases} 1\,, \ x_i \in \Omega_0^j \\ 0\,, \ x_i \in G \backslash \Omega_0^j \end{cases}, \qquad i = 1, ..., N.$$

Then we solve the following *ad hoc* discrete equation,

$$\phi_i = I[\phi](x_i + h_i f(x_i, \widetilde{a}^*(x_i)))\,, \qquad i = 1, ..., N, \tag{13}$$

which is similar to the fixed-point scheme (7) for the main equation. Here $h_i > 0$ is chosen in such a way that $|h_i f(x_i, \widetilde{a}^*(x_i))| = k$. Once the computation is completed, the whole domain will be divided in three zones:

$$\Lambda_1^j = \{x_i : \phi_i = 1\}\,, \quad \Lambda_2^j = \{x_i : \phi_i = 0\}\,, \quad \Lambda_3^j = \{x_i : \phi_i \in (0,1)\}\,,$$

see Fig. 1(b). Note that Λ_3^j will be nonempty because the interpolation operator I in the scheme (13) mixes the values ϕ_i through a convex combination, thus producing values in $[0,1]$ even if the initial datum is in $\{0,1\}$. Since we need a sharp division of the domain, we "project" the colour j into a binary value

$$\widehat{\phi}_i = \begin{cases} 1\,, \ \phi_i \geq \frac{1}{2} \\ 0\,, \ \phi_i < \frac{1}{2} \end{cases}, \qquad i = 1, ..., N \tag{14}$$

and then we define the subdomain $\Omega^j = \{x_i \in G \backslash \Omega_0^j : \widehat{\phi}_i = 1\}$ as the j-th patch, see Fig. 1(c). Once all the patches $j = 1, \ldots, R$ are computed, they are assembled together on the grid G. Thus the grid results to be divided into R patches, each associated to a different colour, as shown in Fig. 1(d). Note that the boundaries of every patch are aligned with the coordinate axes.

The main point here is that the patches Ω^j's are constructed to be invariant with respect to the optimal dynamics, meaning that the solution of the equation

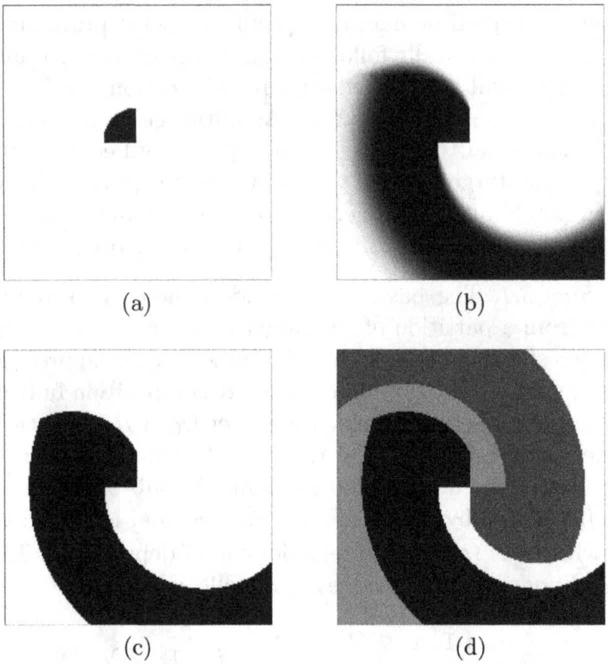

Fig. 1. Creation of patches for a test dynamics, $R = 4$, $\Omega_0 =$ small ball in the centre: (a) Select a subdomain Ω_0^j of the target Ω_0. (b) Find the nodes which depend, at least partially, on Ω_0^j. (c) Define Ω^j projecting the color in a binary value. (d) Assemble all patches.

in each patch will not depend on the solution in other patches. This is equivalent to state that there is no crossing information through the boundaries of the patches.

We stress that Step 3.1 of the algorithm is not expensive, even if it is performed on the fine grid G. The reason for that is the employment of the pre-computed optimal control $\widetilde{a}^*$ in the Eq. (13), which avoids the evaluation of the minimum (see the scheme (8)). Moreover, the stopping rule for the fixed point iterations used to solve (13) can be very rough, since we project the colors at the end and then we do not need precise values.

Numerical examples

We will test the method described above against two minimum time problems of the form (1). The numerical domain is always $\Omega = [-2, 2]^2$.

Test 1 (Eikonal) : $d = 2\,, \quad f(x_1, x_2, a) = a\,, \quad A = B_2(0,1)\,, \quad \Omega_0 = B_2(0, 0.5)$.

Test 2 (Fan) : $d = 2\,, \; f(x_1, x_2, a) = |x_1 + x_2 + 0.1|a\,, \; A = B_2(0,1)\,, \; \Omega_0 = \{x_1 = 0\}$.

In Fig. 2 we show the patchy decomposition for the two dynamics described above in the case $R = 8$, $N_c = 32$, $\widetilde{N} = 50$ and $N = 100$. We also superimpose the optimal vector field $f(x, \widetilde{a}^*)$ to show that patches are (almost) invariant with

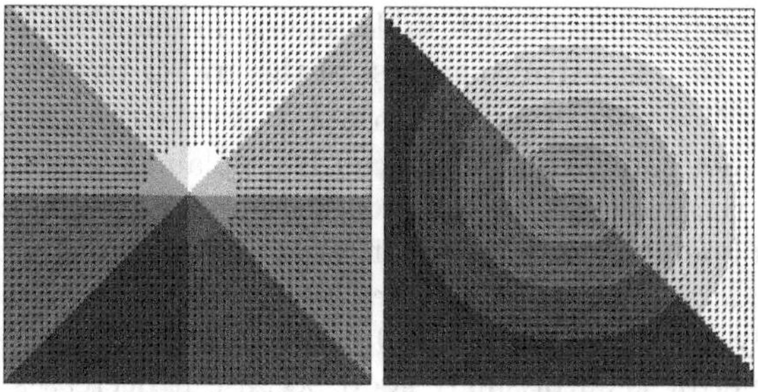

Fig. 2. Patchy decompositions with $R = 8$, $N_c = 32$, $\widetilde{N} = 50$ and $N = 100$. For visualization purposes not all the arrows are shown. (a) Eikonal, (b) Fan.

respect to the optimal dynamics. Indeed, only a few arrows cross from a patch to another. Note that patches cover the whole domain but they are not equivalent in terms of area, even if the target Ω_0 was divided in $R = 8$ equal parts to generate the decomposition.
It is interesting to compare the solution U_P of the patchy algorithm with that of the classical domain decomposition method U_{DD}, both computed on the same fine grid by means of the scheme (8). Let us denote by E_P the difference $E_P := U_P - U_{DD}$ that in the following will be referred to as *patchy error*. In particular we study the norms

$$\|E_P\|_1 := k^d \sum_{i=1}^{N} |E_{P\,i}| \qquad \text{and} \qquad \|E_P\|_\infty := \max_{i=1,\dots,N} |E_{P\,i}|$$

depending on the space steps $\widetilde{k}$ and k. This error is exclusively due to the fact that patches are not completely dynamics-invariant and then it will be considered as a degree of the invariance of the patchy decomposition. Let us stress that we apply state constraint boundary conditions on the patches.

We report the results for $R = 16$, which is the largest number of patches and also the worst case we tested. Indeed, the error E_P necessarily increases as R increases because the number of boundaries increases. We present the results for the Test 2 in Table 1, similar errors appear in other tests.

Table 1. Patchy error $\|E_P\|_1$ ($\|E_P\|_\infty$). Dynamics: Fan, $N_c = 32$, $R = 16$

	$k = 0.08$	$k = 0.04$	$k = 0.02$	$k = 0.01$	$k = 0.005$
$\widetilde{k} = 0.08$	1.393 (3.023)	0.123 (1.507)	0.037 (0.315)	0.017 (0.263)	0.011 (0.263)
$\widetilde{k} = 0.04$	–	0.114 (1.502)	0.032 (0.149)	0.011 (0.095)	0.006 (0.095)
$\widetilde{k} = 0.02$	–	–	0.032 (0.111)	0.011 (0.061)	0.004 (0.037)
$\widetilde{k} = 0.01$	–	–	–	0.011 (0.079)	0.004 (0.037)
$\widetilde{k} = 0.005$	–	–	–	–	0.004 (0.037)

We see that the first line of each table reports in many cases unsatisfactory results, caused by the excessive roughness of the grid $\widetilde{G}$ (see the case $\widetilde{k} = 0.08$, corresponding to $\widetilde{N} = 50$). Even the case $\widetilde{k} = k = 0.08$ (i.e. the grid is not refined at all) is not satisfactory. This can be explained by recalling that the computations on the two grids are not identical because the second one employs state constraints boundary conditions. If the grid is not fine enough, the error due to the boundary conditions is large, and tends to propagate inside each patch. Conversely, if $\widetilde{G}$ has at least 100 nodes per dimension ($\widetilde{k} \leq 0.04$), the behaviour of the error is surprisingly good because it decreases as k decreases (for any fixed $\widetilde{k}$) and $\|E_P\|_1$ is of the same order of k itself. Note that the L^∞ error is always larger than the L^1 error. Quite often we find a very small number of nodes with a large error near the boundaries of the patches, especially at those nodes where two patches and the target meet. This mainly affect the L^∞ error but not the L^1 error.

4 Fast Marching and Fast Sweeping Methods

The second technique which has been proposed to reduce the computational load and memory allocations is based on the localization of the algorithm. At every iteration only a subset of the grid (the active region) is taken into account and the solution is computed just on the nodes belonging to this region. An important feature of this method is the fact that the value at a single node is computed only a finite number of times and this allows to show that the solution is obtained in a finite number of iterations. Here we list and briefly describe some iterative and *single-pass* methods for solving HJ equations.

Let us sketch the Fast Marching Method (FMM) [25,27] introduced as a fast solver for the eikonal equation. Despite the standard global iterative method, the nodes are visited in a solution-dependent order, producing a *single-pass* method: the algorithm itself finds a correct order for processing the grid nodes. The order which is determined satisfies the *causality* principle, i.e. the computation of a node is declared completed only if its value cannot be affected by the future computation. To this end, at each step the grid is divided in three regions: *ACC*, where computation is definitively done, *CONS*, where computation is going on and *FAR*, where computation is not done yet. Then, the node in *CONS* with the minimal value enters *ACC*, its first neighbours enter *CONS* (if not yet in) and are (re)computed.

Following [23], we remark that this *minimum-value rule* corresponds to compute the value function T step by step in the ascending order (i.e., from the simplex containing $-\nabla T$). It follows that *CONS* expands under the gradient flow of the solution itself, which is exactly equivalent to say that *CONS* is, at each step, an approximation of a level set of the value function. In the case of isotropic eikonal Eq. (6), the gradient of the solution coincides with the characteristic field of the HJ equation, hence FMM computes the correct solution. Moreover, FMM still works for problems with mild anisotropy, where gradient lines and characteristics define small angles and lie, at each point, in the same simplex of the underlying grid. On the other hand, when a strong anisotropy comes into play, as for a general anisotropic eikonal equation, FMM fails and there is no way to compute the viscosity solution following its level sets. Finally, we remark that FMM is also a *local* method, since each node is computed by means of first neighbors nodes only and *CONS* is one-cell thick. Moreover, FMM computes the same solution of the global fixed point method (ITM), provided the same scheme is employed.

The *Fast Sweeping Method (FSM)* [24,28] is similar to the global fixed point iteration ITM , but the grid is visited in a multiple-direction predefined order. Usually, a rectangular grid is iteratively swept along four directions: $N \to S$, $E \to W$, $S \to N$, and $W \to E$, where N, S, E, and W stand for North, South, East, and West, respectively. This method has been shown to be much faster than ITM, but (as ITM) it is neither *local* nor *single-pass*. A well known exception is given by the eikonal equation, for which it is proved that only 1 sweep (i.e. four visits of the whole grid) is enough to reach convergence (see [28] for details). FSM computes the same solution of ITM, provided the same scheme and the same stopping rule are employed.

Numerical examples

Let us compare the methods in terms of velocity and accuracy.

Test 1. Let us choose $\mathcal{T} = (0,0)$, $f(x,y,a) \equiv a$. We know the exact solution of the corresponding eikonal equation which is $T(x,y) = \sqrt{(x^2+y^2)}$.

As one can see in Table 2 the two fast marching methods (FM-FD and FM-SL, respectively based on a finite difference and a semi-Lagrangian solver) give a big speed-up in the computation. The fast sweeping method (FS-SL) gives good results but is generally slower than the fast marching methods.

Test 2: state constraint problem. $\mathcal{T} = (-1,-1)$.

$$f(x,y,a) = \begin{cases} 0 \ (x,y) \in ([0,0.5] \times [-2,1.5]) \cup ([1,1.5] \times [-1.5,2]) \\ a \text{ elsewhere.} \end{cases}$$

Table 2. Errors for Test 1.

Method	Δx	L^∞ error	L^1 error	CPU time (s)
FM-FD	0.08	0.0875	0.7807	0.5
FM-SL	0.08	0.0329	0.3757	0.7
SL (46 it)	0.08	0.0329	0.3757	8.4
FS-SL	0.08	0.0329	0.3757	0.8
FM-FD	0.04	0.0526	0.4762	2.1
FM-SL	0.04	0.0204	0.2340	3.1
SL (86 it)	0.04	0.0204	0.2340	60
FS-SL	0.04	0.0204	0.2340	3.2
FM-FD	0.02	0.0309	0.2834	9.4
FM-SL	0.02	0.0122	0.1406	14
SL (162 it)	0.02	0.0122	0.1406	443.7
FS-SL	0.02	0.0122	0.1406	12.5

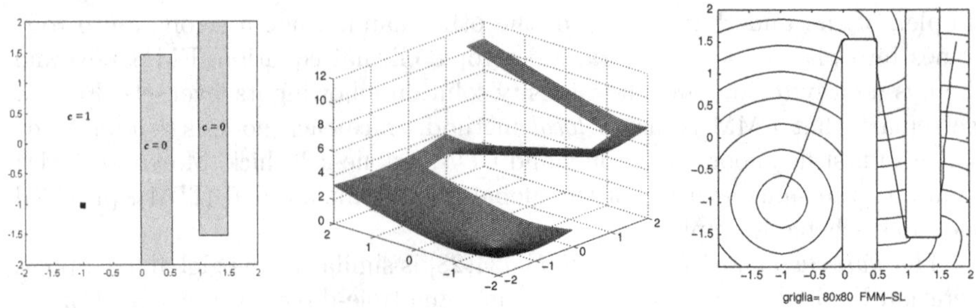

Fig. 3. Domain of the equation (left), value function T (center) and level sets of T with one optimal trajectory (right).

In this test the dynamics has been set to 0 on the obstacles to enforce the state constraint. The results are shown in Fig. 3.

5 An Accelerated Policy Iteration Algorithm with Smart Initialization

Let us conclude with an accelerated iterative algorithm which is constructed upon the building blocks previously introduced in Sect. 2. We aim to an efficient formulation exploiting the main computational features of both value and policy iteration algorithms. As it has been stated in [21], there exists a theoretical equivalence between both algorithms, which guarantees a rather wide convergence framework. However, from a computational perspective, there are significant differences between both implementations. A first key factor can be observed in Fig. 4, which shows, for a two-dimensional minimum time problem, the typical situation arising with the evolution of the error measured with respect to the optimal solution, when comparing value and policy iteration algorithms. To achieve a similar error level, policy iteration requires considerable fewer iterations than the value iteration scheme, as quadratic convergent behavior is reached faster for any number of nodes in the state-space grid. Despite the observed computational evidence, a second issue is observed when examining the policy iteration algorithm in more detail. That is, the sensitivity of the method with respect to the choice of the initial guess of the control field. It can be seen that different initial admissible control fields can lead to radically different convergent behaviors. While some guesses will produce quadratic convergence from the beginning of the iterative procedure, others can lead to an underperformant value iteration-like evolution of the error. This latter is computationally expensive, because it translates into a non-monotone evolution of the subiteration count of the solution of Eq. (2).

A final relevant remark goes back to Fig. 4, where it can be observed that for coarse meshes, the value iteration algorithm generates a fast error decay up to a higher global error. This, combined with the fact that value iteration algorithms are rather insensitive to the choice of the initial guess for the value

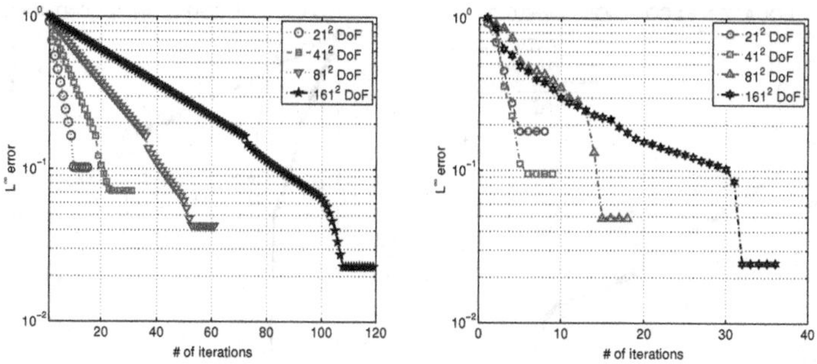

Fig. 4. Error evolution in a 2D problem: value iteration (left) and policy iteration (right).

function, are crucial points for the construction of our accelerated algorithm. The accelerated policy iteration algorithm is based on a robust initialization of the policy iteration procedure via a coarse value iteration which will yield to a good guess of the initial control field (see [1] for details).

Data: Coase mesh G_c and Δt_c , fine mesh G_f and Δt_f, initial coarse guess V_c^0, coarse-mesh tolerance ϵ_c, fine-mesh tolerance ϵ_f.
begin
  Coarse-mesh value iteration step: perform Algorithm 1
  Input: G_c, Δt_c, V_c^0, ϵ_c
  Output: V_c^*
  forall the $x_i \in G_f$ **do**
    $V_f^0(x_i) = I_1[V_c^*](x_i)$
    $A_f^0(x_i) = \underset{a \in A}{argmin}\, \{e^{-\Delta t} I_1[V_f^0](x_i + f(x_i, a)) + \Delta t]$
  end
  Fine-mesh policy iteration step: perform Algorithm 2
  Input: G_f, Δt_f, V_f^0, A_f^0, ϵ_f
  Output: V_f^*
end

Algorithm 3: (API) Accelerated Policy Iteration

Numerical axamples
The next two cases are based on a two-dimensional eikonal equation. For both problems, common settings are given by

$$f(x, y, a) = \begin{pmatrix} \cos(a) \\ \sin(a) \end{pmatrix}, \quad A = [-\pi, \pi], \quad \Delta t = 0.8 \Delta x .$$

What differentiates the problems is the domain and target definitions.

Test 1: $\Omega =]-1, 1[^2$, target $\mathcal{T} = (0, 0)$.

Test 2: $\Omega =]-2, 2[^2$, $\mathcal{T} = \{x \in \mathbb{R}^2 : ||x||_2 \leq 1\}$.

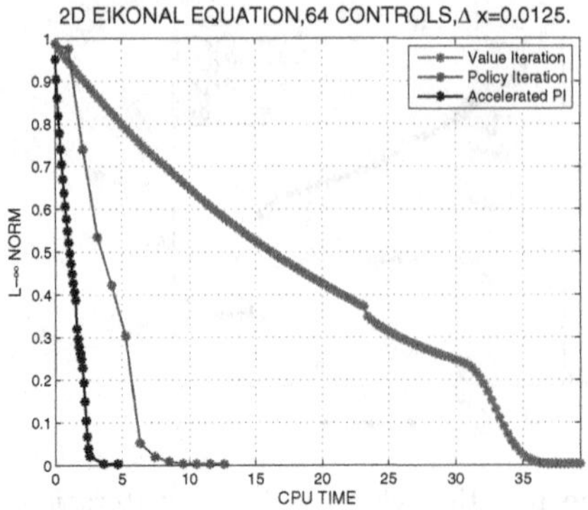

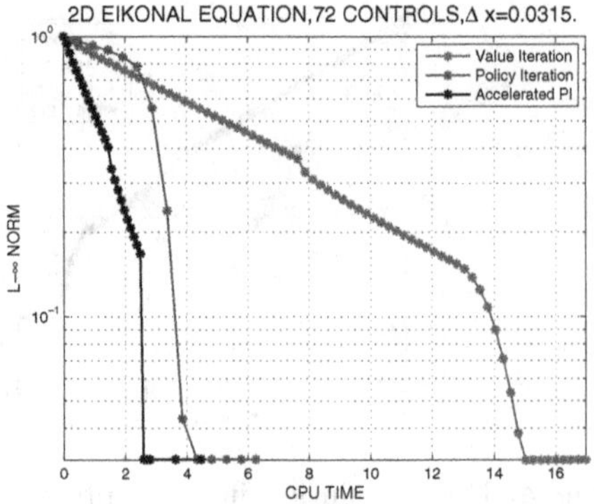

Fig. 5. Error evolution in 2D eikonal equations: Test 1 (left) and Test 2 (right).

Reference solutions are considered to be the distance function to the respective targets, which is an accurate approximation provided that the number of possible control directions is large enough. For Test 1, with a discretization of the control space into a set of 64 equidistant points, it can be seen that API provides a speedup of 8× with respect to VI over fine meshes despite the large set of discrete control points. Figure 5 illustrates, for both problems, the way in which the API idea acts: pre-processing of the initial guess of PI leads to proximity to a "quadratic convergence neighborhood"; fast error decay that coarse mesh VI has in comparison with the fine mesh VI is clearly noticeable.

In Test 2 we have a "fat" target. In general, larger or more complicated targets represent a difficulty in terms of the choice of the minimizing control, which translates into a larger number of iterations. In this case, the CPU time spent in the pre-processing is significant to the overall CPU time, but increasing this ratio in order to reduce its share will lead to an underperformant PI part of the algorithm.

6 Conclusions

We illustrated some recent results concerning the numerical approximation of optimal control problems governed by ordinary differential equations. The above methods can be combined in order to obtain fast algorithms and accurate solutions. Fo example one can use a patchy domain decomposition to set up a parallel algorithm and inside every patch use an Accelerated Policy Iteration (API) or a Fast Marching method. Several open problems still remain. For example, we would like to prove error bounds for the patchy domain decomposition and for the API acceleration method. Moreover, we continue our investigations to extend these methods to differential games and to the control of partial differential equations.

References

1. Alla, A., Falcone, M., Kalise, D.: An efficient policy iteration algorithm for dynamic programming equations. SIAM J. Sci. Comp. (still to appear)
2. Ancona, F., Bressan, A.: Patchy vector fields and asymptotic stabilization. ESAIM: Control Optim. Calc. Var. **4**, 445–471 (1999)
3. Bardi, M., Capuzzo Dolcetta, I.: Optimal Control and Viscosity Solutions of Hamilton-Jacobi-Bellman Equations. Birkhäuser, Boston (1997)
4. Bardi, M., Falcone, M.: An approximation scheme for the minimum time function. SIAM J. Control Optim. **28**, 950–965 (1990)
5. Bardi, M., Falcone, M.: Discrete approximation of the minimal time function for systems with regular optimal trajectories. In: Bensoussan, A., Lions, J.L. (eds.) Analysis and Optimization of Systems. Lecture Notes in Control and Information Sciences, vol. 144, pp. 103–112. Springer, Heidelberg (1990)
6. Bellman, R.: Dynamic Programming. Princeton University Press, Princeton (1957)
7. Cacace, S., Cristiani, E., Falcone, M.: A local ordered upwind method for Hamilton-Jacobi and Isaacs equations. In: Proceedings of the 18th IFAC World Congress, pp. 6800–6805 (2011)
8. Camilli, F., Falcone, M., Lanucara, P., Seghini, A.: A domain decomposition method for Bellman equations. In: Keyes, D.E., Xu, J. (eds.) Domain Decomposition methods in Scientific and Engineering Computing, Contemporary Mathematics, vol. 180, pp. 477–483. AMS, Providence (1994)
9. Cacace, S., Cristiani, E., Falcone, M., Picarelli, A.: A patchy dynamic programming scheme for a class of Hamilton-Jacobi-Bellman equations. SIAM J. Sci. Comp **34**, 2625–2649 (2012)
10. Carlini, E., Falcone, M., Ferretti, R.: An efficient algorithm for Hamilton-Jacobi equations in high dimension. Comput. Vis. Sci. **7**, 15–29 (2004)
11. Carlini, E., Falcone, M., Forcadel, N., Monneau, R.: Convergence of a generalized fast marching method for an Eikonal equation with a velocity changing sign. SIAM J. Numer. Anal. **46**, 2920–2952 (2008)
12. Crandall, M.G., Lions, P.L.: Two approximation of solutions of Hamilton-Jacobi equations. Math. Comput. **43**, 1–19 (1984)
13. Cristiani, E.: A fast marching method for Hamilton-Jacobi equations modeling monotone front propagations. J. Sci. Comput. **39**, 189–205 (2009)
14. Cristiani, E., Falcone, M.: Fast semi-Lagrangian schemes for the Eikonal equation and applications. SIAM J. Numer. Anal. **45**, 1979–2011 (2007)
15. Falcone, M.: Numerical solution of dynamic programming equations, Appendix A in [3].
16. Falcone, M.: Some remarks on the synthesis of feedback controls via numerical methods. In: Menaldi, J.L., Rofman, E., Sulem, A. (eds.), Optimal Control and Partial Differential Equations, pp. 456–465. IOS Press (2001)
17. Falcone, M., Ferretti, R.: Semi-Lagrangian approximation schemes for linear and Hamilton-Jacobi equations. SIAM, Philadelphia (2014)
18. Falcone, M., Lanucara, P., Seghini, A.: A splitting algorithm for Hamilton-Jacobi-Bellman equations. Appl. Numer. Math. **15**, 207–218 (1994)
19. Howard, R.A.: Dynamic programming and Markov processes. Wiley, New York (1960)
20. Navasca, C., Krener, A.J.: Patchy solutions of Hamilton-Jacobi-Bellman partial differential equations. In: Chiuso, A., et al. (eds.) Modeling, Estimation and Control. Lecture Notes in Control and Information Sciences, vol. 364, pp. 251–270. Springer, Heidelberg (2007)

21. Puterman, M.L., Brumelle, S.L.: On the convergence of policy iteration in stationary dynamic programming. Math. Oper. Res. **4**(1), 60–69 (1979)
22. Sethian, J.A.: Level Set Methods and Fast Marching Methods. Cambridge University Press, Cambridge (1999)
23. Sethian, J.A., Vladimirsky, A.: Ordered upwind methods for static Hamilton-Jacobi equations: theory and algorithms. SIAM J. Numer. Anal. **41**, 325–363 (2003)
24. Tsai, Y., Cheng, L., Osher, S., Zhao, H.: Fast sweeping algorithms for a class of Hamilton-Jacobi equations. SIAM J. Numer. Anal. **41**, 673–694 (2004)
25. Tsitsiklis, J.N.: Efficient algorithms for globally optimal trajectories. IEEE Trans. Autom. Control **40**, 1528–1538 (1995)
26. Quarteroni, A., Valli, A.: Domain Decomposition Methods for Partial Differential Equations. Oxford University Press, Oxford (1999)
27. Sethian, J.A.: A fast marching level set method for monotonically advancing fronts. Proc. Nat. Acad. Sci. USA **93**, 1591–1595 (1996)
28. Zhao, H.: A fast sweeping method for Eikonal equations. Math. Comp. **74**, 603–627 (2005)

Development of an Optimization-Based Atomistic-to-Continuum Coupling Method

Derek Olson[1], Pavel Bochev[2](✉), Mitchell Luskin[1], and Alexander V. Shapeev[1]

[1] University of Minnesota, Minneapolis, USA
{olso4056,luskin,ashapeev}@umn.edu
[2] Sandia National Laboratories, Albuquerque, USA
pbboche@sandia.gov

Abstract. Atomistic-to-Continuum (AtC) coupling methods are a novel means of computing the properties of a discrete crystal structure, such as those containing defects, that combine the accuracy of an atomistic (fully discrete) model with the efficiency of a continuum model. In this note we extend the optimization-based AtC, formulated in [17] for linear, one-dimensional problems to multi-dimensional settings and arbitrary interatomic potentials. We conjecture optimal error estimates for the multidimensional AtC, outline an implementation procedure, and provide numerical results to corroborate the conjecture for a 1D Lennard-Jones system with next-nearest neighbor interactions.

1 Introduction

Solid materials have atomic configurations which are arranged as a crystalline lattice, and the properties of these materials are derived from the underlying structure of the lattice. Specifically, defects in the regular, repeating arrangement of atoms such as a dislocation, or an extra plane of atoms, determine fundamental mechanisms such as plastic slip. The presence of defects invalidates the central hypotheses of continuum mechanics so models that recognize the discrete nature of the material on the atomic scale must be used. Such methods can vary in their complexity ranging from quantum mechanical models which incorporate nuclear and electronic forces to empirical potential models that assume the existence of a potential energy which is a function of the nuclear positions only. The latter allows atoms to be considered as classical mechanical particles. Throughout this

DO was supported by the Department of Defense (DoD) through the National Defense Science & Engineering Graduate Fellowship (NDSEG) Program. ML was supported in part by the NSF PIRE Grant OISE-0967140, DOE Award DE-SC0002085, and AFOSR Award FA9550-12-1-0187. AS was supported in part by the DOE Award DE-SC0002085 and AFOSR Award FA9550-12-1-0187.

Sandia National Laboratories is a multi-program laboratory managed and operated by Sandia Corporation, a wholly owned subsidiary of Lockheed Martin Corporation, for the U.S. Department of Energy's National Nuclear Security Administration under contract DE-AC04-94AL85000.

I. Lirkov et al. (Eds.): LSSC 2013, LNCS 8353, pp. 33–44, 2014.
DOI: 10.1007/978-3-662-43880-0_3,

note, we assume that the exact mathematical problem we wish to solve is that of minimizing the global potential energy of a set of N atoms or, equivalently, of equilibrating the internal and external forces on the atoms.

The outstanding issue with empirical atomistic models is the complexity involved in their applications. In even the smallest problems of material interest on the nanoscale, there will be at least 10^9 and up to 10^{15} atoms meaning the number of degrees of freedom in an atomistic model is often far outside the scope of any current computational feasibility. A novel attempt at solving this problem has been to keep the atomistic model only in a small region near the defect, while employing a continuum model such as elasticity in the bulk of the material away from the defect. Continuum models are well understood and can numerically be solved in an efficient manner using finite elements. In effect, the atomistic model provides a constitutive relation near the defect where the constitutive relation of the continuum model fails to hold.

These so called atomistic-to-continuum (AtC) coupling methods have seen a surge of interest in the last two decades, especially with the introduction of the quasicontinuum method in [19]. The problem introduced in these AtC methods is how to combine, or *couple*, the two different models. An informal way of carrying this out is to divide the computational domain, say Ω, into an atomistic region, Ω_{a}, and a continuum region, Ω_{c}. Then, a global hybrid energy or hybrid force field is constructed from the atomistic and continuum models on Ω_{a} and Ω_{c}. The resulting hybrid energy is then minimized, or alternatively, the internal and external forces are equilibrated to find the equilibrium configuration of Ω.

In this note we continue the development of the optimization-based AtC approach commenced in [17]. The core idea is to pose independent atomistic and continuum subproblems on overlapping domains Ω_{a} and Ω_{c} and then couple the models by minimizing an objective functional, which measures the difference between the strains of the atomistic and continuum states on $\Omega_{\mathrm{a}} \cap \Omega_{\mathrm{c}}$. In so doing, our approach combines ideas from *blending AtC* methods [2–4,11–13,15] with the optimization-based domain-decomposition approach for PDEs in [9,10].

The resulting optimization-based AtC method differs substantially from current energy or force-based methods, and to the best of our knowledge [17] is the first instance of using an objective functional of this form to effect atomistic-to-continuum coupling. Conceptually, our AtC approach is similar to the heterogeneous domain decomposition method for PDEs developed in [6] with the important distinction that we couple two fundamentally different material models rather than PDEs.

The main focus of this note is on the formulation of an optimization-based AtC method for modeling material defects in two and three dimensions, while allowing for arbitrary many-body terms in the potential energy. Section 2 quotes the necessary background results and Sect. 3 presents the formulation of the method. Solution of the optimization problem is discussed in Sect. 4. We conjecture error estimates and derive optimal parameters for our algorithm from the complexity analysis of Sect. 5. Finally, Sect. 6 provides numerical evidence in support of these conjectures.

2 Preliminaries

We consider the problem of modeling a crystal occupying the infinite domain, $\mathbb{R}^d$, and take the reference configuration of the atoms to be the integer lattice, $\mathbb{Z}^d$, deformed by the macroscopic deformation gradient F. Deformations of the material are thus described by functions $y : \mathsf{F}\mathbb{Z}^d \to \mathbb{R}^d$. For any deformed configuration, y, of the lattice, we assume the energy due to electronic and nuclear interactions can be described by an empirical site potential $V_\xi(y)$ where V_ξ represents the energy attributable to atom $\xi \in \mathbb{Z}^d$. As usual, we further assume that each ξ interacts with only a finite number of other atoms. The set of atoms that ξ interacts with is given by $\xi + \mathcal{R} \subset \mathbb{Z}^d$ where $\mathcal{R}$ is the interaction neighborhood. The interaction neighborhood can be defined through a cutoff radius, r_{cut}, so that

$$\mathcal{R} = \left\{\xi \in \mathbb{Z}^d \,|\, 0 < |\mathsf{F}\xi| \le r_{\text{cut}}\right\}.$$

Figure 1 depicts $\mathcal{R}$ in $2D$ where F is the identity and $r_{\text{cut}} = 2$. We model point defects in the lattice by allowing V_ξ to depend on ξ while assuming that $V_\xi = V$ when ξ is far from a defect. An evident example is an impurity where atoms of a different species have different interaction laws with the bulk atoms, but these interactions are only limited to small neighborhoods of defect (impure) atoms.

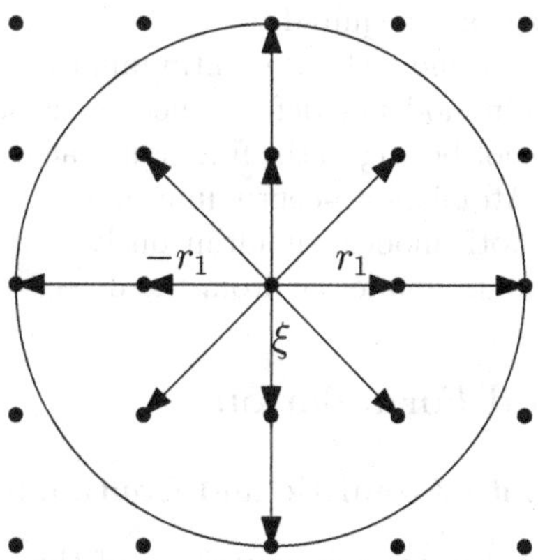

Fig. 1. An atom site ξ and its interaction range $\mathcal{R}$.

The presence of defects in the lattice generates elastic fields causing the atoms to relax. The deformed configurations are generically given by

$$y(\xi) = \mathsf{F}\xi + u(\xi),$$

where $u : \mathbb{Z}^d \to \mathbb{R}^d$ is the displacement field. The energy of the deformed configuration associated to this displacement field is

$$\mathcal{E}(u) := \sum_{\xi \in \mathbb{Z}^d} V_\xi(Du(\xi)) \tag{2.1}$$

where $Du(\xi) := (D_\rho u(\xi))_{\rho \in \mathcal{R}}$ is a collection of finite differences of u and $D_\rho u(\xi) := u(\xi + \rho) - u(\xi)$ defines the finite difference operator. We note that V_ξ implicitly depends on the macroscopic deformation gradient F. Furthermore, without loss of generality, we assume that $V_\xi(0) = 0$ so that the infinite sum in $\mathcal{E}(u)$ is well-defined. For example, in the case of a Lennard-Jones potential, ϕ, with next-nearest neighbor interactions in $1D$, we can define

$$V_\xi(Du(\xi)) = \phi(\mathsf{F} + D_1 u) + \phi(2\mathsf{F} + D_1 u - D_{-1} u) - (\phi(\mathsf{F}) + \phi(2\mathsf{F}))$$

where $\phi(\mathsf{F}) + \phi(2\mathsf{F})$ is subtracted from the usual Lennard-Jones potential (without affecting the computed forces) so that $V_\xi(0) = 0$.

The problem we seek to solve is then

$$\bar{u} \in \arg\min_{u \in \mathcal{U}} \mathcal{E}(u), \tag{2.2}$$

where arg min denotes the set of local minima of a functional and the admissible displacement space is taken to be $\mathcal{U} = \{u : \mathbb{Z}^d \to \mathbb{R}^d\}$. Typically, this energy on an infinite domain is approximated by truncating to a finite domain (the approach taken here) or by imposing periodic boundary conditions. However, the complexity involved in computing the resulting energy may be intractable for current computing capabilities due to the large number of atoms and interactions so a more efficient stratagem is required.

One solution approach would be to use continuum hyperelasticity models, but the elastic fields involved in modeling defects such as dislocations are singular at the defect core and so do not belong to the function spaces required in a standard continuum formulation. Atomistic-to-continuum models seek to overcome these deficiencies by utilizing both models simultaneously: the atomistic model near the defect and the continuum model far from the defect.

3 An AtC Method Formulation

3.1 Decomposition into Atomistic and Continuum Subdomains

Typical AtC methods require the decomposition of the computational domain Ω into atomistic and continuum subdomains, Ω_a and Ω_c, respectively, with a possible blending, or overlap, region $\Omega_o := \Omega_a \cap \Omega_c$. The goal of these methods is to create a *globally* defined hybrid energy or force field derived from using the atomistic model in Ω_a, the continuum model in Ω_c, and some coupling of the two in Ω_o. The distinguishing feature of our algorithm is to pose the atomistic and continuum problems independently on overlapping domains and then couple them by minimizing a suitably defined norm of the difference between the separate atomistic and continuum states that exist simultaneously on the overlap region. As we shall see, some care must be taken in the definitions of Ω_a and Ω_c to account for the interaction range, $\mathcal{R}$, from the previous section.

Truncation of the infinite domain, $\mathbb{R}^d$, to a finite, regular polygonal domain, Ω, inscribed in a sphere of radius R_c, is the first approximation in modeling (2.2).

The boundary of Ω coincides with $\mathsf{F}\mathbb{Z}^d$, and the lattice corresponding to Ω is defined as $\mathcal{L} := \Omega \cap \mathsf{F}\mathbb{Z}^d$. Consequently, we denote the space of admissible displacements which satisfy the far-field boundary condition $u(\xi) = 0$ whenever $\xi \notin \mathcal{L}$ by

$$\mathcal{U}_0 := \{u \in \mathcal{U} \mid u(\xi) = 0 \; \forall \xi \notin \mathcal{L}\}$$

and replace (2.2) by

$$\bar{u} \in \arg\min_{u \in \mathcal{U}_0} \mathcal{E}(u). \tag{3.1}$$

Remark 1. Though we have derived (3.1) with the idea of approximating an infinite domain containing a defect, a second problem of practical interest is minimizing an energy $\mathcal{E}$ on a fixed domain, Ω, subject to some prescribed boundary conditions on $\partial\Omega$ and an imposed external force in Ω. In this case, we typically separate V_ξ into an internal site energy V_ξ^{int} and an external site energy V_ξ^{ext}. Aside from this notational convenience, the formulation of our AtC method is identical for both of these problems.

Remark 2. For any domain, $\Omega_{\mathrm{t}} \subset \mathbb{R}^d$, ($\mathrm{t} = \mathrm{a}, \mathrm{c}, \mathrm{o}$, etc.) we define its (outer) radius, $R_{\mathrm{t}} := \frac{1}{2}\mathrm{diam}(\Omega_{\mathrm{t}})$, and its associated discrete lattice, $\mathcal{L}_{\mathrm{t}} = \Omega_{\mathrm{t}} \cap \mathsf{F}\mathbb{Z}^d$.

We further decompose Ω into overlapping atomistic and continuum subdomains, Ω_{a} and Ω_{c}, as follows. Let $\Omega_{\mathrm{a}} \subset \Omega$ be a regular polytope of radius R_{a} with $R_{\mathrm{a}} \ll R_{\mathrm{c}}$, and take Ω_{core} to be another regular polytope of radius $R_{\mathrm{core}} < R_{\mathrm{a}}$. The continuum subdomain, Ω_{c}, is defined as the closure of $\Omega \backslash \Omega_{\mathrm{core}}$. This decomposition results in an annular overlap region $\Omega_{\mathrm{o}} := \Omega_{\mathrm{a}} \cap \Omega_{\mathrm{c}}$ with width $R_{\mathrm{a}} - R_{\mathrm{core}}$. See Fig. 2 for an illustration in 2D. The atomistic interior of Ω_{a},

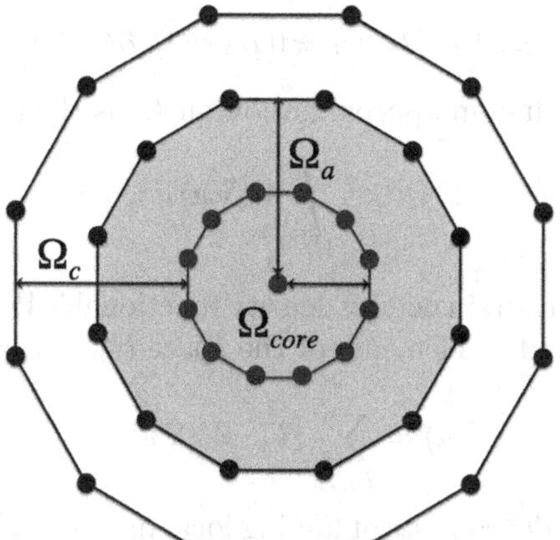

Fig. 2. Decomposition of Ω into atomistic and continuum subdomains.

denoted by Ω_a°, is the set of atoms $\xi \in \Omega_a$ such that all neighbors of ξ are also in Ω_a. Thus

$$\Omega_a^\circ := \{\xi \in \Omega_a \,|\, \xi + \mathcal{R} \subset \Omega_a\} \quad \text{and} \quad \Omega_a^{\circ\circ} := \{\xi \in \Omega_a^\circ \,|\, \xi + \mathcal{R} \subset \Omega_a^\circ\}, \tag{3.2}$$

where $\Omega_a^{\circ\circ}$ can be interpreted as the atomistic interior of Ω_a°.

For these domains, we define the associated displacement spaces

$$\mathcal{U}^a := \{u : \mathcal{L}_a \to \mathbb{R}^n\} \quad \text{and} \quad \mathcal{U}_0^a := \{u \in \mathcal{U}^a \,|\, u = 0 \text{ outside } \Omega_a^{\circ\circ}\}.$$

The energy on these spaces is

$$\mathcal{E}^a(u) := \sum_{\xi \in \mathcal{L}_a^{\circ\circ}} V_\xi(Du(\xi)), \tag{3.3}$$

where $\mathcal{L}_a^{\circ\circ} := \Omega_a^{\circ\circ} \cap \mathcal{L}$. The problem of finding local minima of this energy in the space $\mathcal{U}^a$ subject to some prescribed boundary values on $\mathcal{L}_a \backslash \mathcal{L}_a^{\circ\circ}$ is exactly what has been described as the atomistic model on Ω_a.

Remark 3. As previously mentioned, some care must be exercised in defining Ω_a and Ω_c. Precisely, we must impose the requirement that $\Omega_{\text{core}} \subset \Omega_a^{\circ\circ}$. This ensures that the overlap width is at least twice the size of the interaction range; a necessary condition is $R_a - R_{\text{core}} \geq 2r_{\text{cut}}$.

Our next task is to define a continuum model on Ω_c which is accomplished by defining the Cauchy-Born continuum energy there. We momentarily assume a finite element triangulation, $\mathcal{T}_h$, is given in Ω_c. This triangulation will be explicitly constructed in Sect. 5. Piecewise linear continuous finite elements are employed, and the mesh is fully refined in Ω_o so that a finite element node exists at each $\xi \in \mathcal{L}_o$. We denote the space of finite elements as $\mathcal{U}^c$ while the subspace of $\mathcal{U}^c$ satisfying homogeneous Dirichlet boundary conditions on the "outer" boundary of Ω_c is

$$\mathcal{U}_0^c := \{u \in \mathcal{U}^c \,|\, u = 0 \text{ on } \partial\Omega_c \backslash \partial\Omega_{\text{core}}\}.$$

The Cauchy-Born continuum approximation on Ω_c is then

$$\mathcal{E}^c(u) := \int_{\Omega_c} W(\nabla u)\, dx, \tag{3.4}$$

where the Cauchy-Born strain energy density functional is $W(\mathsf{G}) := V(\mathsf{F}\mathcal{R} + \mathsf{G}\mathcal{R})$. This energy is evaluated at elements of the space $\mathcal{U}_0^c$ so that we may write the continuum energy as

$$\mathcal{E}^c(u) = \sum_{T \in \mathcal{T}_h} |T| \cdot W(\nabla u|_T), \tag{3.5}$$

and the continuum model consists of finding local minima in $\mathcal{U}^c$ of this functional subject to prescribed boundary conditions on $\partial\Omega_c$.[1]

[1] In both the atomistic and continuum model, we have referenced some unknown, prescribed boundary values. These can be interpreted as virtual controls as defined in [7] and discussed in [17].

3.2 Coupling

Having decomposed the computational domain into atomistic and continuum constituencies, we need to provide the mechanism by which these two models are coupled together. This is done by minimizing the energy norm difference between atomistic and continuum states resulting from the atomistic and continuum problems from the spaces $\mathcal{U}^{\mathrm{a}}$ and $\mathcal{U}^{\mathrm{c}}$. Since an atomistic state, $u^{\mathrm{a}} \in \mathcal{U}^{\mathrm{a}}$, is a discrete function defined on a lattice, whereas a continuum state, $u^{\mathrm{c}} \in \mathcal{U}^{\mathrm{c}}$, is a continuous function, we define a continuous, piecewise linear nodal interpolant of u^{a} on $\mathcal{T}_h$ restricted to Ω_{o} by Iu^{a}, which allows us to compare the atomistic and continuum states in the same function space on Ω_{o}.

Our AtC method is to then solve the constrained minimization problem

$$\text{find } (\bar{u}^{\mathrm{a}}, \bar{u}^{\mathrm{c}}) \text{ such that } \|\nabla I u^{\mathrm{a}} - \nabla u^{\mathrm{c}}\|_{L^2(\Omega_{\mathrm{o}})} \text{ is minimized}$$

$$\text{subject to} \begin{cases} \langle \delta\mathcal{E}^{\mathrm{a}}(u^{\mathrm{a}}), v^{\mathrm{a}}\rangle = 0 \; \forall v^{\mathrm{a}} \in \mathcal{U}_0^{\mathrm{a}} \\ \langle \delta\mathcal{E}^{\mathrm{c}}(u^{\mathrm{c}}), v^{\mathrm{c}}\rangle = 0 \; \forall v^{\mathrm{c}} \in \mathcal{U}_0^{\mathrm{c}} \end{cases} \quad \text{and} \quad \int_{\Omega_{\mathrm{o}}} (Iu^{\mathrm{a}} - u^{\mathrm{c}})\, dx = 0 \tag{3.6}$$

The objective in (3.6) ensures that the mismatch between $\bar{u}^{\mathrm{a}}$ and $\bar{u}^{\mathrm{c}}$ over Ω_{o} is as small as possible. The first two constraints in (3.6) imply that $\bar{u}^{\mathrm{a}}$ and $\bar{u}^{\mathrm{c}}$ are equilibria of the atomistic and continuum subproblems defined on Ω_{a} and Ω_{c}. The third constraint is necessary because the objective is a difference of two gradients, and without it the optimal solution would be determined only up to an arbitrary constant[2]. Finally, we define our AtC approximation by

$$\bar{u}^{\mathrm{atc}}(x) = \begin{cases} \bar{u}^{\mathrm{a}}(x), & |x| \le R_{\mathrm{a}}, \\ \bar{u}^{\mathrm{c}}(x), & |x| > R_{\mathrm{a}}. \end{cases} \tag{3.7}$$

4 Solution of the AtC Optimization Problem

The AtC formulation (3.6) is a constrained optimization problem. A standard solution approach for such problems is to recast them into unconstrained optimization problems through the Lagrange multiplier method. Setting the first variations of the resulting Lagrangian with respect to the states and the adjoints to zero yields an optimality system from which we can determine the optimal solution of the original problem. This approach is know as a "one-shot method" [8] because we solve simultaneously for the states, adjoints, and controls.

For the AtC formulation (3.6), we introduce the Lagrange multipliers (adjoint variables) $\lambda_{\mathrm{a}} \in \mathcal{U}_0^{\mathrm{a}}$ and $\lambda_{\mathrm{c}} \in \mathcal{U}_0^{\mathrm{c}}$ for the first two constraints, the multiplier $\eta \in \mathbb{R}$ for the third constraint, and the Lagrangian functional

$$\begin{aligned} \Psi(u^{\mathrm{a}}, u^{\mathrm{c}}, \lambda_{\mathrm{a}}, \lambda_{\mathrm{c}}, \eta) = {} & \frac{1}{2}\|\nabla I u^{\mathrm{a}} - \nabla u^{\mathrm{c}}\|^2_{L^2(\Omega_{\mathrm{o}})} \\ & - \langle \delta\mathcal{E}^{\mathrm{a}}(u^{\mathrm{a}}), \lambda_{\mathrm{a}}\rangle - \langle \delta\mathcal{E}^{\mathrm{c}}(u^{\mathrm{c}}), \lambda_{\mathrm{c}}\rangle - \eta \int_{\Omega_{\mathrm{o}}} (Iu^{\mathrm{a}} - u^{\mathrm{c}})\, dx. \end{aligned} \tag{4.1}$$

[2] In one dimension, or when there are multiple overlap regions associated with modeling multiple defects, a constraint is specified for each individual overlap region.

Setting the first-variations of the Lagrangian to zero yields the optimality system

$$\text{find } u^{\mathrm{a}}, u^{\mathrm{c}}, \lambda_{\mathrm{a}}, \lambda_{\mathrm{c}}, \text{ and } \eta \text{ such that } \nabla\Psi(u^{\mathrm{a}}, u^{\mathrm{c}}, \lambda_{\mathrm{a}}, \lambda_{\mathrm{c}}, \eta) = 0, \tag{4.2}$$

where

$$\nabla\Psi = \left(\frac{\partial\Psi}{\partial u^{\mathrm{a}}}, \frac{\partial\Psi}{\partial u^{\mathrm{c}}}, \frac{\partial\Psi}{\partial\lambda_{\mathrm{a}}}, \frac{\partial\Psi}{\partial\lambda_{\mathrm{c}}}, \frac{\partial\Psi}{\partial\eta}\right)^T \tag{4.3}$$

is the Jacobian[3] of Ψ. The first-order necessary conditions (4.2) are a nonlinear system of equations for the unknowns u^{a}, u^{c}, λ_{a}, λ_{c}, and η. To solve this system, we employ Newton linearization. Specifically, for a given initial guess $\mathbf{z} = [u^{\mathrm{a}},\ u^{\mathrm{c}},\ \lambda_{\mathrm{a}},\ \lambda_{\mathrm{c}},\ \eta]^T$ we solve the linear equation

$$\nabla^2\Psi(\mathbf{z})\mathbf{x} = -\nabla\Psi(\mathbf{z}) \tag{4.4}$$

for the Newton increment $\mathbf{x}$ and set the new iterate to $\mathbf{z} = \mathbf{z} + \mathbf{x}$.

It is not difficult to see that the Hessian $\nabla^2\Psi(z)$ has the form

$$\nabla^2\Psi = \begin{pmatrix} \frac{\partial^2\Psi}{\partial(u^{\mathrm{a}})^2} & \frac{\partial^2\Psi}{\partial u^{\mathrm{c}}\partial u^{\mathrm{a}}} & \frac{\partial^2\Psi}{\partial\lambda_{\mathrm{a}}\partial u^{\mathrm{a}}} & \mathbf{0} & \frac{\partial^2\Psi}{\partial\eta\partial u^{\mathrm{a}}} \\ \frac{\partial^2\Psi}{\partial u^{\mathrm{a}}\partial u^{\mathrm{c}}} & \frac{\partial^2\Psi}{\partial(u^{\mathrm{c}})^2} & \mathbf{0} & \frac{\partial^2\Psi}{\partial\lambda_{\mathrm{c}}\partial u^{\mathrm{c}}} & \frac{\partial^2\Psi}{\partial\eta\partial u^{\mathrm{c}}} \\ \frac{\partial^2\Psi}{\partial u^{\mathrm{a}}\partial\lambda_{\mathrm{a}}} & \mathbf{0} & \mathbf{0} & \mathbf{0} & \mathbf{0} \\ \mathbf{0} & \frac{\partial^2\Psi}{\partial u^{\mathrm{c}}\partial\lambda_{\mathrm{c}}} & \mathbf{0} & \mathbf{0} & \mathbf{0} \\ \frac{\partial^2\Psi}{\partial u^{\mathrm{a}}\partial\eta} & \frac{\partial^2\Psi}{\partial u^{\mathrm{c}}\partial\eta} & \mathbf{0} & \mathbf{0} & \mathbf{0} \end{pmatrix} =: \begin{pmatrix} \boldsymbol{A} & \boldsymbol{B}^T \\ \boldsymbol{B} & \mathbf{0} \end{pmatrix}, \tag{4.5}$$

and so, (4.4) has the typical structure of a saddle-point problem.

5 Formal Error and Complexity Analysis

We measure the error in the energy (semi-)norm, $\|D\bar{u}^{\mathrm{atc}} - D\bar{u}\|^2_{\ell^2(\mathcal{L})}$. Typically, the error has several contributions: (1) the error of truncating the infinite domain, (2) the error of modeling the atomistic interaction with the continuum interaction on a finite element mesh, and (3) an error from coupling the two models. The first error is expected to be $\|Du\|_{\ell^2(\mathbb{Z}^d\setminus\mathcal{L})}$. For P_1 (i.e., piecewise linear) elements, the second error is expected to be $\|hD^2\bar{u}\|_{\ell^2(\mathcal{L}_{\mathrm{c}})}$, where h is the element size and $D^2u := (D_\rho D_\sigma u)_{\rho,\sigma\in\mathcal{R}}$. The third error is usually dominated by the second. For rigorous establishments of similar error estimates, see [5,14,18]. In this note, we conjecture the following result,

Conjecture 1.

$$\|D\bar{u}^{\mathrm{atc}} - D\bar{u}\|^2_{\ell^2(\mathcal{L})} \lesssim \|D\bar{u}\|^2_{\ell^2(\mathbb{Z}^d\setminus\mathcal{L})} + \|hD^2\bar{u}\|^2_{\ell^2(\mathcal{L}_{\mathrm{c}})} =: \mathrm{err}^2, \tag{5.1}$$

where $X \lesssim Y$ indicates that X is less than or equal to Y up to a multiplicative constant (i.e., that $\exists\, c > 0$ such that $X \le cY$).

We note that this is the most "optimistic" conjecture and includes only the error contributions (1) and (2) that cannot be avoided.

[3] The notation $\frac{\partial\Psi}{\partial u^{\mathrm{a}}}$ is used to represent the vector $\frac{\partial\Psi}{\partial u^{\mathrm{a}}_\xi}$ for $\xi \in \mathcal{L}_{\mathrm{a}}$ with analogous definitions for the remaining components.

Optimal Approximation Parameters

A defect can be characterized by a far-field decay rate, $\gamma > 0$, of the elastic displacement or stress fields. That is, we assume that $|D^k \bar{u}(\xi)| \sim |\xi|^{1-k-\gamma}$ (typically, $\gamma = d$ for a point defect and $\gamma = 1$ for a dislocation [5,16]). We further assume a finite element discretization, $\mathcal{T}_h$, with nodes in $\mathcal{L}_\mathrm{c}$ and with a radial mesh size function $h(x) := \mathrm{diam}(T)$ for $x \in T$, which will be chosen to formally optimize the error bound in (5.1) subject to a fixed number of degrees of freedom. We fully resolve the mesh in Ω_o so that each element of $\mathcal{L}_\mathrm{o}$ is taken as a node. The number of remaining degrees of freedom is then given by

$$\#\mathrm{DoF} = \sum_{\substack{T \in \mathcal{T}_h \\ T \cap \Omega_\mathrm{o} = \emptyset}} 1 = \sum_{\substack{T \in \mathcal{T}_h \\ T \cap \Omega_\mathrm{o} = \emptyset}} \frac{|T|}{|T|} \eqsim \int_{R_\mathrm{a}}^{R_\mathrm{c}} \frac{1}{\tilde{h}^d} r^{d-1}\, \mathrm{d}r,$$

where $\tilde{h}(|x|) \eqsim h(x)$ is a mesh size function that depends only on $|x|$ and $X \eqsim Y$ indicates that X and Y are equal up to a multiplicative constant.

Recalling that $|D^k \bar{u}(\xi)| \sim |\xi|^{1-k-\gamma}$, we thus carry out the optimization problem:

$$\begin{aligned} &\text{minimize} \quad \int_{R_\mathrm{a}}^{R_\mathrm{c}} \tilde{h}^2 r^{-2-2\gamma} r^{d-1}\, \mathrm{d}r + \int_{R_\mathrm{c}}^{\infty} r^{-2\gamma} r^{d-1}\, \mathrm{d}r \\ &\text{subject to} \quad \begin{cases} \#\mathrm{DoF} = \int_{R_\mathrm{a}}^{R_\mathrm{c}} \frac{1}{\tilde{h}^d} r^{d-1}\, \mathrm{d}r = C, \\ \tilde{h}(R_\mathrm{a}) = 1 \end{cases} \end{aligned}$$

with respect to $\tilde{h} = \tilde{h}(r)$ and R_c. Notice that we optimize only a part of the error bound, since the remaining contribution $\|hD^2\bar{u}\|_{\ell^2(\mathcal{L}_\mathrm{o})} \eqsim \int_{R_\mathrm{core}}^{R_\mathrm{a}} \tilde{h}^2 r^{-2-2\gamma} r^{d-1}\, \mathrm{d}r$ cannot be optimized after we have fixed the mesh in Ω_o.

Introducing Lagrange multipliers and taking the variation with respect to $\tilde{h}$ we obtain $\tilde{h}(|x|) = c|x|^{\frac{1+\gamma}{1+d/2}}$ for some constant c, and the second constraint, $\tilde{h}(R_\mathrm{a}) = 1$, can then be used to see that $h(x) = \tilde{h}(|x|) = (|x|/R_\mathrm{a})^{\frac{1+\gamma}{1+d/2}}$ (refer to [1] for a related example of mesh optimization for ODEs). Likewise, by differentiating with respect to R_c and using the expression for $\tilde{h}$ we find that $R_\mathrm{c} \eqsim R_\mathrm{a}^{\frac{1+\gamma}{\gamma-d/2}}$, provided $2\gamma - d > 0$. Finally, from the stability condition derived in [17], we should choose $R_\mathrm{a} \eqsim R_\mathrm{core}$.

Since the number of degrees of freedom is

$$\mathrm{DoF} \eqsim R_\mathrm{a}^d + \int_{R_a}^{R_\mathrm{c}} \left((r/R_a)^{\frac{1+\gamma}{1+d/2}} \right)^{-d} r^{d-1} \mathrm{d}r \eqsim R_\mathrm{a}^d, \tag{5.2}$$

we have

$$\mathrm{err}^2 \eqsim (\mathrm{DoF})^{\frac{-2-2\gamma+d}{d}}. \tag{5.3}$$

Remark 4 (Uniform norm). A more involved derivation can be used to optimize the parameters for the conjecture $\|D\bar{u}^{\mathrm{atc}} - D\bar{u}\|_{\ell^\infty(\mathcal{L})} \lesssim \mathrm{errinf}$, where the errors in (5.1) are now measured in the infinity norm. In this case, we would get

$$h(x) = \left(\frac{|x|}{R_{\rm a}}\right)^{1+\gamma}, \qquad R_{\rm c} \precsim R_{\rm a}^{1+\frac{1}{\gamma}}, \qquad \text{errinf} \precsim (\text{DoF})^{-\frac{1+\gamma}{d}}.$$

For a dislocation (i.e., for $\gamma = 1$ and $d = 2$, cf. [5]), the energy norm is infinite so optimizing approximation parameters for err is ill-posed. Nevertheless, optimizing errinf is well-posed.

6 Numerical Experiments

In this section we report the results of numerical experiments conducted in 1D ($d = 1$) using a next-nearest neighbor Lennard-Jones model as the underlying atomistic model. These experiments are analogous to those run for various, popular AtC methods in [14], with the exception that the atomistic model chosen there was the Embedded Atom Method. Numerical experiments for the blended energy and blended force-based quasicontinuum methods using optimal approximation parameters have been presented in [13,15]. Our results provide evidence in support of the estimates conjectured in Sect. 5. We will also show how to incorporate external forces into the model as alluded to in Remark 1. We consider the exact, atomistic energy on the infinite lattice, $\mathbb{Z}$, to be

$$\mathcal{E}^{\rm a}(u) = \sum_{\xi\in\mathbb{Z}} \phi(1+D_1u(\xi))+\phi(2+D_1u(\xi)-D_{-1}u(\xi))-(\phi(1)-\phi(2))-f(\xi)u(\xi),$$

where $f(\xi)$ is an external force at ξ. The Cauchy-Born continuum energy is

$$\mathcal{E}^{\rm c}(u) = \int_{\mathbb{R}} W(\nabla u)\,dx - \int (If)u\,dx, \quad \text{where} \quad W(\mathsf{G}) = \phi(1+\mathsf{G}) + \phi(2+2\mathsf{G}),$$

and If is the continuous linear interpolant of the force. We assume the exact atomistic solution that we wish to approximate is (as in [14])

$$\bar{u}^{\rm a}_\xi = \frac{1}{10}\left(1+\xi^2\right)^{-\gamma/2}\xi.$$

Given this solution, we compute the external forces on an atom ξ to ensure that $\bar{u}^{\rm a}$ is indeed a minimizer of the global atomistic energy. These forces are

$$f_\xi = -\frac{\partial \mathcal{E}^{\rm a}(u)}{\partial u_\xi}\Big|_{u=\bar{u}^{\rm a}}.$$

This implies the Lagrangian from Sect. 4 is

$$\begin{aligned} \Psi(u^{\rm a}, u^{\rm c}, \lambda_{\rm a}, \lambda_{\rm c}, \eta) &= \frac{1}{2}\|\nabla I u^{\rm a} - \nabla u^{\rm c}\|^2_{L^2(\Omega_{\rm o})} + \langle \delta\mathcal{E}^{\rm a}(u^{\rm a}), \lambda_{\rm a}\rangle \\ &+ \langle \delta\mathcal{E}^{\rm c}(u^{\rm c}), \lambda_{\rm c}\rangle + \eta_1 \int_{\Omega_{\rm o}\cap\mathbb{R}^+} (Iu^{\rm a} - u^{\rm c})\,dx + \eta_2 \int_{\Omega_{\rm o}\cap\mathbb{R}^-} (Iu^{\rm a} - u^{\rm c})\,dx \end{aligned} \tag{6.1}$$

where we use the continuous, piecewise linear interpolant of u^{a} in this formulation. (Recall also the need for two Lagrange multipliers to enforce the mean value zero condition on the disconnected overlap region in one dimension.)

We select a value of R_{core} from a range of interest, choose the mesh according to the formal analysis from Sect. 5, and assign the rest of the approximation parameters via the formal derivations of Sect. 5. Namely, we set $R_{\mathrm{a}} = 2R_{\mathrm{core}}$ and recursively construct the nodes, $\mathcal{N}_h$, of the triangulation, $\mathcal{T}_h$, as follows. First, each $\xi \in B_{R_{\mathrm{a}}}(0)$ is chosen as a node. Set $\xi = \max_{\zeta \in \mathcal{N}_h} \zeta$, and sequentially add a new node at $\pm[\xi + h(\xi)]$ where $h(\xi) := \lfloor (\xi/R_{\mathrm{a}})^{\frac{1+\gamma}{1+d/2}} \rfloor$. This is continued until $h(\xi) \approx \xi$, at which point we add two final nodes at $\pm R_C$.

Finally, we take the "defect" approximation parameter to be $\gamma := 3/2$ and employ our optimization-based AtC algorithm to compute u^{atc} for the range of values $R_{\mathrm{core}} \in \{10, 20, 40, 80, 160\}$. According to our estimate (5.3) in Sect. 5, we expect the error to decay as $\mathrm{err}_2 \precsim \mathrm{DoF}^{-2}$. We have plotted the error involved in each of these approximations versus the number of degrees of freedom in Fig. 3. In particular, the error behaves like $(\mathrm{DoFs})^{-2}$, which is truly optimal in the sense of AtC methods because this is the rate of the continuum model. In other words, the error of coupling atomistic and continuum models is dominated by the far field error and the continuum modeling error, as assumed in Conjecture 1.

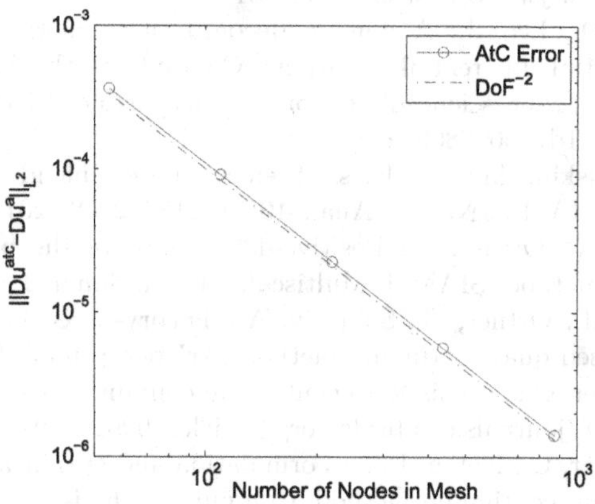

Fig. 3. Error of AtC approximation plotted against number of degrees of freedom.

7 Conclusion

We have formulated a new optimization-based AtC method for arbitrary interatomic potentials in multiple dimensions. Numerical simulations using a next-nearest neighbor Lennard-Jones atomistic model confirm a conjecture that the coupling error is dominated by the modeling and the domain truncation errors, i.e., that our AtC method behaves in an optimal fashion.

References

1. Babuška, I., Rheinboldt, W.: Analysis of optimal finite-element meshes in R^1. Math. Comput. **33**(146), 435–463 (1979)
2. Badia, S., Parks, M., Bochev, P., Gunzburger, M., Lehoucq, R.: On atomistic-to-continuum coupling by blending. Multiscale Model. Simul. **7**(1), 381–406 (2008)
3. Bauman, P., Dhia, H.B., Elkhodja, N., Oden, J., Prudhomme, S.: On the application of the Arlequin method to the coupling of particle and continuum models. Comput. Mech. **42**, 511–530 (2008)
4. Belytschko, T., Xiao, S.: Coupling methods for continuum model with molecular model. Int. J. Multiscale Comput. Eng. **1**, 115–126 (2003)
5. Ehrlacher, V., Ortner, C., Shapeev, A.: Analysis of boundary conditions for crystal defect atomistic simulations (2013). arXiv:1306.5334
6. Gervasio, P., Lions, J.L., Quarteroni, A.: Heterogeneous coupling by virtual control methods. Numer. Math. **90**, 241–264 (2001). http://dx.doi.org/10.1007/s002110100303
7. Gervasio, P., Lions, J., Quarteroni, A.: Heterogeneous coupling by virtual control methods. Numer. Math. **90**(2), 241–264 (2001)
8. Gunzburger, M.D.: Perspectives in Flow Control and Optimization. Society for Industrial and Applied Mathematics, Philadelphia (2002)
9. Gunzburger, M.D., Lee, H.K.: An optimization-based domain decomposition method for the Navier-Stokes equations. SIAM J. Numer. Anal. **37**(5), 1455–1480 (2000). http://www.jstor.org/stable/2587331
10. Gunzburger, M.D., Lee, J.: A domain decomposition method for optimization problems for partial differential equations. Comput. Math. Appl. **40**(2–3), 177–192 (2000). http://www.sciencedirect.com/science/article/B6TYJ-40PR8JD-2/2/b0d35e1928aefcfec1d7550968683a0c
11. Koten, B.V., Luskin, M.: Analysis of energy-based blended quasi-continuum approximations. SIAM J. Numer. Anal. **49**(5), 2182–2209 (2011)
12. Li, X., Luskin, M., Ortner, C.: Positive-definiteness of the blended force-based quasicontinuum method. SIAM J. Multiscale Model. Simul. **10**, 1023–1045 (2012)
13. Li, X., Luskin, M., Ortner, C., Shapeev, A.: Theory-based benchmarking of the blended force-based quasicontinuum method. ArXiv e-prints (2013)
14. Luskin, M., Ortner, C.: Atomistic-to-continuum coupling. Acta Numerica **22**, 397–508 (2013). http://journals.cambridge.org/article_S0962492913000068
15. Luskin, M., Ortner, C., Koten, B.V.: Formulation and optimization of the energy-based blended quasicontinuum method. Comput. Methods Appl. Mech. Eng. **253**, 160–168 (2013). arXiv:1112.2377
16. Mura, T.: Micromechanics of Defects in Solids. Comparative Studies in Overseas History, Dordrecht (1987). http://books.google.com/books?id=N_JmtkfsdZgC
17. Olson, D., Bochev, P., Luskin, M., Shapeev, A.: An optimization-based atomistic-to-continuum coupling method (2013). arXiv:1304.4976
18. Ortner, C., Shapeev, A.V.: Analysis of an energy-based atomistic/continuum approximation of a vacancy in the 2D triangular lattice. Math. Comp. **82**, 2191–2236 (2013)
19. Tadmor, E., Ortiz, M., Phillips, R.: Quasicontinuum analysis of defects in solids. Philos. Mag. A **73**(6), 1529–1563 (1996)

Numerical Modeling of Fluids and Structures

Soliton Solutions as Inverse Problem for Coefficient Identification

Tchavdar T. Marinov[1] and Rossitza Marinova[2(✉)]

[1] Department of Natural Sciences, Southern University at New Orleans, 6801 Press Drive, New Orleans, LA 70126, USA
tmarinov@suno.edu

[2] Department of Mathematical and Computing Sciences, Concordia University College of Alberta, Edmonton, AB T5B 4E4, Canada
rossitza.marinova@concordia.ab.ca

Abstract. We construct an algorithm to investigate numerically non-symmetric solitary wave-like solutions of an ordinary nonlinear differential equation. We reformulate the bifurcation problem, introducing a new parameter; and in such a way we expel the trivial solution of the original problem. The Method of Variational Imbedding (MVI) is used for solving the inverse problem. We illustrate the approach by comparing the numerical solution with a known exact solution of the Boussinesq equation.

1 Introduction

A special numerical technique for identification of symmetric solitary wave solutions of Boussinesq and Korteweg–de Vries equations is proposed in [3]. This method requires the sought solution to be an even function; hence, it can be applied to problems that allow only symmetric solutions. However, not all equations admit only symmetric solutions, and methods for identification of general type solutions are of interest, since they allow researchers to study solitons of any type. The potential impact of results from the proposed work includes a wide class of applications in many disciplines. Insights gained from the application of a general numerical method for the identification of solitons to equations with non-analytic solitary wave solutions could lead to new ideas about describing tsunami waves, nerve signal propagation, and plasma.

2 Problem Formulation

Consider the soliton equation

$$\mathcal{L}(u) = 0, \tag{1}$$

where

$$\mathcal{L}(u) = \mathcal{L}^{\mathrm{KdV}}(u) = u_t + \gamma u_x + 2\alpha u u_x + u_{xxx}, \tag{2}$$

I. Lirkov et al. (Eds.): LSSC 2013, LNCS 8353, pp. 47–54, 2014.
DOI: 10.1007/978-3-662-43880-0_4,

when we deal with the KdV equation, and

$$\mathcal{L}(u) = \mathcal{L}^{\text{Bsq}}(u) = -u_{tt} + \gamma^2 u_{xx} + \alpha(u^2)_{xx} + u_{xxxx}, \tag{3}$$

when we deal with the classical Boussinesq equation. These equations possess special travelling wave solutions called solitary waves. Boussinesqs theory [1] was the first to give a satisfactory and scientific explanation of the phenomenon of solitary waves discovered by Scott Russell [5].

We consider the stationary waves in the moving frame $\xi = x - ct$. After integration (double for the operators (3)) with respect to ξ and taking into account the localized character of the investigated solutions, we obtain the following nonlinear ordinary differential equations

$$\mathcal{L}^{\text{KdV}}(u) = \lambda u + \alpha u^2 + u_{\xi\xi} = 0\,, \qquad \lambda \equiv \gamma - c\,, \tag{4}$$

$$\mathcal{L}^{\text{Bsq}}(u) = \lambda u + \alpha u^2 + u_{\xi\xi} = 0\,, \qquad \lambda \equiv \gamma^2 - c^2\,. \tag{5}$$

We are looking for solutions of the Eq. (5) with $u \to 0$ when $\xi \to \infty$. Then $u^2 \ll u$ in the tails and the linearized version of (5) coincides with its linear part.

2.1 Even and Odd Functions

Every function $u = u(\xi)$, where $\xi \in (-a, a)$, $a > 0$, can be decomposed into a sum of its even part and its odd part

$$u(\xi) = \varphi(\xi) + \psi(\xi), \tag{6}$$

where $\varphi(\xi)$ is an even function and $\psi(\xi)$ is an odd function defined as

$$\varphi(\xi) = \frac{u(\xi) + u(-\xi)}{2}, \qquad \psi(\xi) = \frac{u(\xi) - u(-\xi)}{2}. \tag{7}$$

The following basic facts that apply to even and odd functions are used in the solution process:

- The product of two even functions is even.
- The product of two odd functions is even.
- The product of an even function and an odd function is odd.
- If an even function is differentiable, then its derivative is an odd function.
- If an odd function is differentiable, then its derivative is an even function.

Or, if φ is an even function and ψ is an odd function, both differentiable as many times as necessary, then they satisfy the following conditions

$$\psi(0) = \varphi'(0) = \psi''(0) = \varphi'''(0) = 0. \tag{8}$$

In addition, if a function is zero on a symmetric interval, then its even and odd components must be zero.

2.2 The Inverse Problem

Substituting u, defined with Eq. (6), into Eq. (5), and using the above properties of the even and odd functions, we obtain the following system of two equations for the functions φ and ψ:

$$\lambda\varphi + \alpha\varphi^2 + \alpha\psi^2 + \varphi'' = 0, \tag{9}$$

$$\lambda\psi + 2\alpha\varphi\psi + \psi'' = 0. \tag{10}$$

Because of the symmetry of the two functions φ and ψ, we consider the problem on the half line by adding the following boundary conditions at the point $\xi = 0$:

$$\varphi(0) = \chi, \quad \psi(0) = 0, \quad \varphi'(0) = 0, \quad \psi'(0) = \mu, \tag{11}$$

where χ and μ are unknown constants.

It is convenient to scale the functions φ and ψ by introducing two unknown functions f and g and two unknown constants χ and μ as

$$\varphi(\xi) = \chi f(\xi), \quad \psi(\xi) = \mu g(\xi). \tag{12}$$

Evidently, f is an even function and g is an odd function.

The boundary conditions for the introduced functions f and g are

$$f(0) = 1, \quad g(0) = 0, \quad f'(0) = 0, \quad g'(0) = 1, \tag{13}$$

and

$$\lim_{\xi\to\infty} f(\xi) = \lim_{\xi\to\infty} g(\xi) = 0. \tag{14}$$

Therefore,

$$\lim_{\xi\to\infty} f^{(n)}(\xi) = \lim_{\xi\to\infty} g^{(n)}(\xi) = 0. \tag{15}$$

The system for the functions f and g reads

$$\lambda\chi f + \alpha\chi^2 f^2 + \alpha\mu^2 g^2 + \chi f'' = 0, \tag{16}$$

$$\lambda\mu g + 2\alpha\chi\mu f g + \mu g'' = 0. \tag{17}$$

We divide the Eq. (17) by the unknown constant μ under the condition $\mu \neq 0$. This is a very strong restriction, because $\mu \neq 0$ means that we do not consider even functions $u(\xi)$ as possible solutions of the Eq. (1).

After dividing by μ, the Eq. (17) adopts the form

$$\lambda g + 2\alpha\chi f g + g'' = 0. \tag{18}$$

Thus, we exclude the trivial solution $u(\xi) \equiv 0$. However, we arrive at a problem for coefficient identification from overposed boundary data for solving the system of two second order differential equations (16), (18) under the six boundary condition (13), (14). At the same time, it is necessary to identify the two unknown constants χ and μ.

3 Method for Solving the Inverse Problem

Consider the problem for minimization of the functional

$$I(f, g, \chi, \mu) = \int_0^\infty (A^2 + B^2) d\xi \to \min, \tag{19}$$

where

$$A(\xi) := \lambda\chi f + \alpha\chi^2 f^2 + \alpha\mu^2 g^2 + \chi f'', \tag{20}$$

$$B(\xi) := \lambda g + 2\alpha\chi f g + g''. \tag{21}$$

3.1 Euler-Lagrange Equations for *f* and *g*

The Euler-Lagrange equation for f reads

$$\left(\lambda\chi + 2\alpha\chi^2 f\right) A + 2\alpha\chi g B + \chi A'' = 0, \tag{22}$$

while the Euler-Lagrange equation for g is

$$2\alpha\mu^2 g A + (\lambda + 2\alpha\chi f]\, B + B'' = 0. \tag{23}$$

The system (22), (23) is of the fourth order with respect to each of the functions f and g. There are three boundary conditions for each function from the original problem. To close the system, we also use the following boundary conditions

$$\lim_{\xi\to\infty} f'(\xi) = \lim_{\xi\to\infty} g'(\xi) = 0. \tag{24}$$

3.2 Euler-Lagrange Equations for χ and μ

Since χ and μ are constants, we integrate the functional (19) with respect to ξ. Thus, the problem is to minimize the function

$$\Phi(\chi, \mu) = a_{40}\chi^4 + a_{22}\chi^2\mu^2 + a_{04}\mu^4 + a_{30}\chi^3 + a_{12}\chi\mu^2 + a_{20}\chi^2 + a_{02}\mu^2, \tag{25}$$

with respect to χ and μ, where

$$a_{40} = \int_0^\infty \alpha^2 f^4 dx, \tag{26}$$

$$a_{30} = \int_0^\infty 2\alpha f^2(\lambda f + f'') dx, \tag{27}$$

$$a_{22} = \int_0^\infty 6\alpha^2 f^2 g^2 dx, \tag{28}$$

$$a_{20} = -\int_0^\infty (\lambda f + f'')^2 dx, \tag{29}$$

$$a_{12} = 2\alpha \int_0^\infty g\left[g(3\lambda f + f'') + 2fg''\right] dx, \tag{30}$$

$$a_{04} = -\int_0^\infty \alpha^2 g^4 dx, \tag{31}$$

$$a_{02} = -\int_0^\infty (\lambda g + g'')^2\, dx. \tag{32}$$

The necessary conditions for the minimization of $\Phi(\chi, \mu)$ with respect to χ and μ are

$$\Phi_\chi = 4a_{40}\chi^3 + 3a_{30}\chi^2 + 2a_{22}\chi\mu^2 + 2a_{20}\chi + a_{12}\mu^2 = 0, \tag{33}$$
$$\Phi_\mu = 4a_{04}\mu^3 + 2a_{22}\chi^2\mu + 2a_{12}\chi\mu + 2a_{02}\mu = 0. \tag{34}$$

4 Numerical Scheme

To solve the boundary value problem (13), (14), (22), (23), and (24), we use a finite difference scheme with the second order of approximation of the differential operators and integrals. We use Newton's method to solve the system (33), (34).

4.1 Finite Difference Scheme

The mesh (see Fig. 1) is regular and allows the approximation of all operators using standard central differences.

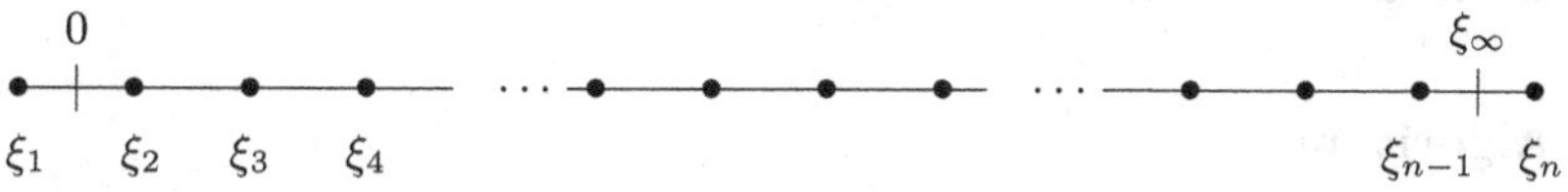

Fig. 1. The mesh

The grid spacing is

$$h \equiv \frac{\xi_\infty}{n-2},$$

where n is the total number of grid points and ξ_∞ is a sufficiently large number, called *'numerical infinity.'* Then, the grid points are defined as follows:

$$\xi_i = (i - 1.5)h \text{ for } i = 1, \ldots, n\,. \tag{35}$$

It is important that the point $\xi = 0$ is the mid-point $\xi_{1\frac{1}{2}}$.

Let us introduce the notation

$$y_i = y(\xi_i), \qquad \text{for} \qquad i = 1, \ldots, n\,. \tag{36}$$

We employ symmetric central differences for approximating the differential operators as follows:

$$y^{(4)}(\xi_i) = \frac{1}{h^4}\left(y_{i-2} - 4y_{i-1} + 6y_i - 4y_{i+1} + y_{i+2}\right) + O(h^2) \tag{37}$$

for $i = 3, \ldots, n-2$, and

$$y''(\xi_i) = \frac{1}{h^2}\left(y_{i-1} - 2y_i + y_{i+1}\right) + O(h^2) \tag{38}$$

for $i = 2, \ldots, n-1$.

The grid also allows the second order approximations of the boundary conditions:

$$y(0) = 1 \Longrightarrow y_1 + y_2 = 2, \tag{39}$$

$$y'(0) = 0 \Longrightarrow y_2 - y_1 = 0, \tag{40}$$

$$y(\xi) \to 0, \quad \xi \to \infty \Longrightarrow y_n + y_{n-1} = 0, \tag{41}$$

$$y'(\xi) \to 0, \quad \xi \to \infty \Longrightarrow y_n - y_{n-1} = 0. \tag{42}$$

4.2 Estimation of χ and μ

We approximate the integrals in Eqs. (33) and (34) for evaluating χ and μ using the so called 'extended midpoint rule,' where the error term is again of the second order. After evaluating the coefficients in the Eqs. (33) and (34), we use Newton's method to solve the system.

4.3 Algorithm

(I) Solve the fourth-order boundary value problem (13), (14), (22), (23), and (24), for the functions f and g with given χ and μ.

(II) With the newly computed f and g, the coefficients χ and μ are evaluated from (33) and (34), respectively. If

$$\left|\chi^{new} - \chi^{old}\right| < \varepsilon, \qquad \left|\mu^{new} - \mu^{old}\right| < \varepsilon, \tag{43}$$

then the calculations are terminated. Otherwise, the index of iterations is stepped up $k := k + 1$ and the algorithm is returned to step **(I)**.

5 Numerical Experiments

Consider the case when $\lambda = -4$ and $\alpha = 6$. Then, the Eqs. (2) and (3) become the following equation

$$\mathcal{L}(u) = u''(x) + 6u^2(x) - 4u(x) = 0. \tag{44}$$

Table 1. Obtained values of $||u^{num} - u^{an}||$, and the rate of convergence for four different values of the mesh spacing.

h	$\lvert\lvert u^{num} - u^{an}\rvert\rvert$	Rate of convergence
1/10	0.0015391100	—
1/20	0.0003865510	1.99337
1/40	0.0001010250	1.93595
1/80	0.0000260587	1.95488

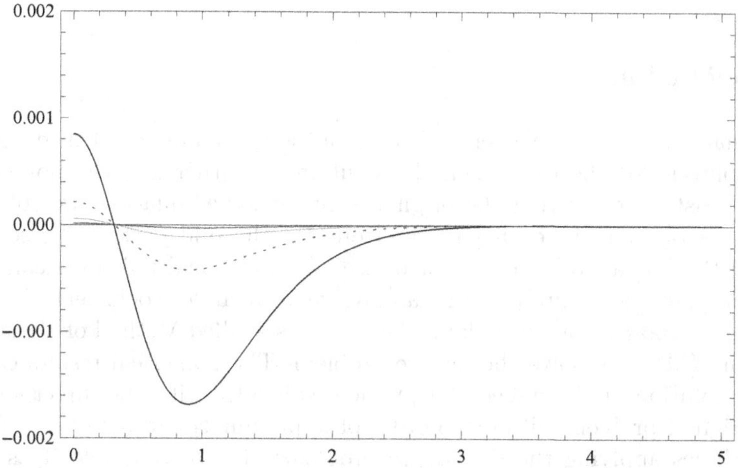

Fig. 2. The shape of the numerical error for four different steps $h = 0.1$, 0.05, 0.025, and 0.0125.

This problem has a non-even analytic solution

$$u(x) = \operatorname{sech}^2(x-1), \tag{45}$$

which can be used for validating the numerical scheme.

For the numerical results described here we use $\xi_\infty = 10$ and the tolerance $\varepsilon = 10^{-8}$.

The discretization error is $O(h^2)$. The numerical results clearly demonstrate these error orders. The values of the l_2 norm of the difference $||u^{num} - u^{an}||$, computed using four different steps are given in Table 1. The rate of convergence, calculated as

$$\text{rate}(u) = \log_2 \frac{||y_{2h}^{num} - y^{an}||}{||u_h^{num} - u^{an}||} \tag{46}$$

is also shown in Table 1.

The distribution of the numerical error is given in Fig. 2. It is seen that the discretization error term is $O(h^2)$, and the total error is $O(h^2)$. The numerical results for the coefficients χ and μ presented in Table 2 demonstrate clearly these error orders.

Table 2. Obtained values of the coefficients χ and μ, and the rate of convergence for four different values of the mesh spacing.

h	χ	Rate	μ	Rate
exact	0.4199743416	—	0.6397000084	—
1/10	0.418352	—	0.637837	—
1/20	0.419562	1.97617	0.639233	1.99611
1/40	0.419866	1.92825	0.639584	2.00922
1/80	0.419946	1.93459	0.639674	2.15718

6 Conclusion

To summarize, we propose a generalization of the procedure for identifying symmetric solitons for the case when the solution is neither an even nor an odd function. Instead of solving the original solitary wave equation, we solve two equations – one for the even part, and one for the odd part of the solution. To avoid the trivial solution, we introduce two additional unknown constants, converting the original problem into an inverse problem for coefficient identification from overposed boundary data. We use the so-called Method of Variational Imbedding (MVI) to solve the inverse problem. The numerical results confirm that the solution of the imbedding problem coincides with the direct solution of the original problem within the order of approximation error $O(h^2)$. Future work includes applying the method to problems of higher complexity such as higher-order and/or two-dimensional equations such as in [2,4].

References

1. Boussinesq, J.: Theorie de l'intumescence liquide appelee onde solitaire ou de translation se propageant dans un canal rectangulare. C. R. Acad. Sci. **72**, 755–759 (1871)
2. Camassa, R., Holm, D.: An integrable shallow water equation with peaked solitons. Phys. Rev. Lett **71**(11), 1661–1664 (1993)
3. Marinov, T.T., Christov, C.I., Marinova, R.S.: Novel numerical approach to solitary-wave solutions identification of Boussinesq and Korteweg-de Vries equations. Int. J. Bifurcat. Chaos **15**(2), 557–565 (2005)
4. Lenells, J.: Traveling wave solutions of the Camassa-Holm equation. J. Differ. Equ. **217**, 393–430 (2005)
5. Russell, J.S.: Report on waves. 14th Meeting of the British Association Report, York (1844)

Control and Uncertain Systems

Improved Error Estimate for an Implicit Discretization Scheme for Linear-Quadratic Control Problems with Bang-Bang Solutions

Walter Alt(✉) and Martin Seydenschwanz

Friedrich-Schiller-Universität Jena, Mathematische Optimierung,
07743 Jena, Germany
{walter.alt,martin.seydenschwanz}@uni-jena.de

Abstract. We analyze an implicit discretization scheme for a class of linear-quadratic optimal control problems without mixed state-control terms. Under the assumption that the optimal control has bang-bang structure we show convergence of the discrete approximation and improve existing error estimates to order $\mathcal{O}(h)$.

1 Introduction

We consider the following linear-quadratic control problem:

$$
\begin{array}{lll}
\text{(OQ)} & \min f(x,u) & \\
 & \text{s.t.} & \\
 & \dot{x}(t) = A(t)x(t) + B(t)u(t) & \text{a.e. on } [0,\mathrm{T}], \\
 & x(0) = a & \\
 & u(t) \in U & \text{a.e. on } [0,\mathrm{T}],
\end{array}
$$

where f is a linear-quadratic cost functional defined by

$$f(x,u) = \left(\tfrac{1}{2}Qx(T) + q\right)^{\mathsf{T}} x(T) + \int_0^T \left(\tfrac{1}{2}W(t)x(t) + w(t)\right)^{\mathsf{T}} x(t) + r(t)^{\mathsf{T}} u(t)\,\mathrm{d}t$$

Here, $u(t) \in \mathbb{R}^m$ is the control, and $x(t) \in \mathbb{R}^n$ is the state of the system at time t. Further $Q \in \mathbb{R}^{n\times n}$, $q \in \mathbb{R}^n$ and the functions $W\colon [0,T] \to \mathbb{R}^{n\times n}$, $w\colon [0,T] \to \mathbb{R}^n$, $r\colon [0,T] \to \mathbb{R}^m$, $A\colon [0,T] \to \mathbb{R}^{n\times n}$, $B\colon [0,T] \to R^{n\times m}$ are Lipschitz continuous. With $b_l, b_u \in \mathbb{R}^m$, $b_l < b_u$ the set $U \subset \mathbb{R}^m$ is defined by lower and upper bounds

$$U = \{u \in \mathbb{R}^m \mid b_l \le u \le b_u\}$$

Moreover we define $\mathcal{U} = \left\{u \in W^1_\infty(0,T;\mathbb{R}^n) \mid u(t) \in U \,\forall' t \in [0,T]\right\}$. The feasible set of (OQ) is

$$\mathcal{F} = \{(x,u) \in X \mid u \in \mathcal{U},\ \dot{x}(t) = A(t)x(t) + B(t)u(t) \text{ a.e. on } [0,T], \quad x(0) = a\}.$$

In [5] we investigated an implicit discretization scheme for problem (OQ) with mixed state-control terms in the cost functional. We showed that this method is

I. Lirkov et al. (Eds.): LSSC 2013, LNCS 8353, pp. 57–65, 2014.
DOI: 10.1007/978-3-662-43880-0_5,

a practically usable, numerically robust alternative to the Euler-discretization, which has severe difficulties in solving stiff problems. We presented error estimates of order $\mathcal{O}(\sqrt{h})$ for the control, state and adjoint state. Here we show that these error estimates can be improved for problems without mixed terms.

The organization of the paper is as follows. In Sect. 2 we recall the implicit discretization scheme and the most important results from [5]. In Sect. 3 we prove our main result.

We use the notations defined in [5]. In view of convexity of Problem (OQ) we assume that the matrices Q and $W(t)$, $t \in [0,T]$, are symmetric and uniformly positive semidefinite. Therefore the cost functional is convex and continuous on $\mathcal{F}$. Moreover the feasible set $\mathcal{F}$ is nonempty, closed, convex and bounded and this is why a minimizer $(x^*, u^*) \in W_2^1(0,T;\mathbb{R}^n) \times L^2(0,T;\mathbb{R}^m)$ of (OQ) exists (see Ekeland/Temam [6], Chap. II, Proposition 1.2). Since $\mathcal{U}$ is bounded we have $(x^*, u^*) \in X = X_1 \times X_2 := W_\infty^1(0,T;\mathbb{R}^n) \times L^\infty(0,T;\mathbb{R}^m)$. Let $(x^*, u^*) \in \mathcal{F}$ be a minimizer of (OQ). Then there exists a function $\lambda \in W_\infty^1(0,T;\mathbb{R}^n)$ such that the adjoint equation

$$-\dot{\lambda}(t) = A(t)^\mathsf{T}\lambda(t) + W(t)x^*(t) + w(t) \text{ a.e. on } [0,T],$$

with end condition

$$\lambda(T) = Qx^*(T) + q\,,$$

and the minimum principle

$$\left[r(t) + B(t)^\mathsf{T}\lambda(t)\right]^\mathsf{T}(u - u^*(t)) \geq 0 \quad \forall u \in U\,,$$

hold a.e. on $[0,T]$. Denoting the *switching function* by

$$\sigma(t) := r(t) + B(t)^\mathsf{T}\lambda(t), \tag{1}$$

it is well-known that for each $i \in \{1,\ldots,m\}$, (1) implies

$$u_i^*(t) = \begin{cases} b_{l,i}, & \text{if } \sigma_i(t) > 0,\\ b_{u,i}, & \text{if } \sigma_i(t) < 0,\\ \text{undetermined}, & \text{if } \sigma_i(t) = 0. \end{cases}$$

2 The Implicit Discretization Scheme

Given a natural number N, let $h_N = T/N$ be the mesh size. We approximate the space X_2 of controls by functions in the subspace $X_{2,N} \subset X_2$ of piecewise constant functions and the state and adjoint state variables by functions in the subspace $X_{1,N} \subset X_1$ of continuous, piecewise linear functions. These approximations are uniquely determined by the function values at the grid points. Moreover we define $\mathcal{U}_N = \mathcal{U} \cap X_{2,N}$ and $X_N = X_{1,N} \times \mathcal{U}_N$. Using a collocation method to approximate the system equation we get the following discretization of (OQ):

$$\text{(OQ)}_\text{N} \quad \min_{(x,u)\in X_{1,N}\times X_{2,N}} \quad f_N(x,u)$$

s.t.

$$\left(I - \tfrac{h_N}{2} A(t_{j+\frac{1}{2}})\right) x_{j+1} = \left(I + \tfrac{h_N}{2} A(t_{j+\frac{1}{2}})\right) x_j + h_N B(t_{j+\frac{1}{2}}) u_j,$$

$$x_0 = a, u_j \in U, \qquad j = 0, 1, \ldots, N-1,$$

where f_N is the linear-quadratic cost functional defined by

$$f_N(x,u) = \frac{1}{2} x_N^\mathsf{T} Q x_N + q^\mathsf{T} x_N + h_N \sum_{j=0}^{N-1} \frac{1}{2} x_{j+\frac{1}{2}}^\mathsf{T} W(t_{j+\frac{1}{2}}) x_{j+\frac{1}{2}} + h_N \sum_{j=0}^{N-1} \left[w(t_{j+\frac{1}{2}})^\mathsf{T} x_{j+\frac{1}{2}} + r(t_{j+\frac{1}{2}})^\mathsf{T} u_j \right].$$

Hereby $u_j = u(t_j)$ and $x_{j+\frac{1}{2}} = \frac{x_{j+1}+x_j}{2} = \frac{x(t_{j+1})+x(t_j)}{2}$. Compactness of U implies that Problem $(\text{OQ})_N$ has a solution (x_h^*, u_h^*), and, for any solution, there exists a continuous, piecewise linear multiplier $\lambda_h \in X_{1,N}$ such that the discrete adjoint equation

$$-\dot{\lambda}_h(t_{j+\frac{1}{2}}) = A(t_{j+\frac{1}{2}})^\mathsf{T} \lambda_h(t_{j+\frac{1}{2}}) + W(t_{j+\frac{1}{2}}) x_{h,j+\frac{1}{2}}^* + w(t_{j+\frac{1}{2}}) + \phi_{\lambda_h,j},$$

with end condition

$$\lambda_{h,N} = Q x_{h,N}^* + q + h_N \phi_{\lambda_h,N},$$

and the discrete minimum principle

$$\left[B(t_{j+\frac{1}{2}})^\mathsf{T} \lambda_{h,j+1} + r(t_{j+\frac{1}{2}}) \right]^\mathsf{T} (u - u_{h,j}^*) \geq 0 \quad \forall u \in U$$

for $j = 0, \ldots, N-1$, are satisfied (compare [12,13]). For further details concerning the perturbations $\phi_{\lambda_h,j}$ see Remark 3 in [5] or [12].

By $\sigma_h \colon [0, t_{N-1}] \to \mathbb{R}^m$ we denote the discrete switching function, which is the continuous and piecewise linear function defined by the values

$$\sigma_h(t_j) := B(t_{j+\frac{1}{2}})^\mathsf{T} \lambda_{h,j+1} + r(t_{j+\frac{1}{2}}), \quad j = 0, \ldots, N-1. \tag{2}$$

From (2) we obtain for $i = 1, \ldots, m$, $j = 0, \ldots, N-1$,

$$u_{h,i}^*(t_j) = \begin{cases} b_{l,i}, & \text{if } \sigma_{h,i}(t_j) > 0, \\ b_{u,i}, & \text{if } \sigma_{h,i}(t_j) < 0, \\ \text{undetermined}, & \text{if } \sigma_{h,i}(t_j) = 0. \end{cases}$$

In order to analyze the implicit scheme, in [5] we made the following assumptions:

(A1) There exists a solution $(x^*, u^*) \in \mathcal{F}$ of (OQ) such that the set Σ of zeros of the components σ_i, $i = 1, \ldots, m$, of the switching function σ defined by (1) is finite and $0, T \notin \Sigma$, i.e. $\Sigma = \{s_1, \ldots, s_l\}$ with $0 < s_1 < \ldots < s_l < T$.

(A2) There exist $\bar{\sigma} > 0$, $\bar{\tau} > 0$ such that $|\sigma_i(\tau)| \geq \bar{\sigma}|\tau - s_j|$ for all $j \in \{1, \ldots, l\}$, $i \in \mathcal{I}(s_j) := \{1 \leq i \leq m : \sigma_i(s_j) = 0\}$ and all $\tau \in [s_j - \bar{\tau}, s_j + \bar{\tau}]$, and moreover it holds $\sigma_i(s_j - \bar{\tau})\sigma_i(s_j + \bar{\tau}) < 0$, i.e. σ_i changes sign in s_j.

Under these assumption, that guarantee the strict bang-bang type of the solution of (OQ), we have proven the following theorem.

Theorem 1. *Let $(x^*, u^*) \in X$ be a solution of* (OQ)*, which satisfies Assumptions* (A1) *and* (A2)*. Then for sufficiently large N and any solution $(x_h^*, u_h^*) \in X_N$ of $(OQ)_N$ we have the error estimates*

$$\|u_h^* - u^*\|_1 \leq c_u h_N^{\frac{1}{2}}, \quad \|x_h^* - x^*\|_\infty \leq c_x h_N^{\frac{1}{2}}$$

with constants c_u and c_x independent of N. For the associated multipliers we have

$$\|\lambda_h - \lambda\|_\infty \leq c_\lambda h_N^{\frac{1}{2}}$$

with some constant c_λ independent of N. ◊

For further details please take a look at [5] and the references specified therein.

3 Structural Stability and Improved Error Estimates

If there is no mixed term in the cost functional, we can improve the error estimates by applying the techniques from [2]. Therefore we define $\hat{u}_h \in X_{2,N}$ due to $\hat{u}_h(t_j) = u^*(t_j)$ for $j = 0, \ldots, N-1$. Now we derive upper and lower bounds for the term

$$J_N = h_N \sum_{j=0}^{N-1} \sigma_h(t_j)^\mathsf{T}(\hat{u}_h(t_j) - u_h^*(t_j))$$

and can deduce convergence of order 1. First we prove a result concerning the derivative of the switching functions analogous to Theorem 5.1. in [2], resp. Theorem 2.8 in [4]. To this end we replace Assumption (A2) by the following slightly stronger assumption:

(A3) The functions r and B are differentiable, $\dot{r}$ and $\dot{B}$ are Lipschitz continuous, and there exists $\bar{\sigma} > 0$ such that

$$\min_{1 \leq j \leq l} \min_{i \in \mathcal{I}(s_j)} \{|\dot{\sigma}_i(s_j)|\} \geq 2\bar{\sigma}.$$

Theorem 2. *Let Assumptions* (A1) *and* (A3) *be satisfied. Let σ be defined by* (1) *and σ_h by* (2)*. Then, for sufficiently large N,*

$$|\dot{\sigma}_h(t) - \dot{\sigma}(t)|_\infty \leq \tilde{c}_\sigma\, h_N^{\frac{1}{2}} \quad \forall t \in [0, t_{N-1}]$$

with a constant $\tilde{c}_\sigma$ independent of N. ◊

Proof. Let $j \in \{0, \ldots, N-2\}$ and $t \in [t_j, t_{j+1})$ be arbitrary. Then we have

$$\dot{\sigma}(t) = \dot{B}(t)^\mathsf{T}\lambda(t) + B(t)^\mathsf{T}\dot{\lambda}(t) + \dot{r}(t)$$

and

$$\dot{\sigma}_h(t) = \frac{B(t_{j+1+\frac{1}{2}})^\mathsf{T}\lambda_{h,j+2} + r(t_{j+1+\frac{1}{2}}) - B(t_{j+\frac{1}{2}})^\mathsf{T}\lambda_{h,j+1} - r(t_{j+\frac{1}{2}})}{h_N}$$

From this we can conclude

$$\begin{aligned}
\dot{\sigma}_h(t) - \dot{\sigma}(t) &= \frac{B(t_{j+1+\frac{1}{2}})^\mathsf{T}\lambda_{h,j+2} + r(t_{j+1+\frac{1}{2}}) - B(t_{j+\frac{1}{2}})^\mathsf{T}\lambda_{h,j+1} - r(t_{j+\frac{1}{2}})}{h_N} \\
&\quad - \dot{B}(t)^\mathsf{T}\lambda(t) - B(t)^\mathsf{T}\dot{\lambda}(t) - \dot{r}(t) \\
&= \frac{r(t_{j+1+\frac{1}{2}}) - r(t_{j+\frac{1}{2}})}{h_N} - \dot{r}(t) \\
&\quad + \left[\frac{B(t_{j+1+\frac{1}{2}})^\mathsf{T} - B(t_{j+\frac{1}{2}})^\mathsf{T}}{h_N} - \dot{B}(t_{j+1})\right]^\mathsf{T} \lambda_{h,j+2} \\
&\quad + \dot{B}(t_{j+1})^\mathsf{T}(\lambda_{h,j+2} - \lambda_h(t)) + \left[\dot{B}(t_{j+1}) - \dot{B}(t)\right]^\mathsf{T} \lambda_h(t) \\
&\quad + \dot{B}(t)^\mathsf{T}(\lambda_h(t) - \lambda(t)) \\
&\quad + B(t_{j+\frac{1}{2}})^\mathsf{T}\frac{\lambda_{h,j+2} - \lambda_{h,j+1}}{h_N} - B(t)^\mathsf{T}\dot{\lambda}(t).
\end{aligned}$$

Plugging in the continuous and discrete adjoint equation leads to

$$\begin{aligned}
\dot{\sigma}_h(t) - \dot{\sigma}(t) &= \frac{r(t_{j+1+\frac{1}{2}}) - r(t_{j+\frac{1}{2}})}{h_N} - \dot{r}(t) + \dot{B}(t)^\mathsf{T}(\lambda_h(t) - \lambda(t)) \\
&\quad + \left[\frac{B(t_{j+1+\frac{1}{2}})^\mathsf{T} - B(t_{j+\frac{1}{2}})^\mathsf{T}}{h_N} - \dot{B}(t_{j+1})\right]^\mathsf{T} \lambda_{h,j+2} \\
&\quad + \dot{B}(t_{j+1})^\mathsf{T}(\lambda_{h,j+2} - \lambda_h(t)) + \left[\dot{B}(t_{j+1}) - \dot{B}(t)\right]^\mathsf{T} \lambda_h(t) \\
&\quad - B(t_{j+\frac{1}{2}})^\mathsf{T}\left[A(t_{j+1+\frac{1}{2}})^\mathsf{T}\lambda_h(t_{j+1+\frac{1}{2}}) + W(t_{j+1+\frac{1}{2}})x_h(t_{j+1+\frac{1}{2}})\right] \\
&\quad - B(t_{j+\frac{1}{2}})^\mathsf{T}\left[w(t_{j+1+\frac{1}{2}}) + \phi_{\lambda_h,j+1}\right] \\
&\quad + B(t)^\mathsf{T}\left[A(t)^\mathsf{T}\lambda(t) + W(t)z(t) + w(t)\right].
\end{aligned}$$

Now we can apply Theorem 1 and use the Lipschitz continuity of the functions $\dot{B}$, $\dot{r}$, A, W, w, x, x_h, λ and λ_h. Moreover without mixed state-control term, it holds $|\phi_{\lambda_h,j}| \le c_\phi h_N$ for $j = 1, \ldots, N-1$, with a constant c_ϕ independent of N (see Remark 3 in [5]). Therefore the assertion follows after some technical transformations. ◊

Theorem 2 guarantees structural stability, i.e. the discrete switching function has the same structure as the original switching function (compare Theorem 5.2 in [2]). In order to get a lower bound for J_N we can therefore follow the argumentation in [2]. After some technical derivations and transformations we obtain

$$\begin{aligned} J_N &\geq \frac{\bar{\sigma}}{8\gamma}\|\hat{u}_h - u_h^*\|_1 \left(\|\hat{u}_h - u_h^*\|_1 - \frac{1}{2}\|\hat{u}_h - u_h^*\|_1 - \gamma h_N \right) \\ &= \frac{\bar{\sigma}}{16\gamma}\|\hat{u}_h - u_h^*\|_1 \left(\|\hat{u}_h - u_h^*\|_1 - 2\gamma h_N \right) , \end{aligned}$$

where $\bar{\sigma}$, γ are constants independent of N. Now we have to consider two cases. If

$$\|\hat{u}_h - u_h^*\|_1 \leq 4\gamma h_N \,, \tag{3}$$

we get a discrete error estimate of order 1. Otherwise, we have

$$2\gamma h_N < \frac{1}{2}\|\hat{u}_h - u_h^*\|_1$$

and therefore

$$J_N \geq \frac{\bar{\sigma}}{32\gamma}\|\hat{u}_h - u_h^*\|_1^2. \tag{4}$$

We can now adapt known proof techniques (see e.g. [1,2,5,9,10]) to derive an upper bound for J_N. By Assumption (A1) the optimal control u^* is piecewise continuous. Therefore the minimum principle (1) holds for all $t \in [0,T]$ (see e.g. [8]). With $t = t_j$ and $u = u_h^*(t_j)$ we obtain

$$\begin{aligned} J_N &= h_N \sum_{j=0}^{N-1} \sigma_h(t_j)^\mathsf{T}(\hat{u}(t_j) - u_h^*(t_j)) \\ &\leq h_N \sum_{j=0}^{N-1} \sigma_h(t_j)^\mathsf{T}(\hat{u}(t_j) - u_h^*(t_j)) + h_N \sum_{j=0}^{N-1} \sigma(t_j)^\mathsf{T}\left(u_h^*(t_j) - \hat{u}(t_j)\right) \\ &= h_N \sum_{j=0}^{N-1} \left(\sigma_h(t_j) - \sigma(t_j)\right)^\mathsf{T}\left(\hat{u}(t_j) - u_h^*(t_j)\right) \\ &= h_N \sum_{j=0}^{N-1} \left(B(t_{j+\frac{1}{2}})^\mathsf{T}\lambda_h^*(t_{j+1}) - B(t_j)^\mathsf{T}\lambda^*(t_j) \right)^\mathsf{T} \left(\hat{u}(t_j) - u_h^*(t_j)\right) \\ &\quad + h_N \sum_{j=0}^{N-1} \left(r(t_{j+\frac{1}{2}}) - r(t_j) \right)^\mathsf{T} \left(\hat{u}(t_j) - u_h^*(t_j)\right). \end{aligned}$$

Some simple transformations, estimates using the Lipschitz continuity of the functions B, r, λ_h^* and $\hat{\lambda}$ and plugging in the discrete system equation lead to

$$\begin{aligned}
J_N \le\ & \frac{h_N}{2}\left(L_r + \|B^\mathsf{T}\|_\infty L_{\lambda_h^*} + L_{B^\mathsf{T}\hat\lambda} + 2c_2\|B^\mathsf{T}\|_\infty\right)\|\hat u - u_h^*\|_1 \\
& + h_N \sum\nolimits_{j=0}^{N-1}\left(B(t_{i+\frac12})^\mathsf{T}\lambda_h^*(t_{i+\frac12}) - B(t_{i+\frac12})^\mathsf{T}\hat\lambda(t_{i+\frac12})\right)^\mathsf{T}(\hat u(t_i) - u_h^*(t_i)) \\
=\ & \frac{h_N}{2}\left(L_r + \|B^\mathsf{T}\|_\infty L_{\lambda_h^*} + L_{B^\mathsf{T}\hat\lambda} + 2c_2\|B^\mathsf{T}\|_\infty\right)\|\hat u - u_h^*\|_1 \\
& - h_N \sum\nolimits_{j=0}^{N-1}\left(\lambda_h^*(t_{j+\frac12}) - \hat\lambda(t_{j+\frac12})\right)^\mathsf{T}\left(A(t_{j+\frac12})(\hat x(t_{j+\frac12}) - x_h^*(t_{j+\frac12}))\right) \\
& + h_N \sum\nolimits_{j=0}^{N-1}\left(\lambda_h^*(t_{j+\frac12}) - \hat\lambda(t_{j+\frac12})\right)^\mathsf{T}\left(\dot{\hat x}(t_{j+\frac12}) - \dot x_h^*(t_{j+\frac12})\right).
\end{aligned}$$

With the help of the discrete adjoint equation we can write

$$\begin{aligned}
J_N \le\ & \frac{h_N}{2}\left(L_r + \|B^\mathsf{T}\|_\infty L_{\lambda_h^*} + L_{B^\mathsf{T}\hat\lambda} + 2c_2\|B^\mathsf{T}\|_\infty\right)\|\hat u - u_h^*\|_1 + c_\phi h_N\|\hat x - x_h^*\|_\infty \\
& - h_N \sum\nolimits_{j=0}^{N-1}\left(\hat x(t_{j+\frac12}) - x_h^*(t_{j+\frac12})\right)^\mathsf{T} W(t_{j+\frac12})\left(\hat x(t_{j+\frac12}) - x_h^*(t_{j+\frac12})\right) \\
& - h_N \sum\nolimits_{j=0}^{N-1}\left(\dot{\hat\lambda}(t_{j+\frac12}) - \dot\lambda_h^*(t_{j+\frac12})\right)^\mathsf{T}\left(\hat x(t_{j+\frac12}) - x_h^*(t_{j+\frac12})\right) \\
& - h_N \sum\nolimits_{j=0}^{N-1}\left(\hat\lambda(t_{j+\frac12}) - \lambda_h^*(t_{j+\frac12})\right)^\mathsf{T}\left(\dot{\hat x}(t_{j+\frac12}) - \dot x_h^*(t_{j+\frac12})\right).
\end{aligned}$$

Now we can use the uniformly positive semidefiniteness of W and some telescoping sums to get

$$\begin{aligned}
J_N \le\ & \frac{h_N}{2}\left(L_r + \|B^\mathsf{T}\|_\infty L_{\lambda_h^*} + L_{B^\mathsf{T}\hat\lambda} + 2c_2\|B^\mathsf{T}\|_\infty\right)\|\hat u - u_h^*\|_1 + c_\phi h_N\|\hat x - x_h^*\|_\infty \\
& - \sum\nolimits_{j=0}^{N-1}\left(\hat\lambda(t_{j+1}) - \lambda_h^*(t_j)\right)^\mathsf{T}(\hat x(t_{j+1}) - x_h^*(t_{j+1})) \\
& - \sum\nolimits_{j=0}^{N-1}\left(\hat\lambda(t_j) - \lambda_h^*(t_j)\right)^\mathsf{T}(x_h^*(t_j) - \hat x(t_j)) \\
=\ & \frac{h_N}{2}\left(L_r + \|B^\mathsf{T}\|_\infty L_{\lambda_h^*} + L_{B^\mathsf{T}\hat\lambda} + 2c_2\|B^\mathsf{T}\|_\infty\right)\|\hat u - u_h^*\|_1 + c_\phi h_N\|\hat x - x_h^*\|_\infty \\
& - (\hat\lambda(T) - \lambda_h^*(T))^\mathsf{T}(\hat x(T) - x_h^*(T)) - (\hat\lambda(0) - \lambda_h^*(0))^\mathsf{T}(x_h^*(0) - \hat x(0)).
\end{aligned}$$

Plugging in the end condition of the discrete adjoint equation and applying some technical transformations and estimates we obtain

$$J_N \le c_u h_N\|\hat u - u_h^*\|_1 + c_x h_N\|\hat x - x_h^*\|_\infty.$$

With Lemma 3.4 in [5] it finally follows

$$J_N \le \tilde c h_N\|\hat u - u_h^*\|_1 \tag{5}$$

with some constant $\tilde c$ independent of N.

We can now state a first order error estimate for the discrete solutions of the implicit discretization and thereby improve the results of Theorem 1 under the slightly stronger Assumption (A3) (compare Theorem 5.3 in [2]).

Theorem 3. *Let* (x^*, u^*) *be a solution of Problem* (OQ) *for which Assumptions* (A1) *and* (A3) *are satisfied. Then, for sufficiently large* N*, any minimizer* (x_h^*, u_h^*) *of Problem* $(\mathrm{OQ})_N$ *and associated multipliers can be estimated by*

$$\|u_h^* - u^*\|_1 + \|x_h^* - x^*\|_\infty + \|\lambda_h^* - \lambda^*\|_\infty \leq c h_N \,, \tag{6}$$

where the constant c *is independent of* N. ◊

Proof. If (3) holds, then, by Theorem 3.3 in [5], we have

$$\|u_h^* - u^*\|_1 \leq \|u_h^* - \hat{u}_h\|_1 + \|\hat{u}_h - u^*\|_1 \leq 4\gamma h_N + h_N \mathrm{V}_0^T u^* \,,$$

i.e. the estimate (6) is satisfied with $c_u = 4\gamma + \mathrm{V}_0^T u^*$. Otherwise, it follows from (4) and (5) that

$$\|\hat{u}_h - u_h^*\|_1^2 \leq \frac{32\gamma}{\bar{\sigma}} J_N \leq \frac{32\gamma}{\bar{\sigma}} h_N \tilde{c}\, \|\hat{u}_h - u_h^*\|_1 .$$

Dividing both sides by $\|\hat{u}_h - u_h^*\|_1$, it follows that the estimate (6) is satisfied with $c_u = \frac{32\gamma}{\bar{\sigma}}\tilde{c} + \mathrm{V}_0^T u^*$. The estimates for x_h^* and λ_h can now be derived as in the proof of Theorem 1. ◊

A numerical example, illustrating the theoretical findings, can be found in [5].

References

1. Alt, W., Bräutigam, N.: Finite-difference discretizations of quadratic control problems governed by ordinary elliptic differential equations. Comp. Optim. Appl. **43**, 133–150 (2009)
2. Alt, W., Baier, R., Gerdts, M., Lempio, F.: Error bounds for Euler approximation of linear-quadratic control problems with bang-bang solutions. Numer. Algebra Control Optim. **2**, 547–570 (2012)
3. Alt, W., Seydenschwanz, M.: Regularization and discretization of linear-quadratic control problems. Control Cybern. **40**(4), 903–921 (2011)
4. Alt, W., Baier, R., Gerdts, M., Lempio, F.: Approximations of linear control problems with bang-bang solutions. Optimization **62**, 9–32 (2013)
5. Alt, W., Seydenschwanz, M.: An implicit discretization scheme for linear-quadratic control problems with bang-bang solutions. Technical Report, Friedrich-Schiller-Universität, Jena (2013). http://users.minet.uni-jena.de/alt/T2.pdf
6. Ekeland, I., Temam, R.: Convex Analysis and Variational Problems. North Holland, Amsterdam-Oxford (1976)
7. Felgenhauer, U.: On stability of bang-bang type controls. SIAM J. Control Optim. **41**, 1843–1867 (2003)
8. Hestenes, M.R.: Calculus of Variations and Optimal Control Theory. Robert E. Krieger Publishing Co., Huntington (1980)

9. Hinze, M.: A variational discretization concept in control constrained optimization: the linear-quadratic case. Comp. Optim. Appl. **30**, 45–61 (2005)
10. Meyer, C., Rösch, A.: Superconvergence properties of optimal control problems. SIAM J. Control Optim. **43**, 970–985 (2004)
11. Sendov, B., Popov, V.A.: The Averaged Moduli of Smoothness. Wiley-Interscience, New York (1988)
12. Seydenschwanz, M.: An implicit discretization scheme for linear-quadratic control problems: optimality conditions. Technical Report, Friedrich-Schiller-Universität, Jena (2012). http://users.minet.uni-jena.de/alt/T1.pdf
13. von Stryk, O.: Numerische Lösung optimaler Steuerungsprobleme: Diskretisierung, Parameteroptimierung und Berechnung der adjungierten Variablen, Fortschr.-Ber. VDI Reihe 8 Nr. 441. Düsseldorf: VDI-Verlag (1995)
14. Veliov, V.M.: Error analysis of discrete approximations to bang-bang optimal control problems: the linear case. Control Cybernet. **34**, 967–982 (2005)

On Optimization Problems for Differential Inclusions with Random Initial Data

Boris I. Ananyev[1,2](✉)

[1] Institute of Mathematics and Mechanics, Kovalevskaya 16, Yekaterinburg, Russia
[2] Ural Federal University, Mira 19, Yekaterinburg, Russia
abi@imm.uran.ru

Abstract. Optimization problems for differential inclusions with convex right-hand side are considered. The initial state of the inclusion being a random vector is unknown, or, in the more general case, is contained in some random closed set. We seek a trajectory of the inclusion which minimizes the expectation of a function of the inclusion's final state, or, in the more general case, the similar trajectory which is a minimizer of an max-(or Choquet) integral. The theorem is proved that gives necessary conditions of optimality for the mentioned problems, conditions which in certain cases are also sufficient.

Keywords: Differential inclusion · Choquet integral · Optimality conditions · Numerical scheme

1 Problems Formulation

Let $\mathcal{Q} = \{q_1, q_2, \ldots\}$ be a countable set in R^n, and $m(\cdot)$ be a Borelian measure with $\mathrm{supp}(m) = \mathcal{Q}$ such that $m(\{q_i\}) = p_i$, $\sum_{i\geq 1} p_i = 1$. Consider the probability space $(R^n, \mathcal{B}(R^n), m)$ and the identity random value q on this space. In this paper, we investigate the following problem:

$$\text{Minimize} \quad E\ell(x(b,q)) = \sum_{i\geq 1} \ell(x(b,q_i))p_i, \tag{1}$$

where E is the mathematical expectation, over all the solutions $x(\cdot, q_i)$ of a differential inclusion

$$\dot{x}(t) \in F(x(t)), \quad t \in \mathcal{T} = [a,b], \tag{2}$$

with initial conditions q_i, i.e. $x(a, q_i) = q_i$. In (2), we have $x(t) \in R^n$, $F(x)$ is a multifunction with convex and compact values. Suppose that $F(x)$ is globally Lipschitzian and bounded, i.e.

$$\exists K > 0, \quad F(x) \subset F(y) + K|x-y|B, \quad B = \{x \mid |x| \leq 1\}, \quad F(x) \subset KB. \tag{3}$$

I. Lirkov et al. (Eds.): LSSC 2013, LNCS 8353, pp. 66–73, 2014.
DOI: 10.1007/978-3-662-43880-0_6,

Hereafter $|\cdot|$ is the Euclidean norm in R^n. An absolutely continuous function $x(\cdot)$ is said to be a solution of (2) if its derivative $\dot{x}(t)$ satisfy (1) almost everywhere (a.e.) on $\mathcal{T}$.

Denote by $\mathcal{S}(F,q) = \{x(\cdot,q)\}$ the set of all solutions of (2) with the initial state q. By $\mathbf{S}(F) = \{x(\cdot,\cdot)\}$ we denote the set of functions of two variables such that $x(\cdot,q) \in \mathcal{S}(F,q)$, $\forall q$, and $x(t,\cdot)$ is Borelian, $\forall t \in \mathcal{T}$. It is well known (see Smirnov [1]) that the set $\mathcal{S}(F,q)$ under conditions (3) is nonempty compact in the space $C(\mathcal{T},R^n)$ and Lipshitzian: $\mathcal{S}(F,q_1) \subset \mathcal{S}(F,q_2) + \exp(K(b-a))|q_1 - q_2|B_C$, where $B_C = \{x(\cdot) \in C(\mathcal{T},R^n) \mid |x(\cdot)|_\infty \leq 1\}$, and $|x(\cdot)|_\infty = \max\{|x(t)| \mid t \in \mathcal{T}\}$. Here $C(\mathcal{T},R^n)$ is the Banach space of all continuous functions from $\mathcal{T}$ into R^n.

In what follows, the function $\ell(\cdot) : R^n \to R$ in (1) is supposed to be globally Lipschitzian. The map $\bar{x}(\cdot,\cdot) \in \mathbf{S}(F)$ is said to be a local minimizer for problem (1) if there exists $\varepsilon > 0$ such that $E\ell(\bar{x}(b,q)) \leq E\ell(x(b,q))$ for all $x(\cdot,\cdot) \in \mathbf{S}(F)$ satisfying the inequality $E|\bar{x}(\cdot,q) - x(\cdot,q)|_\infty \leq \varepsilon$. Our goal is to find the necessary conditions of optimality for the local minimizer.

A generalization of problem (1) consists in the following. Let a multifunction $Q : \Omega \Longrightarrow R^n$ with closed values be given on a probability space $(\Omega, \mathcal{A}, P)$. Suppose that the multifunction is measurable, i.e. the full inverse image $Q^-(B) = \{\omega \mid Q(\omega) \cap B \neq \emptyset\}$ belongs to σ-algebra $\mathcal{A}$ for any Borelian sets B. Such a multifunction is called a closed random set, Molchanov [6]. Let us formulate the problem

$$E \sup_{q \in Q} \ell(x(b,q)) = \int_\Omega \sup\{\ell(x(b,q)) \mid q \in Q(\omega)\} P(d\omega) \to \min_{x(\cdot,\cdot) \in \mathbf{S}(F)} . \tag{4}$$

The left-hand side of (4) has a form of max-(or Choquet) integral with respect to the random set $Q(\omega)$. (See Molchanov [6], Choquet [7], Nguen [8].) We consider below max-integral in more detail and reduce it to the ordinary one with respect to some probability measure. If this measure appears discrete, we can reduce problem (4) to (1).

The problems that we consider are motivated by control problems for an ensemble of trajectories of dynamical systems with uncertain initial data. (See Kurzhanski [2], Schmitendorf [3], Filippova [4].) On the other hand, the necessity of study of control problems with the random data in the form of sets arises under a solution of problems of movement correction, where the random set is naturally appeared at the observation stage (Ananyev [5].)

2 Discrete-Time Inclusions

Let $F : R^n \Longrightarrow R^n$ be a globally Lipschitzian (see (3)) multifunction with compact values and let $\Lambda_k \in R^{n \times kn}$, $k \in 1 : N$, be matrices of the form $\Lambda_k = [\Lambda_{1,k}, \ldots, \Lambda_{k,k}]$, where $\Lambda_{i,k} \in R^{n \times n}$, $i \in 1 : k$. Consider a discrete-time system of inclusions, Smirnov [1],

$$v_{k+1} \in F(\Lambda_{0,k} q + \Lambda_k(v_1, \ldots, v_k)), \quad k \in 0 : N-1, \tag{5}$$

where $\Lambda_k(v_1,\ldots,v_k)=\sum_{i=1}^k \Lambda_{i,k}v_i$, $k\in 0:N$, $\Lambda_{0,k}\in R^n$. Introduce column-vectors $V(q)=[v_1(q);\ldots;,v_N(q)]\in R^{Nn}$, where $\{v_k(q)\}$ is the set of solutions of inclusion (5) with the initial vector q. We can describe the set of all the solutions of system (5) as follows. Let $\mathbf{L}\in R^{2Nn\times Nn}$ and $\mathbf{M}\in R^{2Nn\times n}$ be block matrices of the form

$$\mathbf{L}=[0_{n\times Nn};\Lambda_1,0_{n\times(N-1)n};\ldots;\Lambda_{N-1},0_{n\times n};I_{Nn}],$$
$$\mathbf{M}=[\Lambda_{0,0};\ldots;\Lambda_{0,N-1};0_{Nn\times n}],$$

where I_n is the identity $n\times n$ matrix, $0_{n\times k}$ is the null $n\times k$ matrix, and

$$\mathcal{F}=\{[W;V]\mid v_k\in F(w_k),\ k\in 1:N\}\in R^{2Nn},$$

be the graph of the multifunction described as Cartesian product $F(w_1)\times\cdots\times F(w_N)$. Then system (5) with the initial vector q is equivalent to the inclusion

$$\mathbf{L}V(q)+\mathbf{M}q\in\mathcal{F}. \tag{6}$$

Let $\Phi:R^n\to R$ and $\Psi_k:R^n\to R$, $k\in 1:N$, be globally Lipschitzian functions. As above, let $m(\cdot)$ be the probability discrete measure on R^n with the additional assumption $\sum_{i\geq 1}p_i|q_i|^2<\infty$.

Let us form the functional

$$J(V(\cdot))=\sum_{i\geq 1}p_i(\Phi(\Lambda_{0,N}q_i+\Lambda_N V(q_i))+\Psi(V(q_i))), \tag{7}$$

where $\Psi(V)=\sum_{k=1}^N\Psi_k(v_k)$, and consider the problem

$$J(V(\cdot))\to\min_{V(\cdot)} \tag{8}$$

under constraints (6). We contemplate this problem in Hilbert space $l_2^{Nn}(m)$, where the inner product is $\langle V(\cdot),W(\cdot)\rangle=\sum_{i\geq 1}p_i\langle V(q_i),W(q_i)\rangle$. In order to prove the following theorem let us recall some notions from the theory of nonsmooth analysis [1,9]. Let X, Y be Hilbert spaces, $F:X\Longrightarrow Y$ be a multifunction, and $J:X\to\bar{R}$ be a Lipschitzian function. The conjugate map to a multifunction F is defined as $F^*(x,y)(y^*)=\{x^*\mid (x^*,-y^*)\in N_{\mathrm{gr}F}(x,y)\}$, where gr$F$ is the graph of the multifunction. Hereafter $N_A(x)$ is the basic normal cone to the set A at the point x and $\partial J(x)$ is the Mordukhovich subdifferential at x. In general, these are nonconvex sets. From the other hand, they closely related with convex proximal objects: $N_A(x)=\{w\text{-lim}\,\xi_k\mid \xi_k\in N_A^P(x_k),\ x_k\xrightarrow{A}x\}$, $\partial f(x)=\{w\text{-lim}\,\xi_k\mid\xi_k\in\partial_P f(x_k),\ x_k\xrightarrow{f}x\}$. The definition of proximal objects is given, for example, in [10]. Here w-lim is the weak limit; $x_k\xrightarrow{f}x\Leftrightarrow x_k\to x$, $f(x_k)\to f(x)$.

Theorem 1. *Let the trajectory $\bar{V}(\cdot)$ be a local minimizer for problem* (8) *and $\mathcal{V}(q)=\mathbf{L}^{-1}(\mathcal{F}-\mathbf{M}q)$ be the constraint set for $V(\cdot)$. Then there exist vectors*

$x_k^*(q)$, $s_k(q)$, $y_k^*(q)$, $z^*(q)$ *in* R^n, $k \in 1:N$, $q \in \mathcal{Q}$, *such that*

$$z^*(q) \in \partial\Phi(\Lambda_{0,N}q + \Lambda_N(\bar{v}_1(q),\dots,\bar{v}_N(q))), \quad s_k(q) \in \partial\Psi_k(\bar{v}_k(q)), \tag{9}$$

$$x_k^*(q) \in F^*(\Lambda_{0,k-1}q + \Lambda_{k-1}(\bar{v}_1(q),\dots,\bar{v}_{k-1}(q)),\bar{v}_k(q))(-y_k^*(q)), \tag{10}$$

$$-y_k^*(q) = \Lambda_{k,N}^* z^*(q) + s_k(q) + \sum_{i=k}^{N-1} \Lambda_{k,i}^* x_{i+1}^*(q). \tag{11}$$

Proof. We have to minimize functional (7) under constraints $V(q) \in \mathcal{V}(q) = \mathbf{L}^{-1}(\mathcal{F} - \mathbf{M}q)$, where $\mathbf{L}^{-1}(\mathcal{F}) = \{V \mid \mathbf{L}V \in \mathcal{F}\}$. Let $\mathcal{A} = \{V(\cdot) \in l_2^{Nn}(m) \mid V(q) \in \mathcal{V}(q), q \in \mathcal{Q}\}$. Due to the theory of nonsmooth analysis, there exist $Z^*(\cdot) \in \partial J(\bar{V}(\cdot))$ and $V^*(\cdot) \in N_{\mathcal{A}}(\bar{V}(\cdot))$ such that $0 = Z^*(\cdot) + V^*(\cdot)$. Now we need two lemmas.

Lemma 1. *Let* $J(V(\cdot)) = \sum_{i\geq 1} p_i\Phi(V(q_i))$ *be a functional defined on* $l_2^n(m)$, *where* Φ *is a globally Lipschitzian function on* R^n. *If* $\Xi(\cdot) \in \partial J(V(\cdot))$, *then* $\Xi(q) \in \partial\Phi(V(q))$ *for all* $q \in \mathcal{Q}$.

Proof (of Lemma1). Let $\Xi \in \partial_P J(V(\cdot))$. Then, by definition, there exist $\varepsilon > 0$ and $\sigma > 0$ such that $J(W) \geq J(V) + \langle \Xi, W - V\rangle - \sigma\|W - V\|^2$ for all $W \in V + \epsilon B$, where B is a unit ball in $l_2^n(m)$. It follows from this that for all $q \in \mathcal{Q}$ we have $\Phi(W(q)) \geq \Phi(V(q)) + \langle \Xi(q), W(q) - V(q)\rangle - \sigma|W(q) - V(q)|^2$, if $p_i|W(q) - V(q)|^2 \leq \epsilon^2$, $q = q_i$, i.e. $\Xi(q) \in \partial_P J(V(q))$. Note that $w\text{-}\lim \Xi_k(\cdot) = \Xi(\cdot)$ iff $\Xi_k(q) \to \Xi(q)$, $\forall q \in \mathcal{Q}$, and norms $\|\Xi_k(\cdot)\|$ are bounded. Therefore, if the sequence $\Xi_k(\cdot) \to \Xi(\cdot)$ weakly and $\Xi_k(\cdot) \in \partial_P J(V_k(\cdot))$, $V_k(\cdot) \to V(\cdot)$ strongly, we obtain $\Xi(q) \in \partial\Phi(V(q))$, $\forall q \in \mathcal{Q}$. □

Lemma 2. *Let* $\mathcal{V}(q)$ *be closed sets in* R^n, *and* $\mathcal{A} = \{V(\cdot) \in l_2^n(m) \mid V(q) \in \mathcal{V}(q), q \in \mathcal{Q}\}$. *If* $\Xi(\cdot) \in N_{\mathcal{A}}(V(\cdot))$, *then* $\Xi(q) \in N_{\mathcal{V}(q)}(V(q))$ *for all* $q \in \mathcal{Q}$.

For the proof of Lemma 2, we first note that the inclusion $\Xi(\cdot) \in N_{\mathcal{A}}^P(V(\cdot))$ implies $\Xi(q) \in N_{\mathcal{V}(q)}^P(V(q))$, $\forall q \in \mathcal{Q}$. Then, using the same limiting reasoning as in Lemma 1, we obtain the statement. Due to the lemmas, we have $Z^*(q) \in \partial(\Phi(\Lambda_{0,N}q + \Lambda_N\bar{V}(q)) + \Psi(\bar{V}(q)))$ and $V^*(q) \in N_{\mathcal{V}(q)}(\bar{V}(q))$. As $\partial\mathbf{f}(x) \subset L^*\partial f(Lx)$, where $\mathbf{f}(x) = f(Lx)$, and $\partial(f_1 + f_2)(x) \subset \partial f_1(x) + \partial f_2(x)$, we get the existence of vectors satisfying (9), such that $Z^*(q) = \Lambda_N^* z^*(q) + S(q)$, $S(q) = [s_1(q);\dots;s_N(q)]$. At last, we use the equality $N_{\mathbf{L}^{-1}(\mathcal{F}-\mathbf{M}q)}(\bar{V}(q)) = \mathbf{L}^* N_{\mathcal{F}-\mathbf{M}q}(\mathbf{L}\bar{V}(q)) = \mathbf{L}^* N_{\mathcal{F}}(\mathbf{L}\bar{V}(q) + \mathbf{M}q)$ which holds under condition $\ker\mathbf{L}^* \cap N_{\mathcal{F}-\mathbf{M}q}(\mathbf{L}\bar{V}(q)) = \{0\}$ (See [1, Theorem 3.6].) Thus, we have (10), (11). □

3 Continuous-Time Inclusions

In this section, let us return to problem (1) for continuous inclusions (2). Let $\bar{x}(\cdot,\cdot)$ be a local minimizer and N be a positive integer. Consider the following auxiliary problem: minimize the functional

$$J(V(\cdot)) = \sum_{i\geq 1} p_i \left(\ell\Big(q_i + \delta_N \sum_{k=1}^N v_k(q_i)\Big) + \sum_{k=1}^N \int_{a+(k-1)\delta_N}^{a+k\delta_N} (v_k(q_i) - \dot{\bar{x}}(t,q_i))^2 dt \right),$$

where $\delta_N = (b-a)/N$, over the variables $V(q) = [v_1(q);\ldots;v_N(q)] \in R^{Nn}$ satisfying the inclusions

$$v_k(q) \in F\Big(q + \delta_N \sum_{i=1}^{k-1} v_i(q)\Big), \quad k \in 1:N.$$

Due to global Lipschitzian conditions for ℓ and quadratic growth of integrals in J, we can conclude that auxiliary problem has a minimum, for example, at $\tilde{V}(\cdot)$. Therefore, according to Theorem 1 we obtain the existence of vectors $x_k^*(q)$, $s_k(q)$, $y_k^*(q)$, $z^*(q)$ in R^n, $k \in 1:N$, such that

$$z^*(q) \in \partial\ell\Big(q + \delta_N \sum_{k=1}^{N} \tilde{v}_k(q)\Big), \quad s_k(q) = 2\int_{a+(k-1)\delta_N}^{a+k\delta_N} (\tilde{v}_k(q) - \dot{\bar{x}}(t,q))dt,$$

$$x_k^*(q) \in F^*\Big(q + \delta_N \sum_{i=1}^{k-1} \tilde{v}_i(q), \tilde{v}_k(q)\Big)(-y_k^*(q)),$$

$$-y_k^*(q) = \delta_N\Big(z^*(q) + \sum_{i=k}^{N-1} x_{i+1}^*(q)\Big) + s_k(q).$$

Let $a + k\delta_N = t_k^N$. First, define on $[a,b]$ the piecewise-linear function $\tilde{x}_N(t,q) = q + \delta_N \sum_{i=1}^{k-1} \tilde{v}_i(q) + (t - t_{k-1}^N)\tilde{v}_k(q)$ if $t \in [t_{k-1}^N, t_k^N]$, $k \in 1:N$. Second, define on $[a,b]$ the piecewise-constant function $s_N(t,q) = s_k(q)/\delta_N$ if $t \in (t_{k-1}^N, t_k^N]$, $k \in 1:N$. Put $p_k(q) = -z^*(q) - \sum_{i=k}^{N-1} x_{i+1}^*(q)$ and consider on $[a,b]$ the piecewise-linear function $p_N(t,q) = p_{k-1}(q) + \delta_N^{-1}(p_k(q) - p_{k-1}(q))(t - t_{k-1}^N)$ if $t \in [t_{k-1}^N, t_k^N]$, $k \in 1:N$. Observe that $p_N(\cdot,q)$ satisfies the following inclusions:

$$\begin{gathered} p_N(b,q) \in -\partial\ell(\tilde{x}_N(b,q)), \quad \dot{p}_N(t,q) = \tfrac{p_N(t_k^N,q) - p_N(t_{k-1}^N,q)}{\delta_N} \\ \in F^*(\tilde{x}_N(t_{k-1}^N,q), \dot{\tilde{x}}_N(t,q))\big(-p_N(t_k^N,q) + s_N(t_k^N,q)\big), \quad t \in [t_{k-1}^N, t_k^N]. \end{gathered} \tag{12}$$

Also as in [1, Theorem 7.2] one can prove that the sequence of functions $\dot{\tilde{x}}_N(\cdot,q)$ tends to $\dot{\bar{x}}(\cdot,q)$ in the norm of $L_2^n[a,b]$. In this proof one uses the optimality of $\tilde{V}(\cdot)$. Without loss of generality the functions $|\dot{\tilde{x}}_N(\cdot,q) - \dot{\bar{x}}(\cdot,q)|$ and $|s_N(t,q)|$ tends to zero. From Gronwall inequality, differential inclusions (12), and the Arzela-Ascoli theorem we can conclude that there exists a function $p(\cdot,q)$ for which $p_N(\cdot,q) \to p(\cdot,q)$ in the uniform metric. Using [1, Lemma 4.4] for any $q \in \mathcal{Q}$, we obtain

$$\dot{p}(t,q) \in \mathrm{co}F^*(\bar{x}(t,q), \dot{\bar{x}}(t,q))(-p(t,q)), \quad p(b,q) \in -\partial\ell(\bar{x}(b,q)). \tag{13}$$

From (13) we see that an optimal trajectory satisfies the maximum principle

$$\langle p(t,q), \dot{\bar{x}}(t,q)\rangle = \max\big\{\langle p(t,q), y\rangle : y \in F(\bar{x}(t,q))\big\}, \quad \text{a.e.} \quad t \in [a,b], \tag{14}$$

where $\langle\cdot,\cdot\rangle$ is the inner product in R^n, $q \in \mathcal{Q}$.

Observe some properties of inclusions (14). As F is globally Lipschitzian, we have $\mathrm{co}F^*(x,y)\,(-p) \subset K|p|B$, where K is Lipschitz constant and the set B is defined in (3). Moreover, the multifunction $(x,y,p) \to \mathrm{co}F^*(x,y)(-p)$ is upper semicontinuous. For globally Lipschitzian functions, the subdifferential $\partial\ell(x)$ is bounded and upper semicontinuous in x. Using properties of differential inclusions, we conclude that for optimal solution $\bar{x}(\cdot,q)$ there exists at least one solution $p(t,q)$ of inclusions (14). Thus, summarizing the above consideration, we come to the conclusion.

Theorem 2. *Let $\bar{x}(\cdot,\cdot)$ be a locally optimal trajectory in problem* (1) *under above assumptions about F and ℓ. Then there exists a function $p(t,q)$, absolutely continuous in t, such that it satisfies inclusions (13) and maximum principle (14) for all $q \in \mathcal{Q}$.*

Remark 1. Suppose that the measure $m(\cdot)$ in (1) is not discrete, but absolutely continuous with respect to Lebesgue measure. Then the assertion like Lemma 1 is not valid. Indeed, let $n = 1$ and $J(V(\cdot)) = -\int_0^1 |V(t)|dt$. Then, in the space $L_2[0,1]$, we have $0 \in \partial J(0)$, but $0 \notin \partial\Phi(0) = \{-1,1\}$, where $\Phi(x) = -|x|$. (See [10, P. 156].) Of course, $0 \in \partial_C\Phi(0) = [-1,1]$, where ∂_C is Clarke's subdifferential. Moreover, if $J(V(\cdot)) = \int_{R^n} f(x)\Phi(x)dx$, where $f(\cdot)$ is a probability density function, $\Phi(\cdot)$ is a Lipschitzian function, then $\partial J(V(\cdot)) = \partial_C(V(\cdot)) = \{V(\cdot) \mid V(x) \in \partial_C\Phi(x)\, a.e.\}$. This fact is proved in [10] for the case: $n = 1$, f has a compact support. The general case also can be proved with the help of Aumann's theorem. It will be published.

Remark 2. Suppose, in addition to above assumptions, that the sets $\mathrm{gr}F$, $\mathrm{dom}F$ are convex and closed, and the function ℓ is convex on $\mathrm{dom}F = \mathrm{dom}\ell$. Besides, let the set $\mathcal{S}(F,x) \neq \emptyset$ for any $x \in \mathrm{dom}F$ and the support of measure $m(\cdot)$ be contained in $\mathrm{dom}F$. Then the theorem like Theorem 2 gives sufficient conditions for a global minimum.

Example 1. Let $n = 2$, $a = 0$, $b = 1$, $F(x,y) = \{(y,v) \mid v \in [\exp(-y),1]\}$ when $|x| < \infty$, $y \geq 0$, and $F(x,y) = \emptyset$ when $y < 0$, $\ell(x,y) = x + y$. Then transversality conditions give $p(1) = q(1) = -1$. We have $p(t) \equiv -1$, $\dot{y} = \exp(-y)$, $\dot{q} = 1 + q\dot{y}$ in the domain, where $q(t) < 0$. Then $q(t) = -\big(1/(1+\exp(y_0)) + \log((1+\exp(y_0))\dot{y})\big)/\dot{y}$, $y(t) = \log(t + \exp(y_0))$, $x(t) = x_0 + (t + \exp(y_0))y(t) - (t + \exp(y_0)y_0)$. Therefore, for any probability measure $m(dx_0, dy_0)$ with the support from upper semiplane, for which $E_m(x(1,x_0,y_0) + y(1,y_0))$ is finite, the given trajectory is optimal in problem like (1).

4 The Cost with Choquet Integral

Let $f : R^n \to R \cup \{+\infty\}$ be Borelian function bounded from below. Then for any probability measure m the integral $\int_{R^n} f(x)m(dx)$ is properly defined. Given a closed random set $Q(\omega)$, we can define so called *max*-(or *Choqeut*) integral. First define the function $f_Q(\omega) = \sup\{f(x) \mid x \in Q(\omega)\}$. This function

$f_Q : \Omega \to R\cup\{+\infty\}$ is measurable with respect to σ-algebra $\mathcal{A}$ and bounded from below. Therefore, the value $\int f dQ = Ef_Q$, called max-integral over closed random set $Q(\omega)$, is properly defined for all Borelian functions bounded from below. Note that the original definition consists in the following [7]. Let $f^d(x) = f(x)-d$, where $d = \inf\{f(x) \mid x \in R^n\}$. Then $f_Q^d(\omega) = f_Q(\omega) - d$. Due to the equality $E\xi = \int_0^{+\infty} P\{\xi > t\}dt$ for random value $\xi(\omega)$ (see [8]), we obtain

$$Ef_Q^d = \int_\Omega f_Q^d(\omega)P(d\omega) = \int_0^{+\infty} P\{f_Q^d > t\}dt$$
$$= \int_0^{+\infty} P\{f_Q^{-1}(t+d, +\infty]\}dt = \int_0^{+\infty} P\{\omega \mid Q(\omega) \cap f^{-1}(t+d, +\infty] \neq \emptyset\}dt$$
$$= \int_0^{+\infty} T_Q(\{x \mid f(x) > t+d\})dt = \int_d^{+\infty} T_Q(\{x \mid f(x) > t\})dt.$$

Thus,

$$\int f dQ = Ef_Q = \int_d^{+\infty} T_Q(f^{-1}(t, +\infty])dt + d. \tag{15}$$

Here and above $T_Q(B) = P\{\omega \mid Q(\omega) \cap B \neq \emptyset\}$ is *Choquet capacity* for closed random set $Q(\omega)$. The functional T_Q defined on σ-algebra $\mathcal{B}(R^n)$ is not additive, but monotone: $T_Q(B_1) \leq T_Q(B_2)$, if $B_1 \subset B_2$. Therefore, $T_Q(f^{-1}(t, +\infty])$ is a monotonically nonascending function in t, and integral in (15) is properly defined Riemann integral. Let us accept

Assumption. The set $Q_0(\omega) = \text{Argmax}\{f(x) \mid x \in Q(\omega)\} = \{x \in Q(\omega) \mid f(x) = f_Q(\omega)\} \neq \emptyset$ is Borelian and does not equal empty set almost surely. Besides, the multifunction $Q_0 : \Omega \to \mathcal{B}(R^n)$ is measurable: $Q_0^-(B) \in \mathcal{A}$ for any set $B \in \mathcal{B}(R^n)$. There exists Borelian stochastic kernel $m(dx|\omega)$ that is a measurable probability measure $m : \Omega \to PM(R^n)$, for which $m^{-1}(\mathcal{B}(PM(R^n))) \subset \mathcal{A}$, such that $\int_\Omega m(Q_0(\omega)|\omega)P(d\omega) = 1$. Here $\mathcal{B}(PM(R^n))$ is Borelian σ-algebra on the complete separable metric space $PM(R^n)$ of probability measures with $*$-weak topology [11].

Note that the equality $m(Q_0(\omega)|\omega) = 1$ a.e. follows from the Assumption. The following theorem is valid.

Theorem 3. *Under the Assumption, the equality*

$$\int f dQ = \int_{R^n} f(x)m(dx)$$

takes place for the function f and the closed random set Q, where the probability measure m is defined for any $B \in \mathcal{B}(R^n)$ as $m(B) = \int_\Omega m(B|\omega)P(d\omega)$.

Let there exists a Borelian selection ξ: $\xi(\omega) \in Q_0(\omega)$, $\forall\omega \in \Omega$.

Lemma 3. *If there exists the Borelian selection, as above, the Assumption is valid.*

Proof. As a Borelian stochastic kernel we choose $m(\cdot|\omega) = \delta_{\xi(\omega)}(\cdot)$, that is the composition of Dirac measure and the Borelian selection. All the measurability requirements are fulfilled. □

Example 2. Let $Q(\omega)$ has closed and convex values, and $Q(\omega) \subset K$, where K is a convex compact set in R^n. The function f is strictly concave on K. Then $Q_0(\omega) = \{\xi(\omega)\}$, and the function ξ is measurable due to Filippov lemma. For stochastic kernel $m(\cdot|\omega) = \delta_{\xi(\omega)}(\cdot)$, where $\delta_\xi(\cdot)$ is Dirac measure, our assumptions are fulfilled.

Example 3. Let Ω be the set of all finite sets $\omega \subset \mathcal{Q}$, where $\mathcal{Q}$ is a countable set in R^n. Consider the random set $Q(\omega) = \omega$ with a distribution $\{p(\omega)\}$, such that $\sum_{\omega\in\Omega} p(\omega) = 1$, $p(\emptyset) = 0$. Here, it may be so that $\sum_{i\geq 1} p(\{q_i\}) < 1$. The measure $m(\cdot|\omega)$ on $Q_0(\omega)$ is defined arbitrarily by numbers $q^x_{Q_0(\omega)} \geq 0$, for which $\sum_{x\in Q_0(\omega)} q^x_{Q_0(\omega)} = 1$, $q^x_{Q_0(\omega)} = 0$ if $x \notin Q_0(\omega)$. Then, according to Theorem 3, $m(x) = \sum_{\omega\in\Omega} m(x|\omega)p(\omega)$, $x \in \mathcal{Q}$, and we have $\sum_{\omega\in\Omega} \max\{f(x) \mid x \in \omega\}p(\omega) = \sum_{x\in\mathcal{Q}} f(x)m(x)$.

The situation described in Example 3 is applicable in foregoing Theorem 1 and 2. In practice for the set $\mathcal{Q}$, we can take a countable set $\mathcal{Q} = \{\delta[i_1; \ldots; i_n]\}$, where i_j are integers, $\delta > 0$. Then every compact set will have a finite intersection with $\mathcal{Q}$. This intersection can be taken as an approximation of the compact set.

Acknowledgement. The work is partly supported by RFBR under Project No 13-01-00120 and Program of Presidium RAS "Dynamical Systems and Control Theory" under support of UB RAS (Project 12-P-1-1019).

References

1. Smirnov G.V.: Introduction to the Theory of Differential Inclusions. Graduate Studies in Mathematics, vol. 41, 226 p. AMS (2002)
2. Kurzhanski A.B.: Control and Estimation under Uncertainty. M.: Nauka, 392 p. (1977) (in Russian)
3. Schmitendorf, W.E.: Minimax control of systems with uncertainty in the initial state and in the state equations. IEEE Trans. Automat. Control AC **22**(3), 439–443 (1971)
4. Filippova, T.F.: Control of the system with nonsmooth right-hand side under uncertainty. Diff. Eqn. **19**(10), 1693–1699 (1983). (in Russian)
5. Ananyev, B.I.: Optimal communication channels with noise in problems of estimation and motion correction. Proc. Steklov Inst. Math. **271**(Suppl. 1), 1–17 (2010)
6. Molchanov, I.: Theory of Random Sets, 488 p. Springer, New York (2005)
7. Choquet, G.: Theory of capacities. Ann. Inst. Fourier **5**, 131–295 (1953)
8. Nguyen, H.T.: An Introduction to Random Sets, 257 p. Chapman & Hall, New York (2006)
9. Mordukhovich, B.S.: Variational Analysis and Generalized Differentiation, II. Applications. A Series of Comprehensive Studies in Mathematics, vol. 331, 610 p. Springer, New York (2006)
10. Clarke, F.H., Ledyaev, Y.S., Stern, R.J., Wolenski, P.R.: Nonsmooth Analysis and Control Theory, 276 p. Springer, New York (1998)
11. Bertsecas, D., Shreve, S.: Stochastic Optimal Control. The Discrete-Time Case, 323 p. Athena Scientific, Belmont (1996)

Stability of Switched Systems: An Introduction

Andrea Bacciotti(✉)

Dipartimento di Matematica del Politecnico di Torino,
C.so Duca degli Abruzzi 24, 10129 Torino, Italy
andrea.bacciotti@polito.it

Abstract. In this note we review some problems typically encountered in the theory of stability of switched systems. Moreover, we present some recent achievements on this subject.

Keywords: Switched systems · Stability · Lyapunov functions · Asymptotic controllability

1 Introduction

The study of hybrid systems is one of the most interesting and challenging topic in the modern engineering literature [10]. A *hybrid system* is a system whose evolution is determined by the combination of discrete time and continuous time effects. Hybrid systems may exhibit unusual phenomena, like impulses (discontinuous state evolution) and Zeno behavior (accumulation of discontinuities in finite time). Here we are interested in a special family of hybrid systems, called *switched systems* [13,18]. The evolution of a switched system is described by continuous trajectories: in other words, discontinuities are allowed in the velocity, but not in the state evolution. Moreover, Zeno behavior is excluded by definition. In spite of these simplifications, the dynamical behavior of a switched system can be very complex. In the next section, the notion of switched system is formally introduced. The subsequent sections illustrate some peculiar difficulties arising in the study of stability of switched systems, and survey the way they have been addressed in the recent literature.

2 Switching Signals and Switched Systems

Let $N \geq 2$ be a fixed integer, and let $\mathcal{N} = \{1, \ldots, N\}$ be equipped with the discrete topology. Let $\mathcal{U}_{\mathcal{N}}$ be the set of *switching signals*, that is all the right continuous, piecewise constant functions $\sigma : [0, +\infty) \to \mathcal{N}$. The discontinuity points of a switching signal σ form a finite or countable (possibly empty) subset of the open half line $(0, +\infty)$. They are called *switching times* of σ. Let I_σ be the set whose elements are $t_0 = 0$ and all the switching times t_i of σ, indexed in such a way that $0 = t_0 < t_1 < t_2 < \ldots$. If the set of switching times is infinite, then clearly $\lim_{i \to +\infty} t_i = +\infty$. If it is finite and $\max t_i = t_{i_*}$, then

I. Lirkov et al. (Eds.): LSSC 2013, LNCS 8353, pp. 74–80, 2014.
DOI: 10.1007/978-3-662-43880-0_7,

we set $t_{i_*+1} = +\infty$. The numbers $\theta_i = t_{i+1} - t_i > 0$ are called *durations*. A switching signal σ is said to be *periodic* (of period T) for $\mathcal{F}$ if there exists a string of real numbers $\tau_0, \ldots, \tau_H$ (where H is an integer, $H \geq 1$) and a string of indices $n_1, \ldots, n_H \in \mathcal{N}$ such that: $0 = \tau_0 < \tau_1 < \cdots < \tau_H = T$; $\sigma(t) = n_h$ for $t \in [\tau_{h-1}, \tau_h)$, for each $h = 1, \ldots, H$; and $\sigma(t) = \sigma(t-T)$ for $t \geq T$. The points $\tau_h + mT$, with $h = 1, \ldots, H$, and $m = 0, 1, \ldots$ coincide with the switching times, provided that $H > 1$ and $n_1 \neq n_2, n_2 \neq n_3 \ldots n_H \neq n_1$. Note that σ is constant when $H = 1$.

Let $d \geq 1$ be a fixed integer, and let $\mathcal{F} = \{f_n\}_{n \in \mathcal{N}}$ be a family of vector fields of class C^1 on $\mathbf{R}^d$. For each $n \in \mathcal{N}$, the vector field f_n is called the *n-th component* of $\mathcal{F}$. Assume that for each f_n and each initial state $\bar{x} \in \mathbf{R}^d$ there is a unique differentiable curve $\varphi_n(t, \bar{x}) : [0, +\infty) \to \mathbf{R}^d$ such that $\dot{\varphi}_n(t, \bar{x}) = f_n(\varphi_n(t, \bar{x}))$ for each $t \geq 0$ and $\varphi_n(0, \bar{x}) = \bar{x}$. It is called the (positive) *trajectory* of f_n issued from $\bar{x}$. Analogously, for each $\bar{x}$ and each $\sigma \in \mathcal{U}_{\mathcal{N}}$, there exists a unique continuous curve $\varphi_{\mathcal{F}}(t, \bar{x}, \sigma) : [0, +\infty) \to \mathbf{R}^d$ satisfying the condition $\varphi_{\mathcal{F}}(0, \bar{x}, \sigma) = \bar{x}$, and such that

$$\varphi_{\mathcal{F}}(t, \bar{x}, \sigma) = \varphi_{\sigma(t_i)}(t - t_i, \varphi_{\mathcal{F}}(t_i, \bar{x}, \sigma)) , \qquad \forall t \in [t_i, t_{i+1}) , \quad \forall t_i \in I_\sigma .$$

We say that $\varphi_{\mathcal{F}}(t, \bar{x}, \sigma)$ is the *switched trajectory* of $\mathcal{F}$, issued from the initial state $\bar{x}$, corresponding to the switching signal σ. In the sequel, we will use the simplified notations $\varphi(t, \bar{x}, \sigma)$ or $\varphi(t, \bar{x})$, when the omitted terms are clear from the context.

A *switched system* on $\mathbf{R}^d$ with index set $\mathcal{N}$ is defined by a family $\mathcal{F} = \{f_n\}_{n \in \mathcal{N}}$ of vector fields, together with a set-valued map Σ which assigns a (nonempty) set of switching signals $\Sigma_{\bar{x}} \subset \mathcal{U}_{\mathcal{N}}$ to each point $\bar{x} \in \mathbf{R}^d$, regarded as initial state. A switched system will be denoted by $(\mathcal{F}, \Sigma)$; Σ is called a *switching map*. Roughly speaking, a switching map specifies the set of inputs which is admissible for every initial state, and allows us to take into account possible constraints on the switching signals. A switched system $(\mathcal{F}, \Sigma)$ for which Σ is single-valued and constant i.e., the same switching signal σ is applied for each initial state $\bar{x}$, will be simply denoted by $(\mathcal{F}, \sigma)$. A *periodic switched system* is a pair $(\mathcal{F}, \sigma)$ such that σ is periodic for $\mathcal{F}$.

A switched system $(\mathcal{F}, \Sigma)$ is said to be *linear* if all its components are linear vector fields of $\mathbf{R}^d$, that is $f_n(x) = A_n x$, where A_n is a $d \times d$ real matrix. For a linear switched system, one has $\varphi_n(t, \bar{x}) = e^{tA_n}\bar{x}$ and

$$\varphi(t, \bar{x}, \sigma) = e^{(t-t_i)A_{\sigma(t_i)}} \varphi(t_i, \bar{x}, \sigma) = e^{(t-t_i)A_{\sigma(t_i)}} e^{(t_i - t_{i-1})A_{\sigma(t_{i-1})}} \ldots e^{t_1 A_{\sigma(0)}} \bar{x}$$

for each $t \in [t_i, t_{i+1})$, $t_i \in I_\sigma$, and $\sigma \in \Sigma_{\bar{x}}$. We emphasize that in this case the map $x \mapsto \Phi(t, \sigma)x = \varphi(t, x, \sigma)$ is linear and nonsingular for each $t \geq 0$.

A switched trajectory of a family $\mathcal{F}$ of vector fields can be viewed as a trajectory of a control system of the form $\dot{x} = \sum_{n=1}^{N} u_n f_n(x)$, where the input u is piecewise constant and takes value on the set of the vectors of the canonical basis of $\mathbf{R}^N$. Note that this system is affine, but the control set is not symmetric. A switched trajectory of a family of linear vector fields can be viewed as a trajectory of a bilinear control system. The formalism introduced so far is basically the same used in geometric control theory to represent a control system [12]. However, geometric control theory is especially concerned with the controllability problem, and it is usually assumed $\Sigma_{\bar{x}} = \mathcal{U}_{\mathcal{N}}$ for each $\bar{x} \in \mathbf{R}^d$. Here, we focus on stability and stabilization. Moreover, it is of great interest to consider also cases where $\Sigma_{\bar{x}}$ is a proper subset of $\mathcal{U}_{\mathcal{N}}$. A switched system can be also viewed as a differential inclusion. However, the notion of solution of a differential inclusion is more general than the notion of switched trajectory.

3 Stability

The classical definitions of stability extend to switched systems in a natural way. For simplicity, from now on we consider only switched systems $(\mathcal{F}, \Sigma)$ such that $f_n(0) = 0$ for each $n \in \mathcal{N}$.

Definition 1. *The origin is said to be* stable *for* $(\mathcal{F}, \Sigma)$ *if:*

$$\forall \varepsilon > 0 \ \exists \delta > 0 \text{ s.t.} : |\bar{x}| < \delta \Longrightarrow |\varphi(t, \bar{x}, \sigma)| < \varepsilon \ , \quad \forall t \geq 0 \ , \quad \forall \sigma \in \Sigma_{\bar{x}} \ . \tag{1}$$

The origin is said to be locally attractive *for* $(\mathcal{F}, \Sigma)$ *if:*

$$\exists \delta_0 > 0 \text{ s.t.} : |\bar{x}| < \delta_0 \Longrightarrow \lim_{t \to +\infty} \varphi(t, \bar{x}, \sigma) = 0 \ , \quad \forall \sigma \in \Sigma_{\bar{x}} \ . \tag{2}$$

Moreover, we say that the origin is globally attractive *if* δ_0 *can be taken arbitrarily large;* exponentially attractive *if each switched trajectory decays exponentially to zero when* $t \to +\infty$*;* uniformly attractive *if the limit in (2) is uniform with respect to* σ*. Finally, we say that the origin is* [locally, globally, exponentially, uniformly] asymptotically stable *for the switched system* $(\mathcal{F}, \Sigma)$ *if it is both stable and [locally, globally, exponentially, uniformly] attractive for* $(\mathcal{F}, \Sigma)$.

Even if all the individual components of a family $\mathcal{F}$ are asymptotically stable at the origin, it is sometimes possible to construct switched trajectories of $\mathcal{F}$ which do not converge to the origin. This phenomenon is called *loss of stability*. Loss of stability is a well known, peculiar feature of systems whose behavior is affected by discontinuous or discrete effects. It has been observed in early works [19], and it is the motivation of recent studies [13, 18]. The more popular examples where loss of stability arises, involve families of linear vector fields defined by matrices with complex conjugate eigenvalues: in [2], we give an example where the vector fields have real eigenvalues. These remarks motivate the following definition.

Definition 2. *The family of vector fields* $\mathcal{F}$ *is said to be* stable under arbitrary switching *when the switching system* $(\mathcal{F}, \mathcal{U}_{\mathcal{N}})$ *(i.e., with* $\Sigma_{\bar{x}} = \mathcal{U}_{\mathcal{N}}$ *for each* $\bar{x} \in \mathbf{R}^d$*) is asymptotically stable at the origin.*

4 Lyapunov Functions

A common Lyapunov function for $\mathcal{F}$ is a function $V : \mathbf{R}^d \to \mathbf{R}$ of class C^1 such that $V(0) = 0$, $V(x) > 0$ for $x \neq 0$, and $\nabla V(x) f_n(x) < 0 \; \forall x \neq 0$ and $\forall n \in \mathcal{N}$. A common Lyapunov function is said to be *weak* if the last inequality is replaced by: $\nabla V(x) f_n(x) \leq 0 \; \forall x \in \mathbf{R}^d$ and $\forall n \in \mathcal{N}$.

Theorem 1. *If there exists a common Lyapunov function for $\mathcal{F}$ then the origin is uniformly asymptotically stable for $\mathcal{F}$ under arbitrary switching.*

The existence of a common Lyapunov function implies that each single component is asymptotically stable at the origin. The converse of Theorem 1 holds [14] (see also [9]). However, in practice Theorem 1 does not help very much, due to the difficulty of finding a common Lyapunov function. Another drawback of the common Lyapunov function approach is the following. It is well known that for a single asymptotically stable linear vector field it is always possible to find a quadratic Lyapunov function: but this is not longer true for a family of linear vector fields, asymptotically stable under arbitrary switching [9,13]. We now discuss an extension of the Lyapunov function approach. Let us assume that all the f_n's are asymptotically stable, and let $V_n(x)$ be a Lyapunov function for $f_n(x)$. The family $\{V_n\}$ is called a *multiple Lyapunov function* for $\mathcal{F}$. Of course, the existence of a multiple Lyapunov function is not sufficient to ensure stability of $\mathcal{F}$: one needs to add a compatibility condition, regulating the way the Lyapunov functions of the singular components change their values when a switch arises. The simplest condition of this type is:

$$V_{n_i}(\varphi(t_i, \bar{x}, \sigma)) \leq V_{n_{i-1}}(\varphi(t_i, \bar{x}, \sigma)) \;\; \forall\, \bar{x} \in \mathbf{R}^d,\; \sigma \in \mathcal{U}_{\mathcal{N}},\; t_i \in I_\sigma \,. \tag{3}$$

Theorem 2. *Let $\{V_n\}$ be a multiple Lyapunov function for $\mathcal{F}$. Assume in addition that (3) is fulfilled. Then the origin is asymptotically stable for $\mathcal{F}$ under arbitrary switching.*

This Theorem can be found in [16]. A more sophisticated compatibility condition can be found in [8]. Similar results can be obtained in the context of topological dynamics [4]. The same idea can be used to prove asymptotic stability of a switched system $(\mathcal{F}, \Sigma)$, with $\Sigma_{\bar{x}} \neq \mathcal{U}_{\mathcal{N}}$ for some $\bar{x} \in \mathbf{R}^d$ (constrained switches). To this end, it sufficient to check the compatibility condition for each $\sigma \in \Sigma_{\bar{x}}$ and $\bar{x} \in \mathbf{R}^d$.

5 Invariance Principle

Switched systems share many disadvantages of time-varying systems. In particular, the invariance principle is no more a so powerful tool for proving asymptotic stability of an equilibrium position when only a weak Lyapunov function is available, as in the case of single time-invariant systems [3,11].

Definition 3. *We say that a switched signal σ admits a* dwell time *if the infimum of the durations θ_i is positive. In this case, the number* $\inf \theta_i$ *is called the* dwell time *of σ. The set of all the switched signals $\sigma \in \mathcal{U}_{\mathcal{N}}$ which admit a dwell time (not necessarily the same) is denoted by $\mathcal{D}$.*

If we focus on switching maps satisfying $\Sigma_{\bar{x}} \subset \mathcal{D}$ for each $\bar{x} \in \mathbf{R}^d$, then it is possible to prove extended versions of the Invariance Principle. The following is perhaps the simplest statement in this perspective [3]. The required notion of weakly invariant set is adapted to the switching systems case. More general results can be found in [11,15].

Theorem 3. *Let $V(x) : \mathbf{R}^d \to \mathbf{R}$ be a weak common Lyapunov function for $(\mathcal{F}, \Sigma)$, with $\Sigma_{\bar{x}} \subset \mathcal{D}$ for each $\bar{x} \in \mathbf{R}^d$. Let $l > 0$ and let Ω_l be the connected component of the level set $\{x : V(x) < l\}$ such that $0 \in \Omega_l$. Assume that Ω_l is bounded, and let $Z = \{x : \exists n \in \mathcal{N} \text{ s.t.} : \nabla V(x) f_n(x) = 0\}$. Finally, let M be the union of all the compact, weakly invariant sets which are contained in $Z \cap \Omega_l$. Then every switched trajectory $\varphi(t, \bar{x}, \sigma)$ with $\sigma \in \Sigma_{\bar{x}}$ and $\bar{x} \in \Omega_l$ is attracted by M.*

6 Asymptotic Controllability

The reversed time interpretation of loss of stability shows that even if all the vector fields of a family $\mathcal{F}$ are completely unstable, it is sometimes possible to construct trajectories converging to the origin. Given a family of vector fields $\mathcal{F}$, it is therefore interesting to characterize those switching maps Σ (if any) such that

$$\forall \bar{x} \in \mathbf{R}^d \ \exists \sigma \in \Sigma_{\bar{x}} : \quad \lim_{t \to +\infty} \varphi_{\mathcal{F}}(t, \bar{x}, \sigma) = 0 . \tag{4}$$

Note that as far as we are interested in property (4), we can limit ourselves to single-valued switching maps.

Definition 4. *The family of vector fields $\mathcal{F}$ is said to be* asymptotically controllable *if there exists a single-valued switching map Σ such that (4) holds. The family of vector fields $\mathcal{F}$ is said to be* consistently asymptotically controllable *if there exists a switched signal σ such that (4) holds with $\Sigma_{\bar{x}} \equiv \{\sigma\}$, for each $\bar{x} \in \mathbf{R}^d$.*

The notion of asymptotic controllability is classical: it means that all the initial states can be eventually driven toward the origin, but different switching signals might be required for different initial states. On the contrary, consistent asymptotic controllability means that the same switching signal works for all the initial states. Clearly, if $\mathcal{F}$ is consistently asymptotically controllable then it is asymptotically controllable, but the converse is false in general: an example is given in [18]. Asymptotic controllability has been merely defined here in terms of the attraction property (4). However, if the vector fields of $\mathcal{F}$ are linear,

(4) automatically implies stability. Asymptotic controllability has been deeply investigated for pairs of linear vector fields in $\mathbf{R}^2$. But there is not so many results for more general situations. One of these results reads as follows [5].

Theorem 4. *Let $\mathcal{F} = \{f_n\}$ be any family of vector fields of $\mathbf{R}^d$. Assume that there exist $\alpha_1, \ldots, \alpha_N$ ($\alpha_n \geq 0$, $\sum_{n=1}^{N} \alpha_n = 1$) such that the origin is asymptotically stable for the vector field $f(x) = \sum_{n=1}^{N} \alpha_n f_n(x)$. Then, $\mathcal{F}$ is asymptotically controllable.*

Theorem 4 was originally proved in [20] for families of linear vector fields using a common Lyapunov function approach. In fact, in [20] the authors find out a discontinuous state-dependent switching rule, and introduce hysteresis in order to counteract possible chattering phenomena. Alternatively, the stability could be intended in the sense of Filippov solutions (see [1] and the so called min-projection strategy proposed in [17]). On the other hand, resorting to Filippov solutions has a drawback: indeed in general Filippov solutions cannot be reproduced by means of switched signals [6].

Next we point out that under a mild additional assumption, asymptotic controllability can be achieved by means of switching (open-loop) maps with a very special structure. A switching signal ρ is said to be *eventually periodic* if there exist a periodic switching signal σ and a time $\bar{t} > 0$ such that $\rho(t + \bar{t}) = \sigma(t)$, for all $t \geq 0$. We say that σ is the *periodic part* of ρ and that $\rho\big|_{[0,\bar{t})}$ is the *pre-periodic part* of ρ. We also say that $\bar{t}$ is the *pre-period*. We denote by $\mathcal{E}(\sigma)$ the set of all the eventually periodic switching signals having the same periodic part σ. A single valued switching map Σ is called *eventually periodic* if there exists a periodic switching signal σ such that $\Sigma_{\bar{x}} \in \mathcal{E}(\sigma)$, for each $\bar{x} \in \mathbf{R}^2$. The switched system $(\mathcal{F}, \Sigma)$ is called *eventually periodic* if Σ is eventually periodic. Roughly speaking, an eventually periodic switching map consists of two parts: an initial transient interval (whose length can be predicted) followed by a periodic steady state. During the transient, the control action depends on the initial state and must be operated in open-loop. The advantages of an eventually periodic switching map become evident during the periodic steady state and can be resumed as follows: (1) the periodic steady state is independent of the initial state; (2) the periodic steady state can be constructed according to a systematic procedure; (3) the periodic steady state can be interpreted as a state dependent control law, so that it can be implemented in an automatic way.

To state the main result of this section, we need to introduce a new notion. A family $\mathcal{F}$ of vector fields is said to be *radially controllable* if for each pair of points $x, y \in \mathbf{R}^d$ ($x \neq 0, y \neq 0$) there exist $c > 0$ and $T > 0$ such that cy is reachable in time T from x along a switched trajectory of $\mathcal{F}$. The following result is proven in [6].

Theorem 5. *Let $\mathcal{F}$ be a family of linear vector fields of $\mathbf{R}^d$. Assume that $\mathcal{F}$ is radially controllable and asymptotically controllable. Then there exists an eventually periodic switching map Σ such that the system $(\mathcal{F}, \Sigma)$ is asymptotically stable.*

In general, the radial controllability assumption cannot be dropped out in the previous statement. On the other hand, a family $\mathcal{F}$ can be consistently stabilized by means of a periodic switching rule even if the radial controllability assumption fails. It is reasonable to conjecture that the conclusion of Theorem 5 remains true for homogeneous (not necessarily linear) vector fields.

References

1. Bacciotti, A.: Stabilization by means of state space depending switching rules. Syst. Control Lett. **53**, 195–201 (2004)
2. Bacciotti, A.: A remark about linear switched systems in the plane. In: Proceedings of IFAC-NOLCOS Conference (2013)
3. Bacciotti, A., Mazzi, L.: An invariance principle for nonlinear switched systems. Syst. Control Lett. **54**, 1109–1119 (2005)
4. Bacciotti, A., Mazzi, L.: Stability of dynamical polysystems via families of Lyapunov functions. Nonlin. Anal. TMA **67**, 2167–2179 (2007)
5. Bacciotti, A., Mazzi, L.: Stabilisability of nonlinear systems by means of time-dependent switching rules. Int. J. Control **83**, 810–815 (2010)
6. Bacciotti, A., Mazzi, L.: Asymptotic controllability by means of eventually periodic switching rules. SIAM J. Control Optim. **49**, 476–497 (2011)
7. Boscain, U.: Stability of planar switched systems: the linear single input case. SIAM J. Control Optim. **41**, 89–112 (2002)
8. Branicky, M.S.: Multiple Lyapunov functions and other analysis tools for switched and hybrid systems. IEEE Trans. Autom. Control **43**, 475–482 (1998)
9. Dayawansa, W.P., Martin, C.F.: A converse Lyapunov theorem for a class of dynamical systems which undergo switching. IEEE Trans. Autom. Control **44**, 751–760 (1999)
10. Goebel, R., Sanfelice, R.G., Teel, A.R.: Hybrid Dynamical Systems. Princeton University Press, Princeton (2012)
11. Hespanha, J.P.: Uniform stability of switched linear systems: extensions of LaSalle's invariance principle. IEEE Trans. Autom. Control **49**, 470–482 (2004)
12. Jurdjevic, V.: Geometric Control Theory. Cambridge University Press, Cambridge (1997)
13. Liberzon, D.: Switching in Systems and Control. Birkhäuser, Boston (2003)
14. Mancilla-Aguilar, J.L., Garcia, R.A.: A converse Lyapunov theorem for nonlinear switched systems. Syst. Control Lett. **41**, 67–71 (2000)
15. Mancilla-Aguilar, J.L., Garcia, R.A.: An extension of LaSalle's invariance principle for switched systems. Syst. Control Lett. **55**, 376–384 (2006)
16. Peleties, P., deCarlo, R.A.: Asymptotic stability of m-switched systems using Lyapunov-like functions. In: Proceedings of the American Control Conference, pp. 1679–1684 (1991)
17. Petterson, S., Lennarston, B.: Stabilization of hybrid systems using a min-projection strategy. In: Proceedings of the American Control Conference, pp. 223–228 (2001)
18. Sun, Z., Ge, S.S.: Switched Linear Systems. Springer, London (2005)
19. Utkin, V.I.: Sliding Modes in Control and Optimization. Springer, Berlin (1992)
20. Wicks, M., Peleties, P., DeCarlo, R.A.: Switched controller synthesis for the quadratic stabilization of a pair of unstable linear systems. Eur. J. Control **4**, 140–147 (1998)

Optimal Control of Nonlinear Elliptic PDEs – Theory and Optimization Methods

J. Coletsos(✉) and B. Kokkinis

Department of Mathematics, School of Applied Mathematics and Physics, National Technical University of Athens, Zografou Campus, 15780 Athens, Greece
{coletsos,bkok}@math.ntua.gr

Abstract. We consider an optimal control problem described by a second order elliptic partial differential equation, jointly nonlinear in the state and control with high monotone nonlinearity in the state, with control and state constraints, where the state constraints and the cost functional involve also the state gradient. Since this problem may have no classical solutions, it is also formulated in the relaxed form. The existence of an optimal relaxed control is proved in the relaxed case, without convexity assumptions, and various necessary conditions for optimality are established for the classical and the relaxed problem. For the numerical solution of these problems, we propose a penalized gradient projection method generating classical controls, and a penalized conditional descent method generating relaxed controls. Using also relaxation theory, the behavior in the limit of sequences generated by these methods is examined. Finally, numerical examples are given.

1 Introduction

Relaxation in Optimal Control has been studied by various authors, mainly by Warga [10,11] and Roubíček [9]. It has been introduced, initially, in order to prove existence of optimal controls, then to derive necessary conditions for optimality, and recently for developing optimization and discretization methods. Methods using relaxed controls have been considered in [2,4–6,8,11]. Relaxed controls have been applied to elliptic PDEs in [3,4,7].

2 Classical and Relaxed Problems – Existence and Optimality Conditions

Let Ω be a bounded domain in $\mathbb{R}^d$, with Lipschitz boundary Γ. Consider the nonlinear elliptic state equation

$$Ay + f(x, y(x), w(x)) = 0 \text{ in} \Omega, \quad y(x) = 0 \text{ on} \Gamma,$$

I. Lirkov et al. (Eds.): LSSC 2013, LNCS 8353, pp. 81–89, 2014.
DOI: 10.1007/978-3-662-43880-0_8,

where A is the formal second order elliptic differential operator

$$Ay := -\sum_{i,j=1}^{d} (\partial/\partial x_i)[a_{ij}(x)\partial y/\partial x_j].$$

The state equation will be interpreted in the following weak form

$y \in V := H_0^1(\Omega)$ and $a(y,v) + \int_\Omega f(x, y(x), w(x))v(x)dx = 0, \quad \forall v \in V,$
where $a(\cdot,\cdot)$ denotes the usual bilinear form on $V \times V$ associated with A

$a(y,v) := \sum_{i,j=1}^{d} \int_\Omega a_{ij}(x) \frac{\partial y}{\partial x_i} \frac{\partial v}{\partial x_j} dx.$

We define the set of ***classical controls***

$W := \{w : \Omega \to U |\, w \text{ measurable}\},$

where U is a compact subset of $\mathbb{R}^\nu$, and the functionals

$$G_m(w) := \int_\Omega g_m(x, y(x), \nabla y(x), w(x))dx, \quad m = 0, ..., q.$$

The classical optimal control problem P is to minimize $G_0(w)$ subject to the constraints

$$w \in W, \quad G_m(w) = 0, \quad m = 1, ..., p, \quad G_m(w) \leq 0, \quad m = p+1, ..., q.$$

It is well known that, even if the set U is convex, the problem P may have no solutions. The existence of such a solution is usually proved under strong, often unrealistic for nonlinear systems, convexity assumptions (e.g. Cesari property). Reformulated in the so-called relaxed form, the problem has a solution in a larger space under weaker assumptions.

Next, we define the set of ***relaxed controls*** (see [9,10])

$R := \{r : \Omega \to M_1(U) |\, r \text{ weakly measurable}\} \subset L_w^\infty(\Omega, M(U)) \equiv L^1(\Omega, C(U))^*,$
where $M(U)$ (resp. $M_1(U)$) is the set of Radon (resp. probability) measures on U.

The relaxed controls are thus probability measures defined on U and measurably depended on x. Standard examples of relaxed controls are (i) the Dirac controls and (ii) the probability measures depending on x and defined by a probability distribution function on U.

The set R is endowed with the relative weak star topology, and R is convex, metrizable and compact. If each classical control $w(\cdot)$ is identified with its associated Dirac relaxed control $r(\cdot) := \delta_{w(\cdot)}$, then W may be regarded as a subset of R, and W is thus dense in R. For $\phi \in L^1(\bar{\Omega}; C(U))$ (or $\phi \in B(\bar{\Omega}, U; \mathbb{R})$, where $B(\bar{\Omega}, U; \mathbb{R})$ is the set of Caratheodory functions bounded by an integrable function) and $r \in L_w^\infty(\Omega, M(U))$ (in particular, for $r \in R$), we shall use the simplified notation

$\phi(x, r(x)) := \int_U \phi(x, u) r(x)(du),$

and $\phi(x, r(x))$ is thus linear (under convex combinations, for $r \in R$) in r since the control r acts as an integration with respect to a probability measure on U.

A sequence (r_k) converges to $r \in R$ in R if
$\lim_{k\to\infty} \int_\Omega \phi(x, r_k(x))dx = \int_\Omega \phi(x, r(x))dx$,
for every $\phi \in L^1(\Omega; C(U))$, or $\phi \in B(\bar{\Omega}, U; \mathbb{R})$, or $\phi \in C(\bar{\Omega} \times U)$.

The relaxed optimal control problem $\bar{P}$ is then defined by replacing w by r, with the above notation, and W by R, in the problem P.

Assumptions 1: The coefficients a_{ij} satisfy the ellipticity condition

$$\sum_{i,j=1}^{d} a_{ij}(x) z_i z_j \geq \alpha_0 \sum_{i=1}^{d} z_i^2, \quad \forall z_i, z_j \in \mathbb{R}, \quad x \in \Omega,$$

with $\alpha_0 > 0$, $a_{ij} \in L^\infty(\Omega)$, and that the functions f and f_y are defined on $\Omega \times \mathbb{R} \times U$, measurable for fixed y, u, continuous for fixed x, and satisfy

$$|f(x, 0, u)| \leq \phi_0(x), \quad \forall (x, u) \in \Omega \times U,$$

where $\phi_0 \in L^s(\Omega)$, with $s \geq 2$, $s > d/2$ (e.g. $s = 2$, for $d \leq 3$), and

$$0 \leq f_y(x, y, u) \leq \phi_1(x)\, \eta_1(|y|), \quad \forall (x, y, u) \in \Omega \times \mathbb{R} \times U,$$

where η_1 is an increasing function from $[0, +\infty)$ to $[0, +\infty)$, $\phi_1 \in L^\infty(\Omega)$ if the functionals G_m depend on ∇y, and $\phi_1 \in L^s(\Omega)$ otherwise. The above inequalities remain valid after the relaxation of the data, since the relaxed controls act as a mean value w.r.t. a probability measure.

It follows directly from Theorem 3.1 in [1] that, for every relaxed control $r \in R$, the state equation has a unique solution $y := y_r \in V \cap C^\alpha(\bar{\Omega})$, for some $\alpha \in (0, 1)$. Moreover, there exists constants C, $\bar{C}$ such that
$\|y_r\|_1 + \|y_r\|_\infty \leq C$, $\|y_r\|_{C^\alpha} \leq \bar{C}$, for every $r \in R$.

Assumptions 2: The functions g_m are defined on $\Omega \times \mathbb{R}^{d+1} \times U$, measurable for fixed y, y', u, continuous for fixed x, and satisfy
$|g_m(x, y, y', u)| \leq \psi_{0m}(x) + \beta_{0m}|y'|^2$, $\forall (x, y, y', u) \in \Omega \times \mathbb{R}^{d+1} \times U$ with $|y| \leq C'$, where $C' > C$, $\psi_{0m} \in L^1(\Omega)$, $\beta_{0m} \geq 0$.

Theorem 1. *(i) Under Assumptions 1, the operator $r \mapsto y_r$ (resp. $w \mapsto y_w$), from R (resp. W with the relative topology of $L^2(\Omega; \mathbb{R}^\nu)$, hence of $L^\infty(\Omega; \mathbb{R}^\nu)$) to V, and to $C_0(\bar{\Omega})$, is continuous.*

(ii) Under Assumptions 1 and 2, the functionals $r \mapsto G_m(r)$ on R (resp. $w \mapsto G_m(w)$ on W with the same topologies) are continuous.

(iii) Under Assumptions 1 and 2, if the problem $\bar{P}$ has an admissible control, then it has a solution.

Assumptions 3: The functions f, f_u (resp. $g_m, g_{my}, g_{my'}, g_{mu}$) are defined on $\Omega\times\mathbb{R}\times U'$ (resp. $\Omega\times\mathbb{R}^{d+1}\times U'$), where U' is an open set containing U, measurable on Ω for fixed $(y, u) \in \mathbb{R} \times U$ (resp. $(y, y', u) \in \mathbb{R}^{d+1} \times U$), continuous on $\mathbb{R} \times U$ (resp. $\mathbb{R}^{d+1} \times U$) for fixed $x \in \Omega$, and satisfy

$|f_u(x, y, u)| \leq \phi_2(x)$, $\quad |g_{my}(x, y, y', u)| \leq \psi_{1m}(x) + \beta_{1m}|y'|^{\frac{2(\rho-1)}{\rho}}$,

$|g_{my'}(x, y, y', u)| \leq \psi_{2m}(x) + \beta_{2m}|y'|$, $\quad |g_{mu}(x, y, y', u)| \leq \psi_{3m}(x) + \beta_{3m}|y'|$,

$\forall(x, y, y', u) \in \Omega \times \mathbb{R}^{d+1} \times U$, with $|y| \leq C'$,

where $C' < C$, $\phi_2, \psi_{im} \in L^2(\Omega)$, $\beta_{im} \geq 0$, $2 \leq \rho < \infty$ if $d = 1$ or 2, $2 \leq \rho < \frac{2d}{d-2}$ if $d \geq 3$.

The following Lemma 1 and Theorems 2 and 3 can be proved by using the techniques of [4,9] (the weak relaxed minimum principle in Theorem 2 is proved as in [6]).

Lemma 1. Under Assumptions 1, 2 and 3, where those containing the derivatives in u are omitted, (resp. included) and dropping the index m in g_m, G_m, the directional derivative of the functional G defined on R (resp. W, with U convex) is given by

$DG(r, \bar{r} - r) = \lim\limits_{\alpha \to 0+} \frac{G(r+\alpha(\bar{r}-r))-G(r)}{\alpha}$

$= \int_\Omega H(x, y(x), \nabla y(x), z(x), \bar{r}(x) - r(x))dx$, for $r, \bar{r} \in R$,

(resp. $DG(w, \bar{w} - w) = \lim\limits_{\alpha \to 0+} \frac{G(w+\alpha(\bar{w}-w))-G(w)}{\alpha}$

$= \int_\Omega H_u(x, y(x), \nabla y(x), z(x), w(x))(\bar{w}(x) - w(x))dx$, for $w, \bar{w} \in W$),

where the Hamiltonian is defined by

$H(x, y, y', z, u) := -z\, f(x, y, u) + g(x, y, y', u)$,

and the adjoint state $z := z_r \in V$ (resp. $z := z_w \in V$) satisfies the linear adjoint equation

$a(v, z) + (f_y(y, r)z, v) = (g_y(y, \nabla y, r), v) + (g_{y'}(y, \nabla y, r), \nabla v)$,

(resp. $a(v, z) + (f_y(y, w)z, v) = (g_y(y, \nabla y, w), v) + (g_{y'}(y, \nabla y, w), \nabla v)$),

$\forall v \in V$, with $y := y_r$ (resp. $y := y_w$).

Moreover, the operator $r \mapsto z_r$, from R to V (resp. $w \mapsto z_w$, from W to V), and the functional $(r, \bar{r}) \mapsto DG(r, \bar{r} - r)$, on $R \times R$ (resp. $(w, \bar{w}) \mapsto DG(w, \bar{w} - w)$, on $W \times W$), are continuous.

Theorem 2. *Under Assumptions 1, 2 and 3, where those containing the derivatives in u are omitted, if $r \in R$ is optimal for the problem $\bar{P}$* **or** *the problem P, then r is* ***strongly extremal relaxed****, i.e. there exist multipliers $\lambda_m \in \mathbb{R}$, $m = 0, ..., q$, with $\lambda_0 \geq 0$, $\lambda_m \geq 0$, $m = p+1, ..., q$, $\sum\limits_{m=0}^{q} |\lambda_m| = 1$, such that*

(1) $\sum\limits_{m=0}^{q} \lambda_m DG_m(r, \bar{r} - r) \geq 0$, *for every $\bar{r} \in R$,*

(2) $\lambda_m G_m(r) = 0, \quad m = p+1, ..., q$ *(relaxed transversality conditions).*

The condition (1) is equivalent to the strong relaxed pointwise minimum principle

$H(x, y(x), \nabla y(x), z(x), r(x)) = \min\limits_{u \in U} H(x, y(x), \nabla y(x), z(x), u)$, *a.e. in Ω,*

where the ***complete*** *Hamiltonian and adjoint H, z are defined with* $g := \sum\limits_{m=0}^{q} \lambda_m g_m$.

If U is convex, then this principle implies the weak relaxed pointwise minimum principle

$H_u(x, y, \nabla y(x), z, r(x))r(x) = \min\limits_{\phi} H_u(x, y, \nabla y(x), z, r(x))\phi(x, r(x))$, *a.e.in Ω,*

where the minimum is taken over the set $B(\bar{\Omega}, U; U)$ of Caratheodory functions (see [10]), which in turn implies the global weak relaxed condition

(3) $\int_\Omega H_u(x, y, \nabla y(x), z, r(x))[\phi(x, r(x)) - r(x)]dx \geq 0, \quad \forall \phi \in B(\bar{\Omega}, U; U)$.

A control r satisfying the conditions (3) and (2) is called ***weakly extremal relaxed***. If there are only inequality state constraints and if we impose in addition the regularity condition $G_m(r) + DG_m(r, r' - r) < 0, \quad m = p+1, ..., q$, for some $r' \in R$, then we have $\lambda_0 > 0$.

Theorem 3. *Under Assumptions 1, 2 and 3, with U convex, if $w \in W$ is optimal for the problem P, then w is* ***weakly extremal classical****, i.e. there exist multipliers $\lambda_m \in \mathbb{R}$, $m = 0, ..., q$, with $\lambda_0 \geq 0$, $\lambda_m \geq 0$, $m = p+1, ..., q$, $\sum_{m=0}^{q} |\lambda_m| = 1$, such that*

(4) $\sum_{m=0}^{q} \lambda_m DG_m(w, \bar{w} - w) \geq 0$, *for every* $\bar{w} \in W$,

(5) $\lambda_m G_m(w) = 0, \quad m = p+1, ..., q$ *(classical transversality conditions).*

The condition (4) is equivalent to the weak classical pointwise minimum principle

$$H_u(x, y(x), \nabla y(x), z(x), w(x))w(x) = \min_{u \in U} H_u(x, y(x), \nabla y(x), z(x), w(x))u, \text{ a.e. in } \Omega,$$

where the ***complete*** *Hamiltonian and adjoint H, z are defined with $g := \sum_{m=0}^{q} \lambda_m g_m$.*

3 Classical and Relaxed Optimization Methods

Let (M_m^l), $m = 1, ..., q$, be positive increasing sequences such that $M_m^l \to \infty$ as $l \to \infty$, $\gamma > 0$, $b, c \in (0, 1)$, and (β^l), (ζ_k) positive sequences, with (β^l) decreasing and converging to zero, and $\zeta_k \leq 1$. Define first the penalized functionals on W

$$G^l(w) := G_0(w) + \{\sum_{m=1}^{p} M_m^l [G_m(w)]^2 + \sum_{m=p+1}^{q} M_m^l [\max(0, G_m(w))]^2\}/2.$$

It can be easily shown that the directional derivative of G^l is given by

$$DG^l(w, w' - w) = DG_0(w, w' - w) + \sum_{m=1}^{p} M_m^l G_m(w) DG_m(w, w' - w)$$

$$+ \sum_{m=p+1}^{q} M_m^l \max(0, G_m(w)) DG_m(w, w' - w).$$

The classical penalized gradient projection method is described by the following Algorithm, where U is assumed to be convex.

Algorithm 1.

Step 1. Set $k := 0$, $l := 1$, and choose an initial control $w_0^1 \in W$.

Step 2. Find $v_k^l \in W$ such that

$$e_k := DG^l(w_k^l, v_k^l - w_k^l) + \frac{\gamma}{2}\left\|v_k^l - w_k^l\right\|^2 = \min_{\bar{v} \in W} [DG^l(w_k^l, \bar{v} - w_k^l) + \frac{\gamma}{2}\left\|\bar{v} - w_k^l\right\|^2],$$

and set $d_k := DG^l(w_k^l, v_k^l - w_k^l)$.

Step 3. If $|d_k| \le \beta^l$, set $w^l := w_k^l$, $v^l := v_k^l$, $d^l := d_k$, $e^l := e_k$, $w_k^{l+1} := w_k^l$, $l := l + 1$, and go to step 2.
Step 4. (Modified Armijo Step Search) Find the lowest integer value $s \in \mathrm{Z}$, say $\bar{s}$, such that $\alpha(s) = c^s \zeta_k \in (0, 1]$ and $\alpha(s)$ satisfies
$G^l(w_k^l + \alpha(s)(v_k^l - w_k^l)) - G^l(w_k^l) \le \alpha(s) b d_k$, and then set $\alpha_k := \alpha(\bar{s})$.
Step 5. Set $w_{k+1}^l := w_k^l + \alpha_k(v_k^l - w_k^l)$, $k := k + 1$, and go to step 2.

A (classical or relaxed) extremal (or weakly extremal) control is called ***abnormal*** if there exist multipliers as in the corresponding optimality conditions, with $\lambda_0 = 0$. A control is admissible and abnormal extremal in rather exceptional situations (see [10]).

With w^l as defined in step 3, define the sequences of multipliers
$\lambda_m^l := M_m^l G_m(w^l)$, $m = 1, ..., p$, $\lambda_m^l := M_m^l \max(0, G_m(w^l))$, $m = p+1, ..., q$.

Theorem 4. *We suppose that Assumptions 1, 2 and 3 hold and that U is convex.*

(i) In the presence of state constraints, if the whole sequence $(w_k^{l(k)})_{k \in \mathrm{N}}$ generated by Algorithm 1 converges to some $w \in W$ in L^2 strongly and the sequences (λ_m^l) are bounded, then w is admissible and weakly extremal for the classical problem. In the absence of state constraints, if a subsequence $(w_k)_{k \in K}$ (no index l) converges to some $w \in W$ in L^2 strongly, then w is weakly extremal classical for the problem P.

(ii) In the presence of state constraints, if a subsequence $(w^l)_{l \in L}$ of the sequence generated by Algorithm 1 in step 3, regarded as a sequence of relaxed controls, converges to some r in R, and the sequences $(\lambda_m^l)_{l \in L}$ are bounded, then r is admissible and weakly extremal relaxed for the relaxed problem. In the absence of state constraints, if a subsequence $(w_k)_{k \in K}$ (no index l) converges to some r in R, then r is weakly extremal relaxed for the problem $\bar{P}$.

(iii) In any of the convergence cases (i) or (ii) with state constraints, suppose that the classical, or the relaxed, problem has no admissible, abnormal extremal, controls. If the limit control is admissible, then the sequences (λ_m^l) are bounded, and this control is also extremal as above.

Next, we define the penalized functionals on R
$G^l(r) := G_0(r) + \{\sum_{m=1}^{p} M_m^l [G_m(r)]^2 + \sum_{m=p+1}^{q} M_m^l [\max(0, G_m(r))]^2\}/2.$
The directional derivative of G^l is given by

$$DG^l(r, r' - r) = DG_0(r, r' - r) + \sum_{m=1}^{p} M_m^l G_m(r) DG_m(r, r' - r)$$
$$+ \sum_{m=p+1}^{q} M_m^l \max(0, G_m(r)) DG_m(r, r' - r).$$

The relaxed penalized conditional descent method is described by the following Algorithm, where U is not necessarily convex.

Algorithm 2.
Step 1. Set $k := 0$, $l := 1$, and choose an initial control $r_0^1 \in R$.
Step 2. Find $\bar{r}_k^l \in R$ such that $d_k := DG^l(r_k^l, \bar{r}_k^l - r_k^l) = \min_{r' \in R} DG^l(r_k^l, r' - r_k^l)$.
Step 3. If $|d_k| \leq \beta^l$, set $r^l := r_k^l$, $\bar{r}^l := \bar{r}_k^l$, $d^l := d_k$, $r_k^{l+1} := r_k^l$, $l := l+1$, and go to step 2.
Step 4. Find the lowest integer value $s \in \mathrm{Z}$, say $\bar{s}$, such that $\alpha(s) = c^s \zeta_k \in (0, 1]$ and $\alpha(s)$ satisfies $G^l(r_k^l + \alpha(s)(\bar{r}_k^l - r_k^l)) - G^l(r_k^l) \leq \alpha(s) b d_k$, and then set $\alpha_k := \alpha(\bar{s})$.
Step 5. Choose any $r_{k+1}^l \in R$ such that $G^l(r_{k+1}^l) \leq G^l(r_k^l + \alpha_k(\bar{r}_k^l - r_k^l))$, set $k := k + 1$, and go to step 2.

If the chosen initial control r_0^1 is classical, using Caratheodory's theorem it can be shown by induction that the control r_{k+1}^l in step 5 can be chosen, for each iteration k, to be a Gamkrelidze relaxed control, i.e. a convex combination of a fixed number of classical (Dirac) controls.

With r^l as defined in step 3, define the sequences of multipliers
$\lambda_m^l := M_m^l G_m(r^l), m = 1, ..., p, \lambda_m^l := M_m^l \max(0, G_m(r^l)), m = p + 1, ..., q.$

Theorem 5. *We suppose that Assumptions 1, 2 and 3 hold. The Assumptions containing the derivatives in u are omitted.*

(i) In the presence of state constraints, if a subsequence $(r^l)_{l \in L}$ *of the sequence generated by Algorithm 2 in step 3 converges to some* $r \in R$ *and the sequences* (λ_m^l) *are bounded, then* r *is admissible and strongly extremal relaxed for the relaxed problem. In the absence of state constraints, if a subsequence* $(r_k)_{k \in K}$ *(no index* l*) converges to some* r *in* R*, then* r *is strongly extremal relaxed for the problem* $\bar{P}$.

(ii) In case (i) with state constraints, suppose that the relaxed problem has no admissible, abnormal extremal, controls. If r *is admissible, then the sequences* (λ_m^l) *are bounded and* r *is also strongly extremal relaxed for the problem* $\bar{P}$.

Finally, Gamkrelidze relaxed controls computed by Algorithm 2, can be approximated by classical controls using a standard procedure (see e.g. [9]).

4 Numerical Examples

Example 1. Let $\Omega := (0, 1)^2$. Define the reference control and state
$\bar{u}(x) := \bar{v}(x) := x_1 x_2, \quad \bar{y}(x) := 8x_1 x_2(1 - x_1)(1 - x_2),$
and consider the following classical optimal control problem, with state equation
$-\Delta y + y^3/3 + (1 + u - \bar{u}))y$
$-\bar{y}^3/3 - \bar{y} - 16[x_1(1 - x_1) + x_2(1 - x_2)] - (v - \bar{v}) = 0, \text{ in } \Omega,$
$y(x) = 0 \text{ on } \Gamma,$
control constraints $(u(x), v(x)) \in U := [0, 0.7]^2$, $x \in \Omega$, cost functional
$G_0(u, v) := 0.5 \int_\Omega [(y - \bar{y})^2 + \|\nabla y - \nabla \bar{y}\|_2^2 + (u - \bar{u})^2 + (v - \bar{v})^2] dx$
and state constraint $G_1(u, v) := \int_\Omega (y - 0.22) dx = 0$.
Algorithm 1 was applied to this problem using the finite element method with continuous piecewise linear basis functions on triangular elements (half squares

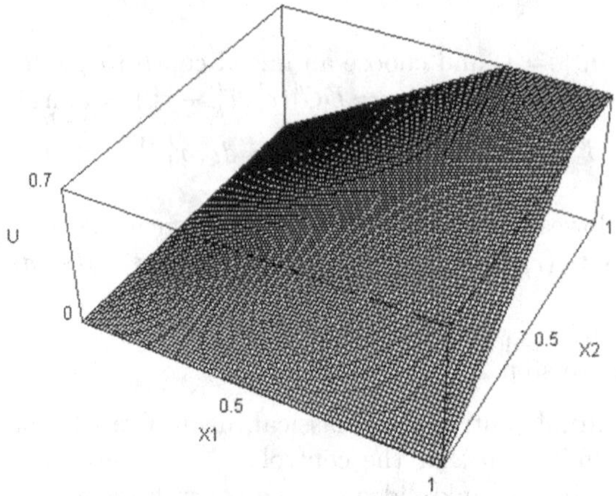

Fig. 1. Last control u

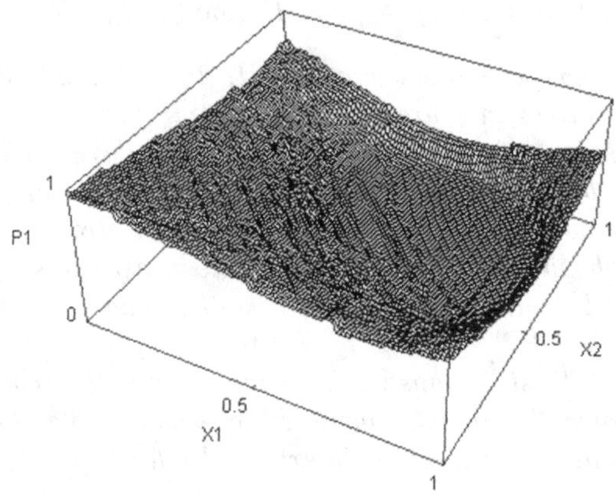

Fig. 2. Last relaxed control probability p_1

of edge size $h = 0.01$) for solving the differential equations, with (not necessarily continuous) elementwise linear classical controls, with $\gamma = 0.5$, Armijo parameters $b = c = 0.5$. After 60 iterations in k we obtained the results
$G_0(u_k^l, v_k^l) = 2.292743040985 \cdot 10^{-3}, G_1(u_k^l, v_k^l) = 8.679 \cdot 10^{-6}, d_k = -4.449 \cdot 10^{-5}$.
The last control u_k is shown in Fig. 1.

Example 2. Let $\Omega := (0,1)^2$ Define the reference control and state

$$\bar{w}(x) := \max(-1, 1 - 1.5(x_1 + x_2)), \quad \bar{y}(x) := 8x_1x_2(1 - x_1)(1 - x_2),$$

and consider the following optimal control problem, with state equation

$$-\Delta y + y^3/3 + (2 + w - \bar{w})y - \bar{y}^3/3 - 2\bar{y} - 16[x_1(1-x_1) + x_2(1-x_2)] = 0 \text{ in } \Omega,$$
$$y(x) = 0 \text{ on } \Gamma,$$

nonconvex control constraint set $U := \{-1\} \cup [0.5, 1]$, nonconvex cost functional

$G_0(w) := \int_\Omega \{0.5\,[(y-\bar{y})^2 + \|\nabla y - \nabla \bar{y}\|_2^2] - w^2 + 1\}dx$
and state constraint $G_1(w) := \int_\Omega (y - 0.22)dx = 0$.

Applying Algorithm 2, we obtained after 200 iterations in k the results $G_0(r_k^l) = 2.600332334904 \cdot 10^{-4}$, $G_1(r_k^l) = 1.552 \cdot 10^{-6}$, $d_k = -2.619 \cdot 10^{-6}$.

The last relaxed control probability function $p_1(x) := r_k^l(x)\{1\}$ is shown in Fig. 2.

References

1. Bonnans, J., Casas, E.: Un principe de Pontryagine pour le contrôle des systèmes semilineaires elliptiques. J. Diff. Equat. **90**, 288–303 (1991)
2. Cartensen, C., Roubíček, T.: Numerical approximation of young measure in non-convex variational problems. Numer. Math. **84**, 395–415 (2000)
3. Casas, E.: The relaxation theory applied to optimal control problems of semilinear elliptic equations. In: Proceedings of the 17th Conference on Systems, Modeling and Optimization, Prague (1995). Chapman and Hall (1996)
4. Chryssoverghi, I., Kokkinis, B.: Discretization of nonlinear elliptic optimal control problems. Syst. Control Lett. **22**, 227–234 (1994)
5. Chryssoverghi, I., Coletsos, J., Kokkinis, B.: Discretization methods for optimal control problems with state constraints. J. Comput. Appl. Math. **191**, 1–31 (2006)
6. Chryssoverghi, I.: Discretization methods for semilinear parabolic optimal control problems. Int. J. Numer. Anal. Modeling **3**, 437–458 (2006)
7. Hongwei, Lou: Existence of optimal controls of semilinear elliptic equations without Cesari type conditions. ANZIAM J. **45**, 115–131 (2003)
8. Mataché, A.-M., Roubíček, T., Schwab, C.: Higher-order convex approximations of Young measures in optimal control. Adv. Comp. Math. **19**, 73–79 (2003)
9. Roubíček, T.: Relaxation in Optimization Theory and Variational Calculus. Walter de Gruyter, Berlin (1997)
10. Warga, J.: Optimal Control of Differential and Functional Equations. Academic Press, New York (1972)
11. Warga, J.: Steepest descent with relaxed controls. SIAM J. Control Optim. **15**, 674–682 (1997)

The Euler Method for Linear Control Systems Revisited

Josef L. Haunschmied[1], Alain Pietrus[2], and Vladimir M. Veliov[1](✉)

[1] Institute of Mathematical Methods in Economics,
Vienna University of Technology, Vienna, Australia
{Josef.Haunschmied,veliov}@tuwien.ac.at
[2] Laboratoire LAMIA, Dépt. de Mathématiques,
Université des Antilles et de la Guyane, Pointe-à-Pitre, Guadeloupe
apietrus@univ-ag.fr

Abstract. Although optimal control problems for linear systems have been profoundly investigated in the past more than 50 years, the issue of numerical approximations and precise error analyses remains challenging due the bang-bang structure of the optimal controls. Based on a recent paper by M. Quincampoix and V.M. Veliov on metric regularity of the optimality conditions for control problems of linear systems the paper presents new error estimates for the Euler discretization scheme applied to such problems. It turns out that the accuracy of the Euler method depends on the "controllability index" associated with the optimal solution, and a sharp error estimate is given in terms of this index. The result extends and strengthens in several directions some recently published ones.

1 Introduction

In this paper we revisit the Euler discretization method applied to the following optimal control problem:

$$\min g(x(T)) \tag{1}$$

subject to

$$\dot{x}(t) = A(t)\, x(t) + B(t)\, u(t), \quad x(0) = x_0, \tag{2}$$

$$u(t) \in U. \tag{3}$$

Here $x \in \mathbf{R}^n$, $u \in U \subset \mathbf{R}^r$, the time interval $[0,T]$ is fixed, $g : \mathbf{R}^n \to \mathbf{R}$, and A and B are matrix functions with appropriate dimensions. The initial state x_0 is given. The control constraining set $U \subset \mathbf{R}^r$ is a convex compact polyhedron. The set of admissible controls, $\mathcal{U}$, consists of all measurable selections of U. The function $x = x[u] : [0,T] \to \mathbf{R}^n$ is a solution of (2) for a given $u \in \mathcal{U}$ if it is absolutely continuous and satisfies (2) for a.e. $t \in [0,T]$.

This research is supported by the Austrian Science Foundation (FWF) under grant No I 476-N13. The paper was written during the visit of the third author at Université des Antilles et de la Guyane, Feb., 2013.

I. Lirkov et al. (Eds.): LSSC 2013, LNCS 8353, pp. 90–97, 2014.
DOI: 10.1007/978-3-662-43880-0_9,

Utilization of the Euler discretization scheme for this problem results in the following discrete-time optimal control problem:

$$\min g(x_N) \tag{4}$$

subject to

$$x_{i+1} = x_i + h\,(A(t_i)\,x_i + B(t_i)\,u_i)\,, \quad x_0 - given, \tag{5}$$

$$u_i \in U, \quad i = 0, \ldots, N-1, \tag{6}$$

where N is a natural number, $h = T/N$, $t_i = ih$. The unknown variables here are $x_1, \ldots, x_N$ and $u_0, \ldots, u_{N-1}$.

The error analysis of the above discretization is burdened by the fact that the optimal control in problem (1)–(3) is typically discontinuous.

In the past few years a number of papers appeared that investigate the accuracy of discrete approximations of optimal control problems with a bang-bang structure of the optimal control. The first one seems to be [10], followed by [1–3]. The main result in the present paper shows that the accuracy of the approximation (measured in the relevant metric defined in the next section) provided by the Euler scheme is $O(h^{1/k})$, where k is the so-called *controllability index* of the optimal solution of problem (1)–(3). A comparison of this result with the abovementioned ones is given in the end of Sect. 3.

We mention also that the error analyses of discrete approximations to control problems for linear systems is facilitated by the recent papers [4–7]. However, our analyses is based on the "companion" paper [9], which extends in an appropriate way the concept of metric regularity of the optimality conditions for optimal control of linear systems.

The organization of the paper is as follows. In the next section we present some material from [9], which is needed for the proof of our main result. The order of accuracy of the Euler scheme is proved in Sect. 3. In the last section we give a numerical example that supports the theoretical result.

2 Assumptions and Preliminaries

In this section we present some necessary preliminary material from the "companion" paper [9]. We begin with some assumptions for problem (1)–(3).

Assumption (A1): The functions $A : [0,T] \to \mathbf{R}^{n\times n}$ and $B : [0,T] \to \mathbf{R}^{n\times r}$ are $\bar{k}$ times, respectively $\bar{k}+1$ times, continuously differentiable (for some natural number $\bar{k}$). Moreover, $g : \mathbf{R}^n \to \mathbf{R}$ is convex and differentiable with a locally Lipschitz derivative.

The reachable set $R = \{x[u](T) : u \in \mathcal{U}\}$ is a convex and compact subset of $\mathbf{R}^n$, hence problem (1)–(3) has at least one solution $(\hat{x}, \hat{u})$.

Define the sequence of matrices

$$B_0(t) = B(t), \quad B_{i+1}(t) = -A(t)B_i(t) + \dot{B}_i(t), \quad i = 0, \ldots, \bar{k}-1. \tag{7}$$

Moreover, denote by E the set of all (non-degenerate) edges of U, and by $\bar{E}$ – the set of all vectors $u_2 - u_1$, where $[u_1, u_2] \in E$.

Assumption (A2): $\text{rank}[B_0(t)\,e, \ldots, B_{\bar{k}}(t)\,e] = n$ for every $e \in \bar{E}$ and every $t \in [0, T]$. Moreover, $\nabla g(x) \neq 0$ for every $x \in R$ (∇g denotes the gradient of g).

The rank condition in the above assumption is the well-known *general position hypotheses* [8]. The second part of the assumption makes the problem meaningful, since it rules out the possibility of infinitely many solutions.

The Pontryagin maximum principle claims that any optimal pair $(\hat{x}, \hat{u})$ together with a corresponding absolutely continuous function $\hat{p} : [0, T] \to \mathbf{R}^n$ satisfies the following (generalized) equations:

$$0 = \dot{x}(t) - A(t)\,x(t) - B(t)\,u(t), \quad x(0) = x_0, \tag{8}$$
$$0 = \dot{p}(t) + A^\top(t)\,p(t), \tag{9}$$
$$0 \in B^\top(t)\,p(t) + N_U(u(t)), \tag{10}$$
$$0 = p(T) - \nabla g(x(T)), \tag{11}$$

where $N_U(u)$ is the normal cone to U at u. Notice that (10) is equivalent to $u(t) \in \underset{w \in U}{\text{Argmin}}\, \langle B^\top(t)\,p(t), w \rangle$.

The following lemma is well-known.

Lemma 1. *Let the matrices A and B be measurable and essentially bounded, and let g be differentiable and convex. Then $(\hat{x}, \hat{u})$ is a solution of problem (1)–(3) if and only if the triple $(\hat{x}, \hat{p}, \hat{u})$ (with an absolutely continuous $\hat{p}$) is a solution of system (8)–(11). If (A1) and (A2) hold, then the solution $(\hat{x}, \hat{u})$ of (1)–(3) is unique, hence that of (8)–(11) is also unique. Moreover, $\hat{u}(t)$ is a vertex of U for a.e. $t \in [0, T]$.*

Let $(\hat{x}, \hat{u})$ be a solution of problem (1)–(3).

Definition 1. *Controllability index* of the solution $(\hat{x}, \hat{u})$ of problem (1)–(3) is the minimal number k such that for every $t \in [0, T]$ and for every $e \in \bar{E}$ at least one of the numbers $\langle B_i^\top(t)\,\hat{p}(t), e \rangle$, $i = 0, \ldots, k$, is not equal to zero. Here $\hat{p}$ is the solution of the equations (9), (11) with $x = \hat{x}$ and $u = \hat{u}$.

Clearly, if (A2) is fulfilled, then the number $k \leq \bar{k}$ exists.

The generalized equations (8)–(11) can be written in the form $0 \in F(x, p, u)$, where

$$F(x, p, u) := \begin{pmatrix} \dot{x} - A\,x - B\,u \\ \dot{p} + A^\top p \\ B^\top p + N_U(u) \\ p(T) - \nabla g(x(T)) \end{pmatrix}. \tag{12}$$

Thus the inclusion $0 \in F(x,p,u)$ is equivalent to our original problem (1)–(3). Namely, under (A1) and (A2) it has a unique solution $(\hat{x},\hat{p},\hat{u})$ and $(\hat{x},\hat{u})$ is the unique solution of problem (1)–(3).

The norms in $L^1(0,T)$ and $L^\infty(0,T)$ are denoted by $\|\cdot\|_1$ and $\|\cdot\|_\infty$, respectively. The notation $W^{1,s} = W^{1,s}([0,T];\mathbf{R}^n)$ (with $s=1$ or $s=\infty$) is used for the space of all absolutely continuous functions $x:[0,T]\to\mathbf{R}^n$ with the derivative $\dot{x}$ belonging to $L^s(0,T)$. The norm in this space is $\|x\|_{1,s} := \|x\|_\infty + \|\dot{x}\|_s$.

The set of admissible controls $\mathcal{U}$ is viewed as a subset of $L^\infty(0,T)$ equipped with the metric

$$d^\#(u_1,u_2) = \text{meas}\,\{t \in [0,T] : u_1(t) \neq u_2(t)\}.$$

This metric is shift-invariant and we shall shorten $d^\#(u_1,u_2) = d^\#(u_1-u_2,0) =: d^\#(u_1-u_2)$. Then the triple (x,p,u) is considered as an element of the (affine) space

$$\mathcal{X} = W^{1,1}_{x_0} \times W^{1,\infty} \times \mathcal{U},$$

where $W^{1,1}_{x_0} = \{x \in W^{1,1} : x(0) = x_0\}$.

The image space of F will be $\mathcal{Y} = L^1 \times L^\infty \times L^\infty \times \mathbf{R}^n$ with the norm

$$\|y\| = \|(\xi,\pi,\rho,\nu)\| := \|\xi\|_1 + \|\pi\|_\infty + \|\rho\|_\infty + |\nu|.$$

We interpret the set $N_U(u)$ in (12) as $\{\rho \in L^\infty : \rho(t) \in N_U(u(t))\ \forall t \in [0,T]\}$ (strictly speaking, we should use the notation $N_{\mathcal{U}}(u)$ instead of the point-wise $N_U(u(t))$, but the overload of the latter does not lead to confusions).

The following is a simplified version of [9, Theorem 2].

Theorem 1. *Let assumptions (A1) and (A2) be fulfilled, let $(\hat{x},\hat{p},\hat{u})$ be a solution of the generalized equation $0 \in F(x,p,u)$ (with F given in (12)) and let k be its controllability index. Then for every number $b > 0$ there exists a number c such that for every $y = (\xi,\pi,\rho,\nu) \in \mathcal{Y}$ with $\|y\| \le b$ and for every solution $(x,p,u) \in \mathcal{X}$ of the inclusion $y \in F(x,p,u)$ it holds that*

$$\|x-\hat{x}\|_{1,1} + \|p-\hat{p}\|_{1,\infty} + \|u-\hat{u}\|_1 \le c\,\|y\|^{\frac{1}{k}}. \tag{13}$$

3 Euler Discretization and Its Accuracy

Consider the discrete-time problem (4)–(6) as introduced in Sect. 1. The maximum principle for discrete-time optimal control problems claims that if $x^N = (x_0,\dots,x_N)$, $u^N = (u_0,\dots,u_{N-1})$ is a solution of problem (4)–(6) then

$$-B(t_i)^\top p_{i+1} \in N_U(u_i), \quad i = 0,\dots,N-1, \tag{14}$$

with $p^N = (p_0,\dots,p_N)$ determined from the equations

$$p_N = \nabla g(x_N), \tag{15}$$

$$p_i = p_{i+1} + hA(t_i)^\top p_{i+1}, \quad i = N-1,\dots,0. \tag{16}$$

We identify any sequence $u^N := (u_0, \ldots, u_{N-1})$ with its piece-wise constant extension: $u^N(t) = u_i$ for $t \in [t_i, t_{i+1})$, $i = 0, \ldots, N-1$. Moreover, we identify any sequence $x^N := (x_0, \ldots, x_N)$ with its piecewise linear interpolation:

$$x^N(t) = x_i + \frac{t - t_i}{h}(x_{i+1} - x_i), \quad t \in [t_i, t_{i+1}), \quad i = 0, \ldots, N-1.$$

Similarly for sequences $p^N := (p_0, \ldots, p_N)$. Then we can view such sequences (x^N, p^N, u^N) as elements of the space $\mathcal{X}$.

The main result in this paper follows.

Theorem 2. *Let assumptions (A1), (A2) be fulfilled and let $(\hat{x}, \hat{u})$ be the unique solution of problem (1)–(3). Let k be the controllability index of this solution. Then there exists a number C such that for any natural number N and corresponding $h = T/N$, and for any solution (x^N, u^N) of the discretized problem (4)–(6) and for the corresponding adjoint functions $\hat{p}$ and p^N (given by equations (9), (11) and (15), (16), respectively) it holds that*

$$\|x^N - \hat{x}\|_{1,1} + \|p^N - \hat{p}\|_{1,\infty} + \|u^N - \hat{u}\|_1 \le C\, h^{1/k}.$$

Moreover, if all u_i, $i = 0, \ldots, N-1$, are vertices of U, then

$$\|x^N - \hat{x}\|_{1,1} + \|p^N - \hat{p}\|_{1,\infty} + d^{\#}(u^N - \hat{u}) \le C\, h^{1/k}.$$

Proof. The proof is simple due to Theorem 1. Essentially, we only have to estimate the residual $y = (\xi, \pi, \rho, \nu)$ of (x^N, p^N, u^N) in the inclusion $0 \in F(x, p, u)$. That is, we have to ensure that there exists $y = (\xi, \pi, \rho, \nu) \in \mathcal{Y}$ such that $y \in F(x^N, p^N, u^N)$, and to estimate its norm.

First we mention that due to the boundedness of U there exists a number M (independent of N) such that any of the numbers $|u|$, $|\hat{x}(t)|$, $|\hat{p}(t)|$, $|x_i|$, $|p_i|$, where $u \in U$, $t \in [0, T]$, $i = 0, \ldots, N$, is smaller than M. Also, let K be such that $|A(t)| \le K$ and $|B(t)| \le K$ for $t \in [0, T]$, where we use the operator norms of matrices. Moreover, we denote by L a Lipschitz constant of A and B.

We define $\xi(t) = \dot{x}^N(t) - A(t)x^N(t) - B(t)u^N(t)$. Clearly $\xi \in L^1$ and for $t \in [t_i, t_{i+1})$ we have

$$\begin{aligned}
|\xi(t)| &= \left| \frac{x_{i+1} - x_i}{h} - A(t)\left(x_i + \frac{t - t_i}{h}(x_{i+1} - x_i)\right) - B(t)u_i \right| \\
&= \left| A(t_i)x_i + B(t_i)u_i - A(t)\left(x_i + \frac{t - t_i}{h}(x_{i+1} - x_i)\right) - B(t)u_i \right| \\
&= \left| (A(t_i) - A(t))x_i + (B(t_i) - B(t))u_i - A(t)\left(\frac{t - t_i}{h}(x_{i+1} - x_i)\right) \right| \\
&\le 2hLM + |A(t)|\,|(x_{i+1} - x_i)| \le 2\,LMh + Kh|A(t_i)x_i + B(t_i)u_i| \\
&\le 2LMh + 2K^2Mh = 2(L + K^2)Mh.
\end{aligned}$$

Hence,

$$\|\xi\|_1 \le 2T(L + K^2)Mh.$$

Now we consider $\pi(t) := \dot{p}^N(t) + A(t)^\top p^N(t)$. Obviously $\pi \in L^\infty$ and using (16) we obtain that for $t \in [t_i, t_{i+1}]$

$$\begin{aligned}
|\pi(t)| &= \left| \frac{p_{i+1} - p_i}{h} + A(t)^\top \left(p_i + \frac{t - t_i}{h}(p_{i+1} - p_i) \right) \right| \\
&= \left| -A(t_i)^\top p_{i+1} + A(t)^\top \left(p_i + \frac{t - t_i}{h}(p_{i+1} - p_i) \right) \right| \\
&\leq \left| -A(t_i)^\top p_{i+1} + A(t)^\top p_i \right| + K|p_{i+1} - p_i| \\
&\leq \left| -A(t_i)^\top p_{i+1} + A(t_i)^\top p_i \right| + \left| -A(t_i)^\top p_i + A(t)^\top p_i \right| + K|p_{i+1} - p_i| \\
&\leq K|p_{i+1} - p_i| + LMh + K|p_{i+1} - p_i| \\
&\leq (2K^2 + L)Mh.
\end{aligned}$$

In order to estimate the residual in (10) we define the function $\rho(t) := B(t)^\top p^N(t) - B(t_i)^\top p^N(t_{i+1})$, $t \in [t_i, t_{i+1})$, which obviously belongs to L^∞. First of all, for $t \in [t_i, t_{i+1})$ we have from (14) that

$$\rho(t) = \rho(t) + 0 \in \rho(t) + B(t_i)^\top p_{i+1} + N_U(u_i) = B(t)^\top p^N(t) + N_U(u^N(t)).$$

We estimate for $t \in [t_i, t_{i+1}]$

$$\begin{aligned}
|\rho(t)| &\leq |(B(t)^\top - B(t_i)^\top)p^N(t)| + |B(t_i)|\,|p^N(t) - p^N_{i+1}| \\
&\leq LMh + K|p^N_i - p^N_{i+1}| \leq LMh + MK^2h = (L + K^2)Mh.
\end{aligned}$$

The fourth residual is $\nu = 0$ since (11) is exactly satisfied by $p^N(T) = p_N$ and $x^N(T) = x_N$ due to (15).

Thus we have obtained so far that

$$\|y\| \leq 2T(L + K^2)Mh + (2K^2 + L)Mh + (L + K^2)Mh \leq (2T + 3)(L + K^2)Mh.$$

Now we apply Theorem 1 with $b = (2T + 3)(L + K^2)MT$ and with the corresponding number c from the formulation of this theorem. It claims that

$$\begin{aligned}
\|x^N - \hat{x}\|_{1,1} + \|p^N - \hat{p}\|_{1,\infty} + \|u^N - \hat{u}\|_1 &\leq c\,\|y\|^{\frac{1}{k}} \leq c((2T + 3)(L + K^2)Mh)^{1/k} \\
&=: c_1 h^{1/k}.
\end{aligned}$$

This proves the first claim of the theorem with $C = c_1$.

To prove the second claim of the theorem we assume that u_i, $i = 0, \ldots, N-1$, are vertices of U. We remind that according to Lemma 1 the values of $\hat{u}$ are also a.e. vertices of U. Then $|u^N(t) - \hat{u}(t)| \geq \eta$ whenever $u^N(t) \neq \hat{u}(t)$, where $\eta > 0$ is the minimal distance between different vertices of U. Then

$$\eta\, d^{\#}(u^N - \hat{u}) \leq \int_0^T |u^N(t) - \hat{u}(t)|\, dt \leq c_1 h^{1/k}.$$

This proves the second claim of the theorem with $C := c_1/\eta$. Q.E.D.

In the rest of this section we compare the above result with those in [1–3, 10].

General Runge-Kutta schemes of (at least) second order global accuracy (third order local consistency) were applied in [10] instead of the Euler scheme. The accuracy of the approximation in the metric in $\mathcal{X}$ is proved to be $O(h^{1/k})$, where k is the controllability index of the optimal solution of problem (1)–(3). In the present paper we show that the same order $h^{1/k}$ is achieved by the Euler discretization (also the assumptions for g are relaxed).

In the recent paper [3] the authors consider linear problems with a linear function g and with controllability index $k = 1$. The result is similar to the one implied by Theorem 2 for the case $k = 1$. The control constraining set U is assumed to be a coordinate box in [3], which is a technical simplification. The linearity of g, however, is a substantial simplification, since the adjoint system (9), (11) is independent of the state x and can be treated by the Euler scheme separately from the rest of the equations in (8)–(11).

Papers [1,2] make a substantial progress by considering a quadratic function g (also a quadratic in x integral term is present there). The $O(h)$ error estimate in this case (again with assumed $k = 1$) becomes nontrivial since the overall interconnected system (8)–(11) has to be investigated. However, its analysis is based on the structural stability of the switching structure of the optimal control (obtained in [4]). Such a stability is no longer valid if $k > 1$, which case is captured by Theorem 2. A different proof is needed in this case and in the present paper it is based on the results of [9].

4 A Numerical Test

The following test example is a slight modification of [3, Example 2.10]. The problem is

$$\begin{aligned} &\min x_3(5) \\ &\dot{x}_1 = -x_2 + u, \quad x_1(0) = 1, \\ &\dot{x}_2 = u, \quad x_2(0) = 1, \\ &\dot{x}_3 = x_1 - 0.5\, u, \quad x_3(0) = 0, \\ &u \in [0, 1]. \end{aligned}$$

The only difference with [3, Example 2.10] is the coefficient -0.5 in the last equation, which equals 4 in the quoted paper. Here assumptions (A1) and (A2) are fulfilled with $\bar{k} = 2$. Therefore, the controllability index of the unique optimal pair (x, u) is $k \leq 2$. In fact, for this example that $k = 2$. The numerical results presented on the table below show that the accuracy of the Euler approximation is $O(h^{1/2})$, indeed.

The optimal control in this problem is easily seen (by applying the maximum principle) to be $\hat{u}(t) \equiv 1$. However, the corresponding "switching function" $\sigma(t) = B(t)^\top p(t) = 5-t-0.5(5-t)^2-0.5$ has a double zero at $t = 4$. Thus $k = 2$ for this problem and the theoretical error estimate is $Err(h) := d^{\#}(u^N - \hat{u}) \leq C\sqrt{h}$. In the table below we show the quantities $Err(h)$, $Err(2h)/Err(h)$, which is expected to be about $C\sqrt{2h}/C\sqrt{h} = \sqrt{2}$, and $Err(h)/\sqrt{h}$, which is expected to be about the constant C.

h	$Err(h) := d^{\#}(u^N - \hat{u})$	$Err(2h)/Err(h)$	$C = Err(h)/\sqrt{h}$
0.01	0.28000000	1.4000	2.8000
$0.01/2$	0.20000000	1.4286	2.8284
$0.01/2^2$	0.14000000	1.4000	2.8000
$0.01/2^3$	0.10000000	1.4286	2.8284
$0.01/2^4$	0.07000000	1.4000	2.8000
$0.01/2^5$	0.05000000	1.4035	2.8284
$0.01/2^6$	0.03562500	1.4250	2.8500
$0.01/2^7$	0.02500000	1.4159	2.8284
$0.01/2^8$	0.01765625	1.4125	2.8250
$0.01/2^9$	0.01250000	$\sqrt{2} \approx 1.4142$	2.8284

References

1. Alt, W., Baier, R., Gerdts, M., Lempio, F.: Error bounds for Euler approximations of linear-quadratic control problems with bang-bang solutions. Numer. Algebra Control Optim. **2**(3), 547–570 (2012)
2. Alt, W., Seydenschwanz, M.: An implicit discretization scheme for linear-quadratic control problems with bang-bang solutions. Optim. Method Softw. **29**(3), 535–560 (2014)
3. Alt, W., Baier, R., Lempio, F., Gerdts, M.: Approximations of linear control problems with bang-bang solutions. Optimization **62**(1), 9–32 (2013)
4. Felgenhauer, U.: On stability of bang-bang type controls. SIAM J. Control Optim. **41**(6), 1843–1867 (2003)
5. Felgenhauer, U., Poggolini, L., Stefani, G.: Optimality and stability result for bang-bang optimal controls with simple and double switch behavior. Control Cybern. **38**(4B), 1305–1325 (2009)
6. Osmolovskii, N.P., Maurer, H.: Equivalence of second order optimality conditions for bang-bang control problems. Part 1: main results. Control Cybern. **34**, 927–950 (2005)
7. Osmolovskii, N.P., Maurer, H.: Equivalence of second order optimality conditions for bang-bang control problems. Part 2: proofs, variational derivatives and representations. Control Cybern. **36**, 5–45 (2007)
8. Pontryagin, L.S., Boltyanskij, V.G., Gamkrelidze, R.V., Mishchenko, E.F.: The Mathematical Theory of Optimal Processes, Fizmatgiz, Moscow, 1961. Pergamon, Oxford (1964)
9. Quincampoix, M., Veliov, V.M.: Metric regularity and stability of optimal control problems for linear systems. SIAM J. Control Optim. **51**(5), 4118–4137 (2013)
10. Veliov, V.M.: Error analysis of discrete approximation to bang-bang optimal control problems: the linear case. Control Cybern. **34**(3), 967–982 (2005)

On Control Synthesis for Uncertain Differential Systems Using a Polyhedral Technique

Elena K. Kostousova(✉)

N.N. Krasovskii Institute of Mathematics and Mechanics,
16, S.Kovalevskaja Street, Ekaterinburg 620990, Russia
kek@imm.uran.ru

Abstract. Problems of feedback terminal target control for linear uncertain systems are considered. We continue the development of polyhedral control synthesis using polyhedral (parallelotope-valued) solvability tubes. New control strategies, which can be calculated on the base of these tubes, are proposed. The cases without uncertainties, with additive parallelotope-valued uncertainties, and also with a bilinear uncertainty (interval uncertainties in coefficients of the system) are considered. Ordinary differential equations, which describe the mentioned tubes, are presented for each of these cases. Numerical results are presented.

Keywords: Differential systems · Uncertain systems · Control synthesis · Polyhedral estimates · Parallelotopes · Interval analysis

1 Introduction

The paper deals with the problems of terminal target feedback control for linear differential uncertain systems. There are known approaches to solving such problems, in particular, based on the notions of the Aumann integral (for the case without uncertainty) and the Pontriagin alternated integral, and on the extremal aiming strategies of N.N. Krasovskii [10]. The problem statement for linear control systems, approaches for solving, and the tight interconnections between solvability tubes (Krasovskii's bridges), the alternated integral, Hamilton-Jacobi-Bellman equations, and funnel equations can be found in [12–14]. Since practical construction of the mentioned tubes may be cumbersome, different numerical methods are devised for this cause, in particular, methods for approximating the set-valued integrals and for numerical solving the mentioned equations, including methods based on approximations of sets by arbitrary polytopes with a large number of vertices, [2,3,17,18] (here and below, we note, as examples, only some references from numerous publications; see also references therein). Such methods are devised to obtain approximations as accurate as possible. But they may require much calculations, especially for large dimensional systems. Other techniques are based on estimates of sets by domains of some fixed shape such as ellipsoids and parallelepipeds, including boxes aligned with coordinate axes as in

I. Lirkov et al. (Eds.): LSSC 2013, LNCS 8353, pp. 98–106, 2014.
DOI: 10.1007/978-3-662-43880-0_10,

the interval analysis [3,5–9,11,12,14–16]. The main advantage of the mentioned techniques is that they enable to obtain approximate/particular solutions using relatively simple tools (up to explicit formulas). More accurate approximations may be obtained by using the whole families (varieties) of such simple estimates (as it was proposed by A.B. Kurzhanski) [7,12,14,15]. In particular, constructive computation schemes for solving the feedback target control problems by ellipsoidal techniques were proposed [12,14] and then expanded to a polyhedral technique [7]. There are also many works devoted to other approaches to solving different control problems under uncertainty, for example, [1,3,4,16].

Here we continue the development of polyhedral control synthesis using polyhedral (parallelotope-valued) solvability tubes. New control strategies, which can be calculated on the base of these tubes, are proposed. In opposite to [7,12,14], they are concretized by explicit formulas when the state belongs to a tube. The cases without uncertainties, with additive uncertainties, and also with a bilinear uncertainty (interval uncertainties in coefficients of the system) are considered. Ordinary differential equations (ODE) for the mentioned tubes are presented. Also polyhedral control synthesis for discrete-time systems is considered. Results of computer simulations are presented too.

The following notation is used below: $\mathbb{R}^n$ is the n-dimensional vector space; $\top$ is the transposition symbol; $\|x\|_2 = (x^\top x)^{1/2}$, $\|x\|_\infty = \max_{1\le i\le n}|x_i|$ are vector norms for $x = (x_1, x_2, \ldots, x_n)^\top \in \mathbb{R}^n$; $\mathrm{e} = (1, 1, \ldots, 1)^\top$; $\mathbb{R}^{n\times m}$ is the space of real $n\times m$-matrices $A = \{a_i^j\} = \{a^j\}$ (with columns a^j); I is the unit matrix; 0 is the zero matrix (vector); $\mathrm{Abs}\, A = \{|a_i^j|\}$ for $A = \{a_i^j\}$; $\mathrm{diag}\,\pi$, $\mathrm{diag}\,\{\pi_i\}$ are the diagonal matrix A with $a_i^i = \pi_i$ (π_i are the components of the vector π); $\det A$ is the determinant of $A \in \mathbb{R}^{n\times n}$; $\mathrm{tr}\, A = \sum_{i=1}^n a_i^i$ is the trace of A; $\mathrm{int}\, \mathcal{X}$ is the set of interior points of the set $\mathcal{X} \subset \mathbb{R}^n$; the notation $k = 1, \ldots, N$ is used instead of $k = 1, 2, \ldots, N$.

2 Problem Formulation

Consider the controlled system with a given terminal set $\mathcal{M}$ ($x \in \mathbb{R}^n$ is the state):

$$\dot{x} = (A(t) + V(t))\,x + u(t) + v(t), \quad t \in T = [0, \theta]. \tag{1}$$

Here $A(t) \in \mathbb{R}^{n\times n}$ is a given matrix function, Lebesgue measurable functions $u(t)$ (a control), $v(t)$ (an unknown but bounded disturbance), $V(t) \in \mathbb{R}^{n\times n}$ (which describes the uncertainty in matrices) are subjected to given set-valued constraints:

$$u(t) \in \mathcal{R}(t), \quad v(t) \in \mathcal{Q}(t), \qquad a.e.\ t \in T, \tag{2}$$

$$V(t) \in \mathcal{V}(t) = \{V \in \mathbb{R}^{n\times n} \,|\, \mathrm{Abs}\,(V - \tilde{V}(t)) \le \hat{V}(t)\}, \qquad \text{a.e. } t \in T. \tag{3}$$

Matrix and vector inequalities ($\le, <, \ge, >$) here and below are understood componentwise. We presume the sets $\mathcal{R}(t)$, $\mathcal{Q}(t)$, and $\mathcal{M}$ to be parallelotopes and a parallelepiped respectively:

$$\mathcal{R}(t) = \mathcal{P}[r(t), \bar{R}(t)],\ \bar{R}(t) \in \mathbb{R}^{n\times n_1}, \quad \mathcal{Q}(t) = \mathcal{P}[q(t), \bar{Q}(t)],\ \bar{Q}(t) \in \mathbb{R}^{n\times n_2},$$
$$\mathcal{M} = \mathcal{P}(p_\theta, P_\theta, \pi_\theta) = \mathcal{P}[p_\theta, \bar{P}_\theta], \quad \bar{P}_\theta \in \mathbb{R}^{n\times n},\ \det \bar{P}_\theta \neq 0, \tag{4}$$

where $r(t)$, $\bar{R}(t)$, $q(t)$, $\bar{Q}(t)$, as well as $A(t)$, $\tilde{V}(t)$, and $\hat{V}(t) \geq 0$ are given continuous vector and matrix functions; the parallelepiped $\mathcal{M}$ is nondegenerate.

By a *parallelepiped* $\mathcal{P}(p, P, \pi) \subset \mathbb{R}^n$ we mean a set such that $\mathcal{P} = \mathcal{P}(p, P, \pi) = \{x \in \mathbb{R}^n \mid x = p + \sum_{i=1}^n p^i \pi_i \xi_i,\ \|\xi\|_\infty \leq 1\}$, where $p \in \mathbb{R}^n$; $P = \{p^i\} \in \mathbb{R}^{n\times n}$ is such that $\det P \neq 0$, $\|p^i\|_2 = 1$[1]; $\pi \in \mathbb{R}^n$, $\pi \geq 0$. It may be said that p determines the center of the parallelepiped, P is the orientation matrix, p^i are the "directions" and π_i are the values of its "semi-axes". We call a parallelepiped *nondegenerate* if $\pi > 0$.

By a *parallelotope* $\mathcal{P}[p, \bar{P}] \subset \mathbb{R}^n$ we mean a set $\mathcal{P} = \mathcal{P}[p, \bar{P}] = \{x \in \mathbb{R}^n \mid x = p + \bar{P}\zeta,\ \|\zeta\|_\infty \leq 1\}$, where $p \in \mathbb{R}^n$ and the matrix $\bar{P} = \{\bar{p}^i\} \in \mathbb{R}^{n\times m}$, $m \leq n$, may be singular. We call a parallelotope $\mathcal{P}$ *nondegenerate* if $m = n$ and $\det \bar{P} \neq 0$.

Each parallelepiped $\mathcal{P}(p, P, \pi)$ is a parallelotope $\mathcal{P}[p, \bar{P}]$ with $\bar{P} = P \operatorname{diag} \pi$; each nondegenerate parallelotope is a parallelepiped with $P = \bar{P} \operatorname{diag}\{\|\bar{p}^i\|_2^{-1}\}$, $\pi_i = \|\bar{p}^i\|_2$ or, in a different way, with $P = \bar{P}$, $\pi = \mathrm{e}$, where $\mathrm{e} = (1, 1, \ldots, 1)^\top$.

For the above system we consider the following cases: (I) *without uncertainty* when v and $V \equiv 0$ are given functions, i.e., $\bar{Q} \equiv 0$, $\tilde{V} \equiv \hat{V} \equiv 0$; (II) *under uncertainty* including the following three subcases: (II, i) *only additive uncertainty* ($V \equiv 0$); (II, ii) *only bilinear uncertainty* ($\bar{\mathcal{Q}} \equiv 0$); (II, iii) *both ones.*

In [12–14], for cases (I) and (II, i) (without bilinear uncertainty), the following problem of *terminal target control synthesis under uncertainty* was investigated.

Problem 1. Specify a *solvability set* $\mathcal{W}(\tau, \theta, \mathcal{M}) = \mathcal{W}(\tau)$ and a set-valued *feedback control strategy*[2] $u = \mathcal{U}(t, x)$, $\mathcal{U}(\cdot,\cdot) \in U_{\mathcal{R}}^{c}$, such that all the solutions to the differential inclusion $\dot{x} \in A(t)x + \mathcal{U}(t, x) + \mathcal{Q}(t)$, $t \in T$, that start from any given position $\{\tau, x_\tau\}$, $x_\tau = x(\tau) \in \mathcal{W}(\tau, \theta, \mathcal{M})$, $\tau \in [0, \theta)$, would reach the terminal set $\mathcal{M}$ at time θ: $x(\theta) \in \mathcal{M}$.

The multivalued function $\mathcal{W}(t)$, $t \in T$, is known as a *solvability tube* $\mathcal{W}(\cdot)$.

The ellipsoidal synthesis was elaborated for solving Problem 1 [12,14]. In [7], the families of external $\mathcal{P}^+(\cdot)$ and internal $\mathcal{P}^-(\cdot)$ parallelepiped-valued and parallelotope-valued (shorter, *polyhedral*) estimates for $\mathcal{W}(\cdot)$ were introduced. The extremal aiming strategies of N.N. Krasovskii were used there (in [7], control strategies were constructed in an analytical form on the base of a solution of some specific mathematical programming problem). Now let us consider Problem 2, which concerns all above cases (I), (II, i), (II, ii), and (II, iii). Unlike Problem 1, it involves single-valued control strategies[3].

---

[1] The normality condition $\|p^i\|_2 = 1$ may be omitted to simplify formulas.

[2] Here the class $U_{\mathcal{R}}^{c}$ of *feasible control strategies* is taken to consist of all convex compact-valued multifunctions $\mathcal{U}(t, x)$ that are measurable in t, upper semicontinuous in x, being restricted by $\mathcal{U}(t, x) \subseteq \mathcal{R}(t)$, $t \in T$. The condition $\mathcal{U}(\cdot,\cdot) \in U_{\mathcal{R}}^{c}$ ensures that the corresponding differential inclusion does have a solution.

[3] This is possible because our strategies will be continuous and even linear with respect to x. Moreover, they will be constructed in an explicit form.

Problem 2. Find a polyhedral tube $\mathcal{P}^-(t) = \mathcal{P}[p^-(t), \bar{P}^-(t)]$, $t \in T$, with $\mathcal{P}^-(\theta) = \mathcal{M}$ and a corresponding feedback control strategy $u = u(t,x)$ such that $u(t,x) \in \mathcal{R}(t)$ for $x \in \mathcal{P}^-(t)$, $t \in T$, and each solution $x(\cdot)$ to the differential equation

$$\dot{x} = (A(t) + V(t))x + u(t,x) + v(t), \qquad t \in T, \tag{5}$$

with $x(0) = x_0 \in \operatorname{int} \mathcal{P}^-(0)$ would be defined on T and would satisfy $x(t) \in \mathcal{P}^-(t)$, $t \in T$, whatever are $v(\cdot)$ and $V(\cdot)$ subjected to (2)–(4). Moreover, introduce a whole family of such tubes $\mathcal{P}^-(\cdot)$.

3 Polyhedral Control Synthesis for Differential Systems

Let us consider the following ODE system for $\mathcal{P}^-(t) = \mathcal{P}[p^-(t), \bar{P}^-(t)]$:

$$\dot{p}^- = (A(t) + \tilde{V}(t))\, p^- + r(t) + q(t), \quad p^-(\theta) = p_\theta; \tag{6}$$

$$\begin{aligned}\dot{\bar{P}}^- &= (A(t)+\tilde{V}(t))\, \bar{P}^- + \bar{P}^- \operatorname{diag} \beta(t, \bar{P}^-) + \bar{R}(t)\, \Gamma(t) + \bar{P}^- \operatorname{diag} \gamma(t, \bar{P}^-),\\ \beta(t, \bar{P}^-) &= \max\{\operatorname{Abs}((\bar{P}^-)^{-1})\, \hat{V}(t) \operatorname{Abs}(p^-(t) + \bar{P}^-\xi) \mid \xi \in \mathbb{E}(\mathcal{C})\},\\ \gamma(t, \bar{P}^-) &= \operatorname{Abs}((\bar{P}^-)^{-1}\bar{Q}(t))\, \mathrm{e}, \qquad \bar{P}^-(\theta) = \bar{P}_\theta.\end{aligned} \tag{7}$$

Here the operation of maximum is understood componentwise, $\mathbb{E}(\mathcal{C})$ denotes the set of all vertices of $\mathcal{C} = \mathcal{P}(0, I, \mathrm{e})$ (i.e., points ξ with $\xi_j \in \{-1, 1\}$); $\Gamma(t) \in \mathbb{R}^{n_1 \times n}$ is an arbitrary Lebesgue measurable matrix function satisfying $\Gamma(t) \in \mathcal{G}$, a.e. $t \in T$, where $\mathcal{G} = \{\Gamma = \{\gamma_i^j\} \in \mathbb{R}^{n_1 \times n} \mid \|\Gamma\|_\infty \leq 1\}$, $\|\Gamma\|_\infty = \max_{1 \leq i \leq n_1} \sum_{j=1}^n |\gamma_i^j|$. Let $\mathbb{G}$ be the set of all such functions $\Gamma(\cdot)$. Let us consider the following control strategy which is connected with the tube $\mathcal{P}^-(\cdot)$ from (6), (7):

$$u(t,x) = r(t) + \bar{R}(t)\Gamma(t)\bar{P}^-(t)^{-1}(x - p^-(t)). \tag{8}$$

Theorem 1. *We consider system (1)–(4), where $\mathcal{M}$ is a nondegenerate parallelepiped. Let $\Gamma(\cdot) \in \mathbb{G}$. Then the system (6), (7) has a unique solution $(p^-(\cdot), \bar{P}^-(\cdot))$ at least on some subinterval $T_1 = [\tau_1, \theta] \subseteq T$, where $0 \leq \tau_1 < \theta$. If $T_1 = T$ and we have $\det \bar{P}^-(t) \neq 0$, $t \in T$, then the tube $\mathcal{P}^-(\cdot)$ and the control strategy (8) give a particular solution to Problem 2; in cases (I) and (II, i), all solutions $x(\cdot)$ to (5) with $x(0) \in \mathcal{P}^-(0)$ (not only with $x(0) \in \operatorname{int} \mathcal{P}^-(0)$) satisfy $x(t) \in \mathcal{P}^-(t)$, $t \in T$.*

Proof. Here we give a sketch. The existence and uniqueness follow from known results similarly to [7,8]. Let $x_0 \in \operatorname{int} \mathcal{P}^-(0)$ ($x_0 \in \mathcal{P}^-(0)$ for cases (I) and (II, i)). Let $x(\cdot)$ be a corresponding solution to (5) with $x(0) = x_0$ (i.e., $x(0) = x_0 = p^-(0) + \bar{P}^-(0)\zeta_0$, where $\|\zeta_0\|_\infty < 1$ (respectively, $\|\zeta_0\|_\infty \leq 1$)), the control $u = u(t,x)$ from (8), and arbitrary admissible functions $v(\cdot)$ (such that $v(t) = q(t) + \bar{Q}(t)\chi(t)$, $\|\chi(t)\|_\infty \leq 1$) and $V(\cdot)$ (which satisfies (3)). Let us represent $x(t) - p^-(t)$ in the form $x(t) - p^-(t) = \bar{P}^-(t)\zeta(t)$. Then we have for the function ζ (here and below we omit functions arguments for shortening): $\frac{d}{dt}\zeta = -(\bar{P}^-)^{-1}(\frac{d}{dt}\bar{P}^-)\zeta + (\bar{P}^-)^{-1}\frac{d}{dt}(x - p^-)$. Taking into account (7) and the relation $\frac{d}{dt}(x - p^-) = (A + \tilde{V})(x - p^-) + (V - \tilde{V})x + u - r + v - q$, which follows from (5), (6), it is not difficult to see that

$$\dot{\zeta} = -(\operatorname{diag}\beta + \operatorname{diag}\gamma)\zeta - (\bar{P}^-)^{-1}\bar{R}\Gamma\zeta + (\bar{P}^-)^{-1}((V - \tilde{V})x + u - r + v - q).$$

Let us denote $b(t) = \beta(t, \bar{P}^-(t)) + \gamma(t, \bar{P}^-(t))$, $c(t,\zeta) = \bar{P}^-(t)^{-1}((V(t) - \tilde{V}(t))\cdot (p^-(t) + \bar{P}^-(t)\zeta) + \bar{Q}(t)\chi(t))$. Then, using (8), (3), (4), we have

$$\dot{\zeta}_i = -b_i(t)\zeta_i + c_i(t,\zeta), \quad i = 1, \ldots, n, \qquad \zeta(0) = \zeta_0; \tag{9}$$

$$b(t) \geq 0, \quad \operatorname{Abs} c(t,\zeta) \leq b(t) \quad \text{for } \zeta \in \mathcal{C} = \mathcal{P}(0, I, \mathrm{e}). \tag{10}$$

It is not difficult to verify that if $\zeta(\cdot)$ satisfies (9), (10) and $\zeta(0) \in \operatorname{int}\mathcal{C}$, then $\zeta(t) \in \operatorname{int}\mathcal{C}$, $t \in T$; if $\zeta(0) \in \mathcal{C}$ and, in addition, $c(t,\zeta) \equiv c(t)$ (i.e., does not depend on ζ), then $\zeta(t) \in \mathcal{C}$, $t \in T$. Thus we obtain $x(t) \in \mathcal{P}^-(t)$, $t \in T$. Also we have $u(t,x) \in \mathcal{R}(t)$ for $x \in \mathcal{P}^-(t)$ because for such x we have $\|\Gamma(t)\bar{P}^-(t)^{-1}(x - p^-(t))\|_\infty \leq \|\Gamma(t)\|_\infty \|\bar{P}^-(t)^{-1}(x - p^-(t))\|_\infty \leq 1 \cdot 1 = 1$ due to $\Gamma(\cdot) \in \mathbb{G}$. □

Theorem 1 describes the whole family of tubes $\mathcal{P}^-(\cdot)$, where $\Gamma(\cdot)$ serves as a parameter. Thus the set $\mathcal{W}^0 = \bigcup\{\operatorname{int}\mathcal{P}^-(0) \mid \Gamma(\cdot) \in \mathbb{G}$ such that $\det\mathcal{P}^-(t) \neq 0$, $t{\in}T\}$ (for cases (I) and (II, i), the analogous set $\mathcal{W}^0 = \bigcup\mathcal{P}^-(0))$ provides the set of initial positions which can be steered to the terminal set $\mathcal{M}$ during the time θ by solving Problem 2. But, generally speaking, it is not true that $\det\mathcal{P}^-(0) \neq 0$ or even $\mathcal{P}^-(0) \neq \emptyset$ for each $\Gamma(\cdot) \in \mathbb{G}$. The attractive property of the control strategies (8) is their explicit form. Generally speaking, the control law depends on the initial state because it depends on $\mathcal{P}^-(\cdot)$, but it is the same for all x_0 from a fixed $\mathcal{P}^-(0)$. For cases (I) and (II, i), the above family of the tubes $\mathcal{P}^-(\cdot)$ coincides with the family of internal estimates for $\mathcal{W}(\cdot)$ introduced in [7]. It follows from [7,9] that for the case (I) we have $T_1 = T$ for each $\Gamma(\cdot) \in \mathbb{G}$ and $\mathcal{W}(0) = \bigcup\{\mathcal{P}^-(0) \mid \Gamma(\cdot) \in \mathbb{G}\}$. But we can not conclude from here that $\mathcal{W}^0 = \mathcal{W}(0)$ because we can not exclude the situation when a cross section of a tube may become a degenerate parallelotope and then (8) is not applicable.

Remark 1. One of the possible heuristic ways to construct a "good" parameter $\Gamma(\cdot)$ is to apply arguments of a "local" volume optimization similarly to [7]. Namely assume, without loss of generality, that $\det\bar{P}_\theta > 0$. Fix a natural number N and introduce a grid T_N of times $\tau_k = kh_N$, $k = 0, \ldots, N$, $h_N = \theta N^{-1}$. Integrating the system (6), (7) from right to left, let us, for each $\tau \in T_N$, solve the optimization problem $\min_\Gamma \operatorname{tr}(\Xi(\tau, \bar{P}^-(\tau))\,\Gamma)$, $\Gamma \in \mathcal{G}$, where $\Xi(\tau, \bar{P}^-) = (\bar{P}^-)^{-1}\bar{R}(\tau)$. This is equivalent to finding the maximal possible velocity of increasing (from right to left) $\det\bar{P}^-(\tau)$ (therefore $\operatorname{vol}\mathcal{P}^-(\tau)$) at time τ, by the choice of the value Γ, when the value $\bar{P}^-(\tau)$ has already been found. Thus we can sequentially construct the piecewise constant function $\Gamma(t) \equiv \Gamma(\tau_k){\in}\operatorname{Argmin}_{\Gamma\in\mathcal{G}}\operatorname{tr}(\Xi(\tau_k, \bar{P}^-(\tau_k))\Gamma)$, $t{\in}(\tau_{k-1}, \tau_k]$, $k = N, \ldots, 1$, and find $\bar{P}^-(\cdot)$.

For the case (I), it is also possible, similarly to [9], to construct $\Gamma(\cdot)$ by minimizing $\mathrm{tr}\,(\Xi(\tau, \bar{P}^-(\tau))\,\Gamma)$ over those Γ that satisfy not only to $\Gamma \in \mathcal{G}$, but also to some additional constraints introduced to produce tight internal estimates $\mathcal{P}^-(t)$ to $\mathcal{W}(t)$, $t \in T$. Note that solutions of both mentioned optimization problems are known in explicit form [7,9].

Now let us briefly consider a problem of control synthesis, similar to Problem 2, for discrete-time systems. This is of independent interest and also may be useful for constructing difference schemes for solving the system (6) and (7).

4 Control Synthesis for Discrete-Time Systems

Consider the controlled discrete-time system with a given terminal set $\mathcal{M}$:

$$\begin{aligned} x[k] &= (A[k] + V[k])\,x[k-1] + u[k] + v[k], \quad k = 1, \ldots, N, \\ x[N] &\in \mathcal{M} = \mathcal{P}[p_\theta, \bar{P}_\theta], \quad \det \bar{P}_\theta \neq 0, \end{aligned} \tag{11}$$

$$u[k] \in \mathcal{R}[k] = \mathcal{P}[r[k], \bar{R}[k]], \quad v[k] \in \mathcal{Q}[k] = \mathcal{P}[q[k], \bar{Q}[k]], \quad k = 1, \ldots, N, \tag{12}$$

$$V[k] \in \mathcal{V}[k] = \{V \in \mathbb{R}^{n \times n} \,|\, \mathrm{Abs}\,(V - \tilde{V}[k]) \leq \hat{V}[k]\}, \quad k = 1, \ldots, N. \tag{13}$$

Problem 3. Find a polyhedral tube $\mathcal{P}^-[k] = \mathcal{P}[p^-[k], \bar{P}^-[k]]$, $k = 0, 1, \ldots, N$, with $\mathcal{P}^-[N] = \mathcal{M}$, and a corresponding feedback control strategy $u = u[k, x]$ such that $u[k, x] \in \mathcal{R}[k]$ for $x \in \mathcal{P}^-[k-1]$, $k = 1, \ldots, N$, and each solution $x[\cdot]$ to the equation $x[k] = (A[k] + V[k])x[k-1] + u[k, x[k-1]] + v[k]$, $k = 1, \ldots, N$, with $x[0] = x_0 \in \mathcal{P}^-[0]$ would satisfy $x[k] \in \mathcal{P}^-[k]$, $k = 1, \ldots, N$, whatever are $v[\cdot]$ and $V[\cdot]$ subjected to (12), (13). Moreover, introduce a whole family of such tubes $\mathcal{P}^-[\cdot]$.

Let us consider the following system of relations for $\mathcal{P}^-[k] = \mathcal{P}[p^-[k], \bar{P}^-[k]]$:

$$p^-[k-1] = B[k]^{-1}(p^-[k] - r[k] - q[k]), \quad B[k] = A[k] + \tilde{V}[k], \quad k = N, \ldots, 1, \tag{14}$$

$$\begin{aligned} \bar{P}^-[k-1] &= P^1[k] - P^0[k]\mathrm{diag}\,\beta[k], \quad P^1[k] = B[k]^{-1}(\bar{P}^-[k]\mathrm{diag}\,(\mathrm{e} - \gamma[k]) - \bar{R}[k]\Gamma[k]), \\ P^0[k] &= B[k]^{-1}\bar{P}^-[k], \quad \gamma[k] = (\mathrm{Abs}\,(\bar{P}^-[k]^{-1}\bar{Q}[k]))\mathrm{e}, \quad k = N, \ldots, 1, \end{aligned} \tag{15}$$

$$p^-[N] = p_\theta, \quad \bar{P}^-[N] = \bar{P}_\theta, \tag{16}$$

$\beta[k]$ satisfy one of the following systems of inequalities or equations or else equalities:

$$\begin{aligned} &H[k, \beta[k]] \leq \beta[k] \leq \mathrm{e} - \gamma[k], \quad k = N, \ldots, 1, \\ &H[k, \beta] = \max_{\xi \in \mathbb{E}(C)} (\mathrm{Abs}\,(\bar{P}^-[k]^{-1}))\hat{V}[k]\,\mathrm{Abs}\,(p^-[k-1] + P^1[k]\xi - P^0[k]\mathrm{diag}\,\xi \cdot \beta); \end{aligned} \tag{17}$$

$$\beta[k] = H[k, \beta[k]], \quad k = N, \ldots, 1; \tag{18}$$

$$\beta[k] = 0, \quad k = N, \ldots, 1. \tag{19}$$

Theorem 2. *We consider the system (11)–(13), where* $\det B[k] \neq 0$, $k = 1, \ldots, N$, *and* $\mathcal{M}$ *is a nondegenerate parallelepiped. Let* $\Gamma[k]$ *be arbitrary matrices such that* $\Gamma[k] \in \mathcal{G}$, $k = N, \ldots, 1$, *and the system (14)–(17) (the system (14)–(16), (18)) has a solution* $(p^-[\cdot], \bar{P}^-[\cdot])$ *such that we obtain* $\det \bar{P}^-[k] \neq 0$ *(and* $\mathrm{e} - \gamma[k] - \beta[k] \geq 0$ *respectively),* $k = N, \ldots, 1$. *Then the tube* $\mathcal{P}^-[\cdot]$ *and the control strategy*

$$u[k, x] = r[k] + \bar{R}[k]\Gamma[k]\bar{P}^-[k-1]^{-1}(x - p^-[k-1]), \quad k = 1, \ldots, N, \tag{20}$$

give a particular solution to Problem 3. For cases (I) and (II, i), similar to ones from Sect. 2, the above is also true if the system (14)–(16), (18) is replaced by the system (14)–(16), (19) of explicit recurrent relations.

Remark 2. If the operator $H[k, \beta]$ is contractive, i.e. $\|H[k, \beta^1] - H[k, \beta^2]\|_\infty \leq L\|\beta^1 - \beta^2\|_\infty$ for any $\beta^1, \beta^2 \in \mathbb{R}^n$, where $L = L[k] \in (0, 1)$, then the equation $\beta = H[k, \beta]$ has a nonnegative solution $\beta = \beta[k]$, which can be found by the simple iteration $\beta^{l+1} = H[k, \beta^l]$, $l = 0, 1, \ldots$, starting from arbitrary β^0; if $\beta^0 = 0$, then we have $\|\beta^l - \beta\|_\infty \leq L^l(1 - L)^{-1}\|(\mathrm{Abs}\,(\bar{P}^-[k]^{-1})\hat{V}[k]\|_\infty(\|p^-[k-1]\|_\infty + \|P^1[k]\|_\infty)$.

Remark 3. Let the system (11)–(13) be obtained by the Euler approximations of (1)–(4) with the same $\mathcal{M}$, $A[k] = I + h_N A(t_{k-1})$, $\mathcal{R}[k] = h_N \mathcal{R}(t_{k-1})$, $\mathcal{Q}[k] = h_N \mathcal{Q}(t_{k-1})$, $\tilde{V}[k] = h_N \tilde{V}(t_{k-1})$, $\hat{V}[k] = h_N \hat{V}(t_{k-1})$, $t_k = k h_N$, $h_N = \theta N^{-1}$. Then it is convenient to use (18) or (19) because the relations $L[k] < 1$, $\gamma[k] + \beta[k] < \mathrm{e}$ for a fixed k are satisfied when $\det \bar{P}^-[k] \neq 0$ and the time step h_N is sufficient small.

5 Examples

We consider some examples, where we use the Euler approximations. Let $\theta = 2$, $A(t) \equiv \begin{bmatrix} 0 & 1 \\ -8 & 0 \end{bmatrix}$, $\tilde{V}(t) \equiv 0$, $\hat{V}(t) \equiv 0$ or $\hat{V}(t) \equiv \begin{bmatrix} 0 & 0 \\ 0.1 & 0 \end{bmatrix}$, $\mathcal{R}(t) \equiv \mathcal{P}(0, I, (0, 1)^\top)$, $\mathcal{Q}(t) \equiv \mathcal{P}(0, I, 0)$ or $\mathcal{Q}(t) \equiv \mathcal{P}(0, I, (0.2, 0)^\top)$, $\mathcal{M} = \mathcal{P}((-0.5, 0)^\top, I, (0.5, 0.5)^\top)$, $N = 200$. We consider 3 cases: (I), (II, i), and (II, iii) (see Fig. 1). The first two examples are the same as in [7], but we construct trajectories of two types, which correspond to controls from [7] (dash lines) and to (8) (solid lines); in the third example we put the "disturbance" $V(t) \equiv \tilde{V}(t) + \hat{V}(t)$. As a rule, if $x_0 \in \mathrm{int}\,\mathcal{P}^-(0)$, then trajectories of the first type turn out to be at the boundary of $\mathcal{P}^-(t)$ starting from some time while the ones of the second type turn out to be nearer to $p^-(t)$.

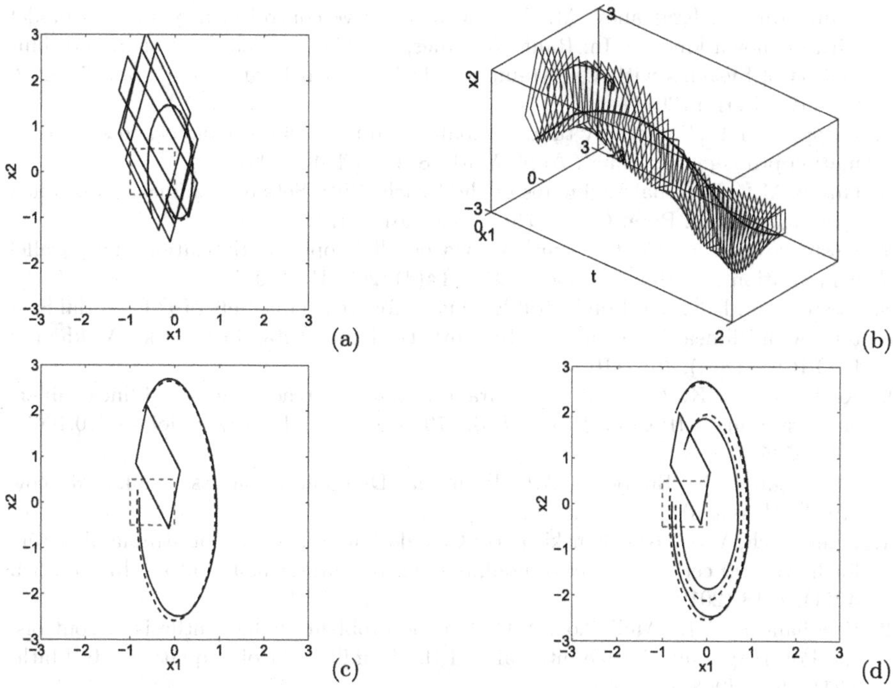

Fig. 1. Examples of polyhedral control synthesis ($n = 2$). (a) Case (I): the set $\mathcal{M}$ (dash line), several parallelotopes $\mathcal{P}^-[0]$ and controlled trajectories for $x_0 = (-0.5, -1)^\top$. (b) Case (I): the suitable tube $\mathcal{P}^-[\cdot]$ and the controlled trajectory. (c) Case (II, i): $\mathcal{M}$, $\mathcal{P}^-[0]$ corresponding to $\Gamma[\cdot]$ from Remark 1, and controlled trajectories for $x_0 = (-0.7, 2)^\top$. (d) Case (II, iii): $\mathcal{M}$, $\mathcal{P}^-[0]$ corresponding to $\Gamma[\cdot]$ from Remark 1, and controlled trajectories for two initial points $x_0 = (-0.7, 2)^\top$ and $x_0 = (-0.5, 1.2)^\top$.

Acknowledgments. The research was supported by the Program of the Presidium of the Russian Academy of Sciences "Dynamic Systems and Control Theory" under support of the Ural Branch of RAS (Project 12-P-1-1019) and by the Russian Foundation for Basic Research (Grants 12-01-00043,13-01-90419).

References

1. Anan'evskii, I.M., Anokhin, N.V., Ovseevich, A.I.: Synthesis of a bounded control for linear dynamical systems using the general Lyapunov function. Dokl. Math. **82**(2), 831–834 (2010)
2. Baier, R., Lempio, F.: Computing Aumann's integral. In: Kurzhanski, A.B., Veliov, V.M. (eds.) Modeling Techniques for Uncertain Systems (Sopron, 1992). Progress in Systems and Control Theory, vol. 18, pp. 71–92. Birkhäuser, Boston (1994)
3. Chernousko, F.L.: State Estimation for Dynamic Systems. CRS Press, Boca Raton (1994)

4. Dimitrova, N., Krastanov, M.: Nonlinear adaptive control of a bioprocess model with unknown kinetics. In: Rauh, A., Auer, E. (eds.) Modeling, Design, and Simulation of Systems with Uncertainties (Mathematical Engineering), pp. 275–292. Springer, Berlin (2011)
5. Filippova, T.F.: Trajectory tubes of nonlinear differential inclusions and state estimation problems. J. Concr. Appl. Math. **8**(3), 454–469 (2010)
6. Gusev, M.I.: External Estimates of the Reachability Sets of Nonlinear Controlled Systems. Autom. Rem. Contr. **73**(3), 450–461 (2012)
7. Kostousova, E.K.: Control synthesis via parallelotopes: optimization and parallel computations. Optim. Methods Softw. **14**(4), 267–310 (2001)
8. Kostousova, E.K.: On Polyhedral Estimates for Reachable Sets of Differential Systems with Bilinear Uncertainty. (Russian). Trudy Instituta Matematiki i Mekhaniki UrO RAN **18**(4), 195–210 (2012)
9. Kostousova, E.K.: On tight polyhedral estimates for reachable sets of linear differential systems. AIP Conf. Proc. **1493**, 579–586 (2012). http://dx.doi.org/10.1063/1.4765545
10. Krasovskii, N.N., Subbotin, A.I.: Positional Differential Games. Nauka, Moscow (1974). (Russian)
11. Kuntsevich, V.M., Kurzhanski, A.B.: Calculation and control of attainability sets for linear and certain classes of nonlinear discrete systems. J. Autom. Inform. Sci. **42**(1), 1–18 (2010)
12. Kurzhanskii, A.B., Mel'nikov, N.B.: On the problem of the synthesis of controls: the Pontryagin alternative integral and the Hamilton-Jacobi equation. Sb. Math. **191**(5–6), 849–881 (2000)
13. Kurzhanski, A.B., Nikonov, O.I.: On the problem of synthesizing control strategies. Evolution equations and set-valued integration. Sov. Math. Dokl. **41**(2), 300–305 (1990)
14. Kurzhanski, A.B., Vályi, I.: Ellipsoidal Calculus for Estimation and Control. Birkhäuser, Boston (1997)
15. Kurzhanski, A.B., Varaiya, P.: On ellipsoidal techniques for reachability analysis. Part I: External approximations. Part II: Internal approximations. Box-valued constraints. Optim. Meth. Softw. **17**(2), 177–237 (2002)
16. Polyak, B.T., Scherbakov, P.S.: Robust Stability and Control. Nauka, Moscow (2002). (Russian)
17. Taras'yev, A.M., Uspenskiy, A.A., Ushakov, V.N.: Approximation schemas and finite-difference operators for constructing generalized solutions of Hamilton-Jacobi equations. J. Comput. Syst. Sci. Int. **33**(6), 127–139 (1995)
18. Veliov, V.M.: Second order discrete approximations to strongly convex differential inclusions. Syst. Contr. Lett. **13**(3), 263–269 (1989)

On the Controllability of a Class of Hybrid Control Systems

Mikhail I. Krastanov[1,2(✉)] and Marc Quincampoix[3]

[1] Faculty of Mathematics and Informatics, Sofia University "St. Kliment Ohridski", James Bourchier Blvd. 5, 1164 Sofia, Bulgaria

[2] Institute of Mathematics and Informatics, Bulgarian Academy of Sciences, Acad. G. Bonchev Str., Block 8, 1113 Sofia, Bulgaria

krastanov@fmi.uni-sofia

[3] Laboratoire de Mathématiques, UMR 6205, Université de Bretagne Occidentale, 6 Avenue Victor Le Gorgeu, 29200 Brest Cedex, France

Marc.Quincampoix@univ-brest.fr

Abstract. The paper considers a class of hybrid control systems with piece-wise linear dynamics and controls which are constrained in convex closed cones. A necessary and sufficient condition for small-time controllability is proved. This result extends the classical Kalman controllability criterion.

1 Introduction

We investigate the controllability problem for piece-wise time-invariant linear systems with controls which are constrained in a cone. These systems belong to a special class of hybrid control systems which comprises a collection of subsystems described by linear dynamics together with a switching rule that specifies the switchings between the subsystems. Such a system can be used to describe a wide range of control systems in practice. Various natural biological and social systems use switch strategies in accordance to environmental changes.

The controllability of linear systems in the presence of state constraints is much less investigated despite the large class of potential applications. There are only a few papers, where the directions of expansion of the reachable set of such kind of control systems are studied (cf., for example, [5–7,14]).

Next we state the problem. Let I be a finite set of indices, C_i, $i \in I$, be convex closed cones in $\mathbb{R}^n$, and L be a linear subspace of $\mathbb{R}^n$ such that

$$\bigcup_{i\in I} C_i = R^n \text{and} \bigcap_{i\in I} C_i = L.$$

This work has been partially supported by the Sofia University "St. Kliment Ohridski" under contract No. 013/09.04.2014, by the Commission of the European Communities under the 7-th Framework Programme Marie Curie Initial Training Networks Project "Deterministic and Stochastic Controlled Systems and Applications" FP7-PEOPLE-2007-1-1-ITN, no. 213841-2 and project SADCO, FP7-PEOPLE-2010-ITN, No 264735. This was also supported partially by the French National Research Agency ANR-10-BLAN 0112.

I. Lirkov et al. (Eds.): LSSC 2013, LNCS 8353, pp. 107–115, 2014.
DOI: 10.1007/978-3-662-43880-0_11,

We assume that the following linear control system is defined on each C_i:

$$\dot{x}(t) = A_i\, x(t) + u(t)$$

with $u(t) \in U_i$, $i \in I$. Here each A_i is an $(n \times n)$-matrix and each U_i is a closed convex cone in $\mathbb{R}^n$.

These linear systems defined on the cones C_i, $i \in I$, determine a discontinuous control system on the whole state space $\mathbb{R}^n$ which can be described in a shorter way (cf., for example, [8]) as follows:

$$\dot{x}(t) \in \text{co} \left\{ A_i\, x(t) + U_i, \begin{array}{c} i \in I \text{and} \\ x(t) \in C_i \end{array} \right\}, \tag{1}$$

where $co(X)$ denotes the closed convex hull spanned by the set X. By definition, a trajectory of (1) defined on $[0,T]$ is any absolutely continuous function $x : [0,T] \to \mathbb{R}^n$ satisfying the inclusion (1) for almost every $t \in [0,T]$. We denote by $R(x,T)$ the reachable set of the system (1) at time T starting from $x \in \mathbb{R}^n$ at $t = 0$, i.e.

$$R(x,T) = \{y \in \mathbb{R}^n : \text{there exists a trajectory } x : [0,T] \to \mathbb{R}^n \text{of (1) such that } x(0) = x \text{ and } x(T) = y\}.$$

In the present paper we study the small-time controllability of (1):

Definition 1. *The control system (1) is called small-time controllable (STC) at the origin if* $R(0,T) \;=\; \mathbb{R}^n$ *for each* $T > 0$.

Remark 1. The small-time local controllability is a local property introduced by H. Sussmann (cf., for example, [10]). Also, the concept for the so called large-time local controllability is defined (cf., for example, [3], where this concept is discussed and a sufficient condition for large-time local controllability is proved).

Remark 2. V. Veliov posed in 1984 the small-time controllability problem for the case of a hybrid control system with piece-wise linear dynamics (cf. [14]), where the case of one hyperplane is analyzed. Later we analyze in [6] a similar controllability problem in the presence of two hyper-planes. Our main result could be viewed as a nontrivial extension of [14]. The corresponding proof is based on the so called differential-geometrical approach (cf., for example, [2,3,9,11,13]). But to treat this more difficult case we need the so called conical inverse mapping theorem (cf., for example, [1,4]).

2 An Approach for Studying Discontinuous Control Systems

The original control system is discontinuous with respect to the state variables. So, even the existence of a local trajectory is not guaranteed. In the present paper we extend the approach proposed in [6,14] as follows: For each index $i \in I$

we project the original system on the linear subspace L by using the admissible velocities $\pm p_{i,j} \in U_i$, $i \in I$. To describe rigorously this projection procedure, we introduce the following notations: if U is a convex closed cone in $\mathbb{R}^n$, we denote by $\mathrm{Rec}(U)$ the recessive subspace of U, i.e. the maximal linear subspace contained in U. Also, $\mathbb{R}^+$ denotes the set of all nonnegative real numbers.

The following assumptions **A** and **B** ensure that the projection procedure is possible:

Assumption A: For each index $i \in I$ there exist a finite index set J_i and admissible velocities $p_{i,j} \in \mathrm{Rec}(U_i)$ and linear function $\alpha_{i,j}$, $j \in J_i$, such that

$$L + \left\{ \sum_{j \in J_i} \alpha_j \, p_{i,j} : \alpha_j \geq 0, \; j \in J_i \right\} \subseteq C_i$$

$\Pi(y) = y$ for each $y \in L$ and

$$\Pi_i(x) := x - \sum_{j \in J_i} \alpha_{i,j}(x) \, p_{i,j} \in L$$

for each point $x \in \mathbb{R}^n$, where $\alpha_{i,j}(x) \geq 0$ for each $x \in C_i$.

Assumption B: The map $\Pi(x) := \{\Pi_i(x) : \; x \in C_i\}$, $x \in \mathbb{R}^n$, is single valued.

Remark 3. Assumption **B** is technical. It ensures that the projection map Π is well defined on the cones C_i, $i \in I$. According to this assumption, Π is a piecewise linear single-valued map, and thus it is Lipschitz continuous. We point out that the map Π is explicitly constructed in [6] for a specific choice of four cones defined by two hyperplanes.

For each index $i \in I$ we define:

(1) the linear operators $B_i : C_i \to L$ as follows:

$$B_i \; x := \Pi(A_i \; x) \, \text{for each} \, x \in C_i, \;\; i \in I;$$

(2) the vectors $r_{i,j} := B_i \; p_{i,j}, j \in J_i$, $i \in I$;
(3) the closed cones

$$V_i := \{\Pi(u) : \;\; u \in U_i\}, \; W_i := V_i + \left\{ \sum_{j \in J_i} \alpha_j \; r_{i,j} : \alpha_j \geq 0, \right\}, \;\; i \in I.$$

One can directly check that W_i, $i \in I$, are convex closed cones whose elements belong to the linear subspace L (remind that $L = \cap_{i \in I} C_i$), and that the linear subspace L is invariant with respect to the linear operator B_i, i.e. $B_i : L \to L$.

Remark 4. We would like to point out that the closed convex cones W_i, $i \in I$, are not just projections of the original cones U_i, $i \in I$, on L. They are enlarged by adding the vectors $r_{i,j}$, $j \in J_i$, $i \in I$, to ensure the best approximation of the original system on L. We discuss below the relation between the trajectories of the original system and the trajectories of the projected system.

If the conditions **A** and **B** hold true, then we can define the following control system

$$\dot{y} \in \text{co}\{B_i\, y + W_i : \quad i \in I\},\ y \in L, \tag{2}$$

on the linear space L.

To show what is the relation between the original system and the projected control system (2) we take an arbitrary trajectory $x(t)$, $t \in [0, T]$, of (1) and set $y(t) := \Pi(x(t))$. According to assumptions **A** and **B**, $y(\cdot)$ is well defined on $[0, T]$. Moreover, it is absolutely continuous on $[0, T]$ and for almost each $t \in [0, T]$ we have that

$$\begin{aligned}
\dot{y}(t) &= \frac{d}{dt}\Pi(x(t)) \\
&\in \text{co}\{\Pi\,(A_i\, x(t) + u_i(t)) : \ u_i(t) \in U_i,\ x(t) \in C_i\} \\
&\subset \text{co}\{B_i\, x(t) + V_i : \ x(t) \in C_i\} \\
&= \text{co}\left\{B_i\left(y(t) + \sum_{j \in J_i} \alpha_{i,j}(x(t))\, p_{i,j}\right) + V_i : \ x(t) \in C_i\right\} \\
&\subset \text{co}\{B_i\, y(t) + \sum_{j \in J_i} \alpha_{i,j}(x(t))\, B_i\, p_{i,j} + V_i : \quad x(t) \in C_i\} \\
&\subset \text{co}\{B_i\, y(t) + \sum_{j \in J_i} \alpha_j\, r_{i,j} + V_i : \ \alpha_j \geq 0,\ x(t) \in C_i\} \\
&\subset \text{co}\{B_i\, y(t) + W_i : \ i \in I\}.
\end{aligned}$$

Hence the projection of each trajectory of the original control system (1) is a trajectory of the projected system (2). This is important for the proof of the necessity of the main result.

3 A Necessary and Sufficient Controllability Condition

Let us remind that $\text{Rec}(V)$ stands for the maximal linear subspace contained in the convex cone $V \subset \mathbb{R}^n$. Also, we introduce notations that will be used later on: cone (S) is the minimal convex cone containing the set S, $\text{relint}_L\ (Z)$ denotes the relative interior of the set S with respect to the linear space L, int $(S) = \text{relint}_{\mathbb{R}^n}(S)$ and $\text{Inv}(S)$ is the minimal linear subspace of $\mathbb{R}^n$ that contains the set S and is invariant with respect to all matrices B_i, $i \in I$. For each subset S of $\mathbb{R}^n$ we define the following nondecreasing sequence of vector subspaces

$$\begin{aligned} L_1(S) &:= \mathrm{Inv}(\mathrm{Rec}(\mathrm{co}(S))), \\ L_2(S) &:= \mathrm{Inv}(\mathrm{Rec}(\mathrm{co}(S \cup L_1(S)))), \\ \dots &\quad \dots\dots\dots \\ L_k(S) &:= \mathrm{Inv}(\mathrm{Rec}(\mathrm{co}(S \cup L_{k-1}(S)))) \\ \dots &\quad \dots\dots\dots \end{aligned}$$

At last we set

$$W := \bigcup_{i \in I} W_i$$

and formulate the main result:

Theorem 1. *The piece-wise control system (1) is small-time controllable at the origin if and only if the following condition holds true*

$$L_{\dim\, L}\,(W) = L. \tag{3}$$

Proof of the necessity. Let us assume that the system Σ is small-time controllable at the origin but the condition (3) does not hold true. Then for each positive T there exist trajectories $x_1(\cdot), x_2(\cdot), \dots, x_d(\cdot)$ of this system such that $x_i(T) \in L$,

$$0 \in \mathrm{relint}_L(co\{x_i(T),\ i = 1, 2 \dots d\}).$$

If we set $y_i(t) := \Pi(x_i(t))$ for $t \in [0, T]$, $i = 1, \dots, d$, then each $y_i(\cdot)$ is a trajectory of the projected system (2) (this is shown in the previous section). Since $y_i(T) := \Pi(x_i(T))$, we obtain that there is no a half-space in L containing all trajectories of the control system (2).

Because we have assumed that the condition (3) is not fulfilled, then there exists $m < \dim L$ such that

$$L_m = L_{m+1} \neq L. \tag{4}$$

Let $\bar{L}$ be the orthogonal complement of L_m with respect to L, and let $\pi : L \to \bar{L}$ be the orthogonal projection onto $\bar{L}$. Since L_m is invariant with respect to all B_i, $i \in I$, we can define the linear operators $\bar{B}_i$ on $\bar{L}$ by

$$\bar{B}_i\ \pi(y) = \pi\left(B_i y\right),\ \ y \in L,\ i \in I,$$

as well as the control systems

$$\dot{\bar{y}}(t)\ \in\ co\left\{\bar{B}_i\ \bar{y}(t) + \bar{W}_i,\ i \in I\right\} \tag{5}$$

where $\bar{W}_i := \pi\left(W_i\right)$. We denote by $\bar{R}(0, T)$ the reachable set of the system (5) at time T and by $\bar{S}$ be the unit sphere in $\bar{L}$.

We claim that

$$0 \notin \mathrm{co}\left(\bigcup_{i\in I} (\bar{W}_i \cap \bar{S})\right). \tag{6}$$

We will argue by contradiction by assuming that

$$0 \in \mathrm{co}\left(\bigcup_{i\in I} (\bar{W}_i \cap \bar{S})\right).$$

This means that there exist $\alpha_i \geq 0$ and $\bar{u}_i \in \bar{W}_i$, $i \in I$, with $\|\bar{u}_i\| = 1$, such that

$$\sum_{i\in I} \alpha_i \bar{u}_i = 0, \quad \sum_{i\in I} \alpha_i = 1. \tag{7}$$

The very definition of $\bar{W}_i = \pi(W_i)$ implies the existence of elements $l_i \in L_m$ such that

$$u_i := \bar{u}_i + l_i \in W_i, \; i \in I.$$

Since $\bar{u}_i = \pi(u_i)$ and $\|\bar{u}_i\| = 1$, it follows that $u_i \in W_i \setminus L_m$. Taking into account (7), we have

$$\sum_{i\in I} \alpha_i u_i = \sum_{i\in I} \alpha_i l_i \in L_m. \tag{8}$$

Let us fix an index i_0 for which $\alpha_{i_0} > 0$. If we set $l := \sum_{i\in I} \alpha_i l_i \in L_m$, then

$$-u_{i_0} = \frac{1}{\alpha_{i_0}}\left(\sum_{i\in I, i\neq i_0} \alpha_i u_i - l\right).$$

Since $u_{i_0} \in W_{i_0} \setminus L_m$, we obtain that $u_{i_0} \in L^{m+1} \setminus L_m$, which contradicts (4). Hence we have proved our claim (6). By the Separation Theorem, we deduce from (6) that there exists a nonvanishing vector $h \in \bar{L}$ and a real number $\alpha > 0$ such that for each $\bar{u} \in \bar{W}_i \cap S$, $i \in I$, we have $\langle h, \bar{u}\rangle \geq \alpha$. This together with the linearity of the considered control systems (cf., also, Theorem 1 in [12]) imply that

$$\langle h, y\rangle \geq 0 \text{ for every } y \in \bar{R}(0,T),$$

whenever $T > 0$ is sufficiently small. So, the system (5) is not small time controllable. Because $\pi(R(0,T)) \subseteq \bar{R}(0,T)$, one can easily deduce that the system (2) is also not small time controllable. This is a contradiction and the proof of the necessity is complete.

Proof of the sufficiency. This part of the proof is based on a general approach for obtaining sufficient controllability conditions (cf. the Appendix at the end of [6]). Let the condition (3) holds true. We have to prove that for each $T > 0$ the reachable set $R(0,T)$ of (1) contains a neighborhood of the origin. For doing this it is enough to show that for every $i \in I$ we have $C_i \subset R(0,T)$: Fix an arbitrary index $i \in I$ and $T > 0$. Let v be an arbitrary element of L. We define the vector field $Z_v : \mathbb{R}^n \to \mathbb{R}^n$ as follows: $Z_v(x) = v$ for each point $x \in \mathbb{R}^n$. We first claim that each vector field Z_v, $v \in L$, is contained in the set of high order variations E_L^+ (for E_L^+ we refer to Definition 2 from the Appendix of [6]). For simplicity of the exposition we just write $L \subset E_L^+$.

The invariance of L with respect to each matrix B_i, $i \in I$, and the inclusion $V_i \subset L$ imply (according to Lemma 3 and Lemma 4 from [6]) that $\text{co}\left(\bigcup_{i\in I} V_i\right)$ (considered as a set of constant vector fields) is also a subset of E_L^+.

Let us fix a positive real $t > 0$, $i \in I$, $j \in J_i$, a point y_0 from L and denote by $y_{i,j}(\cdot)$ the solution of the following system of differential equations

$$\dot{y}_{i,j}(s) = B_i\, y_{i,j}(s) + \alpha(s)p_{i,j},\ y_{i,j}(0) = y_0 \in L,$$

on the interval $[0, 2\sqrt{t}]$, where

$$\alpha(s) = \begin{cases} 1, \text{ if } s \in [0, \sqrt{t}); \\ -1, \text{ if } s \in [\sqrt{t}, 2\sqrt{t}]. \end{cases}$$

According to the definition of the linear operator B_i, we have that

$$B_i\, x = \Pi(A_i\, x) = A_i\, x - \sum_{j\in J_i} \alpha_{i,j}(A_i\, x)\, p_{i,j}.$$

Since the vectors $p_{i,j} \in \text{Rec}\ (U_i)$, $j \in J_i$, we obtain that $y_{i,j}(\cdot)$ is an admissible trajectory for the original system. Moreover, one can directly check that $y_{i,j}(2\sqrt{t}) \in L$.

According to the Campbell-Baker-Hausdorff formula we obtain that $y_{i,j}(2\sqrt{t}) =$

$$\begin{aligned} &= \text{Exp}(\sqrt{t}(B_i(\cdot) + p_{i,j})\text{Exp}(\sqrt{t}(B_i - p_{i,j})) \\ &= \text{Exp}(2\sqrt{t}B_i(\cdot) + tB_i p_{i,j} + o(t)) \\ &= \text{Exp}(2\sqrt{t}B_i(\cdot) + tr_{i,j} + o(t)). \end{aligned}$$

So $r_{i,j} \in E_L^+$. Then Lemma 3 and Lemma 4 from [6] yield

$$\text{co}\left(\bigcup_{i\in I} W_i\right) \subset E_L^+.$$

Applying Lemma 3 from [6] again as well as Lemma 5 from [6], we obtain successively that every subspace L_i, $i = 1, \ldots, \dim L$, is a subset of E_L^+. At last, the very definition of the vectors $p_{i,j}$, $j \in J_i$, $i \in I$, imply that $C_i \subset E_{\mathbb{R}^n}^+$ for each index $i \in I$ and for each $T > 0$. Denote respectively by S and B the unit sphere and the unit ball of $\mathbb{R}^n$, respectively, centered at the origin. Define for any integer $m > 0$ the compact convex set

$$Q_m = \mathrm{co}\left\{ w \in C_i \cap \frac{1}{2}S,\ \mathrm{dist}(w, \mathrm{Lin}(C_i) \setminus C_i) \geq \frac{1}{m} \right\},$$

where Lin (C_i) denotes the linear subspace of $\mathbb{R}^n$ generated by C_i. Also, we denote by R_m denote the convex closed cone generated by the set Q_m. Clearly $R_m \setminus \{0\}$ is contained in the relative interior of C_i. According to the conical inverse mapping theorem (cf. for example Theorem 3.1 of [4] or Theorems 3.3 and 3.2 of [1]), there exists $\delta > 0$ such that

$$R_m \cap \delta B \subset R(0,T).$$

This implies that $Q_m \subset R(0,T)$ because the set $R(0,T)$ is a closed convex cone. Since $C_i \cap \frac{1}{2}S = \bigcup_{m>0} Q_m \subset R(0,T)$ we obtain that $C_i \subset R(0,T)$ for each $i \in I$ (here we use again the fact that $R(0,T)$ is a closed cone). From here we obtain that

$$\mathbb{R}^n = \bigcup_{i \in I} C_i \subset R(0,T)$$

which completes the proof of the theorem.

4 Conclusions

The paper gives a new necessary and sufficient condition for small-time controllability of a discontinuous control system consisting of finite number of linear control systems with control values belonging to convex closed cones. Each of these systems is defined on a convex closed cone. Our approach is based on a projection of the original dynamics on a suitable linear subspace. The obtained necessary and sufficient condition is checkable because it consists in constructing a finite number of linear spaces. Its verification, however, requires determination of the recessive subspace of a convex hull, which goes beyond the pure linear-algebraic considerations typical for the case of unconstrained controls.

References

1. Frankowska, H.: Conical inverse mapping theorems. Bull. Aust. Math. Soc. **45**, 53–60 (1992)
2. Hermes, H.: Lie algebras of vector fields and local approximation of attainable sets. SIAM J. Control Optim. **16**, 715–727 (1978)
3. Hermes, H.: Henry large-time local controllability via homogeneous approximations. SIAM J. Control Optim. **34**(4), 1291–1299 (1996)
4. Kawski, M.: An angular open mapping theorem. In: Bensoussan, A., Lions, J.L. (eds.) Analysis and Optimization of Systems. LNCIS, vol. 111, pp. 361–371. Springer, Heidelberg (1988)

5. Krastanov, M.I.: On the constrained small -time controllability of linear systems. Automatica **44**, 2370–2374 (2008)
6. Krastanov, M.I., Quincampoix, M.: On the small-time controllability of discontinuous piece-wise linear systems. Syst. Control Lett. **62**, 218–223 (2013)
7. Krastanov, M.I., Veliov, V.M.: Local controllability of state constrained linear systems. Acta Universitatis Lodziensis, Folia Mathematica **5**, 103–112 (1992)
8. Krastanov, M.I., Veliov, V.M.: On the controllability of switching linear systems. Automatica **41**, 663–668 (2005)
9. Lobry, C.: Controlabilité des systèmes non linéaires. SIAM J. Control **8**, 573–605 (1970)
10. Sussmann, H.: A general theorem on local controllability. SIAM J. Control Optim. **25**, 158–194 (1987)
11. Sussmann, H.: Small-time local controllability and continuity of the optimal time function for linear systems. J. Optim. Theor. Appl. **53**, 281–296 (1987b)
12. Sussmann, H.: A sufficient condition for local controllability. SIAM J. Control Optim. **16**, 790–802 (1978)
13. Veliov, V.M.: On the controllability of control constrained linear systems. Math. Balkania New Ser. **2**(2–3), 147–155 (1988)
14. Veliov, V.M., Krastanov, M.I.: Controllability of piecewise linear systems. Syst. Control Lett. **7**(5), 335–341 (1986)

BV Regularity and Differentiability Properties of a Class of Upper Semicontinuous Functions

Antonio Marigonda[1(✉)], Khai T. Nguyen[2], and Davide Vittone[2]

[1] Department of Computer Sciences, University of Verona, Strada Le Grazie 15, 37134 Verona, Italy
antonio.marigonda@univr.it

[2] Dipartimento di Matematica, Università di Padova, Via Trieste 63, 35121 Padova, Italy
{khai,vittone}@math.unipd.it

1 Introduction

We study a class of upper semicontinuous functions $f : \mathbb{R}^d \to \mathbb{R}$ whose hypograph $\operatorname{hypo} f$ (see Definition 1) satisfies a geometric regularity property, namely: there exist $c > 0$, $\theta \in]0, 1]$ such that for each P on the boundary of $\operatorname{hypo} f$ there exists a unitary Fréchet (outer) normal $v \in N^F_{\operatorname{hypo} f}(P) \cap \mathbb{S}^d$ to $\operatorname{hypo} f$ with

$$\langle v, P - Q\rangle \leq c\|P - Q\|^{1+\theta} \qquad \text{for every } Q \in \operatorname{hypo} f. \tag{1}$$

Geometrically speaking, this inequality expresses the fact that, in a neighborhood of each point P on the boundary of $\operatorname{hypo} f$, there exists a "subquadratic" smooth hypersurface $\Gamma(P)$ whose intersection with $\operatorname{hypo} f$ reduces to P. One could also say that $\Gamma(P)$ is *supertangent* to $\operatorname{hypo} f$ in a generalized sense. When $\theta = 1$ condition (1) means that the open sphere of center $P - \frac{v}{2c}$ and radius $\frac{1}{2c}$ lies outside $\operatorname{hypo} f$ and touches the boundary of $\operatorname{hypo} f$ at P. This property is also called *exterior sphere condition* and was studied by several authors, mainly in connection with regularity problems arising in the control theory.

If we strenghten the exterior sphere condition by requiring (1) to hold for *every* $v \in N^F_{\operatorname{hypo} f}(P) \cap \mathbb{S}^d$ (while in its formulation this is required just for *at least one* normal) with $\theta = 1$, we are in the class of functions whose hypograph has *positive reach* in the sense of Federer. In finite dimension, sets of positive reach were introduced by Federer in [13] as a generalization of convex sets and sets with C^2-boundary. If moreover we are also allowed to take $c = 0$, then the set is convex.

Upper semicontinuous functions whose hypograph has positive reach share several regularity properties with concave functions: it was proved in [6] that around a.e. points of their domain they are actually Lipschitz continuous, and twice differentiable a.e. In [8,9] and [10] some regularity results were proved for the minimum time function of control problems; under suitable weak controllability assumptions, the latter is proved to have epigraph or hypograph with

I. Lirkov et al. (Eds.): LSSC 2013, LNCS 8353, pp. 116–124, 2014.
DOI: 10.1007/978-3-662-43880-0_12,

locally positive reach, thus generalizing the results of [4] and [5]. Further regularity properties of this class of functions were proved in [7,16], from the nonsmooth analysis and geometric measure theory viewpoints, respectively.

However, it is easy to give examples where the hypograph of the minimum time function does not satisfy an exterior sphere property, so that the results of [9,15] can not be applied. Let us consider the constant control system

$$\begin{cases} x'(t) = 0, \\ y'(t) = u(t) \in [0,1], \\ (x(0), y(0)) = (x_0, y_0) \in \mathbb{R}^2, \end{cases} \tag{2}$$

together with the target $\mathcal{T} = \{(x, \beta) \mid \beta \geq f(x)\}$, where $f(x) = 1$ if $x \leq 0$ and $f(x) = -x^{\frac{2}{3}}$ if $x > 0$.

The minimum time to reach the target $\mathcal{T}$ subject to the above control system is denoted by T. It can be proved (see the Appendix) that hypo T does not satisfy an exterior sphere condition, but still enjoys the weaker uniformity regularity property (1) with $\theta = 1/2$.

The previous considerations motivate us to study the class $\mathscr{F}(\Omega)$ of real functions defined on $\Omega \subset \mathbb{R}^d$ satisfying condition (1) in order to provide a new regularity class which, hopefully, will cover the regularity properties for the minimum time function of certain classes of nonlinear control systems and differential inclusions (see [3]) that do not satisfy an exterior sphere condition. We will refer to this property as *N-regularity* (see Definition 2). We state our first general result, whose main ideas were presented in our recent paper [14], for closed set $K \subset \mathbb{R}^{d+1}$ concerning the structure and dimension of the set $K^{(j)}$ of points on ∂K where the Fréchet normal cone to ∂K has dimension larger than or equal to j. This result generalizes a similar result proved by Federer for sets with positive reach. Indeed, it shows that $K^{(j)}$ can be covered by countably many Lipschitz graphs of $d - j + 1$ variables.

Theorem 1. *Let $K \subseteq \mathbb{R}^{d+1}$ be closed; then $K^{(j)}$ is countably $\mathscr{H}^{d-j+1}$-rectifiable. In particular, also $K_{\pm}^{(j)}$ are countably $\mathscr{H}^{d-j+1}$-rectifiable.*

The sets $K_{\pm}^{(j)}$ are here defined in the same way of $K^{(j)}$ by taking the normal cone to, respectively, K and $\overline{\mathbb{R}^{d+1} \setminus K}$; see Definition 5. Concerning the differentiability properties of functions, we denote by $\mathscr{S}_f$ the set of non-differentiability points of f and prove the following result:

Theorem 2. *Let $\Omega \subseteq \mathbb{R}^d$ be a nonempty open set and $f : \Omega \to \mathbb{R}$ be an upper semicontinuous function with $f \in L^{\infty}_{\mathrm{loc}}(\Omega)$. Assume that the closed set $K :=$ hypo f is N-regular in $\Omega \times \mathbb{R}$. Then $f \in BV_{\mathrm{loc}}(\Omega)$ and $\mathscr{L}^d(\mathscr{S}_f) = 0$. In particular, f is differentiable a.e.*

2 Notation

Let K be a closed subset of $\mathbb{R}^d$, $S \subseteq \mathbb{R}^d$, $x = (x_1, \ldots, x_d) \in K$, $y = (y_1, \ldots, y_d) \in \mathbb{R}^d$, $r > 0$. We denote by $\langle \cdot, \cdot \rangle$, the usual *scalar product* in $\mathbb{R}^d$; ∂S, $\mathrm{int}(S)$, $\overline{S}$, the *topological boundary, interior* and *closure* of S, respectively; $\mathcal{P}(S) := \{B \subseteq \mathbb{R}^d : B \subseteq S\}$, the *power set* of S; $\mathbb{B}^d := \{w \in \mathbb{R}^d : \|w\| < 1\}$, the *unit open ball* (centered at the origin); $\mathbb{S}^{d-1} := \{w \in \mathbb{R}^d : \|w\| = 1\} = \partial\mathbb{B}^d$, the *unit sphere* (centered at the origin); $B(y,r) := \{z \in \mathbb{R}^d : \|z-y\| < r\} = y + r\mathbb{B}^d$, the *open ball* of center y and radius r; $d_K(y) := \mathrm{dist}(y,K) = \min\{\|z-y\| : z \in K\}$, the *distance* of y from K; $\pi_K(y) := \{z \in K : \|z-y\| = d_K(y)\}$, the *set of projections* of y onto K: if $\pi_K(y)$ contains an unique element ξ, we will write $\pi_K(y) = \xi$. $\mathscr{H}^p(S)$ and $\dim_{\mathscr{H}}(S)$, the *p-dimensional Hausdorff measure* and the *Hausdorff dimension* of S. The *characteristic function* $\chi_S : \mathbb{R}^d \to \{0,1\}$ of S is defined as $\chi_S(x) = 1$ if $x \in S$ and $\chi_S(x) = 0$ if $x \notin S$. If $V, W \subseteq \mathbb{R}^d$ are two subset of $\mathbb{R}^d$, we will write $V \subset\subset W$ if V is bounded and $\overline{V} \subseteq W$. Given a set X, $\mathrm{card}(X)$ denotes the number of its elements. The *Fréchet normal cone* and the *Bouligand tangent cone* to K at x are defined respectively by

$$N_K^F(x) := \left\{ v \in \mathbb{R}^d : \limsup_{\substack{y \to x \\ y \in K\setminus\{x\}}} \left\langle v, \frac{y-x}{\|y-x\|} \right\rangle \le 0 \right\};$$

$$T_K^F(x) := \left\{ \lambda\xi \in \mathbb{R}^d : \lambda \ge 0, \exists \{y_n\}_n \subseteq K \setminus \{x\}, y_n \to x \text{ s.t. } \xi = \lim_{n\to\infty} \frac{y_n - x}{\|y_n - x\|} \right\}.$$

Definition 1. *Let $\Omega \subseteq \mathbb{R}^d$ and $f : \Omega \to \mathbb{R} \cup \{\pm\infty\}$ be a function. For $x \in \Omega$ fixed we denote by $\overline{f}(x) := \limsup_{\substack{y \to x \\ y \neq x}} f(y)$; $\widetilde{f}(x) := \limsup_{y \to x} f(y) = \max\{f(x), \overline{f}(x)\}$; $\underline{f}(x) := \liminf_{\substack{y \to x \\ y \neq x}} f(y)$; $\underset{\sim}{f}(x) := \liminf_{y \to x} f(y) = \min\{f(x), \underline{f}(x)\}$; $\mathrm{dom}(f) := \{z \in \Omega : f(z) \in \mathbb{R}\}$, the* domain *of f;* $\mathrm{hypo}\, f := \{(z,\beta) \in \Omega \times \mathbb{R} : \beta \le f(z)\}$, *the* hypograph *of f;* $\mathrm{epi}\, f := \{(z,\alpha) \in \Omega \times \mathbb{R} : \alpha \ge f(z)\}$*, the* epigraph *of f; $\partial^F f(x) := \{v \in \mathbb{R}^d : (-v,1) \in N^F_{\mathrm{hypo}\, f}(x, f(x))\}$; $\partial_F f(x) := \{v \in \mathbb{R}^d : (v,-1) \in N^F_{\mathrm{epi}\, f}(x, f(x))\}$. We say that f is* upper *(respectively,* lower*)* semicontinuous *if $f(x) \ge \overline{f}(x)$ (resp., if $f(x) \le \underline{f}(x)$) for any $x \in \Omega$. The sets $\partial^F f(x)$ and $\partial_F f(x)$ are called respectively the* Fréchet superdifferential *and the* Fréchet subdifferential *of f at x.*

If $\Omega \subseteq \mathbb{R}^d$ is open, we denote by $BV(\Omega)$ the set of *function of bounded variation in* Ω, and if $u \in BV(\Omega)$, we denote by $\|Du\|$ the total variation of the vector-valued measure Du. The *perimeter of* E in Ω is $P(E,\Omega) = \|D\chi_E\|(\Omega)$.

Let $A \subseteq \mathbb{R}^d$ and $0 \le p \le d$. Let $k \in \mathbb{N}$, we say that $A \subseteq \mathbb{R}^d$ is *countably* $\mathscr{H}^k$*-rectifiable* if $A \subseteq \mathcal{N} \cup \bigcup_{i=1}^{\infty} S_i$, where S_i are suitable k-dimensional Lipschitz surfaces and $\mathscr{H}^k(\mathcal{N}) = 0$. For a detailed introduction to BV functions and their properties, we refer to [1,12]

3 Standing Hypothesis and First Consequences

Definition 2. *Let $U \subseteq \mathbb{R}^{d+1}$ be open and $K \subseteq \mathbb{R}^{d+1}$ be nonempty and relatively closed in U. We say that K is* N-regular *in U if there exists an upper* *semicontinuous multifunction* $N : \partial K \cap U \rightrightarrows \mathbb{S}^d$ *such that for every* $x \in \partial K \cap U$ *the following two properties hold:*

(N1) $\emptyset \neq N(x) \subseteq N_K^F(x) \cap \mathbb{S}^d$;

(N2) there exist $\delta_x \in \,]0, dist\,(x, \partial U)[$ *and a continuous function* $\omega_x : \mathbb{R}^+ \to \mathbb{R}^+$ *with* $\lim_{r\to 0^+} \omega_x(r)/r = 0$ *and satisfying the following uniformity property: for every* $y_1 \in \left(x + \delta_x \mathbb{B}^{d+1}\right) \cap \partial K$ *there exists* $\nu(y_1) \in N(y_1)$ *such that*

$$\langle \nu(y_1), y_2 - y_1 \rangle \leq \omega_x(\|y_2 - y_1\|) \text{ for all } y_2 \in \left(x + \delta_x \mathbb{B}^{d+1}\right) \cap K.$$

We will say that $K \subseteq \mathbb{R}^{d+1}$ *is* N-regular *if* K *is* N*-regular in* $\mathbb{R}^{d+1}$. *Possibly replacing the set-valued map* N *with* $x \mapsto \overline{N(x)}$, *when* K *is* N*-regular in* U *we can always assume that* N *has closed graph.*

Example 1. Every set K that is the closure of an open C^1 domain is N-regular, moreover a closed convex set C is N-regular with $N(x) = N_C^F(x) \cap \mathbb{S}^d$

Definition 3. *Let* $U \subseteq \mathbb{R}^{d+1}$ *be open and* $K \subseteq \mathbb{R}^{d+1}$ *be nonempty and relatively closed in* U; *let also* $z \in \partial K \cap U$, $\theta \in]0, 1]$ *and* $C \geq 0$. *We define*

$$\mathscr{N}_K^{C,\theta,U}(z) := \{\zeta \in \mathbb{R}^{d+1} : \langle \zeta, z' - z \rangle \leq C \cdot \|\zeta\| \cdot \|z' - z\|^{1+\theta} \quad \text{for all } z' \in K \cap U\}. \tag{3}$$

If K *is closed,* $U = \mathbb{R}^{d+1}$ *and* $z \in \partial K$ *we will simply write* $\mathscr{N}_K^{C,\theta}(z)$ *instead of* $\mathscr{N}_K^{C,\theta,\mathbb{R}^{d+1}}(z)$. *We notice that if* $\zeta \in \mathscr{N}_K^{C,\theta,U}(x)$, *then* $\mu\zeta \in \mathscr{N}_K^{C,\theta,U}(x)$ *for all* $\mu \geq 0$ *and the multifunction* $\mathscr{N}_K^{C,\theta,U} : \partial K \cap U \rightrightarrows \mathbb{R}^{d+1}$ *has closed graph.*

Let now $\Omega \subseteq \mathbb{R}^d$ *be nonempty and open and* $f : \Omega \to \mathbb{R}$ *be upper semicontinuous. By adapting the previous definition, for* $(x, \beta_x) \in \partial\text{hypo}\, f \cap (\Omega \times \mathbb{R})$ *we define* $\hat{\mathscr{N}}_{\text{hypo}\, f}^{C,\theta}(x, \beta_x)$ *as the set of those* $(v, \lambda) \in \mathbb{R}^d \times \mathbb{R}$ *such that* $\forall (y, \beta) \in \text{hypo}\, f$.

$$\langle (v, \lambda), (y - x, \beta - \beta_x) \rangle \leq C\|(v, \lambda)\| \left(\|y - x\|^{1+\theta} + |\beta - \beta_x|^{1+\theta}\right) \tag{4}$$

We notice that there exist constants $c_1, c_2 > 0$ *depending only on* d *and* θ *such that*

$$\mathscr{N}_{\text{hypo}\, f}^{c_1 C,\theta,\Omega\times\mathbb{R}}(x, \beta_x) \subseteq \hat{\mathscr{N}}_{\text{hypo}\, f}^{C,\theta}(x, \beta_x) \subseteq \mathscr{N}_{\text{hypo}\, f}^{c_2 C,\theta,\Omega\times\mathbb{R}}(x, \beta_x).$$

It is clear from the definition that also $\hat{\mathscr{N}}_{\text{hypo}\, f}^{c,\theta} : \partial\, \text{hypo}\, f \cap (\Omega \times \mathbb{R}) \rightrightarrows \mathbb{R}^{d+1}$ *has closed graph.*

We are ready now to introduce the classes of sets and functions subject of our investigation.

Definition 4. *Let* $U \subseteq \mathbb{R}^{d+1}$ *and* $\Omega \subseteq \mathbb{R}^d$ *be open. We define:*

$$\mathscr{F}^U := \{K \subseteq U : K \text{ is relatively closed in } U \text{ and } \exists C \geq 0,\, 0 < \theta \leq 1 \text{ s.t. } \mathscr{N}_K^{C,\theta,U}(z) \neq \{0\} \text{ for all } z \in \partial K \cap U\}$$

$$\mathscr{F} := \mathscr{F}^{\mathbb{R}^{d+1}}$$

$$\begin{aligned}\mathscr{F}(\Omega) &:= \{f : \Omega \to \mathbb{R} : f \text{ u.s.c., } \text{hypo}\, f \in \mathscr{F}^{\Omega\times\mathbb{R}}\} \\ &= \{f : \Omega \to \mathbb{R} : f \text{ u.s.c., } \exists C \geq 0,\ 0 < \theta \leq 1 \text{ such that} \\ &\qquad \hat{\mathscr{N}}_{\text{hypo}\, f}^{C,\theta}(x, \beta_x) \neq \{0\}\ \forall (x, \beta_x) \in \partial\, \text{hypo}\, f \cap (\Omega \times \mathbb{R})\}.\end{aligned}$$

If $K \in \mathscr{F}^U$, then there exist $C > 0$, $0 < \theta \leq 1$ such that K is N-regular in U with

$$N(x) := \mathscr{N}_K^{C,\theta,U}(x) \cap \mathbb{S}^d \subseteq N_K^F(x), \quad \omega_x(r) := r^{1+\theta} \qquad \forall x \in \partial K \cap U.$$

The upper semicontinuity of N follows from the fact that $\mathscr{N}_K^{C,\theta,U}(x)$ has closed graph.

We refer the reader to [11,13] for a survey of the properties satisfied by sets with positive reach, on which the class $\mathscr{F}$ is modeled.

4 Regularity Results for Sets

In this section we will prove regularity results for the boundary of a closed set $K \subseteq \mathbb{R}^{d+1}$ in a quite general setting. They will be used later to prove fine regularity properties for functions in the class $\mathscr{F}(\Omega)$.

The first result extends an analogous result for the class of sets with positive reach proved by Federer in Remark 4.15 of [13]. Roughly speaking it states that points with *large* normal cone are relatively *few*.

Definition 5. *Let $K \subseteq \mathbb{R}^{d+1}$ be closed; for $j = 1, ..., d+1$ we define*

$$K^{(j)} := \left\{x \in \partial K : \dim\left(N_{\partial K}^F(x)\right) \geq j\right\}, \tag{5}$$

$$K_+^{(j)} := \left\{x \in \partial K : \dim\left(N_K^F(x)\right) \geq j\right\}, \tag{6}$$

$$K_-^{(j)} := \left\{x \in \partial K : \dim\left(N_{\overline{\mathbb{R}^{d+1} \setminus K}}^F(x)\right) \geq j\right\}. \tag{7}$$

We notice that $K^{(j_1)} \supseteq K^{(j_2)}$, $K_+^{(j_1)} \supseteq K_+^{(j_2)}$, $K_-^{(j_1)} \supseteq K_-^{(j_2)}$ if $1 \leq j_1 \leq j_2 \leq d+1$, and that $K_\pm^{(j)} \subseteq K^{(j)}$. Clearly, $K^{(1)} = \{x \in \partial K : N_{\partial K}^F(x) \neq \{0\}\}$.

In order to use local arguments, we will need the following estimate which gives some *uniformity* with respect to the elements of the normal cone, which can be proved exploiting compactness of $N_{\partial K}^F(x) \cap \mathbb{S}^d$: for every $x \in K^{(1)}$ and $0 < \varepsilon \leq 1$ we have

$$\begin{aligned}\delta(x,\varepsilon) := \frac{1}{2}\sup\{\ \delta \in \mathbb{R} :\ & \langle v, y-x\rangle \leq \varepsilon\|y-x\|, \\ & \text{for all } y \in \partial K \cap \left(x + \delta\mathbb{B}^{d+1}\right), v \in N_{\partial K}^F(x) \cap \mathbb{S}^d\right\} > 0\end{aligned} \tag{8}$$

We are now ready to prove the first main result of the paper.

Proof (Proof of Theorem 1.). We begin by constructing a countable covering $\{K_{n,m,h,l}^{(j)}\}_{n,m,h,l \in \mathbb{N}}$ of $K^{(j)}$; we will prove later that each element of the covering is rectifiable and this will establish our result.

Define the function $w : (\mathbb{R}^{d+1})^j \to [0,1]$

$$w(v_1, \ldots, v_j) := \min\left\{\left\|\sum_{i=1}^j \alpha_i v_i\right\| : \alpha_i \in \mathbb{R}, \sum_{i=1}^j |\alpha_i| = 1\right\}.$$

We notice that w is continuous and invariant under permutations of its arguments, so if $V=\{v_1,\dots,v_j\}$ we will write $w(V)$ instead of $w(v_1,...,v_j)$. We have that $w(V)=0$ iff the elements of V are linearly dependent.

Let $\{a_l\}_{l\in\mathbb{N}}$ be a countable dense set in $\mathbb{R}^{d+1}$. For every $x\in K^{(j)}$ choose $V_x\subseteq N^F_{\partial K}(x)\cap\mathbb{S}^d$ with $\operatorname{card}V_x=j$ and $w(V_x)>0$, $V_x=\{v_x^{(i)}\}_{i=1,\dots,j}$. Consider the countable set

$$\mathcal{A}^{(j)}:=\left\{V'\subseteq\mathbb{Q}^{d+1}:\operatorname{card}(V')=j,\,w(V')>0\right\},$$

Being $\mathcal{A}^{(j)}$ countable, we can order its elements and write $\mathcal{A}^{(j)}=\{V'_n\}_{n\in\mathbb{N}}$. We set $V_n^{(j)}=\operatorname{Span}(V'_n)$ and consider the countable set of j-dimensional planes $\mathcal{V}^{(j)}:=\{V_n^{(j)}\}_{n\in\mathbb{N}}$. Define also $W_n^{(j)}:=(V_n^{(j)})^\perp$, $n\in\mathbb{N}$, and $W^{(j)}:=\{W_n^{(j)}\}_{n\in\mathbb{N}}$. Given $n,m,h,l\in\mathbb{N}$, let $v_1,\dots,v_j\in\mathbb{Q}^{d+1}$ be such that $V'_n=\{v_1,\dots,v_j\}$ and set

$$K^{(j)}_{n,m,h,l}:=\left\{x\in K^{(j)}\cap\left(a_l+\tfrac{1}{2(h+1)}\mathbb{B}^{d+1}\right):\begin{array}{l}w(V_x)\geq\frac{1}{m+3},\\ \delta\left(x,\frac{1}{2(m+3)^2}\right)\geq\frac{1}{h+1},\\ \left\|v_x^{(i)}-v_i\right\|\leq\frac{1}{2(m+3)^2}\\ \text{for } i=1,\dots,j\end{array}\right\},$$

where $\delta(x,\frac{1}{2(m+3)^2})$ is as in (8) with $\varepsilon=(2(m+3)^2)^{-1}$.

It turns out that $K^{(j)}\subseteq\bigcup_{n,m,h,l\in\mathbb{N}}K^{(j)}_{n,m,h,l}$: given $x\in K^{(j)}$, we choose in this sequence the indexes: m, n, h, l, to fulfill the properties yielding $x\in K^{(j)}_{n,m,h,l}$.

We prove now that for any $x_1,x_2\in K^{(j)}_{n,m,h,l}$ the orthogonal projection $\pi_{W_n^{(j)}}:K^{(j)}_{n,m,h,l}\to W_n^{(j)}$ satisfies

$$\|\pi_{W_n^{(j)}}(x_2-x_1)\|^2\geq\frac{m+1}{m+3}\|x_2-x_1\|^2. \tag{9}$$

Indeed, we notice that if $V'_n=\{v_1,\dots,v_j\}$, then each v_i is near to a normal vector both at x_1, and at x_2. By exploiting the definition of $K^{(j)}_{n,m,h,l}$, this fact yields:

$$|\langle v_i,x_2-x_1\rangle|\leq\frac{1}{(m+3)^2}\|x_2-x_1\|\quad\text{for every } i=1,...,j.$$

Given $v\in V_n^{(j)}$, $v\neq0$, we can find (in a unique way) $\alpha_i\in\mathbb{R}$, $i=1,\dots,j$ such that $v=\sum_{i=1}^j\alpha_iv_i$; therefore

$$\left|\left\langle\frac{v}{\|v\|},x_2-x_1\right\rangle\right|\leq\frac{\sum_{i=1}^j|\alpha_i|\cdot|\langle v_i,x_2-x_1\rangle|}{\left\|\sum_{i=1}^j\alpha_iv_i\right\|}\leq\frac{\|x_2-x_1\|}{(m+3)^2}\frac{\sum_{i=1}^j|\alpha_i|}{\left\|\sum_{i=1}^j\alpha_iv_i\right\|}.$$

Set $\beta_i := \alpha_i / \sum_{s=1}^{j} |\alpha_s|$; we have $\sum_{i=1}^{j} |\beta_i| = 1$ and thus

$$\left|\left\langle \frac{v}{\|v\|}, x_2 - x_1 \right\rangle\right| \leq \frac{\|x_2 - x_1\|}{(m+3)^2} \frac{1}{\left\|\sum_{i=1}^{j} \beta_i v_i\right\|} \leq \frac{\|x_2 - x_1\|}{(m+3)^2} \frac{1}{w(v_1, \ldots, v_j)}$$
$$\leq \frac{2}{m+3} \|x_2 - x_1\|$$

because $w(v_1, \ldots, v_j) \geq (2(m+3))^{-1}$. Therefore,

$$\|\pi_{W_n^{(j)}}(x_2 - x_1)\|^2 = \|x_2 - x_1\|^2 - \langle \pi_{V_n^{(j)}}(x_2 - x_1), x_2 - x_1 \rangle \geq \frac{m+1}{m+3} \|x_2 - x_1\|^2.$$

By (9), for each n, m, h, l the inverse map $\pi^{-1}_{W_n^{(j)}} : \pi_{W_n^{(j)}}(K^{(j)}_{n,m,h,l}) \to K^{(j)}_{n,m,h,l}$. is Lipschitz continuous and, by Kirszbraun's Theorem, it can be extended to a Lipschitz function defined on the whole $W_n^{(j)}$. This ends the proof.

5 Application to Functions: *BV* Regularity and Structure of Singular Set

In this section we will apply the results obtained in the previous one to closed sets that can be written as hypographs of upper semicontinuous functions possessing at least one normal direction at a.e. point of the boundary of their hypograph; our goal is to obtain regularity results for such functions.

Definition 6. *Let Ω be a nonempty open subset of $\mathbb{R}^d$ and $f : \Omega \to \mathbb{R}$ be a function. For each $x \in \Omega$, we define*

$$J_f := \{x \in \Omega : \widetilde{f}(x) \neq \underset{\sim}{f}(x)\} = \{x \in \Omega : f \text{ is not continuous at } x\},$$
$$S_f := \{x \in \Omega \setminus J_f : (\mathbb{S}^{d-1} \times \{0\}) \cap N^F_{\text{hypo} f}(x, f(x)) \neq \emptyset\},$$
$$\mathscr{S}_f := J_f \cup S_f.$$

We begin with a trivial corollary of Theorem 1, dealing with the singularities corresponding to large dimension of the normal cone.

Corollary 1. *Let Ω be a nonempty open subset of $\mathbb{R}^d$ and $f : \Omega \to \mathbb{R}$ be an upper semicontinuous function. Set $K = \text{hypo} f$ and assume that $N^F_K(x, \beta) \neq \{0\}$ for $\mathscr{H}^d$-a.e. $(x, \beta) \in \partial K \cap (\Omega \times \mathbb{R})$. Then for $\mathscr{L}^d$-almost every $x \in \Omega$ there exists $\zeta_x \in \mathbb{S}^d$ such that $N^F_K(x, \beta) \subseteq \mathbb{R}\zeta_x$ for all β with $(x, \beta) \in \partial K \cap (\Omega \times \mathbb{R})$.*

Proof. By Theorem 1, $K^{(2)}$ is $\mathscr{H}^{d-1}$-rectifiable and hence $\mathscr{H}^d$-negligible. If $\pi : \Omega \times \mathbb{R} \to \Omega$ denotes the canonical projection on Ω, then $\Omega \cap (\pi(\partial K \setminus K^{(1)}) \cup \pi(K^{(2)}))$ is $\mathscr{L}^d$-negligible, hence $E := \Omega \setminus (\pi(\partial K \setminus K^{(1)}) \cup \pi(K^{(2)}))$ has the same measure of Ω. The results follows.

We recall the following result, proved in Theorem 1.2 of [14]:

Theorem 3. *Let Ω be a nonempty open subset of $\mathbb{R}^d$ and let $f \in BV_{\text{loc}}(\Omega)$ be an upper semicontinuous function; set $K := \operatorname{hypo} f$. Assume that for $\mathscr{H}^d$-a.e. $(x, \beta_x) \in \partial K \cap (\Omega \times \mathbb{R})$ it holds $N_K^F(x, \beta_x) \neq \{0\}$. Then $\mathscr{L}^d(\mathscr{S}_f) = 0$.*

We are going to study the regularity properties of upper semicontinuous functions f such that hypo$f$ is N-regular. One of our primary goals is to estimate the size of the singular set $\mathscr{S}_f$; to this aim it will be important to assume that f is of class BV.

We can now prove the second main result of the paper:

Proof (of Theorem 2). Let us prove that $f \in BV(U)$ for any open set U such that $U \subset\subset \Omega$. According to Theorem 1, we have that ∂K is rectifiable, whence $P(K, U \times \mathbb{R}) = P(K, U \times] - 2\|f\|_{L^\infty}, 2\|f\|_{L^\infty}[) < \infty$. According to Theorem 4 in [2], we have that $f \in BV_{\text{loc}}(\Omega)$, so we can apply Theorem 3.

References

1. Ambrosio, L., Fusco, N., Pallara, D.: Functions of Bounded Variation and Free Discontinuity Problems. Oxford Mathematical Monographs. The Clarendon Press Oxford University Press, New York (2000)
2. Chlebík, M., Cianchi, A., Fusco, N.: The perimeter inequality under Steiner symmetrization: cases of equality. Ann. Math. (2) **162**(1), 525–555 (2005)
3. Cannarsa, P., Nguyen, K.T.: Exterior sphere condition and time optimal control for differential inclusions. SIAM J. Control Optim. **49**(6), 2258–2576 (2011)
4. Cannarsa, P., Sinestrari, C.: Convexity properties of the minimum time function. Calc. Var. Partial. Differ. Equ. **3**(3), 273–298 (1995)
5. Cannarsa, P., Sinestrari, C.: Semiconcave Functions, Hamilton-Jacobi Equations, and Optimal Control. Progress in Nonlinear Differential Equations and their Applications, vol. 58. Birkhäuser Boston Inc., Boston (2004)
6. Colombo, G., Marigonda, A.: Differentiability properties for a class of non-convex functions. Calc. Var. Partial. Differ. Equ. **25**(1), 1–31 (2006)
7. Colombo, G., Marigonda, A.: Singularities for a class of non-convex sets and functions, and viscosity solutions of some Hamilton-Jacobi equations. J. Convex Anal. **15**(1), 105–129 (2008)
8. Colombo, G., Marigonda, A., Wolenski, P.R.: Some new regularity properties for the minimal time function. SIAM J. Control Optim. **44**(6), 2285–2299 (2006). (electronic)
9. Colombo, G., Nguyen, K.T.: On the structure of the minimum time function. SIAM J. Control Optim. **48**(7), 4776–4814 (2010)
10. Colombo, G., Nguyen, K.T.: On the minimum time function around the origin. Math. Control Relat. Fields **3**(1), 51–82 (2013)
11. Colombo, G., Thibault, L.: Prox-regular sets and applications. In: Gao, D.Y., Motreanu, D. (eds.) Handbook of Nonconvex Analysis. International Press, Somerville (2010)
12. Evans, L.C., Gariepy, R.F.: Measure Theory and Fine Properties of Functions. Studies in Advanced Mathematics. CRC Press, Boca Raton (1992)
13. Federer, H.: Curvature measures. Trans. Amer. Math. Soc. **93**, 418–491 (1959)
14. Marigonda, A., Nguyen, K.T., Vittone, D.: Some regularity properties for a class of upper semicontinuous functions. Indiana Univ. Math. J. **62**(1), 45–89 (2013)

15. Nguyen, K.T.: Hypographs satisfying an external sphere condition and the regularity of the minimum time function. J. Math. Anal. Appl. **372**, 611–628 (2010)
16. Nguyen, K.T., Vittone, D.: Rectifiability of special singularities of non-Lipschitz functions. J. Convex Anal. **19**(1), 159–170 (2012)

Internal Ellipsoidal Estimates of Reachable Set of Impulsive Control Systems Under Ellipsoidal State Bounds and with Cone Constraint on the Control

Oxana G. Matviychuk[1,2(✉)]

[1] Institute of Mathematics and Mechanics, Russian Academy of Sciences, 16 S. Kovalevskaya str., Ekaterinburg 620990, Russia
[2] Ural Federal University, 19 Mira str., Ekaterinburg 620002, Russia
vog@imm.uran.ru
http://www.imm.uran.ru

Abstract. The problem of estimating reachable sets of linear impulsive dynamical control systems with uncertainty in initial data is considered. It is assumed that the impulsive controls in the dynamical system belong to the intersection of a special cone with a generalized ellipsoid both taken in the space of functions of bounded variation. The algorithms for constructing the ellipsoidal estimates of reachable sets for such control systems are given. Numerical simulation results relating to the proposed procedures are also given.

1 Introduction

We study here the estimation problem for a dynamic process with discontinuous trajectories and generalized controls of impulsive type.

The problem is studied under uncertainty conditions [5–7,13] with set — membership description of uncertain variables, which are taken to be unknown but bounded with given bounds.

The main problem is to estimate the reachable set of the control system. It is assumed that impulsive controls in the dynamical system must belong to the intersection of a special cone with a generalized ellipsoid both taken in the space of functions of bounded variation. The last constraint is motivated by problems of impulsive control theory and by models from applied areas when not every direction of control impulses is acceptable in the system. For example, one can consider implementation of autonomous underwater vehicle systems for oceanographic and environmental field studies [11].

The problem under some more complicated assumption related to the case of state constraints is studied.

Based on the results of estimating the trajectory tubes of ordinary differential systems and using the techniques of ellipsoidal calculus [1,8] we present here

I. Lirkov et al. (Eds.): LSSC 2013, LNCS 8353, pp. 125–132, 2014.
DOI: 10.1007/978-3-662-43880-0_13,

a new state internal estimation method for the studied impulsive control problem. The external ellipsoidal estimates for impulsive systems were considered in [2,3,9,15].

In this paper the reachable set is estimated internally by one ellipsoid [1,8]. In this formulation, the optimality of the internal approximation is understood in terms of the maximum size of the inscribed ellipsoid. The accuracy of the constructed ellipsoidal estimation is not considered for similar problems.

This paper continues previous research [3,4,10]. We present here the state estimation algorithm and illustrate it by an example.

2 Problem Formulation

Let us start by introducing the following basic notations. Let $\mathbb{R}^n$ denotes the n-dimensional Euclidean space. Denote by $\mathbb{R}^{n\times n}$ the set of all $n \times n$ - matrices and by $\widetilde{\mathbb{R}}^{n\times n} \subset \mathbb{R}^{n\times n}$ the set of all symmetric positive definite matrices. Let us denote by the symbol $'$ stands for transposition, $x'y$ the scalar product of vectors $x, y \in \mathbb{R}^n$ and by $\|x\| = (x'x)^{1/2}$ the Euclidean norm of the vector x, and $B(a, r) = \{x \in \mathbb{R}^n : \|x - a\| \leq r\}$. Let $E(y, Y) = \{x \in \mathbb{R}^n : (x - y)'Y^{-1}(x - y) \leq 1\}$ be the ellipsoid in $\mathbb{R}^n$ with center $y \in \mathbb{R}^n$ and a matrix $Y \in \widetilde{\mathbb{R}}^{n\times n}$, and let $I \in \mathbb{R}^{n\times n}$ be the identity matrix. For a set $A \subset \mathbb{R}^n$ we denote its closed convex hull [12] as $\overline{\text{co}}\, A$.

Consider a control system with impulsive control (measure) $u(\cdot)$:

$$dx(t) = A(t)x(t)dt + du(t), \qquad x(t_0 - 0) = x_0, \quad t \in [t_0, T], \tag{1}$$

or in the integral form,

$$x(t) = x(t; u(\cdot), x_0) = \Phi(t, t_0)x_0 + \int_{t_0}^{t} \Phi(t, \tau)du(\tau). \tag{2}$$

Here it is assumed that $A(t)$ is a continuous $n \times n$ – matrix function, $\Phi(t)$ is the Cauchy matrix [6], $\Phi(t, \tau) = \Phi(t)\Phi^{-1}(\tau)$, $u(\cdot) \in \mathbb{V}_p^n$ where $\mathbb{V}_p^n$ $(1 \leq p < \infty)$ means the space of n-vector functions $u(\cdot)$ such that $u(t)$ is continuous from the right on $[t_0, T)$ with $u(t_0 - 0) = 0$ and

$$V_p[u(\cdot)] = \sup_{\{t_i\}} \sum_{i=1}^{k} \|u(t_i) - u(t_{i-1})\|_p \leq \infty, \qquad \|u\|_p = \left(\sum_{i=1}^{n} |u_i|^p\right)^{\frac{1}{p}},$$

where $u = (u_1, \ldots, u_n)$, $t_i : t_0 < \cdots < t_k = T$.

Denote by $\mathbb{C}_q^n$ the space of continuous n-vector functions $y(\cdot)$ with the norm $\|y(\cdot)\|_{\infty,q} = \max_{t_0 \leq t \leq T} \|y(t)\|_q$. It is well known that the space $\mathbb{V}_p^n = \mathbb{C}_q^{n*}$ where $p = 1$ if $q = \infty$, $p = \infty$ if $q = 1$ and $1 < p < \infty$ if $q = (1 - p^{-1})^{-1}$.

Let $E_0 = E(0, Q_0^{-1})$ be an ellipsoid in $\mathbb{R}^n$ with a center at the origin and with $Q_0 \in \widetilde{\mathbb{R}}^{n\times n}$. Consider the so-called generalized "ellipsoid" [12,15] E in $\mathbb{C}_q^n$:

$$E = \{y(\cdot) \in \mathbb{C}_q^n \mid y(t) \in E_0 \ \ \forall t \in [t_0, T]\} \tag{3}$$

and its conjugate ellipsoid [12] E^* in $\mathbb{V}_p^n$ such that

$$E^* = \{u(\cdot) \in \mathbb{V}_p^n \mid \int_{t_0}^{T} y(t)'du(t) \leq 1 \ \ \forall y(\cdot) \in E\}. \tag{4}$$

Let introduce the following cone described as

$$K_0 = \{u \in \mathbb{R}^n \mid u = (u_1, \ldots, u_n), \ \ u_1 \geq 0, \ \ u_i \in \mathbb{R}, \ \ i = 2, \ldots, n\},$$
$$K = \{y(\cdot) \in \mathbb{C}_q^n \mid y(t) \in K_0 \ \ \forall t \in [t_0, T]\},$$

its conjugate cone K^* [12] will have the form

$$K^* = \{u(\cdot) \in \mathbb{V}_p^n \mid \int_{t_0}^{T} y(t)'du(t) \geq 0 \ \ \forall y(\cdot) \in K\}. \tag{5}$$

Definition 1. *The function $u(\cdot) \in \mathbb{V}_p^n$ will be called the admissible control if $u(\cdot) \in U = E^* \cap K^*$.*

Remark 1. Let $u(\cdot) \in \mathbb{V}_p^n$ be a piecewise constant function on $[t_0, T]$ with discontinuity instants $\{t_i \in [t_0, T]\}$ and with "jump" vectors

$$\begin{gathered} \Delta u_i = u(t_{i+1}) - u(t_i) \in E(0, Q_0) \cap K_0 = \\ \{z \in \mathbb{R}^n \mid z'Q_0^{-1}z \leq 1, \ z = (z_1, \ldots, z_n), \ \ z_1 \geq 0, \ z_i \in \mathbb{R}, \ i = 2, \ldots, n\}. \end{gathered} \tag{6}$$

Then $u(\cdot)$ is admissible.

We will assume also that the initial state x_0 for the system (1) is unknown but bounded with a given ellipsoidal bound:

$$x_0 \in X_0 = E(r, R), \tag{7}$$

where $R \in \widetilde{\mathbb{R}}^{n \times n}$ and $r \in \mathbb{R}^n$.

Assume that the state constraint (of terminal type) is imposed, so we have

$$x(T) \in E(y_0, D), \tag{8}$$

where $E(y_0, D)$ is a given ellipsoid $D \in \widetilde{\mathbb{R}}^{n \times n}$, $y_0 \in \mathbb{R}^n$. We will assume that there exists at least one trajectory $x(T)$ satisfying (1), (7), (8) and $u \in U$.

Denote

$$X(T; U, X_0) = \bigcup_{x_0 \in X_0} \bigcup_{u \in U} \{x(T; u(\cdot), x_0)\}, \tag{9}$$

where $x(T; u(\cdot), x_0)$ is defined in (2) under additional assumption (8).

Definition 2. *The set $X(T; U, X_0)$ (9) is called the reachable set of the impulsive differential system (1) from the initial set X_0 (7) at the instant T under controls $u(\cdot) \in U$ and ellipsoidal state bounds (8).*

The main problem of the paper is to find the estimates of ellipsoidal type for the $X(T; U, X_0)$ basing on the special ellipsoidal structure of the data.

3 Main Results

Consider first some auxiliary results. The following theorem is true.

Theorem 1. [3] *The equality holds*

$$X(T;\, U,\, E(r,\, R)) = E(r_1,\, R_1) + X(T;\, U,\, \{0\}), \tag{10}$$

$$X(T;\, U,\, \{0\}) = \overline{\text{co}}\Big(\bigcup_{\tau\in[t_0,\, T]} \Phi(T, \tau)(E(0,\, Q_0) \cap K_0)\Big),$$

$$r_1 = \Phi(T,\, t_0)r, \quad R_1 = \Phi(T,\, t_0)R(\Phi(T,\, t_0))'.$$

Remark 2. From the Theorem 1 we conclude that if we find an ellipsoid E^- such that $E^- \subseteq X(T; U, \{0\})$ then applying well-known formulas for calculating the internal ellipsoidal estimate for the sum of two ellipsoids $E(r_1,\, R_1)$ and E^- [1,8] we can find the resulting internal ellipsoid for $X(T;\, U,\, X_0)$ in (10).

Thus the main difficulty is to construct internal estimates for the $X(T;\, U,\, \{0\})$.

The idea of constructing the internal estimates for $X(T;\, U,\, \{0\})$ is basing on results of ellipsoidal calculus [1,8] and on the procedures of internal approximation of a closed convex hull of the union of some ellipsoids.

First, we find the ellipsoid which is contained in the intersection $E(0,\, Q_0)\cap K_0$.

Theorem 2. *The following internal estimate is true*

$$\begin{aligned} E(a,\, Q) &\subseteq E(0,\, Q_0) \cap K_0, \\ a &= \frac{n}{n+1} \cdot Q_0\nu(\nu' Q_0'\nu)^{-\frac{1}{2}}, \\ Q^{-1} &= (n+1)\cdot(\nu\nu')\cdot(\nu' Q_0'\nu)^{-1} + \frac{n+1}{n}\cdot Q_0^{-1}, \end{aligned} \tag{11}$$

where ν *is the inner normal to the half-space* K_0 *with* $\|\nu\| = 1$ *and the ellipsoid* $E(a,\, Q)$ (11) *has the largest volume among all ellipsoids contained in the* $E(0,\, Q_0) \cap K_0$.

Proof. The idea of the proof is as follows. First, using the affine transformation [1] we transform the ellipsoid $E(0,\, Q_0)$ into the unit ball $B(0,\, 1)$. It is easy to check that under this transformation the cone K_0 is not changed. Now we need to construct an internal ellipsoidal estimate of the spherical segment $B(0,\, 1) \cap K_0$.

This ellipsoidal estimation $E(\bar{a},\, \bar{Q}) \subset B(0,\, 1) \cap K_0$ can be found in the form of the "rotation" ellipsoid [1]:

$$\begin{aligned} E(\bar{a},\, \bar{Q}) = \{x \in \mathbb{R}^n | \; \frac{(y-\xi)^2}{A^2} + \frac{z^2}{B^2} \le 1, \;\; y = \nu' x,\; z = x - y\nu\}, \\ \xi = A = \frac{\sqrt{n}}{n+1}, \quad B = \sqrt{\frac{n}{n+1}}. \end{aligned} \tag{12}$$

Note that the ellipsoid $E(\bar{a}, \bar{Q})$ has a maximal volume among all ellipsoids contained in $B(0, 1) \cap K_0$.

Returning to the initial coordinates [1] we obtain the internal ellipsoidal estimate $E(a, Q)$ for the set $E(0, Q_0) \cap K_0$, where parameters a and Q of the ellipsoid $E(a, Q)$ are defined in (11). This ellipsoid $E(a, Q)$ has the largest volume among all ellipsoids contained in the $E(0, Q_0) \cap K_0$ [1]. □

Consider impulsive control system

$$dx(t) = A(t)x(t)dt + du(t), \quad x(t_0 - 0) = x_0 \in X_0 = E(r, R), \quad t \in [t_0, T], \tag{13}$$

with a new constraint

$$u(\cdot) \in \tilde{U} = \tilde{E}^*, \tag{14}$$

where

$$\tilde{E}^* = \{u(\cdot) \in \mathbb{V}_p^n \mid \int_{t_0}^{T} y(t)'du(t) \le 1 \ \ \forall y(\cdot) \in \tilde{E}\} \supset E^*,$$
$$\tilde{E} = \{y(\cdot) \in \mathbb{C}_q^n \mid y(t) \in E(a, Q) \ \ \forall\, t \in [t_0, T]\} \subset E$$

and parameters a, Q are defined in (11).

We study first the case without state constraint (8). We denote the related the reachable set $X(T; \tilde{U}, \{0\})$ in this case as $\tilde{X}(T; \tilde{U}, \{0\})$.

Consider the following auxiliary problem.

Auxiliary Problem AP. Find the ellipsoid $\tilde{E}^-(\tilde{a}^-, \tilde{Q}^-)$ such that $\tilde{E}^-(\tilde{a}^-, \tilde{Q}^-) \subseteq \tilde{X}(T; \tilde{U}, \{0\})$.

Lemma 1. *The following inclusion holds*

$$\tilde{X}(t; \tilde{U}, \{0\}) \subseteq \tilde{X}(t; U, \{0\}). \tag{15}$$

Proof. The proof of this lemma follows directly from (10). □

For the impulsive control system (13)–(14) and for any $\varepsilon > 0$ there exist $\delta > 0$ and finite set $T_\delta = \{\tau_1, \tau_2, \ldots, \tau_m\} \subset [t_0, T]$ such that the inclusions are true [4]:

$$\begin{gathered}\overline{\text{co}}\Big(\bigcup_{\tau_i \in T_\delta} E(a_{\tau_i}, Q_{\tau_i})\Big) \subseteq \tilde{X}(T; \tilde{U}, \{0\}) \subseteq \overline{\text{co}}\Big(\bigcup_{\tau_i \in T_\delta} E(a_{\tau_i}, Q_{\tau_i})\Big) + B(0, \varepsilon), \\ a_{\tau_i} = \Phi(T, \tau_i)a, \quad Q_{\tau_i} = \Phi(T, \tau_i)Q(\Phi(T, \tau_i))',\end{gathered} \tag{16}$$

Now we need to construct the internal ellipsoidal estimate for the closed convex hull $\overline{\text{co}}\Big(\bigcup_{\tau_i \in T_\delta} E(a_{\tau_i}, Q_{\tau_i})\Big)$.

Here we use the fact that the intersection and the union for overlapping sets are dual operations [12]. For the convex compacts $\mathcal{M}, \mathcal{M}_1, \mathcal{M}_2 \in \mathbb{R}^n$ the following properties holds [14]: if $0 \in \mathcal{M}_1 \cap \mathcal{M}_2$ then $(\overline{\text{co}}(\mathcal{M}_1 \cup \mathcal{M}_2))^* = \mathcal{M}_1^* \cap \mathcal{M}_2^*$ and if $0 \in \mathcal{M}$ then $(\mathcal{M}^*)^* = \mathcal{M}$.

Let us assume that $0 \in \bigcap_{\tau_i \in T_\delta} E(a_{\tau_i}, Q_{\tau_i})$ (if it isn't so we will change the coordinates appropriately). According to the above remark on dual operations we need to construct the upper ellipsoidal bound $\tilde{E}^+(\tilde{a}, \tilde{Q})$ for the intersection of a finite number of transformed ellipsoids $\bigcap_{\tau_i \in T_\delta} E^*(a_{\tau_i}, Q_{\tau_i}) \subseteq \tilde{E}^+(\tilde{a}, \tilde{Q})$.

For the fixed instances $\tau_i \in T_\delta$ and for nondegenerate ellipsoid $E(a_{\tau_i}, Q_{\tau_i})$ the $E^*(a_{\tau_i}, Q_{\tau_i})$ is calculated by the following rule [14]: if $\theta \in \text{int}\, E(a_{\tau_i}, Q_{\tau_i})$ then

$$(E(a_{\tau_i}, Q_{\tau_i}) - \theta)^* = E(a^*, Q^*),$$

$$a^* = -\frac{Q_{\tau_i}^{-1}(a_{\tau_i} - \theta)}{1 - (\theta - a_{\tau_i})' Q_{\tau_i}^{-1}(\theta - a_{\tau_i})}, \quad Q^* = \frac{Q_{\tau_i}^{-1}}{1 - (\theta - a_{\tau_i})' Q_{\tau_i}^{-1}(\theta - a_{\tau_i})} + a^*(a^*)'.$$

Consider first two ellipsoids $E^*(a_{\tau_1}, Q_{\tau_1})$ and $E^*(a_{\tau_2}, Q_{\tau_2})$ $(\tau_1, \tau_2 \in T_\delta)$. There are some approaches for the ellipsoidal estimation of intersection of two ellipsoids, e.g. [1,8,13]. Using one of these methods of estimation for $E^*(a_{\tau_1}, Q_{\tau_1})$ and $E^*(a_{\tau_2}, Q_{\tau_2})$ we calculate the ellipsoid $E_1{}^+(a_1, Q_1)$. Then for a pair of ellipsoids $E_1{}^+(a_1, Q_1)$ and $E^*(a_{\tau_3}, Q_{\tau_3})$, $\tau_3 \in T_\delta$ we construct $E_2{}^+(a_2, Q_2)$, and so on. After a finite number of steps we will find the resulting ellipsoid $\tilde{E}^+(\tilde{a}, \tilde{Q})$.

Then we return to the original coordinates and so we get the internal ellipsoidal estimate $\tilde{E}^-(\tilde{a}^-, \tilde{Q}^-) = \left(\tilde{E}^+(\tilde{a}, \tilde{Q})\right)^*$ of the $\overline{\text{co}}(\bigcup_{\tau_i \in T_\delta} E(a_\tau, Q_\tau))$ and therefore $\tilde{E}^-(\tilde{a}^-, \tilde{Q}^-) \subseteq \tilde{X}(T; \tilde{U}, \{0\})$.

So we have the following theorem.

Theorem 3. *The following internal estimate is true*

$$\tilde{E}^-(\tilde{a}^-, \tilde{Q}^-) = \left(\tilde{E}^+(\tilde{a}, \tilde{Q})\right)^* \subseteq \tilde{X}(T; \tilde{U}, \{0\}).$$

Now we study the case with state constraint (8).

To obtain the required estimate $E^-(a^-, Q^-) \subseteq X(T; U, \{0\})$ we will to construct the internal estimate for the intersection of two ellipsoids: first ellipsoid $\tilde{E}^-(\tilde{a}^-, \tilde{Q}^-)$ is found in Theorem 3 and second ellipsoid $E(y_0, D)$ is given in (8). In order to find the ellipsoid $E^-(a^-, Q^-) \subseteq E(y_0, D) \bigcap \tilde{E}^-(\tilde{a}^-, \tilde{Q}^-)$ we may use standard procedure of ellipsoidal estimates from [4,8,13,14]. At the end we get the Theorem 4.

Theorem 4. *The following internal estimate is true*

$$E^-(a^-, Q^-) \subseteq X(T; \tilde{U}, \{0\}) \subseteq X(T; U, \{0\}).$$

Hence Theorems 1–4 allow us to construct the internal ellipsoidal estimate of reachable sets $X(T; U, X_0)$ of system (1) from the initial set X_0 (7) at the instant T under controls $u(\cdot) \in U$ and with state constraint (8).

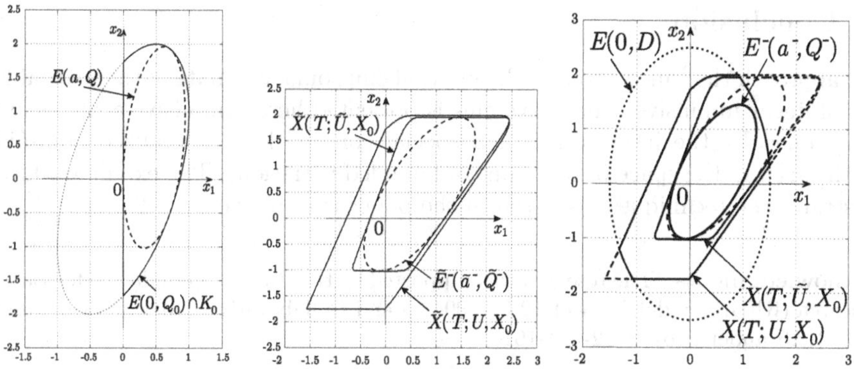

(a) (b) (c)

Fig. 1. (a) Ellipsoidal estimate of the $E(0, Q_0) \cap K_0$. (b) Ellipsoidal estimate of the reachable set $\tilde{X}(T; U, X_0)$. (c) Ellipsoidal estimate of the reachable set $X(T; U, X_0)$.

4 Numerical Simulation: Example

Consider the following control system:

$$\begin{cases} dx_1(t) &= x_2(t)dt + du_1(t), \\ dx_2(t) &= du_2(t), \end{cases} \qquad t \in [0, 1]. \tag{17}$$

We assume that $X_0 = \{0\}$, the set $U = E^* \cap K^*$ and the state constraint is defined as $x(T) \subseteq E(0, D)$, where E^* is generated by of the ellipsoid $E(0, Q_0)$ (see details in formulas (3)–(4))

$$D = \begin{pmatrix} 2.25 & 0 \\ 0 & 6.25 \end{pmatrix}, \qquad Q_0 = \begin{pmatrix} 1 & 1 \\ 1 & 4 \end{pmatrix},$$

and K^* is generated by of the cone $K_0 = \{u \in \mathbb{R}^2 \mid u = (u_1, u_2),\ u_1 \geq 0\}$, according to the formula (5).

Here we have

$$E(a, Q) \subseteq E(0, Q_0) \cap K_0, \tag{18}$$

where parameters a, Q are found according to formulas (11). The inclusion (18) is illustrated at the Fig. 1(a). The ellipsoid $E(a, Q)$ has the largest volume among all ellipsoids contained in the intersection of the ellipsoid $E(0, Q_0)$ and the cone K_0. The cone constraint requires that the ellipsoid $E(a, Q)$ lies in the right half-plane.

The exact reachable sets $\tilde{X}(T; U, X_0)$ and $\tilde{X}(T; \tilde{U}, X_0)$ are presented at the Fig. 1(b), which also shows the internal ellipsoidal estimate $\tilde{E}^-(\tilde{a}^-, \tilde{Q}^-)$ (see Theorem 3). Here the set $\tilde{U} = \tilde{E}^*$ is generated by the ellipsoid $E(a, Q)$. The exact reachable sets $X(T; U, X_0)$ and $X(T; \tilde{U}, X_0)$ with state constraint $E(0, D)$ and resulting internal ellipsoidal estimate $E^-(a^-, Q^-)$ are shown at the Fig. 1(c).

5 Conclusion

The approach that allows to find the internal ellipsoidal estimate of the reachable sets of linear impulsive control systems is presented here. Impulsive controls are constrained by the intersection of a special cone with a generalized ellipsoid (both taken in the space of functions of bounded variation). The example which illustrates the techniques discussed in the paper is also given.

Acknowledgments. The research was supported by the Russian Foundation for Basic Research (RFBR) under Project 12-01-00043-a and Collaborative Program of UB and SB of RAS under Project 12-C-1-1017.

References

1. Chernousko, F.L.: State Estimation for Dynamic Systems. Nauka, Moscow (1988)
2. Filippova, T.F.: Set-valued solutions to impulsive differential inclusions. Math. Comput. Model. Dyn. Syst. **11**, 149–158 (2005)
3. Filippova, T.F., Matviychuk, O.G.: Reachable sets of impulsive control system with cone constraint on the control and their estimates. In: Lirkov, I., Margenov, S., Waśniewski, J. (eds.) LSSC 2011. LNCS, vol. 7116, pp. 123–130. Springer, Heidelberg (2012)
4. Filippova, T.F., Matviychuk, O.G.: Algorithms to estimate the reachability sets of the pulse controlled systems with ellipsoidal phase constraints. Autom. Remote Control **72**(9), 1911–1924 (2011). (Springer)
5. Gusev, M.I.: On optimal control problem for the bundle of trajectories of uncertain system. In: Lirkov, I., Margenov, S., Waśniewski, J. (eds.) LSSC 2009. LNCS, vol. 5910, pp. 286–293. Springer, Heidelberg (2010)
6. Kurzhanski, A.B.: Control and Observation Under Conditions of Uncertainty. Nauka, Moscow (1977)
7. Kurzhanski, A.B., Veliov, V.M. (eds.): Set-Valued Analysis and Differential Inclusions. Progress in Systems and Control Theory, vol. 16. Birkhauser, Boston (1990)
8. Kurzhanski, A.B., Valyi, I.: Ellipsoidal Calculus for Estimation and Control. Birkhauser, Boston (1997)
9. Matviychuk, O.G.: Estimation problem for linear impulsive control systems under uncertainty. In: Proceedings of the 4-rd International Scientific Conference on Physics and Control (PhysCon 2009), pp. 1–6. Universita degli Studi di Catania, Catania (2009). http://lib.physcon.ru/doc?id=c4e6824198c7
10. Matviychuk, O.G.: Estimation problem for impulsive control systems under ellipsoidal state bounds and with cone constraint on the control. AIP Conf. Proc. **1497**, 3–12 (2012)
11. Pereira, F.L., de Sousa, J.B.: Coordinated control of networked vehicles: an autonomous underwater system. Automat. Remote Control **65**(7), 1037–1045 (2004). (Springer)
12. Rockafellar, R.T.: Convex Analysis. Princeton University Press, Princeton (1970)
13. Schweppe, F.C.: Uncertain Dynamical Systems. Prentice-Hall, Englewood Cliffs (1973)
14. Vazhentsev, A.Y.: External ellipsoidal estimation of the union of two concentric ellipsoids an its application. Comput. Math. Model. **15**(2), 110–122 (2004)
15. Vzdornova, O.G., Filippova, T.F.: External ellipsoidal estimates of the attainability sets of differential impulse systems. J. Comput. Syst. Sci. Intern. **45**(1), 34–43 (2006)

Optimal Control Models of Renewable Energy Production Under Fluctuating Supply

Elke Moser[1], Dieter Grass[1], Gernot Tragler[1(✉)], and Alexia Prskawetz[1,2]

[1] Institute for Mathematical Methods in Economics, Vienna University of Technology, 1040 Wien, Austria
{elke.moser,dieter.grass}@tuwien.ac.at, tragler@eos.tuwien.ac.at, afp@econ.tuwien.ac.at

[2] Vienna Institute of Demography (VID), Austrian Academy of Sciences (OeAW), 1040 Wien, Austria

Abstract. The probably biggest challenge for climate change mitigation is to find a secure low-carbon energy supply, which especially is difficult as the supply of renewable sources underlies strong volatility and storage possibilities are limited. We therefore consider the energy sector of a small country that optimizes a portfolio consisting of fossil and/or renewable energy to cover a given energy demand, considering seasonal fluctuations in renewable energy generation. By solving these non-autonomous optimal control models with infinite horizon, we investigate the impact of fossil energy prices on the annual optimal portfolio composition shown by the obtained periodic solutions.

Keywords: Optimal control · Nonlinear dynamical systems · Resources and environment · Renewable energy

1 Introduction

With a constantly increasing world-wide energy demand, the progressively obvious impacts of climate change and the energy sector as the main source of green house gas emission, the possibly biggest challenge of the 21st century is to find a low-carbon, secure and sustainable energy supply. Renewable energy generation is already carried out, but technology and policy efforts are not yet sufficient. Besides the high costs and the limited storage possibilities the possibly biggest problem is the fluctuating supply of renewable sources.

To address this issue we investigate the decision of an energy sector in a small country that optimizes a portfolio consisting of fossil and renewable energy. We assume that this energy sector has full information about the energy demand that has to be covered, which is postulated to be stationary, as done in [3], but instead of assuming that the energy demand is dependent on the GDP of the country (see also [2]) and on the electricity price, we follow [8] and consider the energy demand to be exogenous. Given this demand as well as the mentioned seasonal fluctuations and the fossil energy price, the energy sector optimizes its

I. Lirkov et al. (Eds.): LSSC 2013, LNCS 8353, pp. 133–142, 2014.
DOI: 10.1007/978-3-662-43880-0_14,

portfolio to find the most cost-effective solution. Following [1] we focus especially on solar energy and omit storage completely, so that the generated energy has to be used immediately or is lost.

Due to the seasonal fluctuations this optimal control problem with one state and two controls exhibits a particular mathematical property by being non-autonomous. We solve this problem by applying Pontryagin's Maximum Principle, but instead of the usual steady-state analysis of autonomous approaches we are looking for a periodic solution that solves the non-autonomous canonical system, which makes the problem numerically sophisticated.

2 The Model

While fossil energy is assumed to be constantly available and imported for the price p_F, the supply of renewable energy fluctuates over time but harvesting is for free and the generation is possible within the country. To do so, however, investments for proper energy generation capital are necessary. One important implication of the (small) size of the country is that the energy sector is assumed to be a price taker, which means that its decision does not impact the market prices.

We especially focus in this paper on solar energy as renewable resource. Figure 1a shows the average global radiation per month in Austria. One can clearly observe the seasonal differences underlining the challenge of a constant renewable energy supply over the whole year. To include such seasonal fluctuations[1] in our model, we use a deterministic time-dependent function

$$v_R(t) = \nu \sin(t\pi)^2 + \tau,$$

which can be seen in Fig. 1. The parameter τ defines the minimal supply in winter and ν is the maximal increment during summer. The necessary capital

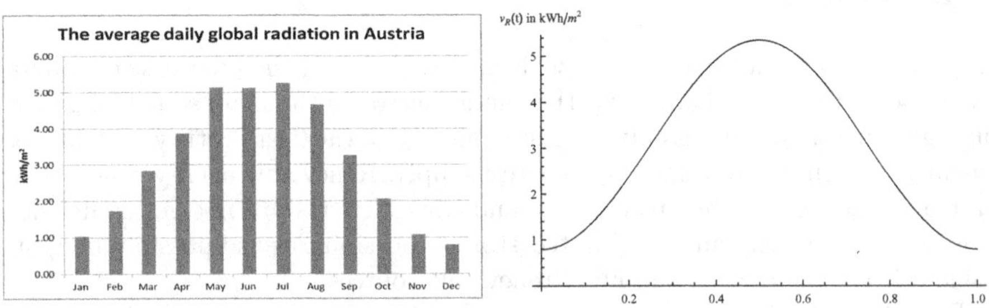

(a) (b)

Fig. 1. (a) Average global radiation per month in Austria. (b) Deterministic function to describe the varying global radiation over one year.

[1] Note that we only consider annual fluctuations and do not include daily fluctuations from day to night nor changes due to weather conditions. To get reasonable parameter values we used Austrian data for the estimation (cf. [9]).

$K_S(t)$ in form of photovoltaic (PV) cells is accumulated by investments $I_S(t)$ and depreciates at a rate δ_S which later on will be set to $\delta_S = 0.03$, implying that a PV cell has a lifetime of about 33 years. With the current capital stock and the given global radiation, renewable energy is generated as in Eq. (1b), where η is the degree of efficiency, which for common PV cells is about 20 %. Note that this function explicitly depends on time t which makes the problem non-autonomous. As the required energy demand E that has to be covered is well known, it is postulated that the demand has to be satisfied completely with the portfolio of fossil, $E_F(t)$, and renewable, $E_S(K_S(t), t)$, energy. This means that shortfalls are not allowed while surpluses are in general possible but are simply lost as saving options do not exist. This balance is included in the model by the mixed path constraint in Eq. (1a). Given this restriction and the current market price for fossil energy, the energy sector searches for the most cost-effective solution by maximizing its profit as shown in Eq. (1), where p is the electricity price. Note that we distinguish between linear investment and quadratic adjustment costs, where the latter arise from installation efforts.

Summing up, we consider a non-autonomous optimal control model with infinite horizon, two controls describing the capital investments and the imported fossil energy, and one state for the capital stock,

$$\max_{E_F(t),\, I_S(t)} \int_0^\infty e^{-rt}\Big(pE - I_S(t)\big(b + cI_S(t)\big) - p_F E_F(t)\Big)\, dt \tag{1}$$

$$\text{s.t.:}\quad \dot{K}_S(t) = I_S(t) - \delta_S K_S(t)$$

$$E_F(t) + E_S\big(K_S(t), t\big) - E \geq 0 \tag{1a}$$

$$E_S\big(K_S(t), t\big) = \big(\nu \sin(t\pi)^2 + \tau\big) K_S(t)\eta \tag{1b}$$

$$E_F(t), I_S(t) \geq 0,$$

where the discount rate r and the parameters b and c are positive constants.

3 Solution

3.1 Canonical System and Necessary First Order Conditions

Let $(K_S^*(t), I_S^*(t), E_F^*(t))$ be an optimal solution of the control problem in Eq. (1), then, according to the maximum principle for infinite time horizon problems (cf. [4]), there exists a continuous and piecewise continuously differentiable function $\lambda(t) \in \mathbb{R}$ satisfying

$$\mathscr{L}(K_S^*(t), I_S^*(t), E_F^*(t), \lambda(t), t) = \max_{I_S(t), E_F(t)} \mathscr{L}(K_S^*(t), I_S(t), E_F(t), \lambda(t), t)$$

where $\mathscr{L}$ defines the Lagrangian which reads as

$$\begin{aligned}\mathscr{L}(K_S(t), I_S(t), E_F(t), \lambda(t), t) = {} & pE - bI_S(t) - cI_S(t)^2 - p_F E_F(t) \\ & + \lambda(t)(I_S(t) - \delta_S K_S(t)) + \mu_1(t)(E_F(t) \\ & + K_S(t)\eta(\nu \sin(t\pi)^2 + \tau) - E) \\ & + \mu_2(t) E_F(t) + \mu_3(t) I_S(t),\end{aligned}$$

with $\mu_1(t), \mu_2(t), \mu_3(t)$ being the Lagrange multipliers for the mixed path constraint and the non-negativity conditions. Further on, at each point where the controls are continuous

$$\dot{\lambda}(t) = r\lambda(t) - \frac{\partial \mathscr{L}(K_S(t), I_S(t), E_F(t), \lambda(t), t)}{\partial K_S}$$

is given and the complementary slackness conditions

$$\begin{aligned} \mu_1(t)\left(E_F^*(t) + E_S^*(K_S^*(t), t) - E\right) &= 0, \quad \mu_1(t) \geq 0, \\ \mu_2(t)E_F^*(t) &= 0, \quad \mu_2(t) \geq 0, \\ \mu_3(t)I_S^*(t) &= 0, \quad \mu_3(t) \geq 0, \end{aligned}$$

have to be satisfied. Hence, the necessary first order conditions and the adjoint equation are given as follows:

$$\begin{aligned} \frac{\partial \mathscr{L}}{\partial E_F(t)} &= -p_F + \mu_1(t) + \mu_2(t) = 0 \\ \frac{\partial \mathscr{L}}{\partial I_S(t)} &= -b - 2cI_S(t) + \lambda(t) + \mu_3(t) = 0 \Leftrightarrow I_S(t) = \frac{\lambda(t) + \mu_3(t) - b}{2c} \\ \dot{\lambda}(t) = r\lambda(t) - \frac{\partial \mathscr{L}}{\partial K_S(t)} &= (r + \delta_S)\lambda(t) - \mu_1(t)\eta(\nu\sin(t\pi)^2 + \tau). \end{aligned}$$

Looking for an interior solution with both controls $I_S(t), E_F(t) > 0$ and the mixed-path constraint of (1a) satisfied with strict inequality, it can be shown that such a solution never can be optimal as costs could be reduced by lowering the amount of fossil energy until the mixed path constraint is satisfied with equality, which makes surpluses in fossil energy inefficient. Hence, we focus for the following analysis on the three boundary cases, which are: the *fossil case* with zero investments[2], $E_F(t) > 0$, $I_S(t) = 0$ and $E_F(t) + E_S(K_S(t), t) - E = 0$; the *mixed case* where both types of energy are used for the coverage, $E_F(t)$, $I_S(t) > 0$ and $E_F(t) + E_S(K_S(t), t) - E = 0$; and finally the *renewable case*, where only renewable energy is used to cover the demand, $E_F(t) = 0$, $I_S(t) > 0$ and $E_S(K_S(t), t) - E > 0$. Inserting the corresponding values of the controls and Lagrange multipliers yields the canonical systems for these boundary cases:

$$\dot{K}_S(t) = A - \delta_S K_S(t), \quad \text{with } A = \begin{cases} 0, & \text{fossil case,} \\ \frac{\lambda(t)-b}{2c}, & \text{mixed and renewable case,} \end{cases} \tag{2}$$

$$\dot{\lambda}(t) = (r + \delta_S)\lambda(t) - B, \quad \text{with } B = \begin{cases} p_F\eta(\nu\sin(t\pi)^2 + \tau), & \text{fossil and mixed case,} \\ 0, & \text{renewable case.} \end{cases} \tag{3}$$

In what follows, we refer to these canonical systems as $\dot{K}_S(t) = f^K(t, K_S(t), \lambda(t))$ and $\dot{\lambda}(t) = f^\lambda(t, \lambda(t))$.

[2] If the initial capital stock along the fossil solution arc is zero, the whole energy demand is covered with fossil energy. If, however, the initial capital stock is positive, also renewable energy contributes to the coverage of the energy demand, nevertheless at a decreasing rate as no investments are done and depreciation reduces the stock.

3.2 Periodic Solution

As the canonical system in (2)–(3) is non-autonomous we have to find a trajectory with the property to be hyperbolic. Detailed theory about the existence, the computation and the manifolds of such distinguished hyperbolic trajectories can be found, e.g., in [5], [7], or [6]. Due to the periodicity of the dynamics candidates for the long-run optimal solution of the problem in (1) are periodic solutions with the period length of one year. In order to find such a periodic solution of the canonical system numerically, we first determine the instantaneous equilibrium points $K_S^{IEP}(t)$ and $\lambda^{IEP}(t)$ (cf. [5]) by setting $(\dot{K}_S, \dot{\lambda})(t) = (0,0)$, and then solve the following boundary value problem using these instantaneous equilibrium points as starting function,

$$\begin{aligned} \dot{K}_S(t) &= f^K(t, K_S(t), \lambda(t)), \quad && \text{with } K_S(0) = K_S^{IEP}(0) \text{ and } K_S(1) = K_S(0), \\ \dot{\lambda}(t) &= f^\lambda(t, \lambda(t)), && \text{with } \lambda(0) = \lambda^{IEP}(0) \text{ and } \lambda(1) = \lambda(0). \end{aligned}$$

Solving this BVP yields the periodic solution $(K_S^*(t), \lambda^*(t))$ that lies completely within one of the three boundary cases. However, it can happen that the solution at some point leaves the current admissible area before the course of the period of one year is completed. In this case one has to switch to the corresponding canonical system to get a periodic solution existing of several arcs. Therefore, a multi-point boundary value problem has to be solved. At each point of time where the constraints of the current region are violated a switch to the proper region happens, meaning that the corresponding canonical system is used to continue the solution. For n switching times $\tau_0 := 0 < \tau_1 < \tau_2 < \cdots < \tau_{n-1} < \tau_n < 1 =: \tau_{n+1}$, one has to calculate $n+1$ arcs, for which the continuity at each switching time has to be guaranteed. We introduce an index $a_i \in \{1,2,3\}$ that distinguishes the canonical systems for the fossil, the mixed and the renewable case, respectively, for each arc i with $i = 1, \dots, n+1$. If n switches are necessary along the periodic solution and we use for simplicity the notation

$$\dot{K}_{S_i}(t) = f_{a_i}^K(t, K_{S_i}(t), \lambda_i(t)), \quad t \in [\tau_{i-1}, \tau_i], \qquad i = 1, \dots, n+1, \tag{4}$$

$$\dot{\lambda}_i(t) = f_{a_i}^\lambda(t, \lambda_i(t)), \quad t \in [\tau_{i-1}, \tau_i], \qquad a_i \in \{1,2,3\}, \tag{5}$$

for the corresponding canonical system at arc i, it has to hold that $a_i \neq a_{i-1}$ and $|a_i - a_{i-1}| = 1$, which means that switches only can happen between fossil/mixed or mixed/renewable cases. For the numerical solution of the system for each arc i we use a time transformation so that it can be solved with fixed time intervals. This means that, in order to solve an equation

$$\dot{x}(t) = f(t, x(t)), \quad t \in [\tau_{i-1}, \tau_i],\ i = 1, \dots, n+1,\ \tau_0 = 0,\ \tau_{n+1} = 1$$

as in (4)–(5), we are looking for a time transformation $t = T(s)$ so that

$$\dot{y}(s) = \tilde{f}(s, y(s)), \quad s \in [i-1, i], \text{ with } y(s) = x(T(s)).$$

It turns out that the linear transformation $T(s) = (\tau_i - \tau_{i-1})(s - i + 1) + \tau_{i-1}$ satisfies the required conditions. Hence, in terms of the original dynamics this yields

$$\dot{x}(s) = \frac{dx(T(s))}{ds} = \frac{dx(T(s))}{dT}\frac{dT(s)}{ds} = f(s, x(s))(\tau_i - \tau_{i-1}).$$

Using this transformation, we have to solve for $i = 1, \ldots, n+1$, $j = 1, \ldots, n$, $s \in [i-1, i]$, $\tau_0 = 0$, $\tau_{n+1} = 1$ the multi-point boundary problem

$$\begin{aligned}
&\dot{K}_{S_i}(s) = (\tau_i - \tau_{i-1}) f^K_{a_i}(T(s), K_{S_i}(s), \lambda_i(s)), \quad \dot{\lambda}_i(s) = (\tau_i - \tau_{i-1}) f^\lambda_{a_i}(T(s), \lambda_i(s)),\\
&(K_{S_j}(\tau_j), \lambda_j(\tau_j)) = (K_{S_{j+1}}(\tau_j), \lambda_{j+1}(\tau_j)), \quad (K_{S_n}(1), \lambda_n(1)) = (K_{S_1}(0), \lambda_1(0)),\\
&(K_{S_1}(0), \lambda_1(0)) = \left(K_S^{IEP}(0), \lambda^{IEP}(0)\right).
\end{aligned} \tag{6}$$

Equation (6) ensures that the continuity in state and controls at each switch is given and, as a periodic solution is calculated, the beginning and the endpoint coincide. The following Eq. (7) finally guarantees the necessary condition that the Lagrangian is continuous as well, which depends on the involved regions as well as on the direction of the switch and is given for $j = 1, \ldots, n$ as

$$0 = c(a_j, a_{j+1}) = \begin{cases} b - \lambda_j(\tau_j), & \text{if } a_j = 1,\ a_{j+1} = 2,\\ \frac{\lambda_j(\tau_j) - b}{2c}, & \text{if } a_j = 2,\ a_{j+1} = 1,\\ E_S(K_{S_j}(\tau_j), \tau_j) - E, & \text{if } a_j = 2,\ a_{j+1} = 3,\\ E_F(\tau_j), & \text{if } a_j = 3,\ a_{j+1} = 2. \end{cases} \tag{7}$$

The periodic solution that solves this BVP then is given as

$$(K_S^*(t), \lambda^*(t)) = \left((K_{S_1}^*(t), \lambda_1^*(t))_{0 \le t < \tau_1}, (K_{S_2}^*(t), \lambda_2^*(t))_{\tau_1 \le t < \tau_2}, \ldots, (K_{S_n}^*(t), \lambda_n^*(t))_{\tau_{n-1} \le t < \tau_n = 1} \right).$$

Calculating the eigenvalues of the monodromy matrix for the obtained periodic solution reflects the stability, which here are given as $e_1 = e^{-\delta_S}$ and $e_2 = e^{r+\delta_S}$. As $\delta_S < 1$ always is satisfied, one can see that $e_1 < 1$ holds. For reasonable values of the discount rate and the depreciation rate it further is supposed that $r+\delta_S<1$ which implies that $e_2 > 1$ in these cases. As the Jacobian is independent of the state and the control variable, this means that every periodic solution that we can find within one of the boundary regions is of saddle-type and, as no eigenvalue $e_i = 1$ occurs, it is a hyperbolic cycle which guarantees that the behavior of the system near this periodic solution can be fully described by its linearisation (see [4]).

Table 1. Parameter values used for the numerical analysis.

Interpretation	Parameter	Value	Interpretation	Parameter	Value
Investment costs	b	0.6	Discount rate	r	0.04
Adjustment costs	c	0.3	Depreciation rate	δ_S	0.03
Energy demand	E	1053.82	Maximal radiation increment	ν	4.56
Electricity price	p	0.1	Degree of efficiency	η	0.2
Fossil energy price	p_F	0.08	Minimal radiation in winter	τ	0.79

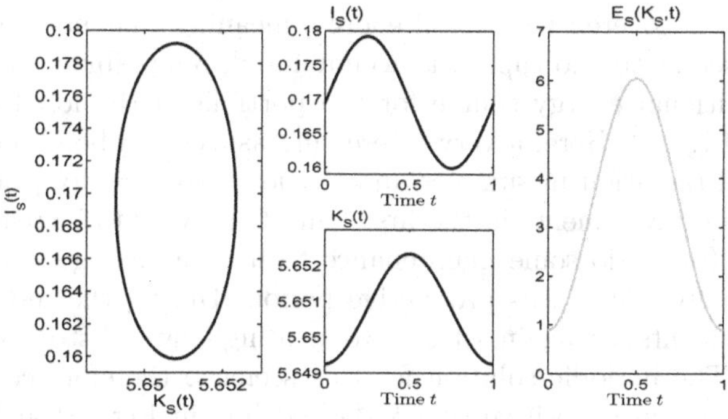

Fig. 2. Periodic solution (left box), time paths for investments and capital over one year (two boxes in the middle) and renewable energy generation (right box) for a fossil energy price of $p_F = 0.08$.

4 Results

For the following numerical analysis, we use the parameter values summarized in Table 1. Figure 2 shows the long-run optimal periodic solution for this parameter value set which corresponds to the mixed case where both types of energy are used. While the initial capital stock in winter is quite low, it increases and peaks during summer due to investments to accumulate new or maintain already existing capital. Note, that the peak is exactly where also the global radiation is maximal and hence, the generation of renewable energy reaches a peak during this time as well. The investments, however, start to decline again even before this period because a further increase of the capital stock in autumn would not be beneficial due to the declining radiation. Therefore, the capital stock decreases again after the summer peak and renewable energy generation goes down. The proportion of renewable energy in this scenario's portfolio with only 0.6 % is very low, but this comes from the fact that fossil energy with $p_F = 0.08$ is really cheap and hence high investments in renewable energy are simply too costly.

5 Sensitivity Analysis

As the previous scenario has shown, not much is invested in renewable energy in case of a low fossil energy price. This aspect raises the question how the portfolio composition will change if fossil energy gets more expensive. We therefore investigate in this section the impact of the fossil energy price on the long-run optimal portfolio solution by increasing the price step by step and then using numerical continuation. Figure 3 shows the results for different values of p_F. The two boxes on the left hand side contain the time paths for investments $I_S(t)$ and capital stock $K_S(t)$, respectively, while the box on the right hand side depicts the composition of the energy portfolio with renewable energy shown as gray line, fossil energy as black line and the energy demand as black dashed line. While for a very low price (below $p_F = 0.06785$) fossil energy is so cheap that the whole energy demand is covered with fossil energy, meaning that no investments are done and, consequently, no capital is accumulated, renewable energy very soon is used as additional energy source for the portfolio if the fossil energy price increases (see Fig. 3a). Here, a very interesting aspect can be observed. Due to the high global radiation in summer and the low fossil energy price, it is only worthwhile to do investments in the first half of the year to increase renewable energy capital (or to do some maintenance to have it in a good condition) in order to optimally utilize this productive period. During the rest of the year, however, investments are again set to zero as a high capital stock would not be cost effective. The periodic solution for this scenario therefore consists of two arcs, the first one with positive (black dashed line in Fig. 3a) and the second one with zero investments (black line in Fig. 3a). Note that the contribution of renewable energy to cover the demand still is very low and hence the line for fossil energy and the energy demand basically coincide. The price interval for which this kind of result can be seen is, however, very small, $p_F \in [0.06785, 0.06897]$. For a higher fossil energy price, investments are done over the whole year but still with a peak before summer, the generation of renewable energy increases and the additional fossil energy amount during the summer period is reduced. During the winter period, however, fossil energy still is required. Figure 3b shows the long-run optimal solution for $p_F = 1.4$, which completely corresponds to the mixed case. At an even higher fossil energy price of $p_F = 2.1025$, the renewable energy generation is so high that it reaches the demand at the peak in summer. This is a certain point of interest because here a switch to the complete renewable case happens. Being only an osculation point at the beginning, it develops to an interval if the price goes further up. In this interval, which always is around the maximum of global radiation, the energy demand can be fully covered with renewable energy and no additional fossil energy is needed. Figure 3c shows this

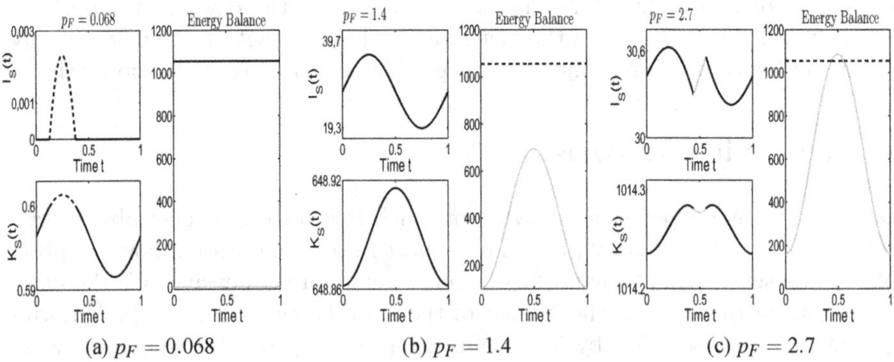

Fig. 3. Periodic solution for (a) $p_F = 0.068$: two arcs given by the mixed solution as dashed line and the fossil solution as solid line, (b) $p_F = 1.4$: mixed solution over the whole year, (c) $p_F = 2.7$: two arcs given by the mixed solution as black solid line and the renewable solution as gray solid line.

scenario for a fossil energy price of $p_F = 2.7$. The energy portfolio in the right box shows that surpluses are generated during summer which are lost as no storage possibilities exist. The periodic solution in this scenario again consists of two arcs, the mixed solution arc (black line) and the renewable solution arc (gray line). The interval in which renewable energy is sufficient to cover the energy demand increases the further the fossil energy price goes up. However, it turns out that this happens at a decreasing speed, and during winter fossil energy still is necessary to cover the shortfalls, even if the fossil energy price is already unreasonably high.

6 Conclusions

We have investigated in this paper the impact of the fossil energy price on the optimal portfolio composition consisting of fossil and renewable (solar) energy in a small country. We postulated that the supply of the renewable source is seasonally varying, the energy demand is well known and constant over the year.

Sensitivity analysis with respect to the fossil energy price p_F showed that a higher fossil energy price indeed is an incentive for more investments in renewable energy capital. However, an autarkic solar energy supply is not possible. While independence of fossil energy can be achieved during some time interval in summer in which global radiation is high and even surpluses can be generated, the shortfalls in winter always have to be covered with fossil energy no matter how high the fossil energy price is.

These results underline that the non-constant supply is one of the major challenges of renewable energy generation. This not only concerns solar energy but also other types of renewable energy like wind and water. Hence, probably a combined portfolio of several types of renewable energy could compensate for the fluctuations of each other and enable a more or less constant supply. Such a portfolio is exactly what we want to study in some model extension in the near future. In addition, also the effect of learning by doing on investment costs and efficiency as well as a time-dependent energy demand will be a special matter of interest.

References

1. Chakravorty, U., Magné, B., Moreaux, M.: A hotelling model with a ceiling on the stock of pollution. IDEI Working Papers 368, Institut d' conomie Industrielle (IDEI), Toulouse (2005). http://ideas.repec.org/p/ide/wpaper/1165.html
2. Chakravorty, U., Magné, B., Moreaux, M.: Resource use under climate stabilization: can nuclear power provide clean energy? J. Public Econ. Theor. **14**(2), 349–389 (2012). http://ideas.repec.org/a/bla/jpbect/v14y2012i2p349-389.html
3. Coulomb, R., Henriet, F.: Carbon price and optimal extraction of a polluting fossil fuel with restricted carbon capture. Working papers 322, Banque de France (2011). http://ideas.repec.org/p/bfr/banfra/322.html

4. Grass, D., Caulkins, J., Feichtinger, G., Tragler, G., Behrens, D.: Optimal Control of Nonlinear Processes: With Applications in Drugs, Corruption, and Terror. Springer, Heidelberg (2008). http://books.google.com/books?id=M7qGPmzrVAkC
5. Ju, N., Small, D., Wiggins, S.: Existence and computation of hyperbolic trajectories of aperiodically time dependent vector fields and their approximations. Int. J. Bifurcat. Chaos **13**(6), 1449–1457 (2003). http://dblp.uni-trier.de/db/journals/ijbc/ijbc13.html#JuSW03d
6. Madrid, J.A.J., Mancho, A.M.: Distinguished trajectories in time dependent vector fields. Chaos **19**(1), 013111-1–013111-18 (2009)
7. Mancho, A.M., Small, D., Wiggins, S.: Computation of hyperbolic trajectories and their stable and unstable manifolds for oceanographic flows represented as data sets. Nonlinear Process. Geophys. **11**(1), 17–33 (2004). http://www.nonlin-processes-geophys.net/11/17/2004/
8. Messner, S.: Endogenized technological learning in an energy systems model. J. Evol. Econ. **7**(3), 291–313 (1997). http://ideas.repec.org/a/spr/joevec/v7y1997i3p291-313.html
9. ZAMG: Klimadaten. Downloaded on 16th of February 2012 (2012). http://www.zamg.ac.at/fix/klima/oe71-00/klima2000/klimadaten_oesterreich_1971_frame1.htm

Pontryagin's Type Optimality Conditions for a Distributed Control Problem Arising in Endogenous Growth Theory

Bernhard Skritek[1], Tsvetomir Tsachev[2(✉)], and Vladimir M. Veliov[1]

[1] Institute of Mathematical Methods in Economics, Vienna University of Technology, Vienna, Austria

[2] Institute of Mathematics and Informatics, Bulgarian Academy of Sciences, Sofia, Bulgaria

tsachev@math.bas.bg

Abstract. The paper presents an optimality condition for a distributed optimal control problem arising in economics, where the (one-dimensional) spatial domain can be influenced by the control. The result enables analytical and numerical investigation of a class of endogenous growth models with investment-dependent technological frontier and heterogeneous capital stock.

1 Introduction

One direction in which the theory of endogenous economic growth develops during the past few years is to take into account the heterogeneity of economic factors, such as productivity, emission levels, consumer preferences, etc., see e.g. [4]. Typically, this requires involvement of specific distributed optimal control models, where the spatial domain changes in accordance with the applied investment policy, viewed as a control variable. A general class of such models was introduced in [3], where additional motivations and references are given.[1] However, the optimality conditions obtained therein are not of Pontryagin's type and are not informative enough to provide a ground for analytic or numerical treatment, unless the optimal control satisfies a certain *regularity condition.* The regularity is not implied by the optimality conditions and turns out to be difficult to ensure a priori. The main goal of the present paper is to verify the required regularity condition for a specific model of endogenous economic growth and to obtain constructive optimality conditions of Pontryagin's type for this model. The result justifies the case study in [3] and opens the door for numerical approaches and further economic analyses.

This research was funded by the Austrian Science Foundation (FWF) under grant No I 476-N13.

[1] The rest of the existing literature on optimal control of heterogeneous systems (see e.g. [6] and the bibliography therein) does not cover problems with endogenously changing spatial domain.

I. Lirkov et al. (Eds.): LSSC 2013, LNCS 8353, pp. 143–151, 2014.
DOI: 10.1007/978-3-662-43880-0_15,

Section 2 gives a precise formulation of the problem. The regularity of the optimal control is proved in Sect. 3, then the Pontryagin's optimality conditions are presented in Sect. 4.

2 Formulation of the Problem and Preliminaries

Here we formulate a particular case, arising in economic growth theory, of the general problem considered in [3]. Let $[0,T]$ be a fixed time-interval and let $[0,\bar{Q}]$ be where the "parameter of heterogeneity", q, takes values ($T>0$ and $\bar{Q}>0$ are given). Denote $D=[0,T]\times[0,\bar{Q}]$. The state variables are

$$x : D \mapsto \mathbf{R}^1, \quad Q : [0,T] \mapsto [0,\bar{Q}], \quad y : [0,T] \mapsto \mathbf{R}^1,$$

while $u : D \mapsto [\underline{u},\overline{u}] \subset \mathbf{R}^1$ is a control function, $\underline{u}>0$, $\overline{u}>0$, being given. For a given $Q : [0,T] \mapsto [0,\bar{Q}]$ we denote $D_Q := \{(t,q) \in D : t \in [0,T], \;\; q \in [0,Q(t)]\}$. The set of admissible controls is $\mathcal{U} = \{u \in L_\infty(D) : u(t,q) \in [\underline{u},\overline{u}] \text{ for a.e. } (t,q) \in D\}$. The optimal control problem we consider is:

$$\max_{u\in\mathcal{U}} \int_0^T e^{-rt} \int_0^{Q(t)} \Big(L(q,Q(t),x(t,q),y(t)) - c(u(t,q))\Big)\, \mathrm{d}q\, \mathrm{d}t, \tag{1}$$

subject to the equations

$$\dot{Q}(t) = g(t,Q(t))y(t), \qquad t \in [0,T], \;\; Q(0)=Q^0, \tag{2}$$

$$y(t) = \int_0^{Q(t)} d(t,q)u(t,q)\, \mathrm{d}q, \qquad t\in[0,T], \tag{3}$$

$$\dot{x}(t,q) = -\delta x(t,q) + u(t,q), \qquad (t,q)\in D_Q \tag{4}$$

$$x(0,q) = x^0(q), \qquad q\in[0,Q^0], \tag{5}$$

$$x(t,Q(t)) = x^b(t), \quad t\in(0,T]. \tag{6}$$

The interpretation is: At time t diverse goods, indexed by $q \in [0,Q(t)]$, are produced using physical capital stock $x(t,q)$. The capital depreciates with rate δ and is replenished by investment, $u(t,q)$. Part of the investment ($y(t)$) leads to the invention of new products, thus increasing the amount of goods, $Q(t)$. The discounted profit is the aggregated revenue $L(q,x,Q,y)$ minus the cost of investment, $c(u)$.
Standing Assumptions.

(i) L and g are differentiable in (Q,x,y), with Lipschitz partial derivatives, uniformly with respect to $(t,q)\in D$; L, L_x and d are continuous with respect to q uniformly in the rest of the variables; g and d are measurable in t, g is locally essentially bounded; L is concave in (x,y) and c is strongly convex and twice continuously differentiable on an open set containing $[\underline{u},\overline{u}]$; $x^b : [0,T] \mapsto \mathbf{R}^1$ is continuously differentiable, $x^0 : [0,Q^0] \mapsto \mathbf{R}^1$ is continuous; $r>0$, $\delta>0$ and $Q^0 \in (0,\bar{Q})$ are given numbers.

(ii) There exist $\overline{g} \geq \underline{g} > 0$ such that $\underline{g} \leq g(t,Q) \leq \overline{g}$ for a.e. $(t,q) \in D$.
(iii) There exist $\overline{d} \geq \underline{d} > 0$ such that $\underline{d} \leq d(t,q) \leq \overline{d}$ for a.e. $(t,q) \in D$.
(iv) For every $u \in \mathcal{U}$ a solution $Q[u]$ of (2) exists on the whole interval $[0,T]$ and takes values in $[Q^0, \bar{Q}]$.

Since in the following equation the right-hand side is locally Lipschitz in Q:

$$\dot{Q}(t) = g(t, Q(t)) \int_0^{Q(t)} d(t,q)u(t,q)\,\mathrm{d}q,$$

$Q[u]$ is unique. Given $u \in \mathcal{U}$, we define for $q \in [0, \bar{Q}]$

$$\theta[u](q) = \begin{cases} 0 & \text{if } q \in [0, Q^0], \\ Q[u]^{-1}(q) & \text{if } q \in (Q^0, Q[u](T)), \\ T & \text{if } q \in [Q[u](T), \bar{Q}]. \end{cases} \tag{7}$$

The definition is correct, since $Q[u]$ is invertible according to Standing Assumptions (ii) and (iii) and its image is $[Q^0, Q[u](T)]$.

We extend the definition of $x^0(q)$ to $[Q^0, \bar{Q}]$ by defining $x^0(q) := x^b(\theta[u](q))$. Then, given $Q(\cdot) = Q[u](\cdot)$ with $u \in \mathcal{U}$, we define the solution x of (4)–(6) as a measurable and bounded function such that for a.e. $q \in [0, Q[u](T)]$ the function $x(\cdot, q)$ is absolutely continuous on $[\theta[u](q), T]$ and satisfies

$$\dot{x}(t,q) = -\delta x(t,q) + u(t,q), \quad t \in [\theta[u](q), T], \tag{8}$$

$$x(\theta[u](q), q) = x^0(q). \tag{9}$$

(We denote by $\dot{x}$ the derivative of x with respect to t.) Thus (8)–(9) is a family of ODEs (one for each $q \in [0, Q[u](T)]$), where the functions $y = y[u]$, $Q = Q[u]$ and $\theta = \theta[u]$ are already defined in (2), (3) and (7) as described above.

The existence and uniqueness of optimal controls is a non-trivial issue and will be considered in a follow-up paper. Below we assume that an optimal control exists.

3 Regularity of the Optimal Control

It was shown in [3, Theorem 2] that Pontryagin's type optimality conditions are valid for any optimal control $\hat{u}$ of problem (1)–(6) under a "regularity condition", namely, that for almost every t the function $\hat{u}(t,\cdot)$ is continuous from the left at $q = Q(t)$. The next proposition implies the required regularity of any optimal $\hat{u}$.

Proposition 1. *Any optimal control $\hat{u}(t,q)$ of problem* (1)–(6) *is continuous in q for a.e. $t \in [0,T]$.*

Let $(\hat{Q}, \hat{x}, \hat{y}, \hat{u})$ be a solution of problem (1)–(6). Before proving the proposition we introduce the following auxiliary problem for $u \in \mathcal{U}$ and x:

$$\max_{u \in \mathcal{U}} \int_0^T e^{-rt} \int_0^{\hat{Q}(t)} \Big(L(q, \hat{Q}(t), x(t,q), \hat{y}(t)) - c(u(t,q)) \Big) \, \mathrm{d}q \, \mathrm{d}t \tag{10}$$

subject to

$$\dot{x}(t,q) = -\delta x(t,q) + u(t,q), \quad (t,q) \in D_{\hat{Q}} \tag{11}$$

$$x(\theta(q), q) = x^0(q), \quad q \in [0, \hat{Q}[u](T)], \tag{12}$$

$$\hat{y}(t) = \int_0^{\hat{Q}(t)} d(t,q)\, u(t,q) \, \mathrm{d}q, \quad t \in [0,T], \tag{13}$$

$$\underline{u} \le u(t,q) \le \overline{u}, \quad (t,q) \in D. \tag{14}$$

This is a reduced form of the original problem, with $\hat{Q}$ and $\hat{y}$ fixed. The meaning of the solution of (11), (12) is the same as of (8), (9). Due to assumption (i), $(\hat{x}, \hat{u})$ is the unique solution of this problem.

We introduce the following notational convention: dependencies on values fixed at the optimal trajectory are suppressed. For example $L(q, x(t,q)) := L(q, \hat{Q}(t), x(t,q), \hat{y}(t))$.

For each $q \in [0, \hat{Q}(T)]$, let $\hat{\lambda}(\cdot, q)$, be the unique solution of the following equation on $[\hat{\theta}(q), T]$:

$$\dot{\hat{\lambda}}(t,q) = \delta \hat{\lambda}(t,q) - e^{-rt} L_x(q), \quad \lambda(T,q) = 0. \tag{15}$$

Lemma 1. *Let $\hat{u}$ be an optimal control of problem* (10)–(14) *and let $\hat{\lambda}$ be the solution to* (15)*. Then for almost every $t \in [0,T]$ the function $\hat{u}(t,\cdot)$ maximizes*

$$\int_0^{\hat{Q}(t)} [\hat{\lambda}(t,q) u(q) - e^{-rt} c(u(q))] \, dq \tag{16}$$

over the set of $u(\cdot) \in L_\infty([0, \hat{Q}(t)])$ satisfying

$$\int_0^{\hat{Q}(t)} d(t,q) u(q) \, dq = \hat{y}(t), \quad \underline{u} \le u(q) \le \overline{u} \quad \textit{for a.e.} \quad q \in [0, \hat{Q}(t)]. \tag{17}$$

Proof. Let $\hat{u}(t,q)$ be an optimal control and $u(t,q)$ be any other control satisfying (13)–(14). Denote by $\hat{x}(t,q)$ and $x(t,q)$ the corresponding trajectories. Further, define $\Delta u(t,q) := u(t,q) - \hat{u}(t,q)$, $\Delta x(t,q) := x(t,q) - \hat{x}(t,q)$ and $\Delta J := J(u) - J(\hat{u})$. Routine calculations lead to

$$\Delta J = \int_0^T e^{-rt} \int_0^{\hat{Q}(t)} [\langle L_x(q), \Delta x(t,q) \rangle - c(u(t,q)) + c(\hat{u}(t,q))] \, \mathrm{d}q \, \mathrm{d}t + e(\Delta u),$$

where

$$e(\Delta u) := \int_0^T e^{-rt} \int_0^{\hat{Q}(t)} \langle L_x(q, \hat{x}(t,q) + s(t,q)\Delta x(t,q)) - L_x(q), \Delta x(t,q) \rangle \, dq \, dt$$

and for some constant M

$$|e(\Delta u)| \le M \, \text{meas} \left(\{ t \in [0,T] : u(t,\cdot) \neq \hat{u}(t,\cdot) \} \right)^2 . \tag{18}$$

The optimality of $\hat{u}$ implies $\Delta J \le 0$. Then standard calculations yield

$$\int_0^T \int_0^{\hat{Q}(t)} [\hat{\lambda}(t,q)(\hat{u}(t,q) - u(t,q)) - e^{-rt}(c(\hat{u}(t,q)) - c(u(t,q)))] \, dq \, dt \ge e(\Delta u). \tag{19}$$

Assume that the assertion of the lemma is not true. Then there exists a subset $A \subset [0,T]$ with positive measure, $u_t \in L_\infty(0, \hat{Q}(t))$, for every $t \in A$ and an $\varepsilon > 0$ such that

$$\int_0^{\hat{Q}(t)} [\hat{\lambda}(t,q)(u_t(q) - \hat{u}(t,q)) - e^{-rt}(c(u_t(q)) - c(\hat{u}(t,q)))] \, dq \ge \varepsilon. \tag{20}$$

The function u_t can be chosen in such a way that $u(t,q) := u_t(q)$ is measurable in (t,q). The proof of this fact is based on [2, Theorem 8.2.9], but we omit the details due to space restrictions.

Now choose m big enough such that $\frac{M}{m} < \varepsilon$ and $\frac{1}{m} < \text{meas}(A)$, then choose a subset A_m of A with $\text{meas}(A_m) = \frac{1}{m}$ and define

$$u_m(t,q) = \begin{cases} u_t(q) & \text{if } t \in A_m \\ \hat{u}(t,q) & \text{if } t \in [0,T] \backslash A_m \end{cases}$$

The so defined control u_m is admissible because $\hat{u}$ and u_t are measurable and satisfy (13)–(14). It differs only on a set of measure $1/m$ from the optimal control and therefore, using (18), $|e(\Delta u_m)| \le M m^{-2}$.

Using (20) and the definition of A_m and u_m, it follows that

$$\begin{aligned} &\int_{A_m} \int_0^{\hat{Q}(t)} [\hat{\lambda}(t,q)\hat{u}(t,q) - e^{-rt} c(\hat{u}(t,q))] \, dq \, dt + \frac{\varepsilon}{m} \\ &\le \int_{A_m} \int_0^{\hat{Q}(t)} [\hat{\lambda}(t,q) u_m(t,q) - e^{-rt} c(u_m(t,q))] \, dq \, dt \\ &< \int_{A_m} \int_0^{\hat{Q}(t)} [\hat{\lambda}(t,q)\hat{u}(t,q) - e^{-rt} c(\hat{u}(t,q))] \, dq \, dt + \frac{\varepsilon}{m}, \end{aligned}$$

where the second inequality comes from (19), the choice of m and the estimation for $e(\Delta u_m)$. This contradiction completes the proof.

Lemma 2. *Let* $\hat{\lambda}$ *be the solution of* (15). *The function*

$$U(t,q,\lambda,\xi) := \begin{cases} \underline{u} & \text{if} \quad e^{rt}(\lambda - \xi\, d(t,q)) < c'(\underline{u}) \\ (c')^{-1}(e^{rt}(\lambda - \xi\, d(t,q))) & \text{if}\, c'(\underline{u}) \le e^{rt}(\lambda - \xi\, d(t,q)) \le c'(\overline{u}) \\ \overline{u} & \text{if} \quad e^{rt}(\lambda - \xi\, d(t,q)) > c'(\overline{u}) \end{cases} \tag{21}$$

is Lipschitz in λ, *and there exists a measurable function* $\hat{\xi}(t)$, $t \in [0,T]$, *such that the optimal control* $\hat{u}(t,q)$ *of problem* (1)–(6) *fulfills* $\hat{u}(t,q) = U(t,q,\hat{\lambda}(t,q),\hat{\xi}(t))$.

Proof. Obviously $(c')^{-1}(e^{rt}(\lambda - \xi d(t,q)))$ is the unique maximizer of the function $u \to \lambda u - e^{-rt}c(u) - \xi d(t,q)$, which is Lipschitz continuous with respect to λ due to the strong convexity of c. On the other hand, $U(t,q,\lambda,\xi)$ is the projection of $(c')^{-1}(e^{rt}(\lambda - \xi d(t,q)))$ on $[\underline{u},\overline{u}]$, therefore it is also Lipschitz continuous in λ.

According to Lemma 1, for almost every $t \in [0,T]$ the optimal control $\hat{u}(t,\cdot)$ maximizes (16) subject to (17). Then for a.e. t there exists a Lagrange multiplier ξ_t (see e.g. the theorem in Sect. 4.2 in [1]) such that $\hat{u}(t,q)$ maximizes in w

$$\hat{\lambda}(t,q)\, w - e^{-rt}c(w) - \xi_t d(t,q)\, w$$

subject to $\underline{u} \le w \le \overline{u}$. In other words, $\hat{u}(t,q) = U(t,q,\hat{\lambda}(t,q),\xi_t)$. Then, due to (17) we have that ξ_t for each $t \in [0,T]$ belongs to the set

$$G(t) := \left\{ \xi \in \mathbf{R} : \int_0^{\hat{Q}(t)} d(t,q) U(t,q,\hat{\lambda}(t,q),\xi)\, \mathrm{d}q - \hat{y}(t) = 0 \right\},$$

therefore $G(t)$ is non-empty. Since $\hat{y}(t)$ is measurable, from [2, Proposition 8.2.9], it follows that $G(t)$ is measurable, therefore it has a measurable selection, $\hat{\xi}(t)$. We shall prove that $U(t,q,\hat{\lambda}(t,q),\hat{\xi}(t)) = U(t,q,\hat{\lambda}(t,q),\xi_t)$ on $D_{\hat{Q}}$, which implies that $U(t,q,\hat{\lambda}(t,q),\hat{\xi}(t)) = \hat{u}(t,q)$ on $D_{\hat{Q}}$ and finalizes the proof.

From the definition of G we have

$$\int_0^{\hat{Q}(t)} d(t,q) U(t,q,\hat{\lambda}(t,q),\hat{\xi}(t))\, \mathrm{d}q = \int_0^{\hat{Q}(t)} d(t,q) U(t,q,\hat{\lambda}(t,q),\xi_t)\, \mathrm{d}q. \tag{22}$$

for $t \in [0,T]$. Obviously $U(t,q,\hat{\lambda}(t,q),\cdot)$ is monotone decreasing in ξ. Therefore, if for some t an inequality

$$U(t,q,\hat{\lambda}(t,q),\hat{\xi}(t)) \; < (>) \; U(t,q,\hat{\lambda}(t,q),\hat{\xi}_t)$$

holds on a subset of $[0,\hat{Q}(t)]$ of positive measure, then $\hat{\xi}(t) > (<) \xi_t$, Since $d(t,q) > 0$, we obtain a contradiction with (22). The proof is complete.

Proof of Proposition 1. Let $\hat{\lambda}(t,q)$ be the solution of (15). Lemma 2 implies $\hat{u}(t,q) = U(t,q,\hat{\lambda}(t,q),\hat{\xi}(t))$, with U and $\hat{\xi}(t)$ being defined as in the Lemma.

Let us consider the following boundary value problem:

$$\dot{x}(t,q) = -\delta x(t,q) + U(t,q,\lambda(t,q),\hat{\xi}(t)), \qquad x(\hat{\theta}(q),q) = x^0(q) \tag{23}$$

$$\dot{\lambda}(t,q) = \delta\lambda(t,q) - e^{-rt}L_x(t,q,x(t,q)), \qquad \lambda(T,q) = 0. \tag{24}$$

Obviously $(\hat{x},\hat{\lambda})$ is a solution of this system. Our goal below will be to prove that it is the only solution and that it depends continuously on q, hence also $\hat{u}$ depends continuously on q.

Consider the initial value problem (23)–(24) with $\lambda(\hat{\theta}(q),q) = p$ instead of the end point condition for λ. Due to the standing assumptions and Lemma 2, the right-hand side of the differential system in (23), (24) is Lipschitz continuous in (λ, x). Then for every q the initial value problem has a unique solution $(x(t,q;p),\lambda(t,q;p))$ on $[\hat{\theta}(q),T]$. Let us fix q and suppress it, as well as $\hat{\xi}(t)$, in the notations below.

To prove uniqueness of the solution of (23)–(24), assume that there are two solutions, (x_1,λ_1) and (x_2,λ_2). If $\lambda_1(\hat{\theta}) = \lambda_2(\hat{\theta})$, then both solutions coincide with $(x(t;p),\lambda(t;p))$ for $p = \lambda_1(\hat{\theta})$. Therefore, let us assume that $\lambda_2(\hat{\theta})-\lambda_1(\hat{\theta}) > \varepsilon$ for some $\varepsilon > 0$. Let τ be the maximal number in $[\hat{\theta},T]$ such that $\lambda_2(t)-\lambda_1(t) \geq \varepsilon$ for all $t \in [\hat{\theta},\tau]$. Using that the function $\lambda \mapsto U(t,\lambda)$ is non-decreasing due to its definition in (21) and the convexity of c, we obtain that for $t \in [\hat{\theta},\tau]$

$$\dot{x}_2(t) - \dot{x}_1(t) = -\delta(x_2(t)-x_1(t)) + U(t,\lambda_2(t)) - U(t,\lambda_1(t)) \geq -\delta(x_2(t)-x_1(t)).$$

Since $x_1(\hat{\theta}) = x_2(\hat{\theta})$, the above inequality implies $x_2(t) - x_1(t) \geq 0$ for all $t \in [\hat{\theta},\tau]$. Using the last inequality and the fact that the function $x \mapsto L_x(t,x)$ is non-increasing due to the concavity of L, we obtain

$$\begin{aligned}\dot{\lambda}_2(t) - \dot{\lambda}_1(t) &= \delta(\lambda_2(t)-\lambda_1(t)) - e^{-rt}\left(L_x(x_2(t)) - L_x(x_1(t))\right)\\ &\geq \delta(\lambda_2(t)-\lambda_1(t)) \geq 0, \quad t \in [\hat{\theta},\tau],\end{aligned}$$

which implies $\lambda_2(\tau) - \lambda_1(\tau) > \varepsilon$. This means that $\tau = T$ and, in particular, $\lambda_2(T)-\lambda_1(T) \geq \varepsilon$. This contradicts the boundary condition in (24), and implies that the solution of (23), (24) is unique (namely, $(\hat{x}(\cdot,q),\hat{\lambda}(\cdot,q))$).

Next, we shall prove the continuity of $\hat{\lambda}$ with respect to q. Due to the boundedness of $\hat{u}$ it is easy to verify that there is a compact interval $P \subset \mathbf{R}$ containing all values $\hat{\lambda}(\hat{\theta}(q),q)$, $q \in [0,\hat{Q}(T)]$. Let us prove the following property (P): for every $\varepsilon > 0$ there exists $\delta > 0$ such that for any q_1, $q_2 \in [0,\hat{Q}(T)]$ and p_1, $p_2 \in P$ satisfying $|q_1 - q_2| < \delta$ and $\lambda(T,q_1;p_1) = \lambda(T,q_2;p_2) = 0$, it holds that $|p_1 - p_2| \leq \varepsilon$.

According to the continuous dependence of the solution of an ODE with Lipschitz continuous right-hand side with respect to initial data and parameter (see e.g. [5, Theorem 2], where the required continuity in t is not necessary.) the mapping $[0,\hat{Q}(T)] \times \mathbf{R} \ni (q,p) \to \lambda(\cdot,q;p) \in C([\hat{\theta}(q),T])$ is continuous, hence it is uniformly continuous on $[0,\hat{Q}(T)] \times P$.

Assume that property (P) does not hold. Then there exists $\varepsilon > 0$ such that for every $\delta > 0$ there exist q_1, $q_2 \in [0,\hat{Q}(T)]$ and p_1, $p_2 \in P$ such that $|q_1 - q_2| < \delta$,

$\lambda(T,q_1;p_1) = \lambda(T,q_2;p_2) = 0$, and $p_2 - p_1 > \varepsilon$. Due to the (uniform) continuous dependence we may choose $\delta > 0$ so small that

$$|\lambda(T,q_2;p_2) - \lambda(T,q_1;p_2)| \le \varepsilon/2.$$

We have

$$\lambda(\hat\theta(q_1),q_1;p_2)-\lambda(\hat\theta(q_1),q_1;p_1) = p_2-p_1 \ge \varepsilon, \quad x(\hat\theta(q_1),q_1;p_2) = x(\hat\theta(q_1),q_1;p_1).$$

Then we can prove in exactly the same way as a few paragraphs above that $\lambda(t,q_1;p_2) - \lambda(t,q_1;p_1) \ge \varepsilon$ for all $t \in [\hat\theta(q_1),T]$. Hence

$$\lambda(T,q_2;p_2) - \lambda(T,q_1;p_1) \ge \lambda(T,q_1;p_2) - \lambda(T,q_1;p_1) - \varepsilon/2 \ge \varepsilon/2.$$

This contradicts the equality $\lambda(T,q_1;p_1) = \lambda(T,q_2;p_2)$ and proves property (P).

Applying property (P) for $p_i = \hat\lambda(\hat\theta(q_i),q_i)$, $i = 1, 2$, we obtain that for every $\varepsilon > 0$ there exists $\delta > 0$ such that $|q_1 - q_2| < \delta$ implies $|\hat\lambda(\hat\theta(q_1),q_1) - \hat\lambda(\hat\theta(q_2),q_2)| < \varepsilon$. Then using again the continuous dependence of the solution of ODEs we conclude that $q \mapsto \hat\lambda(\cdot,q)$ is continuous. From the equality $\hat u(t,q) = U(t,q,\hat\lambda(t,q),\hat\xi(t))$ and the continuity of $d(t,\cdot)$ we obtain the desired continuity of $\hat u(t,\cdot)$. Q.E.D.

4 Pontryagin Maximum Principle

Theorem 2 in [3] claims that under the continuity of the optimal control $\hat u$ with respect to q (proved in Proposition 1) Pontryagin type optimality conditions hold. Below we formulate these conditions for the particular problem considered in the present paper. We use the so-called current-value adjoint variables and current-value Hamiltonian, which differ from the ones used above by a multiplier e^{rt}.

Let us introduce the following *adjoint system* for the variables $(\lambda(t,q), \mu(t), \nu(t))$, where the meaning of the solution is as that of (4)–(6) in inverse time:

$$\dot\lambda(t,q) = (\delta + r)\lambda(t,q) - L_x(q) \text{ for a.e. } (t,q) \in D_{\hat Q},\ \lambda(T,q) = 0 \text{ on } [0,\bar Q] \quad (25)$$

$$\begin{aligned}\dot\mu(t) = {}& \mu(t)\Big(- g_{_Q}(t)\, y(t) + r\Big) - L(\hat Q(t)) + c(u(t,\hat Q(t))) \qquad (26)\\ & - \lambda(t,\hat Q(t))[-\delta x^b(t) - \dot x^b(t) + u(t,\hat Q(t))] - \nu(t)d(t,\hat Q(t))u(t,\hat Q(t))\\ & - \int_0^{\hat Q(t)} L_{_Q}(q)\,\mathrm{d}q \quad \text{for a.e. } t \in [0,T], \qquad \mu(T) = 0,\end{aligned}$$

$$\nu(t) = \mu(t)\, g(t) + \int_0^{\hat Q(t)} L_y(q)\,\mathrm{d}q. \quad (27)$$

Define $H : D \times \mathbf{R} \times [0,\bar Q] \times \mathbf{R} \times [\underline{u},\overline{u}] \times \mathbf{R} \times \mathbf{R} \times \mathbf{R} \mapsto \mathbf{R}$ as

$$\begin{aligned}H(t,q,x,Q,y,u,\lambda,\mu,\nu) = {}& L(q,x,Q,y) - c(u) + \lambda(-\delta x + u) \qquad (28)\\ & + \mu\, g(t,Q)\, y + \nu\, d(t,q)u.\end{aligned}$$

Theorem 1. *Let $\hat{u} \in L_\infty(D)$ be an optimal control for the problem* (1)–(6) *and denote by $\hat{z} := (\hat{x}, \hat{Q}, \hat{y})$ the corresponding trajectory. Then the adjoint system* (25)–(27) *has a unique solution $\hat{\pi} := (\hat{\lambda}, \hat{\mu}, \hat{\nu})$ and for a.e. $(t,q) \in D_{\hat{Q}}$ it holds that*

$$H(t,q,\hat{z}(t,q),\hat{u}(t,q),\hat{\pi}(t,q)) = \max_{u \in [\underline{u},\overline{u}]} H(t,q,\hat{z}(t,q),u,\hat{\pi}(t,q)).$$

The maximization condition in the theorem means that knowing the adjoint function $\hat{\lambda}$ one can obtain the optimal control $\hat{u}$ by solving for $(t,q) \in D_{\hat{Q}(\cdot)}$

$$\max_{u \in [\underline{u},\overline{u}]} \left\{ -c(u) + [\hat{\lambda}(t,q) + \hat{\nu}(t)\, d(t,q)]\, u \right\}.$$

References

1. Alekseev, V.M., Tikhomirov, V.M., Fomin, S.V.: Optimal Control. Plenum, New York (1987)
2. Aubin, J.-P., Frankowska, H.: Set-Valued Analysis. Birkhäuser, Boston (1990)
3. Belyakov, A., Tsachev, Ts., Veliov, V.M.: Optimal control of heterogeneous systems with endogenous domain of heterogeneity. Appl. Math. Optim. **64**(2), 287–311 (2011)
4. Sorger, G.: Horizontal innovations with endogenous quality choice. Economica **78**, 697–722 (2011)
5. Strauss, A.: Continuous dependence of solutions of ordinary differential equations. Am. Math. Mon. **71**(6), 649–652 (1964)
6. Veliov, V.M.: Optimal control of heterogeneous systems: basic theory. J. Math. Anal. Appl. **346**, 227–242 (2008)

Invariance Property in Approaching Problem on a Finite Time Interval

Vladimir Ushakov[1(✉)], Alexander Matviychuk[1,2], and Sergey Brykalov[1]

[1] Institute of Mathematics and Mechanics, Russian Academy of Sciences, 16 S. Kovalevskaya str., Ekaterinburg 620990, Russia
{ushak,brykalov}@imm.uran.ru

[2] Ural Federal University, 19 S. Mira, Ekaterinburg 620002, Russia
matv@uran.ru

Abstract. The problem of approaching a target set in the phase state space by controlled system at the fixed time moment is under consideration. Algorithm for solving this problem is described in the paper. This method is based on weak invariance property of problem solvability sets.

1 Introduction

In this paper the problem of approaching a compact target set in the Euclidean space by controlled system at the fixed time moment is under consideration. The method used for solving this problem is based on the weakly invariance property of approaching problem solvability sets [1–5] and constructions with a guide [3,5]. Here the guide is a finite set of points in a phase state space which unfolds along the time and arrives to the target set at the fixed time moment. The subject of this paper is closely connected with results provided in [6–15].

2 Problem Formulation

We consider a controlled system

$$\frac{dx}{dt} = f(t, x, u), \quad x \in \mathbb{R}^n, \quad u \in P; \tag{1}$$

on an interval $[t_0, \vartheta]$, $t_0 < \vartheta < \infty$; here P is a compactum in the Euclidean space $\mathbb{R}^p$.

The right-hand side of the system (1) satisfies the assumptions

Assumption A.1. *The vector function $f(t, x, u)$ is defined and continuous in t, x, u and for any bounded and closed domain $D \subset [t_0, \vartheta] \times \mathbb{R}^n$ there exists a number $L = L(D) \in [0, \infty)$ such that $\|f(t, x^{(1)}, u) - f(t, x^{(2)}, u)\| \leqslant L\|x^{(1)} - x^{(2)}\|$, $(t, x^{(i)}, u) \in D \times P$, $i = 1, 2$.*

Assumption A.2. *A number $\gamma \in (0, \infty)$ exists such that $\|f(t, x, u)\| \leqslant \gamma(1 + \|x\|)$, $(t, x, u) \in [t_0, \vartheta] \times \mathbb{R}^n \times P$.*

I. Lirkov et al. (Eds.): LSSC 2013, LNCS 8353, pp. 152–158, 2014.
DOI: 10.1007/978-3-662-43880-0_16,

Here $\|\cdot\|$ is the norm in $\mathbb{R}^n$.

Together with system (1), the following differential inclusion is considered:

$$\frac{dx}{dt} \in F(t,x) = co\{f(t,x,u) : u \in P\}. \tag{2}$$

Let $t_0 \leqslant t_* \leqslant t^* \leqslant \vartheta$ and $X_* \subset \mathbb{R}^n$. We will consider motions $x(t)$ of system (1) generated by the control $u(t)$ on $[t_0, \vartheta]$. The motions $x(t)$ of inclusion (2) on $[t_0, \vartheta]$ will also be considered.

Definition 1. *The attainability set $X(t^*, t_*, x_*)$ $(Y(t^*, t_*, x_*))$ of controlled system (1) (differential inclusion (2)) is the set of all $x^* \in \mathbb{R}^n$ such that $x^* = x(t^*)$ for some motion $x(t)$, $x(t_*) = x_* \in X_*$ of system (1) (differential inclusion (2)).*

Definition 2. *The set $X(t_*, X_*)$ $(Y(t_*, X_*))$ is called integral funnel of controlled system (1) (differential inclusion (2)) of the sets $X(t_*, X_*) = \bigcup\limits_{t^* \in [t_*, \vartheta]} (t^*, X(t^*, t_*, X_*))$ $(Y(t_*, X_*) = \bigcup\limits_{t^* \in [t_*, \vartheta]} (t^*, Y(t^*, t_*, X_*)))$ on $[t_*, \vartheta] \times \mathbb{R}^n$; here $(t^*, X^*) = \{(t^*, x^*) :\ x^* \in X^*\}$.*

We give definitions of invariant and weak invariant sets with respect to controlled system (1) (differential inclusion (2)).

Let Φ be a nonempty closed set in $[t_0, \vartheta] \times \mathbb{R}^n$.

Definition 3. *A set Φ is called invariant with respect to system (1) (inclusion (2)) if for any $(t_*, x_*) \in \Phi$ and $t_* \in [t_0, \vartheta]$ one has $X(t_*, x_*) \subset \Phi$ $(Y(t_*, x_*) \subset \Phi)$.*

Definition 4. *A set Φ is called weakly invariant with respect to system (1) (inclusion (2)) if for any $(t_*, x_*) \in \Phi$ there exists a solution $x(t)$, $x(t_*) = x_*$ of system (1) (inclusion (2)) on $[t_*, \vartheta]$ that satisfies the inclusion $(t, x(t)) \in \Phi$, $t \in [t_*, \vartheta]$.*

Question about (exact) calculation of attainability sets and integral funnels or their effective analytical description often arises. At the same time the determination of analytical description can be an overwhelming affair. Therefore the task of approximate calculation of attainability sets is especially relevant. There exist number of methods and approaches aimed at effective solving the problem of approximate calculation of the attainability sets [5–15]. These methods can be used for solving the control problems and, in particular, for solving the approaching of a target set problem.

3 Particular Case of Solution for Problem of Approaching a Compactum in $\mathbb{R}^n$ by System (1)

We formulate a problem of approaching a compact target set in the space $\mathbb{R}^n$ by system (1) at the time moment ϑ. We will discuss method of solving this problem based on employing weakly invariant sets.

Let M be a compactum in $\mathbb{R}^n$, and $x_0 \in \mathbb{R}^n$.

Problem 1. Find an admissible control $u^*(t)$ on $[t_0, \vartheta]$, that generates a motion $x^*(t)$, $x^*(t_0) = x_0$ of system (1) on the interval $[t_0, \vartheta]$, such that $x^*(\vartheta) \in M$.

There exist points $x_0 \in \mathbb{R}^n$ for which problem 1 is not solvable. Moreover for some $x_0 \in \mathbb{R}^n$ it may be that $M \bigcap X(\vartheta, t_0, x_0) \neq \varnothing$ and $M \bigcap Y(\vartheta, t_0, x_0) \neq \varnothing$. In this situation there is no control $u^*(t)$ solving the problem 1 but there is control $u^*(t)$ on $[t_0, \vartheta]$ that solves a problem of approaching of system (1) with preassigned ε-neighborhood M_ε of the set M.

Taking into account this circumstance we will formulate a problem of approaching a set M by system (1) as the problem of finding an admissible control $u^*(t)$ on $[t_0, \vartheta]$ that brings motion $x^*(t)$ of system (1) into a preassigned ε-neighborhood M_ε (i.e. $x^*(\vartheta) \in M_\varepsilon$).

Let us assume that initial position (t_0, x_0), initial set M and a moment ϑ are such that $M^* = M \bigcap Y(\vartheta, t_0, x_0) \neq \varnothing$.

We associate the direct time interval $t \in [t_0, \vartheta]$ with the reverse time interval $\tau \in [t_0, \vartheta] : \tau = t_0 + \vartheta - t$. We also associate differential inclusion (2) with a differential inclusion

$$\frac{dz}{d\tau} = F^0(\tau, z), \quad \tau \in [t_0, \vartheta]; \tag{3}$$

here $F^0(\tau, z) = co\{f^0(\tau, z, v) : v \in P\}$, $f^0(\tau, z, v) = -f(t_0+\vartheta-\tau, z, v)$, $(\tau, z, v) \in [t_0, \vartheta] \times \mathbb{R}^n \times P$.

Motions of differential inclusions (2) and (3) are dual, i.e. any movement $x(t)$ of differential inclusion (2) may be associated with the movement $z(\tau)$ of differential inclusion (3) in such way: $z(\tau) = x(t)$, $\tau = t_0 + \vartheta - t$, $t \in [t_0, \vartheta]$.

We associate integral funnel $Z = Z(t_0, M)$ of inclusion (3) with a set $W \in D$ by the formula $W(t) = Z(\tau)$, $t = t_0 + \vartheta - \tau$, $\tau \in [t_*, \vartheta]$.

The closed set W is the solvability set for inclusion (2) in problem of approaching the set M by the inclusion movements at time moment ϑ.

The set W is weakly invariant with respect to inclusion (2) and $(t_0, x_0) \in W$. Let us use the weak invariance property of W in the construction of control $u^*(t)$ on $[t_*, \vartheta]$ that brings movement of system (1) from x_0 into the preassigned ε-neighborhood M_ε of the set M.

Let us describe step-by-step control procedure with a guide. This procedure implements control $u^*(t)$ on $[t_*, \vartheta]$.

On the axis t (on the axis τ) consider a finite mesh $\Gamma = \{t_0, t_1, \ldots, t_j, \ldots, t_N = \vartheta\}$ ($\Gamma = \{\tau_0 = t_0, \tau_1, \ldots, \tau_i, \ldots, \tau_N = \vartheta\}$) with equal steps $\Delta_i = t_{i+1} - t_i = \Delta > 0$.

Assume that we can calculate the sets $W(t_j) = Z(\tau_i)$, $i = \overline{0 - N}$, and they are already calculated (precisely). But we can calculate $Y(t_{j+1}, t_j, x^{(j)})$, $x^{(j)} \in \mathbb{R}^n$ only approximately as

$$\tilde{Y}(t_{j+1}, t_j, x^{(j)}) = x^{(j)} + \Delta F(t_j, x^{(j)});$$

here $x^{(j)} + \Delta F(t_j, x^{(j)}) = \{x^{(j)} + \Delta f : f \in F(t_j, x^{(j)})\}$.

Under these assumptions let us begin the description of a step-by-step procedure for control with the guide.

Assume that we have started at the time moment t_0 from a point $x_0 \in W(t_0)$ and have implemented construction of movement $x^*(t)$ of system (1) using step-by-step control procedure with the guide up to the time moment $t_j \in \Gamma$, $t_j < \vartheta$.

Suppose that at the time moment t_j the system (1) is located at the point $x^*(t_j)$. Also there are a guide's point $\tilde{x}^{(j)}$ that corresponds to the time moment t_j and point $x^{(j)}$ on $W(t_j)$ nearest to $\tilde{x}^{(j)}$.

The function $\omega^*(\rho) > 0$, $\rho \in (0, \infty)$ ($\omega^*(\rho) \downarrow 0$ as $\rho \downarrow 0$) such that

$$d\big(F(t_*, x_*), F(t^*, x^*)\big) \leq \omega^*(|t_* - t^*| + \|x_* - x^*\|) \text{ as } (t_*, x_*) \text{ and } (t^*, x^*) \text{ from } D_* \tag{4}$$

is used in considerations of paper [7].

Since W is weakly invariant with respect to inclusion (2) and taking into account the inclusion $x^{(j)} \in W(t_j)$ and inequality (4) we have

$$W(t_{j+1})_{\omega(\Delta)} \bigcap \tilde{Y}(t_{j+1}, t_j, x^{(j)}) \neq \varnothing,$$

from which it follows that there exists a vector $f^{(j)} \in F(t_j, x^{(j)})$ such that $x^{(j+1)} = x^{(j)} + \Delta f^{(j)} \in W(t_{j+1})_{\omega(\Delta)}$.

The relation holds

$$f^{(j)} = \sum_{k=1}^{n+1} \alpha_k f(t_j, x^{(j)}, u^{(k)}), \tag{5}$$

where $\alpha_k \geqslant 0$, $u^{(k)} \in P$ for $k = \overline{1, n+1}$, $\sum\limits_{k=1}^{n+1} \alpha_k = 1$.

The vector $f^{(j)} \in F(t_j, x^{(j)})$ can be calculated according to the following scheme:

1. Find the set $W(t_{j+1})_{\omega(\Delta)} \bigcap \tilde{Y}(t_{j+1}, t_j, x^{(j)})$;
2. Choose a point $\tilde{x}^{(j)} \in W(t_{j+1})_{\omega(\Delta)} \bigcap \tilde{Y}(t_{j+1}, t_j, x^{(j)})$;
3. Calculate the vector $f^{(j)}$ according to formula

$$f^{(j)} = \Delta^{-1}(\tilde{x}^{(j+1)} - x^{(j)}). \tag{6}$$

For a given $f^{(j)}$ (6), we can consider (5) as an equation with respect to α_k, $u^{(k)}$, $k = \overline{1, n+1}$, which satisfy also the conditions ($\alpha_k \geqslant 0$, $u^{(k)} \in P$ for $k = \overline{1, n+1}$ and $\sum\limits_{k=1}^{n+1} \alpha_k = 1$).

We suppose that we can solve equation (5) approximately, that is, we have found β_k, $v^{(k)}$, $k = \overline{1, n+1}$ ($\beta_k \geqslant 0$, $v^{(k)} \in P$, $k = \overline{1, n+1}$ and $\sum\limits_{k=1}^{n+1} \beta_k = 1$) so

$$\tilde{f}^{(j)} = \sum_{k=1}^{n+1} \beta_k f(t_j, x^{(j)}, v^{(k)}), \quad \|\tilde{f}^{(j)} - f^{(j)}\| \leqslant æ^{(j)}, \tag{7}$$

where $æ^{(j)} = \omega^*((1 + K)\Delta)$.

We introduce the numbers $\Delta_k = \beta_k \Delta$, $k = \overline{1, n+1}$ and a mesh $\Gamma^{(j)} = \{t_1^{(j)} = t_j, t_2^{(j)}, \ldots, t_k^{(j)}, t_{k+1}^{(j)}, \ldots, t_{n+1}^{(j)}, t_{n+2}^{(j)} = t_{j+1}\}$ in the segment $[t_j, t_{j+1}]$; here $t_{k+1}^{(j)} = t_k^{(j)} + \Delta_k$, $k = \overline{1, n+1}$.

Consider also a control $u^*(t)$, $t \in [t_j, t_{j+1})$, that is constant on the segments $[t_k^{(j)}, t_{k+1}^{(j)})$ of the mesh $\Gamma^{(j)}$

$$u^*(t) = v^{(k)} \text{ for } t \in [t_k^{(j)}, t_{k+1}^{(j)}), k = \overline{1, n+1}. \tag{8}$$

Let $x^*(t_0) = x^{(0)} = x_0$ and we introduce the notation $\rho_j = \|x^*(t_j) - x^{(j)}\|$, $j = \overline{0, N}$ so that $\rho_0 = 0$ holds.

The recurrent estimate holds

$$\rho_{j+1} \leqslant e^{L\Delta} \rho_j + 3\omega(\Delta), \quad j = \overline{0, N-1}; \tag{9}$$

here $\omega(\Delta) = \Delta\omega^*((1+K)\Delta)$, $\Delta \in (0, \infty)$.

We derive from (9)

$$\rho(x^*(\vartheta), M) \leqslant \rho_N \leqslant 3e^{L(\vartheta - t_0)}(\vartheta - t_0)\omega^*((1+K)\Delta). \tag{10}$$

Theorem takes place

Theorem. *Let system* (1) *satisfies assumptions A.1, A.2 and problem 1 be solvable from the initial position* (t_0, x_0) *of system* (1). *Then for any* $\varepsilon > 0$, *one can find a finite uniform mesh* Γ *on the segment* $[t_0, \vartheta]$ *and an admissible finite-constant control* $u^*(t)$ on $[t_0, \vartheta]$ *corresponding to the mesh* Γ *such that the motion* $x^*(t)$, $x^*(t_0) = x_0$ *of system* (1) *generated by this control satisfies the relation* $x^*(\vartheta) \in M_\varepsilon$.

4 Example: Inverted Pendulum System

We consider inverted pendulum fixed on a train. The train can move backward and forward on a horizontal plane (Fig. 1).

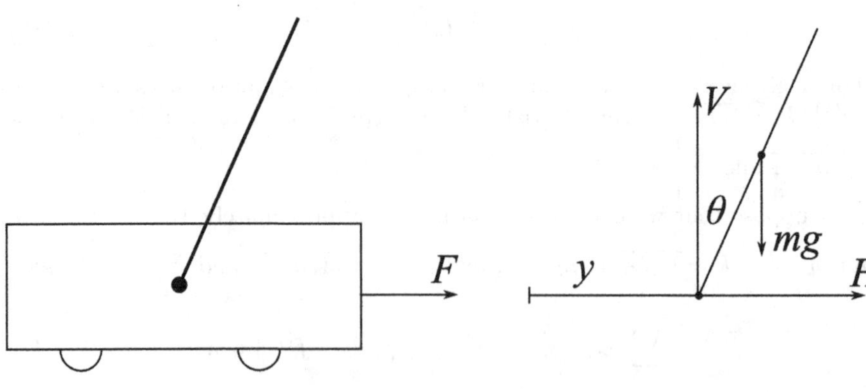

Fig. 1. Inverted pendulum fixed on the train

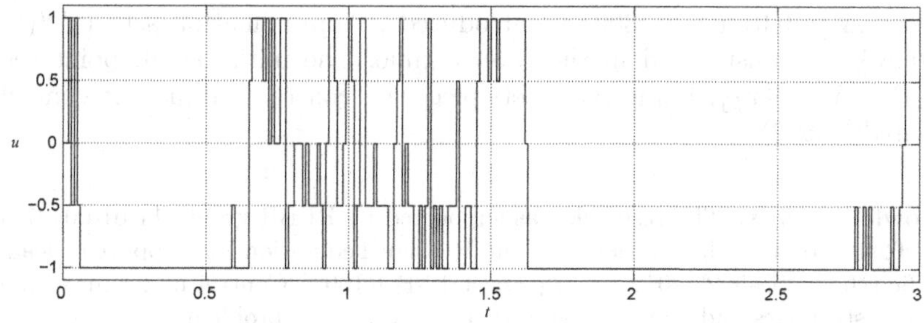

Fig. 2. Control $u^*(t)$

The train is affected by traction force $\overrightarrow{F}$. The pendulum is affected by gravitation $\overrightarrow{mg}$ applied at the pendulum center of gravity. Also horizontal force $\overrightarrow{H}$ and vertical force $\overrightarrow{V}$, which are the forces of reaction, affect the pendulum. m— mass of the pendulum g— gravitational constant. L— distance between pendulum center of gravity and support point, y— support point offset, ϑ— pendulum angle, I— moment of inertia in relation to center of gravity, M— mass of the train, k— friction coefficient.

Let us introduce variables $x_1 = \vartheta$, $x_2 = \dot{\vartheta}$, $y_1 = y$, $y_2 = \dot{y}$. Using these variables we can write down the system of equations

$$\begin{cases} \dot{x}_1 = x_2 \\ \dot{x}_2 = -\dfrac{1}{\Delta(x_1)}\{(m+M)mgL\sin x_1 - mL\cos x_1(F + mLx_2^2 - ky_2)\} \\ \dot{y}_1 = y_2 \\ \dot{y}_2 = -\dfrac{1}{\Delta(x_1)}\{-mL\cos x_1 mgL\sin x_1 + (I+mL^2)(F + mLx_2^2\sin x_1 - ky_2\} \end{cases}$$

here $\Delta(x_1) = (I+mL^2)(m+M) - m^2L^2\cos x_1$.

For this mechanical system $u = F$ is a control.

Let $X = \begin{pmatrix} x_1 \\ x_2 \\ x_3 \\ x_4 \end{pmatrix}$ — be the phase state vector of the system.

Example 1. We have interval $[0;T] = [0;3]$, start point $x_0 = \begin{pmatrix} 1.582 \\ 2.006 \\ -0.469 \\ -0.403 \end{pmatrix}$ and finish point $x_f = \begin{pmatrix} 0 \\ 0 \\ 1 \\ 0 \end{pmatrix}$, which is a target set M. It is necessary to construct admissible control $u^*(t)$ that brings motion $x^*(t)$ of the system to $x^*(T) = x_f$.

The solution of this problem is based on the scheme from **3**. According to this scheme time interval partition $\Gamma = \{t_0 = 0, t_1, \ldots, t_j, \ldots, t_N = \vartheta = 3\}$ with diameter $\Delta = 0.01$ $(N = 300)$ is introduced . Approximation sets $\{\mathcal{W}^a(t_j) : j = \overline{0, N}\}$ are constructed in the reverse time. The initial single-point set is $\mathcal{W}^a(t_N) = M = \{x_f\}$. Using these sets piecewise constant admissible control is calculated (Fig. 2).

Acknowledgments. The research was supported by RFBR grants 11-01-00427 and 12-01-00290, grant of the President of the Russian Federation for support of leading scientific schools SS-5927.2012.1 and project 12-P-1-1002 "Control under uncertainty. Positional strategies and Hamiltonian structures in control problems".

References

1. Krasovskii, N.N.: Control of Dynamical System, 520 p. Nauka, Moscow (1985)
2. Krasovskii, N.N.: The basic game problem of approach. Target absorption. Extremal strategy. Lectures on control theory. 4 Uralstate university, 96 p. Sverdlovsk (1970)
3. Krasovskii, N.N., Subbotin, A.I.: Positional Differential Games, 456 p. Nauka, Moscow (1974)
4. Pontryagin, L.S.: On linear differential games. I. Dokl. Acad. Nauk USSR **156**(4), 738–741 (1964)
5. Ushakov, V.N., Matviychuk, A.R., Ushakov, A.V., Parshikov, G.V.: Invariance of sets in approach problem solution construction. Proc. IMM Ur. Br. RAS **19**(1), 134–145 (2013)
6. Guseinov, KhG, Moiseev, A.N., Ushakov, V.N.: On control systems attainability sets approximation. Prikl. Math. Mech. **62**(2), 179–187 (1998)
7. Ushakov, V.N., Matviychuk, A.R., Ushakov, A.V.: Approximation of attainability sets and integral funnels. Vestn. Udm. Univ. (Math., Mech., Comp. Sci.) **4**, 23–39 (2011)
8. Nikolskii, M.S.: On differential inclusion attainability set approximation. Vestn. Mos. Univ. (Ser. 15. Vychisl. Mat. Kibernet) **4**, 31–34 (1987)
9. Kurzhanski, A.B., Valyi, I.: Ellipsoidal Calculus for Estimation and Control. Birkhauser, Boston (1997)
10. Cardaliaguet, P., Quincampoix, M., Saint-Pierre, P.: Numerical schemes for discontinuous value functions of optimal control. Set-valued Anal. **8**(1–2), 111–126 (2000)
11. Falcone, M., Saint-Pierre, P.: Algorithms for constrained optimal control problem: a link between viability and dynamic programming. University of Rome (1995) (Preprint)
12. Chernousko, F.L.: Dynamical System Phase State Estimation: Method of Ellipsoids. Nauka, Moscow (1988)
13. Gusev, M.I.: Estimation of attainability sets of multidimensional control systems with nonlinear crossing connections. Proc. IMM Ur. Br. RAS **15**(4), 82–94 (2009)
14. Filippova, T.F.: Differential equations for ellipsoidal estimates of attainability sets of nonlinear dynamical controlled system. Proc. IMM Ur. Br. RAS **16**(1), 223–232 (2010)
15. Lempio, F., Veliov, M.V.: Discrete approximations of differential inclusions. Bayereuther Math. Schr. **54**, 149–232 (1998)

Monte Carlo Methods: Theory, Applications and Distributed Computing

Hybrid Monte Carlo CT Simulation on GPU

Gábor Jakab[1,2] and László Szirmay-Kalos[1(✉)]

[1] Budapest University of Technology and Economics, Budapest, Hungary
szirmay@iit.bme.hu
http://cg.iit.bme.hu

[2] Mediso Medical Equipment Developing and Service Ltd., Budapest, Hungary
research@mediso.hu
http://www.mediso.hu

Abstract. Developing image reconstruction algorithms for diagnostic medical devices requires physically accurate and effective simulation tools. In this paper we present a hybrid Monte Carlo (MC) particle simulation method for Computed Tomography (CT) scanners. To meet the performance requirements, we combine several variance reduction techniques and tailor the algorithms for effective GPU execution. Variance reduction methods include main part separation, sample weighting, reuse, forced collision, next event estimation and table driven importance sampling. We show that the resulting method can deliver accurate simulations orders of magnitude faster than direct physical simulation.

Keywords: GPU · CT · Image reconstruction · Photon transport

1 Introduction

In medical imaging, the 3D density field is generated with the reconstruction algorithm from the measured data acquired by the scanner. MC particle simulation has various applications in the context of medical imaging. First, during the development of new equipment and reconstruction algorithms, we need to simulate the transport process to obtain controllable "measured data". On the other hand, iterative reconstruction schemes involve a transport simulation and an update of the model, so MC simulation is a part of an iterative process. Finally, MC simulation can also be used to estimate the radiation dose imposed on the patient.

In Computer Tomography (CT) an X-ray source emits photons in a spectrum of energy levels, i.e. frequencies. Photons are scattered and absorbed in the examined object. Some of the emitted photons arrive at detectors, generating hit events that are the input of the reconstruction algorithm.

From mathematical point of view, we need to solve a Fredholm type integral equation. Along a ray of direction $\boldsymbol{\omega}$ at point $\boldsymbol{x}$ the intensity $I(\boldsymbol{x}, \boldsymbol{\omega}, E)$ of particle

I. Lirkov et al. (Eds.): LSSC 2013, LNCS 8353, pp. 161–169, 2014.
DOI: 10.1007/978-3-662-43880-0_17,

flow at energy level E satisfies

$$\boldsymbol{\omega}\cdot\boldsymbol{\nabla}I(\boldsymbol{x},\boldsymbol{\omega},E) = -(\sigma_a(\boldsymbol{x},E)+\sigma_s(\boldsymbol{x},E))I(\boldsymbol{x},\boldsymbol{\omega},E) + \int_{\Omega} I(\boldsymbol{x},\boldsymbol{\omega}',E')\sigma_s(\boldsymbol{x},E')P(\boldsymbol{\omega}'\cdot\boldsymbol{\omega},E')\mathrm{d}\omega', \quad (1)$$

where σ_a is the absorption cross section, $\sigma_s = \sigma_c + \sigma_i$ is the scattering cross section that can be further decomposed into coherent and incoherent scattering, $\sigma_t = \sigma_a + \sigma_s$ is the total cross section, Ω is the directional sphere, E' and E are the incident and scattered photon energies, respectively, and $P(\boldsymbol{\omega}'\cdot\boldsymbol{\omega},E')$ is the phase function, i.e. the probability density of the scatter direction. Energy level E' is unambiguously determined by the scattered energy E and the angle between incident direction $\boldsymbol{\omega}'$ and scattered direction $\boldsymbol{\omega}$. The boundary condition is given by a point source at $\boldsymbol{s}$ of a known directional and spectral characteristic $\Phi(\boldsymbol{\omega},E)$, which is the source intensity on energy level E.

We are interested in the measured value of detectors, where each detector d is associated with a measuring function $M_d(\boldsymbol{y},E)$ that is non-zero if point $\boldsymbol{y}$ is on the surface A_d of the detector and can be a non-linear function of photon energy E. Thus, we need to determine a large number of measured values

$$m_d = \int_{E_{\min}}^{E_{\max}} \int_{A_d} \int_{\Omega} M_d(\boldsymbol{y},E)I(\boldsymbol{y},\boldsymbol{\omega},E)\mathrm{d}\omega\mathrm{d}\boldsymbol{y}\mathrm{d}E.$$

The most straightforward way is the direct simulation of physical effects, i.e. following the life cycle of photons from the source to the detectors [2,3]. As physical processes describing photon–matter interaction are inherently random, MC simulation mimics the phenomena of real life, including coherent, incoherent scatter and photoelectric absorption. To obtain an accurate enough CT simulation in this way, we need about 10^{12} photons. The industry standard MC simulators such as GATE or GEANT (http://geant4.cern.ch/), can only handle 10^6 particles per second on a desktop computer, which means that such simulations may require supercomputers to get the results in reasonable time.

To attack this problem, we exploit the massively parallel architecture of graphics cards (GPU), and get rid of the concept of direct physical simulation to allow the application of different variance reduction techniques. GPUs are designed to solve data parallel problems, therefore they have substantially more processing cores then CPUs. These cores are grouped into Streaming Multiprocessors (SMX) which can be considered as SIMD processors, so each core in one SMX executes the same instruction, but on different data. The Monte Carlo simulation tracks the particles individually, so it can be distributed into thousands of threads. On the other hand, the algorithm contains a lot of conditional statements and this is not optimal for GPUs. The classic ray marching algorithm is not only data parallel, but is free from conditional statements. The idea is to combine these different approaches into one algorithm.

2 CT Simulation with the MC Method

In direct physical simulation, we generate photons from the X-ray source and track them individually. A photon life cycle starts with sampling the initial photon energy E and direction $\boldsymbol{\omega}$ by mimicking the power spectrum of the source $\Phi(\boldsymbol{\omega}, E)$. Then, simulation continues with a sequence of free path travel and scattering steps, and finishes either with absorption in the phantom or in the detector, or with recognizing that the photon has left the volume of interest.

Generating a single step of the random path involves the sampling of the free path traveled by the photon before scattering, deciding whether scattering or absorption happens, and finally sampling the new scattering direction. The cumulative probability density of the free path length L along a ray of origin $\boldsymbol{x}$ and direction $\boldsymbol{\omega}$ is

$$P(L) = 1 - \exp\left(-\int_0^L \sigma_t(\boldsymbol{x} + \boldsymbol{\omega} l, E)\mathrm{d}l\right).$$

Thus, sampling length L with this distribution means the solution of the following *sampling equation* for L:

$$\mathrm{rnd} = 1 - \exp\left(-\int_0^L \sigma_t(\boldsymbol{x} + \boldsymbol{\omega} l, E)\mathrm{d}l\right) \implies -\log(1 - \mathrm{rnd}) = \int_0^L \sigma_t(\boldsymbol{x} + \boldsymbol{\omega} l, E)\mathrm{d}l \tag{2}$$

where rnd is a random number uniformly distributed in the unit interval. One option is *ray marching* that approximates the integral by a Riemann sum and finds $L = n\Delta l$ where a running sum exceeds $-\log(1 - \mathrm{rnd})$. The other popular free sampling method is the *Woodcock tracking* [4,6] which advances in the media with random length steps based on the maximum cross section $\sigma_{\max}$ to get tentative interaction points:

$$L_t = \frac{-\log(1 - \mathrm{rnd})}{\sigma_{\max}}. \tag{3}$$

Tentative interaction points are either accepted or rejected with probability $\sigma_t/\sigma_{\max}$ and $1 - \sigma_t/\sigma_{\max}$. In case of rejection, and the same sampling step is repeated from there. If the interaction point is accepted, then we identify the type of interaction (absorption, coherent (aka Rayleigh) and incoherent (aka Compton) scattering randomly proportionally to their cross sections.

In coherent scattering the photon keeps its original energy, and the Rayleigh phase function is:

$$P_{\mathrm{Rayleigh}}(\boldsymbol{\omega}) = \frac{3}{16\pi}\left(1 + (\boldsymbol{\omega}' \cdot \boldsymbol{\omega})^2\right). \tag{4}$$

In incoherent scattering, the energy of the scattered photon is determined by the Compton law:

$$E_i(E, \boldsymbol{\omega} \cdot \boldsymbol{\omega}') = \frac{E}{1 + \frac{E}{m_e c^2}(1 - \boldsymbol{\omega} \cdot \boldsymbol{\omega}')}, \tag{5}$$

where E_i is the scattered energy, E is the incident energy, and $m_e c^2$ is the energy of the electron, $\boldsymbol{\omega}$ is the scatter direction, and $\boldsymbol{\omega}'$ is the incident direction. The phase function is given by the Klein-Nishina formula [7]:

$$P_{\mathrm{KN}}(\boldsymbol{\omega}) \propto E_i(E, \boldsymbol{\omega} \cdot \boldsymbol{\omega}') + E_i^3(E, \boldsymbol{\omega} \cdot \boldsymbol{\omega}') - E_i^2(E, \boldsymbol{\omega} \cdot \boldsymbol{\omega}')(1 - (\boldsymbol{\omega} \cdot \boldsymbol{\omega}')^2). \quad (6)$$

To sample the scatter direction with these phase functions, we used the idea of [5], and calculate the solution of the sampling equation for many random numbers and energy levels and store the results in two dimensional texture in the GPU memory. During simulation when the random number and the energy are available, the random scattering angle can be obtained by a texture lookup.

If a photon leaves the bounding box of the measured object, no more interaction will be calculated. If the ray intersects the detector, a measurement function is evaluated to determine the weight of the sample.

We implemented this algorithm and examined two different CT setups: a preclinical one used for small animal imaging (e.g.: pharmacy industry), and a clinical CT for human diagnostics. We found that the scattering is negligible for preclinical solutions, but it can be significant for the clinical case. Despite multi-GPU implementation generating a series of images with a noise statistic similar to a real acquisition took too much time, which can be explained with several problems. The detectors in a CT occupy just a smaller solid angle, so photons shot from the source do not necessarily hit them. This is true even for unscattered photons and becomes crucial for scattered photons. This means that a detector gets just small number of photons, and consequently the variance of its detected value will be high. The *efficiency*, i.e. the fraction of non-zero contribution samples is rather law. The second problem is that — similarly to nature — all photons are simulated independently, which means that we cannot reuse knowledge gathered when other photons are traced. For example, the simulation starts with the identification of the energy level of the source photon since material properties like cross sections depend on this value. Thus the generated path of this photon will correspond to only this initial photon energy, and when another photon of different energy is born, its path should be generated from scratch.

3 Hybrid Simulation

In order to speed up the physically motivated MC algorithm and improve its efficiency, our *hybrid simulation* uses different variance reduction techniques, which are discussed in the following subsections.

3.1 Main Part Separation

A significant part of detected values comes from the contribution of *direct*, i.e. unscattered photons. These direct photons travel along a linear path between the source and the detector and the probability that an emitted photon remains to be direct, i.e. it is neither absorbed nor scattered, can be expressed by an analytical formula:

$$m_d^{\text{direct}} = \int_{E_{\min}}^{E_{\max}} \int_{A_d} M_d(\boldsymbol{y}, E) \exp\left(- \int_{\boldsymbol{s}}^{\boldsymbol{y}} \sigma_t(\boldsymbol{l}, E) \mathrm{d}l \right) \Phi(\boldsymbol{\omega}_{\boldsymbol{s}\to\boldsymbol{y}}, E) \frac{\cos\theta_{\boldsymbol{s}\to\boldsymbol{y}}}{|\boldsymbol{s}-\boldsymbol{y}|^2} \mathrm{d}\boldsymbol{y}\mathrm{d}E,$$

where $\boldsymbol{\omega}_{\boldsymbol{s}\to\boldsymbol{y}}$ is the direction vector from source $\boldsymbol{s}$ to point $\boldsymbol{y}$ on the detector, and $\theta_{\boldsymbol{s}\to\boldsymbol{y}}$ is the angle between this direction and the surface normal of the detector. The integral is calculated with ray-marching. After separating the direct contribution, MC simulation needs to concentrate only on scattered contribution.

3.2 Forced Interaction

A photon flying not into the direction of the detectors or leaving the volume of interest without scattering is a loss from the point of view of efficiency of scattered contribution estimation. Our random sampler should guarantee that no such photon is generated, while the correct expectation is maintained by weighting. This modification keeps the sampling unbiased but the variance is significantly reduced. Interaction can be enforced by the modification of the free path sampling (Eq. 2). Knowing the initial position and direction of the photon, the maximum length $L_{\max}$ the photon can travel in the volume of interest can be determined by simple geometric calculations. This maximum traveling distance corresponds to a maximum random value $r_{\max}$ in Eq. 2:

$$r_{\max} = 1 - \exp\left(- \int_0^{L_{\max}} \sigma_t(\boldsymbol{x} + \boldsymbol{\omega} l, E) \mathrm{d}l \right)$$

Random values that are greater than $r_{\max}$ correspond to samples where the photon leaves the space without interaction. The probability of this is $1 - r_{\max}$. So, efficiency can be increased to 100% by modifying the sampling equation to

$$r_{\max} \cdot \text{rnd} = 1 - \exp\left(- \int_0^{L} \sigma_t(\boldsymbol{x} + \boldsymbol{\omega} l, E) \mathrm{d}l \right),$$

and weighting the contribution of each photon by $r_{\max}$. If the photon is already close to the boundary of the volume of interest, the weight of this method can be close to zero. Such cases can be handled with *next event estimation*, which means that a detector point is sampled and the sample point is deterministically connected to the interaction point. If the detector area is A_d, then the probability density of finding a single point $\boldsymbol{y}$ with uniform distribution is $1/A_d$, thus the probability density of direction $\boldsymbol{\omega}_{\boldsymbol{x}\to\boldsymbol{y}}$ is

$$p(\boldsymbol{\omega}_{\boldsymbol{x}\to\boldsymbol{y}}) = \frac{|\boldsymbol{x}-\boldsymbol{y}|^2}{A_d \cos\theta_{\boldsymbol{x}\to\boldsymbol{y}}}.$$

3.3 Absorption Handling with Weighting

When a photon interacts with the material, it can get absorbed with probability σ_a/σ_t. In case of absorption, the sample gets lost. The efficiency can be improved if absorption is not sampled but the photon is weighted with $1 - \sigma_a/\sigma_t$ at each interaction.

3.4 Polychromatic Particles

For a polychromatic X-ray source we should sample the spectrum of the source to obtain the initial energy of photons because cross sections and phase functions depend on this energy. However, when a complex particle path is established, it is worth reusing this path for other energy levels as well without starting the simulation from scratch. The possibility of reuse is provided by that cross sections can be factorized to a material dependent but energy independent factor and a material independent but energy dependent factor:

$$\sigma(\boldsymbol{x}, E) = \sigma(\boldsymbol{x}, E_r) \cdot \nu(E)$$

where E_r is an appropriate reference energy level. For example, the probability of the absorption due to the photoelectric effect is inversely proportional to the cube of the photon energy:

$$\sigma_a(\boldsymbol{x}, E) = \frac{\sigma_a(\boldsymbol{x}, E_r)}{(E/E_r)^3}.$$

The energy dependence of the incoherent scattering cross section can be computed from the scaling factor in the Klein-Nishina formula:

$$\sigma_i(\boldsymbol{x}, E) = \sigma_i(\boldsymbol{x}, E_r) \cdot \frac{\int_{-1}^{1} E_i(E,c) + E_i^3(E,c) - E_i^2(E,c)(1-c^2)\mathrm{d}c}{\int_{-1}^{1} E_i(E_r,c) + E_i^3(E_r,c) - E_i^2(E_r,c)(1-c^2)\mathrm{d}c}$$

where $c = \cos\theta = \boldsymbol{\omega} \cdot \boldsymbol{\omega}'$.

During the simulation of direct and scattered paths, we use ray marching to obtain the attenuation along line segments of a path. The attenuation is an exponent of a line integral:

$$A(E) = \exp\left(-\int \sigma(\boldsymbol{l}, E)\mathrm{d}\boldsymbol{l}\right) = \exp\left(-\nu(E)\int \sigma(\boldsymbol{l}, E_r)\mathrm{d}\boldsymbol{l}\right) = [A(E_r)]^{\nu(E)}.$$

This means that computing the attenuation separately for absorption, coherent and incoherent scatter on the reference energy level, the results can be transformed to arbitrary energy levels without the lengthy ray marching process.

4 Results and Future Work

We implemented the algorithms in CUDA, and used a GeForce GTX-590 in dual GPU setup in performance measurements. One GPU thread tracked a large amount of particles at the same time, and at least 512 threads are executed in parallel. The cross section tables were stored in GPU shared memory, the precalculated interaction tables were represented in 2D textures. The density and material distribution were stored in 3D textures, the calculated projection images and dose distribution were kept in GPU global memory.

For the preclinical scanner, the simulated phantom object was a 4 cm diameter, homogeneous water cylinder. We generated 180 projection images at 256×512 resolution. For the clinical study, we used the Zubal[1] phantom. We computed 180 projection images at 128×1024 resolution.

This new combined method uses significantly less particles in the Monte Carlo simulation, and executes ray marching where it is efficient on the GPU. We achieved 11 times speed-up for the preclinical scanner and 43 times acceleration for the human scanner. Figure 1 shows a slice from original Zubal phantom, a simulated projection, and also the dose distribution. The reconstructed slices are in Fig. 2.

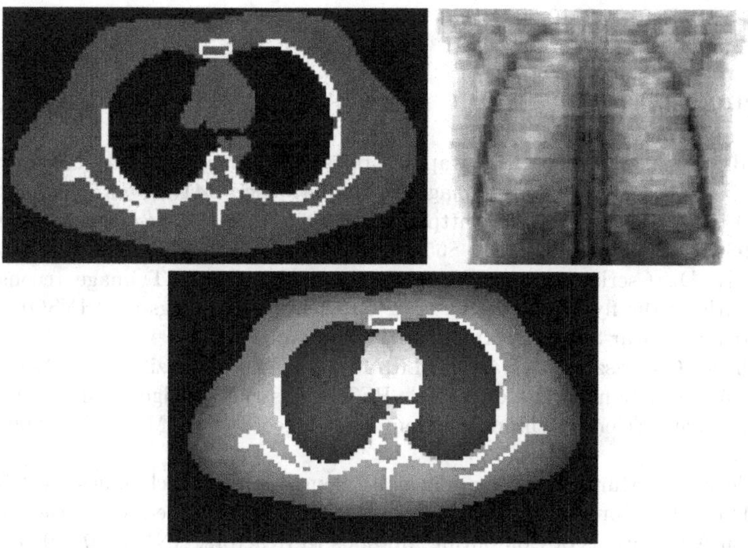

Fig. 1. Original Zubal phantom (left upper), a simulated projection (right upper), and simulated dose distribution (lower).

[1] http://noodle.med.yale.edu/zubal/data.htm

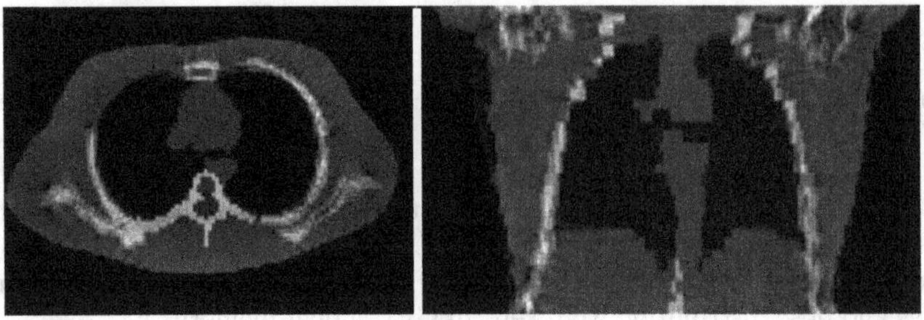

Fig. 2. Reconstructed slices

5 Conclusion

This paper proposed a hybrid MC simulation algorithm for particle transport, taking into account the special requirements of Computer Tomographs. Using various variance reduction techniques, we could significantly increase the efficiency of the algorithm.

Acknowledgement. This work has been supported by the OTKA K-104476 and by TÁMOP -4.2.2.B-10/1–2010-0009.

References

1. Euclid, S.: Computed tomography: physical principles and recent technical advances. J. Med. Imaging Radiat. Sci. **41**(2), 87–109 (2010). doi:10.1016/j.jmir.2010.04.001. http://www.sciencedirect.com/science/article/pii/S1939865410000317. ISSN 1939–8654
2. Légrády, D., Cserkaszky, A., Wirth A., Domonkos B.: PET image reconstruction with on the fly Monte Carlo using GPU. In: Proceedings of PHYSOR 2010, American Nuclear Society, Pittsburgh, Pennsylvania (2010)
3. Wirth, A., Cserkaszky, A., Kári, B., Légrády, D., Fehér, S., Czifrus, S., Domonkos, B.: Implementation of 3D Monte Carlo PET reconstruction algorithm on GPU. In: IEEE Nuclear Science Symposium Conference Record (NSS/MIC), pp. 4106–4109 (2009)
4. Woodcock E., Murphy T., Hemmings P., Longworth S.: Techniques used in the GEM code for Monte Carlo neutronics calculation. In: Proceedings of the Conference on Applications of Computing Methods to Reactors, ANL-7050 (1965)
5. Szirmay-Kalos, L., Tóth, B., Magdics, M., Légrády, D., Penzov, A.: Gamma photon transport on the GPU for PET. In: Lirkov, I., Margenov, S., Waśniewski, J. (eds.) LSSC 2009. LNCS, vol. 5910, pp. 435–442. Springer, Heidelberg (2010)
6. Szirmay-Kalos, L., Tóth, B., Magdics, M.: Free path sampling in high resolution inhomogeneous participating media. Comput. Graph. Forum **30**(1), 85–97 (2011)
7. Yang, C.N.: The Klein-Nishina formula and quantum electrodynamics. In: Yang, C.N. (ed.) Nishina Memorial Lectures. Lecture Notes in Physics, pp. 393–397. Springer, Berlin (2008)

8. Jakab, G., Rácz, A., Nagy, K.: High quality cone-beam CT reconstruction on the GPU. In: 8th KÉPAF Conference, Budapest, Hungary (2011)
9. Magdics, M., Szirmay-Kalos, L., Tóth, B., Csendesi, Á., Penzov, A.: Scatter estimation for PET reconstruction. In: Dimov, I., Dimova, S., Kolkovska, N. (eds.) NMA 2010. LNCS, vol. 6046, pp. 77–86. Springer, Heidelberg (2011)
10. Jakab, G., Huszár T., Csébfalvi, B.: Iterative CT Reconstruction on the GPU. In: Sixth Hungarian Conference on Computer Graphics and Geometry, Budapest (2012)

Analysis and Control of the Accuracy and Convergence of the ML-EM Iteration

Milán Magdics[1,2], László Szirmay-Kalos[2(✉)], Balázs Tóth[2], and Anton Penzov[3]

[1] University of Girona, Girona, Spain
[2] Budapest University of Technology, Budapest, Hungary
szirmay@iit.bme.hu
[3] Bulgarian Academy of Sciences, Sofia, Bulgaria

Abstract. In inverse problems like tomography reconstruction we need to solve an over-determined linear system corrupted with noise. The ML-EM algorithm finds the solution for Poisson noise as the fixed point of iterating a forward projection and a non-linear back projection. In tomography we have several hundred million equations and unknowns. The elements of the huge matrix are high-dimensional integrals, which cannot be stored, but must be re-computed with Monte Carlo (MC) quadrature when needed. In this paper we address the problems of how the quadrature error affects the accuracy of the reconstruction, whether it is possible to modify the back projection to speed up convergence without compromising the accuracy, and whether we should always take the same MC estimate or modify it in every projection.

1 Introduction

In Positron Emission Tomography (PET) we need to find the spatial density of radioactive tracer materials [4]. The tracer density is computed from the statistics of detected hits, which is the *inverse problem* of particle transport in scattering and absorbing media. Inverse problems are usually solved iteratively, by alternating the simulation of the forward problem and a correction step.

The output of the reconstruction is the activity density that is defined on a 3D voxel grid $\mathbf{x} = (x_1, x_2, \ldots, x_{N_{voxel}})$. The inputs of the reconstruction algorithm are the measured coincident photon hits in detector pairs, called LORs: $\mathbf{y} = (y_1, y_2, \ldots, y_{N_{LOR}})$. Using *maximum likelihood estimation* (ML-EM), vector $\mathbf{x}$ is found by maximizing the probability of the actually measured data $\mathbf{y}$ [5], alternating forward projection and back projection that together update estimate $\mathbf{x}^{(n)}$:

$$\text{Forward: } \tilde{\mathbf{y}} = \mathbf{A} \cdot \mathbf{x}^{(n)}, \qquad \text{Back: } \frac{\mathbf{x}^{(n+1)}}{\mathbf{x}^{(n)}} = \frac{\mathbf{A}^T \cdot \frac{\mathbf{y}}{\tilde{\mathbf{y}}}}{\mathbf{A}^T \cdot \mathbf{1}}$$

where vector division is defined in an element-wise manner, $\mathbf{A}_{LV}$ or $\mathbf{A}$ in short is the *System Matrix* (*SM*), which is the probability that a photon pair born in voxel V is detected by LOR L of expected value $\tilde{y}_L$.

I. Lirkov et al. (Eds.): LSSC 2013, LNCS 8353, pp. 170–177, 2014.
DOI: 10.1007/978-3-662-43880-0_18,

The true solution $\mathbf{x}^*$ of the reconstruction is the fixed point of this scheme, which satisfies:

$$\mathbf{A}^T \cdot \frac{\mathbf{y}}{\mathbf{A} \cdot \mathbf{x}^*} = \mathbf{A}^T \cdot \mathbf{1}. \quad (1)$$

In tomography we have several hundred million LORs and voxels, thus an SM may have more than 10^{16} elements. To handle the huge SM, it can be *factored* [3], and simpler physical phenomena may be obtained by on-the-fly analytic approximations. However, as these approximations are part of an iteration process, even a small error can accumulate unacceptably. Accurate and consistent estimations can be obtained with MC quadrature, but its computational burden is high [1]. There is an important difference between applying MC for estimating a quadrature and using MC as a part of an iteration process [6]. While the goal is an integral quadrature, the convergence rate is known and the error can be minimized by variance reduction techniques and increasing the number of samples. When MC is applied in an iteration, the accuracy of a single estimate is not so relevant since later iteration steps may correct the error of an earlier estimate. However, decreasing the samples in a single step means that we can make more iterations under the given budget of samples or computation time.

This paper examines the process of iteration with random MC estimates. Furthermore, we also investigate the potential of using simplified back projection matrices to speed up the projection.

2 Error and Convergence Analysis

SM estimations may be different in forward projection and back projection, and due to the numerical errors both differ from the exact matrix. Let us denote the forward projection SM by $\mathbf{F} = \mathbf{A} + \mathbf{\Delta F}$ and the back projection estimation by $\mathbf{B} = \mathbf{A} + \mathbf{\Delta B}$. We use the following notations for the normalized back projectors

$$\bar{\mathbf{A}}_{LV} = \frac{\mathbf{A}_{LV}}{\sum_{L'} \mathbf{B}_{L'V}}, \; \bar{\mathbf{B}}_{LV} = \frac{\mathbf{B}_{LV}}{\sum_{L'} \mathbf{B}_{L'V}} \implies \bar{\mathbf{B}} = \bar{\mathbf{A}} + \mathbf{\Delta}\bar{\mathbf{B}} \;\text{ and }\; \mathbf{\Delta}\bar{\mathbf{B}} \cdot \mathbf{1} = \mathbf{0}.$$

Note that $\mathbf{\Delta}\bar{\mathbf{B}} \cdot \mathbf{1} = \mathbf{0}$ is the consequence of the normalization of matrix $\mathbf{\Delta}\bar{\mathbf{B}}$, i.e. each element is divided by the row sum.

The question is how these approximations modify the convergence and the fixed point of the iteration scheme. Let us express the activity estimate in step n as $\mathbf{x}^{(n)} = \mathbf{x}^* + \mathbf{\Delta x}^{(n)}$. Substituting this into the iteration formula and replacing the terms by first order Taylor's approximations we obtain:

$$\mathbf{\Delta} x^{(n+1)} \approx \left(\mathbf{1} - \langle x_V^* \rangle \cdot \bar{\mathbf{B}}^T \cdot \langle \frac{y_L}{\tilde{y}_L^2} \rangle \cdot \mathbf{F}\right) \cdot \mathbf{\Delta} x^{(n)} + \langle x_V^* \rangle \cdot \bar{\mathbf{B}}^T \cdot \langle \frac{y_L}{\tilde{y}_L} \rangle \cdot \frac{\mathbf{\Delta}\tilde{\mathbf{y}}}{\tilde{\mathbf{y}}} - \mathbf{\Delta}\bar{B}^T \cdot \frac{\mathbf{y}}{\tilde{\mathbf{y}}}.$$

where $\langle x_V^* \rangle$ is an N_{voxel}^2 element diagonal matrix of true voxel values, $\langle \frac{y_L}{\tilde{y}_L^2} \rangle$ is an N_{LOR}^2 element diagonal matrix of ratios $\frac{y_L}{\tilde{y}_L^\alpha}$, and $\mathbf{\Delta}\tilde{\mathbf{y}} = \mathbf{\Delta F} \cdot \mathbf{x}$ is the error of the expected LOR hits made in the forward projection. Note that Taylor's approximation is acceptable only if function $1/y$ can be well approximated by a line in $\tilde{y}_L \pm \Delta\tilde{y}_L$. The iteration is convergent if

$$\mathbf{T} = \mathbf{1} - \langle x_V^* \rangle \cdot \bar{\mathbf{B}}^T \cdot \langle \frac{y_L}{\tilde{y}_L^2} \rangle \cdot \mathbf{F}$$

is a *contraction* after certain number of iterations (note that $\mathbf{T}$ is not constant but depends on $\mathbf{x}^{(n)}$ via $\tilde{y}_L$). Even for convergent iteration, the limiting value will be different from $\mathbf{x}^*$ due to the errors of the forward and back projections:

$$\Delta\mathbf{x}^{(\infty)} = \mathbf{S} \cdot \left(\Delta\bar{\mathbf{B}}^T \cdot \frac{\mathbf{y}}{\tilde{\mathbf{y}}} - \mathbf{A}^T \cdot \langle \frac{y_L}{\tilde{y}_L} \rangle \cdot \frac{\mathbf{d}\tilde{\mathbf{y}}}{\tilde{\mathbf{y}}} \right) \quad \text{where} \quad \mathbf{S} = \left(\mathbf{A}^T \cdot \langle \frac{y_L}{\tilde{y}_L^2} \rangle \cdot \mathbf{A} \right)^{-1}. \tag{2}$$

We can make several observations examining these formulae:

1. As measured hits y_L are Poisson distributed with expectations $\tilde{y}_L$, ratios $y_L/\tilde{y}_L$ have expected value 1 and variance $1/\tilde{y}_L$, thus $E[\Delta\bar{\mathbf{B}}^T \cdot \mathbf{y}/\tilde{\mathbf{y}}] = \mathbf{0}$ and even the variance caused by the back projector error diminishes when the measurement is high dose and thus the result is statistically well defined. Thus, for high dose measurement, the error made in forward projection is mainly responsible for the accuracy of the reconstruction, which adds the following error in each iteration step:

$$\langle x_V^* \rangle \cdot \bar{\mathbf{B}}^T \cdot \langle \frac{y_L}{\tilde{y}_L} \rangle \cdot \frac{\Delta\tilde{\mathbf{y}}}{\tilde{\mathbf{y}}} = \langle x_V^* \rangle \cdot \bar{\mathbf{B}}^T \cdot \langle \frac{y_L}{\tilde{y}_L} \rangle \cdot \frac{\Delta\mathbf{F} \cdot \mathbf{x}}{\tilde{\mathbf{y}}} \tag{3}$$

2. If the back projection accuracy is not so important, it is worth using a modified normalized SM $\bar{\mathbf{B}}$ to increase the contraction of $\mathbf{T}$ and thus speeding up the iteration.

3 ML-EM Iteration Using MC Quadrature

In tomography the size of the SM is enormous, thus matrix elements cannot be pre-computed and stored, but must be re-computed each time with MC quadrature when a matrix element is needed. It means that forward projector $\mathbf{F}$ and back projector $\bar{\mathbf{B}}$ are random variables. We use unbiased MC estimates, i.e.

$$E[\mathbf{F}] = \mathbf{A}, \quad E[\bar{\mathbf{B}}] = \bar{\mathbf{A}}.$$

As these estimates are re-made in every iteration, we can choose whether the same random estimate is used in all iterations, the estimate is modified in each iteration, or even between the forward projection and back projection. Note that as we have to re-compute the matrix elements anyway, the computation costs of different options are the same, the algorithms differ only in whether or not the seed of the random number generator is reset.

The contribution to the error of a single iteration is defined by Eq. 3. Errors of different iteration steps accumulate. However, the accuracy can be improved if

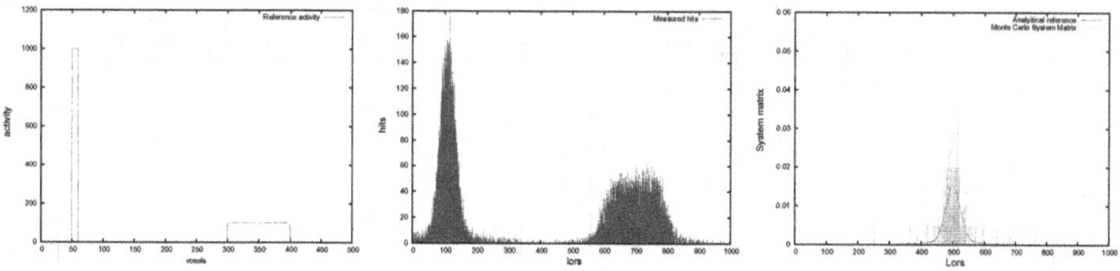

ref. activity | measured values | 1 column of the SM

Fig. 1. The measured function (left), the distribution of the hits in different LORs for the high dose case (middle), and the MC estimate of the SM (1 column is shown) obtained with 10^5 random samples.

we use an MC estimation where the expectation value of this contribution is zero since it means that the error contributions of different iteration steps compensate each other and we may get a precise reconstruction even with inaccurate SM estimates. So, our goal is to guarantee that

$$E\left[\langle x_V^*\rangle \cdot \bar{\mathbf{B}}^T \cdot \langle \frac{y_L}{\tilde{y}_L}\rangle \cdot \frac{\mathbf{\Delta F}\cdot \mathbf{x}}{\tilde{\mathbf{y}}}\right] = E\left[\langle x_V^*\rangle \cdot (\bar{\mathbf{A}}^T + \mathbf{\Delta}\bar{B}^T) \cdot \langle \frac{y_L}{\tilde{y}_L}\rangle \cdot \frac{\mathbf{\Delta F}\cdot \mathbf{x}}{\tilde{\mathbf{y}}}\right] = 0$$

which, taking into account that both the forward projector and the back projector are unbiased estimators, is held if

$$E\left[\mathbf{\Delta}\bar{\mathbf{B}}^T \cdot \langle \frac{y_L}{\tilde{y}_L}\rangle \cdot \mathbf{\Delta F}\right] = 0.$$

Note that this is true if the forward projector is statistically independent from the back projector, but is false when they are correlated. This means that it is worth using independent random samples in each iteration and re-sampling even between forward projection and back projections.

To demonstrate this, we analyze a simple analytical problem, when an SM of dimensions $N_{LOR} = 1000$ and $N_{voxel} = 500$ is defined as the sum of two Gaussian density functions.

The ground truth activity is another simple function of the left of Fig. 1. The measured values are obtained by sampling Poisson distributed random variables setting their means to the product of the SM and the reference activity. We examined a high dose and a low dose case, which differ in a factor of 10 of their activities. The middle of Fig. 1 shows the measurement of the high activity case. The error of the reconstruction is tested with random SM approximations, which are obtained by replacing the $5 \cdot 10^5$ analytical SM elements by unbiased MC estimates calculated with 10^4, 10^5, and 10^6 discrete samples in total, respectively.

In the first set of experiments we examine the L2 error of the reconstruction process of the *fixed case*, i.e. when the same SM approximation is used in all iteration steps (Fig. 2). These results indicate that working with the same MC estimate during an EM iteration is generally a bad idea. Reconstructing with a modified SM means that we altered the physical model, so the EM iteration

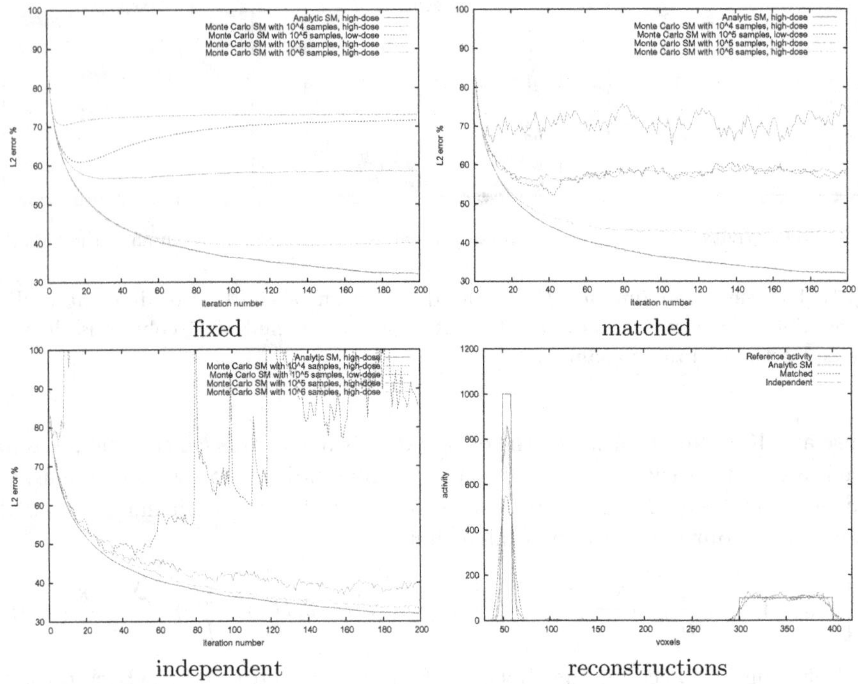

Fig. 2. L2 error curves of different sampling strategies and the reconstructed results. Fixed sampling takes the same samples in every projection. Matched sampling keeps the samples of forward even for back projection of the same iteration step. Independent sampling uses different random numbers in all projections.

converges to a different solution. *Matched sampling* takes the same samples in the forward and back projections of a single iteration but regenerates samples for each iteration. Matched sampling does not help, the error curves are quite similar to those of generated with fixed SM.

Independent sampling, where samples of forward projection are independent of the samples in back projection, has advantages and disadvantages as well. If the sample number is small, then the error curves are strongly fluctuating. The explanation is that matrix $\mathbf{T}$ is just probably a contraction, so the iteration have convergent and divergent stages. If the number of samples is higher, then the iteration gets similar to iterating with the analytic SM.

4 Speeding up the Convergence with Simplified Back Projectors

We concluded that the reconstruction accuracy of high dose measurements is just slightly affected by the accuracy of the back projector. In a special case when $\mathbf{B} = \mathbf{A} \cdot \mathbf{P}$ where $\mathbf{P}$ is an invertible square matrix of N_{voxel}^2 elements,

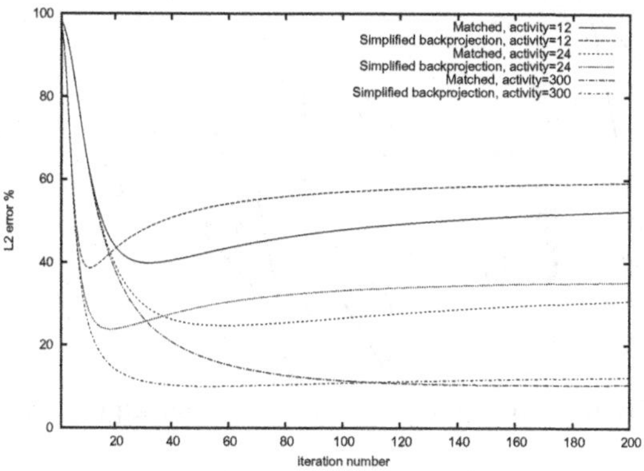

Fig. 3. Convergence in L2 for matched and simplified back projectors for different activities.

the fixed point is preserved, which can be seen if both sides of Eq. 1 are multiplied with matrix **P**. The convergence speed depends on the contraction of matrix **T**, which is strong if

$$\langle x_V^* \rangle \cdot \tilde{\mathbf{B}}^T \cdot \langle \frac{y_L}{\tilde{y}_L^2} \rangle \cdot \mathbf{A}$$

is close to the identity matrix. We need to find matrix **P** so that for every voxel V just the most significant $\mathbf{A}_{LV}$ elements are kept while others are replaced by zero during the multiplication with **P**. As the SM represents a sequence of physical phenomena, this means ignoring voxel space blurring effects.

Using the example of the previous section, we examined the convergence of the reconstruction for different activity levels (recall that back projection accuracy becomes important only for low dose measurements).

The results are shown by Fig. 3. Note that simplified and original back projectors converge to the same result, the approximation is more accurate when the measurement is of high dose. The initial convergence of the simplified back projector is much faster and it becomes poorer only when the iteration overfits the result and therefore the iteration is worth stopping anyway (such overfitting may be avoided with regularization).

5 Application in 3D Positron Emission Tomography

To test the presented method with a realistic 3D PET reconstruction, we took a LOR-centric, i.e. ray-based forward projection and a voxel-based back projection (Fig. 4). The forward projection samples are multiple rays or line segments connecting two uniformly distributed points on the two crystals' surfaces of the LOR. The line integral is evaluated between the two endpoints by sampling points being equal size but having a random starting offset. In back projection, a discrete point is sampled in each voxel and the solid angles subtended from this point by the two crystals' surfaces of each LOR are randomly sampled

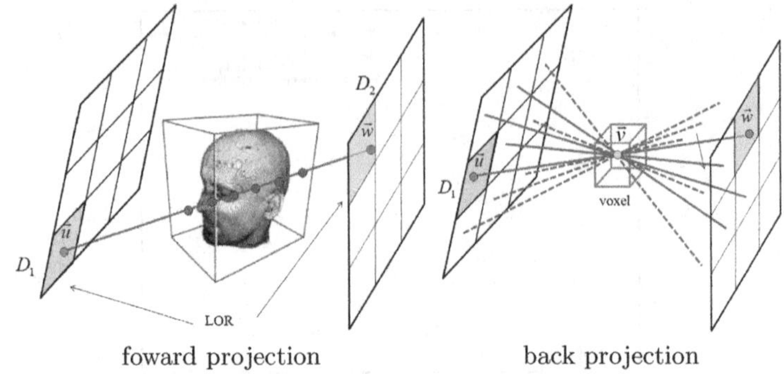

Fig. 4. Forward projection samples are line segments connecting uniformly distributed sample points on the crystals. Back projection samples are points in voxels and then points of detectors.

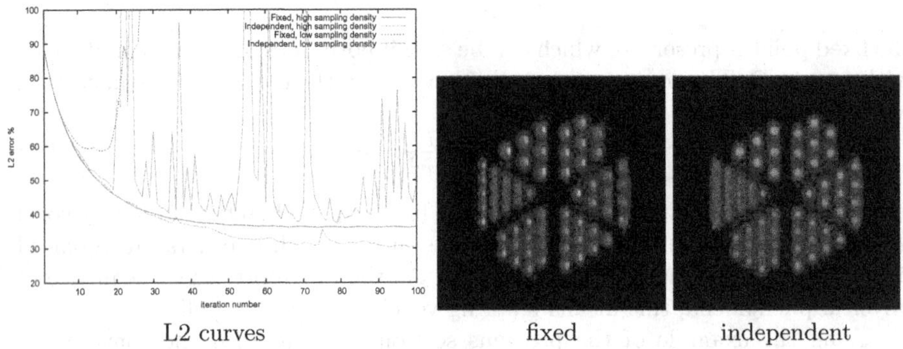

Fig. 5. L2 error curves and reconstructions of the Derenzo phantom.

by a line. The SM element can then be computed from the solid angle, and the total attenuation of the line between the two detectors.

We modeled the Mediso's small animal *nanoScan PET/CT* [2], which consists of twelve detector modules of 81×39 crystals detectors of surface size $1.12 \times 1.12\,\text{mm}^2$, thus the total number of LORs is 180 million when crystals of a module are connected by LORs to crystals of three opposite modules. We examined the *Micro Derenzo phantom* with rod diameters $1.0, 1.1, \ldots, 1.5\,\text{mm}$ in different segments. The Derenzo is virtually filled with 1.6 MBq activity and we simulated a 1000 s measurement.

The error curves and slices of reconstructions when the random number generator is reset in each iteration and when independent samples are generated are shown by Fig. 5. We considered the cases when integrals are estimated with many and with fewer samples. Note that for low sampling density fixed iteration diverges, while independent sampling oscillates. For higher sampling density, both of them are stable and independent sampling has better results.

Examining the reconstruction results we can observe that fixed sampling distorts the uniform activity distribution in rods, while independent sampling better preserves the ground truth.

6 Conclusion

This paper proposed the application of independent sampling and simplified back projector in inverse problems when elements are re-computed in each iteration step. The independent re-sampling has the advantage that it can gather more information about the system, probably not in a single step but as iteration proceeds. This additional information helps increase the accuracy. Independent sampling in forward and back projectors has a drawback that solution oscillates if the sample density is low, so sample numbers should be carefully selected. We also shown that if back projector is properly simplified, then not only its computation can be speeded up, but also the iteration can be made faster.

Acknowledgement. This work has been supported by the OTKA K-104476 and by TÁMOP - 4.2.2.B-10/1–2010-0009. The GATE simulation of the Derenzo phantom has been executed by Gergely Patay.

References

1. Buvat, I., Castiglioni, I.: Monte Carlo simulations in SPET and PET. J. Nucl. Med. **46**(1), 48–61 (2002)
2. nanoscan PET/CT. http://www.mediso.com/products.php?fid=2,11&pid=86
3. Qi, J., Leahy, R.M., Cherry, S.R., Chatziioannou, A., Farquhar, T.H.: High-resolution 3D Bayesian image reconstruction using the microPET small-animal scanner. Phys. Med. Biol. **43**(4), 1001 (1998)
4. Reader, A.J., Zaidi, H.: Advances in PET image reconstruction. PET Clin. **2**(2), 173–190 (2007)
5. Shepp, L., Vardi, Y.: Maximum likelihood reconstruction for emission tomography. IEEE Trans. Med. Imaging **1**, 113–122 (1982)
6. Szirmay-Kalos, L., Magdics, M., Tóth, B., Bükki, T.: Averaging and Metropolis iterations for positron emission tomography. IEEE Trans. Med. Imaging **32**(3), 589–600 (2013)

Stochastic Formulation of Newton's Acceleration

P. Schwaha[1,2], M. Nedjalkov[1], S. Selberherr[1], J.M. Sellier[3], I. Dimov[3], and R. Georgieva[3(✉)]

[1] Institute for Microelectronics, TU Wien Gußhausstraße 27-29/E360, 1040 Vienna, Austria
[2] AVL List GmbH, Hans-List-Platz 1, 8020 Graz, Austria
[3] Institute for Parallel Processing, Bulgarian Academy of Sciences, Acad. G.Bontchev str. Bl25A, 1113 Sofia, Bulgaria
rayna@parallel.bas.bg

Abstract. The theoretical equivalence of the Wigner and ballistic Boltzmann equations for up to quadratic electric potentials provides the convenient opportunity to evaluate stochastic algorithms for the solution of the former equation with the analytic solutions of the latter equation - Liouville trajectories corresponding to acceleration due to a constant electric field. The direct application of this idea is impeded by the fact that the analytic transformation of the first equation into the second involves generalized functions. In particular, the Wigner potential acts as a derivative of the delta function which gives rise to a Newtonian accelerating force. The second problem is related to the discrete nature of the Wigner momentum space. These peculiarities incorporate unphysical effects in the approximate Wigner solution, which tends to the Boltzmann counterpart in a limiting case only.

Operator mechanics are the established representation of quantum mechanics, where the evolution of expectation values of physical quantities are given by operators $\hat{A}$ along with a commutator bracket and an Hamiltonian operator. This is a departure from the classical descriptions of phase spaces where the Hamiltonian and the Poisson bracket impress the space's geometry on the equations of motion. The Wigner formalism [1] is a return to a phase space description of quantum systems and their evolution. In the case of quantum systems the phase space accommodates features not found in the classical case. Where the Liouville component of the Boltzmann equation is governed by the first derivative of the electric potential - the electric field, quantum evolution is determined by the Wigner integral, which accounts for the entire potential in a nonlocal manner. By performing a Taylor expansion of the Wigner integral it is possible to link derivatives of the potential to powers of $\hbar$. Classical systems then appear by a limit of $\hbar \to 0$, which in this case causes only the linear component, the electric field, to remain. This also means that in the case of a linear potential the Wigner equation reduces to the ballistic Boltzmann equation and the nature, classical or quantum, is determined purely by the initial condition. From the multitude of purely mathematically available solutions only a subset is physically viable. In classical systems this requires all states to be nonnegative, which also allows

I. Lirkov et al. (Eds.): LSSC 2013, LNCS 8353, pp. 178–185, 2014.
DOI: 10.1007/978-3-662-43880-0_19,

for a direct interpretation as densities. In the case of a quantum system, on the other hand, this means that a legitimate quantum state must conform to the uncertainty relation [2,3].

Since the nature only depends on the initial state, it offers a test facility where simulations for quantum simulations may be examined and tested. The generalized functions required for the treatment of the Wigner transport have made application difficult in direct numerical treatment.

Here, an ensemble particle algorithm for general transport regimes determined by initial and boundary conditions and transients is presented. It uses annihilation of indistinguishable particles at consecutive time steps and is rooted in the use of a quantized momentum space. Force effects are introduced exclusively using the Wigner potential, so that individual particles are unaccelerated as they evolve according to the fieldless Liouville operator of the Wigner equation.

1 Monte Carlo Algorithm

The foundation of the algorithm is the reformulation of the problem as a Fredholm integral equation of the second kind, which can be solved by a Neumann series. The series is evaluated using a Monte Carlo approach. Newton trajectories link the individual terms of the series, where the integral kernel is applied repeatedly. Thus the scheme can be presented as comprised by the two major components

- Evolve along a Newton trajectory
- Apply the kernel as a scattering event
- Record

The Newton trajectories used are exactly the same as in a purely classical setup without force.

A representation of the kernel responsible for the scattering transitions, is needed. We employ a discretized version [4] of the originally continuous Wigner potential. Choosing a finite coherence length L also fixes a finite delta in momentum space proportional $\sim 1/L$. When using wave numbers to represent momentum space, as is customary in the field of solid state physics, this yields $\Delta k = \pi/L$. This discrete approach allows for the identification of momenta with integers. The scattering introduced in this manner deviates greatly from classical transport simulations. Where the kernel in classical transport descriptions is positive definite, this is no longer the case in the quantum setting. This requires the introduction of opposing signs for the particles to accommodate the action of the kernel on a particle. Where in a classical case the kernel will act on any given particle and simply may change its state in a possibly discontinuous manner, the Wigner kernel will spawn a pair of new particles from the initial particle. The interaction with the Wigner potential occurs after traversing the trajectory for a certain time. The interaction can be expressed as:

$$\Gamma(\mathbf{r}, \mathbf{m}, \mathbf{m}') = V_W^+(\mathbf{r}, \mathbf{m} - \mathbf{m}') - V_W^-(\mathbf{r}, \mathbf{m}' - \mathbf{m}) + \gamma(\mathbf{r})\delta_{\mathbf{m},\mathbf{m}'} \qquad (1)$$

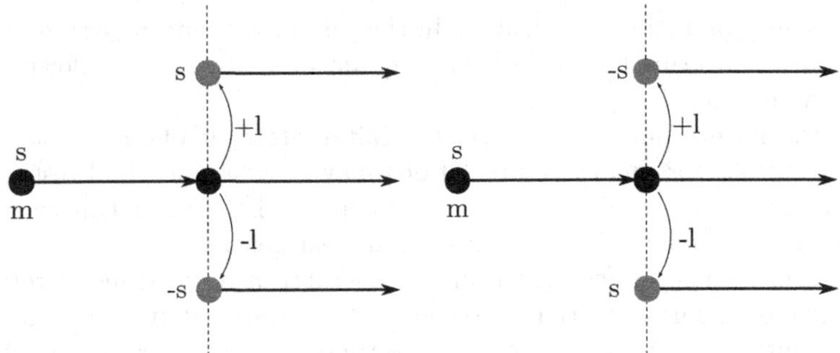

Fig. 1. The signs of the generated particles depend not only on the sign of the original particle, but also on the sign of the Wigner potential at the generating location.

m and **m**′ are integers representing the initial and final nodes in momentum space, respectively. The antisymmetry of the Wigner integral, which acts as the scattering source, enforces

$$V_w^+ = max(V_w, 0), \qquad V_W^-(\mathbf{m}) = V_W^+(-\mathbf{m}) \tag{2}$$

so that the generation of the two new particles is actually linked. A single choice l remains, which is the offset of the new states from the original momentum node m. When choosing the signs of the generated particles, the sign of V_W must be considered. In case V_W is positive, the particle at the position of the final node $m + l$ retains the sign of the generating particle, while the particle at $m - l$ is constructed with the inverted sign. In case the Wigner potential is negative, the signs of the newly spawned particles are flipped. This process of generation is depicted in Fig. 1. The left side shows the case of $V_W > 0$, while the flip of the generated signs is shown in the right part of the figure.

The particles are of opposite signs and each moved in momentum space from the original particle's momentum. In addition to the two newly spawned particles due to interaction with the Wigner potential, the original particle continues along the original trajectory unperturbed, due to the δ function in Eq. 1, as is also depicted in Fig. 1. Thus, after such an scattering event, instead of the single original particle, now three particles must be processed, each of which not only needs to be processed further but can also generate new particle in the same manner. Thus the total number of particles increases exponentially. This makes it essential to have a means of reducing the number of particles again.

The mechanism counteracting the generation of particles employed in the presented algorithm is annihilation at the time of recording; which marks the end of any chosen time step. Two particles at the same position at the same time but of opposite sign not only have no net contribution to the value of a recording estimator, they also annihilate each other. This means that neither of the two opposing particles will continue to evolve. Thus the number of overall particles is reduced by two. Since it is necessary for two particles to be at the same place at the same time, the phase space must be subdivided into cells in order to make annihilation feasible, as otherwise the probability of two particles meeting would be zero. The discrete momentum space is already inherently subdivided into a finite set of cells identifiable by the integer indexed nodes. The number of nodes in the momentum component is linked to the resolution selected in space. The number of nodes required to fill the characteristic length L used to obtain V_W is identical to the number of nodes required for momentum quantization.

2 Numerical Analysis

The outlined algorithm for quantum transport is applied to a test configuration consisting of a single peak in the centre of the phase space. It thus is a discrete and finite model of a delta function. From a physical point of view this setting violates the uncertainty relation inherent to quantum phenomena, but since the setting is such that the nature of the system is determined entirely by the condition placed within it, it is expected that this classical initial state should also yield classical results, even as it is subjected to quantum evolution.

Figure 2 shows how the number of particles depends on the length of the time step and how particles are generated not only from the initial particles. The initial particles, comprising Generation 0, create an avalanche of subsequent particles. As the time step is increased the number of generated particles and with it the computational burden increases drastically. This can be attributed not only to the fact, that for a fixed probability of interaction with V_W, more particles will be spawned by the primary particles, but also to the circumstance, that the generated particles themselves have a long time span to again generate new particles. The maximum of particles is reached in the 7th generation, after which the number of new particles declines, since the average time remaining until the end of the time step makes generation less probable.

Since the computational burden increases so dramatically when extending the time step, the question arises, if calculations using a series of several shorter time steps will produce results matching a single longer one. As can be seen in Fig. 3, the agreement between the different strategies to reach an absolute time is excellent.

This indicates that by substituting one long time step by several shorter ones it is possible to save considerable computational effort, as after each time step the number of particles is reduced by annihilation.

Figure 3 also shows oscillations of the distribution including negative values. This nonphysical behaviour is attributed to the fact that the initial condition used here is in violation of the uncertainty principal required in the quantum setting.

Furthermore, Fig. 3 also shows the process of transition from the initially occupied node at 0 to the node at 1. The transition is worth examining in more detail, since it reveals that the transition now occurs as in cellular automata [5].

Figure 4 shows a particle's transition from one node to another node. The intermediate time regime, where the initial peak has already decreased, while

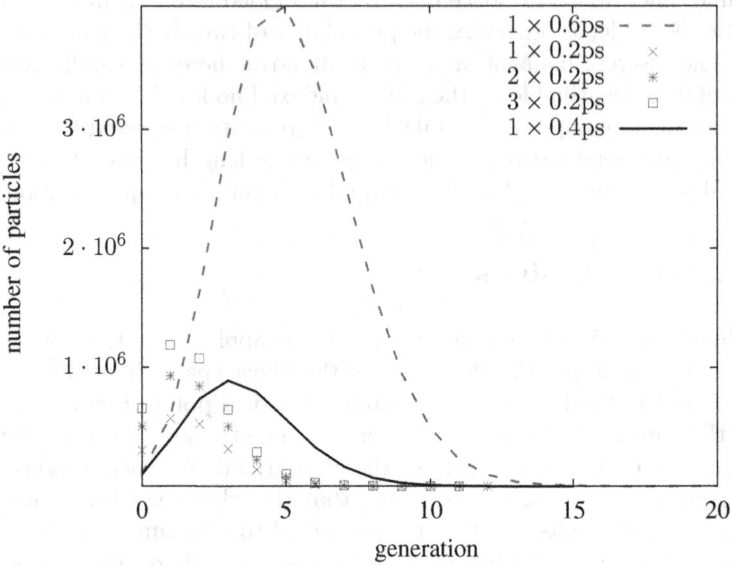

Fig. 2. The length of a time step determines the number of generation before annihilation. When tripling the length of the time step, the shorter time step's number appears almost negligible.

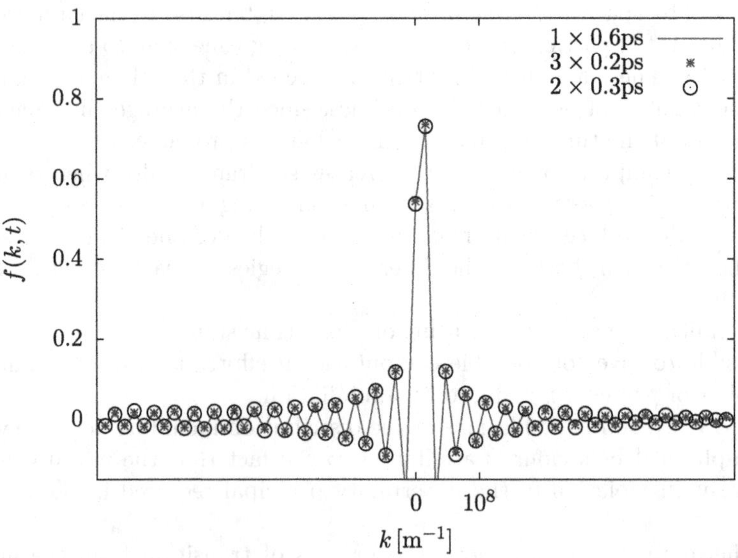

Fig. 3. Choosing a single long time step yields the same results as choosing several shorter time steps, as can be observed for the case of a single 0.6 ps time step vs two 0.3 ps and three 0.2 ps time steps.

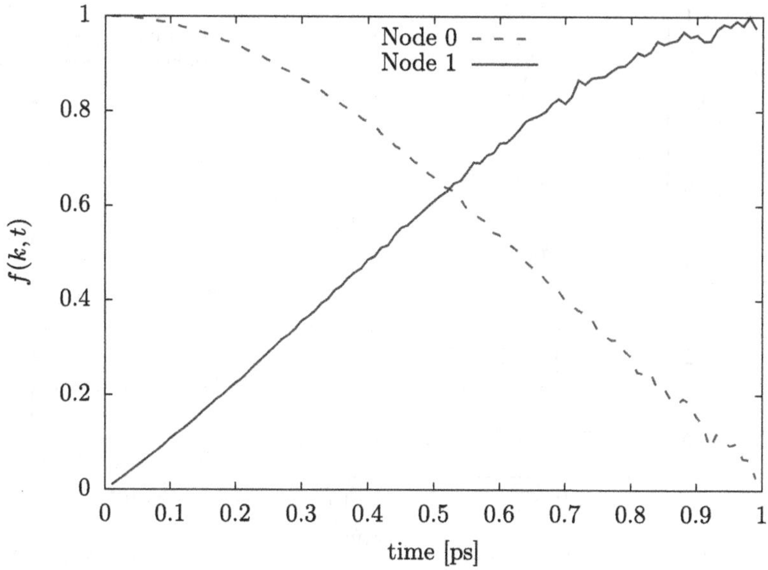

Fig. 4. The densities at the Node 0, which holds the initial condition, and Node 1, which is the first to be occupied. Particles are *not* transferred continuously from node to node until all reach the destination node. Instead the node occupancy is controlled by a generation of positive and negative particles.

the target has not yet fully formed, is entirely controlled by generation of positive and negative particles and a subsequent annihilation. An analysis of this process will be presented in the sequel. We now point to Fig. 5 which shows the reconstitution of the initial distribution at the target node: the momentum of the particles increases, which corresponds to acceleration but this time without an explicite action of the field. Another interesting physical aspect of the density is the substantial reduction of the spurious oscillations observed in Fig. 3. The quantum system is closest to the classical counterpart at dicrete points in time and momentum. At the limit $\Delta k \to 0$ which corresponds to infinite L and thus the continuous case the behaviour becomes classical.

Investigating the manner in which the algorithm moves particles in more detail it is possible to elucidate how the force term is accommodated by purely relying on the mechanism of the Wigner potential V_W. Given a number of particles N_0 located at a given node of a phase space grid a certain number will be scattered as they evolve along a Newton trajectory. Even without knowing this number it is possible to examine the ratios of how they will be distributed if we know V_W. In the case under investigation, V_W at the nodes was calculated for a constant electric field to the form of:

$$V_W(n) = \frac{(-1)^{n+1}}{n} \quad \forall n \neq 0, \qquad 0 \quad n = 0 \tag{3}$$

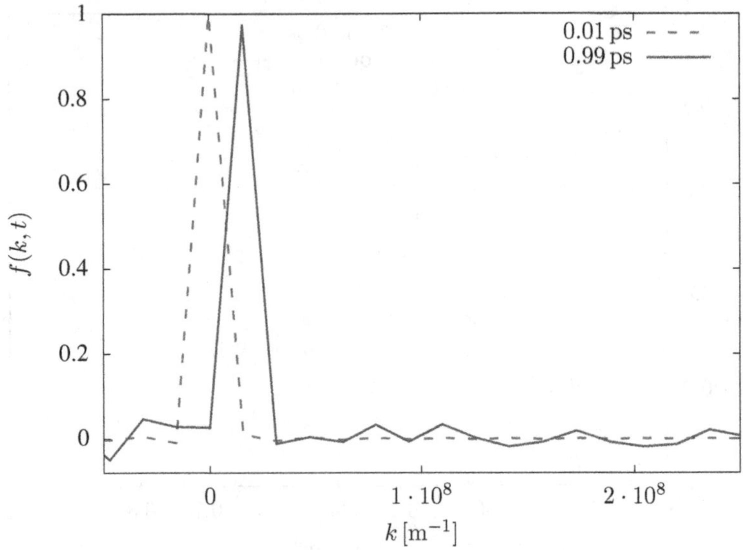

Fig. 5. At times corresponding to Newton's law, the peaks not only reappear but also the nonphysical oscillations are dampened to a minimum.

By following the described algorithm the following table is obtained, which shows how many particles are assigned to which node. The common factors are denoted by N_x, where x gives the generation of the particle. The sign of the factor indicates the signs assigned to the particles generated for the particular node. The table reveals several peculiarities: The signs of the contribution to the

0	0	0	N_0	0	0	0
$-\frac{1}{3}N_1$	$\frac{1}{2}N_1$	$-N_1$	x	N_1	$-\frac{1}{2}N_1$	$\frac{1}{3}N_1$
$-\frac{1}{2}N_2$	N_2	x	$-N_2$	$\frac{1}{2}N_2$	$-\frac{1}{3}N_2$	$\frac{1}{4}N_2$
$\frac{1}{4}N_2$	$-\frac{1}{3}N_2$	$\frac{1}{2}N_2$	$-N_2$	x	N_2	$-\frac{1}{2}N_2$
$-\frac{1}{2}N_2$	x	$\frac{1}{2}N_2$	$-\frac{1}{2}\frac{1}{2}N_2$	$\frac{1}{3}\frac{1}{2}N_2$	$-\frac{1}{4}\frac{1}{2}N_2$	$\frac{1}{5}\frac{1}{2}N_2$
$-\frac{1}{5}\frac{1}{2}N_2$	$-\frac{1}{4}\frac{1}{2}N_2$	$\frac{1}{3}\frac{1}{2}N_2$	$-\frac{1}{2}\frac{1}{2}N_2$	$\frac{1}{2}N_2$	x	$-\frac{1}{2}N_2$
x	$-\frac{1}{3}N_2$	$\frac{1}{2}\frac{1}{3}N_2$	$-\frac{1}{3}\frac{1}{3}N_2$	$\frac{1}{4}\frac{1}{3}N_2$	$-\frac{1}{5}\frac{1}{3}N_2$	$\frac{1}{6}\frac{1}{3}N_2$
$\frac{1}{6}\frac{1}{3}N_2$	$-\frac{1}{5}\frac{1}{3}N_2$	$\frac{1}{4}\frac{1}{3}N_2$	$-\frac{1}{3}\frac{1}{3}N_2$	$\frac{1}{2}\frac{1}{3}N_2$	$-\frac{1}{3}N_2$	x

originating node are negative, while they are all positive for the first node to the right. For the remainder of the nodes, the signs are mixed. This supports the conjecture that the algorithm indeed allows to model the effects of force by purely relying on the interaction with V_W. The initial peak is moved by being annihilated by the particles of opposing sign and reconstructed at the neighbouring node.

3 Conclusion

An algorithm for quantum transport has been presented. Its main features include the use of a quantized momentum space and discrete selection rules for the scattering. The discrete nature of the momentum component works very well in conjunction with the employed annihilation scheme, which helps to reduce the number of generated signed particles.

It was shown numerically that it is possible to utilize short time steps to iterate to a longer duration in a stable manner. This is important due to the significant increase of particle generation with the extension of the time step.

Furthermore, an explanation has been provided, how this algorithm accelerates particles without explicitly incorporating a force term.

Finally an interesting mixture of quantum and classical phenomena have been observed in the behavior of the modeled transport process.

Acknowledgements. This work has been supported by the Austrian Science Fund Project FWF-P21685-N22, the EC FP7 Project AComIn (FP7-REGPOT-2012-2013-1), and Bulgarian NSF Grants DTK 02/44/2009.

References

1. Wigner, E., Margenau, H.: Symmetries and reflections. Sci. Essays Am. J. Phys. **35**(12), 1169–1170 (1967)
2. Tatarskii, V.I.: The Wigner representation of quantum mechanics. Sov. Phys. Usp. **26**, 311–327 (1983)
3. Dias, N.C., Prata, J.N.: Admissible states in quantum phase space. Ann. Phys. **313**, 110–146 (2004)
4. Nedjalkov, M., Vasileska, D.: Semi-discrete 2d Wigner-particle approach. J. Comput. Electron. **7**, 222–225 (2008)
5. Zandler, G., Di Carlo, A., Krometer, K., Lugli, P., Vogl, P., Gornik, E.: A comparison of Monte Carlo and cellular automata approaches for semiconductor device simulation. IEEE Electron Dev. Lett. **14**(2), 77–79 (1993)

The Role of Annihilation in a Wigner Monte Carlo Approach

Jean Michel Sellier[1], Mihail Nedjalkov[2],
Ivan Dimov[1(✉)], and Siegfried Selberherr[2]

[1] Institute for Parallel Processing, Bulgarian Academy of Sciences,
Acad. G.Bontchev, Bl 25A, 1113 Sofia, Bulgaria
ivdimov@bas.bg

[2] Institute for Microelectronics, TU Wien Gußhausstraße 27-29/E360,
1040 Vienna, Austria

Abstract. The Wigner equation provides an interesting mathematical limit, which recovers the constant field, ballistic Boltzmann equation. The peculiarities of a recently proposed Monte Carlo approach for solving the transient Wigner problem, based on generation and annihilation of particles are summarized. The annihilation process can be implemented at consecutive time steps to improve the Monte Carlo resolution. We analyze theoretically and numerically this process applied to the simulation of important quantum phenomena, such as time-dependent tunneling of a wave packet through potential barriers.

1 Introduction

In the theory of carrier transport involving the concept of phase-space, there are strong similarities between classical and quantum regimes. From this perspective, the Wigner theory is a promising approach for the simulation of fully quantum transport phenomena in semiconductor devices. Investigations of this approach have been carried out in the recent past. Efforts have been performed to reuse successful ideas of the semi-classical transport regime.

Eventually, two particle models were derived. The first model, an ensemble Monte Carlo (MC) technique based on particles endowed with an affinity, has proved to be a reliable method. Unfortunately it needs heavy computational resources [1]. The second model is a single particle MC approach, based on the ergodicity of the system and thus restricted to stationary regimes determined by the boundary conditions [2]. As compared to the previous one, it has very different attributes, related to the generation of particles endowed with a sign. Particles are created in the phase space and are consecutively evolved to the boundary.

A generalization of the second approach has been recently developed. This new method exploits the concepts of momentum quantization and indistinguishable particles. These concepts treat properly quantum mechanics, entangled with the notions of classical trajectories, particle ensemble, and particle-with-sign

I. Lirkov et al. (Eds.): LSSC 2013, LNCS 8353, pp. 186–193, 2014.
DOI: 10.1007/978-3-662-43880-0_20,

generation giving rise to a time-dependent, fully quantum transport model which naturally includes both open and closed boundary conditions along with general initial conditions.

Focusing on this model we introduce a recently developed technique, a time-dependent renormalization by means of particles annihilation, which allows the simulation of time-dependent quantum phenomena. We apply this novel technique to the tunneling of a wave packet through a barrier.

2 Stochastic Aspects

The developed Monte Carlo algorithm aims to evaluate the expectation value $\langle A\rangle(t)$ of a generic physical quantity A at given evolution time t.

2.1 Monte Carlo Algorithm

By reformulating the semi-discrete Wigner equation as a Fredholm integral equation of the second kind one can derive a proper adjoint equation, which allows to express $\langle A\rangle$ as a series of terms:

$$\langle A\rangle(t) = \int_0^{\infty} dt' \int \mathrm{d}x_i \sum_{m'=-\infty}^{\infty} f_i(x_i, m') e^{-\int_0^{t'} \gamma(x_i(y))dy} g(x_i(t'), m', t') \quad (1)$$

where $x'(y) = x_i(y)$ is the Newtonian trajectory of a field-less particle, $g(x, m, t)$ is solution of the adjoint equation, represented by its resolvent series [2]:

$$\begin{aligned} g(x', m', t') = {} & A_\tau(x', m', t') + \\ & \int_{t'}^{\infty} \mathrm{d}t \sum_{m=-\infty}^{\infty} g(x'(t), m, t) \Gamma(x', m, m') e^{-\int_{t'}^{t} \gamma(x'(y))dy} \end{aligned} \quad (2)$$

with

$$\begin{aligned} A_\tau(x, m, t) &= A(x, m)\delta(t-\tau) \\ \gamma(r) &= \sum_{m=-\infty}^{\infty} V_w^+(r, m) \\ \Gamma(r, m, m') &= V_w^+(r, m-m') - V_w^+(r, m'-m) + \gamma(r)\delta_{m,m'} \\ V_w^+ &= \begin{cases} V_w & \text{if } V_w > 0 \\ 0 & \text{otherwise.} \end{cases} \end{aligned}$$

The approach is a generalization of the stationary counterpart [2] for general transient transport problems. The terms in the resolvent expansion of (1) are ordered by consecutive applications of the kernel in (2), which is used to construct the transition probability for the numerical Monte Carlo trajectories. The latter consist of pieces of Newton trajectories linked by a change of the momentum number from m to m' according to Γ. Thus a numerical trajectory may be associated with a moving particle which undergoes scattering events. The exponent gives the probability that the particle remains on its Newton trajectory provided that the scattering rate is γ. If not scattered until time τ, the particle contributes to $< A >_0 (\tau)$ with the value $f_i(x_i, m')g(x_i(\tau), m')$. If scattered, the particle contributes to a next term in the expansion. It may be shown that a single particle contributes to one and only one term in the expansion and thus is an independent realization of the random variable sampling the whole sum: $< A > (t)$ is then estimated by averaging over N particles.

It is possible to give another interpretation to the equations due to the special appearance of Γ: after any free flight the trajectory forks into three contributive terms, the initial trajectory and two new trajectories with wave-vector offset equal to $l = m - m'$ and $-l$ around the initial wave number. Thus, a particle picture of a Monte Carlo branching process can be associated.

The process of creation of new couples is exponential [2]. The upper part of Fig. 1 shows what happens for one particle which moves in a constant field. The field magnitude is chosen to be 10339.2 V/m, so that the particle goes to the next k-space cell after 1 ps, and the total length of the domain is $1\,\mu m$. The following interpretation can be given: one initial particle evolves until, at some given random time, it generates a couple of positive and negative particles (recorded at 200 fs). Those particles evolve along with their parent particle, until they also generate couples. In turn, couples generate other couples until the final time is reached. This process triggers an avalanche of particle creations. From one particle, one ends up with 111 new particles at a recording time of 650 fs. We chose this final time since it is of the same order of typical final times to reach stationary regimes of practical nanodevices. Now, realistic simulations also involve several milions of initial particles. They rapidly generate a number of particles, which is of the order of several billions, a daunting numerical task.

2.2 The Annihilation Technique

The renormalization technique by annihilation described below represents a way to avoid this situation.

The renormalization technique is based on the fact that, in the Wigner formalism, particles are now mathematical objects deprived of any physical meaning. They are independent realizations of certain probability distributions related to the time-dependent solution of the Wigner equation. The unknown (and main object) of our problem is the Wigner quasi-distribution function. Furthermore, particles having the same wave number and position (i.e. being in the same cell of the phase-space), with opposite signs, do not contribute in the calculation of the average value for a macroscopic variable. They simply cancel the contribution of each other reciprocally or, in other words, annihilate.

These observations highlight the possibility of removing, periodically, the particles which do not contribute to the calculation of the Wigner function, i.e. one can apply a renormalization of the numerical average of the Wigner quasi-distribution by means of a particles annihilation process. This is in accordance to the Markovian character of the evolution to progress at consecutive time steps so that the final solution at a given time step becomes the initial condition for the next step. This method can be implemented as follows: one fixes a recording time at which the annihilating particles are checked. Particles found to annihilate are immediately removed from the simulation. This considerably reduces the number of particles at every recording time, as shown in the bottom of Fig. 1. In this case, the simulation ends up with 35 particles instead of 111 particles. This technique is advantageous since the reduction of the number of particles by means of the renormalization process allows the simulation of time-dependent technologically relevant cases with the Wigner formalism. Indeed, it is known

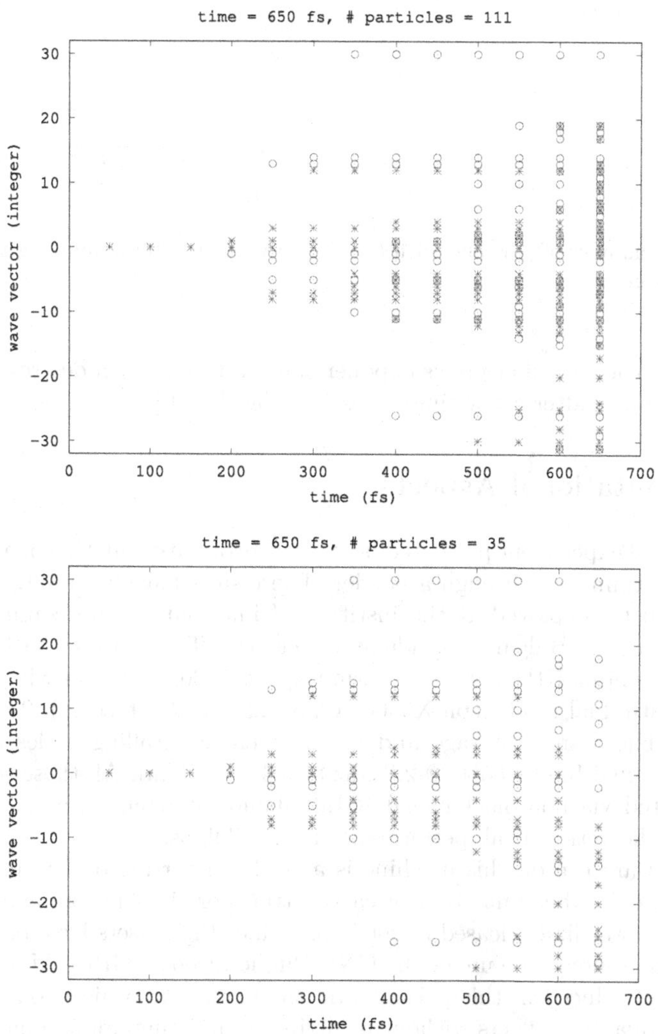

Fig. 1. Particle creation avalanche process with (bottom) and without (top) the annihilation technique. * = positive sign particles, o = negative sign particles.

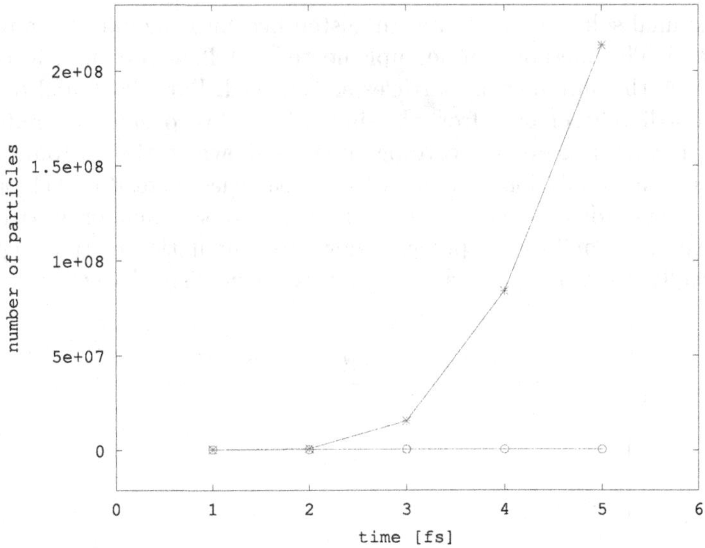

Fig. 2. Total number of particles with (dots) and without (stars) applying the annihilation technique.

that the particle's number grows exponentially in time, which destroys the simulation feasibility after a few time steps (see Fig. 2 and [2]).

3 Computational Aspects

The numerical experiment presented here consists of the simulation of a Gaussian wave packet tunneling through a barrier. The results have been obtained using the HPC cluster deployed at the institute of information and communication technologies of the Bulgarian academy of sciences. This cluster consists of two racks which contain HP Cluster Platform Express 7000 enclosures with 36 blades BL 280 C with dual Intel Xeon X5560 @ 2.8 Ghz (total 576 cores), 24 GB RAM per blade. There are 8 storage and management controlling nodes 8 HP DL 380 G6 with dual Intel X5560 @ 2.8 Ghz and 32 GB RAM. All these servers are interconnected via non-blocking DDR Infiniband interconnect at 20 Gbps line speed. The theoretical peak performance is 3.23 Tflops.

The software run on this machine is a modified version of Archimedes, the GNU package for the simulation of carrier transport in semiconductor devices [3]. This code was first released in 2005, and, since then, users have been able to download the source code under the GNU Public License (GPL). Many features have been introduced in this package. In this particular project, our aim is to include the quantum effects without recurring to quantum approximations such as effective potentials which are not satisfying when applied to nanodevices. The code is entirely developed in C and optimized to get the best performance from

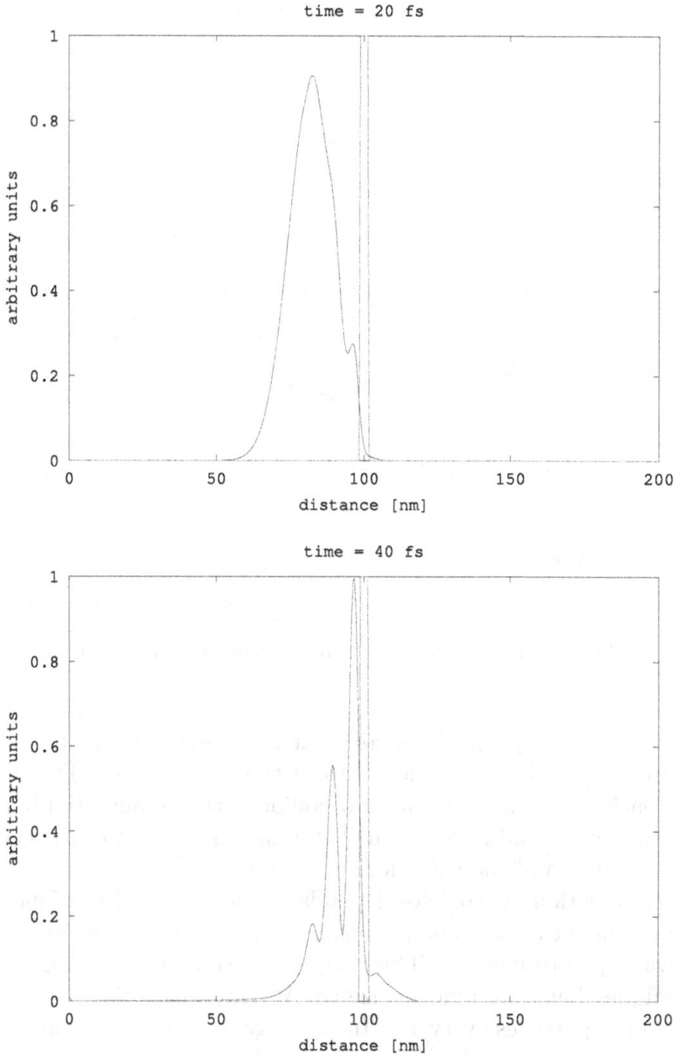

Fig. 3. Evolution at different times of a wave packet in proximity of a potential barrier.

the hardware [4]. The results of the new version will be, eventually, deployed on nano-archimedes website, dedicated to the simulation of nanodevices (see [5]).

The wave packet moves in a domain that has a potential barrier in the center. The domain is 200 nm long, and the barrier is 3 nm thick with an energy equal to 0.2 eV. The corresponding initial Wigner function reads:

$$f_W^0(r,n) = Ne^{-\frac{(r-r_0)^2}{\sigma^2}} e^{-(n\Delta k - k_0)^2 \sigma^2} \quad (3)$$

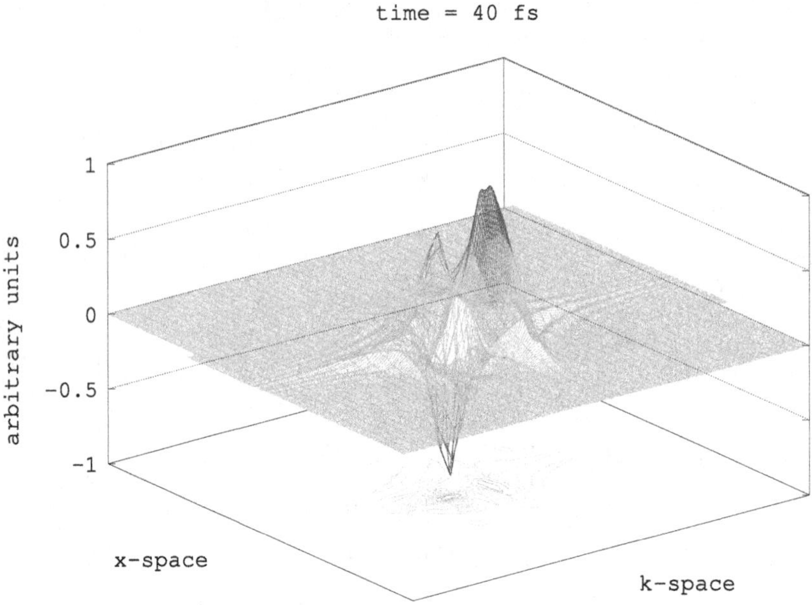

Fig. 4. Wigner distribution of a wave packet after 40 fs evolution.

N, k_0, r_0 and σ are, respectively, a constant of normalization, the initial wave vector, the initial position, and the width of the wave packet. The parameters defining the packet are chosen to make it collide with the potential barrier. The initial wave vector is equal to $4 \cdot 10^8\, m^{-1}$, the initial position of the center wave packet is at 68.5 nm, and the value for σ is 10 nm.

The rate of creation of particles depends on the shape of the function $\gamma(x)$. For the simulation of this particular transport problem, one needs at least 40 fs to observe relevant quantum effects. This is achievable only by applying a renormalization technique. Thus, we renormalize the Wigner quasi-distribution function by annihilation of particles every 1 fs. In this way one can reach long final times even equal to 80 fs.

Figures 3 and 4 show the wave packet at different evolution times, and the Wigner distribution function, respectively. The density obtained from the Wigner equation ressembles the time-dependent Schroedinger counterpart. A thorough comparison between Wigner and Schroedinger models will be presented somewhere else. Furthermore, the smoothness of the solution indicates the low stochastic noise of the method. Finally, Fig. 4 shows negative values in the calculated solution in the proximity of the barrier. This is a manifest of the quantum nature of the Wigner function as compared to the non-negative semi-classical distribution function.

4 Conclusions

We presented a renormalization technique for the Wigner quasi-distribution function, which involves the annihilation of particles at chosen recording times. We applied the method to the simulation of the tunneling process, where a Gaussian wave packet interacts with an energetic barrier. We have shown that, due to this renormalization technique, it is possible to calculate the time-dependent solution of the Wigner equation in realistic transport problems governed by quantum effects.

Acknowledgment. This work has been supported by the the project EC AComIn (FP7-REGPOT-2012-2013-1), and Bulgarian NSF Grants DTK 02/44/2009 and Austrian Science Fund Project FWF-P21685-N22.

References

1. Querlioz, D., Dollfus, P.: The Wigner Monte Carlo Method for Nanoelectronic Devices - A Particle Description of Quantum Transport and Decoherence. ISTE-Wiley, London (2010)
2. Nedjalkov, M., Kosina, H., Selberherr, S., Ringhofer, C., Ferry, D.K.: Unified particle approach to Wigner-Boltzmann transport in small semiconductor devices. Phys. Rev. B **70**, 115319 (2004)
3. Sellier, J.M.: www.gnu.org/software/archimedes
4. Sellier, J.M.: Archimedes, the Free Monte Carlo simulator (2012). http://arxiv.org/abs/1207.6575arXiv:1207.6575
5. Sellier, J.M.: www.nano-archimedes.com

Theoretical and Algorithmic Advances in Transport Problems

The Reference Solution Approach to Hp-Adaptivity in Finite Element Flux-Corrected Transport Algorithms

Melanie Bittl(✉) and Dmitri Kuzmin(✉)

Institute of Applied Mathematics LS III, Dortmund University of Technology,
Vogelpothsweg 87, 44227 Dortmund, Germany
melanie.bittl@math.tu-dortmund.de,
kuzmin@math.uni-dortmund.de

Abstract. This paper presents an *hp*-adaptive flux-corrected transport algorithm based on the reference solution approach. It features a finite element approximation with unconstrained high-order elements in smooth regions and constrained $Q1$ elements in the neighborhood of steep fronts. The difference between the reference solution and its projection into the current (coarse) space is used as an error indicator to determine the local mesh size h and polynomial degree p. The reference space is created by increasing the polynomial degree p in smooth elements and h-refining the mesh in nonsmooth elements. The smoothness is determined by a hierarchical regularity estimator based on discontinuous higher-order reconstructions of the solution and its derivatives. The discrete maximum principle for linear/bilinear finite elements is enforced using a linearized flux-corrected transport (FCT) scheme. p-refinement is performed by enriching a continuous bilinear approximation with continuous or discontinuous basis functions of polynomial degree $p \geq 2$. The algorithm is implemented in the open-source software package HERMES. The use of hierarchical data structures that support arbitrary level hanging nodes makes the extension of FCT to *hp*-FEM relatively straightforward. The accuracy of the proposed methodology is illustrated by a numerical example for a two-dimensional benchmark problem with a known exact solution.

Keywords: Hp-adaptation · Flux-corrected transport · Finite elements · Maximum principles

1 Introduction

The standard Galerkin finite element discretization of convection-dominated problems is known to produce nonphysical oscillations in the neighborhood of steep fronts. In the case of linear/bilinear finite elements the discrete maximum principle can be enforced by using an algebraic flux correction scheme [4,5]. This method has its origin in the multidimensional flux-corrected transport (FCT)

I. Lirkov et al. (Eds.): LSSC 2013, LNCS 8353, pp. 197–204, 2014.
DOI: 10.1007/978-3-662-43880-0_21,

algorithm [12] and limits the coefficients of the discrete transport operator in order to make it local extremum diminishing. Since the design of FCT-like limiters for higher-order elements is complicated [6], these elements may only be used in smooth regions where no oscillations can occur and so the use of the standard Galerkin method is safe.

In the present paper we construct the reference space by increasing p in smooth elements of the coarse space and h-refining all others. Algebraic flux correction of FCT type is only applied in matrix blocks corresponding to $Q1$ elements. The regularity of the solution and its derivatives is estimated using a parameter-free smoothness indicator [3] which regards the p-th derivative as smooth if the (discontinuous) high-order reconstruction is bounded by the original values at the centers of surrounding elements. The difference between the computed reference solution and its projection into the coarse space is used as an error indicator for the hp-refinement of the coarse space.

The algorithm is implemented in the hp-adaptive finite element library HERMES [9] which provides hierarchical basis functions and supports the use of irregular meshes with arbitrary-level hanging-nodes [10,11]. A 2D test problem is solved using hp-adaptivity with continuous and discontinuous p-enrichment. It is shown that the proposed framework leads to a sharp resolution of steep fronts and preserves the optimal order of accuracy at smooth peaks.

2 Algebraic Flux Correction

We consider the following unsteady linear convection equation

$$\frac{\partial u}{\partial t} + \nabla \cdot (\mathbf{v}u) = 0 \qquad \text{in } \Omega, \tag{1}$$

where u is the concentration of a conserved quantity, $\mathbf{v}$ is a given velocity and Ω is a bounded domain. This equation (1) is of hyperbolic type and is endowed with suitable initial and boundary conditions.

For the discretization in space we use the (continuous) Galerkin finite element method. This yields a system of equations which can be written in generic form

$$M_C \frac{\mathrm{d}u}{\mathrm{d}t} = Ku, \tag{2}$$

where u is the vector of unknowns, $M_C = \{m_{ij}\}$ is the consistent mass matrix and $K = \{k_{ij}\}$ is the discrete transport operator.

Since the standard Galerkin discretization can produce non-physical oscillations we enforce the discrete maximum principle by using algebraic flux correction [4]. For this reason, we replace the matrix M_C with its lumped counterpart

$$M_L := \operatorname{diag}\{m_i\}, \qquad m_i = \sum_j m_{ij}. \tag{3}$$

Next, we fix K by adding a discrete diffusion operator $D = \{d_{ij}\}$ with [4,5]

$$d_{ii} := -\sum_{j\neq i} d_{ij}, \qquad d_{ij} = \max\{-k_{ij}, 0, -k_{ji}\} = d_{ji}, \forall j \neq i \tag{4}$$

so that $K + D$ has no negative off-diagonal entries and D has zero row sums.

Using (3) and (4), we can split the semi-discrete Galerkin scheme (2) as follows:

$$M_L \frac{\mathrm{d}u}{\mathrm{d}t} = (K + D)u + f(u), \tag{5}$$

where $f(u)$ is the sum of antidiffusive terms that may cause over- and undershoots

$$f(u) = (M_L - M_C)\frac{\mathrm{d}u}{\mathrm{d}t} - Du. \tag{6}$$

It can be shown [4] that each component of (6) admits a flux decomposition of the form

$$f_i = \sum_{j\neq i} f_{ij}, \qquad f_{ij} = \left(m_{ij}\frac{\mathrm{d}}{\mathrm{d}t} + d_{ij}\right)(u_i - u_j), \qquad \forall j \neq i. \tag{7}$$

Some fluxes may create undershoots or overshoots in proximity to troubled cells. For this reason the contribution of these fluxes must be limited in order to make the antidiffusive term local extremum diminishing. After this correction, the generic form of the semi-discrete problem becomes

$$M_L \frac{\mathrm{d}u}{\mathrm{d}t} = (K + D)u + \bar{f}(u), \tag{8}$$

where $\bar{f}(u)$ is a vector containing the sums of limited antidiffusive fluxes

$$\bar{f}_i = \sum_{j\neq i} \alpha_{ij} f_{ij}, \qquad 0 \leq \alpha_{ij} \leq 1. \tag{9}$$

A well-designed flux limiter produces $\alpha_{ij} \approx 1$ in smooth regions and $\alpha_{ij} = 0$ in troubled cells. For our computations we use a nonclipping version of Zalesak's limiter [4,12].

If we discretize (8) in time using the Crank-Nicolson method, we obtain a nonlinear algebraic system

$$Au^{n+1} = Bu^n + \bar{f}, \tag{10}$$

where $\bar{f}$ is the fully discrete counterpart of the limited antidiffusive term,

$$A = \frac{1}{\Delta t}M_L - \frac{1}{2}(K + D) \quad \text{and} \quad B = \frac{1}{\Delta t}M_L + \frac{1}{2}(K + D). \tag{11}$$

Since the implicit part of $\bar{f}$ depends on the unknown solution u^{n+1}, it must be linearized or calculated in an iterative way. In this paper, we use a flux-corrected transport (FCT) algorithm [5] in which the raw antidiffusive fluxes

$$f_{ij} = \left(\frac{m_{ij}}{\Delta t} + \frac{d_{ij}}{2}\right)(u_i^H - u_j^H) - \left(\frac{m_{ij}}{\Delta t} - \frac{d_{ij}}{2}\right)(u_i^n - u_j^n) \tag{12}$$

are evaluated using the unconstrained Galerkin solution u^H.

3 Regularity Estimator

The restriction of u_h to a single element K of the computational mesh $\mathcal{T}_h$ is given by a linear or multilinear shape function $u_h|_K$. To estimate the smoothness of u_h in a neighborhood of cell K, we consider a linear approximation of the form

$$\hat{u}_h(\mathbf{x}) = u_h(\mathbf{x}_c) + R_h u_h(\mathbf{x}_c) \cdot (\mathbf{x} - \mathbf{x}_c), \tag{13}$$

where $\mathbf{x}_c$ denotes the center of K and $R_h : V_h \to V_h \times V_h$ is a gradient recovery operator. In contrast to ∇u_h, the reconstructed gradient is continuous, and $R_h u_h(\mathbf{x}_c)$ depends on the data in all elements that share a vertex with K. In this paper, we construct $R_h u_h = (R_h^1 u_h, R_h^2 u_h)^T$ using an L^2 projection.

The shape functions given by (13) define a **discontinuous** piecewise-linear approximation $\hat{u}_h$. The difference between u_h and $\hat{u}_h$ may serve as a smoothness indicator. We regard the solution on cell K as smooth if the value of $\hat{u}_h$ at each vertex $\mathbf{x}_i \in K$ is bounded by the values of u_h at the centers of surrounding elements

$$u_i^{\min} < \hat{u}_h(\mathbf{x}_i) < u_i^{\max}, \qquad \forall \mathbf{x}_i \in K, \tag{14}$$

where

$$u_i^{\max} := \max\{u_h(\mathbf{x}_c) \mid \exists K \in \mathcal{T}_h : \mathbf{x}_i, \mathbf{x}_c \in K\}, \tag{15}$$

$$u_i^{\min} := \min\{u_h(\mathbf{x}_c) \mid \exists K \in \mathcal{T}_h : \mathbf{x}_i, \mathbf{x}_c \in K\}. \tag{16}$$

Note that the inequalities in (14) are strict, which implies that a constant function is not regarded as smooth.

Since conditions (14)–(16) are violated at the local maxima and minima of u_h, all cells containing these extrema are marked as "troubled" [15]. To distinguish between smooth peaks and spurious undershoots/overshoots, the regularity estimator must be applied to each component of the gradient $\nabla u_h = (u_x, u_y)^T$.

Building on the analogy with [13,14], we use the derivatives of the recovered gradient $R_h u_h = (R_h^1 u_h, R_h^2 u_h)^T$ to define the linear reconstructions

$$\hat{g}_h^1(\mathbf{x}) = \frac{\partial u_h}{\partial x}(\mathbf{x}_c) + \nabla(R_h^1 u_h)(\mathbf{x}_c) \cdot (\mathbf{x} - \mathbf{x}_c), \tag{17}$$

$$\hat{g}_h^2(\mathbf{x}) = \frac{\partial u_h}{\partial y}(\mathbf{x}_c) + \nabla(R_h^2 u_h)(\mathbf{x}_c) \cdot (\mathbf{x} - \mathbf{x}_c). \tag{18}$$

Similarly to (14)–(16), the gradient is regarded as smooth if the values of $\hat{g}_h^1$ and $\hat{g}_h^2$ at all vertices of K are bounded by the centroid values of $\frac{\partial u_h}{\partial x}$ and $\frac{\partial u_h}{\partial y}$, respectively.

As shown in [13,14], no shock capturing is required if the finite element solution u_h and/or both components of its gradient are found to be smooth.

4 Reference Solution Approach

The reference solution approach [9–11] is based on the assumption that the reference solution is a better approximation to the exact solution than the solution in the current space. The reference space/mesh is usually created by increasing the polynomial degree and refining the mesh size of the current (coarse) space. Since the FCT limiter works only for linear/bilinear ($P1/Q1$) finite elements we restrict the p-enrichment to smooth elements only and h-refine all non-smooth elements. Here the smoothness is determined by the regularity estimator from the last section which labels an element as smooth if the solution or both components of its gradient are found to be smooth on this element. For the computation of the reference solution we apply FCT only to matrix blocks associated with the $P1/Q1$ approximation.

The algorithm works iteratively and we have to update an initial/coarse space and the numerical solution in each time step in the following way [1]:

1. Adaptivity loop:
   (a) Construct the reference space by increasing p in smooth elements and h-refining non-smooth elements.
   (b) Project the old solution u^n into the reference space and compute the reference solution u^{ref}.
   (c) Project the reference solution into the coarse space and calculate the difference between the reference solution and its projection.
   (d) Adjust the local mesh size and polynomial degree of the coarse mesh/space according to the error indicator in (c). (Details can be found in [9]).
2. Set $u^{n+1} = u^{ref}$.

Note that this algorithm doesn't coarsen the mesh, so after a certain number of time steps the coarse space is reset to the initial space. Furthermore we use a constrained L^2-projection [1,7] to transfer the previous timestep solution to the reference space.

5 Numerical Examples

In our numerical study, we consider the *solid body rotation* problem [2,8]. We solve equation (1) with the incompressible velocity field $\mathbf{v}(x, y) = (0.5-y, x-0.5)$ that describes a counterclockwise rotation about the center of $\Omega = (0,1)\times(0,1)$.

The exact solution to the solid body rotation problem reproduces the initial state u_0 exactly after each full revolution ($t = 2\pi k,\ k \in \mathbb{N}$). Hence, the challenge of this test is to preserve the shape of u_0. Following LeVeque [8], we consider a slotted cylinder, a sharp cone, and a smooth hump.

Figure 1 shows the *hp*-adaptive solution at $t = 2\pi$ calculated using the reference solution approach. The regularity estimator labeled most elements of the hump and the cone as smooth, so that p was increased in these elements. Note that the maximal polynomial degree $p = 2$ was limited by the error indicator which means that no higher polynomial degree was found necessary for a higher accuracy. The elements at the top of the cone and inside the cylinder were handled as non-smooth using $p = 1$.

Figure 2 compares the *hp*-adaptive solution with an h-adaptive solution at $t = 2\pi$. Both solutions exhibit a similar resolution of the cylinder. In both cases

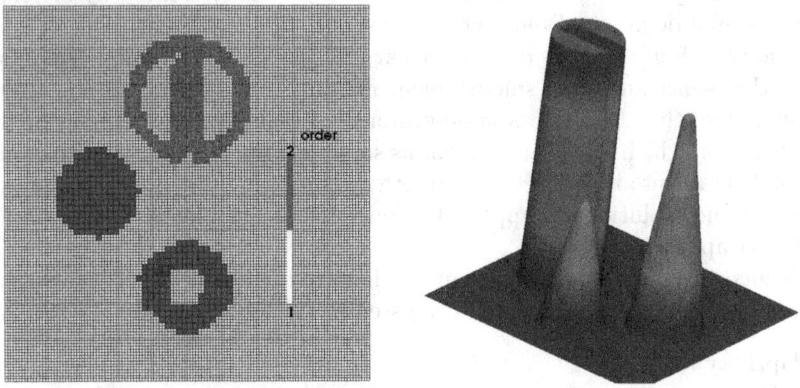

Fig. 1. Solid body rotation: reference solution at $t = 2\pi$

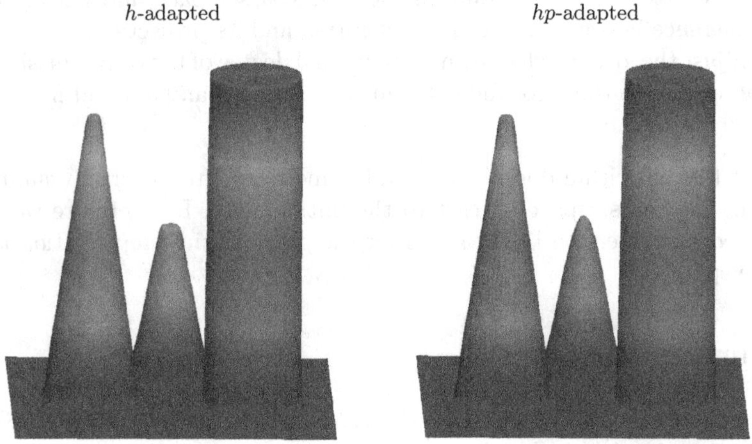

Fig. 2. Solid body rotation: solution at $t = 2\pi$

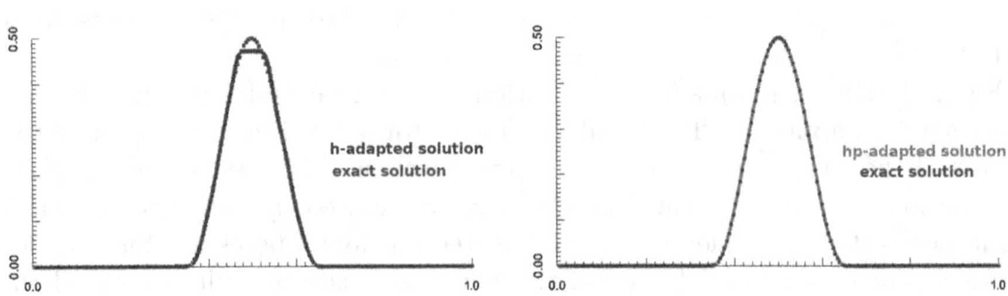

Fig. 3. Solid body rotation: comparison of exact and numerical solution at $x = 0.25$

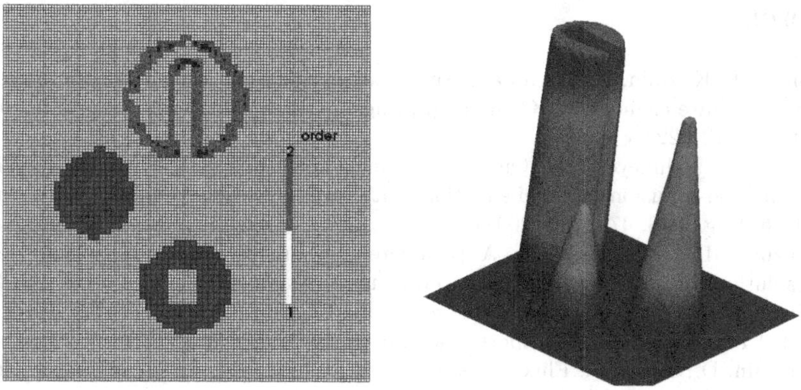

Fig. 4. Solid body rotation: reference solution at $t = 2\pi$ with discontinuous edge functions

the peak of the cone is smeared. Since the FCT limiter is deactivated in $Q2$ elements, no peak clipping is visible at the hump of the *hp*-solution. This can be seen in more detail in Fig. 3 where the profiles of the numerical and exact solutions along the line $x = 0.25$ are presented.

Figure 4 shows the *hp*-adaptive reference solution at $t = 2\pi$ using a new approach for high-order elements (CG1-DG2) where we enrich continuous linear finite elements with discontinuous higher-order basis functions. This leads to discontinuities across the edges of $Q2$ elements but continuity is preserved at the vertices. The result is comparable to the pure continuous solution of Fig. 1.

6 Conclusion

In this paper, we combined algebraic flux correction of FCT type with *hp*-adaptivity for finite element approximations to convection-dominated transport problems. In particular, we presented an *hp*-adaptivity algorithm based on the reference solution approach. The proposed scheme enables *hp*-adaptivity in smooth elements and *h*-refinement in non-smooth elements. The FCT limiter is applied in low-order ($Q1$) elements, whereas the unconstrained Galerkin approximation is used in high-order elements. The presented numerical results illustrate the benefits of this approach in the context of a linear convection equation. The possibility of enriching linear finite elements with discontinuous higher-order basis functions was explored. The CG1-DG2 approach was found to be as accurate as CG1-CG2. Future work will focus on the numerical analysis of this promising new approach to *hp*-adaptivity.

Acknowledgements. The authors would like to thank Pavel Solin (University of Nevada, Reno) for inspiring discussions. This research was supported by the German Research Association (DFG) under grant KU 1530/6-1.

References

1. Bittl, M., Kuzmin, D.: An hp-adaptive flux-corrected transport algorithm for continuous finite elements. J. Comput. (Springer) (2012). http://dx.doi.org/10.1007/s00607-012-0223-y
2. John, V., Schmeyer, E.: On finite element methods for 3D time-dependent convection-diffusion-reaction equations with small diffusion. Comput. Meth. Appl. Mech. Eng. **198**, 475–494 (2008)
3. Kuzmin, D., Schieweck, F.: A parameter-free smoothness indicator for high-resolution finite element schemes. Cent. Eur. J. Math. **11**(8), 1478–1488 (2013)
4. Kuzmin, D.: Algebraic flux correction I. In: Kuzmin, D., Löhner, R., Turek, S. (eds.) Flux-Corrected Transport, 2nd edn. Springer, Berlin (2012)
5. Kuzmin, D., Turek, S.: Flux correction tools for finite elements. J. Comput. Phys. **175**, 525–668 (2002)
6. Kuzmin, D.: On the design of algebraic flux correction schemes for quadratic finite elements. Comput. Appl. Math. **218**(1), 79–87 (2008)
7. Kuzmin, D., Möller, M., Shadid, J.N., Shashkov, M.: Failsafe flux limiting and constrained data projections for equations of gas dynamics. J. Comput. Phys. **229**, 8766–8779 (2010)
8. LeVeque, R.J.: High-resolution conservative algorithms for advection in incompressible flow. SIAM J. Numer. Anal. **33**, 627–665 (1996)
9. Solin, P., et al.: Hermes - Higher-Order Modular Finite Element System (User's Guide). http://hpfem.org/
10. Solin, P., Segeth, K., Dolezel, I.: Higher-Order Finite Element Methods. Chapman and Hall/CRC Press, London (2003)
11. Solin, P., Cerveny, J.: Automatic hp-adaptivity with arbitrary-level hanging nodes. Research report, University of Texas at El Paso (2006)
12. Zalesak, S.T.: Fully multidimensional flux-corrected transport algorithms for fluids. J. Comput. Phys. **31**, 335–362 (1979)
13. Kuzmin, D.: A vertex-based hierarchical slope limiter for p-adaptive discontinuous Galerkin methods. J. Comput. Appl. Math. **233**, 2077–3085 (2010)
14. Kuzmin, D.: Slope limiting for discontinuous Galerkin approximations with a possibly non-orthogonal Taylor basis. Int. J. Numer. Methods Fluids **71**(9), 1178–1190 (2013)
15. Michoski, C., Mirabito, C., Dawson, C., Wirasaet, D., Kubatko, E.J., Westerink, J.J.: Adaptive hierarchic transformations for dynamically p-enriched slope-limiting over discontinuous Galerkin systems of generalized equations. J. Comput. Phys. **230**, 8028–8056 (2010)

Optimization-Based Conservative Transport on the Cubed-Sphere Grid

Kara Peterson[1(✉)], Pavel Bochev[1], and Denis Ridzal[2]

[1] Numerical Analysis and Applications, Sandia National Laboratories, Albuquerque, NM 87185-1320, USA
[2] Optimization and Uncertainty Quantification, Sandia National Laboratories, Albuquerque, NM 87185-1320, USA
{kjpeter,pbboche,dridzal}@sandia.gov

Abstract. Transport algorithms are highly important for dynamical modeling of the atmosphere, where it is critical that scalar tracer species are conserved and satisfy physical bounds. We present an optimization-based algorithm for the conservative transport of scalar quantities (i.e. mass) on the cubed sphere grid, which preserves local solution bounds without the use of flux limiters. The optimization variables are the net mass updates to the cell, the objective is to minimize the discrepancy between these variables and suitable high-order cell mass update (the "target"), and the constraints are derived from the local solution bounds and the conservation of the total mass. The resulting robust and efficient algorithm for conservative and local bound-preserving transport on the sphere further demonstrates the flexibility and scope of the recently developed optimization-based modeling approach [1,2].

1 Introduction

We present a conservative, and monotone optimization-based transport algorithm and its application to a cubed sphere grid. The method is based on an incremental remap approach [6] with an optimization-based remap step at the core. The efficient mass variable mass target (MVMT) algorithm [5] is used for the remap step. In this approach a high-order mass update is used as the target for the optimization and local solution bounds and mass conservation are guaranteed through the constraints.

Numerical results are shown for standard transport tests on the sphere. A similar incremental remap transport algorithm in which the remap step is implemented using the flux-corrected remap (FCR) [9] provides a benchmark for the numerical studies. The studies show that the optimization-based algorithm is computationally competitive with the benchmark and is more robust in the case of complex flows.

Sandia National Laboratories is a multi-program laboratory managed and operated by Sandia Corporation, a wholly owned subsidiary of Lockheed Martin Corporation, for the U.S. Department of Energy's National Nuclear Security Administration under contract DE-AC04-94AL85000.

I. Lirkov et al. (Eds.): LSSC 2013, LNCS 8353, pp. 205–212, 2014.
DOI: 10.1007/978-3-662-43880-0_22, © Springer-Verlag Berlin Heidelberg 2014

2 Optimization-Based MVMT Transport

We briefly review the MVMT optimization-based transport algorithm for the scalar transport problem

$$\frac{\partial \rho}{\partial t} + \nabla \cdot \rho \mathbf{v} = 0 \quad \text{on} \quad \Omega \times [0, T] \qquad \text{and} \quad \rho(\mathbf{x}, 0) = \rho^0(\mathbf{x}). \tag{1}$$

Our approach combines the incremental remapping idea [6] with an optimization-based remap step [3–5].

Consider a partition $C(\Omega)$ of Ω into cells $c_i, i = 1, \ldots, C$. Let $m_i = \int_{c_i} \rho(\mathbf{x}, t)$ dV, $\mu_i = \int_{c_i} dV$, and $\rho_i = m_i/\mu_i$ denote the cell mass, the cell volume, and the mean cell density, respectively. The algorithm is motivated by the fact that mass is conserved within Lagrangian volumes and cell average density depends only on the constant mass and the updated Lagrangian volume. Given a grid configuration $C(\Omega(t))$, cells masses $(m_i(t))$, cell volumes $(\mu_i(t))$, and cell average densities $(\rho_i(t))$ at time t, the incremental remapping algorithm consists of three steps:

1. Project the departure grid at time t to an arrival grid at time $t + \Delta t$ using the velocity field: $C(\Omega(t)) \mapsto C(\Omega(t + \Delta t))$;
2. Update mass and cell average density on the arrival grid: $m_i(t+\Delta t) = m_i(t)$, $\rho_i(t + \Delta t) = m_i(t)/\mu_i(t + \Delta t)$ for $i = 1, \ldots, C$;
3. Remap mean cell density back to departure grid $\widetilde{C} = C(\Omega(t))$: $m(t+\Delta t) \mapsto \widetilde{m}$ and $\rho(t + \Delta t) \mapsto \tilde{\rho}$, for $i = 1, \ldots, C$.

In the final remap step the mean density values on the arrival grid $\rho_i(t + \Delta t)$ are used to find approximations of the new masses $\widetilde{m}_i$ and mean densities $\widetilde{\rho}_i$ on the departure grid. To formulate the remap as an optimization problem we write the remapped mass in cell c_i as

$$\widetilde{m}_i = \int_{\widetilde{c}_i} \rho(\mathbf{x}) dV = \int_{c_i} \rho(\mathbf{x}) dV + \left(\int_{\widetilde{c}_i} \rho(\mathbf{x}) dV - \int_{c_i} \rho(\mathbf{x}) dV \right).$$

The quantity in the parentheses is the incremental mass update $(\widehat{u}_i)$ on c_i, which is the optimization variable. The approximate mass update

$$u_i^{\mathsf{T}} := \int_{\widetilde{c}_i} \rho^h(\mathbf{x}) dV - \int_{c_i} \rho^h(\mathbf{x}) dV; \; i = 1, \ldots, C,$$

defined using a mean-preserving linear density reconstruction

$$\rho^h(\mathbf{x})|_{c_i} = \rho_i + \mathbf{g}_i \cdot (\mathbf{x} - \mathbf{b}_i), \quad \mathbf{b}_i \text{ - barycenter of } c_i \tag{2}$$

provides the optimization *target*. The conditions that the remapped values satisfy conservation of mass and local bounds define the optimization constraints. Succinctly, we require that $\sum_{i=1}^{C} \widehat{u}_i = 0$ and $\widetilde{m}_i^{\min} \leq m_i + \widehat{u}_i \leq \widetilde{m}_i^{\max}$ for $i = 1, \ldots, C$, where $\widetilde{m}_i^{min/max} = \rho_i^{min/max}\widetilde{\mu}_i$. Thus, we have the following MVMT optimization formulation of remap:

$$\begin{cases} \text{minimize} \quad \frac{1}{2}\|\widehat{u} - u^{\mathsf{T}}\|_{\ell_2}^2 \qquad \text{subject to} \\ \sum_{i=1}^{C} \widehat{u}_i = 0 \quad \text{and} \quad \widetilde{m}_i^{\min} \leq m_i + \widehat{u}_i \leq \widetilde{m}_i^{\max} \quad i = 1, \dots, C. \end{cases} \tag{3}$$

The inequality constraints guarantee that the remapped mean cell density satisfies the local bounds $\rho_i^{min} \leq \widetilde{\rho}_i \leq \rho_i^{max}$. These bounds are sufficient to ensure monotone solutions provided each arrival grid cell remains in the immediate neighborhood of its departure grid parent. Note that the global inequality constraints are separable, which allows for an efficient implementation of the MVMT algorithm. The approximation $\mathbf{g}_i$ to the gradient of the density in (2) is obtained by a least-squares fit from the mean cell densities in neighboring cells. The target mass increment is computed by integrating the density reconstruction over the intersections of the arrival mesh with the departure mesh. Rather than computing exact intersections we use a swept region approximation [3,4].

3 Extension of MVMT Transport to Cubed Sphere Grid

The cubed sphere grid, originally introduced by Sadournay [12], consists of six faces or panels of a cube that are projected onto the surface of a sphere. This configuration avoids the pole singularity that plagues latitude/longitude grids and is in increasing use among the climate community. To define the grid partition we use an equiangular gnomonic projection where $\alpha, \beta \in [-\pi/4, \pi/4]$ are central angles, which can be related to the local panel coordinates x_p, y_p by

$$x_p = a \tan \alpha \qquad y_p = a \tan \beta \quad p = 1, \dots, 6,$$

where $a = R/\sqrt{3}$ and R is the radius of the sphere. Figure 1 shows a plot of the six cube panels and an example cubed sphere grid. For mappings between latitude/longitude coordinates and the cubed sphere coordinates we refer to [11].

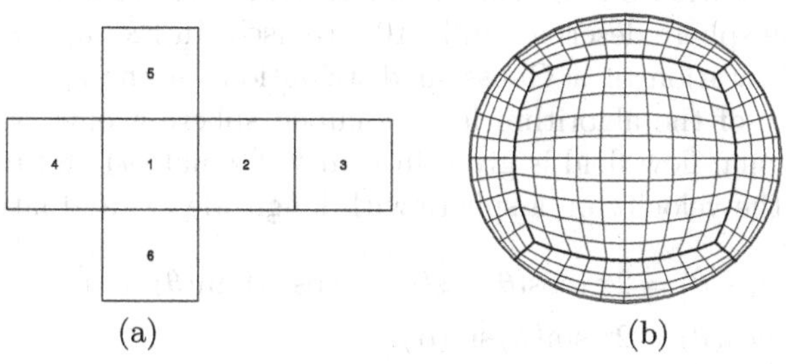

Fig. 1. (a) The six cube panels. (b) A cubed-sphere grid with 10×10 elements per panel.

Extension of the MVMT algorithm to a cubed-sphere grid requires modifications to the target computation, but does not affect the constraints or the optimization algorithm. This is one of the key advantages of the optimization-based approach, which has been exploited in [5] to formulate remap and transport algorithms in spherical coordinates and to explore adaptive target definitions.

To compute the target mass increment the area integrals and linear density reconstruction must be reformulated for the cubed sphere curvilinear coordinates. In the incremental remap approach the area integrals are generally converted to line integrals via Green's theorem. Using Green's theorem with the nonorthogonal curvilinear cubed sphere panel coordinates, the area integral over a cell can be expressed as

$$\mu_i = \int_{c_i} dV = -\int_{\partial c_i} \frac{y_p}{r_p(1+x_p^2)} dx_p \,,$$

for $r_p = \sqrt{1 + x_p^2 + y_p^2}$. The linear density reconstruction additionally requires barycenters of cells, which can be similarly written as

$$b_{xi} = \frac{1}{\mu_i} \int_{c_i} x_p dV = -\frac{1}{\mu_i} \int_{\partial c_i} \frac{x_p y_p}{r_p(1+x_p^2)} dx_p \,,$$

and

$$b_{y_i} = \frac{1}{\mu_i} \int_{c_i} y_p dV = -\frac{1}{\mu_i} \int_{\partial c_i} \frac{1}{r_p} dx_p \,.$$

Using these expressions, the mean preserving density reconstruction on the cubed sphere grid for a position $\mathbf{s}$ on a panel p is

$$\rho^h(\mathbf{s})|_{c_i} = \rho_i + g_i^{x_p}(x_p - b_{xi}) + g_i^{y_p}(y_p - b_{y_i}).$$

Once this density reconstruction is determined the MVMT algorithm as described in [3,5] can be applied.

4 Results

To test the formulation on the cubed sphere grid, two standard test cases for transport on the sphere described in [8,10] are used. In Example 1, we compute the solid body rotation of a Gaussian distribution on the sphere to test the convergence rate of the algorithm for the cubed sphere geometry. The temporally constant zonal flow field is given in terms of zonal (u) and meridional (v) components of the velocity on a sphere with longitude (λ) and latitude (θ) as

$$u(\lambda, \theta) = 2\pi\left(\cos(\theta)\cos(\alpha) + \cos(\lambda)\sin(\theta)\sin(\alpha)\right)$$
$$v(\lambda, \theta) = 2\pi\sin(\lambda)\sin(\alpha).$$

The rotation angle α provides the orientation of the flow. For this test α is taken to be $\pi/4$, which is the most demanding orientation for the cubed-sphere

Table 1. (1) Comparison of the computational costs of FCR and MVMT as measured by Matlab™ wall-clock times in seconds, on a single Intel Xeon X5450 3.0 GHz processor, for the slotted-cylinder translation test on the sphere. (2) Comparison of the L_1 errors with respect to the initial condition.

Solid-body translation on the sphere (timings and L_1 error)

Mesh	# steps	FCR Time (s)	MVMT Time (s)	FCR L_1 error	Rate	MVMT L_1 error	Rate
3°	240	18.1	17.6	1.33e-2	–	1.49e-2	–
1.5°	480	108.5	109.3	2.43e-3	2.45	2.65e-3	2.50
0.75°	960	816.5	811.0	5.17e-4	2.34	5.44e-4	2.39

geometry because the density distribution is transported over four of the corners of the cubed-sphere grid. The smooth Gaussian density distribution is initially centered at $(\lambda_1, \theta_1) = (3\pi/2, 0)$ and is defined in terms of three-dimensional Cartesian coordinates (X, Y, Z) as

$$\rho(\lambda, \theta) = \exp(-5((X - X_1)^2 + (Y - Y_1)^2 + (Z - Z_1)^2)) \tag{4}$$

where $X_1 = \cos\lambda_1 \cos\theta_1$, $Y_1 = \sin\lambda_1 \cos\theta_1$, and $Z_1 = \sin\theta_1$.

Three grids are used with $30 \times 30, 60 \times 60$, and 120×120 elements per panel corresponding to resolutions of 3°, 1.5°, and 0.75° along the equator. Results are computed using the incremental remapping approach discussed in Sect. 2 with the MVMT algorithm used for the remap step. For comparison, results from the FCR algorithm are also given.

At the final time the density distribution returns to the initial position, which allows for an error analysis. L_1 errors are computed as in [8]. Timings as well as L_1 errors and rates for the MVMT and FCR solutions are given in Table 1. For this simple translation of a smooth density distribution it is expected that MVMT and FCR perform similarly. Slightly better than second-order convergence is seen for both methods and the absolute errors are comparable. The

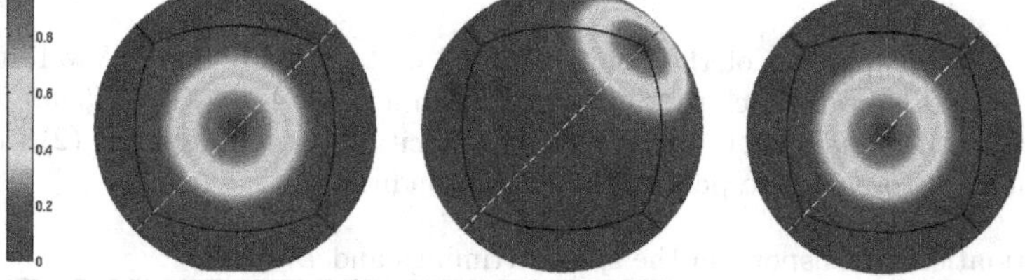

Fig. 2. Transport results for the solid-body rotation test on the sphere at the time the center of the density distribution passes over a cubed sphere corner (t = 0.15) and at the final time (t = 1) after one revolution (960 time steps) on a mesh with 120 × 120 elements per panel. The rotation angle of $\pi/4$ determines the trajectory shown on the plots as a white dashed line.

computational costs of MVMT and FCR are virtually identical, owing to the efficiency of the MVMT optimization scheme. Plots of the density distribution at the initial time, at time $t = 0.15$ as the Gaussian hill is passing over a cubed sphere corner, and at the final time $t = 1$ are shown in Fig. 2 for the MVMT algorithm. Results for FCR are visually very similar and thus not shown here.

Example 2 is more demanding with an initial density distribution consisting of two notched cylinders with radius $r = 1/2$, height $h = 1$, and initial positions $(\lambda_1, \theta_1) = (5\pi/6, 0)$ and $(\lambda_2, \theta_2) = (7\pi/6, 0)$. Given the great circle distance between an arbitrary point (λ, θ) and a cylinder center (λ_i, θ_i)

$$r_i(\lambda, \theta) = \arccos\left(\sin\theta_i \sin\theta + \cos\theta_i \cos\theta \cos(\lambda - \lambda_i)\right),$$

the initial configuration of the notched cylinders may be expressed in latitude-longitude coordinates as

$$\rho(\lambda, \theta) = \begin{cases} h \text{ if } r_i < r \text{ and } |\lambda - \lambda_i| \geq r/6 \text{ for } i = 1,2 \\ h \text{ if } r_1 < r \text{ and } |\lambda - \lambda_0| < r/6 \text{ and } \theta - \theta_0 < -5r/12 \\ h \text{ if } r_2 < r \text{ and } |\lambda - \lambda_1| < r/6 \text{ and } \theta - \theta_1 > 5r/12 \\ 0 \text{ otherwise.} \end{cases}$$

The cylinders are transported in the following deformational flow field with superimposed rotation

$$\begin{aligned} u(\lambda, \theta, t) &= 2\sin^2(\lambda - 2\pi t/T)\sin(2\theta)\cos(\pi t/T) + 2\pi\cos(\theta)/T \\ v(\lambda, \theta, t) &= 2\sin\bigl(2(\lambda - 2\pi t/T)\bigr)\cos(\theta)\cos(\pi t/T) \end{aligned}$$

where the period T is set to 5.

Timings as well as L_1 errors and rates for the MVMT and FCR solutions are given in Table 2 and plots of the density distribution at the initial time, at a time of maximum deformation $t = 2.5$ and a final time $t = 5$ are shown in Fig. 3. Second-order convergence is not seen for either method in this case due to the discontinuous density field, but the errors and convergence rates appear comparable. Note, however that for this case with 2400 time steps and a maximum Courant-Friedrichs-Lewy (CFL) number of 0.677 the FCR solution

Table 2. (1) Comparison of the computational costs of FCR and MVMT as measured by Matlab™ wall-clock times in seconds, on a single Intel Xeon X5450 3.0 GHz processor, for the nondivergent deformational velocity test on the sphere. (2) Comparison of the L_1 errors with respect to the initial condition.

Deformational transport on the sphere (timings and L_1 error)

Mesh	# steps	FCR Time (s)	MVMT Time (s)	FCR L_1 error	Rate	MVMT L_1 error	Rate
3°	600	45.9	45.0	9.38e-1	–	9.54e-1	–
1.5°	1200	274.1	277.5	6.17e-1	0.60	6.53e-1	0.55
0.75°	2400	2081.0	2071.5	4.16e-1	0.59	4.45e-1	0.55

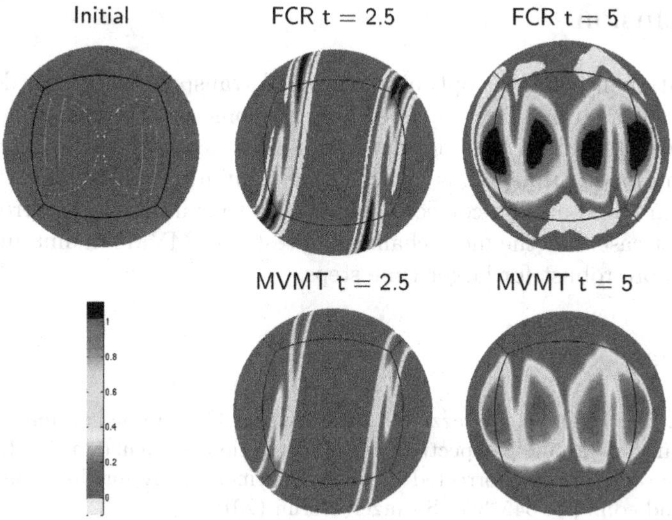

Fig. 3. Transport results for the nondivergent deformational flow test on the sphere, shown at the time of maximum deformation (t = 2.5) and at the final time (t = 5) for a total of 2400 time steps on a mesh with 120 × 120 elements per panel.

has a minimum value of −0.0639 and a maximum value of 1.075, while the MVMT solution remains within the physical bounds [0, 1]. This case illustrates that fact that the low-order fluxes used in the FCR method are not guaranteed to be monotone for relatively high CFL numbers if exact cell intersections are not used. When the number of time steps is decreased to 1650, which corresponds to a maximum CFL number of 0.985, the FCR solution blows up, but the MVMT solution still remains monotone and appears reasonable visually (Fig. 4).

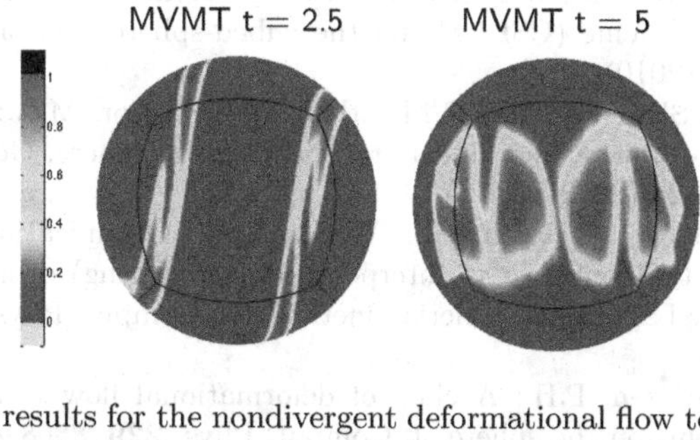

Fig. 4. Transport results for the nondivergent deformational flow test with rotation on the sphere, shown at the time of maximum deformation (t = 2.5) and at the final time (t = 5) for a total of 1650 time steps for a maximum CFL number of 0.985 on a mesh with 120 × 120 elements per panel. The FCR solution blows up for this long time step case and is therefore not shown.

5 Conclusion

A computationally efficient optimization-based transport algorithm detailed in [3,5] has been modified for the cubed sphere geometry. The resulting formulation has been tested on two standard transport cases for the sphere [7]. The optimization-based transport is shown to be computationally competitive with an algorithm based on flux-corrected remap and to exhibit similar errors for the simplest test case. For the more challenging test case MVMT maintains positivity and is more robust for larger time steps.

References

1. Bochev, P., Ridzal, D., Scovazzi, G., Shashkov, M.: Constrained optimization based data transfer - a new perspective on flux correction. In: Kuzmin, D., Lohner, R., Turek, S. (eds.) Flux-Corrected Transport. Principles, Algorithms and Applications, 2nd edn., pp. 345–398. Springer, Berlin (2012)
2. Bochev, P., Ridzal, D., Young, J.: Optimization-based modeling with applications to transport: Part 1. Abstract formulation. In: Lirkov, I., Margenov, S., Waśniewski, J. (eds.) LSSC 2011. LNCS, vol. 7116, pp. 63–71. Springer, Heidelberg (2012)
3. Bochev, P., Ridzal, D., Peterson, K.: Optimization-based remap and transport: a divide and conquer strategy for feature-preserving discretizations. J. Comput. Phys. (2013, accepted)
4. Bochev, P., Ridzal, D., Scovazzi, G., Shashkov, M.: Formulation, analysis and numerical study of an optimization-based conservative interpolation (remap) of scalar fields for arbitrary Lagrangian-Eulerian methods. J. Comput. Phys. **230**(13), 5199–5225 (2011)
5. Bochev, P., Ridzal, D., Shashkov, M.: Fast optimization-based conservative remap of scalar fields through aggregate mass transfer. J. Comput. Phys. (2012, accepted)
6. Dukowicz, J.K., Baumgardner, J.R.: Incremental remapping as a transport/advection algorithm. J. Comput. Phys. **160**(1), 318–335 (2000)
7. Lauritzen, P.H., Nair, R.D., Ullrich, P.A.: A conservative semi-Lagrangian multi-tracer transport scheme (CSLAM) on the cubed-sphere grid. J. Comput. Phys. **229**, 1401–1424 (2010)
8. Lauritzen, P.H., Skamarock, W.C., Prather, M.J., Taylor, M.A.: A standard test case suite for two-dimensional linear transport on the sphere. Geosci. Model Dev. Discuss. **5**, 189–228 (2012)
9. Liska, R., Shashkov, M., Váchal, P., Wendroff, B.: Optimization-based synchronized flux-corrected conservative interpolation (remapping) of mass and momentum for arbitrary Lagrangian-Eulerian methods. J. Comput. Phys. **229**, 1467–1497 (2010)
10. Nair, R.D., Lauritzen, P.H.: A class of deformational flow test cases for linear transport problems on the sphere. J. Comput. Phys. **229**, 8868–8887 (2010)
11. Nair, R.D., Thomas, S.J., Loft, R.D.: A discontinuous Galerkin transport scheme on the cubed sphere. Mon. Weather Rev. **133**, 814–828 (2005)
12. Sadournay, R.: Conservative finite-differencing approximations of the primitive equations on quasi-uniform spherical grids. Mon. Weather Rev. **100**, 136–144 (1972)

Applications of Metaheuristics to Large-Scale Problems

Application of Metaheuristics to Large-Scale Transportation Problems

Luca D'Acierno[1(✉)], Mariano Gallo[2], and Bruno Montella[1]

[1] Department of Civil, Architectural and Environmental Engineering, 'Federico II' University of Naples, Naples, Italy
{luca.dacierno,bruno.montella}@unina.it
[2] Department of Engineering, University of Sannio, Benevento, Italy
gallo@unisannio.it

Abstract. In this paper we propose a general model for solving the Transportation Network Design Problem (TNDP). Since in real-scale networks the number of feasible solutions to be examined does not allow an exhaustive search and objective functions are not convex, it is necessary to adopt metaheuristic algorithms to obtain sub-optimal solutions within suitable calculation times. Hence, we show and analyse some algorithms proposed in the literature for solving TNDPs both in urban and extra-urban contexts in order to highlight the importance of metaheuristic algorithms in large-scale transportation problems.

Keywords: Transportation · Network design · Metaheuristic algorithms · Real-scale networks

1 Introduction

The *Transportation Network Design Problem* (TNDP) consists in optimising the features of a transportation network (such as timing of traffic lights or public transport frequencies) so as to minimise the value of an objective function, taking user behaviour and several constraints into account. It is worth noting that, in the case of real-scale networks, the number of decision variables and their values are such that the number of feasible solutions has an order of magnitude equal to 10^{10}–10^{60}. Hence, since the use of an exhaustive search has to be excluded and objective functions are generally not convex, it is necessary to adopt or develop metaheuristic algorithms in order to obtain sub-optimal solutions within suitable calculation times.

In previous years, by means of metaheuristic algorithms, the authors proposed the solution of several transportation problems in the case of real-scale networks (see papers [1–4]). The aim of this paper is not to provide a comparison among metaheuristic algorithms in order to identify the best but to show the feasibility of adopting such algorithms for solving large-scale problems. Hence, we propose an overview of the literature, recommending the above papers for in-depth analysis. However, an extensive state-of-the-art review of the TNDP

I. Lirkov et al. (Eds.): LSSC 2013, LNCS 8353, pp. 215–222, 2014.
DOI: 10.1007/978-3-662-43880-0_23,

can be found in [5,6]. Generally, the TNDP can be decomposed into a *Road Network Design Problem* (RNDP) or a *Mass Transit Network Design Problem* (MTNDP). The RNDP, which consists of designing variables of the private car system, has been widely studied elsewhere. The proposed models may be classified into discrete variable models, continuous variable models and mixed variable models. Likewise, the MTNDP, which consists of designing variables of the public transport system, has been explored by [7,8]. The models in question can be classified into service frequency optimisation, joint design of frequencies and routes or elastic demand approach models.

To show the validity of metaheuristic algorithms in solving large-scale transportation problems, we compare algorithm frameworks and numerical results obtained from the above literature. The paper is structured as follows: Sect. 2 provides the theoretical formulation of the TNDP; Sect. 3 describes the metaheuristic algorithms adopted and Sect. 4 shows numerical applications; finally, Sect. 5 summarises the conclusions and outlines research prospects.

2 The Transportation Network Design Problem

In general, the TNDP can be formulated as:

$$\hat{\boldsymbol{y}} = \arg \min_{\boldsymbol{y} \in \boldsymbol{S}_{\boldsymbol{y}}} Z\left(\boldsymbol{y}, \boldsymbol{f}^*\right) \tag{1}$$

subject to:

$$\boldsymbol{f}^* = \boldsymbol{\Lambda}\left(\boldsymbol{f}^*, \boldsymbol{y}, \boldsymbol{d}\right) \tag{2}$$

$$\boldsymbol{\lambda}\left(\boldsymbol{y}, \boldsymbol{f}^*\right) \in \boldsymbol{S}_{\boldsymbol{\lambda}} \tag{3}$$

where $\boldsymbol{y}$ is the vector of decision variables to be optimised; $\hat{\boldsymbol{y}}$ is the optimal value of $\boldsymbol{y}$; $\boldsymbol{S}_{\boldsymbol{y}}$ is the feasibility set of vector $\boldsymbol{y}$; $Z\left(\cdot\right)$ is the objective function to be minimised; $\boldsymbol{f}^*$ is the vector of equilibrium flows to be calculated by means of (2); $\boldsymbol{\Lambda}\left(\cdot\right)$ is the assignment function; $\boldsymbol{d}$ is the vector of travel demand; $\boldsymbol{\lambda}\left(\cdot\right)$ is a function which expresses transportation system features; $\boldsymbol{S}_{\boldsymbol{\lambda}}$ is the feasibility set of function $\boldsymbol{\lambda}\left(\cdot\right)$.

In particular, constraint (2) represents the assignment constraint which provides equilibrium flows $\boldsymbol{f}^*$ as a function of equilibrium flows $\boldsymbol{f}^*$, decisional variables $\boldsymbol{y}$ and travel demand $\boldsymbol{d}$. Indeed, transportation network performance generally depends on the number of users of transportation systems and on the features of the considered project expressed by means of the decisional variable value, that is:

$$\boldsymbol{C} = \tilde{\boldsymbol{C}}\left(\boldsymbol{f}, \boldsymbol{y}\right) \tag{4}$$

where $\boldsymbol{C}$ represents the vector of network performance, generally indicated as generalised cost; $\tilde{\boldsymbol{C}}\left(\cdot\right)$ represents the generalised cost function; $\boldsymbol{f}$ the vector of generic transportation network flows.

Likewise, user behaviour is affected by transportation systems performance and travel demand, that is:

$$\boldsymbol{f} = \tilde{\boldsymbol{f}}(\boldsymbol{C}, \boldsymbol{d}) \tag{5}$$

where $\tilde{\boldsymbol{f}}(\cdot)$ represents the network loading function, i.e. a function which describes user behaviour in terms of network flows.

By combining (4) and (5), we obtain a fixed-point formulation, that is:

$$\boldsymbol{f} = \tilde{\boldsymbol{f}}\left(\tilde{\boldsymbol{C}}(\boldsymbol{f}, \boldsymbol{y}), \boldsymbol{d}\right) \tag{6}$$

whose aim is to determine a flow vector, indicated as equilibrium flow vector $\boldsymbol{f}^*$, which, together with the value of design variables $\boldsymbol{y}$, yields network performance $\boldsymbol{C}$ which, jointly with travel demand $\boldsymbol{d}$, affects user behaviour such that they generate flows equal to equilibrium flow vector $\boldsymbol{f}^*$, that is:

$$\boldsymbol{f}^* = \tilde{\boldsymbol{f}}\left(\tilde{\boldsymbol{C}}(\boldsymbol{f}^*, \boldsymbol{y}), \boldsymbol{d}\right) \tag{7}$$

Analytically, (7) represents the constraint (2) expressed in terms of network loading and generalised cost functions, whose theoretical properties were analysed by [9,10], and relative solution algorithms were proposed by [9–13]. In particular, [12,13] are based on Ant Colony Optimisation.

Constraint (3) indicates that some transportation system features, whose values depend on decisional variables $\boldsymbol{y}$ and equilibrium flows $\boldsymbol{f}^*$, may have to satisfy some conditions such as budget or technical constraints described by $\boldsymbol{S}_{\boldsymbol{\lambda}}$.

It is worth noting that the TNDP can be classified as a monomodal or a multimodal problem according to the assumptions on analysed transportation systems. Indeed, as shown by [14], if the designed intervention does not provide relevant effects on all transportation systems, we may analyse only a single mode by a monomodal approach; otherwise it is worth considering all influenced modes by a multimodal approach. Moreover, if the modal share of travel demand can be assumed independent from the analysed intervention scenarios, we may assume that the travel demand for each analysed transportation system is constant and the assignment model (2) is indicated as a rigid demand model. Otherwise the assignment model (2) is indicated as an elastic demand model.

3 Metaheuristic Solution Algorithms

As shown in the introduction, some TNDPs have been analysed in the case of large-scale networks by proposing and/or adopting the following solution algorithms: the *Neighbourhood Search Algorithm* (NSA), a *Heuristic Local Search Algorithm* (HLSA), the *Scatter Search* (SS) and the *Genetic Algorithm* (GA).

In this section the main features of the above algorithms are provided. However, for a more detailed description of these algorithms, we suggest reading [1–4].

3.1 Neighbourhood Search Algorithm (NSA)

The NSA is a heuristic algorithm for solving discrete optimisation problems. Each vector $\boldsymbol{y}$ has an associated set of vectors $\boldsymbol{N}(\boldsymbol{y}) \subseteq \boldsymbol{S}_{\boldsymbol{y}}$, called neighbourhood of $\boldsymbol{y}$, where the generic element $\boldsymbol{y}' \in \boldsymbol{N}(\boldsymbol{y})$ is obtained from solution $\boldsymbol{y}$ by an operation consisting in changing only one component of vector $\boldsymbol{y}$.

In the NSA, one of the most commonly adopted rules for generating the next solution is the *Steepest Descent Method* (SDM) consisting in examining all elements of the neighbourhood and identifying the solution with the best objective function value. In managing large-scale networks, calculation times may be reduced by adopting a *Random Descent Method* (RDM). It consists in extracting randomly a solution from the neighbourhood and determining its objective function value: if the new solution is better than the current one, it becomes the current solution; otherwise, another neighbourhood solution is randomly extracted until a better solution is found.

Obviously, the use of random draws, especially in the case of non-convex functions, could provide different results both in terms of objective function improvement (i.e. we may obtain different local optima) and calculation time requirements.

In both approaches, the algorithm ends when a solution is a local optimum. We used the NSA mainly as a subroutine of more complex algorithms.

3.2 Heuristic Local Search Algorithm (HLSA)

The HLSA is a metaheuristic algorithm consisting of five phases. In the first phase, each component of vector $\boldsymbol{y}$ is optimised, assuming that the values of other components are constant. This phase may be developed according to two approaches: an exhaustive or a monodimensional NSA approach. In this phase, constraint (3) is neglected. The second phase consists in determining the first starting solution by setting each component of vector $\boldsymbol{y}$ at the optimal value calculated in the previous phase. The third consists in performing an NSA with an SDM or RDM approach. Also in this phase, constraint (3) is neglected. The fourth phase entails analysing all solutions generated in the previous phases, selecting the one that minimises the objective function and jointly satisfies constraint (3). Finally, the last phase performs NSA with an SDM approach by considering constraint (3).

3.3 Scatter Search (SS)

The SS, as shown by [15], is a metaheuristic algorithm for solving complex combinatorial optimisation problems which consists of five phases. The first phase consists in generating a set of solutions which should cover different parts of the solution space. The second entails applying for each element of the current reference set an improvement method for generating improved solutions. The improvement method could be, for instance, the NSA or HLSA. In the third phase, a reference set is generated by selecting all improved solutions generated in the previous phase or, if they are too numerous, only part of them. In the fourth phase, some solution subsets are generated, consisting of some solutions belonging to the reference set, which will be combined in the following phase to generate other solutions. Finally, the solutions of each subset are combined by associating a score (depending on the objective function value and the times that the specific value is assumed by the variable in all solutions belonging to the subset) to each value that can be assumed by a component of vector $\boldsymbol{y}$. The combined solution obtained from the subset will be that in which every variable assumes the value with the best score.

The solutions obtained in the last phase are improved (phase 2), generating a new reference set. The procedure ends when the reference sets in two successive iterations are equal or when a fixed a priori number of iterations is reached.

3.4 Genetic Algorithm (GA)

The genetic algorithm is a metaheuristic evolutionary technique which explores the solution space by mimicking natural evolution [16,17]. Each solution $\boldsymbol{y}$ is indicated as a chromosome and each solution component as a gene. The main phases of the algorithm are: initialisation, selection, reproduction and termination.

The first phase of the algorithm consists in defining the starting population, that is an initial set of solutions. In the second phase, for each element of the population the objective function and the related fitness function are calculated. Moreover, in this phase some pairs of members of the population are extracted (*parent selection*), for instance, by adopting a roulette wheel selection scheme. Once two elements have been selected as parents, the reproduction phase is performed by means of two sub-phases: crossover and mutation. *Crossover* consists of extracting randomly an integer number, x, in the interval $[1; n_y]$, where n_y is the dimension of vector $\boldsymbol{y}$. The first offspring will have the first x genes which are the same as the father's and the others $(n_y - x)$ identical to the mother's. Likewise, it is possible to generate complementary offspring which have the first x genes like the mother's and the others $(n_y - x)$ just like the father's. The *mutation* consists in randomly selecting a gene for each offspring and randomly selecting a number in the set of feasible values of that gene. The offspring are added to the best solution in the previous population and the selection phase will be once again performed. The procedure ends when the optimal values of objective function in two successive iterations are equal or when a fixed a priori number of iterations is reached.

4 Application to Real Scale Networks

In this section we describe some applications of the above-mentioned algorithms proposed in the literature for solving the following TNDPs: the *Urban Road Network Design Problem* (URNDP) [1], where link directions were optimised with a rigid demand monomodal approach; the *Extra-urban Road Network*

Table 1. Application contexts

Analysed problem	Real-scale network	Population	Area (km^2)
URNDP	Benevento (Italy)	61,700	130
ERNDP	Campania (Italy)	6,075,000	13,600
UTNDP	Salerno (Italy)	132,000	59
ETNDP	Campania (Italy)	6,075,000	13,600

Table 2. Variables and exhaustive approach features

Analysed problem	Number of variables	Type	Feasible values	Feasible solutions	Exhaustive calculation times (years)
URNDP	12 + 129	Integer	2/3	1.45e+65	3.57e+58
ERNDP	102	Binary	2	5.07e+30	4.18e+24
UTNDP	40	Integer	15	1.11e+47	4.82e+41
ETNDP	14	Integer	10	1.00e+14	3.30e+08

Design Problem (ERNDP) [2], where roads to be improved were optimised with a rigid demand monomodal approach; the *Urban Transit Network Design Problem* (UTNDP) [3], where bus line frequencies were optimised with an elastic multimodal approach; the *Extra-urban Transit Network Design Problem* (ETNDP) [4], where rail line frequencies were optimised with an elastic multimodal approach.

Table 1 describes some features of the application contexts. Table 2 provides details on the design variable numbers (12 + 129 and 2/3 means that there are 12 variables with 2 feasible values and 129 variables with 3 values) and an estimation of calculation times in the case of an exhaustive approach.

Table 3 provides numerical results obtained by testing different algorithms on large-scale networks. In particular, SSn-k indicates the use of an SS algorithm implemented n times with the use of algorithm k as an improvement method; NSA(x) indicates the use of an NSA implemented with the x approach; and HLSA(x) indicates the use of an HLSA algorithm obtained by implementing the NSA phases with an x approach. Finally, the last column expresses the improvement in objective function value obtained with the best solution with respect to the initial solution.

Further details concerning the analysed application can be found in [1–4]. However, in terms of application results, it may be concluded that the adoption of analysed metaheuristic algorithms allows sub-optimal solutions to be obtained within suitable calculation times.

In detail, we may state that the use of an NSA with an RDM approach provides similar results but with lower calculation times than the use of an SDM approach (see URNDP, ERNDP and UTNDP). However, since the NSA is based on random draws, results could be affected by the sequence of draws. Hence, it may be useful, in terms of future research, to provide an analysis of RDM approaches in the case of different draws.

Likewise, in the case of SS, we may highlight that an increase in the number of implementations could provide a better value of objective function (see URNDP).

Table 3. Numerical results

Analysed problem	Implemented algorithms	Number of analysed solutions	Calculation times (h)	Objective function improvements (%)
URNDP	SS1-NSA(SDM)	171,751	370	13.95
URNDP	SS1-NSA(RDM)	26,851	58	13.82
URNDP	SS2-NSA(RDM)	52,735	114	15.05
ERNDP	NSA(SDM)	5,918	9	19.85
ERNDP	NSA(RDM)	660	1	20.02
UTNDP	HLSA(SDM)	2,946	113	1.93
UTNDP	HLSA(RDM)	1,106	42	1.62
ETNDP	HLSA(RDM)	1,075	31	1.42
ETNDP	SS2-HLSA(RDM)	4,645	133	1.43
ETNDP	GA	4,645	134	1.34

Hence, in terms of future research, it could be useful to provide an analysis of the best compromise between the increase in calculation times and the improvement in solutions.

Finally, we also provide a comparison between HLSA, SS, and GA (see ETNDP). Obviously the analysis could be expanded by comparing described algorithms in the case of different parameters such as population size for GA or reference set for SS.

5 Conclusions and Research Prospects

Analysis of a large-scale transportation network requires the implementation of a bi-level optimisation problem where the lower level consists in solving a fixed-point problem. The huge number of feasible solutions and the non-convexity of the objective function necessarily requires the adoption of metaheuristic algorithms.

The paper proposed a brief description of algorithms applied by the authors in previous years for solving transportation problems. In particular, initial results show that the use of metaheuristic algorithms is actually one of the few approaches for managing real-scale problems.

Future research will focus on comparing all described algorithms in all analysed networks in order to provide a homogeneous field of analysis, comparing each algorithm by varying implementation parameters and implementing other metaheuristics in order to explore and show the actual power of metaheuristics.

References

1. Gallo, M., D'Acierno, L., Montella, B.: A meta-heuristic approach for solving the Urban Network Design Problem. Eur. J. Oper. Res. **201**, 144–157 (2010)
2. Gallo, M., D'Acierno, L., Montella, B.: A meta-heuristic algorithm for solving the road network design problem in regional contexts. Procedia Soc. Behav. Sci. **54**, 84–95 (2012)

3. Gallo, M., Montella, B., D'Acierno, L.: The transit network design problem with elastic demand and internalisation of external costs: an application to rail frequency optimisation. Transp. Res. C-Emer. **19**, 1276–1305 (2011)
4. Gallo, M., D'Acierno, L., Montella, B.: A multimodal approach to bus frequency design. WIT Trans. Built Env. **116**, 193–204 (2011)
5. Yang, H., Bell, M.G.H.: Models and algorithms for road network design: a review and some new developments. Transp. Rev. **18**, 257–278 (1998)
6. Feremans, C., Labb, M., Laporte, G.: Generalized network design problems. Eur. J. Oper. Res. **148**, 1–13 (2003)
7. Guihaire, V., Hao, J.K.: Transit network design and scheduling: a global review. Transp. Res. A-Pol. **42**, 1251–1273 (2008)
8. Kepaptsoglou, K., Karlaftis, M.: Transit route network design problem: review. J. Transp. Eng. **135**, 491–505 (2009)
9. Cantarella, G.E.: A general fixed-point approach to multimodal multi-user equilibrium assignment with elastic demand. Transp. Sci. **31**, 107–128 (1997)
10. Cascetta, E.: Transportation Systems Analysis: Models and Applications. Springer, New York (2009)
11. Sheffi, Y., Powell, W.B.: A comparison of stochastic and deterministic traffic assignment over congested networks. Transp. Res. B-Meth. **15**, 53–64 (1981)
12. D'Acierno, L., Montella, B., De Lucia, F.: A stochastic traffic assignment algorithm based on Ant Colony Optimisation. In: Dorigo, M., Gambardella, L.M., Birattari, M., Martinoli, A., Poli, R., Stützle, T. (eds.) ANTS 2006. LNCS, vol. 4150, pp. 25–36. Springer, Heidelberg (2006)
13. D'Acierno, L., Gallo, M., Montella, B.: Ant Colony Optimisation approaches for the transportation assignment problem. WIT Trans. Built Env. **111**, 37–48 (2010)
14. Montella, B., Gallo, M., D'Acierno, L.: Multimodal network design problems. Adv. Transp. **5**, 405–414 (2000)
15. Glover, F., Laguna, M., Martì, R.: Scatter search. In: Ghosh, A., Tsutsui, S. (eds.) Advances in Evolutionary Computation: Theory and Applications, pp. 519–537. Springer, New York (2003)
16. Holland, J.H.: Adaptation in Natural and Artificial Systems. The University of Michigan Press, Ann Arbor (1975)
17. Goldberg, D.E.: Genetic Algorithms in Search. Optimization and Machine Learning. Kluwer Academic Publishers, Boston (1989)

Genetic Operators Significance Assessment in Simple Genetic Algorithm

Maria Angelova(✉) and Tania Pencheva

Institute of Biophysics and Biomedical Engineering,
Bulgarian Academy of Sciences, 105 Acad. G. Bonchev Str., 1113 Sofia, Bulgaria
{maria.angelova,tania.pencheva}@biomed.bas.bg

Abstract. Genetic algorithms, proved as successful alternative to conventional optimization methods for the purposes of parameter identification of fermentation process models, search for a global optimal solution via three main genetic operators, namely selection, crossover, and mutation. In order to determine their importance for finding the solution, a procedure for significance assessment of genetic algorithms operators has been developed. The workability of newly elaborated procedure has been tested when simple genetic algorithm is applied to parameter identification of *S. cerevisiae* fed-batch cultivation. According to obtained results the most significant genetic operator has been distinguished and its influence for finding the global optimal solution has been evaluated.

Keywords: Simple genetic algorithm · Genetic operators · Parameter identification · *S. cerevisiae* fed-batch cultivation

1 Introduction

Genetic algorithms (GA) [1] are a metaheuristic method based on biological evolution. Some properties such as hard problems solving, noise tolerance, easiness to interface and hybridize make GA a suitable and quite workable tool especially for incompletely determined tasks. Such a task and a real challenge for researchers is the parameter identification of fermentation processes (FP) models [2–6]. FP are known as complex, dynamic systems with interdependent and time-varying process variables, and their modeling is a specific task, rather difficult to be solved. Failure of conventional optimization methods to reach to a satisfactory solution for parameters identification of FP models [6] provokes idea as an alternative technique genetic algorithms to be tested.

Inspired by natural genetics, Goldberg [1] initially presents the standard single-population genetic algorithm (SGA) that searches a global optimal solution using three main genetic operators in a sequence selection, crossover and mutation. When GA are applied for the purposes of model parameter identification, there are many operators, functions, parameters and settings that may vary depending on the considered problems [1,7]. In [7] three of the main GA parameters, namely generation gap (GGAP), crossover (XOVR) and mutation

I. Lirkov et al. (Eds.): LSSC 2013, LNCS 8353, pp. 223–231, 2014.
DOI: 10.1007/978-3-662-43880-0_24,

Table 1. Range of investigated genetic algorithms parameters

GGAP	XOVR	MUTR
0.5	0.65	0.02
0.67	0.75	0.05
0.8	0.85	0.08
0.9	0.95	0.1

(MUTR) rates have been investigated towards model accuracy and algorithm convergence time with values shown in Table 1, according to some statements [8]. Among them three, GGAP has been distinguished as the most sensitive GA parameter. Up to almost 40 % of the algorithm calculation time can be saved in the case of one of the considered in [7] SGA using GGAP = 0.5 instead of 0.9. Exploring different values of XOVR no such time saving was realized but it was pointed that value of 0.85 can be assumed as more appropriate. Only in MUTR no tendency of influence was drawn, but for the same algorithm it was shown that using MUTR = 0.1 instead of 0.05 leads to save up to 20 % of convergence time without loss of model accuracy.

In general, the quality of GA performance might be assessed by some representative criteria such as objective function value and algorithm convergence time. But from biological, and even biotechnological, point of view it is valuable to be known how the main GA parameters influent to model parameters. Going further, such analysis might be worth to assess the GA operators significance. As an alternative for such a purpose, intuitionistic fuzzy logic might be applied.

The aim of this study is to apply intuitionistic fuzzy estimations for assessing the influence of the three main GA operators, namely selection, crossover and mutation. For that purpose, the three main GA parameters - GGAP, XOVR and MUTR are going to be evaluated towards the values of model parameters when standard SGA is implemented to parameter identification of *S. cerevisiae* fed-batch cultivation.

2 Background

2.1 Mathematical Model of *S. cerevisiae* Fed-Batch Cultivation

Experimental data of *S. cerevisiae* fed-batch cultivation is obtained in the *Institute of Technical Chemistry - University of Hannover, Germany* [6]. The cultivation of the yeast *S. cerevisiae* is performed in 1.5 l reactor, using a Schatzmann medium. Glucose in feeding solution is 50 g/l. The temperature was controlled at 30 °C, the pH at 5.7. The stirrer speed was set to 500 rpm.

Mathematical model of *S. cerevisiae* fed-batch cultivation is commonly described as follows, according to the mass balance [6]:

$$\frac{dX}{dt} = \mu X - \frac{F}{V}X \tag{1}$$

$$\frac{dS}{dt} = -q_S X + \frac{F}{V}(S_{in} - S) \tag{2}$$

$$\frac{dE}{dt} = q_E X - \frac{F}{V} E \tag{3}$$

$$dt = -q_{O_2} X + k_L^{O_2} a (O_2^* - O_2) \tag{4}$$

$$\frac{dV}{dt} = F \tag{5}$$

where X is the concentration of biomass, [g/l]; S - concentration of substrate (glucose), [g/l]; E - concentration of ethanol, [g/l]; O_2 - concentration of oxygen, [%]; O_2^* - dissolved oxygen saturation concentration, [%]; F - feeding rate, [l/h]; V - volume of bioreactor, [l]; $k_L^{O_2}a$ - volumetric oxygen transfer coefficient, [1/h]; S_{in} - glucose concentration in the feeding solution, [g/l]; μ, q_S, q_E, q_{O_2} - specific growth/utilization rates of biomass, substrate, ethanol and dissolved oxygen, [1/h]. All functions are continuous and differentiable.

The fed-batch cultivation of *S. cerevisiae* considered here is characterized by keeping glucose concentration equal to or below to its critical level, sufficient dissolved oxygen and availability of ethanol in the broth. This state corresponds to the so called mixed oxidative state (FS II) according to functional state modeling approach [6]. Hence, specific rates in Eqs. (1)–(5) are:

$$\mu = \mu_{2S}\frac{S}{S + k_S} + \mu_{2E}\frac{E}{E + k_E}, \quad q_S = \frac{\mu_{2S}}{Y_{SX}}\frac{S}{S + k_S}$$
$$q_E = -\frac{\mu_{2E}}{Y_{EX}}\frac{E}{E + k_E}, \quad q_{O_2} = q_E Y_{OE} + q_S Y_{OS} \tag{6}$$

where μ_{2S}, μ_{2E} are the maximum growth rates of substrate and ethanol, [1/h]; k_S, k_E - saturation constants of substrate and ethanol, [g/l]; Y_{ij} - yield coefficients, [g/g]; and all model parameters fulfill the non-zero division requirement.

As an optimization criterion, mean square deviation between the model output and the experimental data obtained during cultivation has been used:

$$J = \sum (Y - Y^*)^2 \rightarrow min, \tag{7}$$

where Y is the experimental data, Y^* - model predicted data, $Y = [X, S, E, O_2]$.

2.2 Intuitionistic Fuzzy Estimations

In intuitionistic fuzzy logic (IFL) [9,10] if p is a variable then its truth-value is represented by the ordered couple

$$V(p) = \langle M(p), N(p) \rangle \tag{8}$$

so that $M(p), N(p), M(p) + N(p) \in [0, 1]$, where $M(p)$ and $N(p)$ are degrees of validity and of non-validity of p. These values can be obtained applying different formula depending on the problem considered.

For the purpose of this investigation the degrees of validity/non-validity can be obtained, e.g., by the following formula:

$$M(p) = \frac{m}{u}, \quad N(p) = 1 - \frac{n}{u}, \tag{9}$$

where m is the lower boundary of the "narrow" range; u - the upper boundary of the "broad" range; n - the upper boundary of the "narrow" range.

If there is a database collected having elements with the form $< p, M(p), N(p) >$, different new values for the variables can be obtained. In case of two records in the database, the new values might be as follows:

$$V_{opt} = \langle \max(M_1(p), M_2(p)), \min(N_1(p), N_2(p)) \rangle \,, \tag{10}$$

$$V_{aver} = \langle (M_1(p) + M_2(p))/2, (N_1(p) + N_2(p))/2 \rangle \,, \tag{11}$$

$$V_{pes} = \langle \min(M_1(p), M_2(p)), \max(N_1(p), N_2(p)) \rangle \,, \tag{12}$$

Therefore, for each p: $V_{pes}(p) \leq V_{aver}(p) \leq V_{opt}(p)$.

3 Procedure for Significance Assessment of Genetic Algorithms Operators Applying IFL

The implementation of IFL for assessment the significance of GA operators steps on the construction of validity and non-validity degrees. For that purpose, it is required the algorithm to be performed in two different intervals of model parameters variation. One interval could be determined as so-called "broad" range known from the literature [4]. The other one, called "narrow" range, is user-defined and might be obtained using different criteria — e.g. based on the minimum and maximum values, or on the average ones, or some other.

In this study a procedure for significance assessment of GA operators applying IFL is proposed: at the beginning, a number of runs for each value of the GA parameters, object of the investigation, have to be performed in both "broad" and "narrow" ranges of model parameters. In this study, the "narrow" range is constructed based on the average values of model parameters for the concrete value of GA parameter. Further, the degrees of validity/non-validity are determined according to (9) for each value of GA parameters. Then, according to (10)–(12), *optimistic*, *average* and *pessimistic* values are defined for each one of the model parameters. Next, determined in such way values are assigned to each of the model parameters for each of the ranges for each of GA parameters. Finally, based on these assigns, the significance of GA operators is assessed.

4 Significance of Genetic Algorithms Operators in Standard SGA

Standard SGA [1] has been applied to parameter identification of *S. cerevisiae* fed-batch cultivation. Following model (1)–(6) of *S. cerevisiae* fed-batch cultivation, nine model parameters have been estimated altogether. When one of GA parameters GGAP, XOVR or MUTR is investigated according to Table 1, the basic values for the other two are as follows, according to some statements [8]: GGAP = 0.8, XOVR = 0.95 and MUTR = 0.05. These values are considered as referent points. Based on the thorough investigations done in [7], the following values of three main GA parameters are suggested as more appropriate: GGAP = 0.5, XOVR = 0.85 and MUTR = 0.1. These values are considered as optimal. The type of genetic operators are as tuned in [7]. GA is terminated when a certain number of generations is fulfilled, in this case 100. Parameter identification of the model (1)–(6) has been performed using *Genetic Algorithm Toolbox* [11] in *Matlab 7* environment. All the computations are performed using a PC Intel Pentium 4 (2.4 GHz) platform running Windows XP.

Table 2 presents previously used "broad" range with low (LB) and up (UB) boundaries for each model parameter according to [4] as well as new "narrow" range, proposed as presented above, alltogether for three GA parameters. Additionally, Table 2 consists of degrees of validity/non-validity, obtained by (9).

Table 3 presents the boundaries for the *optimistic*, *average* and *pessimistic* prognoses for model parameters at different values of GA parameters, obtained based on (10)–(12).

Since the average values of objective functions and convergence time have been presented in [7], Table 4 lists only the average values of model parameters when SGA has been executed at different values of GA parameters.

If one would like to go in details in the data presented in Table 4, it is worth to note that there are few parameters that differ more than accepted as refer point of 5 % deviation when model parameter values corresponding to referent and optimal values have been compared. If one has a look at the GGAP, there are only 2 parameters - $k_L^{O_2}a$ and Y_{OS} that differ with, respectively, 11.96 % and 11.29 % when GGAP = 0.5 (optimal value) is used instead of GGAP = 0.8 (referent value). Altogether 6 parameters differ more than 5 % when the referent value of XOVR has been replaced with the optimal one: μ_{2E}, k_S and Y_{OE} increase with, respectively, 5.50 %, 7.45 % and 72.92 %, while μ_{2S}, Y_{EX} and Y_{OS} decrease with, respectively, 11 %, 13.15 % and 5.97 %. In the case of MUTR only Y_{OE} increases with 6.51 %. Nevertheless, all parameters are within acceptable from biotechnological point of view limits.

Table 5 lists the number and type of the estimations assigned to model parameters when SGA is applied at referent and optimal values of GA parameters.

As seen from Table 5, there are 3 absolutely "winners" with only *optimistic* prognoses - at optimal value of GGAP and both cases of MUTR - at optimal and at referent values. But if anyone consideres Table 5 parameter by parameter, when GGAP is changed from referent to optimal value, only 2 parameters change with more than 5 % (see above) but namely they both "collapse" to the

Table 2. Model parameters boundaries

SGA			μ_{2S}	μ_{2E}	k_S	k_E	Y_{SX}	Y_{EX}	$k_L^{O_2}a$	Y_{OS}	Y_{OE}
previously used		LB	0.9	0.05	0.08	0.5	0.3	1	0.001	0.001	0.001
		UB	1	0.15	0.15	0.8	10	10	300	1000	1000
GGAP 0.8	advisable after procedure application	LB	0.92	0.1	0.11	0.77	0.4	1.38	88	700	150
		UB	0.98	0.12	0.13	0.8	0.43	1.47	95	750	160
	degrees of validity of *p*	$M_1(p)$	0.92	0.67	0.73	0.96	0.04	0.14	0.29	0.70	0.15
	degree of non-validity of *p*	$N_1(p)$	0.02	0.20	0.13	0.00	0.96	0.85	0.68	0.25	0.84
GGAP 0.5	advisable after procedure application	LB	0.92	0.1	0.11	0.76	0.4	1.33	99	780	150
		UB	0.98	0.12	0.13	0.8	0.43	1.43	106	830	165
	degrees of validity of *p*	$M_2(p)$	0.92	0.67	0.73	0.95	0.04	0.13	0.33	0.78	0.15
	degree of non-validity of *p*	$N_2(p)$	0.02	0.20	0.13	0.00	0.96	0.86	0.65	0.17	0.84
XOVR 0.95	advisable after procedure application	LB	0.9	0.12	0.11	0.77	0.39	1.6	85	680	80
		UB	0.96	0.14	0.13	0.8	0.43	1.8	92	730	90
	degrees of validity of *p*	$M_1(p)$	0.90	0.80	0.73	0.96	0.04	0.16	0.28	0.68	0.08
	degree of non-validity of *p*	$N_1(p)$	0.04	0.07	0.13	0.00	0.96	0.82	0.69	0.27	0.91
XOVR 0.85	advisable after procedure application	LB	0.94	0.11	0.12	0.77	0.4	1.4	80	640	140
		UB	1	0.12	0.14	0.8	0.43	1.55	90	690	155
	degrees of validity of *p*	$M_2(p)$	0.94	0.73	0.80	0.96	0.04	0.14	0.27	0.64	0.14
	degree of non-validity of *p*	$N_2(p)$	0.00	0.20	0.07	0.00	0.96	0.85	0.70	0.31	0.85
MUTR 0.05	advisable after procedure application	LB	0.94	0.11	0.12	0.77	0.4	1.5	90	720	130
		UB	1	0.13	0.14	0.8	0.43	1.7	100	780	150
	degrees of validity of *p*	$M_1(p)$	0.94	0.73	0.80	0.96	0.04	0.15	0.30	0.72	0.13
	degree of non-validity of *p*	$N_1(p)$	0.00	0.13	0.07	0.00	0.96	0.83	0.67	0.22	0.85
MUTR 0.1	advisable after procedure application	LB	0.91	0.12	0.12	0.77	0.39	1.5	90	720	140
		UB	0.98	0.14	0.14	0.8	0.42	1.7	100	780	160
	degrees of validity of *p*	$M_2(p)$	0.91	0.80	0.80	0.96	0.04	0.15	0.30	0.72	0.14
	degree of non-validity of *p*	$N_2(p)$	0.02	0.07	0.07	0.00	0.96	0.83	0.67	0.22	0.84

Table 3. Prognoses for SGA performance

		μ_{2S}		μ_{2E}		k_S		k_E		Y_{SX}		Y_{EX}		$k_L^{O_2}a$		Y_{OS}		Y_{OE}	
		LB	UB	LB	UB	LB	UB	LB	UB	LB	UB	LB	UB	LB	UB	LB	UB	LB	UB
GGAP	V_{opt}	0.92	0.98	0.10	0.12	0.11	0.13	0.77	0.80	0.40	0.43	1.38	1.47	99	106	780	830	150	165
	V_{aver}	0.92	0.98	0.10	0.12	0.11	0.13	0.77	0.80	0.40	0.43	1.36	1.45	94	101	740	790	150	163
	V_{pes}	0.92	0.98	0.10	0.12	0.11	0.13	0.76	0.80	0.40	0.43	1.33	1.43	88	95	700	750	150	160
XOVR	V_{opt}	0.94	1.00	0.12	0.14	0.12	0.14	0.77	0.80	0.40	0.43	1.60	1.80	85	92	680	730	140	155
	V_{aver}	0.92	0.98	0.12	0.13	0.12	0.14	0.77	0.80	0.40	0.43	1.50	1.68	82.5	91	660	710	110	123
	V_{pes}	0.90	0.96	0.11	0.12	0.11	0.13	0.77	0.80	0.39	0.43	1.40	1.55	80	90	640	690	80	90
MUTR	V_{opt}	0.94	1.00	0.12	0.14	0.12	0.14	0.77	0.80	0.40	0.43	1.50	1.70	90	100	720	780	140	160
	V_{aver}	0.93	0.99	0.12	0.14	0.12	0.14	0.77	0.80	0.40	0.43	1.50	1.70	90	100	720	780	135	155
	V_{pes}	0.91	0.98	0.11	0.13	0.12	0.14	0.77	0.80	0.39	0.42	1.50	1.70	90	100	720	780	130	150

pessimistic prognoses and this is the case with the most *pessimistic* prognoses. When XOVR is changed from referent to optimal value, altogether 6 parameters change with more than 5 % (see above) but results in both cases might be considered as similar: model parameters at the optimal value "lose" only one of the *optimistic* prognosis that becomes *average* one. When MUTR is changed

Table 4. Results from model parameter identification

Parameter/ Value	GGAP		XOVR		MUTR	
	ref 0.8	opt 0.5	ref 0.95	opt 0.85	ref 0.05	opt 0.1
μ_{2S}, 1/h	0.95	0.95	0.92	0.97	0.97	0.94
μ_{2E}, 1/h	0.11	0.11	0.13	0.12	0.12	0.13
k_S, g/l	0.12	0.12	0.12	0.13	0.13	0.13
k_E, g/l	0.80	0.78	0.80	0.80	0.80	0.79
Y_{SX}, g/g	0.42	0.41	0.41	0.41	0.41	0.41
Y_{EX}, g/g	1.43	1.38	1.69	1.47	1.56	1.64
$k_L^{O_2}a$, 1/h	91.56	102.51	88.56	84.96	96.18	94.63
Y_{OS}, g/g	722.58	804.16	703.55	661.53	748.61	752.30
Y_{OE}, g/g	154.84	156.64	85.29	147.49	142.62	151.90

Table 5. Model parameters estimations

	GGAP		XOVR		MUTR	
	ref 0.8	opt 0.5	ref 0.95	opt 0.85	ref 0.05	opt 0.1
opt	7	9	7	6	9	9
aver	0	0	1	2	0	0
pes	2	0	1	1	0	0

from referent to optimal value, only one parameter changes with more than 5 % (see above) but results remain only with the *optimistic* prognoses. That means that the MUTR is with the less "sensitivity" to the changes. Thus, based on the intuitionistic fuzzy estimations of the model parameters and further constructed prognoses, GGAP is again distinguished as the most sensitive GA parameter, due to even small changes (only 2 parameters exceeding 5 %) lead to significant difference in the final outcome, namely reliable values of model parameters. GGAP is followed by XOVR with more or less comparable results for referent and optimal values and - at last - MUTR with no changes observed. In general, these results confirm once again those reported in [7] towards model accuracy and algorithm convergence time: GGAP is the most sensitive GA parameter, followed by XOVR and MUTR. Since each of investigated here three GA parameters are closely associated to GA operators, it could be inferred that selection is the operator with the most significant influence both on the model accuracy and algorithm convergence time, as well as on the model parameters values.

Figure 1 shows results from experimental data and model prediction, respectively, for biomass (top left), ethanol (top right), substrate (bottom left) and dissolved oxygen (bottom right) when SGA has been applied with the optimal values of three investigated GA parameters, namely GGAP = 0.5, XOVR = 0.85 and MUTR = 0.1.

The obtained results show the effectiveness of SGA applied with optimal values for three investigated here GA parameters.

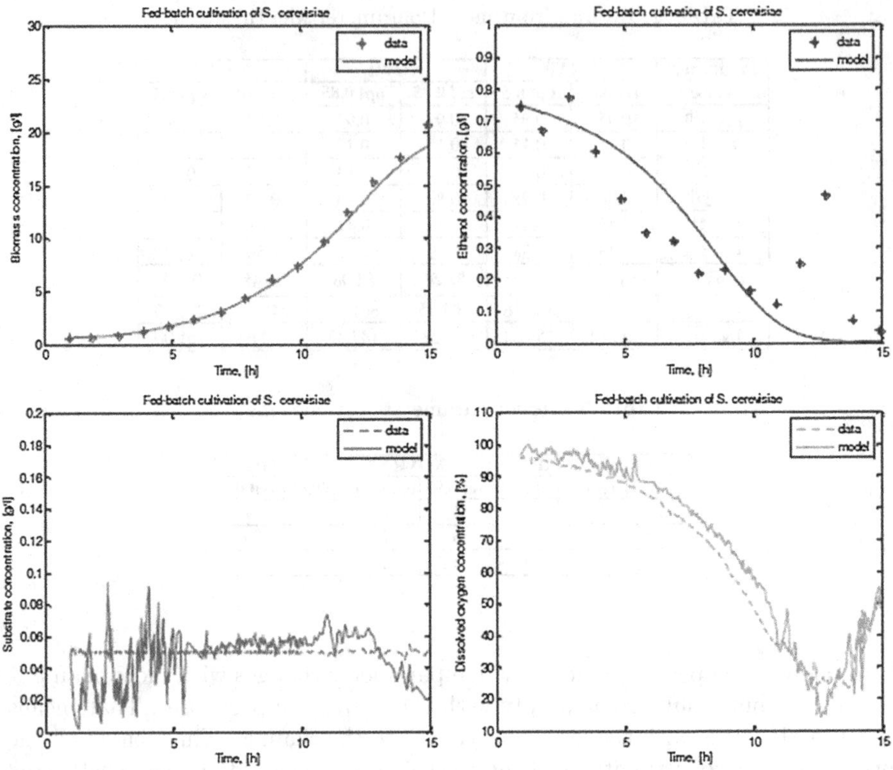

Fig. 1. Model prediction compared to experimental data when SGA has been applied with the optimal values

5 Conclusions

In this investigation intuitionistic fuzzy logic has been implemented in order to assess the significance of GA operators for the purposes of parameter identification of *S. cerevisiae* fed-batch cultivation. For that aim a procedure has been developed and further applied to SGA for three main GA operators, namely selection, crossover and mutation. After the procedure implementation and based on the constructed *optimistic*, *average* and *pessimistic* prognoses for model parameters, results have been compared. This study confirms once again the statement previously reported by authors: GGAP is again distinguished as the most sensitive GA parameter, since even small changes lead to significant difference in the results. Changes of XOVR do not have as strong influence and results are more or less comparable, while changes of MUTR even do not reflect. Since the three investigated GA parameters are closely related to GA operators, it could be inferred that selection is the operator with the most significant influence.

Presented here "cross-evaluation" based on IFL appears as an appropriate tool and might be applied for reliable assessment of other GA parameters, different optimization algorithms as well as to various objects of parameter identification.

Acknowledgements. This work is partially supported by National Science Fund of Bulgaria, grants DID 02-29 and DMU 03-38.

References

1. Goldberg, D.: Genetic Algorithms in Search, Optimization and Machine Learning. Addison-Wiley Publishing Company, Massachusetts (1989)
2. Jones, K.: Comparison of genetic algorithms and particle swarm optimization for fermentation feed profile determination. In: Proceedings of the CompSysTech'2006, Veliko Tarnovo, Bulgaria, pp. IIIB.8-1–IIIB.8-7, 15–16 June 2006
3. Adeyemo, J., Enitian, A.: Optimization of fermentation processes using evolutionary algorithms - a review. Sci. Res. Essays **6**(7), 1464–1472 (2011)
4. Schuegerl, K., Bellgardt, K.-H. (eds.): Bioreaction Engineering, Modeling and Control. Springer, Berlin (2000)
5. Roeva, O. (ed.): Real-World Application of Genetic Algorithms. In Tech, Rijeka (2012)
6. Pencheva, T., Roeva, O., Hristozov, I.: Functional State Approach to Fermentation Processes Modelling. Prof. Marin Drinov Academic Publishing House, Sofia (2006)
7. Angelova, M., Pencheva, T.: Tuning genetic algorithm parameters to improve convergence time. Int. J. Chem. Eng., Article ID 646917 (2011)
8. Obittko, M.: Genetic algorithms (2005). http://cs.felk.cvut.cz/~xobitko/ga/main.html
9. Atanassov, K.: Intuitionistic Fuzzy Sets. Springer, Heidelberg (1999)
10. Atanassov, K.: Intuitionistic Fuzzy Sets Theory. Springer, Berlin (2012)
11. Chipperfield, A.J., Fleming, P., Pohlheim, H., Fonseca, C.M.: Genetic algorithm toolbox for use with MATLAB, User's guide, version 1.2. Department of Automatic Control and System Engineering, University of Sheffield, UK (1994)

Influence of the Number of Ants on Multi-objective Ant Colony Optimization Algorithm for Wireless Sensor Network Layout

Stefka Fidanova[1(✉)], Pencho Marinov[1], and Marcin Paprzycki[2,3]

[1] Institute of Information and Communication Technologies, Bulgarian Academy of Sciences, Acad. G. Bonchev Street Bl25A, 1113 Sofia, Bulgaria
{stefka,pencho}@parallel.bas.bg
[2] Systems Research Institute, Polish Academy of Sciences, Warsaw, Poland
[3] Management Academy, Warsaw, Poland
marcin.paprzycki@ibspan.waw.pl

Abstract. Wireless sensor networks monitor physical or environmental conditions. One of key objectives during their deployment is full coverage of the monitoring region with a minimal number of sensors and minimized energy consumption of the network. The problem is hard from the computational point of view. Thus, the most appropriate approach to solve it is application of some metaheuristics. In this paper we apply multi-objective Ant Colony Optimization to solve this important telecommunication problem. The aim is to study the influence of the number of the ants on the algorithm performance.

1 Introduction

A sensor is a device which can collect and transmit data. First the wireless sensor networks were used by the military for reconnaissance and surveillance [2]. Examples of possible applications are forest fire prevention, volcano eruption study [14], health data monitoring [16], civil engineering [12], and others. Sensor networks depend on deployment of sensors. The sensors can sense any various phenomena or material such as temperature, voltage, or chemical substances. A Wireless Sensor Network (WSN) allows automatic monitoring.

The energy for collecting data and its transmission comes from the battery of a node. In battery-powered systems, higher data rates and more frequent radio use consume more power. One of the nodes of the WSN has special role. It is a High Energy Communication Node (HECN), which collects data from across the network and transmits it to the main computer to be processed. The sensors transmit their data to the HECN, either directly or via hops, using closest sensors as communication relays. When deploying a WSN, the positioning of the sensor nodes becomes one of major concerns. The coverage obtained with the network and the economic cost of the network depends directly on it. Note that, the WSN can have large numbers of nodes, and therefore the task of selecting the geographical positions of the nodes for an optimally designed network can be very

I. Lirkov et al. (Eds.): LSSC 2013, LNCS 8353, pp. 232–239, 2014.
DOI: 10.1007/978-3-662-43880-0_25,

complex. Thus, it is unpractical to solve the problem with traditional numerical methods. In this case, one of the best choices is to apply some metaheuristic method.

The problem is multi-objective with two objective functions. They are (1) minimizing the energy consumption of the nodes in the network, and (2) minimizing the number of the nodes. The full coverage of the network and connectivity are considered as constraints. It is an NP-hard multi-objective problem. We propose a multi-objective ant (ACO) algorithm, which solves the WSN layout problem. Our aim is to study the influence of the number of ants on the algorithm performance and quality of the achieved solutions and to find the minimal number of ants which are enough to achieve good solutions.

Jourdan [8] solved an instance of the WSN layout using a multi-objective genetic algorithm. In their formulation, a fixed number of sensors had to be placed in order to maximize the coverage. In some applications the most important is the network energy. In this context, in [7] an ACO algorithm was proposed, while in [15] an evolutionary algorithm was applied to this variant of the problem. In [4] an ACO algorithm was investigated that took into account only the number of the sensors. In [10] several evolutionary algorithms to solve the problem were proposed. Finally, in [9] a genetic algorithm, which achieves similar solutions as the algorithms in [10] was studied, but tested on small test problems.

The paper is organized as follows. In Sect. 2 the WSN is introduced and the layout problem is formulated. Section 3 presents the ACO algorithm. In Sect. 4 we show the experimental results. Finally, Sect. 5 contains concluding remarks.

2 Problem Formulation

A wireless sensor network consists of spatially distributed autonomous sensors that cooperatively monitor physical or environmental conditions, such as temperature, sound, vibration, pressure, motion, or pollutants. The development of wireless sensor networks was motivated by military applications such as battlefield surveillance, and are now used in many industrial and civilian application areas, including industrial process monitoring and control, machine health monitoring, environment and habitat monitoring, health-care applications, home automation, and traffic control, etc.

Each node in a sensor network is equipped with wireless communications device and an energy source, usually a battery. A sensor node might vary in size and cost. Each sensor node sens an area around itself. The sensing radius determines the sensing area of the node. The nodes communicate among themselves using wireless communication links, determined by a communication radius. The HECN is responsible for the external access to the network. Therefore, every sensor node in the network must have communication with the HECN. Since the communication radius is often much smaller than the network size, direct links are not possible for the peripheral nodes. A multi-hop communication path is then established for those nodes that are far from the HECN. Overall, the quantity of the transmitted data defines the used energy. The node with the highest energy defines the energy of the network. Note that an unspecified number of sensor nodes has to be placed in a terrain to provide full coverage. Therefore, the objectives are to construct a network, with minimal number of sensors (cheapest for construction) and with minimal energy (cheapest for exploitation), while keeping the connectivity of the network.

3 Multi-objective ACO for WSN Layout

Multi-Objective Optimization (MOP) has his roots in the nineteenth century in the work in economics, of Edgeworth and Pareto [11]. The optimal solution for MOP is not a single solution as for mono-objective optimization problems, but a set of solutions defined as Pareto optimal solutions. A solution is Pareto optimal if it is not possible to improve a given objective without deteriorating at least another objective. The main goal of the resolution of a multi-objective problem is to obtain the Pareto optimal set and consequently the Pareto front. One solution dominates another if minimum one of its component is better than the same component of other solutions and other components are not worse. The Pareto front is the set of non-dominated solutions. When metaheuristics are applied, the goal becomes to obtain solutions close to the Pareto front.

We apply multi-objective ant colony optimization to solve the problem. The idea for ant algorithm comes from the real ant behavior. When walking, they put on the ground chemical substance called pheromone. The ants smell the pheromone and follow the path with a stronger pheromone concentration. Thus they find shorter path between the nest and the food. The ACO algorithm uses a colony of artificial ants that behave as cooperating agents. With the help of the pheromone they try to construct better solutions and to find the optimal ones. The problem is represented by a graph and the solution is represented by a path in the graph or by tree in the graph. Ants start from random nodes and construct feasible solutions. When all ants construct their solution we update the pheromone. Ants compute a set of feasible moves and select the best one, according to the transition probability rule. The transition probability p_{ij}, to chose the node j when the current node is i, is based on the heuristic information η_{ij} and on the pheromone level τ_{ij} of the move, where $i, j = 1, \ldots, n$.

$$p_{ij} = \frac{\tau_{ij}^{\alpha} \eta_{ij}^{\beta}}{\sum\limits_{k \in \{allowed\}} \tau_{ik}^{\alpha} \eta_{ik}^{\beta}} \qquad (1)$$

The ant selects the move with highest probability. The initial pheromone is set to a small positive value τ_0 and then ants update this value after completing the construction stage [1,5]. In our implementation we use the MAX-MIN Ant System (MMAS) [3,13], which is one of the most successful ant approach. The main feature of the MMAS is using a fixed upper bound τ_{max} and a lower bound τ_{min} of the pheromone. Thus the accumulation of big amounts of pheromone by part of the possible movements and repetition of same solutions is partially prevented. In our case the graph of the problem is represented by a square grid. The ants will deposit their pheromone on the nodes of the grid. We will deposit the sensors on the nodes of the grid. The solution is represented by tree starting by the high energy communication node. An ant starts to create the rest of the solution from a random node, which communicates with the HECN. Using transition probability (Eq. 1), the ant chooses the next node to visit. If there is more than one node with the same probability, the ant chooses one of them randomly. Construction of the heuristic information is a crucial point in ant algorithms. Our heuristic information is a product of three values (Eq. 2).

$$\eta_{ij}(t) = s_{ij} l_{ij} (1 - b_{ij}), \tag{2}$$

where s_{ij} is the number of the new points which the sensor will cover, and

$$l_{ij} = \begin{cases} 1 \text{ if communication exists;} \\ 0 \text{ if there is not communication,} \end{cases} \tag{3}$$

b is the solution matrix and the matrix element $b_{ij} = 1$ when there is sensor on this position otherwise $b_{ij} = 0$. With s_{ij} we try to increase the number of points covered by one sensor and thus to decrease the number of sensors we need. With l_{ij} we guarantee that all sensors will be connected. The search stops when $p_{ij} = 0$ for all values of i and j.

The pheromone trail update rule is given by:

$$\tau_{ij} \leftarrow \rho\tau_{ij} + \Delta\tau_{ij}, \tag{4}$$

$$\Delta\tau_{ij} = \begin{cases} 1/F(k) \text{ if } (i,j) \in \text{non-dominated solution constructed by ant } k, \\ 0 \qquad \text{otherwise}. \end{cases}$$

We decrease the pheromone with a parameter $\rho \in [0, 1]$. This parameter models evaporation in the nature and decreases the influence of old information in the search process. After that, we add the new pheromone, which is proportional to the value of the fitness function. If the pheromone of some node becomes less than the lower bound of the pheromone we put it to be equal to the lower bound and thus we prevent the pheromone of some nodes to become very low close to 0 (and to be undesirable). It is a kind of diversification of the search. The F is the fitness function. The role of the fitness function is to estimate the achieved solutions. The aim is to add more pheromone on non-dominated solutions and thus to force the ants to search around them for new non-dominated solutions. The fitness function is constructed as follows:

$$F(k) = \frac{f_1(k)}{\max_i f_1(i)} + \frac{f_2(k)}{\max_i f_2(i)} \tag{5}$$

Where $f_1(k)$ is the number of sensors achieved by the kth ant and $f_2(k)$ is the energy of the solution of the kth ant. These are also the objective functions of the WSN layout problem. We normalize the values of two objective functions with their maximal achieved values from the first iteration.

4 Experimental Results

Every ant start to create its solution from random point. In our case it is such point, which communicates with the HECN. Thus the ant algorithm uses small number of agents (ants). Smaller number of ants means less memory, which is important when we solve large problems. The aim of this work is to learn the influence of the number of the ants on quality of the solution.

We have created a software which realizes our ant algorithm. Our software can solve the problem at any rectangular area, the communication and the coverage radius can be different and can have any positive value. The HECN can be fixed in any point in the area. The program was written in C language and the tests were run on computer with Intel Pentium 2.8 GHz processor. In our tests we use an example where the area is square and consists of 500 points in every side. The coverage and communication radii cover 30 points. The HECN is fixed in the center of the area. We use this example for comparison, because other authors use the same. We apply our algorithm on smaller test problem too. The area consists of 350×350 points. The HECN is fixed in the center of the area, the coverage and communication radii are as in a previous case.

In our previous work [6], we showed that our ant algorithm outperforms existing algorithms for this problem. There, after several runs of the algorithm we specify the most appropriate values of its parameters. We apply MAX-MIN ant algorithm with the following parameters: $\alpha = \beta = 1$, $\rho = 0.5$. In the ACO, if we fix the number of iterations and double the number of ants the execution time will be doubled. We study the influence of the number of ants on the quality of the solutions. We fixed the number of the iterations to be 60 (H ant) and the number of ants to have following values $\{1, 2, 3, 4, 5, 6, 7, 8, 9, 10\}$.

We run our ACO algorithm 30 times for each number of ants. We extract the Pareto front from the solutions of these 30 runs. In Tables 1 and 2 we show the achieved non dominated solutions (Pareto fronts) for case 500×500 and 350×350 respectively. In the left column are the number of sensors and in other columns is the energy corresponding to this number of sensors and the number of ants. Analyzing the Table 1 (case 500×500) we observe that the Pareto front achieved by 6 ants dominates the Pareto fronts achieved by 1, 2, 3, 4 and 5 ants. The is not dominance between Pareto fronts achieved by 6, 7, 8, 9 and 10 ants and we cannot say which of them is better. Analyzing the Table 2 (case 350×350) we observe that the Pareto front achieved by 3 ants is dominated by other Pareto fronts. The Pareto fronts achieved by 1, 2, 4, 5, 6 and 9 ants are part of the Pareto front achieved by 7, 8 and 10 ants. More ants leads to more computational time. Thus the best Pareto front in the case 350×350 is achieved by 7 ants.

We prepare a Pareto front achieved by all runs of the algorithm with any number of ants (from 6 to 10) and we call it a common Pareto front. In the case 500×500 the common Pareto front is $\{(232, 48), (230, 52), (228, 54), (226, 56), (224, 57), (223, 81)\}$ and for the case 350×350 it is $\{(111, 30), (113, 28), (114, 26), (116, 25)\}$. Let us have a set of number of sensors from 223 to 244 for the case 500×500 and 111 to 116 for the case 350×350 respectively. If for

Table 1. Pareto fronts, example 500 × 500

Sensors	Ants									
	1	2	3	4	5	6	7	8	9	10
244					52					
243										
242										
241										
240	53	53								
239	56			50						
238			53							
237										
236										
235			54						50	
234					53			48	53	
233							51			
232			55	51	54	50	52	51		48
231		55			55		53			
230	57					52		54		
229	58			55				56		56
228									54	
227		57	57		57	56		57		
226	59	95	73	57	59	57	56			
225				58	60	58	57	58		57
224	61				88	65	61	59	57	71
223					89	81				

some number of sensors there is not corresponding energy in the common Pareto front, we put the energy to be equal to the point of the front with lesser number of sensors. We can do this because, if we take some solution and if we include a sensor close to the HECN it will not increase the value of the energy and will increase by 1 only the number of the sensors. Thus, there is corresponding energy to any number of nodes. This front we will call the Extended front. In the case 500 × 500 the Extended front is {(234, 48), (233, 48), (232, 48), (231, 52), (230, 52), (229, 54), (228, 54), (227, 56), (226, 56), (225, 57), (224, 57), (223, 81)}. In the case 350 × 350 the Extended front is {(111, 30), (112, 30), (113, 28), (114, 26), (115, 26), (116, 25)}.

We have included additional criteria to decide which Pareto front is better in the case when there are not dominance between Pareto fronts. We calculated the distance between a Pareto front and the Extended front. To calculate the distance, we extend every element of Pareto fronts in a similar way as the Extended front. The distance between a Pareto front and the Extended front is the sum of distances between the points with a same number of sensors, or it is the difference between their energy. These distances are always positive because the Extended front dominates the Pareto fronts. Thus, by this criteria, the best Pareto front will be the closest to the Extended front.

Table 2. Pareto fronts, example 350 × 350

Sensors	Ants									
	1	2	3	4	5	6	7	8	9	10
111				30	30	30	30	30	30	30
112										
113	28	35	28				28	28	28	28
114	26	26	26	26	26	26	26	26	26	26
115										
116							25	25		25

Table 3. Distances from extended front case 500 × 500

Ants	6	7	8	9	10
Distance	20	23	21	22	29

In Table 3 we show the distances between the Extended front and the Pareto fronts achieved by 6, 7, 8, 9, and 10 ants. Analyzing the Table 3 we conclude that the distance between the Extended front and the Pareto front achieved by 6 ants is the shortest. Thus, by our criteria, the Pareto front (solutions) achieved by 6 ants in the case 500 × 500 is better.

5 Conclusion

In this paper we studied the influence of the number of ants on the performance of the ACO algorithm, applied to the wireless sensor network. Smaller number of ants leads to the shorter running time and minimizes memory use, which is important for complex / large cases. We varied the number of ants, while fixing the number of iterations. Furthermore, we included the concept of an Extended front, as an additional tool to compare Pareto fronts that do not dominate each other. The best Pareto front and the best performance were achieved when the number of ants was equal in case 500 × 500 in the case 350 × 350.

Acknowledgments. This work has been partially supported by the Bulgarian National Scientific Fund under the grants DID 02/29 and DTK 02/44. It is a part of the Poland-Bulgaria bilateral grant "Parallel and distributed computing practices".

References

1. Bonabeau, E., Dorigo, M., Theraulaz, G.: Swarm Intelligence: From Natural to Artificial Systems. Oxford University Press, New York (1999)
2. Deb, K., Pratap, A., Agrawal, S., Meyarivan, T.: A fast and elitist multi-objective genetic algorithm: NSGA-II. IEEE Trans. Evol. Comput. **6**(2), 182–197 (2002)
3. Dorigo, M., Stutzle, T.: Ant Colony Optimization. MIT Press, Cambridge (2004)

4. Fidanova, S., Marinov, P., Alba, E.: Ant algorithm for optimal sensor deployment. In: Madani, K., Correia, A.D., Rosa, A., Filipe, J. (eds.) Studies of Computational Inteligence: Computational Inteligence, vol. 399, pp. 21–29. Springer, Berlin (2012)
5. Fidanova, S., Atanasov, K.: Generalized net model for the process of hybrid ant colony optimization. C. R. Acad. Bulg. Sci. **62**(3), 315–322 (2009)
6. Fidanova, S., Shindarov, M., Marinov, P.: Multi-objective ant algorithm for wireless sensor network positioning. C. R. Acad. Bulg. Sci. **66**(3), 353–360 (2013). ISSN 1310-1331
7. Hernandez, H., Blum, C.: Minimum energy broadcasting in wireless sensor networks: an ant colony optimization approach for a realistic antenna model. J. Appl. Soft Comput. **11**(8), 5684–5694 (2011)
8. Jourdan, D.B.: Wireless sensor network planning with application to UWB localization in GPS-denied environments. Ph.D. Thesis Massachusets Institute of Technology (2000)
9. Konstantinidis, A., Yang, K., Zhang, Q., Zainalipour-Yazti, D.: A multi-objective evolutionary algorithm for the deployment and power assignment problem in wireless sensor networks. J. Comput. Netw. **54**(6), 960–976 (2010)
10. Molina, G., Alba, E., Talbi, E.-G.: Optimal sensor network layout using multi-objective metaheuristics. Univ. Comput. Sci. **14**(15), 2549–2565 (2008)
11. Mathur, V.K.: How well do we know pareto optimality? J. Econ. Educ. **22**(2), 172–178 (1991)
12. Paek, J., Kothari, N., Chintalapudi, K., Rangwala, S., Govindan, R.: The performance of a wireless sensor network for structural health monitoring. In: Proceedings of 2nd European Workshop on Wireless Sensor Networks, Istanbul, Turkey (2005)
13. Stutzle, T., Hoos, H.H.: MAX-MIN ant system. Future Gener. Comput. Syst. **16**, 889–914 (2000)
14. Werner-Allen, G., Lorinez, K., Welsh, M., Marcillo, O., Jonson, J., Ruiz, M., Lees, J.: Deploying a wireless sensor network on an active volcano. IEEE Internet Comput. **10**(2), 18–25 (2006)
15. Wolf, S., Merz, P.: Evolutionary local search for the minimum energy broadcast problem. In: van Hemert, J., Cotta, C. (eds.) EvoCOP 2008. LNCS, vol. 4972, pp. 61–72. Springer, Heidelberg (2008)
16. Yuce, M.R., Ng, S.W., Myo, N.L., Khan, J.Y., Liu, W.: Wireless body sensor network using medical implant band. Med. Syst. **31**(6), 467–474 (2007)
17. Zitzler, E., Thiele, L.: Multiobjective evolutionary algorithms: a comparative case study and the strength pareto approach. IEEE Trans. Evol. Comput. **3**(4), 257–271 (1999)

Dynamic Differential Evolution Algorithm for Clustering Temporal Data

Kristina S. Georgieva(✉) and Andries P. Engelbrecht

Department of Computer Science,
University of Pretoria, Pretoria, South Africa
kristina.s.georgieva@gmail.com,
engel@cs.up.ac.za

Abstract. Temporal data clustering is the process of grouping similar patterns in a dataset together when the patterns change with time. This change in patterns introduces the issue of loss of diversity in differential evolution algorithms. The lack of re-diversification of the population limits the exploration ability of differential evolution algorithms resulting in early convergence around stale solutions. This paper describes and evaluates three algorithms that were applied to a temporal data clustering problem, namely the standard data clustering DE, the reinitialising data clustering DE, and the data clustering DynDE.

1 Introduction

Data clustering [8] refers to the process of grouping similar data patterns together, where the data can be either static or dynamic. Static data refers to already existing data which does not change, while dynamic data refers to frequently changing data [7]. Temporal data [13], which refers to data that changes at some frequency of time, can therefore be considered as a type of dynamic data due to its frequent changes.

A clustering dataset may change in various ways and combinations [5]. Patterns from one cluster may move to another cluster, clusters themselves may also move around the search space as a whole and, lastly, old clusters may disappear and new ones may appear due to migrating, disappearing or appearing patterns.

The differential evolution [16] algorithm is a population-based evolutionary algorithm that uses selection, mutation, and crossover to adapt the individuals of its population. The clustering of data using differential evolution has been previously evaluated by a number of researchers [1,3,9,14,15,17].

When clustering dynamic data, the differential evolution algorithm suffers from loss of diversity [12]. This occurs as individuals begin to converge. The problem emerges when data is dynamic, as exploration of the search space is necessary to evolve to the correct positions and the population needs to be re-diversified in order to explore the search space. Unlike particle swarm optimization algorithms, however, DEs do not have the issue of outdated memory

I. Lirkov et al. (Eds.): LSSC 2013, LNCS 8353, pp. 240–247, 2014.
DOI: 10.1007/978-3-662-43880-0_26,

[2] as they do not store any information about the population's previous state. This may make them more appropriate algorithms for clustering dynamic data.

This paper evaluates the temporal data clustering ability of the data clustering DE proposed in [6]. The paper also adapts this algorithm to re-initialise part of the population if a change in the data has occurred. Lastly, it adapts the DynDE proposed in [10] to perform the clustering process. The three DEs are compared using three measures [5], namely the inter-cluster distance, the intra-cluster distance, and the Ray-Turi validity index. These results are then ranked using the Friedman test.

2 Differential Evolution

The differential evolution [16] algorithm is a population-based evolutionary algorithm guided by distance information about the current population [4]. Each possible solution in the population is a vector, called an individual. Individuals adapt using three evolutionary processes, namely selection, mutation, and crossover, where the mutation process is the one that mainly differentiates the DE from all other evolutionary algorithms.

The mutation process uses a target vector selected from the population at random, as well as distance information from the population to generate a new vector called the trial vector. This is done by adding a scaled difference between two random individuals of the population to a target vector selected at random. The trial vector is created using [4]

$$\mathbf{u}_i(t) = \mathbf{x}_{i_1}(t) + \beta(\mathbf{x}_{i_2}(t) - \mathbf{x}_{i_3}(t)) \tag{1}$$

$\mathbf{x}_{i_1}(t)$ is the target vector, β is the scaling factor, $\mathbf{x}_{i_2}(t)$ is a randomly selected individual from the population that is not $\mathbf{x}_{i_1}(t)$ and $\mathbf{x}_{i_3}(t)$ is a randomly selected individual from the population that is not $\mathbf{x}_{i_1}(t)$ or $\mathbf{x}_{i_2}(t)$.

The current individual being adapted, $\mathbf{x}_i$, is then used by the crossover process to produce a single offspring. The offspring is created by [4]

$$x'_{ij}(t) = \begin{cases} u_{ij} & \text{if } j \text{ is} \in J \\ x_{ij}(t) & \text{if } j \text{ is} \notin J \end{cases} \tag{2}$$

where $\mathbf{u}_i$ is the trial vector, $\mathbf{x}_i$ is the parent vector, and J is a set of crossover points, which are determined by random selection. This paper uses binomial crossover where a crossover probability is used to determine which dimensions of the offspring will become part of the new solutions and which dimensions of the parent vector will remain as part of the solution, ensuring that one randomly selected dimension is forced to be within the set J.

Lastly, a selection algorithm is applied in order to determine which offspring and which parents survive to the next generation. This paper uses an elitist selection strategy, where the individual with the best fitness value between the offspring and the parent survives to the next generation.

3 Data Clustering DE and Its Reinitialising Alternative

The solution to a data clustering problem consists of K vectors, where K is the total number of clusters and each vector represents the position of the center of one cluster. Due to the solution no longer being one vector, the representation of individuals in a clustering DE needs to change. The clustering DE, therefore, represents each solution as a set of centroid positions, where each dimension of the solution is a vector holding the position of a cluster centroid.

The dataset has an influence on determining whether one solution is better or worse than another solution. For this reason, the dataset needs to be used when the fitness of an individual is calculated. This is done by using the quantization error [11] as a fitness measure for the clustering problem. The quantization error is minimised and it is calculated using

$$J_e = \frac{\sum_{k=1}^{K} \frac{\sum_{\forall \mathbf{z_p} \in C_k} d(\mathbf{z_p}, \mathbf{c_k})}{|C_k|}}{K} \tag{3}$$

where $\mathbf{z}$ is a data pattern, C_k is the set of data patterns assigned to the cluster, $d(\mathbf{z_p}, \mathbf{c_k})$ is the Euclidean distance between data pattern $\mathbf{z_p}$ and cluster centroid $\mathbf{c_k}$, $|C_k|$ is the total number of patterns assigned to the centroid, and K is the total number of centroids. This measure can divide by zero if a centroid is positioned far from all patterns leading to no patterns being assigned to it, making $|C_k|$ zero. To remedy this, if $|C_k|$ was zero, the fitness value was approximated to infinity, such bad fitness will, with time, filter out the bad solution.

The data clustering DE [6] performs the selection, mutation, and crossover processes in the same manner as the DE described in Sect. 2. A target vector and two difference vectors are selected at random, a trial vector is then created using (1) on each cluster centroid position, the offspring and the parent individual are then crossed over using (2) on each cluster centroid, and then the individual with the best fitness is selected to survive to the next generation. The complete algorithm is shown in Algorithm 1.

The reinitialising data clustering DE adapts the data clustering DE in an attempt to re-diversify the population and overcome the loss of diversity problem that the DE experiences. This adaptation is the addition of a re-initialisation step, where part of the population is re-initialised if a change in the dataset has been detected.

4 DynDE for Data Clustering

The DynDE algorithm proposed by Mendes and Mohais [10] adapts the original DE described in Sect. 3 by adding an element of exclusion and Brownian individuals in order to increase diversity. It is a multi-population algorithm where each population searches for the optimal solution to a problem in a different area of the search space. If two populations are optimising the same area of the search space, the population with the worst global best solution is re-initialised.

Brownian individuals are individuals that are not adapted using the standard DE strategies discussed, but instead their new positions are generated based on the global best position of their population and Gaussian noise. Gaussian noise is added to the position of the global best individual of a sub-swarm using $\mathbf{b}_i = \mathbf{y}_i + \mathbf{N}(0, \sigma)$ where $\mathbf{b}_i$ is the new value for dimension i of the Brownian individual, $\mathbf{y}_i$ the value for dimension i of the global best individual and $\mathbf{N}(0, \sigma)$ is Gaussian noise with mean 0 and standard deviation σ. Mendes and Mohais [10] evaluated quantum individuals and entropic differential evolution, but concluded that the Brownian individuals displayed the most successful results.

Algorithm 1. Data Clustering Differential Evolution

1: Initialize a population of N individuals with K centroid positions each
2: **while** stopping condition has not been reached **do**
3: **for** each individual x_i **do**
4: **for** each data pattern $\mathbf{z_p}$ **do**
5: Calculate the Euclidean distance $d(\mathbf{z_p}, \mathbf{x_{ij}})$ to all clusters
6: Assign pattern $\mathbf{z_p}$ to cluster C_{ij}, such that $d(\mathbf{z_p}, \mathbf{x_{ij}}) = min_{\forall k=1 \ldots N_k} \{d(\mathbf{z_p}, \mathbf{x_{ik}})\}$
7: **end for**
8: calculate the fitness using equation (3)
9: Select a target entity and two difference entities
10: Generate a trial entity $u_i(t)$ using equation (1)
11: Generate the offspring x_i' using equation (2)
12: **if** $f(x_i')$ is better than $f(x_i)$ **then**
13: add x_i' to the next population
14: **else**
15: add x_i to the next population
16: **end if**
17: **end for**
18: **end while**

Algorithm 2. DynDE

1: Initialise M populations of individuals
2: Assign a percentage of each population to be Brownian Individuals
3: Evaluate each population by performing the steps described in lines 4-9 of Algorithm 1
4: Compare the global best individuals from each population to each other
5: **if** One or more centroids of best individuals of two populations are within an exclusion radius r_{excl} of each other **then**
6: re-initialise the population with the worst global best
7: **else**
8: **for** each population **do**
9: **for** each individual x_i **do**
10: **if** x_i is a normal individual **then**
11: Update x_i using lines 10-12 of Algorithm 1
12: **else if** x_i is a Brownian Individual **then**
13: Update using $b_i = y_i + N(0, \sigma)$
14: **end if**
15: **end for**
16: **end for**
17: Make the old Brownian Individuals normal Individuals
18: Make a percentage of the weakest individuals Brownian Individuals
19: **end if**

The DynDE algorithm begins by initialising the sub-populations, making a percentage of each sub-population Brownian individuals. It then calculates the fitness of all the individuals in each sub-population. If two populations are within an exclusion radius r_{excl} from each other, the entire population with the worst global best solution is re-initialised to random positions within the search space. The populations that were not re-initialised update their normal individuals by using the standard DE update methods described in Sect. 2 and their Brownian individuals using the Brownian individual update. For this paper an asynchronous approach was taken when updating the DynDE individuals, where, instead of adding survivors to the next population, survivors replace their parents in the current population.

What changes when the DynDE algorithm is used for data clustering is simple. The representation of individuals described for the data clustering DE is used and the fitness calculation takes the dataset into consideration. Inherently, the data clustering DE updates are used instead of the DE updates described in Sect. 3. This algorithm is shown in Algorithm 2, where the number of clusters is known.

5 Experimental Procedure

The migrating patterns cluster datasets used in [5] were used as the clustering problems in this paper. Datasets had 8 clusters, 80 patterns per timestep, 100 timesteps and combinations of severities of 1, 2, 3, 4, and 5, frequencies of 1, 2, 3, 4, and 5, and dimensions 3, 8, and 15. A window of 80 patterns was used to slide from one timestep to the next. Higher severities refer to larger changes while higher frequencies refer to less intervals of change, where the $iterationOfChange = \frac{f}{10} * totalIterations$.

Fifty individuals were used and were initialised within the bounds of the dataset. The averages of 30 individual simulations are reported in this paper, where each simulation ran for 1000 generations with 8 populations. For the reinitialising DE, 10 % of the population was re-initialised when a change took place. A boundary constraint was also used, where individuals that pass the boundaries were reset to stay on the boundary. Lastly, a scaling factor and crossover probability of 0.5 were used.

6 Results and Discussion

The inter-cluster distance, intra-cluster distance and Ray-Turi validity index were used as measures to compare the clustering ability of each of the three DEs. Table 1 shows the average values and standard deviations for each of these

Table 1. Averages, standard deviation and chi-square values for inter-cluster distance, intra-cluster distance and Ray-Turi validity for each algorithm's last iteration. Where the Chi-square was calculated with $n = 75, k = 3, df = 2, \alpha = 95\,\%$, showing a statistical significant difference for all three measures where $\chi^2 > 5.991$

Algorithm	Inter-cluster distance	Intra-cluster distance	Ray-Turi validity
Standard	9.029 ± 2.700	21.769 ± 1.191	1.084 ± 0.166
Standard Reinitialising	9.020 ± 2.688	21.772 ± 1.188	1.0716 ± 0.168
DynDE	3.262 ± 1.214	20.074 ± 1.316	0.902 ± 0.388
Chi-Square value χ^2	112.6666667	21.94666667	10.90666667

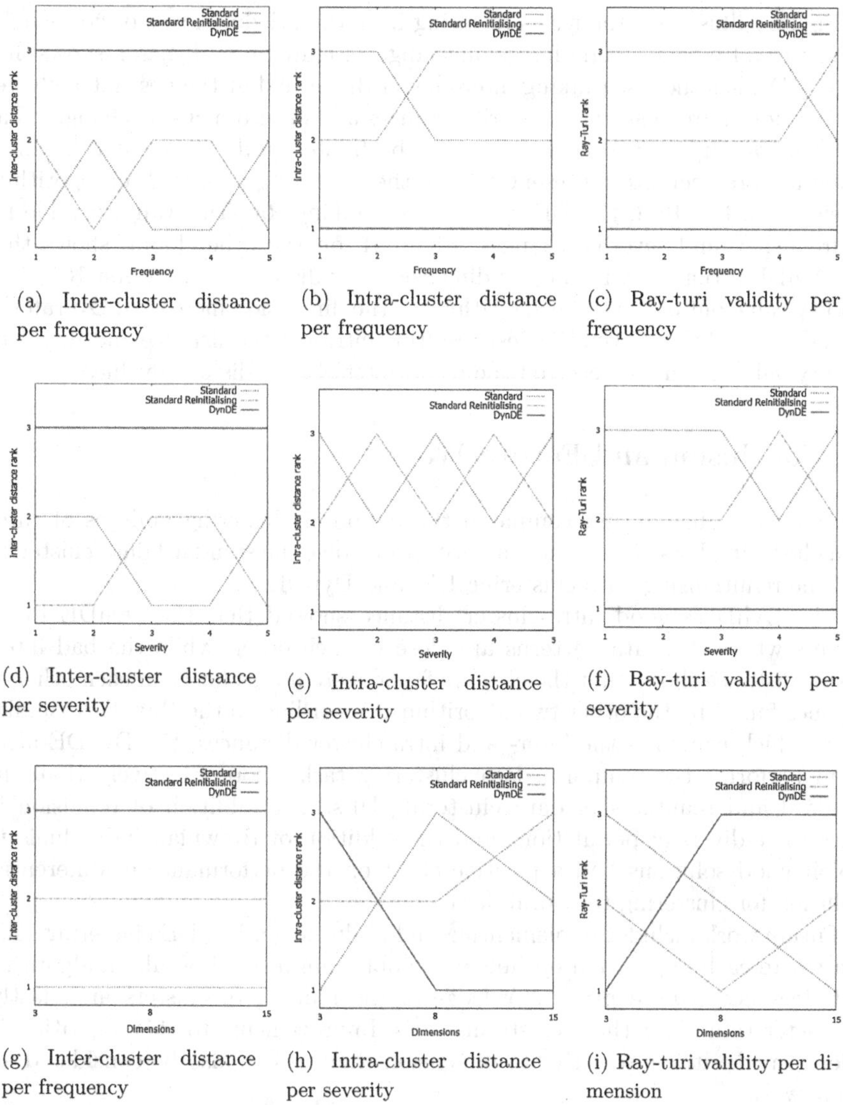

(a) Inter-cluster distance per frequency

(b) Intra-cluster distance per frequency

(c) Ray-turi validity per frequency

(d) Inter-cluster distance per severity

(e) Intra-cluster distance per severity

(f) Ray-turi validity per severity

(g) Inter-cluster distance per frequency

(h) Intra-cluster distance per severity

(i) Ray-turi validity per dimension

Fig. 1. Ranks of Inter-cluster distance, Intra-cluster distance and Ray-Turi validity index per frequency, severity and dimension

three measures. On average, the standard and reinitialising data clustering DEs have the best inter-cluster distances, showing that the clusters found by these algorithms are further apart than the ones found by the DynDE. The DynDE, however, has a more optimal average intra-cluster distance, showing that the clusters found by the algorithm were more compact. Lastly, the DynDE's Ray-Turi validity value shows that the overall clustering capability of the DynDE surpassed that of the standard and reinitialising DEs.

Figure 1 illustrates the ranks resulting from the Friedman test performed for the severity, frequency and dimension changes. Figure 1a, b, d, e, g, h shows that the DynDE has the last-ranking inter-cluster distance, but the best intra-cluster distance for all frequency and severity changes and most dimension changes. The Ray-Turi validity value takes into account both inter- and intra-cluster distances giving a more accurate representation of the clustering ability of an algorithm. As shown in Fig. 1c, f, DynDE has the first-ranking Ray-Turi validity value for all frequency and severity changes. Figure 1i, on the other hand, shows that the DynDE struggled with higher dimensions by displaying the worst Ray-Turi validity value out of the three algorithms as the dimension increased. Overall the DynDE algorithm received the best resulting intra-cluster distance and Ray-Turi validity values, while it received the worst inter-cluster distance values.

7 Conclusion and Future Work

This paper evaluated and compared the temporal clustering abilities of three data clustering DEs. These three algorithms involve the standard data clustering DE, the reinitialising data clustering DE, and DynDE.

The DynDE's good intra-cluster distance showed that the DynDE found clusters where the data patterns are close to each other, while the bad inter-cluster distance shows that the clusters found were not as far from each other as the ones found by the other two algorithms. According to the Ray-Turi validity value, which combines the inter- and intra-cluster distances, the DynDE algorithm performs the temporal data clustering tasks more effectively than the standard and reinitialising data clustering DEs. The addition of repulsion in order to re-diversify populations and the addition of Brownian individuals to exploit good solutions had a positive effect on the performance of differential evolution for clustering temporal data.

Future work includes implementing and evaluating a DynDE clustering solution where each population optimises one optimum instead of all, an algorithm that does not require prior knowledge of the number of clusters and, lastly, parameter tuning for these clustering DEs. Improvements to this algorithm in order to make it less sensitive to dimension changes can also be considered for future work.

References

1. Abraham, A., Das, S., Konar, A.: Document clustering using differential evolution. In: Congress on Evolutionary Computation, pp. 1784–1791 (2006)
2. Blackwell, T.: Particle swarm optimization in dynamic environments. In: Yang, S., et al. (eds.) Evolutionary Computation in Dynamic and Uncertain Environments, pp. 29–49. Springer, Berlin (2007)
3. Das, S., Abraham, A., Konar, A.: Automatic clustering using an improved differential evolution algorithm. IEEE Trans. Syst. Man Cybern. Part A Syst. Humans **38**(1), 218–237 (2008)

4. Engelbrecht, A.P.: Computational Intelligence: An Introduction, Chap. 1. Wiley, New York (2007)
5. Graaff, A.: A local network neighbourhood artificial immune system. Ph.D. thesis, University of Pretoria, June 2011
6. Hanuman, S., Babu, V., Govardhan, A., Satapathy, S.: Data clustering using almost parameter free differential evolution technique. Int. J. Comput. Appl. **8**(13), 1–7 (2010)
7. Iyengar, A., Ramaswamy, L., Schroeder, B.: Techniques for efficiently serving and caching dynamic web content. In: Tang, X., et al. (eds.) Web Content Delivery. Web Information Systems Engineering and Internet Technologies Book Series, vol. 2, Chap. 5, pp. 101–130. Springer, New York (2005)
8. Jain, A.K.: Data clustering: 50 years beyond k-means. Pattern Recogn. Lett. **31**, 651–666 (2009)
9. Maulik, U., Saha, I.: Modified differential evolution based fuzzy clustering for pixel classification in remote sensing imaging. Pattern Recogn. **42**(9), 2135–2149 (2009)
10. Mendes, R., Mohais, A.S.: Dynde: a differential evolution for dynamic optimization problems. In: The 2005 IEEE Congress on Evolutionary Computation, vol. 3, pp. 2808–2815, Sept 2005
11. van der Merwe, D.W., Engelbrecht, A.: Data clustering using particle swarm optimization. In: The 2003 Congress on Evolutionary Computation, CEC '03, vol. 1, pp. 215–220 (2003)
12. Pant, M., Thangaraj, R., Abraham, A., Grosan, C.: Differential evolution with laplace mutation operator. In: Congress on Evolutionary Computation, pp. 2841–2849 (2009)
13. Patel, J.: Temporal database system. Master's thesis, Department of Computing, Imperial College, University of London, June 2003
14. Paterlini, S., Krink, T.: High performance clustering with differential evolution. In: Congress on Evolutionary Computation, vol. 2, pp. 2004–2011 (2004)
15. Paterlini, S., Krink, T.: Differential evolution and particle swarm optimisation in partitional clustering. Comput. Stat. Data Anal. **50**(5), 1200–1247 (2006)
16. Price, K., Storn, R., Lampinen, J.: Differential Evolution: A Practical Approach to Global Optimization. Springer, Berlin (2005)
17. Tian, Y., Liu, D., Qi, H.: K-harmonic means data clustering with differential evolution. In: Conference on Future BioMedical Information Engineering, pp. 369–372 (2009)

Adaptive Critic Design and Heuristic Search for Optimization

Petia Koprinkova-Hristova(✉)

Institute of Information and Communication Technologies,
Bulgarian Academy of Sciences, Acad. G. Bonchev, bl. 25A, 1113 Sofia, Bulgaria
pkoprinkova@bas.bg

Abstract. The main aim of the present work is to combine Adaptive Critic Design (ACD) approach that falls with the gradient optimization techniques with the associative learning that is heuristic search algorithm. The relatively new neural network structure — Echo state network (ESN) — is used as critic network within ACD scheme. It is trained minimizing temporal difference error using Recursive Least Squares (RLS) algorithm. The actor in ACD scheme is trained by associative learning. The proposed approach is tested on optimization of a complex nonlinear process for biopolymer production. The obtained previously results using gradient descent algorithm for actor training are compared with those obtained using heuristic search algorithm and the advantages and shortcomings of both methods are discussed.

Keywords: Reinforcement learning · Hebbian learning · Adaptive critic design · Echo state networks

1 Introduction

Reinforcement learning (RL) is introduced as a method of artificial neural network training "by experience", rather than "by examples". Created initially to mimic animal behavior in an attempt to explain Pavlovian conditioning, RL is also recognized as an approximation of Bellman's dynamic programming method [2] that is well known in the control community. During the last thirty years theoretical developments in this field (a very exhaustive retrospective can be found in [9]) have lead to methodologies known as neuro-dynamic programming [3] and adaptive critic designs (ACD) [12] also commonly known as Adaptive Dynamic Programming. The core of the methods is the approximation of Bellman's equation or value function (which is the discounted sum of future rewards) using neural networks (also called "heuristic adaptive critic"). Having such well-trained critic networks allows solving dynamic programming or RL tasks in a forward manner. Different training schemes for adaptive critic design depend on the presence or absence of a model of the environment [12]. In both cases the critic's training is done using temporal difference (TD) error [18] thereby mimicking the brain's ability to learn how to predict future outcomes on the basis

I. Lirkov et al. (Eds.): LSSC 2013, LNCS 8353, pp. 248–255, 2014.
DOI: 10.1007/978-3-662-43880-0_27, © Springer-Verlag Berlin Heidelberg 2014

of previous experience without awaiting the final results from future actions. The key component of ACD training and solving the optimization task is the backpropagation method that is gradient method based on the chain rule of derivative calculation [20]. Usually the critic is trained off-line since it needs a collection of a variety of data from the beginning to the end of several process runs. Combination between off-line and on-line learning is also considered [13]. True on-line applications of ACD approaches, however, needs very fast training algorithms [14]. In highly non-linear environments the necessity for additional feedback connections arises, which further complicates the on-line training. In such cases the application of backpropagation trough time (BPTT) [20] is an alternative. However, it is impossible to be used in an on-line mode. Instead of that the Extended Kalman Filter (EKF) method [4] is usually applied, which is more complicated and resource demanding. Hence it is crucial to work towards finding simply trainable recurrent network structures for ACD schemes.

The recently proposed ESN structure [4,5,10] incorporates a dynamic reservoir generated randomly and easily trainable output neurons. The less complex and much faster Recursive Least Square method (RLS) [4] can be applied for their on-line training. Moreover, the derivative calculation with respect to the ESN inputs (that is needed for gradient descent), requires much less computational effort, because of the ESN structure that naturally separates the reservoir from its input and output connections. In our previous investigations we applied this approach to a robot control task for obstacle avoidance [8]. In [6] on-line training of ESN critic for solving the optimization task of a complex nonlinear process of biopolymer production is investigated. From biological point of view however, the gradient learning is considered as non-plausible. It is claimed that associative learning algorithms like Hebbian law are closer to the biological neurons behavior. In fact the ACD originate from the first actor-critic scheme [1] that uses associative learning algorithms for both critic and actor. In [11] it was proven that heuristic search algorithms arising from animal behavior such as ant colony optimization can be considered as analog to the stochastic gradient descent algorithms and to reinforcement learning techniques.

In the present study combination between non-associative training of critic network and associative training of the actor is proposed. It combines the advantage of on-line training of ESN critic using RLS and exploits biologically plausible Hebbian law for the actor training. The results are compared with previous work [6] using the gradient training of actor.

2 Problem Statement

2.1 ACD Approach

The ACD approach also called neural dynamic programming or heuristic dynamic programming [3,12] is an approximation of the classical dynamic programming in which the Bellman equation is approximated by a neural network that is then used to predict the future utility function to be minimized by adjusting control actions. The scheme for on-line training of ACD without known process

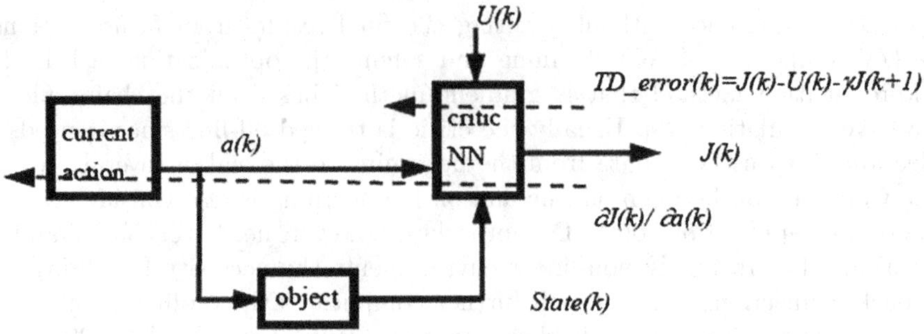

Fig. 1. ACD scheme. Dashed lines represent the training cycle.

model (that is analog to the RL task) adopted with some changes from [16] is given on Fig. 1. The vector $State(k)$ represents the object state vector, $a(k)$ is the control variable. The critic NN has to be trained to predict the utility function $U(k)$ by approximating Bellman's equation as follows:

$$J(State(k), a(k)) = \sum_{t=0}^{k} \gamma^t U(State(t), a(t)) \tag{1}$$

where γ is discount factor taking values between 0 and 1.

The value of control variable is adjusted by gradient descent algorithm as follows:

$$a_i(k) = a_{i-1}(k) \pm \alpha \frac{\partial J_i(k)}{\partial a_i(k)} \tag{2}$$

Here i denotes the iteration number and $0 < \alpha < 1$ is learning rate. The sign ($\pm$) in Eq. (2) depends on whether the optimization task is to maximize or to minimize the utility function.

2.2 Associative Learning Approach

The fundamental law of Hebb states that stable pairing of pre- and postsynaptic activity strengthens the weight of corresponding connection between neurons. Based on it associative learning rule include product of presynaptic (or input $I(k)$) and postsynaptic (or output $O(k)$) activities instead of the gradient and by analogy with the Eq. (2), the Hebbian learning rule for a given connection weight w at i-th training iteration is as follows:

$$w_i(k) = w_{i-1}(k) \pm \alpha I_i(k) O_i(k) \tag{3}$$

This learning rule is used also in the first actor/critic adaptive algorithm [1]. Here the training algorithm for the Adaptive Search Element (ASE) is applied for action training as follows:

$$a_i(k) = a_{i-1}(k) \pm \alpha J_i(k) e_i(k) \tag{4}$$

where $e_i(k)$ denotes the eligibility trace of action. According to [1] and accounting for specificity of the ACD scheme used here, the eligibility trace for control action becomes:

$$e_i(k) = \delta e_{i-1}(k) + (1-\delta)a_i(k) \tag{5}$$

where $0 < \delta < 1$ is decay rate of the trace.

2.3 Echo State Networks

ESNs are a kind of recurrent neural networks that arise from so called "reservoir computing approaches" [10]. The basic ESN structure is shown in Fig. 2. The ESN output vector denoted here by $J(k)$ (since the ESN will be in the place of critic network from Fig. 1) for the current time instance k is usually a linear function of its input and current state:

$$J(k) = f^{out}(W^{out}[in(k), X(k)]) \tag{6}$$

Here, $in(k)$ is a vector of network inputs and $X(k)$ a vector composed of the reservoir neuron states; f^{out} is a linear function (usually the identity), W^{out} is a $n_J \times (n_{in} + n_X)$ trainable matrix (here n_J, n_{in}, and n_X are the sizes of the corresponding vectors J, in, and X). The neurons in the reservoir have a simple sigmoid output function f^{res} (usually hyperbolic tangent) that depends on both the ESN input $in(k)$ and the previous reservoir state $X(k-1)$:

$$X(k) = f^{res}(W^{in}in(k) + W^{res}X(k-1)) \tag{7}$$

Here W^{in} and W^{res} are $n_{in} \times n_X$ and $n_X \times n_X$ matrices that are randomly generated and are not trainable. There are different approaches for reservoir parameter production [10]. A recent approach used in the present investigation is proposed in [15]. It is called intrinsic plasticity (IP) and suggests initial adjustment of these matrices, aiming at increasing the entropy of the reservoir neurons outputs. For on-line training, the RLS algorithm [5] was used.

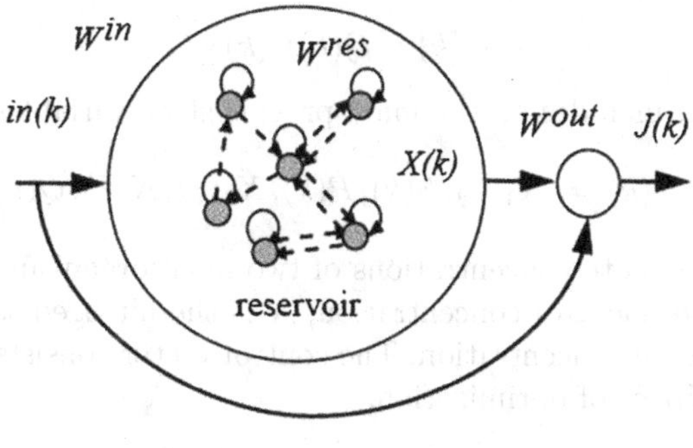

Fig. 2. Echo state network structure.

2.4 PHB Production Process

The object under consideration here (PHB production process) is a kind of mixed culture cultivation biotechnological process. Mixed culture systems are quite common in nature: the human body, waste water treatment, ecosystems are some well known examples. In such systems one microorganism assimilates substrate A and converts it to metabolite B which is converted by another microorganism to metabolite C. Since the change in culture conditions affects all microorganisms differently it is difficult to control them in an optimal way. That is why they are extremely difficult for dynamic analysis and control strategies in typical industrial applications. A mixed culture system where sugars (glucose) are converted to lactate by the microorganism *L.delbrueckii* and then the lactate is converted to PHB (poly-β-hydroxybutyrate) by the microorganism *R.euthropha* is the subject of optimization here. The main product obtained — PHB — is a biodegradable polymer used as thermoplastic in food and drug industry. The main purpose of the process control strategy is to maximize the final product of the process (PHB) accounting for the needs and mutual relations of both microorganisms in the culture. Several approaches to this problem are known by now. In [19] quite a complete mathematical model of the process has been developed and different control strategies were exploited separately or in combination. The PHB production process was modeled by seven nonlinear ordinary differential equations. The model details can be found in [7,19]. In present study the aim was to optimize set point time profiles of all three variables used as control inputs — dissolved oxygen (DO^*), glucose (S^*), and nitrogen source (N^*) concentrations. The previously developed model is used as process simulator as it is difficult to make multiple real on-line experiments with such kind of processes.

3 Results and Discussion

In the present investigation the action dependent heuristic dynamic programming is applied. The main goal is to maximize the process outcome, i.e. the target product PHB (denoted by Q here) by the end of the process. The utility function is:

$$U(k) = Q(k)V(k) \tag{8}$$

Vector $State(k)$ includes all the main process state variables, i.e.:

$$State(k) = (X_1(k), S(k), P(k), X_2(k), N(k), Q(k)) \tag{9}$$

where X_1 and X_2 denote concentrations of two microorganisms; P is the intermediate metabolite (lactate) concentration; N is the nitrogen source concentration; S is sugar source concentration. The control vector consists of the three set points that are subject of optimization:

$$a(k) = (S^*(k), N^*(k), DO^*(k)) \tag{10}$$

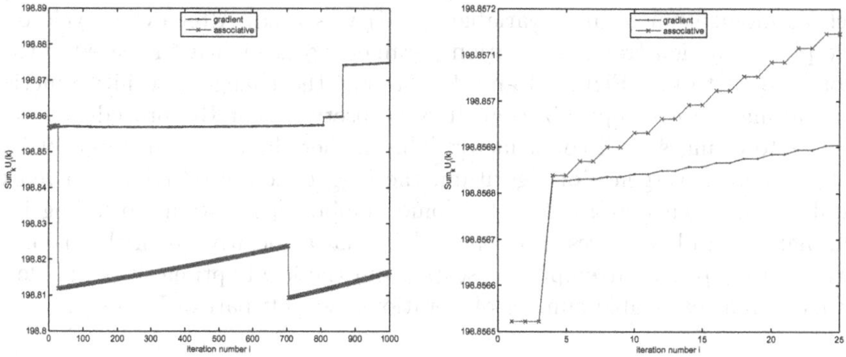

Fig. 3. Change of utility function value during iterative optimization. The discount factor γ is increased by 0.001 each critic/actor training iteration. On the left is whole run of 1000 iterations; on the right is a zoom of first 25 iterations.

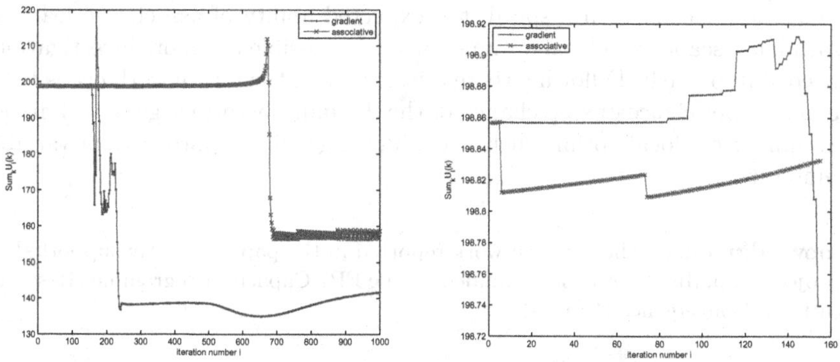

Fig. 4. Change of utility function value during iterative optimization. The discount factor γ is increased by 0.01 each critic/actor training iteration. On the left is whole run of 1000 iterations; on the right is a zoom of first 155 iterations.

Here DO is dissolved oxygen concentration in the cultural broth.

We suppose that all concentration controllers work properly and are able to follow their set points. The applied control scheme is described in more detail in [7,19]. For the ESN critic training and simulation a Matlab toolbox from [17] with our improvements for IP training as in [15] was used. The critic network has 9 inputs, 10 reservoir neurons and 1 output. The reservoir neurons have hyperbolic tangent output function. Instead of a complex action network here we have only time profiles of the set points of the sugar, nitrogen, and dissolved oxygen concentrations that have to be adjusted during the training phase. The initial set point profiles were taken from [7]. Detailed optimization algorithm can be found in [6]. It consists of consecutive critic and actor training iterations. Here for comparative purpose simple gradient algorithm without any improvement (such as momentum term or variable speed) was used. After every cycle of a critic plus an action training iteration parameter γ is slightly increased until it become equal to 0.5. Figures 3 and 4 represent the change of utility function value during iterative optimization. It was observed that the procedure is too sensitive to changes in discount factor. The smaller change (with step of 0.001) led to slower convergence of algorithm; the bigger step of 0.01 however led to unstable convergence and even after some number of iterations to falling in a local maxima. In both cases it becomes clear that associative algorithm is much slower and in spite of attempts to escape from the local optima, it is unable to do this within reasonable number of iterations (see left part of Fig. 4).

4 Conclusions

The carried out initial investigations led to the following conclusions: the combination of ACD approach with the associative learning of actor is a possible alternative to the purely gradient ACD scheme; however, as it was expected, it has much lower convergence speed; the expected ability of associative learning algorithm to escape from local optima obviously possible but more investigations are needed to prove it. Following these initial results, the future work needs to be done on: proper choice of step change of the discount factor; usage of techniques for escaping from local optima such as conjugate gradient approaches or variable learning rate.

Acknowledgments. The research work reported in the paper is partly supported by the project AComIn, grant 316087, funded by the FP7 Capacity Programme (Research Potential of Convergence Regions).

References

1. Barto, A.G., Sutton, R.S., Anderson, C.W.: Neuronlike adaptive elements that can solve difficult learning control problems. IEEE Trans. Syst. Man Cybern. **13**(5), 834–846 (1983)
2. Bellman, R.E.: Dynamic Programming. Princeton Univ. Press, Princeton (1957)
3. Bertsekas, D.P., Tsitsiklis, J.N.: Neuro-Dymanic Programming. Athena Scientific, Belmont (1996)
4. Jaeger, H.: Tutorial on training recurrent neural networks, covering BPPT, RTRL, EKF and the "echo state network approach. GMD Report 159, German National Research Center for Information Technology (2002)
5. Jaeger, H.: Adaptive nonlinear system identification with echo state networks. In: Advances in Neural Information Processing Systems 15 (NIPS 2002), pp. 593–600. MIT Press, Cambridge (2003)
6. Koprinkova-Hristova, P., Palm, G.: Adaptive critic design with ESN critic for bioprocess optimization. In: Diamantaras, K., Duch, W., Iliadis, L.S. (eds.) ICANN 2010, Part II. LNCS, vol. 6353, pp. 438–447. Springer, Heidelberg (2010)
7. Koprinkova-Hristova, P.: Knowledge-based approach to control of mixed culture cultivation for PHB production process. Biotechnol. Biotechnol. Equip. **22**(4), 964–967 (2008)

8. Koprinkova-Hristova, P., Oubbati, M., Palm, G.: Adaptive critic design with echo state network. In: Proceedings of 2010 IEEE International Conference on Systems, Man and Cybernetics, Istanbul, Turkey, pp. 1010–1015, 10–13 October 2010
9. Lenardis, G.G.: A retrospective on adaptive dynamic programming for control. In: Proceedings of Interenational Joint Conference on Neural Networks, Atlanta, GA, USA, pp. 1750–1757, 14–19 June 2009
10. Lukosevicius, M., Jaeger, H.: Reservoir computing approaches to recurrent neural network training. Comput. Sci. Revi. **3**, 127–149 (2009)
11. Meuleau, N., Dorigo, M.: Ant colony optimization and stochastic gradient descent. Artif. Life **8**, 103–121 (2002)
12. Prokhorov, D.V.: Adaptive critic designs and their applications. Ph.D. dissertation. Department of Electrical Engineering, Texas Tech University (1997)
13. Prokhorov, D.: Toward effective combination of off-line and on-line training in ADP framework. In: Proceedings of the 2007 IEEE Symposium on Approximate Dynamic Programming and Reinforcement Learning (ADPRL 2007), pp. 268–271 (2007)
14. Prokhorov, D.: Training recurrent neurocontrollers for real-time applications. IEEE Trans. Neural Netw. **18**(4), 1003–1015 (2007)
15. Schrauwen, B., Wandermann, M., Verstraeten, D., Steil, J.J.: Improving reservoirs using intrinsic plasticity. Neurocomputing **71**, 1159–1171 (2008)
16. Si, J., Wang, Y.-T.: On-line learning control by association and reinforcement. IEEE Trans. Neural Netw. **12**(2), 264–276 (2001)
17. Simple and very simple Matlab toolbox for Echo State Networks by H. Jaeger and group members. http://www.reservoir-computing.org/software
18. Sutton, R.S.: Learning to predict by methods of temporal differences. Mach. Learn. **3**, 9–44 (1988)
19. Tohyama, M., Patarinska, T., Qiang, Z., Shimizu, K.: Modeling of the mixed culture and periodic control for PHB production. Biochem. Eng. J. **10**, 157–173 (2002)
20. Werbos, P.J.: Backpropagation through time: What it does and how to do it. Proc. IEEE **78**(10), 1550–1560 (1990)

Using Self-Adaptive Evolutionary Algorithms to Evolve Dynamism-Oriented Maps for a Real Time Strategy Game

Raúl Lara-Cabrera(✉), Carlos Cotta, and Antonio J. Fernández-Leiva

Department "Lenguajes Y Ciencias de la Computación", ETSI Informática, University of Málaga, Campus de Teatinos, 29071 Málaga, Spain
{raul,ccottap,afdez}@lcc.uma.es

Abstract. This work presents a procedural content generation system that uses an evolutionary algorithm in order to generate interesting maps for a real-time strategy game, called *Planet Wars*. Interestingness is here captured by the dynamism of games (i.e., the extent to which they are action-packed). We consider two different approaches to measure the dynamism of the games resulting from these generated maps, one based on fluctuations in the resources controlled by either player and another one based on their confrontations. Both approaches rely on conducting several games on the map under scrutiny using top artificial intelligence (AI) bots for the game. Statistic gathered during these games are then transferred to a fuzzy system that determines the map's level of dynamism. We use an evolutionary algorithm featuring self-adaptation of mutation parameters and variable-length chromosomes (which means maps of different sizes) to produce increasingly dynamic maps.

1 Introduction

Videogames, with a total consumer spent of 24.75 billion US dollars in 2011 [1], is a very important pillar of the entertainment industry. Until the last decade, the graphical quality of a game determined its quality but, since then, the attractiveness of video-games has fallen on additional features, such as music, interesting stories and the player immersion into the game. It is difficult to measure how much fun a game is since it depends on each player; however it is related to the player satisfaction: the higher the satisfaction, the higher the fun.

This high satisfaction can be achieved via the automated adaptation of the game in response to the player's needs [7] using computational intelligence (CI) techniques. Traditionally, CI has been applied to generate strategies that define the behaviour of the non-player characters (NPC), but it can be also applied to many other aspects of game development such as computational narratives, player modelling, learning in games, intelligent camera control, and procedure content generation (PCG), among other – see [6].

PCG involves algorithms and techniques devoted to create game content automatically, providing several advantages to game developers, such as reduced

I. Lirkov et al. (Eds.): LSSC 2013, LNCS 8353, pp. 256–263, 2014.
DOI: 10.1007/978-3-662-43880-0_28,

memory consumption, the possibility of create endless video-games (i.e. the game changes every time a new game is started) and a reduction in the expense of creating the game content. This work focuses in PCG in the context of the real-time strategy (RTS) game *Planet Wars* by means of evolutionary algorithms (EAs).

Planet Wars is a real-time strategy game based on *Galcon* and used in the *Google AI Challenge 2010*. The objective is to conquer all the planets on the map or eliminate every opponent. Every game takes place on a map on which several planets are scattered. These planets are able to host ships and they can be controlled by any player or remain neutral if no player conquer them. Moreover, planets have different sizes, a property that defines their growth rate (i.e., the number of new ships created every time step, as long as the planet belongs to some player). Players send fleets of ships from controlled planets to other ones. If the player owns the target planet the number of fleet's ships is added to the number of ships on that planet, otherwise a battle takes place in the target planet: ships of both sides destroy each other so the player with the highest number of ships owns the planet (with a number of ships determined by the difference between the initial number of ships). The distance between the planets affects the required time for a fleet to arrive to her destination, which is fixed during the flight (i.e., it is not possible to redirect a fleet while it is flying).

PCG for *Planet Wars* involves in this case generating the maps on which the game takes place. The particular structure of these maps can lead to games exhibiting specific features. In previous work [3,4] we focused on achieving balanced games, i.e., games in which none of the players strongly dominates her opponent. Such balanced games can be of little interest though, due to the lack of action. For this reason, we turn our attention to the evolution of maps resulting in interesting, action-packed games. We use the label dynamism to refer to this property of games. Next section is devoted to analyse the evolution of dynamism-oriented games.

2 Evolution of Maps with Dynamism

To study the evolution of *Planet Wars* maps leading to dynamic games, let us firstly analyse how to capture dynamism within an objective function. Subsequently, we focus on an evolutionary approach optimizing this objective function.

2.1 Capturing Dynamism

In order to evaluate the dynamism of the generated maps we had to specify which are the characteristics that define a dynamic game. To do so, we consider two groups of indicators. The first group reflects dynamism from a resource-based perspective (i.e., we try to relate dynamism with the variation in the amount of resources owned by either player); the second group focuses on confrontations between the players (i.e., dynamism is tried to be captured by the extent to which the players repeatedly clash). More precisely, the indicators for a game i are the following:

- Resource-based:
  - *Game length* T_i: this is the ratio of the maximum number of turns allowed $\tau_{\max}$ that have been played in the current game: $T_i = \tau_i/\tau_{\max}$.
  - *Conquering rate* K_i: this is the ratio of planets which are not neutral at the end of the game.
  - *Reconquering rate* Z_i: Let ζ_{ij} be the number of planets that were owned by a player in turn $j-1$ and conquered by the other player in turn j. Then $Z_i = \frac{1}{\tau_i}\sum_{j=1}^{\tau_i}\zeta_{ij}/n_p$, where n_p is the total number of planets.
  - *Peak difference*: this is a family of variables measuring the maximal amplitude of the variation in any of the resources accounted for, namely planets (π), growth capacity (γ), and ships (ξ). Let $\phi_{ij}^{(a)}$ be the amount of resource ϕ owned by player a in the j-th turn of the i-th game, we record the two points in which the relative difference is best for one player and the other one and sum both quantities, i.e., $\Delta_i^{\phi} = \max_{1\leqslant j\leqslant\tau_i}\{(\phi_{ij}^{(1)} - \phi_{ij}^{(2)})/(\phi_{ij}^{(1)} + \phi_{ij}^{(2)})\} - \min_{1\leqslant j\leqslant\tau_i}\{(\phi_{ij}^{(1)} - \phi_{ij}^{(2)})/(\phi_{ij}^{(1)} + \phi_{ij}^{(2)})\}$
- Confrontation-based:
  - *Battle rate:* B_i this is the ratio of planets under attack throughout the game. Let β_{ij} be the number of planets that were under attack during the j-th turn, then $B_i = \frac{1}{\tau_i}\sum_{j=1}^{\tau_i}\beta_{ij}/n_p$.
  - *Destroyed ships* S_i: this is the ratio of the generated ships that have been destroyed throughout the game. Let χ_i be the number of destroyed ships and ψ_i the number of created ships, then $S_i = \chi_i/\psi_i$.

Since we use a tournament system whereby a number of bots are paired and compete on the map under evaluation, the consider the average value of the above indicators across the N_g total games. We drop the sub-index to denote this average quantity. Subsequently, we have defined a set of fuzzy rules to express dynamism as a function of these indicators. The fuzzy rule base is depicted in Fig. 1. In general the underlying fuzzy sets (LO and HI) are defined so as to hit a maximum at the corresponding end of the value range of the variable under consideration, and decrease linearly towards the other end. The exceptions are Z and B whose usual values are far from the theoretical maximal value 1.0. in this case, we saturate the HI value to 1 when they reach the value 0.1 and 0.35 respectively (these values were empirically determined). For the output variable *dyn* a middle triangular fuzzy set MED is defined, hitting a maximum at 0.5 and linearly decreasing towards both ends. In this case, both HI and LO reach their minimum at this 0.5 value.

2.2 Evolutionary Approach

We have used a self-adaptive evolutionary approach in order to optimize the dynamism of the generated maps. These maps have been encoded in mixed real-integer vectors which define the characteristics of the planets (i.e. position, size and number of ships) in a way that each gene represents a planet. The mutation operator depends on the parameter's type: Gaussian mutation for the real-valued parameters and a method that generates suitable integer mutations [5,8] for the

Resource-based:
1. **if** K **is** HI **and** Z **is** HI **then** *dyn* **is** HI
2. **if** Δ^π **is** HI **and** Δ^γ **is** HI **and** Δ^ξ **is** HI **then** *dyn* **is** HI
3. **if** Δ^π **is** HI **and** (Δ^γ **is** LO **or** Δ^ξ **is** LO) **then** *dyn* **is** MED
4. **if** Δ^γ **is** HI **and** (Δ^π **is** LO **or** Δ^ξ **is** LO) **then** *dyn* **is** MED
5. **if** Δ^ξ **is** HI **and** (Δ^γ **is** LO **or** Δ^π **is** LO) **then** *dyn* **is** MED
6. **if** Δ^π **is** LO **and** Δ^γ **is** LO **and** Δ^ξ **is** LO **then** *dyn* **is** LO
7. **if** K **is** LO **or** Z **is** LO **or** T **is very** LO **then** *dyn* **is** LO

Confrontation-based:
1. **if** B **is** HI **and** S **is** HI **then** *dyn* **is** HI
2. **if** (B **is** HI **and** S **is** LO) **or** (B **is** LO **and** S **is** HI) **then** *dyn* **is** MED
3. **if** B **is** LO **and** S **is** LO **then** *dyn* **is** LO

Fig. 1. Fuzzy rule bases for dynamism.

rest of the parameters. The parameters of these operators have been included into the solutions, thus providing the means for self-adapting them. More precisely, in the case of real-valued parameters $\langle x_1, ..., x_n \rangle$ mutation is done by having $\sigma'_i = \sigma_i \cdot \exp(\tau' \cdot N(0,1) + \tau \cdot N_i(0,1))$ and $x'_i = x_i + \sigma_i \cdot N_i(0,1)$ where σ_i is the mutation parameter for x_i, $\tau' = 1/\sqrt{2n}$, and $\tau = 1/\sqrt{2\sqrt{n}}$. Likewise, integer-valued parameters $\langle z_1, ..., z_m \rangle$ are mutated by having $\varsigma'_i = \max(1, \varsigma_i \cdot \exp(\tau \cdot N(0,1) + \tau' \cdot N(0,1)))$, $\psi_i = 1 - (\varsigma'_i/m)(1 + \sqrt{1 + (\varsigma'_i/m)^2})^{-1}$ and $z'_i = z_i + \lfloor \ln(1 - U(0,1))/\ln(1 - \psi_i) \rfloor - \lfloor \ln(1 - U(0,1))/\ln(1 - \psi_i) \rfloor$ where ς_i is the mutation parameter for z_i, $\tau = 1/\sqrt{2m}$ and $\tau' = 1/\sqrt{2\sqrt{m}}$, U(0,1) is a function that returns a random number drawn from a uniform distribution in (0,1).

As for recombination, we have used a "cut and splice" operator that recombines two individuals by swapping cut pieces with different sizes that provides new maps which contain a different number of planets in relation to their parents, hence adding again additional self-adapting capabilities to the algorithm.

3 Experimental Results

We have used the DEAP library [2] to implement the EA described previously. The algorithm has employed a population size of 100 individuals and a $(\mu + \lambda)$ generational scheme, with $\mu = 10$, $\lambda = 100$. The bots used to evaluate the individuals were obtained from the Google AI Challenge competition- These bots (*Manwe*[1], *Flagscapper's bot*[2] and *fglider's bot*[3]) ranked in the top 100 (there were over 4600 participants) and have their source code available. The duration of the games was limited to $\tau_{\max} = 400$ turns. Regarding the evaluation of fuzzy rules, we have used the min t-norm, the max t-conorm, and the centre of mass as defuzzification method.

[1] https://github.com/Manwe56/Manwe56-ai-contest-planet-wars
[2] http://flagcapper.com/?c1
[3] http://planetwars.aichallenge.org/profile.php?user_id=8490

We have run two sets of experiments focusing on the behaviour of the algorithm when optimizing dynamism using either a confrontation-based (CB) or a resource-based (RB) approach. The results are shown in Fig. 2. Let us firstly focus on the middle and rightmost sub-figures which provide an indication on the correlation of both objective functions when one of them is being optimized. It is clear from these plots that both measures are fairly orthogonal, in the sense that when one of them is optimized the other one follows a rather flat trajectory. Thus they can be seen as truly complementary views on game dynamism. Notice also that CB fitness seems to converge faster, likely indicating that confrontations can be induced more easily than wide resource fluctuations by adjusting the map. Another interesting fact is shown in the leftmost sub-figure. Therein we show how balanced are the games when either objective function is being optimized. The measure of balanced was defined elsewhere aiming to analyse its trade-off with a RB version of dynamism, and essentially amounts to measure how the three resources (planets, growth capacity and ships) remain balanced (i.e., their absolute difference is small) for the two players throughout the game. While balance seems to follow a flat trajectory when RB dynamism is being optimized, there is an increasing trend in the case of optimizing CB dynamism. We believe this can be due to the fact that continuous battles prevent the accumulation of resources by either party and push towards their mutual cancellation.

Figure 3 shows the progress of the variables used to measure the dynamism of the maps during evolution. Unsurprisingly, variables used in the function under optimization in either case exhibit an increasing trend in general. It is more interesting to note some cross-relationships. Firstly, the conquering rate

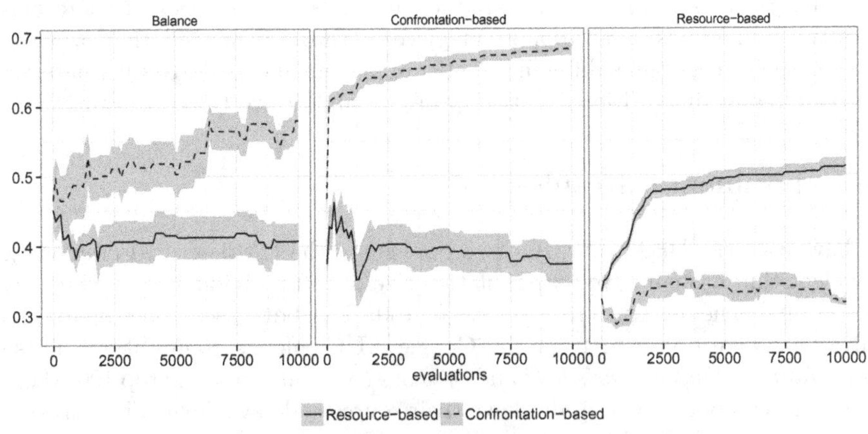

Fig. 2. Evolution of the different objective functions. In the middle and rightmost graph we depict the evolution of dynamism measured in both ways when one of them is subject to optimization (the one indicated in the sub-graph title). The leftmost graph indicates the evolution of balance when either objective function is being optimized. Each line represents the average of 10 runs of the evolutionary algorithm and the shaded area indicates the standard error of the mean.

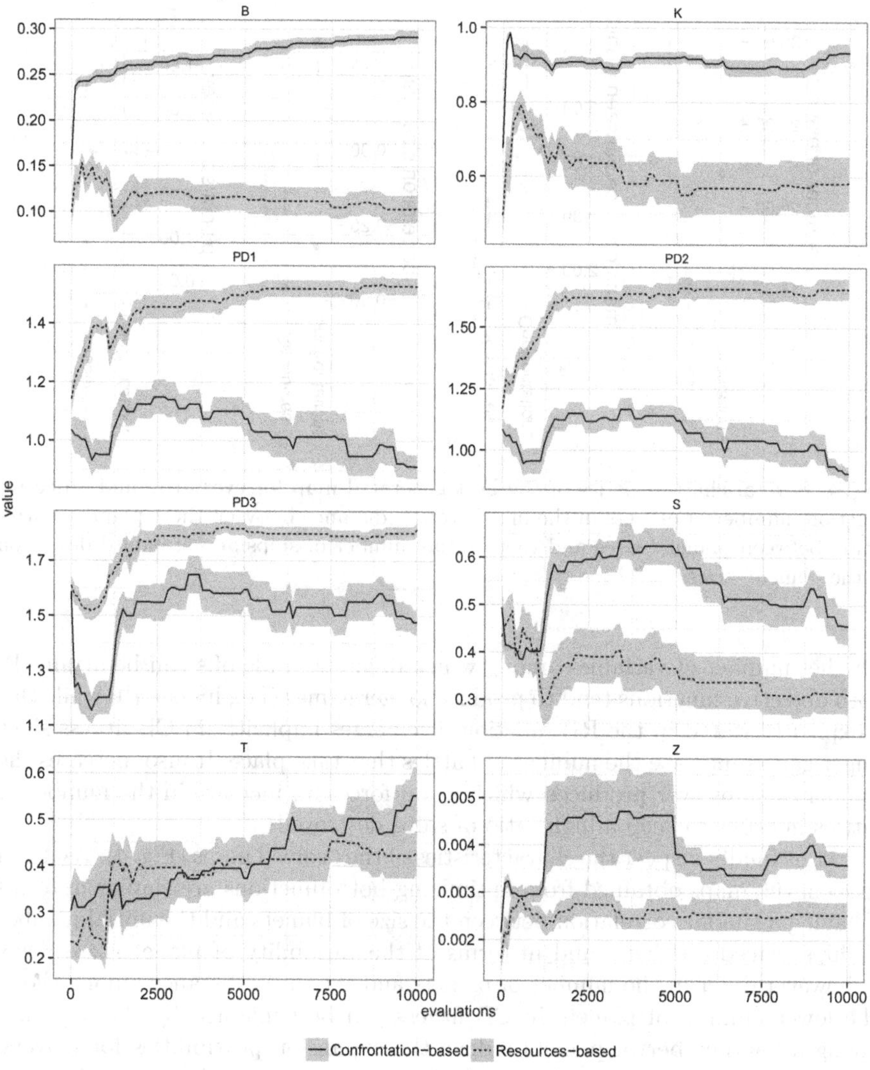

Fig. 3. Evolution of the different variables involved in the rules of the fuzzy sets.

K grows higher in the case of CB fitness than in RB fitness, despite the fact it is only explicitly included in the latter. This is side effect of the optimization of the battle rate B: in order to conquer a planet for the first time it must be placed under attack; hence an increase in the number of conquered plants implies another increase in the number of battles, a fact exploited by the evolutionary algorithm. Likewise the reconquering rate is also higher in CB fitness since a high number of battles can eventually lead to numerous planets changing hands (note also that CB fitness heavily revolves around B whereas RB fitness involves

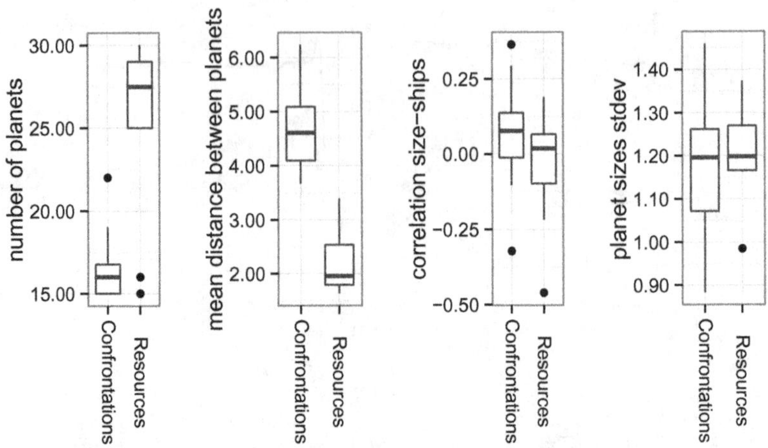

Fig. 4. Several characteristics of the best generated map for every run and objective function: number of planets in the map, average distance between these planets, correlation between planets' sizes and their initial number of ships and standard deviation of the planets' sizes.

a higher number of variables among which different trade-offs can be attained). Both objective functions tend to produce longer games (i.e., higher T); while this is explicitly stated in the RB function, it emerges implicitly in CB fitness since longer games increase the number of battles that take place. It also increases the number of ships ever produced which again forces an increase in the number of battles in order to keep a high ratio of ships destroyed.

Let us finally inspect the characteristics of the evolved maps (Fig. 4). As it can be seen, the maps obtained from optimizing both functions are similar in terms of having a similar correlation between the size of planets and the initial number of ships placed on them, and in terms of the variability of planet sizes. They do however differ in the number of planets and the mean distance among them. The lower number of planets in CB fitness can be explained by the fact that having a low number of planets reduce the expansion possibilities for players, thus forcing them to focus on the same planets more often and hence leading to a higher number of battles. Having a higher distance among planets has the effect of increasing the time-lag between the moment decisions are taken (i.e., ships are dispatched to a target) and the moment ships arrive to their destination. We hypothesize that this larger time-lag introduces a factor of instability by making it more difficult to hold positions and increasing the time of reaction upon attacks, thus promoting more battles to regain control of planets.

4 Conclusion and Future Work

This paper presents a PCG method that is capable of generating maps for the RTS game *Planet Wars*. These maps should fulfil some desirable requirements in terms of dynamism in order to obtain interesting and attractive games. This dynamism has been tackled from two different approaches: a RB approach that looks for a high level of dynamism in players' resources, and a CB approach, that focuses on battles and lost ships. Both approaches have been shown to be orthogonal, thus suggesting their joint optimization (either in a single- or a multi-objective scenario) as a potential line of future research. It is interesting to note the higher correlation of CB fitness with balance. We plan to analyse further this connection by introducing subjective evaluation of the generated maps in a future work, so as to analyse the attractiveness of games for human players.

Acknowledgements. This work is partially supported by Spanish MICINN under project ANYSELF (TIN2011-28627-C04-01), and by Junta de Andalucía under project P10-TIC-6083 (DNEMESIS).

References

1. Entertainment Software Association: Essential facts about the computer and video game industry (2012). http://www.theesa.com/facts/pdfs/esa_ef_2012.pdf
2. Fortin, F.A., Rainville, F.M.D., Gardner, M.A., Parizeau, M., Gagné, C.: DEAP: evolutionary algorithms made easy. J. Mach. Learn. Res. **13**, 2171–2175 (2012)
3. Lara-Cabrera, R., Cotta, C., Fernández-Leiva, A.J.: Procedural map generation for a RTS game. In: Leiva, A.F., et al. (eds.) 13th International GAME-ON Conference on Intelligent Games and Simulation, Eurosis, Malaga (Spain), pp. 53–58 (2012)
4. Lara-Cabrera, R., Cotta, C., Fernández-Leiva, A.J.: A procedural balanced map generator with self-adaptive complexity for the real-time strategy game planet wars. In: Esparcia-Alcázar, A.I. (ed.) EvoApplications 2013. LNCS, vol. 7835, pp. 274–283. Springer, Heidelberg (2013)
5. Li, R.: Mixed-integer evolution strategies for parameter optimization and their applications to medical image analysis. Ph.D. Thesis, Leiden University (2009)
6. Lucas, S.M., Mateas, M., Preuss, M., Spronck, P., Togelius, J.: Artificial and computational intelligence in games (Dagstuhl Seminar 12191). Dagstuhl Rep. **2**(5), 43–70 (2012)
7. Nogueira, M., Cotta, C., Fernández-Leiva, A.J.: On modeling, evaluating and increasing players' satisfaction quantitatively: steps towards a taxonomy. In: Di Chio, C., et al. (eds.) EvoApplications 2012. LNCS, vol. 7248, pp. 245–254. Springer, Heidelberg (2012)
8. Rudolph, G.: An evolutionary algorithm for integer programming. In: Davidor, Y., Männer, R., Schwefel, H.-P. (eds.) PPSN 1994. LNCS, vol. 866, pp. 139–148. Springer, Heidelberg (1994)

Simple Iterative Heuristics for Correlation Clustering

Andrzej Lingas[1](✉) and Mia Persson[2]

[1] Department of Computer Science, Lund University, Lund, Sweden
Andrzej.Lingas@cs.lth.se

[2] Department of Computer Science, Malmö University, Malmö, Sweden
mia.persson@mah.se

Abstract. A straightforward natural iterative heuristic for correlation clustering in the general setting is to start from singleton clusters and whenever merging two clusters improves the current quality score merge them into a single cluster. We analyze the approximation and complexity aspects of this heuristic and its randomized variant where two clusters to merge are chosen uniformly at random among cluster pairs amenable to merge.

1 Introduction

Correlation clustering is nowadays a well known technique of partitioning documents, e.g., web pages [3], into groups on the base of their similarity and/or dissimilarity. It has several useful applications, e.g., in combining the output of different machine learning classifiers [2]. Correlation clustering can be also viewed as an agnostic learning problem [3,7]. In contrast to such classic clustering problems as k-means, k-sum, and k-center, in case of correlation clustering the number of clusters does not have to be specified. One assumes that there is a classifier function that assigns to pairs of documents + if they are similar and − if they are dissimilar. It can be modeled by a graph whose vertices correspond to the documents and edges are labeled by + or −. In general setting, the classifier function may be partial and real weights may be assigned to the + and − labels so the corresponding graph is edge weighted and non-necessarily complete.

In their pioneering work [2], Bansal et al. introduced two objectives of correlation clustering: minimizing the number of disagreements and maximizing the number of agreements. A disagreement is a − edge within a cluster or a + edge between clusters while an agreement is a + edge within a cluster or a − edge between clusters. They showed both problems to be NP-hard even if the underlying graph is unweighted and complete (in fact both problems are equivalent in the exact setting but different in the approximation setting). Bansal et al. also provided a polynomial-time approximation scheme (PTAS) for the maximization problem and a constant factor approximation for the minimization problem, in both cases for complete unweighted graphs. This constant factor approximation

I. Lirkov et al. (Eds.): LSSC 2013, LNCS 8353, pp. 264–271, 2014.
DOI: 10.1007/978-3-662-43880-0_29,

was later improved by Charikar et al. [4], where a factor 4 approximation algorithm is given for complete graphs based on linear programming relaxation. The latter problem was also proved to be APX-hard.

The problems of maximizing agreements and minimizing disagreements were subsequently studied in the general setting of non-necessarily complete graphs with edge weights in [4–6,8]. More recently, Ailon et al. [1] have provided a randomized expected 3-approximation algorithm for minimizing disagreements.

In our paper, we study simple iterative heuristics for correlation clustering for unweighted graphs. A straightforward natural iterative heuristic for correlation clustering in the general setting is to start from singleton clusters and whenever merging two clusters improves the current quality score (i.e., the number of agreements or disagreements, respectively) merge them into a single cluster. The heuristic has several refinements depending on the way in which two clusters to merge are chosen. For example, the choice can be based on the degree of the improvement of the current score and/or be random. We analyze the approximation and complexity aspects of the basic variant, termed as Simple Merging heuristic (SM for short) and the random variant, termed as Random Merging heuristic (RM for short), where two clusters to merge are chosen uniformly at random among cluster pairs amenable to merge.

To begin with, we observe that the difference between the number of agreements and the number of disagreements in the clustering of a graph with n vertices and each edge labeled either with $+$ or $-$ produced by SM or RM is at least n minus the number of clusters produced.

Next, we show that the clustering produced by SM and RM run for maximizing agreements on such a graph achieves an approximation factor strictly smaller than 2. As for minimizing disagreements, we provide an upper bound $O(nq)$ on the approximation factor of SM and RM, where n is the number of vertices of the input graph and q is the maximum number of vertices in a cluster of the resulting SM or RM clustering.

On the other hand, we exhibit an infinite family of graphs for which the clustering produced by SM run for maximizing agreements can be $\approx$1.172414 times worse than the optimum. We also present another infinite family of graphs for which the clustering produced by SM run for minimizing disagreements can be $\frac{n-3}{2}$ times worse than the optimum. Furthermore, we show that SM in the maximization case (the minimization case, respectively) can produce a clustering with the number of agreements additively smaller by (the number of disagreements larger by, respectively) $\Omega(n^2)$.

Finally, we observe that the SM and RM heuristics, run on an n-vertex graph, admit a simple/practical $O(n^2)$-time implementation.

2 Preliminaries

Merging two disjoint clusters C_i and C_j increases the number of agreements or equivalently decreases the number of disagreements if and only if the number of agreements between pairs of vertices in $C_i \times C_j$ is greater than the number of disagreements between pairs of vertices in $C_i \times C_j$. Hence, we can define more formally the SM and RM heuristics as follows.

Definition 1. Let G be a graph with each edge labeled either with $+$ or with $-$.

Simple Merging Heuristic (SM): Put all vertices of G into singleton clusters; while there are two clusters C_i, C_j such that the number of $+$ edges in $C_i \times C_j$ is greater than the number of $-$ edges in $C_i \times C_j$ then merge C_i with C_j.

Random Merging Heuristic (RM): Put all vertices of G into singleton clusters; while there are two clusters C_i, C_j such that the number of $+$ edges in $C_i \times C_j$ is greater than the number of $-$ edges in $C_i \times C_j$ then pick such a pair uniformly at random and merge it.

3 Upper Bounds

Lemma 1. *Let G be a graph with n vertices and each edge labeled either with $+$ or $-$. Next, let C be the clustering resulting from running SM or RM on G. Let t be the number of clusters in C. The number of $+$ edges whose endpoints belong to different clusters in C is not greater than the number of $-$ edges whose endpoints are in different clusters in C. Also, the number of $-$ edges whose endpoints are in the same cluster of C is smaller than the number of $+$ edges with both endpoints in the same cluster of C by $n - t$.*

Proof. For each pair C_i, C_j of clusters in C, let $E^+_{i,j}$ ($E^-_{i,j}$, respectively) be the set of $+$ edges ($-$ edges, respectively) connecting a vertex in C_i with a vertex in C_j. Since the clusters C_i, C_j cannot be merged by SM or RM, we have $|E^+_{i,j}| \leq |E^-_{i,j}|$. By taking the summation over all pairs of clusters in C, we obtain the first part of lemma.

We prove the second part by induction on the number of merging steps in SM or RM, respectively. The induction hypothesis holds after step 0, where there are only singleton clusters. Let $0 \leq s \leq n-1$. Suppose that the hypothesis holds for the clustering into $s+1$ clusters resulting from the $n-s-1$ merging steps, and C_i and C_j are the clusters to be merged in the $(n-s)$-th step. The number of $+$ edges with one endpoint in C_i and the other in C_j has to be larger than the corresponding number of $-$ edges. This combined with the inductive hypothesis yields the second part. □

Lemma 1 immediately yields the following theorem.

Theorem 1. *Let G be a graph with n vertices and each edge labeled either with $+$ or $-$. Next, let C be the clustering resulting from running SM or RM on G, and let t be the number of clusters in C. The difference between the number of agreements and the number of disagreements in C is at least $n - t$.*

Theorem 2. *Let G be a graph with e^+ + edges and e^- − edges. Next, let m^+ be the cardinality of maximum matching in the subgraph G^+ of G induced by the + edges. The clustering resulting from running SM or RM for maximizing agreements on G is within $\min\{\frac{e^++e^-}{e^+}, \frac{e^++e^-}{e^-+\lceil m^+/2\rceil}\}$ of an optimal clustering of G.*

Proof. The start clustering into singletons already achieves e^- agreements. Consider an edge d of the maximum matching of G^+. Note that its two endpoints cannot belong to two singleton clusters in the clustering C produced by SM or RM since such clusters would be merged. Hence, at least one of the two singleton clusters for the endpoints is merged with another cluster and we can assign at least $\frac{1}{2}$ agreement increase to d. We conclude that C achieves at least $e^- + \lceil\frac{m^+}{2}\rceil$ agreements which yields the approximation factor $\frac{e^++e^-}{e^-+\lceil m^+/2\rceil}$. On the other hand, it follows from the first part of Lemma 1 that the resulting clustering has at least e^+ agreements which yields the approximation factor $\frac{e^++e^-}{e^+}$. □

Let us assume the notation of Theorem 2.

Corollary 1. *The clustering resulting from running SM or RM for maximizing agreements on G achieves an approximation strictly smaller than 2.*

Proof. If $e^+ = 0$ then the start singleton clustering is optimal. Hence, we may assume w.l.o.g. $e^+ > 0$ and consequently $\lceil m^+/2\rceil > 0$. If $e^- + \lceil m^+/2\rceil > e^+$ then we obtain an approximation factor strictly smaller than 2 by Theorem 2. Otherwise, e^- is strictly smaller than e^+, and we again obtain an approximation factor strictly smaller than 2 by Theorem 2. □

We can also derive a parametrized upper bound on the approximation factor of the SM or RM clustering in the minimization case.

A *bad cycle* (or, an erroneous cycle [6]) in a graph with all edges labeled either with + or − is a simple cycle that has exactly one − edge. Note that a bad cycle incurs at least one disagreement in any clustering of the graph. Hence, the cardinality of a set of pairwise edge-disjoint bad cycles is a lower bound on the number of disagreements in an optimal clustering of the graph.

Lemma 2. *For each cluster in the SM or RM clustering of a graph with all edges labeled either with + or −, the subgraph induced by the + edges within the cluster is connected and spans all vertices in the cluster.*

Proof. The proof is by a straightforward induction on the number of merging steps of SM or RM, respectively. It is sufficient to observe that two clusters to merge have to have at least one + edge connecting them and then use the inductive hypothesis. □

Lemma 3. *Let G be a graph with each edge labeled either with + or −. In the SM and RM clustering of G, each disagreement occurs on a bad cycle in G lying within a subgraph of G induced by vertices of at most two clusters and using at most two edges between a pair of clusters.*

Proof. If a disagreement is caused by an − edge whose both endpoints are in the same cluster then the endpoints are connected by a path composed of + edges within the cluster by Lemma 2 and jointly with the − edge form a bad cycle. Otherwise, if the disagreement is caused by a + edge between two clusters then there must be at least one − edge between the clusters and again by Lemma 2, paths composed of + edges within each of the two clusters respectively connect the endpoint of the − edge with the endpoint of the + edge sharing the same cluster. □

Theorem 3. *Let G be a graph with each edge labeled either with + or −. Suppose that the maximum number of vertices in a cluster in the SM (RM, respectively) clustering of G is q. The SM (RM, respectively) clustering of G has at most $O(nq)$ times more disagreements than the optimal clustering.*

Proof. Let us assign to each disagreement a bad cycle containing it according to the thesis of Lemma 3. Let B be the multiset of the assigned bad cycles. Each cycle b in B lies within a subgraph of G induced by at most two clusters C_i, C_j by Lemma 3. Note that a cluster cannot contain more than $\binom{q}{2}$ disagreements on − edges and between two clusters there can be at most $\frac{1}{2}q^2$ disagreements on + edges. It follows that the part of b within C_i as well as the part of b within C_j can overlap with at most $O(\frac{n}{q} \times q^2)$ bad cycles in B. Finally, if b contains two edges between C_i and C_j, these edges can overlap only with the $O(q^2)$ bad cycles in B that are contained in the subgraph of G induced by C_i and C_j. We conclude that b can overlap with at most $O(nq)$ bad cycles in B. Hence, the optimal clustering has $\Omega(\frac{|B|}{nq})$ disagreements. □

4 Lower Bounds

Theorem 4. *Let G be a graph with n vertices 1, …, n and each edge labeled either with + or − such that vertex 1 is connected by + edges with vertices 2 and 3 and the + edges on vertices 2 through n form a $(n-1)$-clique, see Fig. 1. Next, let C be the clustering resulting from running SM on G. The ratio between the number of disagreements in C and that in an optimal clustering can be at least $\frac{n-3}{2}$.*

Proof. SM starting from singleton clusters can produce the cluster $\{1, \ldots, i+1\}$ in the i-th iteration. Simply, merging $\{i+1\}$ with $\{1, \ldots, i\}$ decreases the number of disagreements at least by one. Hence, the resulting clustering C can be the singleton cluster containing all the n vertices. Thus, the number of disagreements can be as large as $n-3$. On the other hand, the clustering consisting of $\{1\}$ and $\{2, \ldots, n\}$ has only 2 disagreements. □

We can also run SM for maximizing agreements on the graph G defined as in Theorem 4. The ratio between the number of agreements in an optimal clustering and that resulting from running SM can be even $\frac{17}{15} \approx 1.1333$. However, we can subsume this lower bound by pruning the set of + edges in G appropriately.

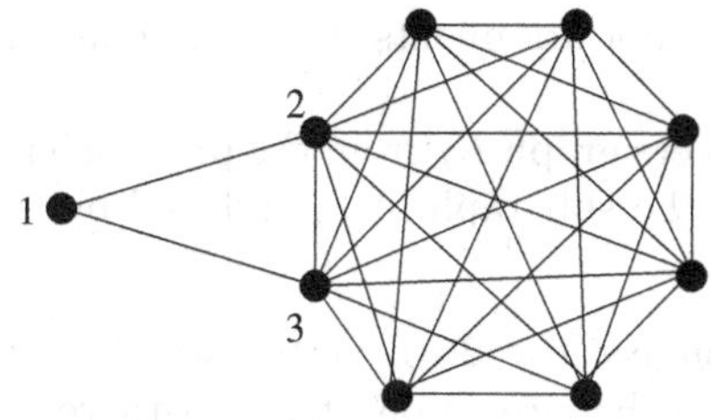

Fig. 1. The subgraph of the graph from Theorem 4 induced by + edges.

Theorem 5. *There is a graph with each edge labeled either with + or with − such that the ratio between the number of agreements in an optimal clustering of the graph and that number in the clustering of the graph produced by SM is at least* $\approx$1.172414.

Proof. We refine the construction of the graph G by removing some + edges. We assume that the constructed graph has $2k$ vertices and denote it by H_k. For $l = 1, \ldots, k$, we iterate the following two steps in order to specify the subgraph of H_k induced by the + edges:

1. connect the $2l$-th vertex with l vertices of numbers smaller than $2l$, avoiding connection to vertex 1 for $l > 1$;
2. if $l < k$ then connect the $(2l + 1)$-th vertex with $l + 1$ vertices of numbers smaller than $2l + 1$, avoiding connection to vertex 1 for $l > 1$.

All pairs of vertices not connected by + edges are connected by − edges.

Note that by the construction, SM can consecutively produce clusters $\{1, \ldots, i\}$ for $i = 1, \ldots, 2k$, since merging i into $\{1, \ldots, i-1\}$ for $i > 1$ will always increase the number of agreements by 1. The final cluster $\{1, \ldots, 2k\}$ produced by SM achieves $2(k(k+1)/2) + k - (k+1) = k(k+1) - 1$ agreements. On the other hand, the two clustering $\{1\}$, $\{2, \ldots, 2k\}$ has $k(k+1) - 1 - 2$ agreements on + edges and $2k - 3$ agreements on − edges, totally $k(k+3) - 6$ agreements. For $k = 5$, the ratio becomes at least $\frac{34}{29} \approx 1.172414$. □

In fact, we can modify the graph construction from the proof of Theorem 5 in order to show that SM can produce a clustering with an additive error quadratic in the number of vertices.

Theorem 6. *There is a graph with n vertices and each edge labeled either with + or with − such that:*

- *the number of disagreements in the clustering of the graph produced by SM is larger than that in an optimal clustering by* $\Omega(n^2)$, *or equivalently,*
- *the number of agreements in the clustering of the graph produced by SM is smaller than that in an optimal clustering by* $\Omega(n^2)$.

Proof. To simplify the proof arguments, let us assume that n is divisible by 2 and 3. Let $n = 2k$.

We modify the complete graph H_k from the proof of Theorem 5 to a complete graph F_k by specifying the subgraph of F_k induced by the + edges as follows. Let $l = 1, \ldots, k$.

1. if $l \leq 2k/3$ then connect the $2l$-th vertex with l vertices of largest numbers smaller than $2l$, otherwise connect the $2l$-th vertex with all vertices in $\{2k/3, 2k/3+1, \ldots, 2l-1\}$;
2. if $l \leq 2k/3$ then connect the $(2l+1)$-th vertex with $l+1$ vertices of largest numbers smaller than $2l+1$, otherwise, if $l < k$ then connect the $(2l+1)$-th vertex with all vertices in $\{2k/3, 2k/3+1, \ldots, 2l\}$

It is easy to see that SM can cluster all vertices of F_k into one cluster.

The number of disagreements in the one clustering is at least $(n/3-1)^2 + (n/6-1)^2 + D$, where D is the total number of $-$ edges with both endpoints in $\{1, \ldots, n/3\}$ or both endpoints in $\{n/3+1, \ldots, n\}$. The first component of the sum follows from the absence of + edges between the first $n/3-1$ vertices and the last $n/3-1$ vertices of F_k. The second component follows from the absence of + edges between the first $n/6-1$ vertices and the vertices $3n/6+1, \ldots, 4n/6$. On the other hand, consider the two clustering $\{1, \ldots, n/3\}$, $\{n/3+1, \ldots, 2k\}$. The number of disagreements in the two clustering is not larger than $(n/3)^2 + D$.

The second part of the theorem follows from the fact that the number of agreements in a clustering of the complete graph F_k is equal to $\binom{n}{2}$ decreased by the number of disagreements in the clustering. □

5 Time Complexity of SM and RM

To implement the iterative steps of SM or RM, we shall maintain a simple data structure keeping the number of + edges and $-$ edges between pairs of current clusters and listing candidate pairs of current clusters suitable for merging.

Theorem 7. *The SM and RM heuristics run on an n-vertex graph with each edge labeled with either + or $-$ can be implemented in $O(n^2)$ time.*

Proof. Let G be the input n-vertex graph. For each pair C_i, C_j of clusters in the current clustering of G, we maintain the number $e^+_{i,j}$ of + edges in G as well as the number $e^-_{i,j}$ of $-$ edges in G, connecting pairs of vertices in $C_i \times C_j$. We also maintain the set of candidate pairs of current clusters suitable for merging in the next step, by using an $0-1$ $n \times n$ array.

The initialization in both cases takes $O(n^2)$ time. When two clusters C_i and C_j are merged, say to form a new cluster C_i, we simply update the aforementioned numbers by setting $e^+_{i,k}$ to $e^+_{i,k} + e^+_{j,k}$ and $e^-_{i,k}$ to $e^-_{i,k} + e^-_{j,k}$ for all other clusters C_k. This takes $O(n)$ time. Also, forming a list or a $0-1$ vector representing the new cluster can be done in $O(n)$ time.

Having the numbers $e^+_{i,k}$ and $e^-_{i,k}$ updated, we can easily update the set of candidate pairs of current clusters suitable for merging, in $O(n)$ time.

Since the number of merging steps does not exceed n the upper bound stated in the theorem follows. □

6 Final Remarks

Among other things, we have demonstrated that the SM heuristic for maximizing agreements behaves reasonably well (Theorem 2, Corollary 1) while the SM heuristic for minimizing disagreements can be extremely bad (Theorem 4). It is easy to observe that the RM heuristic for minimizing disagreements run on the graphs from the proof of Theorem 4 would produce the optimal two clustering with very high probability. Unfortunately, the RM heuristic seems much harder to analyze generally than the randomized pivot heuristic of Ailon et al. shown to achieve expected 3-approximation [1]. It is a challenging open problem to prove a non-trivial approximation factor for the RM heuristic for minimizing disagreements. The simplicity of implementations of the SM and RM heuristics makes that especially the RM heuristic might be an interesting alternative to more sophisticated approximation algorithms for correlation clustering with provable low approximation factors.

References

1. Ailon, N., Charikar, M., Newman, A.: Aggregating inconsistent information: ranking and clustering. In: Proceedings of the 37th Annual ACM Symposium on Theory of Computing (STOC 2005), pp. 684–693 (2005)
2. Bansal, N., Blum, A., Chawla, S.: Correlation clustering. Mach. Learn. **56**(1–3), 89–113 (2004)
3. Becker, H.: A survey of correlation clustering. In: COMS E6998: Advanced Topics in Computational Learning Theory, pp. 1–10 (2005)
4. Charikar, M., Guruswami, V., Wirth, A.: Clustering with qualitative information. In: Proceedings of the 44th Annual Symposium on Foundations of Computer Science (FOCS 2003), pp. 524–533 (2003)
5. Demaine, E.D., Immorlica, N.: Correlation clustering with partial information. In: Arora, S., Jansen, K., Rolim, J.D.P., Sahai, A. (eds.) RANDOM 2003 and APPROX 2003. LNCS, vol. 2764, pp. 1–13. Springer, Heidelberg (2003)
6. Emanuel, D., Fiat, A.: Correlation clustering – minimizing disagreements on arbitrary weighted graphs. In: Di Battista, G., Zwick, U. (eds.) ESA 2003. LNCS, vol. 2832, pp. 208–220. Springer, Heidelberg (2003)
7. Kearns, M.J., Shapire, R.E., Sellie, L.M.: Toward efficient agnostic learning. Mach. Learn. **17**(2/3), 115–142 (1994)
8. Swamy, C.: Correlation clustering: maximizing agreements via semidefinite programming. In: Proceedings of the 15th Annual ACM-SIAM Symposium on Discrete Algorithms (SODA 2004), pp. 526–527 (2004)

Evolutionary Estimation of Parameters in Computational Models of Thymocyte Dynamics

Lavinia Moatar-Moleriu(✉), Viorel Negru, and Daniela Zaharie

Department of Computer Science, West University of Timişoara, Timişoara, Romania
{lmoatar,vnegru,dzaharie}@info.uvt.ro

Abstract. This paper presents an evolutionary-based parameter estimation procedure able to deal with the particularities of the constraints arising in mathematical models of biological systems. A measure of the constraint satisfaction degree and several feasibility-based ranking rules are proposed and comparatively analyzed for the problem of estimating the parameters involved in a model describing the dynamics of thymocytes. The numerical results illustrate the effectiveness of the procedure in inferring models which fit well the experimental data and also satisfy the biological constraints.

Keywords: Population dynamics · Parameter estimation · Constrained · Optimization · Differential evolution

1 Introduction

Inferring models from experimental data is an important task in computational biology and usually lead to difficult constrained optimization problems. As the estimation of the model quality can involve simulation of complex systems and the apriori knowledge on the parameters could be limited, local optimization methods involving gradient computation are inapplicable. Therefore, metaheuristics proved to be viable methods for parameter estimation of such models [7]. In this paper we focus on the problem of identifying computational models able to simulate the dynamics of thymocyte populations taking place in the thymus, as part of the complex process through which the organisms defend against infections [2]. Aiming to model transient perturbations of the normal dynamics of thymocytes we arrived to the problem of estimating several dozens of parameters such that some biologically motivated constraints are satisfied. As these constraints are related to properties of some time-dependent functions they require specific handling techniques. The paper is organized as follows. Section 2 presents the mathematical model and the components of the constrained optimization problem. Section 3 shortly reviews evolutionary constraint optimization while the proposed parameter estimation procedure, including the specific constraint handling variants, is presented in Sect. 4. Results of a comparative

I. Lirkov et al. (Eds.): LSSC 2013, LNCS 8353, pp. 272–280, 2014.
DOI: 10.1007/978-3-662-43880-0_30,

analysis and the numerical validation of the estimation procedure are presented in Sect. 5. Finally, Sect. 6 concludes the paper.

2 The Mathematical Model and the Parameter Estimation Problem

One of the simplest models describing the dynamics of thymocyte populations, proposed in [2], consists of four coupled differential equations (Eqs. (1)). Each equation describes the evolution of the number of cells in the corresponding population, controlled by proliferation, death, and transfer rates (denoted by r, d, and s, respectively). Besides these rates, the model contains other three parameters: b (inflow of progenitor cells), K and K_n (carrying capacities).

$$\begin{aligned}
\dot{N}(t) &= r_n N(t)(1 - N(t)/K_n) - d_n N(t) - s_n N(t) + b(1 - N(t)/K_n) \\
\dot{P}(t) &= r_p P(t)(1 - Z(t)/K) - d_p P(t) - (s_4 + s_8)P(t) + s_n N(t) \\
\dot{M_4}(t) &= r_4 M_4(t)(1 - Z(t)/K) - d_4 M_4(t) - s_{o4} M_4(t) + s_4 P(t) \\
\dot{M_8}(t) &= r_8 M_8(t)(1 - Z(t)/K) - d_8 M_8(t) - s_{o8} M_8(t) + s_8 P(t) \\
Z(t) &= N(t) + P(t) + M_4(t) + M_8(t)
\end{aligned} \tag{1}$$

These equations proved to be appropriate in modelling the thymocyte dynamics in a normal thymus [2]. However various pathological situations or the administration of some substances can perturb the normal dynamics by inducing a significant involution followed by a regeneration of the thymocyte populations. Such a dynamics can be simulated by replacing the constant rate parameters in Eqs. (1) with variable rates obtained by adding to the initial rates a time-depending function which model the transient perturbation. A family of functions appropriate to model a perturbation starting from a zero value at an initial time moment and approaching again zero after a time interval is described in Eq. (2), where $C = \{c_1, c_2, c_3, c_4, c_5\}$ denotes a set of positive parameters.

$$\xi(C;t) = \frac{c_1}{t^{c_3} + c_2} - \frac{c_1 c_4 / c_2}{t^{c_5} + c_4} \tag{2}$$

By replacing each constant rate r with $r + \xi(C;t)$, five new parameters are introduced for each of the thirteen rates, leading to a set of $k = 71$ parameters in the model. Estimating the parameters values means finding $x^* \in \mathbf{R}^k$ which minimizes the mean squared error described in Eq. (3) and satisfies constraints related to the positivity of all perturbed rates and the vanishing of the perturbation.

$$MSE(x) = \frac{1}{4n} \sum_{\pi \in \{N,P,M_4,M_8\}} \left(\frac{1}{\max_{j=\overline{1,n}}\{\bar{\pi}_j^2\}} \sum_{j=1}^{n} (\pi(x;t_j) - \bar{\pi}_j)^2 \right) \tag{3}$$

In Eq. (3) n denotes the number of experimental values available for each thymocyte population, $\bar{\pi}$ denotes experimental values corresponding to each of the four populations and $\pi(x;t)$ denote numerically estimated solutions corresponding to the given set of parameters and to the time moments of the experimental measurements. These estimated solutions are obtained by numerically solving Eqs. (1) for initial values compatible with the experimental data. The division of the error terms corresponding to each population by the maximal measured value ensures the balance between the errors corresponding to different thymocyte populations and avoid the bias in the estimation process toward parameters of the dominant population. For each perturbed rate the two constraints to be satisfied are described in Eq. (4) where t_a denotes the time moment when the perturbation starts and t_f the time moment when it should be small enough (e.g. smaller than a given value $\epsilon_f > 0$).

$$r + \xi(C;t) \geq 0 \text{ for all } t \in [t_a, t_f]; \quad |\xi(C;t_f)| < \epsilon_f \tag{4}$$

3 Evolutionary Constrained Optimization

A constrained optimization problem usually requires to find $x^* \in \mathbf{R}^k$ which minimizes an objective function $f : \mathbf{R}^k \to \mathbf{R}$ and satisfies $g_j(x^*) \geq 0$ for each $j \in \{1, \ldots, q\}$. Evolutionary constrained optimization relies on using a constraint handling technique when applying the basic evolutionary operators (variation and/or selection). Most approaches interferes with the selection process by changing either the fitness value computation (penalty method) or the comparison rule between two candidate solutions (feasibility rules, stochastic ranking) [3]. Despite the differences between them, all these techniques uses the so-called constraint violation amount which for a constraint $g_j(x) \geq 0$ is defined by $\phi_j(x) = \min(0, g_j(x))^2$. The overall violation of the constraints is defined as the sum $\phi(x) = \phi_1(x) + \ldots + \phi_q(x)$. While in the penalty function technique the objective function and the constraint violation function are combined, in the feasibility based rules they are separately used. The classical Deb's rule [1] specifies that a candidate solution x is better than x' if one of following conditions is satisfied: *(i)* x is feasible and x' is not feasible; *(ii)* both x and x' are feasible and $f(x) < f(x')$; *(iii)* both x and x' are infeasible and $\phi(x) < \phi(x')$.

Using the objective function as comparison criterion only in the case of feasible solutions can lead to premature convergence [3]. Two variants which enlarges the set of cases when the objective function is used as optimization criterion are ϵ-feasibility [6] and stochastic ranking [5]. The ϵ-feasibility rule is based on a relaxation of the feasibility concept, i.e. x is considered better than x' if one of following conditions is satisfied: *(i)* $\phi(x) \leq \epsilon$ and $\phi(x') \leq \epsilon$ (x and x' are almost feasible) and $f(x) < f(x')$; *(ii)* $\phi(x) = \phi(x')$ (same constraints violation) and $f(x) < f(x')$; *(iii)* $\phi(x) < \phi(x')$. On the other hand, the stochastic ranking enhances the role of the objective function by involving it in the decision rule not only when the solutions are feasible but also when a random event occurs. However all these feasibility based rules use, when deciding if a solution is better than another, either the objective function or the constraints violation function but not both of them.

4 The Proposed Parameter Estimation Procedure

The parameters of the model described in Sect. 2 which should be estimated can be grouped in several sets: $(C_0; C_1; \ldots; C_q)$. C_0 denotes the parameters involved in Eqs. (1) (i.e. $C_0 = (r_n, r_p, r_4, r_8, d_n, d_p, d_4, d_8, s_n, s_4, s_8, s_{o4}, s_{o8}, b, K_n, K)$). The sets C_j correspond to the parameters involved in the functions used to perturb the rates specified in C_0 and should satisfy the constraints described in Eq. (4). The first constraint is particularly difficult to check as, in the general case, it requires the analysis of the values of $r + \xi(C; t)$ over a continuous time interval. As a consequence it is neither easy to decide if a solution is feasible nor to compute the constraint violation amount. However for smooth continuous functions $\xi(C; t)$ one can estimate the degree of constraint satisfaction by sampling the time interval and computing the proportion of cases when the constraint is satisfied. More specifically, by considering an uniform discretization $T_h = \{t_a, t_a + h, t_a + 2h, \ldots, t_f\}$ of $[t_a, t_f]$ we can compute an estimation of the constraint satisfaction degree as given in Eq. (5).

$$S_p^j(C_j) = \frac{\text{card}\{t \in T_h | r_j + \xi(C_j; t) > 0\} - \delta}{\text{card}(T_h)} \tag{5}$$

The constant $\delta > 0$ in Eq. (5) has a small value and is used only to discriminate the cases when the constraint satisfaction can be mathematically proved. For instance in the particular case of constraints given in Eq. (4) a sufficient condition ensuring the positivity is $r \geq \max\{c_1/(c_2 + c_2^2), c_1c_4/(c_2 + c_2c_4)\}$. Therefore this condition is first checked and if it is satisfied then the positivity constraint is considered satisfied and S_p is set to 1. Otherwise S_p is computed using Eq. (5).

For the second type of constraints the satisfaction degree can be computed following a standard approach which leads to a value S_v as defined in Eq. (6).

$$S_v^j(C_j) = \begin{cases} 1 & \text{if } |\xi(C_j; t_f)| \leq \epsilon_f \\ 1 - \min\{1, |\xi(C_j; t_f)|\} & \text{otherwise} \end{cases} \tag{6}$$

The overall degree of constraints satisfaction, $S \in [0, 1]$ is defined as $S(C) = \prod_{j=1}^{q} S_p^j(C_j) S_v^j(C_j)$. The values of S can be interpreted as follows: *(i)* if $S(C) = 1$ then C is surely feasible; *(ii)* if $S(C) \geq 1 - \delta/\text{card}(T_f)$ then C is probably feasible (there is neither evidence that the first constraint is violated nor guarantees that it is satisfied); *(iii)* if $S(C) = 0$ then at least one of the constraints is severely violated (at least one perturbation is too large or for at least one perturbed rate all sampled points are negative); *(iv)* in all other cases, the value of S offers information about the degree of constraint satisfaction.

Examples of several cases of perturbed rates satisfying or violating the constraints and the corresponding S values are illustrated in Fig. 1. Having a value in $[0, 1]$, S can be used to penalize the value of the objective function or as acceptance probability of infeasible configurations.

Constraints handling. There are several ways to use the satisfaction degree S and the MSE value in order to decide which of two candidate solutions is

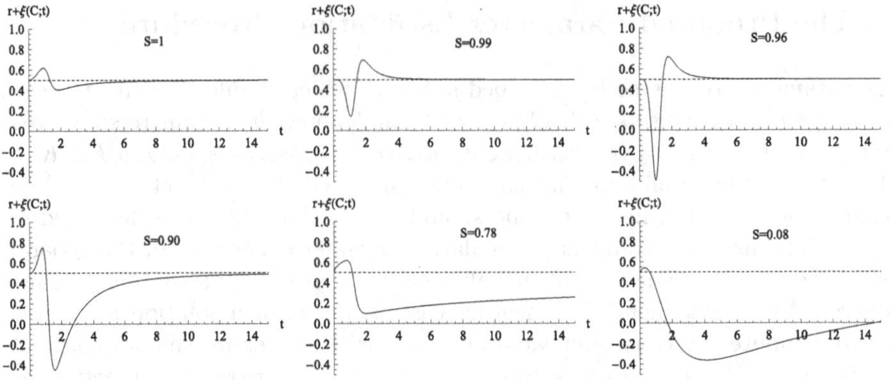

Fig. 1. Illustration of the relationship between the properties of the perturbed rates and the values of the constraint satisfaction degree, S. Continuous line: perturbed rate, dashed line: initial value of the rate.

better. Starting from the existing feasibility and ranking rules [3] and using the properties of S we identified several variants which we comparatively analyzed with respect to their effectiveness in solving the addressed parameter estimation problem.

Ranking rule A. Using the assumption that $S(x) \geq \theta$ suggests that x is feasible, while $S(x) < \theta$ means that it is infeasible (for a given threshold θ), the Deb's feasibility rule can be rewritten as follows. A candidate solution x is better than x' if one of the following conditions is satisfied: *(i)* $S(x) \geq \theta$ and $S(x') < \theta$; *(ii)* $S(x) \geq \theta$ and $S(x') \geq \theta$ and $MSE(x) \leq MSE(x')$; *(iii)* $S(x) < \theta$ and $S(x') < \theta$ and $S(x) \geq S(x')$.

Ranking rule B. One of the particularities of the previous rule is that the objective function and the constraint satisfaction degree are used in a decoupled way. A variant which aggregates MSE and S states that x is better than x' if one of the conditions is satisfied (checked in this specific order): *(i)* $S(x) \geq \theta$ and $S(x') < \theta$; *(ii)* $S(x)S(x') = 0$ and $MSE(x) \leq MSE(x')$; *(iii)* $S(x) \neq 0$ and $S(x') \neq 0$ and $MSE(x)/S(x) \leq MSE(x')/S(x')$.

Ranking rule C. The first two rules analyze first the cases when the constraints are satisfied or close to be satisfied. By ruling out first the cases when the constraints are severely violated one obtains a slightly different variant when x is better than x' if: *(i)* $S(x) > 0$ and $S(x') = 0$; *(ii)* $S(x) = 0$ and $S(x') = 0$ and $MSE(x) \leq MSE(x')$; *(iii)* $S(x) \neq 0$ and $S(x') \neq 0$ and $MSE(x)/S(x) \leq MSE(x')/S(x')$.

Ranking rule D. Instead of inferring the feasibility in a deterministic way from the value of S one can do it in a probabilistic manner. In this case $S(x)$ is interpreted as a probability that x, if selected, can lead to a feasible configuration. Thus, by denoting with U_1 and U_2 two independent random values uniformly selected from $[0,1]$ one can say that x is better than x' if one of the following conditions is satisfied: *(i)* $U_1 \leq S(x)$ and $U_2 > S(x')$; *(ii)* $U_1 > S(x)$ and $U_2 > S(x')$ and $MSE(x) \leq MSE(x')$; *(iii)* $U_1 \leq S(x)$ and $U_2 \leq S(x')$ and $MSE(x)/S(x) \leq MSE(x')/S(x')$.

Ranking rule E. Starting from the idea of stochastic ranking [5], which allows in a probabilistic manner to use the objective function as comparison criterion, even for infeasible solutions, we arrived at the following rule which states that x is better than x' by sequentially checking the following conditions: *(i)* $S(x) \geq \theta$, $S(x') \geq \theta$ and $MSE(x) < MSE(x')$; *(ii)* $U < P_f$, $S(x)S(x') \neq 0$ and $MSE(x)/S(x) < MSE(x')/S(x')$; *(iii)* $S(x) \geq S(x')$.

Search method. As stated in [3] one of the most competitive evolutionary metaheuristic in solving constrained optimization problems seems to be Differential Evolution (DE). On the other hand the effectiveness of DE for parameter estimation in biological systems was reported in several comparative studies [7]. This motivated us to use JADE, an adaptive DE variant introduced in [8]. The JADE overall structure is described in Algorithm 1 and its main features are: (i) the elements used in the recombination rule described in Eq. (7) are chosen such that a new candidate is created in a neighborhood of a good element but away from a worse one (x_{rbest} is selected from the $p\,\%$ elites of the current population and x_{r2} is one of the inferior elements which were discarded in a previous selection step and was stored in an archive); (ii) the scale factor (F) and the crossover probability (CR) are generated for each element of the population using a probability distribution (Gaussian and Cauchy, respectively) whose mean is recomputed at each generation using information from successful elements.

$$z_i^l = \begin{cases} x_i^l + F_i \cdot (x_{rbest}^l - x_i^l) + F_i \cdot (x_{r1}^l - x_{r2}^l) & \text{if } rand() \leq CR_i \\ x_i^l & \text{otherwise} \end{cases}, \quad i = \overline{1,m},\ l = \overline{1,k} \tag{7}$$

The constraint handling techniques interfere with two of the JADE components: (i) the ranking process used to select the top $p\,\%$ elements; (ii) the selection of the survivor between the parent and the trial element.

Algorithm 1. JADE overall structure

```
Initialization step (population, control parameters, archive)
while ⟨the stopping condition is false⟩ do
  Rank the population and identify the top p% elements
  for i = 1,m do
    Construct z_i using Eq. (7); Choose the best between z_i and x_i
  end for
  Update the control parameters and the archive
end while
```

Table 1. Quality of fit (MSE), constraints satisfaction (S) and feasibility probability ($FP, \theta_1 = 0.999927$ and $\theta_2 = 0.99$) for JADE combined with the proposed ranking rules.

Ranking rule	MSE	S	$FP(\theta_1)$	$FP(\theta_2)$
Rule A ($\theta = 1$)	0.0338 ± 0.0012	1 ± 0	1	1
Rule A ($\theta = 0.99$)	0.0270 ± 0.0010	0.9966 ± 0.0033	0.5	1
Rule B ($\theta = 0.99$)	0.0268 ± 0.0014	$0.9999 \pm 5 \cdot 10^{-6}$	1	1
Rule C	0.0261 ± 0.0009	0.9878 ± 0.0119	0.45	0.45
Rule D	0.0290 ± 0.0017	$0.9999 \pm 3 \cdot 10^{-6}$	1	1
Rule E ($P_f = 0.45$)	0.0250 ± 0.0005	0.9935 ± 0.0011	0.03	1
Unconstrained	0.0208 ± 0.0022	0.0468 ± 0.0776	0	0

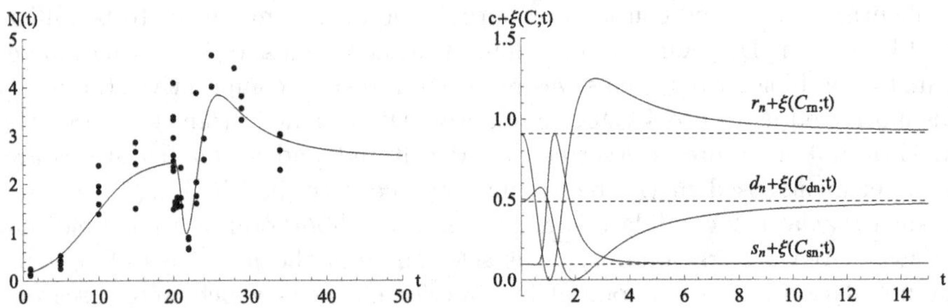

Fig. 2. Left: Experimental data and simulated dynamics of N ($MSE = 0.023, S = 0.993, t_a = 20, t_f = 35$ days). Right: initial rates (dashed lines) and perturbed rates (continuous lines) for N.

5 Numerical Results

The used experimental dataset consists of 232 estimates of the number of cells in each of the four thymocyte populations collected from young and adult mice thymus either before or after a treatment administration. Each of the five ranking rules proposed in the previous section was combined with JADE leading to a specific procedure to estimate all $k = 71$ parameters of the model which satisfies $q = 26$ constraints. In each case, results from 30 independent runs were collected. The results reported in Table 1 have been obtained by using populations of $m = 20$ elements, 5000 generations, a percent $p = 10\,\%$ in defining the set of top elements and 0.5 as initial mean of distributions used to provide values for F and CR. A preliminary numerical study using several population sizes (e.g. 20, 50 and 100) suggested that $m = 20$ leads to the best quality/cost trade-off. For the variant inspired by stochastic ranking, the value of P_f was set to 0.45, as suggested in [5]. The numerical solutions of system Eq. (1) required for MSE estimation were obtained using the ODE solver from Mathematica 7.0.

Table 1 presents statistical values of MSE, the constrained satisfaction degree S (with $\epsilon_f = \delta \doteq 0.001$) and the feasibility probability (FP) defined as the ratio between the number of runs when the solution can be considered feasible (i.e. $S \geq \theta$) and the total number of runs [4]. As feasibility threshold, any value θ larger than 0.99 proved to lead to solutions satisfying the positivity constraint. The value of the threshold corresponding to the cases when all sampled values of the perturbed rates are positive is $1 - \delta/\text{card}(T_h) = 0.999927$ (for $h = 0.1$ and $\delta = 0.001$). The results in Table 1 show that the analyzed ranking rules are characterized through different quality of fit vs. constraint satisfaction trade-offs. The best trade-off is obtained by the variant inspired from stochastic ranking (with respect to MSE it is superior to other ranking rules, as a Mann-Whitney statistical test returns p-values less than 10^{-5} when rule E is compared with any of the other ones). With respect to the constraint satisfaction the most effective ones are rules B and D. Best behavior was observed for the ranking rules using an aggregation of MSE and S.

The ability of the proposed procedure to lead to a model which fits well to the data and satisfies the constraints on rates is illustrated (for one of the four populations) in Fig. 2.

6 Conclusions

By combining an evolutionary algorithm with an appropriate constraint handling technique we succeeded in inferring a model of the perturbed thymus dynamics which is in accordance with the experimental data. The constraint satisfaction degree and the proposed ranking rules, characterized by aggregating the quality of fit measure and the constraint satisfaction degree, can be used for other optimization problems involving constraints which can be only partially checked.

Acknowledgments. This work was supported by grant no. PN-II-ID-PCE-2011-3-0571 and by the strategic grant POSDRU/CPP107/DMI1.5/S/78421, Project ID 78421 (2010), co-financed by the European Social Fund Investing in People, within the Sectoral Operational Programme Human Resources Development 2007–2013. The authors thank Dr. Felix Mic (University of Medicine and Pharmacy, Timisoara) for providing the experimental data.

References

1. Deb, K.: An efficient constraint handling method for genetic algorithms. Comput. Methods Appl. Mech. Eng. **186**, 311–338 (2000)
2. Mehr, R., Globerson, A., Perelson, A.S.: Modeling positive and negative selection and differentiation processes in the thymus. J. Theor. Biol. **175**, 103–126 (1995)
3. Mezura-Montes, E., Coello Coello, C.A.: Constraint-handling in nature-inspired numerical optimization: past, present and future. Swarm Evol. Comput. **1**, 173–194 (2011)
4. Mezura-Montes, E., Miranda-Varela, M.E., Gomez-Ramon, R.C.: Differential evolution in constrained numerical optimization: an empirical study. Inf. Sci. **180**, 4223–4262 (2010)
5. Runarsson, T.P., Yao, X.: Stochastic ranking for constrained evolutionary optimization. IEEE Trans. Evol. Comput. **4**, 284–294 (2000)

6. Takahama, T., Sakai, S.: Constrained optimization by ϵ-constrained particle swarm optimizer with ϵ-level control. In: Abraham, A., Dote, Y., Furuhashi, T., Köppen, M., Ohuchi, A., Ohsawa, Y. (eds.) WSTST 2005. Advances in Soft Computing, vol. 29, pp. 1019–1029. Springer, Heidelberg (2005)
7. Tashkova, K., Korošek, P., Šilc, J., Todorovski, L., Džeroski, S.: Parameter estimation with bio-inspired meta-heuristic optimization: modeling the dynamics of endocytosis. BMC Syst. Biol. **5**, 159 (2011). doi:10.1186/1752-0509-5-159
8. Zhang, J., Sanderson, A.C.: JADE: adaptive differential evolution with optional external archive. IEEE Trans. Evol. Comput. **13**(5), 945–958 (2009)

Micro Differential Evolution Performance Empirical Study for High Dimensional Optimization Problems

Mauricio Olguin-Carbajal[1](✉), J. Carlos Herrera-Lozada[1], Javier Arellano-Verdejo[2], Ricardo Barron-Fernandez[2], and Hind Taud[1]

[1] Centro de Innovación y Desarrollo Tecnológico en Cómputo, IPN, Mexico City, México
molguinc@ipn.mx
[2] Centro de Investigación en Computación, IPN, Mexico City, México

Abstract. This paper presents an empirical study of a micro Differential Evolution algorithm (micro-DE) performance versus a canonical Differential Evolution (DE) algorithm performance. Micro-DE is a DE algorithm with reduced population and some other differences. This paper's objective is to show that our micro-DE outperforms the canonical DE for large scale optimization problems by using a test bed consisting of 20 complex functions with high dimensionality for a performance comparison between the algorithms. The results show two important points; first, the relevance of an accurate set of the optimization algorithms parameters regarding the problem itself. Second, we demonstrate the superior performance of our micro-DE with respect to DE in 19 out 20 tested functions. In some functions, the difference is up to seven orders of magnitude. Also, we show that micro-DE is better statistically than a simple DE and an adjusted DE for high dimensionality. In several problems where DE is used, micro-DE is highly recommended, as it achieves better results and statistic behavior without much change in code.

Keywords: Micro-algorithm · Reduced population · Differential evolution · High dimensionality

1 Introduction

A Differential Evolution (DE) algorithm using one hundred individuals over a thousand generations requires (at least) one hundred thousand (1.0e+5) calls to the objective function. For this reason, it is worth seeking new strategies to achieve better performance with less labor. One of these strategies is a DE algorithm with reduced population, with fewer individuals and fewer function evaluations (FE's) for similar or better results. Also, more objective function evaluations, per individual, could lead to better performance for the same FE's than algorithms with bigger populations.

I. Lirkov et al. (Eds.): LSSC 2013, LNCS 8353, pp. 281–288, 2014.
DOI: 10.1007/978-3-662-43880-0_31,

In this study, we show the performance of our micro Differential Evolution (micro-DE) algorithm with respect to a Differential Evolution algorithm in high dimensional problems.

The strategy of small populations begins with the work of Goldberg [1]. In his work, Goldberg presented experiments with different population sizes for Genetic Algorithms and shows a relationship between size and errors. Later, Krishnakumar used a population of five individuals and adopted an elitist strategy [3]. Krishnakumar reported better and faster results for his micro-GA (micro Genetic Algorithm) when tested as two static functions and two real world control problems.

Rahnamayan and Tizhoosh [5] developed a micro-DE and a micro-ODE for image threshold where the micro-ODE outperformed the micro-DE as well a traditional Kittler threshold method. Rahnamayan and Tizhoosh algorithms (micro-DE and micro-ODE) used a five individual population, and test their algorithms with sixteen challenging test images. Micro-DE achieves ten best results out of sixteen test images.

Recently, Parsopoulos [4] developed a Cooperative micro-Differential Evolution (COMDE) and proposed an approach that employs small cooperative sub-populations to detect subcomponents of the original problem's solution concurrently. In his experiments, COMDE achieves, for all operators, better results than a classical DE algorithm. However, he only used five functions for his experiments with dimensions: 300, 600, 900, and 1200. The functions used were: Sphere, Generalized Rosenbrock, Rastrigin, Griewank, and Ackley. We intend to go further with a greater number of complex functions for high dimensional problems.

In this paper, we carry out an empirical study of the performance of our micro-DE regards to DE by using a benchmark consisting of twenty functions for high dimensionality, such as the special session LSGO (Large Scale Global Optimization) from CEC2012. By using this benchmark and comparing the performance of micro-DE against DE we have obtained remarkable results.

2 Micro Differential Evolution Algorithm (Micro-DE)

In this section we explain the basis of DE as background for, later, explain our micro-DE algorithm.

2.1 Differential Evolution Algorithm (DE)

The original DE algorithm proposed by Storn and Price [6] starts by generating a group of vectors with random values (Eq. 1), called initial population (Eq. 2).

$$\boldsymbol{x}_{i,G} = 1, 2, \ldots, D \tag{1}$$

$$P_0 = \boldsymbol{x}_1, \boldsymbol{x}_2, \ldots, \boldsymbol{x}_{Np} \tag{2}$$

Each P_0 individual is evaluated with respect to its performance of the function to be optimized. In the next stage (mutation), with the aid of scale factor (F), it proceeds to perform a noisy vector generation,

$$\boldsymbol{v} = \boldsymbol{x}_{r_1,G} + F \cdot (\boldsymbol{x}_{r_2,G} - \boldsymbol{x}_{r_3,G}), \tag{3}$$

with $I, r_1, r_2, r_3 \in [1, Np], I \neq r_1 \neq r_2 \neq r_3$ the recombination operator is applied to vector $\boldsymbol{v}$ and the population selected vectors (r_1, r_2, r_3), obtaining a trial vector $\boldsymbol{u}_{i,G}$,

$$\boldsymbol{u} = (u_0, u_1, \ldots, u_{D-1}) \tag{4}$$

with

$$u_{ij,G+1} = \begin{cases} v_{ij,G+1}, \; if \; R_j \leq CR \text{ or } \quad j = r_k \\ x_{ij,G}, \quad if \; R_j > CR \text{ and } j \neq r_k \end{cases} \tag{5}$$

where $CR \in [0, 1]$ is a crossover constant for the $u_{i,G}$ trial vector generation; $j = 1, 2, \ldots, n$; R_j is the *j-th* evaluation of a uniform random number generator [0,1]; and $r_k \in [1, n]$ is the random individual index. Storn and Price highlighted some common variants of DE algorithm; the most common used variant, as well as the used in all our algorithms, is DE/rand/1/bin (for further reference see [6]).

2.2 Micro Differential Evolution Algorithm (Micro-DE)

To be fully functional, a DE algorithm must have at least four individuals (see Eq. 3). However, in our tests we find that the best results for our micro-DE were achieved by using five individuals.

Algorithm 1. Pseudocode for micro Differential Evolution

$P_0 \leftarrow GenerateRandomlyInitialPopulation \quad i = 1 \ldots Np$
$Ext_{Lim} \leftarrow FunctionEvaluationsLimit$
$Int_{Lim} \leftarrow InternalGenerationsLimit$
$CR \leftarrow 0.001$
for $E_G = 1$ to Ext_{Lim} **do**
  $P_G \leftarrow$GenerateWorkPopulation(P_{G-1});
  EvaluateWorkPopulation(P_G);
  for $I_G = 1$ to Int_{Lim} **do**
    for $n = 1$ to Np **do**
      Randomly select $i \neq r1 \neq r2 \neq r3$
      $\boldsymbol{u}_{i,G} \leftarrow$ CalculateTrialVector$(\boldsymbol{x}_{i,G}, \boldsymbol{x}_{r_1}, \boldsymbol{x}_{r_2}, \boldsymbol{x}_{r_3})$;
      $\boldsymbol{x}_{i,G+1} \leftarrow$ TestTrialVector$(\boldsymbol{x}_{i,G}, \boldsymbol{u}_{i,G})$;
    end for
    $I_G \leftarrow I_G + 1$
  end for
  $S_{ellite} \leftarrow$SelectEllitist(P_G);
  $E_G \leftarrow E_G + 1$
end for

Our micro-DE proposal uses two loops (one internal and one external). The external loop runs until it reaches the stop criterion (Function Evaluations limit), Algorithm 1, line 2. The internal loop has a generational stop criterion (five generations), line 3. The work population generation, in the first generation, is constructed by vectors with random values, but subsequently integrated with *elite* and *remaining* individuals, line 7.

The internal loop is a DE algorithm of five generations, using five individuals, lines 8–15. Here we obtain a nominal convergence and pass resulting population to the *SelectEllitist* procedure, line 16.

The *elitist* individuals (best fitness) from the previous inner cycle are preserved to pass to the external cycle. The *remaining* individuals (worst fitness) are generated with random values to maintain diversity in the *GenerateWorkPopulation* procedure, line 6. In our micro-DE diversity is achieved, with the phases of mutation and recombination, as well as the *remaining* individuals. The algorithms go this way until the stop criterion is achieved.

3 Experiments

This section presents the experimental methodology used to evaluate and compare the different versions of Differential Evolution. In this study there are three: a canonical DE with $CR = 0.9$, an Adjusted DE with CR parameter for high dimensionality (ADE), and our micro-DE with CR parameter for high dimensionality. In the functions test bed used for this study, we can highlight their heterogeneous nature as some of the most notable features including: separable, not separable, rotated, and shifted. The detailed description of these functions can be found in [2].

3.1 Experimental Configuration

The settings used for each algorithm are: same variant for all (DE/rand/1/bin), population (60 individual for DE, ADE and 5 for micro-DE), as well as F parameter (the same for the all algorithms), CR parameter (0.9 for DE) and for high dimensionality (0.001 for ADE and micro-DE), as shown in Table 1.

DE, ADE and micro-DE were programmed in C++ language and were compiled by using g++ as well as the CEC2012 C++ libraries given by Tang et al.

Table 1. Parameter settings

Parameter	DE	ADE	micro-DE
Max runs number	100	100	100
FE's	3 000 000	3 000 000	3 000 000
Population size	60	60	5
F	[0.0, 1.0]	[0.0, 1.0]	[0.0, 1.0]
CR	0.9	0.001	0.001

Table 2. Median obtained for DE, ADE, and micro-DE

Function	DE	ADE	micro-DE
F1	4,52E+10	1,06E+08	**3,62E+06**
F2	1,48E+04	6,02E+02	**1,85E+02**
F3	2,09E+01	2,00E+01	**1,08E+00**
F4	2,51E+14	1,09E+14	**4,15E+13**
F5	4,37E+08	5,97E+08	**3,36E+08**
F6	**1,52E+07**	2,10E+07	1,93E+07
F7	6,28E+10	8,54E+10	**3,57E+10**
F8	2,48E+13	3,50E+10	**8,85E+08**
F9	5,09E+10	1,89E+09	**3,94E+08**
F10	1,54E+04	8,41E+03	**3,86E+03**
F11	2,28E+02	2,32E+02	**1,95E+02**
F12	7,92E+06	1,28E+06	**3,06E+05**
F13	3,31E+11	1,59E+06	**1,42E+04**
F14	6,19E+10	3,85E+09	**7,11E+08**
F15	1,58E+04	1,69E+04	**7,42E+03**
F16	4,18E+02	4,24E+02	**3,87E+02**
F17	1,40E+07	**2,71E+06**	4,03E+06
F18	1,01E+12	4,70E+06	**3,10E+04**
F19	3,56E+07	1,20E+07	**5,81E+06**
F20	1,03E+12	4,86E+06	**2,61E+04**

in [7]. We proceeded to perform 100 independent runs of each function, and a stop condition of 3 million function evaluations, (3.00e+06 FEs) for each algorithm and for each function.

4 Results

This section reports the micro-DE algorithm performance with regards the other two versions of DE, a canonical DE with a *CR* of 0.9, and a canonical DE with adjusted *CR* of 0.001 for high dimensionality (named ADE).

The first set of results shows that ADE outperformed DE, having better results than the classical DE only for the *CR* set to 0.001. The adjustment improves the performance of the DE in 14 functions. In eleven out of twenty functions, this improvement is significant by one or more orders of magnitude (F1, F2, F8, F9, F10, F13, F14, F15, F17, F18, and F20). It is noteworthy that in some cases performance improved by several orders of magnitude, as in the case of F8 with 3 orders, F13, F18, and F20 with five orders of magnitude (see Table 2).

With regard to our micro-DE, the best performance is obtained with four *elitist* individuals and one *remaining* individual. By observing the behavior of our micro-DE, one can see better performance in **nineteen** out of twenty functions, relative to the ADE algorithm for dimensionality with *CR* for one thousand variables. Our micro-DE performs even better, in terms of orders of magnitude,

Table 3. Average rankings of the algorithms by applying the Friedman procedure

Algorithm	Ranking
DE	2.65
ADE	2.25
micro-DE	1.10

with respect to the DE with *CR* 0.9, also with **nineteen** out of twenty functions (see Table 2) micro-DE column. One can see that our micro-DE is better than the enhanced feature set ADE in **nineteen** out of twenty functions, in some cases by two orders of magnitude: F1, F8, F13, F18 and F20. Against DE, our micro-DE showed improvements of up to **eight** orders of magnitude, as seen in F18 and F20.

5 Analysis of Results

In this section we use a non-parametrical statistical test (Friedman) to verify whether or not there is difference between algorithms (see [8]). As we can see in Table 3, there is no significant difference between DE and ADE algorithms. However, between DE and ADE relative to micro-DE, there is a significant difference. As we know, this result only means there is a statistical difference between algorithms, but it does not tell us if micro-DE is better or not.

Once the Friedman test showed that there are statistical differences between the algorithms studied, we performed a post-hoc test, to analyse in more detail the behavior of the algorithms. So we started from the fact that the null hypothesis indicates that the behaviour of the algorithm is not similar.

5.1 Post Hoc Comparisons

As Table 4 shows, the null hypothesis is rejected for all cases. This means all the algorithms are different. Nevertheless, DE and ADE algorithm comparison shows that, although the null hypothesis is rejected, the behavior of the two algorithms is statistically similar; so there is no difference observed when using DE or ADE. However, when the comparison is between DE and ADE relative to micro-DE, the difference between the algorithms is very significant. This confirms

Table 4. *p*-values table for $\alpha = 0.05$ and $\alpha = 0.10$

i	Algorithms	$z = (R_0 - R_i)/SE$	p	$Holm_{\alpha=0.05}$	$Holm_{\alpha=0.10}$
3	DE vs. micro-DE	4.901530	0.000001	0.017	0.033
2	ADE vs. micro-DE	3.636619	0.000276	0.025	0.050
1	DE vs. ADE	1.264911	0.205903	0.050	0.100

Table 5. Adjusted *p*-values

i	Hypothesis	Unadjusted p	p_{Holm}
1	DE vs. micro-DE	0.000001	0.000003
2	ADE vs. micro-DE	0.000276	0.000552
3	DE vs. ADE	0.205903	0.205903

the observations of the Friedman test, and allows us to conclude that, the micro-DE algorithm is clearly superior to its two competitors.

As shown in Table 5, the behavior of the DE and ADE algorithms is statistically similar, so there is no advantage to indiscriminate use of either, based on the adjusted p-values. However, comparison between DE and micro-DE, as well as ADE and micro-DE, shows that micro-DE is statistically better than canonical DE and Adjusted DE (DE and ADE). The main result is that micro-DE is a better option for optimization for the tested functions for high dimensionality.

6 Conclusions

We have demonstrated that our micro-DE outperforms a canonical DE and a high dimensional Adjusted DE in the Large Scale benchmark used. Also, we know that the DE algorithm is a robust and fast algorithm; however, if it is used without adjusted parameters (for the problem of optimization) it will perform poorly, as we have demonstrated. Recently, DE algorithm has been used as a reference (CEC2008), and got poor results relative to new algorithms, as it uses a *CR* of 0.9. When we use a cross rate of 0.001 the results are vastly different in orders of magnitude.

Moreover, our micro-DE provides superior performance than the classical and adjusted DE. So it is likely to be improved in many ways in which the canonical DE has been improved. This can impact its performance compared to other state of the art algorithms. Also, our micro-DE algorithm behave differently than canonical DE, and it is in our interest to investigate these issues and find improvements for this algorithm (micro-DE) using strategies previously used in DE, as well as other improvements and algorithms that we will develop.

Acknowledgments. Authors would like to thank, for their economic support:

– "Instituto Politécnico Nacional", CONACyT (register number 175589 and 290674), SNI, COFAA (register number SeAca/COTEPABE/79/12), Academic Secretary, Postgraduate and Research Secretary.

– The project roadMe: Fundamentals for Real World Applications of Metaheuristics: The Vehicular Network Case TIN2011-28194 (2012-2014).

References

1. Goldberg, D.E.: Genetic algorithms, noise, and the sizing of populations. In: Schaffer, J.D. (ed.) Proceedings of the Third International Conference on Genetic Algorithms, San Mateo, California, pp. 70–79. Morgan Kaufmann Pub (1989)
2. Herrera, F., Lozano, M., Molina, D.: Test suite for the special issue of soft computing on scalability of evolutionary algorithms and other meta-heuristics for large scale continuous optimization problems. Technical report, SCI2S, University of Granada, Spain (2010)
3. Krishnakumar, K.: Micro-genetic algorithms for stationary and non-stationary function optimization. In: SPIE Proceedings: Intelligent Control and Adaptive Systems, Philadelphia, PA, vol. 1196, pp. 289–296 (1989)
4. Parsopoulos, K.E.: Cooperative micro-differential evolution for high-dimensional problems. In: Proceedings of the GECCO'09, Montréal Québec, Canada, pp. 531–538. ACM (2009). ISBN 978-1-60558-325-9/09/07
5. Rahnamayan, S., Tizhoosh, H.R.: Image thresholding using micro opposition-based differential evolution (Micro-ODE). In: Proceedings of the IEEE Congress on Evolutionary Computation (CEC 2008), pp. 1409–1416 (2008)
6. Storn, R., Price, K.: Differential evolution - a simple and efficient adaptive scheme for global optimization over continuous spaces. Technical report, TR-95-012, ICSI (1995). ftp://ftp.icsi.berkeley.edu
7. Tang, K., Li, X., Suganthan, P.N., Yang, Z., Weise, T.: Benchmark functions for the CEC'2010 special session and competition on large scale global optimization. Technical report, NICAL, USTC, Hefei, Anhui, China (2009)
8. Friedman, M.: A comparison of alternative tests of significance for the problem of m rankings. Ann. Math. Stat. **1**, 86–92 (1940)

Free Search in Multidimensional Space

Kalin Penev(✉)

Maritime and Technology Faculty, Technology School,
Southampton Solent University, East Park Terrace, Southampton SO14 0YN, UK
kalin.penev@solent.ac.uk

Abstract. One of the challenges for modern search methods is resolving multidimensional tasks where optimization parameters are hundreds, thousands and more. Many evolutionary, swarm and adaptive methods, which perform well on numerical test with up to 10 dimensions are suffering insuperable stagnation when are applied to the same tests extended to 50, 100 and more dimensions. This article presents an original investigation on Free Search, Differential Evolution and Particle Swarm Optimization applied to multidimensional versions of several heterogeneous real-value numerical tests. The aim is to identify how dimensionality reflects on the search space complexity, in particular to evaluate relation between tasks' dimensions' number and corresponding iterations' number required by used methods for reaching acceptable solution with non-zero probability. Experimental results are presented and analyzed.

1 Introduction

This study focuses on multidimensional optimisation, where tasks parameters are within the range of hundreds. Explored are real coded optimisation algorithms Free Search [4], Differential Evolution [7] and Particle Swarm Optimisation [2]. In contrast to combinatorial optimisation the number of potential solutions in real coded tasks could tend to infinity. Combined with multiple dimensions this makes task difficult even for clarification of the optimal value with low precision. Considerable research efforts are directed towards evaluation and improvement of existing and design of new methods capable of resolving multidimensional tasks [1,5,6,8–11].

According to the literature existing methods perform well on variety of problems. However when applied to multidimensional tasks with many parameters at the range of hundreds variables and more well-known methods face difficulties such as: - need for large number of evaluations per iteration - large populations size; - need for large computational resources; - need for large period of time for calculations; - inability to identify optimal solution with appropriate level of precision. In summary identification of optimal solutions with acceptable level of precision and within acceptable period of time seems a challenge.

The aim of this investigation is to evaluate Free Search, Differential Evolution and Particle Swarm Optimisation abilities to avoid stagnation and trapping in local suboptimal solution, to identify minimal number of iterations required to

I. Lirkov et al. (Eds.): LSSC 2013, LNCS 8353, pp. 289–296, 2014.
DOI: 10.1007/978-3-662-43880-0_32,

resolve multidimensional tasks with acceptable precision. For this purpose three global numerical test are used - Bump test function [3], Michalewicz test function [12] and Norwegian test function [13].

2 Test Problems

Criteria for tests selection for this investigation are: - must be for global optimisation with many local suboptimal solutions; - must not provide initial knowledge for optimal solution value and location; - optimal solution must be dependent on dimensions number. Chosen for this study numerical test, which meet the above criteria are - Bump test function, Michalewicz test function and Norwegian test function. From available publications is visible that existing optimisation methods faced difficulties to reach optimal solutions with appropriate level of precision.

Michalewicz and Norwegian tests are explored for 100 dimensions by Free Search, Differential Evolution and Particle Swarm Optimisation. Free Search is tested additionally on Bump test for 50, 100, 200, 500, and 1000 dimensions.

2.1 Bump Test Function

This is hard constrained global optimisation problem [3] transformed in this study for maximisation.

$$f(x_i) = \left| \sum_{i=1}^{n} \cos^4(x_i) - 2 \prod_{i=1}^{n} cos^2(x_i) \right| / \sqrt{\sum_{i=1}^{n} i x_i^2}$$

subject to: $\prod_{i=1}^{n} x_i > 0.75$ and $\sum_{i=1}^{n} x_i < 15n/2$
for $0 < x_i < 10$, and $i = 1, \ldots, n$ start from $x_i = 5$, $i = 1, \ldots, n$, where x_i are the variables (expressed in radians) and n is the number of dimensions.

2.2 Michalewicz Test Function

The Michalewicz test function is described in the domain of Kyoto University [12] as global optimisation problem. In this study test function is transformed for maximization.

$$f(x_i) = \sum_{i=1}^{n} \sin(x_i)(sin(ix_i^2/\pi))^{2m}$$

where search space is defined as $0 \leq x_i \leq \pi$, $i = 1, \ldots, n$.

2.3 Norwegian Test Function

Norwegian test function is global test problem.

$$f(x_i) = \prod_{i=1}^{n} \left(\cos(\pi x_i^3) \left(\frac{99 + x_i}{100} \right) \right)$$

where search space borders are defined by $-1.1 < x_i < 1.1$, $i = 1, \ldots, n$.

3 Optimization Methods

In this study three optimization methods are used - Free Search, Differential Evolution and Particle Swarm Optimisation.

3.1 Free Search

Free Search is adaptive heuristic method [4] for real coded optimisation. It is based on a conceptual model, which is different from other evolutionary methods. Free Search simulates behaviour of animals in nature. In Free Search animals (individuals) explore continuous search space by taking exploration walks within their continuous neighbouring area.

In contrast to other optimisation methods optimisation process in Free Search is organised in explorations. For one exploration each individual performs certain number of steps. In this study steps number per exploration is five. In order to guarantee equal number of objective function calculations Free Search explorations are limited to 20000, 200000, and 2000000, multiplied to 5 steps this correspond to 100000, 1000000, and 10000000 iterations for other methods. Further in this article limitations are indicated in iterations.

Free Search is implemented with a population of 10 individuals and the explorations are 5 steps, for all experiments. The sense is random in the highest 10 % of the sensibility, and the neighbouring space varies from 0.5 to 1.5 with step 0.1.

3.2 Differential Evolution

A main feature of Differential Evolution is the concept for generation of new individuals. Individuals in DE are called vectors. DE selects from the current population target, donor and differential vectors. From these vectors DE generates a new trial vector, which replaces the target vector, if it is better, in the new population. The authors proposed several strategies for generation of a trail vector [7].

Differential Evolution is implemented with a population of 10 individuals and explored with strategy 5 [7]. All individuals from the population are involved in modification and replacement. The crossover probability is 0.5. Differential factor [7] varies from 0.5 to 1.5 with step 0.1.

3.3 Particle Swarm Optimisation

The Particle Swarm Optimisation is motivated from the simulation of social behaviour of the group of individuals [2]. An original feature of PSO is generation of new population. PSO generates new values for all particles (individuals) in the swarm (population). It memorises the previous individual and social (swarm) experience and it uses them for generation of new particles. This concept for generation of new individuals differentiates PSO from other population-based algorithms.

The modification strategy of Particle Swarm Optimisation has been improved by use of the original concept for the so called inertia parameter that increases the overall performance of PSO [14].

Particle Swarm Optimisation is implemented and explored with the inertia parameter. The inertia parameter varies from 0.5 to 1.5 with step 0.1. The implemented PSO has a population of 10 individuals for all experiments. The individual and the social learning factors are 2 for all experiments.

4 Experimental Results

Experimental methodology is organized to identify minimal number of iterations required to achieve optimal result with acceptable level of precision. Norwegian and Michalewicz test functions are implemented for 100 dimensions and evaluated for three series of 320 experiments, with start from different random locations, limited to 100000, 1000000, and 10000000 iterations for all three methods Free Search, Differential Evolution and Particle Swarm Optimisation. Achieved experimental results are analysed and compared for maximal achieved result.

Free Search additionally is tested on Bump test function for 50, 100, 200, 500, and 1000 dimensions. Experimental results are summarised and compared to results earlier published in the literature.

Achieved maximal results from Free Search (FS), Differential Evolution (DE) and Particle Swarm Optimisation (PSO) experimental results on Norwegian (denoted as F1) and Michalewicz test functions (denoted as F2) are presented in Table 1.

Table 1. Maximal results on Norwegian and Michalewicz test functions

	Iterations	FS	DE	PSO
F1	100 000	0.750627	0.448729	0.220553
F1	1 000 000	0.967082	0.448729	0.224411
F1	10 000 000	1.00004	0.490885	0.225525
F2	100 000	99.5808	82.1164	79.2948
F2	1 000 000	99.6157	87.4321	79.2948
F2	10 000 000	99.6191	88.2164	79.2948

Free Search is tested additionally on Bump test function for 50, 100, 200, 500, and 1000 dimensions. Achieved results are presented in Table 4.

The variables values which correspond to the results in Table 2 are available on request for evaluation or comparison. For experiments is used desktop PC with motherboard ASUS Rampage VI, processor Intel i7 3960x overclocked to 4.895 GHz and memory G.Skill TridentX at 1866 MHz.

Table 2. Mean results on Norwegian and Michalewicz test functions

	Iterations	FS	DE	PSO
F1	100 000	0.69120580	0.12018652	0.00747217
F1	1 000 000	0.92401156	0.20175355	0.00798572
F1	10 000 000	0.98937421	0.2493765	0.00836006
F2	100 000	99.5021065	52.1924515	31.9071906
F2	1 000 000	99.6109537	57.4708753	33.0173021
F2	10 000 000	99.618175	59.863695	34.2029145

Table 3. Standard deviation on Norwegian and Michalewicz test functions

	Iterations	FS	DE	PSO
F1	100 000	0.02712148	0.18597598	0.03091758
F1	1 000 000	0.01853559	0.20359603	0.03013784
F1	10 000 000	0.00843936	0.21503870	0.03148975
F2	100 000	0.11389434	19.591272	20.5197324
F2	1 000 000	0.00266902	19.2625445	20.9821918
F2	10 000 000	0.00048003	15.9760239	22.3544008

Table 4. Standard deviation on Bump test functions

Dimension	50	100	200	500	1000
FS	0.835262348358115	0.8456854	0.8506636	0.8512628	0.8514553

5 Discussion

Analysis of experimental results suggests that for Norwegian test function used implementations of DE and PSO stagnate in suboptimal solutions for all experiments limited to 100000, 1000000, and 10000000 iterations. Taking into that DE and PSO had difficulties on 2 dimensional versions of Norwegian test [15] reasons for this stagnation could be a subject of further research. In contrast FS confirms its abilities to avoid stagnation and escape from trapping in suboptimal areas. For the first two series of 320 tests on Norwegian test function limited to 100000 and 1000000 iterations FS does not reach optimal solution. However for tests limited to 10000000 iterations from 320 runs with different start locations FS reaches 117 times optimal solutions with acceptable precision (above 1.00002).

This is a good illustration of the effectiveness of FS modification strategy, which guarantees non-zero probability for access to the whole search space during entire optimization process. On Michalewicz test function DE and PSO stagnate in suboptimal solutions for all experiments limited to 100000, 1000000, and 10000000 iterations. Reasons for this stagnation could be a subject of further research. On Michalewicz test function FS achieves optimal solution for all experiments limited to 100000, 1000000, and 10000000 iterations. Solving 100 dimensional Michalewicz test function for each run confirms good exploration abilities of FS. It indirectly suggests that this task could be resolved within less number of iterations, which could be a subject of further research. FS results on Bump test function are different from achieved by other methods and published in the literature [1,5,6]. Table 3 compares FS and Asynchronous Parallel Evolutionary Algorithm (APEA) [5,6] for 50, 100, 200, and 500 dimensions.

Table 5. Comparison of FS and APEA on Bump test function

Dimension	50	100	200	500
FS	0.835262348	0.8456854	0.8506636	0.8512628
APEA	0.8352620	0.8448539	0.8468442	0.8504975

Table 6. Period of time in minutes on Norwegian and Michalewicz test functions

	Iterations	FS (min)	DE (min)	PSO (min)
F1	10 000 000	4	15	35
F2	10 000 000	19	31	55

In order to clarify comparison of the results in Table 5 should be noted that FS achieves these results running in single thread on single processor desktop PC. Other essential issue is a period of time required for completion of optimization task. For experiments limited to 10 000 000 iterations average period of time in minutes, from 320 experiments, required for completion of one experiment on Norwegian and Michalewicz test is presented in Table 6.

Table 7. Variables for Bump test Fmax50 = 0.835262348358115

x0	6.2835790261751	x17	2.9524114588141508	x34	0.46001886537952297
x1	3.169937677750014	x18	2.9379900975870918	x35	0.45827603976373454
x2	3.156074749723996	x19	2.9232836057447771	x36	0.45656222211455522
x3	3.1423609878041932	x20	0.48823744173286926	x37	0.45487684374388804
x4	3.1287695107543283	x21	0.48593392519529544	x38	0.45321821190663503
x5	3.1152747643628493	x22	0.4836826364813	x39	0.45158651639926856
x6	3.1018528645810473	x23	0.48148246973308972	x40	0.449980222959319
x7	3.088480538514534	x24	0.47932981475899472	x41	0.44839856026986158
x8	3.0751349167360189	x25	0.47722236395944401	x42	0.44684046137453542
x9	3.0617943894947892	x26	0.47515900821764157	x43	0.44530576420283136
x10	3.0484368235138755	x27	0.4731373982323247	x44	0.44379365517281105
x11	3.0350390366956366	x28	0.47115575813689387	x45	0.44230323452275311
x12	3.0215778555508499	x29	0.46921217893825617	x46	0.44083365158292653
x13	3.0080295243393778	x30	0.46730534146908231	x47	0.43938498943233062
x14	2.9943676920815716	x31	0.46543440391236818	x48	0.43795641918838452
x15	2.9805647610851183	x32	0.46359705504397125	x49	0.43654683784954496
x16	2.9665903794608957	x33	0.46179196398120126		

Time periods in Table 6 is measured on processor Intel i7 3960x overclocked to 4.895 GHz and memory G.Skill TridentX at 1866 MHz, motherboard ASUS Rampage VI and solid state disk - SanDisk Extreme SSD SATA III. Experiments are completed simultaneously in hyper-treading processor mode. For verification of archived optimal results and for comparison to other methods variables for maximal value on 50 dimensional bum test are presented in Table 7. These variables could be used for start location for search of potential better value.

6 Conclusion

This article presents experimental evaluation of Free Search, Differential Evolution and Particle Swarm Optimization on hard global multidimensional tests. Identified are required number of iterations for which selected test could be resolved with high probability. Experimental results are also summarized and compared to results published in the literature. Further investigation could focus on evaluation and measure of time and computational resources sufficient for completion of 200 dimensional and other multidimensional tasks. Algorithms analysis and improvement could be also subject of future research.

Acknowledgements. I would like to thank to my students Asim Al Nashwan, Dimitrios Kalfas, Georgius Haritonidis, and Michael Borg for the design, implementation and overclocking of desktop PC used for completion of the experiments presented in this article.

References

1. MacNish, C., Yao, X.: Direction matters in high-dimensional optimisation. In: IEEE Congress on Evolutionary Computation, pp. 2372–2379 (2008)
2. Eberhart R., Kennedy J.: Particle swarm optimisation. In: Proceedings of the 1995 IEEE International Conference on Neural Networks, vol. 4, pp. 1942–1948. IEEE Press (1995)
3. Keane, A.J.: A brief comparison of some evolutionary optimization methods. In: Rayward-Smith, V.J., Osman, I.H., Reeves, C.R., Smith, G.D. (eds.) Modern Heuristic Search Methods, pp. 255–272. John Wiley, Chichester (1996)
4. Penev, K.: Free search of real value or how to make computers think. St. Qu, UK (2008). ISBN 978-0-9558948-0-0
5. Liu, P., Lau, F., Lewis, M.J., Wang, C.: A new asynchronous parallel evolutionary algorithm for function optimization. In: Guervós, J.J.M., Adamidis, P.A., Beyer, H.-G., Fernández-Villacañas, J.-L., Schwefel, H.-P. (eds.) PPSN 2002. LNCS, vol. 2439, pp. 401–410. Springer, Heidelberg (2002)
6. Liu, P., Lewis, M.J.: Communication aspects of an asynchronous parallel evolutionary algorithm In: Proceedings of the Third International Conference on Communications in Computing, Las Vegas, NV, 24–27 June 2002, pp. 190–195
7. Storn, R.: Constrained optimisation. Dr. Dobb's J. pp. 119–123 (1994)
8. Yanga, Z., Tanga, K., Yaoa, X.: Large scale evolutionary optimization using cooperative coevolution. Inf. Sci. **178**(15), 2985–2999 (2008)

9. Yang, Z., Tang, K., Yao, X.: Differential evolution for high-dimensional function optimization. In: IEEE Congress on Evolutionary Computation (2007)
10. Noman, N., Iba, H.: Enhancing differential evolution performance with local search for high dimensional function optimization. In: Proceedings of the 2005 Conference on Genetic and Evolutionary Computation, pp. 967–974 (2005)
11. Hendtlass, T.: Particle swarm optimization and high dimensional problem spaces. In: IEEE Congress on Evolutionary Computation (CEC 2009), pp. 1988–1994 (2009)
12. Hedar, A.-R.: Test functions for unconstrained global optimization. http://www-optima.amp.i.kyoto-u.ac.jp/member/student/hedar/Hedar_files/TestGO_files/Page2376.htm. Accessed 4 April 2013
13. Vasileva, V., Penev, K.: Free search and particle swarm optimization applied to non-constrained test. In: Proceedings of Optimization of Mobile Communications Networks, pp. 20–27 (2013). ISBN 978-0-9563140-4-8
14. Eberhart, R., Shi, Y.: Comparing inertia weights and construction factors in particle swarm optimization. In: Proceedings of the 2000 Congress on Evolutionary Computation, pp. 84–89 (2000)
15. Penev, K.: Adaptive intelligence - essential aspects. J. Inf. Technol. Control **VII**(4), 8–17 (2009). ISSN 1312–2622

Scale Multi-commodity Flow Handling on Dynamic Networks

Alain Quilliot[(✉)], Heito Liberalino, and Benoit Bernay

Labex IMOBS3 LIMOS, UMR CNRS 6158 Bat. ISIMA, Université Blaise Pascal, Campus des Cézeaux, BP 125, 63173 Aubiere, France
alain.quilliot@isima.fr

Abstract. We present here an experimental Large Scale Flow/Multi-commodity model for a routing and scheduling Transportation Problem. We deal with this model through a mix of a GRASP scheme and of Hierarchical Decomposition/Agregation techniques.

Keywords: Large scale dynamic networks · Multi-commodity flows

1 Introduction

We were asked to propose I.T tools for the management of a shuttle fleet which operates on several sites inside a median suburban zone and help workers in moving between their home and their working place, while also performing internal mail transportation. Of course, users ask this service to induce short transportation times, while managers expect low running costs. Using I.T in order to optimize such a system means dealing with some class of *Pick Up and Delivery* problem (see [6,8,9]).

We modeled this problem as a flow/multi-commodity problem defined on a specific dynamic network (see [2]), that is a network with time indexed vertices, which helped us in handling temporal constraints. While this **FMS** model is close to **CFA** (*Capacitated Flow Assignment*) models (see [1,4,5,7]) used in telecommunication network design, we had to cope with the large scale of the related dynamic network. So we handled this problem in a heuristic way, by mixing a hierarchical decomposition/aggregation process with a GRASP scheme.

2 The FMS Model

Main definitions/notations: A network G, with vertex set X and arc set E, is denoted by $G = (X, E)$. A flow vector defined on G is an arc indexed vector with values in **Q** (rational numbers) or **Z** (integral numbers), such that for every vectex x, we have $\sum_{e\ enter\ into\ x} f_e = \sum_{e\ comes\ out\ x} f_e$ (Kirshoff Law). A *multi-commodity flow* vector f, defined on G, is a flow vector collection $f = \{f(k), k \in K\}$ whose components $f(k), k \in K$ are flow vectors. We denote by $Sum(f)$ the *Aggregated Flow* $Sum(f) = \sum_{k \in K} f(k)$.

I. Lirkov et al. (Eds.): LSSC 2013, LNCS 8353, pp. 297–304, 2014.
DOI: 10.1007/978-3-662-43880-0_33,

The "Shuttle" Problem: The urban space is represented with a *Urban Area* network $H = (Z, U)$: nodes mean either production sites, labeled as specific nodes $y_1, \ldots, y_m$ ($m = 7$ in our application), or a *residential* area, and every arc means an elementary connection. A demand $D_k, k \in K$, is a 4-uple (o_k: *origin node*, d_k: *destination node*, L_k: *Load*, t_k: *deadline*): L_k users have to be transported from o_k to d_k (at least one of both nodes being an production site) and arrive (or start) before (after) time t_k. *Quality of Service* (QoS) requires the related trip not to last more than T_k time units. Users move by alternatively walking and using the *shuttle* system; so, every arc e of H is endowed with a *walking* length $l_p(e)$, and with a *vehicle* length $l_\nu(e)$. Our goal is to design and schedule routes for the shuttle fleet which meet the demands, under a minimal cost and a good QoS. *Route preemption* is allowed, which means that several vehicles may be involved in meeting a given demand. For the sake of simplicity we suppose that: every tour starts and ends into a unique *Depot* node; every vehicle has capacity 1; coefficients L_k are rational; the economical cost linearly depends on the number of vehicles (*Fixed Investment Cost*) and on their running times (*Running Cost*).

The Dynamic Network *H-Dyn* = *(X, E)*: we build it from H by associating, with any node x of $Z, (N+1)$ copies of x, indexed from 0 to N, which represent the state of x at the instants $0, \delta, \ldots, N\delta; \delta$ is an elementary time unit, chosen between 3 mn and 6 mn in our application; N is a parameter which fixes the planning period (between 2 and 3 h). We add a ficticious vertex DP and set $X = \{x_r, x \in Z, r \in 0, \ldots, N\} \cup \{DP\}$. As for the arc set E, we round modulo δ the *vehicle* and *walking* lengthes of any arc u in U by setting: $l_p^*(u) = \lceil l_p(u)/\delta \rceil$, $l_v^*(u) = \lceil l_v(u)/\delta \rceil$, and, for any k in $K: t_k^* = \lceil t_k/\delta \rceil$, and we define the labeled arc family E as containing:

- arcs $(x_r, x_{r+1}), x \in Z, r \in 0, \ldots, N-1$: such an arc is considered twice, with *walk* and *vehicle* labels;
- arcs $(DP, Depot_r), (Depot_r, DP), r \in 0, \ldots, N$, with *vehicle* labels.
- arcs $(x_r, z_r + l_v^*(u)), u = (x, z) \in U, r$ such that $0 \leq r \leq N - l_v^*(u)$, with *vehicle* labels;
- arcs $(x_r, z_r + l_p^*(u)), u = (x, z) \in U$, r such that $0 \leq r \leq N - l_p^*(u)$, with *walk* label.

We denote by A the subset of E defined by all the *vehicle* arcs and call it *Vehicle Set*. We provide any arc e in E, with an *Economical Cost* c_e and a *QoS Cost* p_e as follows:

- if e is a *vehicle* arc $(x_r, z_r + l_v^*(u))$ associated with an arc $u = (x, y)$ of the network H, then $c_e = l_v(u)$ and $p_e = S.c_e$, where S is a fixed scaling coefficient;
- if e is a *walk* arc $(x_r, z_r + l_v^*(u))$, associated with $u = (x, z)$ of the network H, then $c_e = 0$ and $p_e = S.l_p(u)$;
- if e is any *vehicle* arc $(x_r, x_r + 1)$ then c_e = a fixed *waiting* cost μ and $p_e = \gamma + S.\delta$ where γ is a fixed cost;
- if e is any *walk* arc $(x_r, x_r + 1)$, then $c_e = 0$ and $p_e = S.\delta$

- if e is an arc $(DP, Depot_r)$, then c_e = a fixed cost α and $p_e = 0$;
- p_e and c_e are equal to 0 elsewhere.

Size of the dynamic network: As a matter of fact, we need, in order to avoid negative rounding effects, to consider as an arc of the *Urban Area* network H any pair (x, y) such that a vehicle may move from x to y during a discrete time unit δ. So, in the case of a 200 node *Urban Area* network, Card(U) may vary between 2000 and 3000 and the dynamic network *H-Dyn* may contain up to 100000 arcs.

The Flow/Multi-commodity Shuttle Model (FMS): We must compute the routes of the vehicles and users, together with their schedules. Aggregating vehicle routes yields an **integral flow vector** F, while user's routes may be represented as a **rational multi-commodity flow** $f = \{f(k), k \in K\} \geq 0$, in such a way that: **(FMS)**

- F is null on the *walk* arcs;
- $f(k)$ expresses the moves of L_k users from o_k to d_k before (or after) time t_k;
- $Sum(f)_e = \sum_{k,e} f(k)_e \leq F_e$, for any arc of *H-Dyn* with *vehicle* label;
- The *global cost* $Cost(F) + QoS(f) = c.F + p.Sum(f) = \sum_{e \ in \ E} c_e.F_e + \sum_{e \ in \ E} p_e.Sum(f)_e$ is the smallest possible.

One easily checks that if F and f satisfy **FMS** constraints, then F may be disaggregated into a solution of the *Shuttle* Problem. This flow model allows us to cast *temporal* and *synchronization constraints* into the construction of the network *H-Dyn*.

Size of the FMS Model: In the case of our application, the number of origin/destination pairs could vary between 200 and 300. So, the size of multi-commodity flow vector f varies accordingly between 2.10^6 and 3.10^6, and the resulting **FMS** model is a large scale Mixed Linear Programming problem. One easily checks that **FMS** is NP-Hard and that relaxing the integrality constraint on F yields poor bounds, since it means providing every user with a fraction of vehicle for a shortest path trip.

3 The FMS Problem: General Resolution Scheme

Because of the size of the problem, we proceed in a heuristic way and consider the aggregated flow *Sum(f)* as a Master object, which we handle through simple path and cycle local transformation procedures. We say that a flow vector g defined on the network *H-Dyn* is *D-decomposable* in relation to the demand set $D = \{D_k, k \in K\}$ if there exists a multi-commodity flow $f = \{f(k), k \in K\}$, such that:

- for every $k \in K$, $f(k)$ routes the load L_k from o_k to d_k before (after) time t_k;
- $g = Sum(f)$.

Then we restate the ***FMS*** Problem as follows *(Aggregated Reformulation)*:

{*Compute an integral flow vector* $F \geq 0$, *and a rational flow vector* $g \geq 0$, *such that:*

- g is *D-decomposable*;
- $F_A \geq \lceil g_A \rceil$;
- *the cost* $c.F + p.g$ *is the smallest possible*}

Given a *D-decomposable* flow vector g, we denote by $\mathbf{FMS_g}$ the restriction of **FMS** obtained by fixing g. The min cost integral flow problem $\mathbf{FMS_g}$ is polynomial.

3.1 A Randomized Initialization Procedure FMS-INIT

Let us recall that if γ is a cycle of *H-Dyn*, a *cycle flow* vector f^γ is defined by $f_e^\gamma = 1$ if arc e of γ is oriented as γ, $f_e^\gamma = -1$ if e is oriented opposite to γ, and $f_e^\gamma = 0$ if e is not in γ. So **FMS-INIT** works in a greedy way, by picking up demands D_k, $k \in K$ according to a randomly generated order, and inserting them into the current aggregated flow g in such a way it induces the lowest possible increase of the related cost $c.\lceil g \rceil + p.g$:

FMS-INIT:
K-In ← *Nil*; *K-Out* ← K;
While *K-Out* ≠ Nil do
  Randomly pick up k_0 in *K-Out*; Insert it in *K-In* and withdraw it from *K-Out*;
  Route D_{k0} along a shortest path Γ for the length function h defined by:
  - If arc e is oriented as Γ, then $h(e) = L_{k0}.p_e + c_e.Sup(0, (\lceil (f_e + L_{k0}) \rceil - \lceil f_e \rceil)$;
  - If e is opposite to Γ, then $h(e) = -L_{k0}.p_e + c_e.Sup(0, (\lceil f_e \rceil - \lceil (f_e - L_{k0}) \rceil)$;
  Add $L_{k0}.f_\Gamma$ to g (we identify path Γ and its natural completion as a cycle).

3.2 A Hierarchical Decomposition Scheme for FMS

Let *(F, g)* be some feasible solution of **FMS**, such that F is an optimal solution of FMS_g. We may associate with (F, g) a dual solution $(\mu = (\mu_x, x \in X), \alpha = (\alpha_e, e \in E) \geq 0, \lambda = (\lambda_e, e \in A) \geq 0)$. Then, improving the pair (F, g) means modifying g in such a way that it remains *D-decomposable* and that $\lambda.\lceil g \rceil + p.g$ decreases. So we set:

FMS-Aux(λ) Problem: {*Compute a flow* $g \geq 0$, *which is D-decomposable and such that* $p.g + \lambda.(\lceil g \rceil)$ *is the smallest possible.*}

Next Sect. 3.3 will describe a *local search* procedure **P-FMS-Aux**, which deals with **FMS-Aux**. Being provided with this procedure leads us to the following general GRASP resolution scheme **GRASP-MSD** (*Master/Slave Decomposition*):

Algorithmic Scheme GRASP-MSD (n: Number of Replications)
For $i = 1..n$ **do**
  Initialize g through **FMS-INIT**; Not *Stop*;
  While not *Stop* **do**
    Solve $\mathbf{FMS_g}$ and **compute** the related dual component λ;

Starting from g, **perform** iterations of **P-FMS-Aux** until *Failure* or the $\lambda.\lceil g \rceil + p.g$ decreases; (I1)
If *Failure* **then** *Stop* Else Solve $\mathbf{FMS_g}$; Update *Stop*;
Keep the best solution (F, g) ever obtained; Update *Stop*;

Remark: The *While* Loop is a kind of *inverse* Benders scheme, which uses $\mathbf{FMS_g}$ polynomiality, since Benders scheme usually sees integral vector as the master object.

3.3 Handling the FMS-Aux Problem

Key questions are here about the non linear $p.g + \lambda.\lceil g \rceil$. cost and about the *Decomposability Constraint.*

Handling $\mathbf{p.g + \lambda.\lceil g \rceil}$**: extended cycles.** We deal with g through cycle procedures. So, g being given, we search for a circuit γ and a number q in $]0, 1]$ such that turning g into $g + f^{\gamma}$ improves g, which means that if we set, for any arc e:

- $D_g(e, q) = q.p_e + \lambda_e.(\lceil (g)_e + q \rceil - \lceil (g)_e \rceil)$;
- $D_g^*(e, q) = -q.p_e - \lambda_e.(\lceil (g)_e \rceil - \lceil (g)_e - q \rceil)$.

then:

- length$(\gamma) \geq 3$; (*we do not want to use the same arc forth and back*)
- $\sum_{e \text{ oriented as } \gamma} D_g(e, q) + \sum_{e \text{ oriented opposite to } \gamma} D_g^*(e, q) < 0$.

Such a pair (γ, q) is called a *good extended* cycle for the current flow vector g.
One easily checks that the search for a *good extended cycle* is time-polynomial: only a finite set Q of q values has to be tested, and for every value q, searching for a related γ means searching for a negative circuit in some *ad hoc* network. Still, practically, the search for a *good extended cycle* remains here a difficult problem, due to the *scale* of Q and *H-Dyn.* So we deal with it through sampling, by iteratively randomly picking up a source node x_0 and performing a local BFS (*Breadth First Search*) $CYGEN_{x_0}$ process from x_0 onto a restricted area: at any time during this BFS, a visited node x is provided with a boolean value *Mark(x)*, a list $L(x) = (q_1..q_s)$ of increasing q values, and a list $\Gamma(x) = (\Gamma_1..\Gamma_s)$ of paths from x_0 to x. Then a node *x-pivot* is chosen, such that $Mark(x) = 0$ and $\sum_i (q_i - q_{i-1}).\Pi_i$ be minimal, where Π_i is the length of Γ_i in the sense of $D_g(., q_i)$. *Mark(x-pivot)* is set to 1 and *x-pivot* is expanded, as in BFS Dijsktra Algorithm, which means that pairs $(L(x), \Gamma(x))$, x such that $Mark(x) = 0$, are updated and that the existence of y in $\Gamma(x - pivot)$ and q_i in $L(x - pivot)$ which would induce a *good extended cycle* is tested.

$CYGEN_{x_0}$ is tried N times (N = parameter), with N distinct starting nodes x_0. It is run as a *local* process, which means that, x_0 being given, no more than M nodes (M is a parameter) are going to be visited by $CYGEN_{x_0}$. This stems from the knowledge that, practically, the length of a *good extended cycle* is usually rather small.

CYGEN Algorithmic Scheme (*N, M: Integer*)
Not *Success: Counter* $\leftarrow$ 0;
While Not *Success* and *Counter* $\leq N$ **do**
  Randomly pick up a source node x_0; **Apply** the BFS $CYGEN_{x_0}(M)$ process;
  If $CYGEN_{x_0}(M)$ yields a *good extended cycle* (γ, q)
  then *Success* **else** *Counter* $\leftarrow$ *Counter* + 1;

Handling the Decomposability Constraint: Every time we modify $g = Sum(f)$ through **CYGEN**, we apply some update process in order to maintain the D-decomposability of g. We do it by imposing that all the flow vectors $f(k), k \in K$, be *path flow* vectors, which means that demand D_k is routed from o_k to d_k along a single path, and by making in such a way that at any iteration of the **P-FMS-Aux** iterative resolution process, we are provided with a current set Δ of *metric cuts* (see [3]) $MCUT(Z), Z \subseteq X$:

$$\sum\nolimits_{e\ has\ its\ origin\ in\ Z\ and\ destination\ in\ X-Z} g_e \geq \sum\nolimits_{k\ in\ OD(Z)} D_k,$$

where $OD(Z) = k \in K$ *such that* $o_k \in Z$ *and* $d_k \in X - Z$. So, **P-FMS-Aux** involves two steps:

- Step 1: compute a *good extended cycle* (γ, q), such $g + q.f_\gamma$ does not violate any constraint of Δ.
- Step 2: in case of success, replace g by $g + q.f_\gamma$ and redirect some of the paths $(f(k), k \in K)$ in order to minimize $|Sum(f) - g + q.f_\gamma|$.

We may summarize this into the following procedure **P-FMS-Aux**:

P-FMS-AuxAlgorithmic Scheme:
Initialize g, current decomposition $\Sigma = \Gamma_k, k \in K$ and Δ through **FMS-INIT** (in case we start from nothing) or with the current g; Not *Stop;*
While Not *Stop* **do**
  Apply *CYGEN* to g and Δ;
  If *CYGEN* yields such a *good extended cycle* (γ, q) such that replacing g by $g + q.f_\gamma$ does then violate any cut in Δ, then *POSSIBLE;* $h \leftarrow g + q.f_\gamma$;
    While *POSSIBLE* and $(h \neq g)$ **do** (*Update current decomposition Σ*):
      $\delta := Sup_{e \in E}|(g - h)_e|$; Look for $k \in K$ and for a path Γ from o_k to d_k such that replacing Γ_k by Γ in Σ induces a strict diminution of the quantity $Sup_{e \in E}|(g - h)_e|$;
      If k and Γ exist **then** replace Γ_k by Γ in Σ and modify g accordingly
        Else Not *POSSIBLE*; Deduce a *metric cut* related to the failure of the computation of Γ;
      Update *Stop;*
      Keep the best aggregated flow vector g which appeared during this "While" loop;

4 Numerical Experiments

We present here 3 experiments, performed on a LINUX server CentOS 5.4, Processor Intel Xeon 3.6 GHZ, with help of the CPLEX 12 library, on both small instances, in order to evaluate approximation levels, and on large scale instances with sparse structures (Tables 1, 2).

4.1 First Experiment: The CYGEN Procedure on FMS-Aux Instances

Vector g being given, we compare failure ratio which derive from the replacement of a *good extended cycle* **EXACT** procedure by the fast search algorithm **CYGEN**.

In every case, we also compare related CPU times. *NA* is the number of arcs of the involved *H-Dyn*, *EXACT-CPU* and *CYGEN-CPU* are the respective mean CPU times for **EXACT** and **CYGEN**, and *FAILURE* is the mean percentage of failure induced by the use of **CYGEN** instead of **EXACT-CYCLE**. We get (in average):

Table 1. CYGEN Evaluation.

NA	*FAILURE*	*EXACT-CPU (s)*	*CYGEN-CPU (s)*
148	0.2 %	0.04	0.01
1128	1.1 %	1.9	0.07
2354	2.7 %	5.2	0.15

Comment: as expected, *FAILURE* increases with the size *NA*.

4.2 Second Experiment: The Ṕ-FMS-Aux Procedure

Small instances: *NOD* is the number of origin/destination pairs. *GAP* (in %) = $(VAL - OPT)/OPT$ is the gap between the optimal cost value *OPT* computed by CPLEX and *VAL* obtained through **P-FMS-Aux.**

Table 2. P-FMS-Aux Evaluation, Small Instances.

Instance	*NA*	*NOD*	*GAP* (%)
1	80	5	5.7
2	128	10	10.5
3	156	5	7.6
4	204	20	6.7

Large instances: we test the relation between the size of the instances and the related running costs (Tables 3, 4, and 5).

Table 3. P-FMS-Aux Evaluation, Large Instances.

Instance	*NA*	*NOD*	*CPU (s)*
1	8088	30	36
2	20404	60	204
3	66256	100	1025
4	98468	250	59 mn

Table 4. MSD Evaluation, Small Instances.

Instance	*NA*	*OD*	*MSD-GAP*
1	80	5	4.5
2	128	10	10.3
3	156	5	7.4
4	204	20	11.2

Table 5. MSD Evaluation, Large instances.

Instance	*NA*	*OD*	*MSD-Time*
1	8088	30	98
2	20404	60	678
3	66256	100	1635
4	98468	250	1 h 29

4.3 Third Experiment: Evaluation of the Global Scheme MSD

Small instances: same small networks as in 4.2; $MSD\text{-}GAP = (MSD\text{-}VAL - OPT)/OPT$, where $MSD\text{-}VAL$ is the cost value of the solution (F, f) induced by the **MSD** scheme and OPT the optimal value. We get:

Large instances: As in 4.2; *MSD-Time* denotes induced CPU times. We get:

Comment: the approximation level of **MSD** is not so high, if we refer to real context requirements, and it enables us to efficiently manage the large scale feature.

References

1. Ahuja, R.K., Magnanti, T.L., Orlin, J.B., Reddy, M.R.: Applications of network optimization; Chap. 1 Network Models, Handbook O.R & Manag. Sci. 7, 1–83 (1995)
2. Aronson, J.E.: A survey on dynamic network flows. Ann. Op. Res. **20**, 1–66 (1989)
3. Barahona, F.: Network design using cut inequalities. SIAM J. Optim. **6**, 822–837 (1995)
4. Benchakroun, A., Ferland, J., Gascon, V.: Benders decomposition for network design problems with underlying tree structure. Invest. Operativa **6**, 165–180 (1997)
5. Bienstock, D., Günlük, O.: Capacitated network design polyhedral structure and computation. INFORMS J. Comput. **8**, 243–259 (1996)
6. Cordeau, J.P., Toth, P., Vigo, D.: A survey of optimization models for train routing and scheduling. Transp. Sci. **32**, 380–404 (1998)
7. Crainic, T., Gendreau, M., Farvolden, J.M.: A simplex based Tabu search method for capacitated network design. INFORMS J. Comput. **12**, 223–236 (2000)
8. Crainic, T., Frangioni, A., Gendron, B.: Bundle based relaxation methods for multicommodity capacitated fixed charge network design. DAM **112**, 73–99 (2001)
9. Magnanti, T., Wong, R.T.: Network design and transportation planning models and algorithms. Trans. Sci. **18**, 1–5 (1984)

Hybrid Genetic Algorithm for Allocation Mapping in Processor Array Design

Piotr Ratuszniak(✉)

Koszalin University of Technology,
ul. Śniadeckich 2, 75-453 Koszalin, Poland
ratusz@ie.tu.koszalin.pl

Abstract. In this paper a hybrid genetic method for processor arrays design dedicated to realization of linear algebra algorithm for banded matrices is presented. The proposed method is a modification of previous genetic algorithm which is characterized by few important advantages relative to well-known linear projection methods. The main disadvantages of this previous method are: long program runtime and problems with obtaining acceptable allocation mapping results for huge information dependency graphs. Linear projection methods don't allow obtaining better allocation mapping solutions but are characterized by shorter program runtime. New hybrid algorithm combines these both linear and genetic methods and merges their advantages. Summarizing, this new proposed method is characterized by: a shorter program runtime, better allocation mapping results in comparison with both previous methods, possibly allocation mapping for large input linear algebra banded matrices and possibility of defining the designed processor array structure.

1 Introduction

In recent years, a high popularity increase of using parallel architectures for computation acceleration is still observed. The most popular platforms for parallel computation include multicore PC processors and General Purpose Graphical Processor Units. Another one, still undervalued hardware platform, are FPGA devices. In comparison to alternative parallel hardware platforms FPGAs allow adaptation of designed system architecture to the algorithm [1], are not expensive, consume less power [2,3], support any data representation and thanks to this provides higher calculations accuracy. Moreover, modern FPGA devices contain huge number of Configurable Logic Blocks, built-in DSP and memory blocks that allow implement whole complex systems in the form of System on Chip. Thanks to this FPGAs are used for implementation of mobile system for example, which are limited by physical dimensions and low power consumption. Unfortunately, for efficient system design good knowledge of one of the hardware description languages (HDL such as VHDL or Verilog) and design experience are needed. For this reason and in order to increase the effectiveness of the design process, with the rule "right first time", complete functional block generators, called intellectual property cores (IPCore), are used [4]. Presented processor array design

I. Lirkov et al. (Eds.): LSSC 2013, LNCS 8353, pp. 305–312, 2014.
DOI: 10.1007/978-3-662-43880-0_34,

method, based on the hybrid genetic algorithm, is the part of original IPCore generator for linear algebra hardware accelerators and is a new modification of previous design method presented in the following paper [5].

At present, there are few translators available from the C language to one of the HDL languages, such as ImpuleC. Unfortunately, the architectures obtained from these translators are not optimal [1]. There are also few well-known methods [6–9] or whole environment [10,11] for processor array design dedicated to realization of recursive algorithms. For example, the method [6] is very fast because involves projection matrices multiplication but generates a lot of non-operation instructions for linear projection which cause longer runtime needed for designed architectures. Only proposed method [5] provides exact definition of the designed structure by a user. Moreover, the realization runtime needed for designed architectures is often smaller in comparison with similar architectures obtained with the use of linear and nonlinear design methods. The main disadvantage of method described in the paper [5] is problem with finding permissible solutions of allocation mapping, which keeps connection locality constraint, in a short program runtime (under 30 min) for large information dependency graphs (with above 3000 nodes). For this reason the parallel realization of genetic algorithm and information dependency graphs decomposition were applied. In this method the solution "grows up" during the runtime of the designed method and genetic operators worked at the next assumed limits of a position in a chromosome. The method presented in this paper combines the previous genetic method and linear projection from method [6] for linear algebra algorithms for banded matrices.

2 New Allocation Mapping Method for Processor Array Design

In the allocation mapping for each node one processor was assigned from an array. The graph projection process has one main limitation — a connection locality. A condition is necessary for a local connection between the processors in a complete processor array. The connection locality means that each graph edge can connect two nodes only from neighbor processors. Thus, whole processor array can work with a high clock frequency after an implementation in FPGA device. After the allocation mapping process, the tact number for each operation in a whole processor element is estimated. In the optimization process, the program minimized the maximum tact number needed for realization of the complete given linear algebra algorithm.

In the first stage of proposed method the allocation mapping are computed by genetic algorithm for a representative subgraph from information dependency graph for linear algebra algorithms for banded matrices. The node number for a pattern subgraph and the stage runtime are defined by a program user. After the allocation mapping computation, in the second program stage, the computed projection for a pattern subgraph is duplicated for the rest nodes from a whole graph without nodes from a graph tail, which is characterized by another shape.

In the third stage the allocation mapping for graph tail nodes is computed in the runtime also defined by a program user. In the last, fourth stage the connection locality constraint is checked for a whole information dependency graph. The schedule mapping is computed then, which define a time projection in the form of a tact number for all given algorithm operations (graph nodes) assigned to the proper Processor Units from the designed array. The allocation mapping for a pattern subgraph is computed with using a bigger computation subgraph (usually the bigger subgraph additionally contains nodes from one addition graph layer). The bigger subgraph is necessary because of the need to check connection locality constrain for the nodes from the current and next pattern subgraph in allocation mapping duplication process. Figure 1 presents the input information dependency graph for Gauss LU decomposition with numbered nodes, where the input matrix dimension N = 8 and band width L = 7.

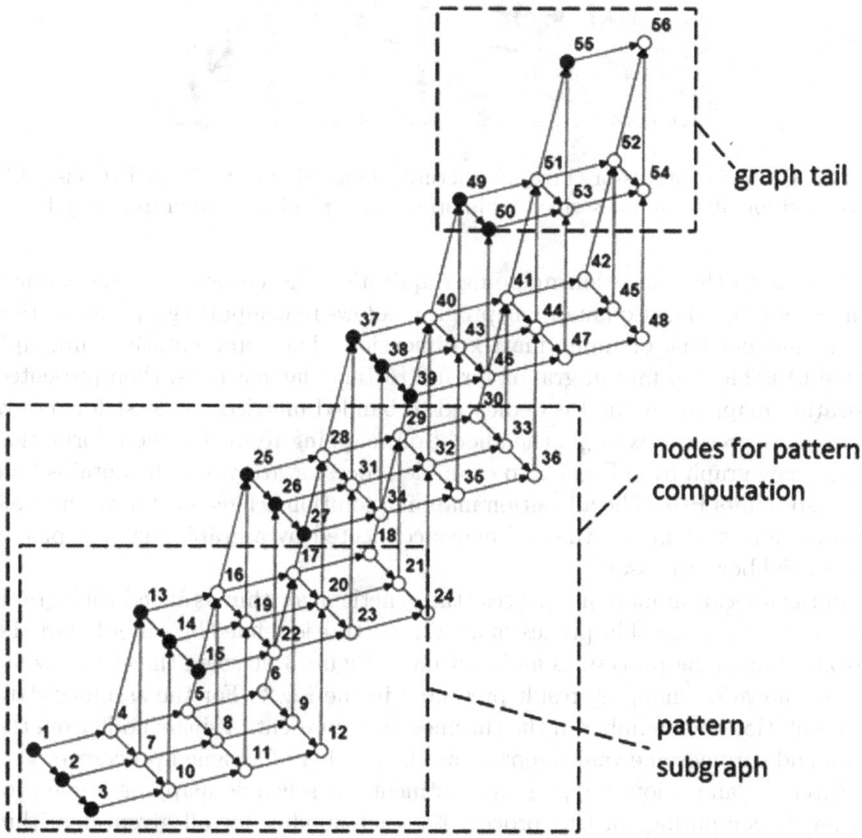

Fig. 1. Information dependency graph for example Gauss LU decomposition algorithm (dimension N = 8, band width L = 7) with their marked subgraph.

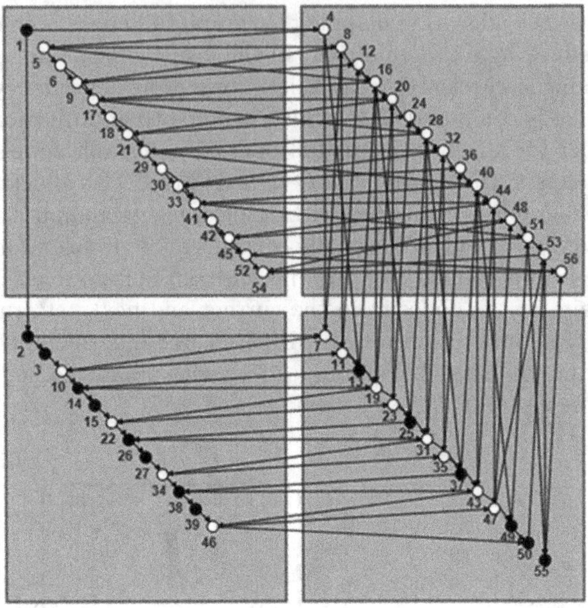

Fig. 2. Example processor array structure designed to realization Gauss LU decomposition algorithm described by information dependency graph from Fig. 1.

Thanks to the allocation mapping duplication the needed runtime is much shorter and the allocation mapping program allows to computes graph projection almost independent of input matrix dimension. The representative subgraph (pattern) is hard to find in graphs for full rectangular matrices, then presented allocation mapping method is dedicated to banded matrices, at least for today.

Figure 2 presents example the allocation mapping result for the information dependency graph from Fig. 1 into example processor array, which contains four Processor Units PUs. The allocation mapping solution achieves connection locality constrain, that means that all nodes connected by a graph edge are placed into a neighbor processor.

In the allocation mapping process the genetic algorithm assigned each graph node to one processor. The processor array structure is defined by a user before the allocation mapping process as an input data. Figure 3 presents the chromosome for the allocation mapping result presented in the Fig. 2. For the assumed data representation each number in the chromosome represents a single node from the graph and chromosome values represents the number of assigned processors.

After the allocation mapping appointment the schedule mapping (time projection) is computing. In this process the tact number for all given algorithm operations (graph nodes) are computed by the iterative method. The iterative method is characterized by a large computation complexity for huge graphs but supports minimal values for a computed tact number. Figure 4 presents schedule mapping computations result for example processor array from Fig. 2.

genBest->[0,2,2,1,0,0,3,1,0,2,3,1,3,2,2,1,0,0,3,1,0,2,3,1,3,2,2,1,
0,0,3,1,0,2,3,1,3,2,2,1,0,0,3,1,0,2,3,1,3,3,1,0,1,0,3,1]

Fig. 3. Genetic algorithm chromosome which represents the allocation mapping solution from Fig. 2.

Tacts-> [1,2,3,2,3,4,3,4,5,4,5,6,4,5,6,5,6,7,6,7,8,7,8,9,7,8,9,8,9,10,9,10,11,10,
11,12,10,11,12,11,12,13,12,13,14,13,14,15,13,15,14,15,16,17,17,18]

Fig. 4. Schedule mapping result for processor array from Fig. 2.

3 Genetic Algorithm for Allocation Mapping Method

There are several well-known methods for a processor array design [6–10]. These methods are not automatic and only experienced designers can use them. For methods which use linear or non-linear functions of allocation mapping, it is difficult or impossible to design a processor array with a structure exactly defined by a user. For most of these methods designer defines several projection directions and choose one more interesting architecture from the obtained solutions. Some methods use graph transformations, but these transformations are specific to the given input algorithm [11]. Proposed method and their previous version [5] allows for the exact definition of the structure of the designed processor array before the allocation mapping process. Thanks to use of genetic algorithm any definitions of projection functions are not required and proposed method can be used by designers without a strong experience in the parallel architectures domain.

The main problem with the use of the genetic algorithm for a whole information dependency graph projection is the graph size for larger input matrices [5]. For this reason the information dependency graph decomposition and parallel realization of the genetic algorithm were proposed. This allowed to obtain a permissible solution of the graphs projection for larger input matrices dimensions but the necessary runtime was proportional to the graph nodes. For example graphs with nodes number under 1500 in the 30 min runtime program has found permissible projection with accuracy near 50 percent. The proposed new approach merges linear projection speed with advantages of the developed genetic algorithm. The method proposed in this paper uses the genetic algorithm in the allocation mapping process for pattern and tail subgraphs. The genetic algorithm from the presented method is characterized by the same parameters like the algorithm described in the paper [5] except runtime which for experiment results presented in the next part of this article was equal to 1 or 3 min. The main optimization goal is the tact number minimization necessary for realization a given input linear algebra algorithm in designed processor array. In the genetic algorithm a division of groups coding by using numbers in a chromosome is used, which means that each number in the chromosome represents a single node from the information dependencies graph. The chromosome values represent the number of assigned processors in the array. The genetic algorithm population contains 100 chromosomes and the initial population was generated randomly. A standard one-point crossover operator with a fixed probability and a mutation operator with a variable probability were used. After many experiments with several kinds and dimensions of input graphs, the value of crossover probability was experimentally chosen at the 0.2 level and a mutation operator with a variable probability was used. Before the genetic computation the value of the mutation probability was calculated, that exactly one position in a chromosome was mutated. This caused a constant number of mutations for different graph sizes. The described genetic algorithm, like in the previous version, was operated in two stages: before and after finding a permissible solution. In the first stage, before finding a permissible solution, objective function F1 depended on the space projection errors. These errors depended on unfulfilled connection locality conditions in the form of non local connections between the processors. The objective function F1 in the first stage of the algorithm calculation is presented in the Eq. (1)

$$F1 = 1 + EN * (EN - SE) \tag{1}$$

where: EN - edges number, SE - space projection errors.

In the second stage of the algorithm computations, after finding a permissible solution, the value of the objective function was calculated using two formulas. For the permissible solutions the value of the objective function F2 was additionally dependent on the number of the clock cycles necessary to realization of the current pattern or tail sugraph. For the other solutions the objective function F3 gives much lower values (penalty function). The detailed formulas for the objective function in the second stage of the algorithm computations are presented in the Eqs. (2) and (3).

$$F2 = 1 + EN^2 + NN - T \tag{2}$$

$$F3 = 1 + EN - SE \tag{3}$$

where additionally: NN - nodes number, T - number of clock cycles.

In the proposed genetic algorithm, the standard elitist selection model was used because of strict runtime limits and because of the mutation process, which for this data representation, could easily change the chromosome into a non-permissible solution. The same way as in the previous version of the genetic algorithm the parallel fitness function computations were implemented with use Microsoft ParallelFX extension, which allowed using all cores of PC processor for allocation mapping computations.

4 Experiments. Designed Processors Arrays Parameters

For all the computational experiments the strict limits for described program runtimes were assumed (1 or 3 min) and for all set of algorithms parameters

Table 1. Processor arrays (2×2) parameters designed for realization of Gauss LU algorithm for banded matrices (bandwidth = 9) obtained for previous and current design method.

Matrix size		**30**	**40**	**50**	**60**	**70**	**80**	**90**	**100**
Graph nodes number		**520**	**720**	**920**	**1120**	**1320**	**1520**	**1720**	**1920**
Graph edges number		**1213**	**1683**	**2153**	**2623**	**3093**	**3563**	**4033**	**4503**
previous method (30 min runtime)	avg tacts	258	305	384	498	576	-	-	-
	best	245	294	375	478	576	-	-	-
	find	100%	100%	60%	80%	20%	-	-	-
	avg PU load	50%	59%	60%	56%	57%	-	-	-
current method (1 min runtime)	avg tacts	184	264	307	401	464	530	608	664
	best	149	222	258	340	370	424	523	580
	find	100%	100%	100%	100%	100%	100%	100%	100%
	avg PU load	70%	68%	74%	69%	71%	71%	70%	72%
current method (3 min runtime)	avg tacts	169	238	305	378	421	504	569	631
	best	150	220	261	316	370	423	521	581
	find	100%	100%	100%	100%	100%	100%	100%	100%
	avg PU load	76%	75%	75%	74%	78%	75%	75%	76%

the program was running 10 times. Table 1 presents the parameters of processor arrays, designed to realization the Gauss LU algorithm, like: a best and average value of a tact number necessary to realization the all operations from the given algorithm, an average Processor Unit loads and a percent value, which describes algorithm runtimes number when the permissible solution was found.

Based on the results presented in Table 1, one can conclude that the new proposed design method allows obtaining processor array architecture characterized by better quality parameters in a much shorter program runtime. Additionally, for the new proposed approach the program always generates permissible solutions, even for large information dependency graphs, which contain over few thousand nodes.

5 Conclusions and Future Research

In this paper, the new approach of genetic algorithm usage for a processor array design is presented. The allocation mapping was calculated for a representative subgraph and in the next stage this mapping was duplicated for the rest of graph without a graph tail. In the last stage the allocation mapping for the tail was calculated. Thanks to this, the new method merges advantages of the genetic allocation mapping with a linear projection speed. Additionally, the design processor array architectures are characterized by a higher average processors load and a shorter necessary runtime for realization of all input operations. Moreover, the new approach allows designing processor arrays for huge information dependency graphs.

In the future research the application of mix a selection operator [12] is considered for steering of the genetic algorithm selection pressure which allows controlling diversity index for genetic algorithm population.

References

1. Peterson, M.: FPGA acceleration for outstanding performance. Challenges and opportunities. In: Parallel Processing and Applied Mathematics, Wroclaw, Poland (2009)
2. Kestur, S., Davis, J., Williams, O.: BLAS comparision on FPGA, CPU and GPU. In: IEEE Computer Society Symposium on VLSI, pp. 288–293 (2010)
3. Williams, J., George, A.D., Richardson, J., Gosrani, K., Suresh, S.: Computational density of fixed and reconfigurable multi-core devices for application acceleration. In: Proceedings of the 4th Reconfigurable Systems Summer Institute, Nat'l Center for Supercomp. App., Illinois (2008)
4. Chen, Y.K., Kung, S.Y.: Trend and challenge on system-on-a-chip designs. J. Signal Proc. Syst. **53**, 217–229 (2008). (Springer)
5. Ratuszniak, P.: Processor array design with the use of genetic algorithm. In: Lirkov, I., Margenov, S., Waśniewski, J. (eds.) LSSC 2011. LNCS, vol. 7116, pp. 238–246. Springer, Heidelberg (2012)
6. Kung, S.Y.: VLSI Array Processors. Prentice Hall, Englewood Cliffs (1988)
7. Quinton, P., Robert, Y.: Systolic Algorithms and Architectures. Prentice Hall, Hertfordshire (1991)
8. Sergyienko A., Kaniewski, J., Maslennikow, O., Wyrzykowski, R.: A metod for mapping DSP algorithm into application specific processor. In: Proceedings of the 24th Euromicro Conference on Parallel and Distributed Processing, Vasteras, Sweden, vol. 1. IEEE Comp. Soc. Press (1998)
9. Fimmel, D., Merker, R.: Design of processor arrays for reconfigurable architectures. J. Supercomput. **19**, 41–56 (2001). (Kluwer Academic Publishers)
10. Le Verge, H., Mauras, C., Quinton, P.: The ALPHA language and its use for the design arrays. J. VLSI Signal Proc. **3**, 173–182 (1991)
11. Wilde D.K., Sie O.: Regular array synthesis using ALPHA. In: International Conference on Application Specific Array Processors (1994)
12. Słowik, A.: Steering of balance between exploration and exploitation properties of evolutionary algorithms - mix selection. In: Rutkowski, L., Scherer, R., Tadeusiewicz, R., Zadeh, L.A., Zurada, J.M. (eds.) ICAISC 2010, Part II. LNCS, vol. 6114, pp. 213–220. Springer, Heidelberg (2010)

Hybrid ACO-GA for Parameter Identification of an *E. coli* Cultivation Process Model

Olympia Roeva[1(✉)], Stefka Fidanova[2], and Vassia Atanassova[1,2]

[1] Institute of Biophysics and Biomedical Engineering, BAS, Acad. G. Bonchev Str., bl. 105, 1113 Sofia, Bulgaria
olympia@biomed.bas.bg, vassia.atanassova@gmail.com

[2] Institute of Information and Communication Technologies, BAS, Acad. G. Bonchev Str., bl. 25A, 1113 Sofia, Bulgaria
stefka@parallel.bas.bg

Abstract. The present work offers a novel approach to parameter identification of an *E. coli* cultivation process model, using a hybrid of two metaheuristics, namely Ant Colony Optimization (ACO) and Genetic Algorithms (GAs). Our basic idea is to generate initial solutions by the ACO method, and then serve these solutions to the GA as its initial population of individuals. Thus, the GA will start with a population, which is not randomly generated, as in the general case, but one rather closer to an optimal solution. The motivation behind this hybridization is to combine the benefits of both approaches, aimed at achieving commensurate calculations precision with less computation resources, in terms of time and memory. The proposed method is approbated with the estimation of the parameters of a real *E. coli* fed-batch cultivation process model. The presented results are affirmative of our goal to yield better performance of the hybrid algorithm: almost twice less computational time and approximately five times smaller populations needed, compared to both ACO and GAs, as taken separately.

1 Introduction

Modeling approaches are central in system biology and provide new ways towards the analysis and understanding of cells and organisms. A common approach to model cellular dynamics is by using sets of nonlinear differential equations. Real parameter optimization of cellular dynamics models has become a research field of particularly great interest. A major deficiency is the lack of applicable methods. Since the considered problem has been known to be NP-complete, using metaheuristic techniques can solve this problem more efficiently than exact or traditional methods whose can not solve the problem with reasonable computational effort. Metaheuristics has become increasingly popular in different research areas and industry. Most of them mimic natural metaphors to solve complex optimization problems (evolution, ant colony, particle swarm, immune system). In contrast to many classical methods, metaheuristics do not build a model of the tackled optimization problem, but treat the problem as it is (black-box optimization). Therefore, they are directly applicable to complex real-world

I. Lirkov et al. (Eds.): LSSC 2013, LNCS 8353, pp. 313–320, 2014.
DOI: 10.1007/978-3-662-43880-0_35,

problems with relatively few modifications. Two of the most successfully performing metaheuristics are the Genetic Algorithms (GAs) and the Ant Colony Optimization (ACO) [5]. Their effectiveness have been already demonstrated for model parameter identification of bioprocesses [1,6,12,13].

In recent years, it has become evident that the concentration on a sole metaheuristic is rather restrictive. A skilled combination of a metaheuristic with other optimization techniques, so called hybrid metaheuristic, can provide more efficient behavior and higher flexibility when dealing with real-world and large-scale problems. In general, hybrid metaheuristic approaches can be classified as either "collaborative combinations" or "integrative combinations" [2].

Most of the authors combine some metaheuristics with local search procedure or with some exact method. There are some applications of the hybrid algorithms between different metaheuristics. For example, the superiority of the hybrid algorithms between metaheuristics ACO and GA is shown in applications in different areas and problems [3,8,10,11]. In these papers first GA is applied and when it stagnates the ACO is used to go out of the stagnation and the algorithm continue with GA.

In this paper, based on the idea in [2], a hybrid metaheuristics ACO-GA is realized and applied for parameter identification of a cultivation process of the bacteria *E. coli* model. Cultivation of recombinant micro-organisms e.g. *E. coli*, in many cases is the only economical way to produce pharmaceutical biochemicals such as interleukins, insulin, interferons, enzymes, and growth factors.

A system of ordinary nonlinear differential equations is proposed to model *E. coli* biomass growth and substrate utilization. Model parameters optimization is performed using real experimental data set from an *E. coli* MC4110 fed-batch cultivation process [14].

The paper is organized as follows. In Sect. 2, after a brief description of ACO and GA, the hybrid metaheuristics ACO-GA is introduced. Section 3 presents the problem definition. The numerical results and a discussion are presented in Sect. 4. Section 5 provides some conclusion remarks and ideas for further research.

2 Hybridization of Ant Colony Optimization Method and Genetic Algorithms

ACO is a stochastic optimization method that mimics the social behavior of real ants colonies, which manage to establish the shortest route to feeding sources and backwards to the nest [5]. Real ants foraging for food lay down quantities of pheromone (chemical cues) marking the path that they follow. An isolated ant moves essentially at random but an ant encountering a previously laid pheromone will detect it and decide to follow it with high probability, thereby reinforcing it with a further quantity of pheromone. The repetition of the above mechanism represents the auto-catalytic behavior of a real ant colony where the more the ants follow a trail, the more attractive that trail becomes. The original idea comes from observing the exploitation of food resources among ants, in which

```
Ant Colony Optimization
Initialize number of ants;
Initialize the ACO parameters;
while not end-condition do
    for k = 0 to number of ants
        ant k choses start node;
        while solution is not constructed do
            ant k selects higher probability node;
        end while
    end for
    Update-pheromone-trails;
end while
```

Fig. 1. Pseudocode for ACO

ants' individually limited cognitive abilities have collectively been able to find the shortest path between a food source and the nest.

The structure of the ACO algorithm is shown by the pseudo-code in Fig. 1.

GAs originated from the studies of cellular automata, conducted by John Holland and his colleagues at the University of Michigan [9]. The GA is a model of machine learning which derives its behavior from a metaphor of the processes of evolution in nature [7]. This is done by the creation within a machine of a population of individuals represented by chromosomes. Depending on the specific problem, a chromosome could be an array of real numbers, a binary string, a list of components in a database. The GAs are highly relevant to industrial applications, because they are capable of handling problems with non-linear constraints, multiple objectives, and dynamic components – properties that frequently appear in the real-world problems [7]. Since their introduction and subsequent popularization [9], the GAs have been frequently used as an alternative optimization tool to the conventional methods [7] and have been successfully applied to a variety of areas, and enjoy increasing acceptance.

The structure of the GA is shown by the pseudo-code in Fig. 2.

The population at time t is represented by the time-dependent variable P, with the initial population of random estimates being $P(0)$. Here, each decision variable in the parameter set is encoded as a binary string (with precision of binary representation). The initial population is generated using a random number generator that uniformly distributes numbers in the desired range. The objective function (see Eq. (4)) is used to provide a measure of how individuals have performed in the problem domain.

Our idea is to combine two metaheuristics, ACO and GA. ACO is a constructive method which does not need initial solutions. GA is a population-based method and in traditional GA's initial population is randomly generated. In this random generation the initial solutions can be very far from the optimal solutions and may need a lot of iterations to draw close to it. Therefore, our idea is to generate initial solutions by ACO and then use them as an initial population in GA. Thus, the GA will start with a population, which is closer to optimal

```
begin
    i = 0
    Initial population P(0)
    Evaluate P(0)
    while (not done) do (test for termination criterion)
    begin
        i = i + 1
        Select P(i) from P(i - 1)
        Recombine P(i)
        Mutate P(i)
        Evaluate P(i)
    end
end
```

Fig. 2. Pseudocode for GA

solution. The best model parameters vector will be obtained by the genetic evolution of ant colony.

3 Problem Formulation

The mathematical model is a tool that allows to be investigated the static and dynamic behavior of the process without doing (or at least reducing) the number of practical experiments. In practice, an experimental approach often has serious limitations that make it necessary to work with mathematical models instead. Development of adequate models is an important step for process optimization and high-quality control. Application of the general state space dynamical model [4] to the *E. coli* cultivation fed-batch process leads to the following nonlinear differential equation system [6]:

$$\frac{dX}{dt} = \mu_{max} \frac{S}{k_S + S} X - \frac{F_{in}}{V} X \tag{1}$$

$$\frac{dS}{dt} = -\frac{1}{Y_{S/X}} \mu_{max} \frac{S}{k_S + S} X + \frac{F_{in}}{V} (S_{in} - S) \tag{2}$$

$$\frac{dV}{dt} = F_{in} \tag{3}$$

where X is biomass concentration, [g/l]; S is substrate concentration, [g/l]; F_{in} is feeding rate, [l/h]; V is bioreactor volume, [l]; S_{in} is substrate concentration in the feeding solution, [g/l]; μ_{max} is maximum value of the specific growth rate, [h^{-1}]; k_S is saturation constant, [g/l]; $Y_{S/X}$ is yield coefficient, [-].

For the model parameter identification problem the objective function is presented as a minimization of a distance measure J between experimental and model predicted values of state variables (X and S):

$$J = \sum_{i=1}^{m} (X_{\exp}(i) - X_{\mathrm{mod}}(i))^2 + \sum_{i=1}^{m} (S_{\exp}(i) - S_{\mathrm{mod}}(i))^2 \rightarrow \min \tag{4}$$

where m is the number of experimental data; $X_{\exp}$ and $S_{\exp}$ are the known experimental data for biomass and substrate; X_{mod} and S_{mod} are the model predictions for biomass and substrate with a given set of parameters (μ_{max}, k_S and $Y_{S/X}$).

4 Numerical Results and Discussion

The theoretical background of the GA and ACO is presented in [13]. Parameters of the GA and ACO were tuned based on several pre-tests according considered here optimization problem. After tuning procedures the main algorithm parameters are set to the optimal settings. The basic operators and parameters in GA are summarized in Table 1. The parameter setting for ACO is shown in Table 2.

Table 1. Operators and parameters of GA

Operator	Type	Parameter	Value
Fitness function	Linear ranking	ggap	0.97
Selection function	Roulette wheel selection	xovr	0.75
Crossover function	Simple crossover	mutr	0.01
Mutation function	Binary mutation	maxgen	200
Reinsertion	Fitness-based	nind	180

We perform 30 independent runs of the three metaheuristics: ACO, GA, and hybrid ACO-GA. Computer specification to run all identification procedures are Intel Core i5-2329 3.0 GHz, 8 GB Memory, Windows 7 (64bit) operating system and Matlab 7.5 environment. For comparison of hybrid performance pure GA and ACO are run with parameters shown in Tables 1 and 2. Hybrid ACO-GA starts with 5 ants for 10 generation. To form population of chromosomes for further improvement from GA, pure ACO repeat 30 times. We take the best ACO solution from every one of the runs to form population. The obtained population with 30 chromosomes (ACO best solutions) is used from GA to obtain the best model parameters vector by the genetic evolution for 100 generations. The main numerical results are summarized in Table 3. The obtained average values of the model parameters (μ_{max}, k_S, and $Y_{S/X}$) are summarized in Table 4.

Table 2. Parameters of ACO algorithm

Parameter	Value
Number of ants	20
Initial pheromone	0.5
Evaporation	0.1
Generations	200

Table 3. Experimental results

Value	Algorithm	Algorithm performance	
		Time [s]	J
Best	GA	67.5172	4.4396
	ACO	67.3456	4.9190
	ACO-GA	35.5212	4.4903
Worst	GA	66.5968	4.6920
	ACO	66.6280	6.6774
	ACO-GA	35.3498	4.6865
Average	GA	67.1370	4.5341
	ACO	69.5379	5.5903
	ACO-GA	36.1313	4.5765

Table 4. Model parameters' estimations

Value	Algorithm	Model parameters		
		μ_{max}	k_S	$1/Y_{S/X}$
Average	GA	0.4857	0.0115	2.0215
	ACO	0.5154	0.0151	2.0220
	ACO-GA	0.4976	0.0135	2.0221

Table 3 shows that our ACO-GA algorithm achieves similar to pure ACO and pure GA solutions, but the running time is twice less. The pure GA algorithm starts from randomly generated initial solutions (population) which can be very fare from the optimal one. The ACO is a constructive method, which does not need any initial solution. ACO can faster find solutions which are not very fare from the optimal one. In our hybrid ACO-GA algorithm we explore the advantages of the both ACO and GA. We run the ACO for several iterations only and thus we generate initial solutions for GA which are closer to the optimal. The GA starts from solutions which are not fare from the optimal and thus we increase the convergence of the GA. More over in ACO-GA the population is very small, only 5 ants for ACO and only 30 individuals for GA (vs. 20 ants and 180 individuals), which decreases the used memory. Thus our hybrid algorithm has two advantages - less running time and less memory.

5 Conclusion

In this paper, a hybrid metaheuristic approach, which is a combination between two metaheuristics, Ant Colony Optimization and Genetic Algorithms is applied to the problem of parameter identification of an *E. coli* fed-batch cultivation process model.

Combining the advantages of both approaches, better performance of the hybrid algorithm was achieved in terms of computational time and memory, yet preserving the precision of the calculation. As shown in the detailed comparison above, in the hybrid algorithm almost twice less computational time was consumed. Moreover, approximately four times (in case of ACO – 20 vs. 5 individuals)) and six times (in case of GA – 180 vs. 30 individuals)) smaller populations were required to achieve results, as compared to the ACO and GA approaches, taken separately. Thus, the hybrid ACO-GA uses in times less memory for the computation.

As a next step of this research, we would further elaborate the hybrid ACO-GA algorithm, taking into consideration the possibility to have the GA stagnating after a number of iterations. For this reason, we intend to return the obtained GA solutions back to the ACO algorithm, and then run the ACO with the accordingly updated pheromone, thus generating a new population for further GA execution. In this way, a bidirectional hybridization will take place: GA is hybridized with ACO, and ACO is hybridized with GA. It is noteworthy that we can use any variant of both ACO and GA methods, depending on the specific problem being solved.

Acknowledgements. This work has been partially supported by the Bulgarian National Scientific Fund under the grants DID 02/29, DMU 02/4 and EU Project AComIn: Advanced Computing for Innovation, FP7 Capacity Programme, Research Potential of Convergence Regions.

References

1. Angelova, M., Pencheva, T.: Algorithms improving convergence time in parameter identification of fed-batch cultivation. Comptes Rendus de L'Academie Bulgare des Sciences **65**(3), 299–306 (2012)
2. Atanassova, V., Fidanova, S., Popchev, I., Chountas, P.: Generalized nets, ACO algorithms and genetic algorithms, monte carlo methods and applications. In: Proceedings of the 8th IMACS Seminar on Monte Carlo Methods, August 29 – September 2, 2011, Borovets, Bulgaria. Sabelfeld, K.K., Dimov, I. (eds.) De Gruyter Proceedings in Mathematics, pp. 39–46 (2012)
3. Basiri, M.E., Nemati, S.: A novel hybrid ACO-GA algorithm for text feature selection. In: IEEE Congress on Evolutionary Computation, CEC '09, pp. 2561–2568 (2009)
4. Bastin, G., Dochain, D.: On-line Estimation and Adaptive Control of Bioreactors. Els. Sc. Publ, Amsterdam (1991)
5. Dorigo, M., Stutzle, T.: Ant Colony Optimization. MIT Press, Cambridge (2004)
6. Fidanova, S., Roeva, O., Ganzha, M.: ACO for parameter settings of E. coli cultivation model. In: Proceedings of the Federated Conference on Computer Science and Information Systems, WCO 2012, Poland, pp. 407–414 (2012)
7. Goldberg, D.E.: Genetic Algorithms in Search, Optimization and Machine Learning. Addison Wesley Longman, London (2006)
8. Guangdong, H., Qun, W.: A hybrid ACO-GA on sports competition scheduling. In: Ostfeld, A. (ed.) Ant Colony Optimization - Methods and Applications, pp. 89–100. InTech (2011)
9. Holland, J.H.: Adaptation in Natural and Artificial Systems, 2nd edn. MIT Press, Cambridge (1992)

10. Li, N., Wang, S., Li, Y.: A hybrid approach of GA and ACO for VRP. J. Comput. Inf. Syst. **7**(13), 4939–4946 (2011)
11. Nemati, S., Basiri, M.E., Ghasem-Aghaee, N., Aghdam, M.H.: A novel ACO-GA hybrid algorithm for feature selection in protein function prediction. J. Expert Syst. Appl.: An Int. J. Arch. **36**(10), 12086–12094 (2009)
12. Roeva, O.: Improvement of Genetic Algorithm Performance for Identification of Cultivation Process Models, Advances Topics on Evolutionary Computing. Artificial Intelligence Series - WSEAS 2008, pp. 34–39 (2008)
13. Roeva, O., Fidanova, S.: A comparison of genetic algorithms and ant colony optimization for modeling of E. coli cultivation process. In: Roeva, O. (ed.) Real-World Application of Genetic Algorithms, ch. 13, pp. 261–282. InTech (2012)
14. Roeva, O., Pencheva, T., Hitzmann, B., Tzonkov, S.: A genetic algorithms based approach for identification of Escherichia coli fed-batch fermentation. Int. J. Bioautomation **1**, 30–41 (2004)

Modeling Forest Fire Spread Through a Game Method for Modeling Based on Hexagonal Cells

Evdokia Sotirova[1], Emilia Velizarova[2], Stefka Fidanova[3](✉), and Krassimir Atanassov[4]

[1] Prof. Asen Zlatarov University, 8000 Bourgas, Bulgaria
esotirova@btu.bg
[2] FRI – BAS, St. Kl. Ohridski Blvd. 132, 1756 Sofia, Bulgaria
velizars@abv.bg
[3] IICT – BAS, Acad. G. Bonchev Str. Bl25A, 1113 Sofia, Bulgaria
stefka@parallel.bas.bg
[4] IBPhBME – BAS, Acad. G. Bonchev Str., Block 105, 1113 Sofia, Bulgaria
krat@bas.bg

Abstract. A mathematical model for predicting the spread of a fire front in homogeneous and inhomogeneous forest is presented. It is based on the application of the game method for modeling with hexagonal lattice. The results of the modeling show the advantage of using hexagonal cells for the forest areas considered, thus avoiding the limitations of spurious symmetries of the square cells used in previous studies. Further validation of the model developed with a real case of forest fire is foreseen.

Keywords: Game method · Modeling · Forest fire spread

1 Introduction

In a series of papers, the Game Method for Modeling (GMM) is described and some of its applications are discussed. Some of these papers are related to fire front in homogeneous forest (see, e.g. [4–6]). All these models are based on square lattices. Here, for a first time, we use hexagonal lattice.

In Sect. 2, short remarks on GMM, following [1], are given and the main results are described in Sect. 3.

In Sect. 4 we compare the models of homogeneous forest (with initial values "9") when the lattice is square and hexagonal.

Short discussion for the plans for future research are given in the Conclusion.

2 Short Remarks on GMM

Conway's Game of Life (CGL, see, e.g. [2,3]), introduced by John Horton Conway has a "universe", which is an infinite two-dimensional orthogonal grid of square

I. Lirkov et al. (Eds.): LSSC 2013, LNCS 8353, pp. 321–328, 2014.
DOI: 10.1007/978-3-662-43880-0_36,

cells, each of which is in one of two possible states, alive or dead, or (as an equivalent definition) in the square there is an asterisk, or there is not. The first situation corresponds to the case when the cell is alive and the second corresponds to the case when the cell is dead.

One of the modifications of the CGL is called "Game Method for Modeling" (GMM) [1]. In this section, we give its short description.

Let us consider a set of symbols S and an n-dimensional simplex comprising of n-dimensional cubes (when $n = 2$, a two-dimensional grid of squares).

We assume that material points (referred in brief as objects) can be found in some of the centers of the n-dimensional cells. The GMM-grid can be either finite or infinite. In the first case, for i-th dimension of the grid there is natural number g_i that corresponds to the number of the sequential cells of the grid in the present dimension. Therefore, when the n-dimensional GMM-grid is finite, there is a vector $\langle g_1, g_2, ..., g_n \rangle$ of the lengths of its sides. Here, we use finite grids. We also consider a set of rules $\mathcal{A}$ as follows:

1. rules for the motion of the objects along the vertices of the simplex;
2. rules for the interactions among the objects, e.g., when they are collected in one cell.

Let the rules from the i-th type be denoted as i-rules, where $i = 1, 2$.

When $S = \{*\}$, we obtain the standard CGL.

We can call an *initial configuration* every set of (ordered) $(n+2)$-tuples with an initial component being the number of the object; the second, third, etc. until the $(n+1)$-st – its coordinates; and the $(n+2)$-nd – its corresponding symbol from S.

We can call a *final configuration* the ordered set of $(n+2)$-tuples having the above form and being a result of the modifications that occurred during a certain number of applications of the rules from $\mathcal{A}$ over a (fixed) initial configuration.

The single application of a rule from $\mathcal{A}$ over a given configuration K is called an *elementary step* in the transformation of the model and is denoted by $A_1(K)$.

When we have some initial configuration, we obtain new configurations in a stepwise manner. We must determine some constructive criteria for stopping the process. For example, such a condition may be the following.

1. The rules of the GMM are applied over the initial configuration and its derivatives for exactly n iterations.
2. A predefined configuration is obtained on the GMM-grid. For example, if we model a process of interaction between some objects, the process should stop when the grid contains only one of these objects.
3. A previous configuration is obtained on the GMM-grid, i.e., the process oscillates. This criterion is applicable for deterministic processes.

Let us consider a rule P which juxtaposes to a set of configurations M a single configuration $P(M)$ being the mean of the given ones. We will call this rule a *concentrate rule*. The concentration can be made either over the values of the symbols from S for the objects, or over their coordinates (not over both of them simultaneously). In [1], different formulas for P are given. Here, we suppose that as a result of applying the rule P over the set of configurations M, we will obtain a new configuration $P(M)$, for which the $(i_k; j_k)$-th place is occupied by a digit calculated as:

$$d_{i,j} = \left[\frac{1}{s} \sum_{k=1}^{s} d_{i,j}^{k} \right],$$

where for any real number $x = a + \alpha$, where a is a natural number and $\alpha \in [0, 1)$: $[x] = a$.

If K is an initial configuration, let $A_1(K)$ be the configuration, obtained in a result of application of the rules from A over K (for one step). Let $A_{n+1}(K) = A_1(A_n(K))$ for every natural number $n \geq 1$.

Let $\mathcal{B}$ be a criterion derived from physical or mathematical considerations. For two given configurations K_1 and K_2, it answers to the question whether they are close enough to each other or not. For two configurations K_1 and K_2 lying in a planar rectangle with lengths p and q, we can use the following criterion:

$$\mathcal{B}(K_1, K_2) = \frac{1}{p.q} \sum_{i=1}^{p} \sum_{j=1}^{q} |d_{i,j}^1 - d_{i,j}^2| < C,$$

where C is a predefined constant. For the set of configurations M and the set of rules $\mathcal{A}$, we define the set of configurations

$$\mathcal{A}(M) = \{L | (\exists K \in M)(L = A(K))\}.$$

The rules $\mathcal{A}$ are called statistically correct, if for a large enough (from a statistical point of view) natural number N:

$$(\forall n > N)(\forall M = \{K_1, K_2, ..., K_n\})(\forall m \geq 1)$$

$$(\mathcal{B}(A_m(P(M)), P(\{L_i | L_i = A_m(K_i), 1 \leq i \leq n\})) < C).$$

3 GMM-Model of Forest Fire

We describe a finite grid, having the form of a hexagonal lattice, with size 11×11 in which we check the development of forest fire processes. We assume that in the field there is a river (its territory being marked by letter R), stones (their territory being marked by S) and on the rest part of the field there is a homogeneous forest. The digits correspond to the wood mass per one unit square. These digits are specific for different types of forests, but here the forest is homogeneous and initially, the digits are only "9". After the beginning of the fire, the digits will decrease by specific rules, described below.

If some rule determines that symbol Y must be changed with symbol Z, let us denote this fact by $Y \rightarrow Z$.

Let everywhere $r \in [0, 1]$ be a random number that is generated for the current procedure.

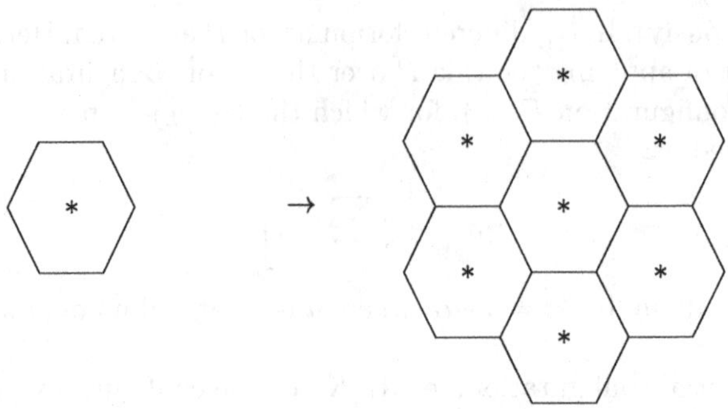

Fig. 1. Hexagonal sells

In the present research, we discuss a scenario without a wind.

In our model, the fire will develop in concentric circles (see Fig. 1).

The rules for the GMM are the following.

1. $R \to R$;
2. $S \to S$;
3. $0 \to 0$;
4. In the initial time-step, the fire starts from a fixed cell containing digit 9 that represents the density of the trees in that cell of the forest. On the second time-step, for the same cell $9 \to 8$. On the third time-step, for the same cell $8 \to 7$. In the same moment, all neighboring cells of the cell with the fire change their digits with the previous digit.
5. On the next time-steps, for the cells with fire

$$n \to \begin{cases} 0, & \text{if } n = 1 \\ n-1, & \text{if } n > 1 \end{cases}$$

6. The process continues until all cells in the region contain only digit 0. In the opposite case, go to 5.

The development of a forest fire is shown in Fig. 2.

4 Comparison of the Models of the Fire-Processes in Square and Hexagonal Lattices

On Fig. 3, a diagram for the fire-processes in a hexagonal lattice is shown for 18 time-steps. On Fig. 4, the diagram corresponds to the case of a square lattice for the same 18 time-steps.

It is obvious that both processes flow analogously and this is a non-formal proof that the GMM gives correct results.

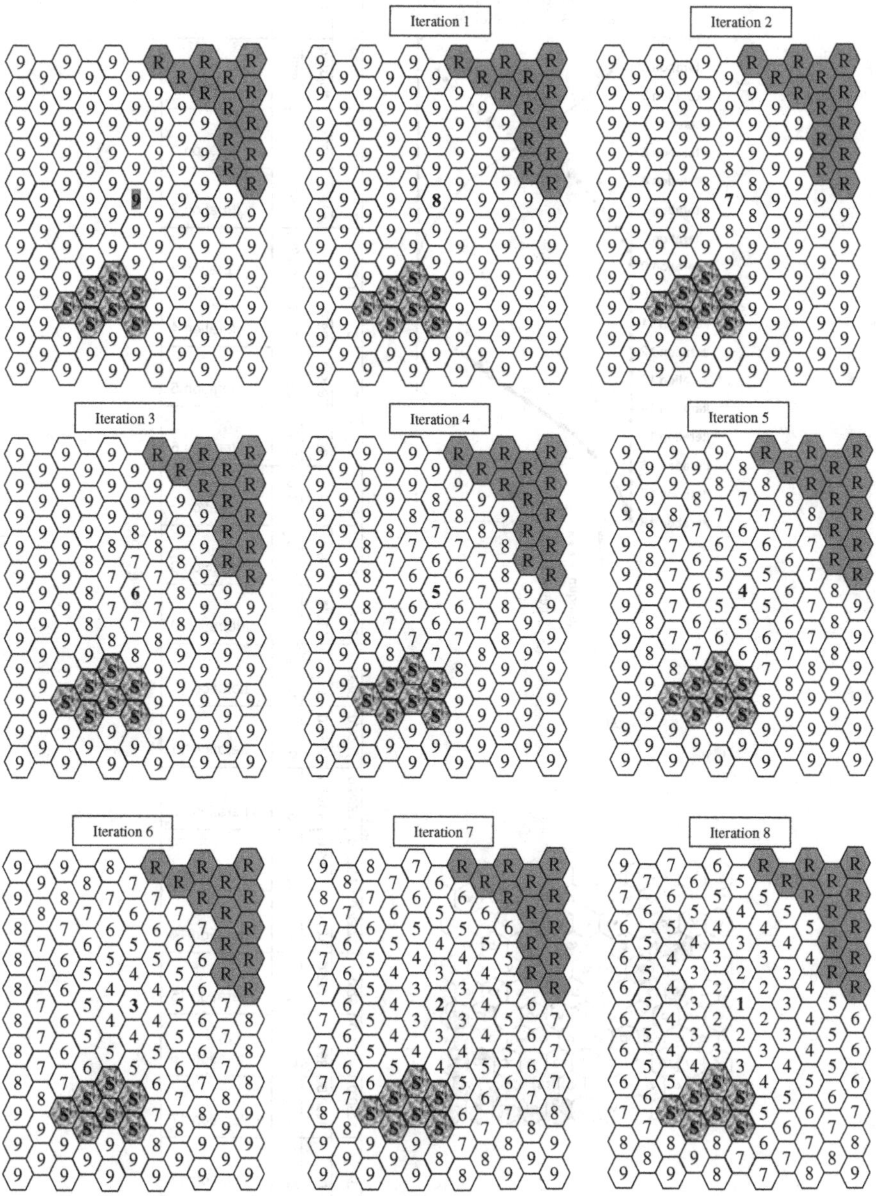

Fig. 2. Fire spread from time step 0 to time step 8

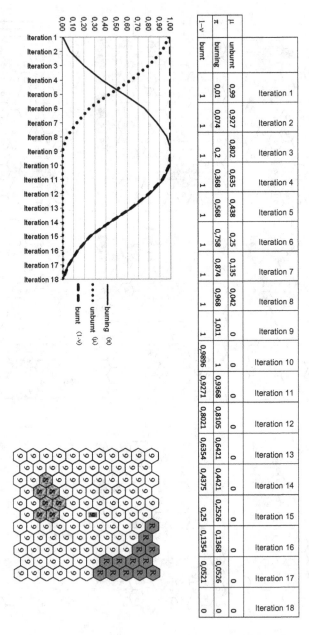

		Iteration 1	Iteration 2	Iteration 3	Iteration 4	Iteration 5	Iteration 6	Iteration 7	Iteration 8	Iteration 9	Iteration 10	Iteration 11	Iteration 12	Iteration 13	Iteration 14	Iteration 15	Iteration 16	Iteration 17	Iteration 18
μ	unburnt	0,99	0,927	0,802	0,635	0,438	0,25	0,135	0,042	0	0	0	0	0	0	0	0	0	0
π	burning	0,01	0,074	0,2	0,368	0,568	0,758	0,874	0,968	1,011	1	0,9368	0,8105	0,6421	0,4421	0,2526	0,1368	0,0526	0
1–ν	burnt	1	1	1	1	1	1	1	1	1	0,9896	0,9271	0,8021	0,6354	0,4375	0,25	0,1354	0,0521	0

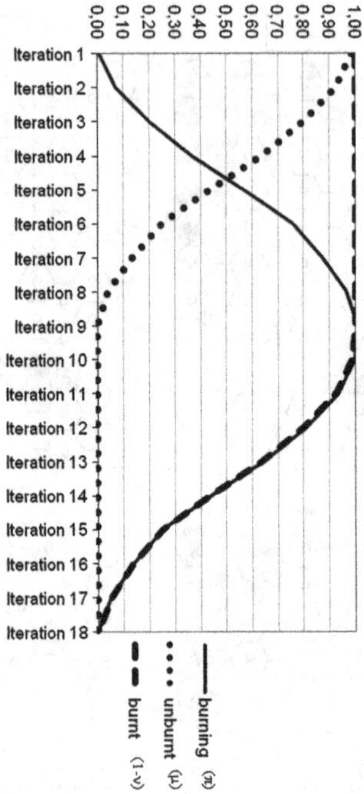

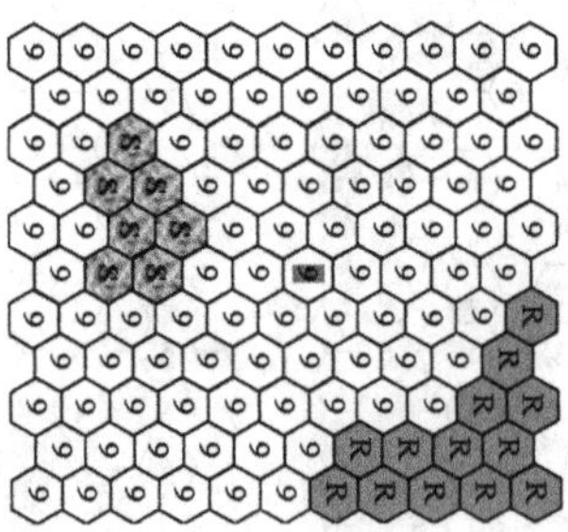

Fig. 3. Fire-processes in a hexagonal lattice

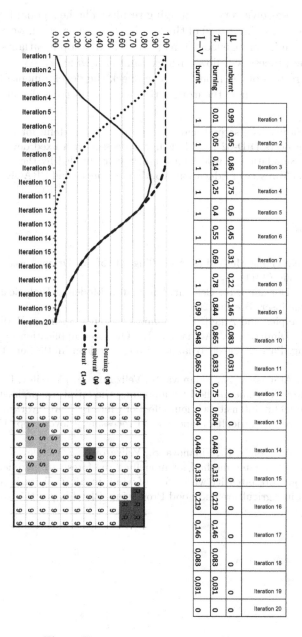

		Iteration 1	Iteration 2	Iteration 3	Iteration 4	Iteration 5	Iteration 6	Iteration 7	Iteration 8	Iteration 9	Iteration 10	Iteration 11	Iteration 12	Iteration 13	Iteration 14	Iteration 15	Iteration 16	Iteration 17	Iteration 18	Iteration 19	Iteration 20
μ	unburnt	0,99	0,95	0,86	0,75	0,6	0,45	0,31	0,22	0,146	0,083	0,031	0	0	0	0	0	0	0	0	0
π	burning	0,01	0,05	0,14	0,25	0,4	0,55	0,69	0,78	0,844	0,865	0,833	0,75	0,604	0,448	0,313	0,219	0,146	0,083	0,031	0
$1-\nu$	burnt	1	1	1	1	1	1	1	1	0,99	0,948	0,865	0,75	0,604	0,448	0,313	0,219	0,146	0,083	0,031	0

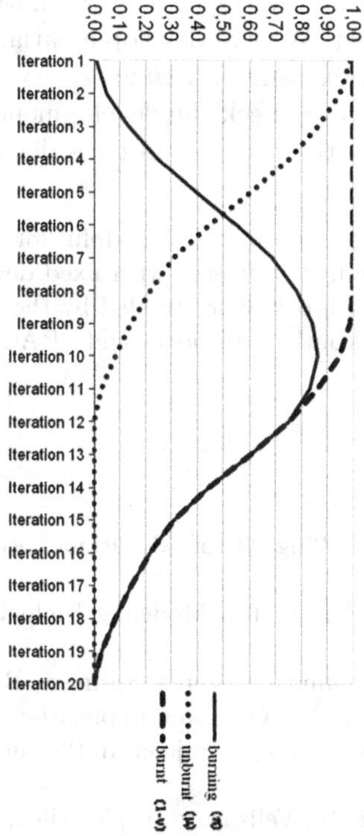

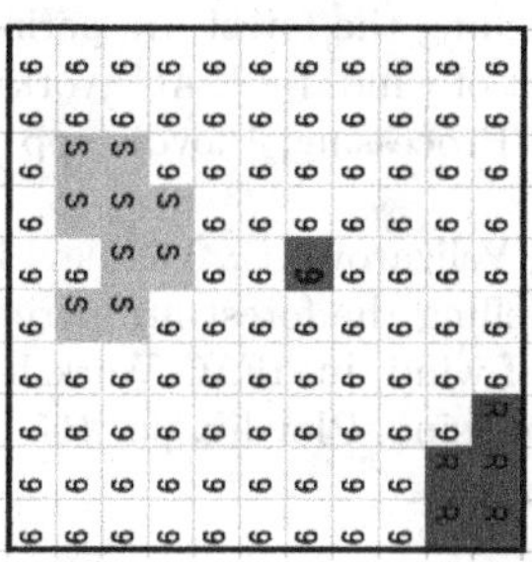

Fig. 4. Fire-processes in a square lattice

5 Conclusion

On this work we receive very encouraging results. The hexagonal lattice presents in more realistic way the spread of the fire. In the next research we will model the fire-process in an inhomogeneous forest. We will develop a variants of scenarios. After that, the process will flow with existing of a wind.

Using the results of the model, we can check the development of a real forest fire, the occupation of new territories, the decrease in trees density, etc.

Acknowledgment. The second and third authors are grateful for the support provided by the projects DID-02-29 "Modelling processes with fixed development rules" and I01/0006 "Simulation of wild-land fire behavior" funded by the National Science Fund, Bulgarian Ministry of Education, Youth and Science and PRACE project of FP7 of EC.

References

1. Atanassov, K.: Game Method for Modelling, "Prof. M. Drinov" Academic Publishing House, Sofia (2011)
2. Deutsch, A., Dormann, S.: Cellular Automaton Modeling Biological Pattern Formation. Birkhäuser, Boston (2005)
3. Gros, C.: Complex and Adaptive Dynamical Systems. Springer, Berlin (2011)
4. Sotirova, E., Dimitrov, D., Atanassov, K.: On some application of game-method for modelling. Part. 1: forest dynamics. Proc. Jangjeon Math. Soc. **15**(2), 115–123 (2012)
5. Sotirova, E., Atanassov, K., Fidanova, S., Velizarova, E., Vassilev, P., Shannon, A.: Application of the game method for modelling the forest fire perimeter expansion. Part 1: a model fire intensity without effect of wind. In: IFAC Workshop on Dynamics and Control in Agriculture and Food Processing, Plovdiv, pp. 159–163, 13–16 June 2012
6. Sotirova, E., Atanassov, K., Fidanova, S., Velizarova, E., Vassilev, P., Shannon, A.: Application of the game method for modelling the forest fire perimeter expansion. Part 2: a model fire intensity with effect of wind. In: IFAC Workshop on Dynamics and Control in Agriculture and Food Processing, Plovdiv, pp. 165–169, 13–16 June 2012

Modeling and Numerical Simulation of Processes in Highly Heterogeneous Media

Mixed FEM for Second Order Elliptic Problems on Polygonal Meshes with BEM-Based Spaces

Yalchin Efendiev[1], Juan Galvis[2], Raytcho Lazarov[1], and Steffen Weißer[3](✉)

[1] Texas A and M University, College Station, TX 77843, USA
{efendiev,lazarov}@math.tamu.edu
[2] Matematicas, Universidad Nacional de Colombia,
Carrera 45 No 26-85 - Edificio Uriel Gutierréz, Bogotá D.C., Colombia
jcgalvisa@unal.edu.co
[3] Saarland University, Saarbrücken, Germany
weisser@num.uni-sb.de

Abstract. We present a Boundary Element Method (BEM)-based FEM for mixed formulations of second order elliptic problems in two dimensions. The challenge, we would like to address, is a proper construction of $\mathbf{H}(\text{div})$–conforming vector valued trial functions on arbitrary polygonal partitions of the domain. The proposed construction generates trial functions on polygonal elements which inherit some of the properties of the unknown solution. In the numerical realization, the relevant local problems are treated by means of boundary integral formulations. We test the accuracy of the method on two model problems.

Keywords: Mixed formulation · BEM-based FEM · Polygonal mesh

1 Introduction

Mixed Finite Element Methods (FEM) have been instrumental in the development of flexible and accurate approximations of elliptic problems with heterogeneous coefficient on triangular and rectangular grids. Recent strategies, like the BEM-based FEM [5,10,12], aim at extending classical Finite Element Methods to polygonal and polyhedral meshes. Such general cells are very desirable in many applications, e.g. flows in heterogeneous porous media as models in hydrology and reservoir simulation. Therefore, a variety of approximation and solution methods on general grids, such as Mixed Finite Element Methods [8], Mimetic Finite Difference Methods [3] and the Virtual Element Methods [1], have been considered, studied, and tested in the last decade.

The goal of this note is to introduce a mixed formulation for the BEM-based Finite Element Method which until now has been studied only for the primal formulation of boundary value problems. The key idea is to construct a finite dimensional approximation space by implicitly defined trial functions which fulfill certain boundary value problems on a local, element-by-element-wise level. These problems are treated by means of boundary integral formulations which are discretized by Boundary Element Methods.

I. Lirkov et al. (Eds.): LSSC 2013, LNCS 8353, pp. 331–338, 2014.
DOI: 10.1007/978-3-662-43880-0_37,

Since these ideas are applied to the mixed formulation of the problem, we need a suitable discretization of the Sobolev space $\mathbf{H}(\mathrm{div})$ on polygonal meshes. This is done by implicitly generating trial functions by the BEM. A construction of suitable trial function for the mixed FEM on polygonal meshes was done by Kuznetsov and Repin in [8] by using subdivision of the polygonal cell into triangular elements and subsequently generating the test functions locally by mixed FEM. Also similar ideas were implemented in the Mixed Multiscale Finite Element Method [4,7]. The novelty in our approach is that instead of treating the local problem by the classical mixed FEM (as in [8]) or by the multiscale FEM (as in [4]) the local problems are treated by Boundary Element Methods. Thus, we avoid an additional triangulation of the polygonal elements. The Boundary Element Method works directly on the discretization of the element boundaries given naturally by their polygonal shapes. Therefore, the spacial dimension of the local problems is reduced by one and the effort for local meshing is not needed. The BEM-based FEM benefits from an elegant formulation and an efficient handling of local boundary value problems, which could be implemented as a stand-alone procedure.

2 Mixed Formulation

Let $\Omega \subset \mathbb{R}^2$ be a convex polygonal domain which is bounded, and let n_Ω be the outer unit normal vector to its boundary $\Gamma = \partial\Omega$. The boundary $\Gamma = \Gamma_D \cup \Gamma_N$ is divided into Γ_D (with non vanishing length) and Γ_N where Dirichlet and Neumann data is prescribed, respectively. We consider the problem

$$-\mathrm{div}(a\nabla p) = f \quad \text{in } \Omega, \qquad n_\Omega \cdot a\nabla p = 0 \quad \text{on } \Gamma_N, \qquad p = g_D \quad \text{on } \Gamma_D \tag{1}$$

with right hand side $f \in L^2(\Omega)$, Dirichlet data $g_D \in H^{1/2}(\Gamma_D)$ and material coefficient $a \in L^\infty(\Omega)$ which, for some positive constants $a_{\min}$ and $a_{\max}$, fulfills

$$0 < a_{\min} \leq a \leq a_{\max} \quad \text{almost everywhere in } \Omega.$$

The usual notation for Sobolev spaces $H^s(D)$, $s \geq 0$ and the Lebesgue space $L^2(D)$ is utilized for any domain $D \subset \Omega$. Vector valued spaces are indicated by bold face. We further assume that every interior angle at any transient point between the boundary Γ_D and Γ_N is less than π, so that the solution of (1) with $a = 1$, $f = 0$ and $g_D = 0$ is in the space $H^s(\Omega)$, $s > \frac{3}{2}$.

Next, a new unknown flux variable $u = a\nabla p$ is introduced and the boundary value problem is presented in a mixed form: Find $(u, p) \in \mathbf{H}_0(\mathrm{div}, \Omega) \times L^2(\Omega)$ such that

$$\begin{aligned} a(u,v) + b(v,p) &= (n_\Omega \cdot v, g_D)_{L^2(\Gamma_D)} && \forall v \in \mathbf{H}(\mathrm{div}, \Omega), \\ b(u,q) &= -(f,q)_{L^2(\Omega)} && \forall q \in L^2(\Omega), \end{aligned} \tag{2}$$

where

$$a(u,v) = (a^{-1}u, v)_{\mathbf{L}^2(\Omega)} \quad \text{and} \quad b(v,q) = (\mathrm{div}\, v, q)_{L^2(\Omega)}$$

and

$$\mathbf{H}_0(\mathrm{div}, \Omega) = \{v \in \mathbf{L}^2(\Omega) : \mathrm{div}\, v \in L^2(\Omega) \text{ and } n_\Omega \cdot v = 0 \text{ on } \Gamma_N\}.$$

The space $\mathbf{H}_0(\mathrm{div}, \Omega)$ is equipped with the norm

$$\|v\|^2_{\mathbf{H}(\mathrm{div},\Omega)} = \|v\|^2_{\mathbf{L}^2(\Omega)} + \|\mathrm{div}\, v\|^2_{L^2(\Omega)}.$$

It is known that the bilinear forms $a(\cdot,\cdot)$ and $b(\cdot,\cdot)$ are bounded, i.e.

$$\begin{aligned} |a(u,v)| &\le \varrho_1 \|u\|_{\mathbf{H}(\mathrm{div},\Omega)} \|v\|_{\mathbf{H}(\mathrm{div},\Omega)} && \text{for } u, v \in \mathbf{H}_0(\mathrm{div}, \Omega), \\ |b(v,q)| &\le \varrho_2 \|v\|_{\mathbf{H}(\mathrm{div},\Omega)} \|q\|_{L^2(\Omega)} && \text{for } v \in \mathbf{H}_0(\mathrm{div}, \Omega), q \in L^2(\Omega), \end{aligned}$$

the form $b(\cdot,\cdot)$ satisfies the *inf-sup* condition, the form $a(\cdot,\cdot)$ is coercive in $\mathbf{H}_0(\mathrm{div}, \Omega)$ on $\mathbf{Z} = \{v \in \mathbf{H}_0(\mathrm{div}, \Omega) : \; b(v,q) = 0 \; \forall q \in L^2(\Omega)\}$ and thus, the existence of a unique solution follows from the Babuska-Brezzi theory [2].

For the numerical treatment of problem (2) we shall need a splitting of Ω into finite elements as well as finite dimensional subspaces of $\mathbf{H}_0(\mathrm{div}, \Omega)$ and $L^2(\Omega)$. For this we use a family of polygonal meshes $\{\mathcal{K}_h\}$ which satisfy the regularity conditions specified below. Denote by $\mathcal{E}_h$ all edges of the elements in $\mathcal{K}_h$ which are in the interior of Ω or belong to Γ_D, and let h_K and h_E be the diameter of the element K and the edge length of E, respectively. We introduce the diameter ρ_K of the largest circle inscribed in K with center z_K. If z_K is not unique an arbitrary but fixed one is chosen. Following [9], we call the mesh *regular* if

1. All elements $K \in \mathcal{K}_h$ are convex polygons;
2. The aspect ratio is uniformly bounded from above by σ,
   i.e. for all $K \in \mathcal{K}_h$ we have $h_K/\rho_K < \sigma$;
3. There is a constant $c_1 > 0$ such that for all elements $K \in \mathcal{K}_h$ and all its edges $E \subset \partial K$ we have $h_K \le c_1 h_E$.

When studying convergence, the constants σ and c_1 have to be uniform over the hole sequence of meshes and they do not depend on the mesh size.

The finite dimensional subspaces of $\mathbf{H}_0(\mathrm{div}, \Omega)$ are defined through their basis functions ψ_E associated with edges $E \in \mathcal{E}_h$. For any $E \in \mathcal{E}_h$, let n_E be a unit normal vector, which in the sequel is considered to be fixed. Furthermore, let K_1 and K_2 be the two adjacent elements sharing the common edge E with the outer normal vectors n_{K_1} and n_{K_2}, respectively. The functions ϕ_E are defined implicitly as solution of the following local boundary value problems

$$\begin{aligned} \mathrm{div}(a\nabla\phi_E) &= \kappa_E(K)/|K| \quad \text{in } K \in \{K_1, K_2\}, \\ n_E \cdot a\nabla\phi_E &= \begin{cases} h_E^{-1} & \text{on } E, \\ 0 & \text{on all other edges,} \end{cases} \end{aligned} \tag{3}$$

see Fig. 1. Here, $\kappa_E(K) = n_E \cdot n_K = \pm 1$, so that the solvability condition for the Neumann problem is satisfied and (3) has a weak solution $\phi_E \in H^1(\Omega)$ which is unique up to an additive constant. Now, for $E \subset \overline{K}_1 \cap \overline{K}_2$, we define

$$\psi_E(x) = \begin{cases} a\nabla\phi_E(x) & \text{for } \; x \in \overline{K}_1 \cup \overline{K}_2, \\ 0 & \text{for } \; x \; \text{ elsewhere.} \end{cases} \tag{4}$$

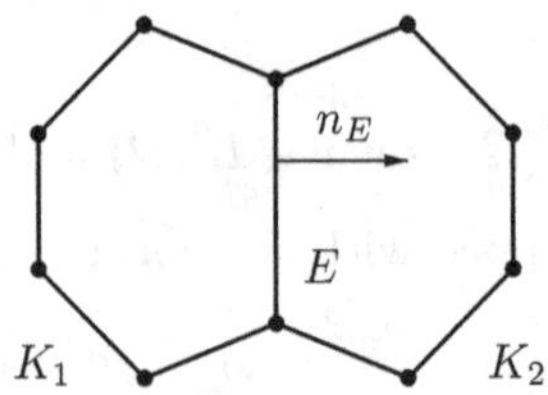

Fig. 1. Adjacent elements to E for the definition of ψ_E.

By construction, ψ_E has continuous normal flux across E and zero normal flux along all other internal edges of Ω so that $\psi_E \in \mathbf{H}_0(\text{div}, \Omega)$. An edge $E \in \Gamma_D$ has only one neighboring element K, and therefore the basis function is constructed in the same way by considering problem (3) solely on K.

We now define $\mathbf{X}_h$, a finite dimensional subspace of $\mathbf{H}_0(\text{div}, \Omega)$, as

$$\mathbf{X}_h = \text{span}\{\psi_E : E \in \mathcal{E}_h\}. \tag{5}$$

In the standard finite element terminology the "degrees of freedom" in $\mathbf{X}_h$ are associated with the edges $E \in \mathcal{E}_h$ and defined by

$$w_E = \int_E n_E \cdot w \; ds, \quad E \in \mathcal{E}_h.$$

Then, the corresponding interpolation operator $\pi_h : \mathbf{H}_0(\text{div}, \Omega) \to \mathbf{X}_h$ is

$$\pi_h w = \sum_{E \in \mathcal{E}_h} w_E \; \psi_E. \tag{6}$$

Further, we introduce the approximation space for the pressure p as

$$M_h = \{q \in L^2(\Omega) : \quad q|_K = const, \quad K \in \mathcal{K}_h\}. \tag{7}$$

By the use of the previously introduced spaces, the discrete version of the variational formulation (2) reads: Find $(u_h, p_h) \in \mathbf{X}_h \times M_h$ such that

$$\begin{aligned} a(u_h, v_h) + b(v_h, p_h) &= (n_\Omega \cdot v_h, g_D)_{L^2(\Gamma_D)} \qquad && \forall v_h \in \mathbf{X}_h, \\ b(u_h, q_h) &= -(f, q_h)_{L^2(\Omega)} && \forall q_h \in M_h. \end{aligned} \tag{8}$$

To prove unique solvability of the discrete problem, we use a fundamental theorem in the mixed finite element analysis, see [2]. This theory relies on the space

$$\mathbf{Z}_h = \{v_h \in \mathbf{X}_h : b(v_h, q_h) = 0 \quad \forall q_h \in M_h\}$$

and the following two assumptions:

Assumption [A1] There exists a constant $\alpha^* > 0$ such that

$$a(v_h, v_h) \geq \alpha^* \|v_h\|^2_{\mathbf{H}(\text{div},\Omega)} \quad \text{for } v_h \in \mathbf{Z}_h.$$

Assumption [A2] There exists a constant $\beta^* > 0$ such that

$$\inf_{q_h \in M_h} \sup_{v_h \in \mathbf{X}_h} \frac{b(v_h, q_h)}{\|v_h\|_{\mathbf{H}(\mathrm{div},\Omega)} \|q_h\|_{L^2(\Omega)}} \geq \beta^*.$$

The continuity of the bilinear forms $a(\cdot,\cdot)$ on $\mathbf{X}_h \times \mathbf{X}_h$ and $b(\cdot,\cdot)$ on $\mathbf{X}_h \times M_h$ and the assumptions A1 and A2 are sufficient for existence and uniqueness of the solution of the discrete problem (8), cf. [2]. In order to use this theory we have to show that Assumptions 1 and 2 are satisfied. This can be done by using the projection operator π_h introduced in (6) through showing its stability in $\mathbf{H}_0(\mathrm{div},\Omega)$–norm, see [6]. Then, using Babuska-Brezzi theory the main result of this Section is proven.

Theorem 1. *The problem* (8) *with* $\mathbf{X}_h$ *defined by* (5) *and* M_h *defined by* (7) *has unique solution* $(u_h, p_h) \in \mathbf{X}_h \times M_h$. *Furthermore, there exists a constant* c_2 *depending only upon* α^*, β^*, ϱ_1 *and* ϱ_2 *such that*

$$\|u - u_h\|_{\mathbf{H}(\mathrm{div},\Omega)} + \|p - p_h\|_{L^2(\Omega)} \leq c_2 \left\{ \inf_{v_h \in \mathbf{X}_h} \|u - v_h\|_{\mathbf{H}(\mathrm{div},\Omega)} + \inf_{q_h \in M_h} \|p - q_h\|_{L^2(\Omega)} \right\}.$$

Analysing the approximation properties of the discrete space $\mathbf{X}_h$ and the interpolation operator π_h, the following estimate for the interpolation error can be obtained.

Lemma 1. *Let* $\mathcal{K}_h$ *be a regular mesh and* $v \in \mathbf{H}^s(\Omega)$, $1 \geq s > \frac{1}{2}$. *Then the following estimate is valid*

$$\|v - \pi_h v\|_{\mathbf{H}(\mathrm{div},\Omega)} \leq c_3 h^s \, |v|_{\mathbf{H}^s(\Omega)} + \inf_{q_h \in M_h} \|\mathrm{div}\, v - q_h\|_{L^2(\Omega)}$$

with $h = \max\{h_K : K \in \mathcal{K}_h\}$.

3 BEM Approximation of Trial Space

In Sect. 2, we have introduced a conforming subspace $\mathbf{X}_h$ of the Sobolev space $\mathbf{H}_0(\mathrm{div},\Omega)$. However, the trial functions ψ_E rely on solutions of local boundary value problems and thus, it is almost impossible to give an analytic formula for these functions on arbitrary polygonal elements. To make this approach feasible, we make use of the Boundary Element Method. Due to this choice, we have two additional assumptions. First, we assume that the diffusion coefficient is piecewise constant such that $a(x) = a_K$ for $x \in K$ $\forall K \in \mathcal{K}_h$, and second, we assume that all elements have a diameter less than one. The second assumption does not represent a serious restriction of the method, since it can always be fulfilled by proper rescaling of the domain.

In the definition of ψ_E, the boundary value problem (3) has to be solved for ϕ_E. Due to the piecewise constant diffusion coefficient, it is possible to rewrite this problem as

$$-\Delta\phi_E = f_E \quad \text{in } K, \qquad n_E \cdot \nabla\phi_E = g_E \quad \text{on } \partial K$$

with appropriate data f_E and g_E. Furthermore, since f_E is constant on K, there is a polynomial $\phi_{E,p}$ with $-\Delta\phi_{E,p} = f_E$ and we write $\phi_E = \phi_{E,0} + \phi_{E,p}$, where $\phi_{E,0}$ is obviously the solution of

$$-\Delta\phi_{E,0} = 0 \quad \text{in } K, \qquad n_E \cdot \nabla\phi_{E,0} = g_{E,0} \quad \text{on } \partial K$$

for some data $g_{E,0}$. It is known, cf. [11], that the weak solution of this problem in $H^1(K)$, which is unique up to an additive constant, can be expressed as

$$\phi_{E,0}(x) = \int_{\partial K} U^*(x,y) g_{E,0}(y)\, ds_y - \int_{\partial K} \gamma_{1,y}^K U^*(x,y) \gamma_0^K \phi_{E,0}(x)\, ds_y \tag{9}$$

for $x \in K$. Here, $U^*(x,y) = -\frac{1}{2\pi} \ln|x-y|$ for $x, y \in \mathbb{R}^2$ is the so called fundamental solution and $\gamma_0^K : H^1(K) \to H^{1/2}(\partial K)$ denotes the usual trace operator. If the trace $\gamma_0^K \phi_{E,0}$ is known, the representation formula (9) can be used to evaluate $\phi_{E,0}$ and thus ϕ_E inside of K. It is known, cf. [11], that this trace fulfills the following variational formulation on the boundary of the element K: Find $\gamma_0^K \phi_{E,0} \in H^{1/2}(\partial K)$ such that

$$\left(\widetilde{\mathbf{D}}_K \gamma_0^K \phi_{E,0}, \vartheta\right)_{L^2(\partial K)} = \left((\tfrac{1}{2}\mathbf{I} - \mathbf{K}_K')g, \vartheta\right)_{L^2(\partial K)} \quad \forall \vartheta \in H^{1/2}(\partial K), \tag{10}$$

where

$$\left(\widetilde{\mathbf{D}}_K \theta, \vartheta\right)_{L^2(\partial K)} = (\mathbf{D}_K \theta, \vartheta)_{L^2(\partial K)} + (\theta, 1)_{L^2(\partial K)} (\vartheta, 1)_{L^2(\partial K)},$$

and $\mathbf{D}_K$ and $\mathbf{K}_K'$ are the well studied hypersingular integral operator and the adjoint double layer potential operator, respectively. The operator $\widetilde{\mathbf{D}}_K$ is bounded and elliptic on $H^{1/2}(\partial K)$ and thus, the variational formulation (10) admits a unique solution $\gamma_0^K \phi_{E,0} \in H^{1/2}(\partial K)$.

For the computer implementation, the space $H^{1/2}(\partial K)$ is discretized by piecewise linear and continuous functions over ∂K and a discrete Galerkin formulation for (10) is used to approximate the trace $\gamma_0^K \phi_{E,0}$. Inserting this approximation into the representation formula (9), we obtain an approximation $\widetilde{\phi}_{E,0}$ of $\phi_{E,0}$ and therefore an approximation $\widetilde{\phi}_E$ of ϕ_E. Finally, we replace ϕ_E by its approximation in (4) to get an approximated trial function $\widetilde{\psi}_E$. This gives an approximated trial space

$$\widetilde{\mathbf{X}}_h = \operatorname{span}\{\widetilde{\psi}_E : E \in \mathcal{E}_h\}.$$

Due to the approximation errors coming from the Boundary Element Method, the space $\widetilde{\mathbf{X}}_h$ is not conforming any more. Therefore, the discrete version of the initial problem now reads: Find $(\widetilde{u}_h, \widetilde{p}_h) \in \widetilde{\mathbf{X}}_h \times M_h$ such that

$$\begin{aligned} a_h(\widetilde{u}_h, \widetilde{v}_h) + b_h(\widetilde{v}_h, \widetilde{p}_h) &= (n_\Omega \cdot \widetilde{v}_h, g_D)_{L^2(\Gamma_D)} && \forall \widetilde{v}_h \in \widetilde{\mathbf{X}}_h, \\ b_h(\widetilde{u}_h, \widetilde{q}_h) &= -(f, \widetilde{q}_h)_{L^2(\Omega)} && \forall \widetilde{q}_h \in M_h, \end{aligned}$$

where $a_h(u,v) = \sum_{K \in \mathcal{K}_h} (a^{-1}u, v)_{\mathbf{L}^2(K)}$ and $b_h(v,q) = \sum_{K \in \mathcal{K}_h} (\operatorname{div} v, q)_{L^2(K)}$. The stability, error estimates, and number of numerical experiments for this method are presented in our work [6].

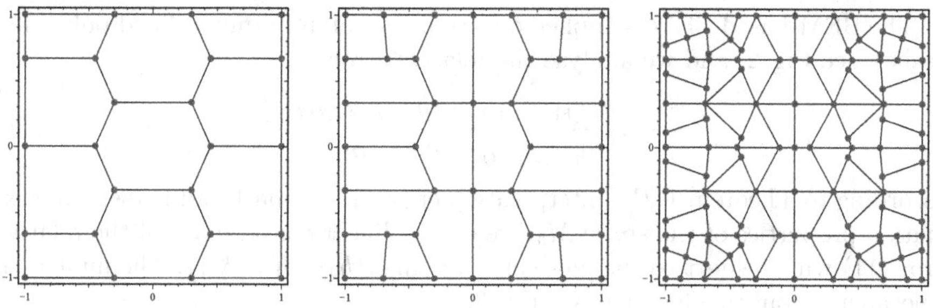

Fig. 2. Initial mesh (left), refined mesh after one step (middle), refined mesh after three steps (right)

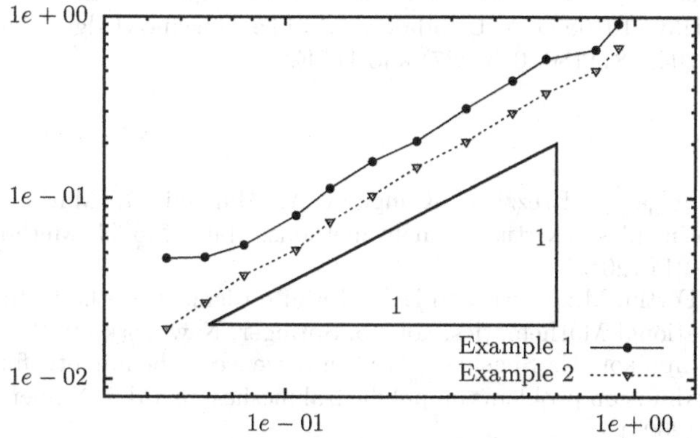

Fig. 3. Relative error (11) with respect to h in logarithmic scale

4 Numerical Experiments

To validate our theoretical results, we give the first numerical experiments for the mixed formulation of the BEM-based Finite Element Method. Two model problems are posed on the domain $\Omega = (-1,1)^2$ and we decompose its boundary into $\Gamma_D = \{(x_1,-1)^\top : -1 \leq x_1 \leq 1\}$ and $\Gamma_N = \partial\Omega \setminus \Gamma_D$.

In the first example, we choose the data g_D and g_N in such a way that the smooth function $p(x) = \exp(2\pi(x_1 - 0.3)) \cos(2\pi(x_2 - 0.3))$, $x \in \mathbb{R}^2$ is the exact solution of

$$-\Delta p = 0 \quad \text{in } \Omega, \qquad n_\Omega \cdot \nabla p = g_N \quad \text{on } \Gamma_N, \qquad p = g_D \quad \text{on } \Gamma_D.$$

Thus, (u,p) with $u = \nabla p$ solves the corresponding mixed formulation (2). For the second example, we take $p(x) = \sin(\pi x_1)\sin(\pi x_2)$, $x \in \mathbb{R}^2$ as solution of

$$-\Delta p = f \quad \text{in } \Omega, \qquad n_\Omega \cdot \nabla p = g_N \quad \text{on } \Gamma_N, \qquad p = 0 \quad \text{on } \Gamma_D$$

with corresponding data f and g_N.

The BEM-based FEM is applied on a sequence of uniformly refined polygonal meshes, see Fig. 2, and we analyse the relative error

$$\frac{\|u - u_h\|_{\mathbf{H}(\mathrm{div},\Omega)} + \|p - p_h\|_{L^2(\Omega)}}{\|u\|_{\mathbf{H}(\mathrm{div},\Omega)} + \|p\|_{L^2(\Omega)}}. \tag{11}$$

According to Theorem 1, the interpolation error in Lemma 1 and known approximation properties of the space M_h, we expect linear convergence of the relative error (11) with respect to the mesh size $h = \max\{h_K : K \in \mathcal{K}_h\}$. The numerical experiments confirm this fact, see Fig. 3.

Acknowledgments. The research of Y. Efendiev, J. Galvis, and R. Lazarov has been supported in parts by award KUS-C1-016-04, made by King Abdullah University of Science and Technology (KAUST). R. Lazarov is also supported in part by the award made by NSF DMS-1016525. Y. Efendiev would like to acknowledge a partial support from NSF (724704, 0811180, 0934837) and DOE.

References

1. Beirão da Veiga, L., Brezzi, F., Cangiani, A., Manzini, G., Marini, L.D., Russo, A.: Basic principles of virtual element methods. Math. Models Methods Appl. Sci. **23**(1), 199–214 (2013)
2. Brezzi, F., Fortin, M.: Mixed and Hybrid Finite Element Methods. Springer Series in Computational Mathematics, vol. 15. Springer, New York (1991)
3. Brezzi, F., Lipnikov, K., Shashkov, M.: Convergence of the mimetic finite difference method for diffusion problems on polyhedral meshes. SIAM J. Numer. Anal. **43**(5), 1872–1896 (2005)
4. Chen, Z., Hou, T.Y.: A mixed multiscale finite element method for elliptic problems with oscillating coefficients. Math. Comp. **72**(242), 541–576 (2003)
5. Copeland D., Langer U., Pusch D.: From the boundary element domain decomposition methods to local Trefftz finite element methods on polyhedral meshes. In: Bercovier M., Gander M., Kornhuber R., Widlund O. (eds) Domain Decomposition Methods in Science and Engineering XVIII. Lecture Notes in Computational Science and Engineering (LNCSE), pp. 315–322. Springer, Heidelberg (2009)
6. Efendiev Y., Galvis J., Lazarov R.D., Weißer S.: Mixed formulation of BEM-based FEM for second order elliptic problems on general polygonal meshes (in progress)
7. Efendiev, Y., Hou, T.Y.: Multiscale Finite Element Methods. Theory and Applications. Surveys and Tutorials in the Appl. Math. Sci. Springer, New York (2009)
8. Kuznetsov, Y., Repin, S.: New mixed finite element method on polygonal and polyhedral meshes. Russian J. Numer. Anal. Math. Model. **18**(3), 261–278 (2003)
9. Rjasanow, S., Weißer, S.: Developments in BEM-based finite element methods on polygonal and polyhedral meshes. In: Eberhardsteiner, J., Böhm, H.J., Rammerstorfer, F.G. (eds.) ECCOMAS 2012, e-Book Full Papers, pp. 1421–1431. Vienna University of Technology, Austria (2012)
10. Rjasanow, S., Weißer, S.: Higher order BEM-based FEM on polygonal meshes. SIAM J. Numer. Anal. **50**(5), 2357–2378 (2012)
11. Steinbach, O.: Numerical Approximation Methods for Elliptic Boundary Value Problems: Finite and Boundary Elements. Springer, New York (2008)
12. Weißer, S.: Residual error estimate for BEM-based FEM on polygonal meshes. Numer. Math. **118**(4), 765–788 (2011)

Topology Optimization Using Multiscale Finite Element Method for High-Contrast Media

Boyan S. Lazarov(✉)

Department of Mechanical Engineering, Solid Mechanics,
Technical University of Denmark, Nils Koppels Alle B. 404, 2800 Lyngby, Denmark
bsl@mek.dtu.dk

Abstract. The focus of this paper is on the applicability of multiscale finite element coarse spaces for reducing the computational burden in topology optimization. The coarse spaces are obtained by solving a set of local eigenvalue problems on overlapping patches covering the computational domain. The approach is relatively easy for parallelization, due to the complete independence of the subproblems, and ensures contrast independent convergence of the iterative state problem solvers. Several modifications for reducing the computational cost in connection to topology optimization are discussed in details. The method is exemplified in minimum compliance designs for linear elasticity.

Keywords: Topology optimization · Multiscale finite element method · High contrast media

1 Introduction

The aim of this work is to investigate and to demonstrate the applicability of the recently developed multiscale finite element method (MsFEM) with local spectral basis functions [8] in topology optimization. Topology optimization [4] is an iterative process which finds a material distribution in a given design domain by minimizing an objective and fulfilling a set of predefined constraints. The material distribution is represented by a density field which takes value one, if the material point is occupied with material, and zero if the material point is void. In order to utilize gradient based optimization, the problem is relaxed to take intermediate values between zero and one. The optimization algorithm consists of alternating finite element analyzes, gradient evaluations, regularization steps and math programming updates. Most of the computational efforts are spent on solving the discretized physical problem and wider industrial adoption requires reduction of the solution time. A promising direction is the development of new scalable algorithms and codes for proper utilization of the modern parallel machines [1]. Here, the MsFEM with local spectral basis functions is adopted for achieving this goal.

This work was financially supported by Villum Fonden and by a grant from the Danish Center of Scientific Computing (DCSC).

I. Lirkov et al. (Eds.): LSSC 2013, LNCS 8353, pp. 339–346, 2014.
DOI: 10.1007/978-3-662-43880-0_38,

The main idea behind the MsFEM approach is to construct basis functions which provide a good approximation of the solution on a coarse grid [9]. The method has been applied mainly to scalar elliptic problems, and recently applications in linear elasticity [5] have been demonstrated as well. The original MsFEM approach [9] constructs a single shape function per coarse node and the error in the solution is controlled by the coarse mesh size. The MsFEM for high contrast problems [8] constructs several basis functions per coarse node which are capable of representing the important features of the solution. The accuracy of the approximation is controlled by the dimension of the coarse space and the approach is briefly presented below.

2 Multiscale Finite Element Method - Linear Elasticity

The considered mechanical system is linear elastic with system response governed by the Navier-Cauchy partial differential equation given as

$$\begin{aligned} -\mathrm{div}\boldsymbol{\sigma}\left(\boldsymbol{u}\right) &= \boldsymbol{f} \text{ in } \Omega \\ \boldsymbol{\sigma}\left(\boldsymbol{u}\right) &= \boldsymbol{C} : \boldsymbol{\varepsilon}\left(\boldsymbol{u}\right) \text{ in } \Omega \end{aligned} \tag{1}$$

where $\boldsymbol{\sigma}$ is the stress tensor, $\boldsymbol{\varepsilon}$ is the strain tensor, $\boldsymbol{C}$ is an elastic material properties tensor, $\boldsymbol{u}$ denotes the displacement field and $\boldsymbol{f}$ is the input supplied to the system, i.e., distributed and concentrated forces. The mechanical system occupies the bounded physical domain $\Omega \subset \mathbb{R}^2$. The boundary $\Gamma = \overline{\Gamma_{D_i} \cup \Gamma_{N_i}}, i = 1,2$, is decomposed into two disjoint subsets for each component $i = 1,2$, Γ_{D_i} with prescribed displacements $u_i = 0$, and Γ_{N_i} with prescribed traction t_i. The material properties tensor is assumed to be isotropic and has the following form $\boldsymbol{C}(\boldsymbol{x}) = E(\boldsymbol{x})\boldsymbol{C}_0$, where $\boldsymbol{C}_0$ is a constant elasticity tensor obtained for predefined Poisson ratio ν and modulus of elasticity one, and $E(\boldsymbol{x})$ is a spatially varying modulus of elasticity $E(x) \in [E_{\min}, E_{\max}]$. The variational formulation of (1) is given as

$$a\left(\boldsymbol{u},\boldsymbol{v}\right) = l\left(\boldsymbol{v}\right) \text{ for all } \boldsymbol{v} \in V_0 \tag{2}$$

with bilinear form a and linear form l defined by

$$\begin{aligned} a\left(\boldsymbol{u},\boldsymbol{v}\right) &= \int_{\Omega} \left(C : \boldsymbol{\varepsilon}\left(\boldsymbol{u}\right)\right) : \boldsymbol{\varepsilon}\left(\boldsymbol{v}\right) \mathrm{d}\boldsymbol{x} \text{ for all } u, v \in V_0 \\ l\left(\boldsymbol{v}\right) &= \int_{\omega} \left(\boldsymbol{f}\cdot\boldsymbol{v}\right) \mathrm{d}\boldsymbol{x} + \int_{\Gamma_N} \left(\boldsymbol{t}\cdot\boldsymbol{v}\right) \mathrm{d}s \text{ for all } v \in V_0 \end{aligned}$$

where $V_0 = \left\{\boldsymbol{v} \in \left[H^1(\Omega)\right]^2 : v_i = 0 \text{ on } \Gamma_{\mathrm{D_i}}, \mathrm{i} = 1,2\right\} \subset V = \left[H^1(\Omega)\right]^2$. The weak formulation is discretized using finite element space $V_h \subset V_0$ with vector valued shape function defined on a uniform rectangular mesh $\mathcal{T}^h$. Each basis function is a scalar linear Lagrange function in one of the components and zero in the other. This leads to a linear system of equations in the form

$$\boldsymbol{K}\boldsymbol{u} = \boldsymbol{f} \tag{3}$$

where $\boldsymbol{K}$ is the so-called stiffness matrix of the system, $\boldsymbol{u}$ is the vector with nodal displacements and $\boldsymbol{f}$ is the vector with the external forces.

In topology optimization the linear system (3) is solved on each iteration step. Therefore, the solution cost of the optimization algorithm depends strongly on the time necessary for obtaining the solution of the state equations. Solving the linear system of equations can account for more than 99 % of the total computational time [1]. As the discretization of realistic 3D industrial problems often leads to linear system of equations with several million degrees of freedom, the utilization of direct solvers becomes prohibitive and the state solution is usually obtained by iterative methods. For multi-material physical systems with high contrast between the material parameters arising in topology optimization and classical preconditioning, the number of iterations increases with increasing the contrast, e.g. [1].

In [8] it is demonstrated that the solution cost for the diffusion equation with heterogeneous coefficients can be significantly reduced by using coarse space approximations which contain important features of the solution. Here the same approach is followed for the linear elastic case and the main steps are briefly outlined. The mesh $\mathcal{T}^h$ is assumed to be obtained as a refinement of a coarser one $\mathcal{T}^H = \{K\}$, and K denotes a coarse grid block. The nodes of the coarse mesh are denoted as $\{\boldsymbol{y}_i\}_{i=1}^{N_c}$ and the neighborhood of node $\boldsymbol{y}_i$ is defined as

$$\omega_i = \bigcup \{\overline{K}_j \in \mathcal{T}^H; \boldsymbol{y}_i \in \overline{K}_j\} \tag{4}$$

A set of coarse basis functions $\{\boldsymbol{\phi}_{i,j}, j = 1 \ldots N_i\}$, defined w.r.t. $\mathcal{T}^h$, with support on ω_i is introduced for each node $\boldsymbol{y}_i$ in the coarse mesh. The solution in the coarse space is sought as $\boldsymbol{u}_a = \sum_{i,j} c_{i,j}\boldsymbol{\phi}_{i,j}$. A coarse discretization of the variational formulation is given as $\boldsymbol{K}_c\boldsymbol{u}_c = \boldsymbol{f}_c$, where the coarse stiffness matrix $\boldsymbol{K}_c$ and the coarse right hand side $\boldsymbol{f}_c$ are obtained as

$$\boldsymbol{K}_c = \boldsymbol{R}_c\boldsymbol{K}\boldsymbol{R}_c^{\mathsf{T}}, \quad \boldsymbol{f}_c = \boldsymbol{R}_c\boldsymbol{f} \tag{5}$$

The vector $\boldsymbol{u}_c$ contains the coefficients $c_{i,j}$ and $\boldsymbol{R}_c^{\mathsf{T}} = [\boldsymbol{\phi}_1, \boldsymbol{\phi}_2, \ldots, \boldsymbol{\phi}_{N_t}], N_t = \sum_{i=1}^{N_c} N_i$ is a matrix which describes mapping from the coarse to the fine space, and consists of the nodal values of the coarse basis functions in the fine space. Approximation of the nodal solution in the fine space is obtained as $\boldsymbol{u}_a = \boldsymbol{R}_c^{\mathsf{T}}\boldsymbol{u}_c$.

Following [10], the coarse basis functions are obtained by solving the following eigenvalue problem

$$-\mathrm{div}\left(\boldsymbol{C}\left(\boldsymbol{x}\right) : \boldsymbol{\varepsilon}\left(\boldsymbol{u}\right)\right) = \lambda \mathrm{E}\left(\boldsymbol{x}\right)\boldsymbol{u}, \quad \boldsymbol{x} \in \omega_i \tag{6}$$

for each neighborhood ω_i of node $\boldsymbol{y}_i$ with the same boundary conditions applied to (1) on $\partial\omega_i \cap \Gamma \neq \emptyset$, and homogeneous Neumann ($\boldsymbol{t} = 0$) otherwise. By using the subspace $V_h(\omega_i) = \{\boldsymbol{v}_h \in V_h : \mathrm{supp}\boldsymbol{v}_\mathrm{h} \subset \omega_\mathrm{i}\}$, the eigenvalue problem in discrete form is written as

$$\boldsymbol{K}^{\omega_i}\boldsymbol{\psi}^{\omega_i} = \lambda\boldsymbol{M}^{\omega_i}\boldsymbol{\psi}^{\omega_i} \tag{7}$$

The eigenvalues and the eigenvectors are denoted as $\{\lambda_l^{\omega_i}\}$ and $\{\psi_l^{\omega_i}\}$ respectively, and the eigenvalues are ordered as $\lambda_1^{\omega_i} \leq \lambda_2^{\omega_i} \leq \lambda_3^{\omega_i} \leq \cdots \leq \lambda_j^{\omega_i} \leq \ldots$.

The set of coarse base functions $\{\boldsymbol{\phi}_{i,j}\}$ associated with node $\boldsymbol{y}_i$ is defined as the fine space finite element interpolant of $\left\{\chi_i\left(\boldsymbol{x}\right)\boldsymbol{\psi}_j^{\omega_i}\left(\boldsymbol{x}\right), j = 1 \ldots N_i\right\}$, e.g. [6]. Where, $\{\chi_i\}_{i=1}^{N_c}$ is a partition of unity subordinated to $\{\omega_i\}$, such that $\chi_i \in H^1\left(\Omega\right)$ and $|\nabla\chi_i| \leq 1/H, i = 1, \ldots, N_c$, and H is the characteristic length of a coarse element K. For each ω_i, N_i is determined as the number of the eigenvalues smaller than a globally selected threshold λ_Ω.

The above procedure is a direct generalization of the coarse space construction procedure for scalar diffusion problem with high contrast coefficients [10]. Coarse spaces can be constructed by replacing the mass matrix in (7) with the diagonal of $\boldsymbol{K}^{\omega_i}$ as proposed in [7]. An alternative for general positive definite bi-linear forms is analyzed in [6]. Numerical experiments performed on the high contrast test problems considered here reveal that both alternatives lead to similar convergence rate with respect to the number of coarse degrees of freedom and, due to the limits on the size, they are not presented in the current work.

3 Topology Optimization

Topology optimization [4] is an iterative process which minimizes a predefined objective function, e.g. structural weight or compliance, by distributing material in a given design domain. The material distribution fulfills a predefined set of constraints, e.g. volume or stress constraints. The design domain is discretized using cells (voxels in 3D) and the design field is parametrized using constant design variables associated with each cell. The discretization coincides with the finite element mesh. The existence of the solution is ensured by filtering (see references in [4]). The compliance minimization problem in discrete form is given as:

$$\begin{aligned} \min_{\boldsymbol{\rho}} &: c(\boldsymbol{\rho}), \\ \text{s.t.} &: \boldsymbol{K}\left(\tilde{\boldsymbol{\rho}}\left(\boldsymbol{\rho}\right)\right)\boldsymbol{u} = \boldsymbol{f}, \\ &: V\left(\tilde{\boldsymbol{\rho}}\left(\boldsymbol{\rho}\right)\right) \leq V^*, \\ &: 0 \leq \boldsymbol{\rho} \leq 1 \end{aligned} \tag{8}$$

where $\boldsymbol{\rho}$ is a vector with design variables, $\tilde{\boldsymbol{\rho}}$ is filtered density vector with elements $\tilde{\rho}_e\left(\boldsymbol{\rho}\right)$ computed as weighted average of the design variables ρ_j in the elements within radius R_f from the center of element e, e.g. [3]. The objective is given as $c(\boldsymbol{\rho}) = \boldsymbol{f}^\mathsf{T}\boldsymbol{u}$ and the material properties for each element e are computed as

$$E_e = E_{\min} + \tilde{\rho}_e^p\left(E_{\max} - E_{\min}\right) \tag{9}$$

where $p = 3$ penalizes the intermediate densities. The optimization problem is non-convex and the design converges to a local minimum. In order to simplify the notations the explicit dependence of $\boldsymbol{\rho}$ and $\tilde{\boldsymbol{\rho}}$ will be omitted in the following text.

The gradients of the objective with respect to the filtered design variables are obtained by adjoint sensitivity analysis, and with respect to the design variables by employing the chain rule [4]. The sensitivities in discrete form with respect to the filtered design variables $\tilde{\rho}_e, e = 1, \ldots, N_e$ are given as

$$\frac{\partial c}{\partial \tilde{\rho}_e} = -\boldsymbol{u}^{\mathsf{T}} \frac{\partial \boldsymbol{K}}{\partial \tilde{\rho}_e} \boldsymbol{u} = -\boldsymbol{u}_e^{\mathsf{T}} \frac{\partial \boldsymbol{K}_e}{\partial \tilde{\rho}_e} \boldsymbol{u}_e \tag{10}$$

where N_e is the number of elements in the fine mesh. Following the same approach, the sensitivities for the reduced coarse model are given as

$$\frac{\partial c}{\partial \tilde{\rho}_e} = -\boldsymbol{u}_c^{\mathsf{T}} \frac{\partial \boldsymbol{K}_c}{\partial \tilde{\rho}_e} \boldsymbol{u}_c = -\boldsymbol{u}_c^{\mathsf{T}} \frac{\boldsymbol{R}_c}{\partial \tilde{\rho}_e} \boldsymbol{K} \boldsymbol{R}_c^{\mathsf{T}} \boldsymbol{u}_c - \boldsymbol{u}_c^{\mathsf{T}} \boldsymbol{R}_c \frac{\boldsymbol{K}}{\partial \tilde{\rho}_e} \boldsymbol{R}_c^{\mathsf{T}} \boldsymbol{u}_c - \boldsymbol{u}_c^{\mathsf{T}} \boldsymbol{R}_c \boldsymbol{K} \frac{\boldsymbol{R}_c^{\mathsf{T}}}{\partial \tilde{\rho}_e} \boldsymbol{u}_c \tag{11}$$

where the matrices $\boldsymbol{K}$ and $\boldsymbol{R}_c$ are functions of the filtered densities $\tilde{\boldsymbol{\rho}}$. The gradients of all basis functions with support $\omega_i \cap K_e = \emptyset$ are zero. Therefore, the rows of $\frac{\partial \boldsymbol{R}_c}{\partial \rho_e}$ are zero everywhere except for the rows with indices corresponding to the basis functions with support $\omega_i \cap K_e \neq \emptyset$. The gradients of an eigenvector ψ can be obtained by adjoint sensitivity analysis [11]. For a large number of shape functions/eigenvectors the computational cost becomes significant, therefore, in the numerical examples presented later in the paper the optimization process is driven by approximate sensitivities obtained by neglecting the first and last term in (11). Numerical experiments, as well as analysis of the gradient expression, reveal that the convergence rate of the gradient vector is slower than the energy norm of the state problem solution. However, in-line with the observation in [2], the larger relative difference has little effect on the optimization process.

4 Numerical Examples

The first set of numerical experiments demonstrates the convergence of MsFEM for high contrast problems in linear elasticity. The test example is shown in Fig. 1. The computational domain has dimensions $1\,\mathrm{m} \times 3\,\mathrm{m}$. It represents half of the so-called MBB (Messerschmitt–Bölkow–Blohm) beam problem and therefore symmetric boundary conditions are applied to the left edge of the model as shown in Fig. 1. The elastic modulus is set to $E_{\max} = 1.0\mathrm{GPa}$ and unit force is applied to the upper left corner. The test design is obtained by filtering 0/1 black and white design. The filter radius is set to $R_f = 0.015\,\mathrm{m}$. The reference solution is obtained with a regular quadrilateral finite element mesh consisting of 320×960 elements, which corresponds to the fine mesh $\mathcal{T}_h$. Convergence of the solution obtained using the presented spectral multiscale method to the reference solution is shown in Fig. 2. The coarse discretization consists of 5×15 coarse elements for all experiments. The partition of unity is obtained using standard finite element interpolation on the internal coarse elements, linear interpolation along the element edges aligned with the boundary of the computational domain and constant value for the corner elements. It should be noted that the coarse subdomains are not aligned with the repetitive cells of the test structure. Spectral

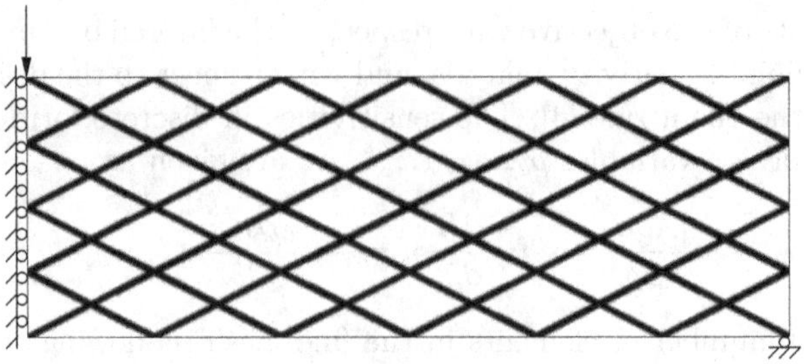

Fig. 1. Design domain and boundary conditions. The sample design is used as a base for testing the convergence of MsFEM for high contrast problems. Black corresponds to density one and white to $E_{\min}$.

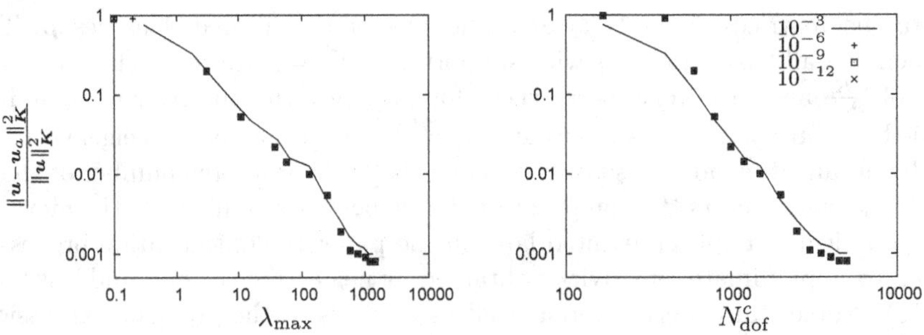

Fig. 2. MsFEM convergence rate for different contrast values $E_{\min}/E_{\max} = 10^{-3},\ldots,10^{-12}$, different thresholds $\lambda_{\max}$ (left) and different coarse dofs numbers N_{dof}^c (right).

convergence can be observed with respect to both the eigenvalue threshold, and to the number of coarse degrees of freedom. Furthermore, the convergence rate is independent on the contrast ratio $E_{\min}/E_{\max}$.

The method is tested in Matlab environment utilizing a modified version of the assembly process presented in [3]. The implementation of the MsFEM solver for the 2D test problem is slower than the implementation based on the standard Matlab sparse direct solver. The time consuming part in the MsFEM algorithm is the construction of the coarse basis. Several techniques for reducing the time spent on the small eigenvalue problems, like partial homogenization, multilevel and multigrid coarsening, have been tested and have resulted in significant reduction in both problem size and computational time. Efficient implementation requires full vectorization in Matlab or implementation in languages like Fortran/C/C++, therefore, a detailed analysis, discussion and comparison are left for future papers.

The topology optimization process is based on repetitive design updates and solutions of the state equation (3). After the first several iterations the design changes are localized which is utilized here for further reduction of the computational effort. The coarse bases are updated only for agglomerates which exhibit changes in the design. The update is triggered when a point of the local design field becomes larger or smaller than a prescribed upper $\min\{\rho+\Delta\rho, 1\}$ or lower bounds $\max\{\rho-\Delta\rho, 0\}$, with $\Delta\rho = 0.1$. For optimization steps within the limits the changes in the design are taken into account by solving an eigenvalue problem obtained by projection of (7) on reduced basis. The reduced basis for neighborhood ω_i consists of three sets of n eigenvectors. The first set is from the latest update $\boldsymbol{\psi}_i = \{\psi_l^{\omega_i}\}_{l=1\ldots n}$. The second set $\boldsymbol{\psi}_d$ is orthogonal to $\boldsymbol{\psi}_i$ and obtained from (7) written for the dilated bound, and the third set $\boldsymbol{\psi}_e$ is orthogonal to the previous two sets, and obtained by (7) written for the eroded bound. This modification results in significant speed-up which for the 2D optimization example presented here is faster than a standard implementation with direct sparse solver [3].

An optimized topology obtained after 500 iterations using the above algorithm is shown in Fig. 3. The initial design is uniform and the volume constraint is set to 50 % of the total volume. The design is updated using the method of moving asymptotes (MMA) [12]. In order to avoid early convergence to a local minimum, the penalization parameter p in (9) is increased every 20 iterations during the first 300 iterations, from one to its maximum $p = 3$. The number of the coarse degrees of freedom varies from 4000 to 6000 during the optimization, which corresponds to an average of 71 to 108 dofs per coarse node. The maximal eigenvalue is set to be smaller than 800. Reducing the size of the basis (decreasing $\lambda_{\max}$) results in a design with disconnected bars, isolated material islands and checkerboard-like patterns, due to the stiffness averaging of the coarse shape functions and the lack of uniqueness of the associated microstructure. The reduction of the solution basis as well as the utilization of inexact objective gradients, provides additional regularization to the optimization problem. The effect is pronounced in the early stages when the initial design is formed and the basis consists mainly of smooth solutions of the eigenvalue problem. For topologies with good contrast the reduction does not have any significant influence on the optimization process. The regularization effect can be removed by utilizing the coarse solver as a preconditioner. Preconditioned GMRES applied to (3) with multigrid like preconditioner and one post-smoothing Gauss-Seidel step converges with relative error 10^{-6} in 12 to 15 iterations (independent of the contrast) with negligible increase of the computational time.

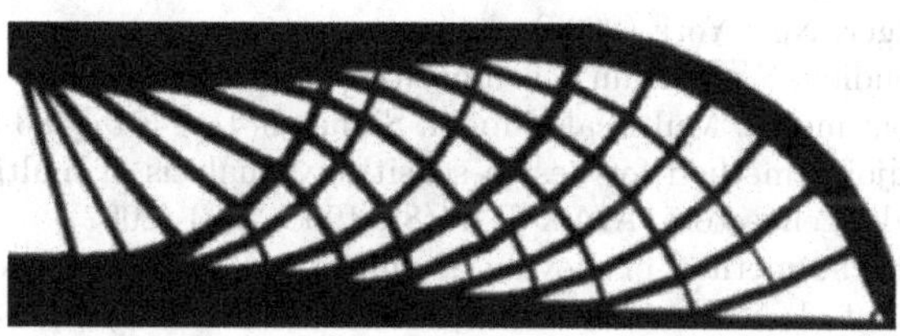

Fig. 3. Optimized MBB design with compliance $c = 197.633$ and mesh size 160×480.

5 Conclusions

MsFEM with local spectral basis provides a good alternative to the complex re-meshing approaches used for reducing the computational burden in topology optimization. It ensures good contrast independent convergence on a fixed mesh. For large 3D examples the method is perfectly scalable as it relies on the solution of a set of completely independent coarse grained tasks and it will excel in realistic industrial problems. Reduction of the basis update frequency is a key for decreasing the computational time. The multiscale coarse bases, in combination with iterative solvers, provide the most robust and computationally effective approach for topology optimization. A detailed study will be the subject of future research.

References

1. Aage, N., Lazarov, B.: Parallel framework for topology optimization using the method of moving asymptotes. Struct. Multi. Optim. **47**(4), 493–505 (2013)
2. Amir, O., Sigmund, O., Lazarov, B.S., Schevenels, M.: Efficient reanalysis techniques for robust topology optimization. Comput. Meth. Appl. Mech. Eng. **245–246**, 217–231 (2012)
3. Andreassen, E., Clausen, A., Schevenels, M., Lazarov, B.S., Sigmund, O.: Efficient topology optimization in matlab using 88 lines of code. Struct. Multi. Optim. **43**, 1–16 (2011)
4. Bendsoe, M.P., Sigmund, O.: Topology Optimization - Theory, Methods and Applications. Springer, Heidelberg (2003)
5. Buck, M., Iliev, O., Andrä, H.: Multiscale finite element coarse spaces for the application to linear elasticity. Cent. Eur. J. Math. **11**(4), 680–701 (2013)
6. Efendiev, Y., Galvis, J., Lazarov, R., Willems, J.: Robust domain decomposition preconditioners for abstract symmetric positive definite bilinear forms. ESAIM: Math. Model. Numer. Anal. **46**, 1175–1199 (2012)
7. Efendiev, Y., Galvis, J., Vassilevski, P.: Spectral element agglomerate algebraic multigrid methods for elliptic problems with high-contrast coefficients. In: Huang, Y., Kornhuber, R., Widlund, O., Xu, J. (eds.) Domain Decomposition Methods in Science and Engineering XIX. Lecture Notes in Computational Science and Engineering, vol. 78, pp. 407–414. Springer, Heidelberg (2011)
8. Efendiev, Y., Galvis, J., Wu, X.H.: Multiscale finite element methods for high-contrast problems using local spectral basis functions. J. Comput. Phys. **230**(4), 937–955 (2011)
9. Efendiev, Y., Hou, T.Y.: Multiscale Finite Element Methods: Theory and Applications. Springer, New York (2009)
10. Galvis, J., Efendiev, Y.: Domain decomposition preconditioners for multiscale flows in high-contrast media. Multiscale Model. Simul. **8**(4), 1461–1483 (2010)
11. Lee, T.H.: Adjoint method for design sensitivity analysis of multiple eigenvalues and associated eigenvectors. AIAA J. **45**(8), 1998–2004 (2007)
12. Svanberg, K.: The method of moving asymptotes - a new method for structural optimization. Int. J. Numer. Meth. Eng. **24**, 359–373 (1987)

Numerical Homogenization of Heterogeneous Anisotropic Linear Elastic Materials

S. Margenov, S. Stoykov, and Y. Vutov(✉)

Institute of Information and Communication Technologies,
Bulgarian Academy of Sciences, Sofia, Bulgaria
yavor@parallel.bas.bg

Abstract. The numerical homogenization of anisotropic linear elastic materials with strongly heterogeneous microstructure is studied. The developed algorithm is applied to the case of trabecular bone tissue. In our previous work [1], the orthotropic case was considered. The homogenized anisotropic tensor is transformed according to the principle directions of anisotropy (PDA). This provides opportunities for better interpretation of the results as well as for classification of the material properties.

The upscaling procedure is described in terms of six auxiliary elastic problems for the reference volume element (RVE). Rotated trilinear Rannacher-Turek finite elements are used for discretization of the involved subproblems. A parallel PCG method is implemented for efficient solution of the arising large-scale systems with sparse, symmetric, and positive semidefinite matrices. Then, the bulk modulus tensor is computed from the upscaled stiffness tensor and its eigenvectors are used to define the transformation matrix. The stiffness tensor of the material is transformed with respect to the PDA which gives a canonical (unique) representation of the material properties.

Numerical experiments for two different RVEs from the trabecular part of human bones are presented.

1 Introduction

Many materials, including the human bone have a complex microstructure. In recent years micro computed tomography (μCT) and micro finite element method (μFEM) analysis proved to be a valuable tool for analyzing bone properties, see e.g. [2]. The macro level material properties strongly depend on their microstructure. Nevertheless, the overall mechanical responses can be described using multilevel techniques that are built upon basic conservation principles at the micro level.

In our previous work [1], we studied a numerical homogenization algorithm for computing the upscaled orthotropic stiffness tensor. This approach is further developed to the general case of anisotropic materials. Here we obtain an effective stiffness tensors of a reference volume element (RVE).

I. Lirkov et al. (Eds.): LSSC 2013, LNCS 8353, pp. 347–354, 2014.
DOI: 10.1007/978-3-662-43880-0_39,

The trabecular bone is a strongly heterogeneous composition of solid and fluid phases. Its voxel representation obtained from μCT images is used to formulate the problem. Our goal is to obtain upscaled material properties of trabecular bone tissue. In this work, only the mechanical response of the solid phase is considered. To this purpose a fictitious domain approach is used.

This paper is organized as follows. The applied numerical homogenization scheme is described in Sect. 2. In Sect. 3 transformation to the principal directions of anisotropy (PDA) is recalled. And finally the upscaled and transformed tensors are presented and discussed in the last section.

2 Homogenization Technique

Let Ω be a parallelepipedal domain representing our reference volume element (RVE) and $\mathbf{u} = (u_1, u_2, u_3)$ be the displacements vector in Ω. Here, components of the small strain tensor [3] are:

$$\varepsilon_{ij}(\mathbf{u}(\mathbf{x})) = \frac{1}{2}\left(\frac{\partial u_i(\mathbf{x})}{\partial x_j} + \frac{\partial u_j(\mathbf{x})}{\partial x_i}\right) \tag{1}$$

We assume that Hooke's law holds. The stress tensor σ is expressed in the form

$$\sigma_{ij} = s_{ijkl}\varepsilon_{kl}, \tag{2}$$

where summation over repeating indexes is assumed. The forth-order tensor s is called the stiffness tensor, and has the following symmetry [4]:

$$s_{ijkl} = s_{jikl} = s_{ijlk} = s_{klij}. \tag{3}$$

Often, the Hooke's law is written in matrix form:

$$\begin{bmatrix} \sigma_{11} \\ \sigma_{22} \\ \sigma_{33} \\ \sigma_{23} \\ \sigma_{13} \\ \sigma_{12} \end{bmatrix} = \begin{bmatrix} s_{1111} & s_{1122} & s_{1133} & s_{1123} & s_{1113} & s_{1112} \\ s_{2211} & s_{2222} & s_{2233} & s_{2223} & s_{2213} & s_{2212} \\ s_{3311} & s_{3322} & s_{3333} & s_{3323} & s_{3313} & s_{3312} \\ s_{2311} & s_{2322} & s_{2333} & s_{2323} & s_{2313} & s_{2312} \\ s_{1311} & s_{1322} & s_{1333} & s_{1323} & s_{1313} & s_{1312} \\ s_{1211} & s_{1222} & s_{1233} & s_{1223} & s_{1213} & s_{1212} \end{bmatrix} \begin{bmatrix} \varepsilon_{11} \\ \varepsilon_{22} \\ \varepsilon_{33} \\ 2\varepsilon_{23} \\ 2\varepsilon_{13} \\ 2\varepsilon_{12} \end{bmatrix}. \tag{4}$$

The symmetric 6×6 matrix in (4) is denoted with S and. is called also the stiffness matrix. For an isotropic material matrix S, and the tensor s have only two independent degrees of freedom. For orthotropic materials (materials containing three orthogonal planes of symmetry), the matrix S has nine independent degrees of freedom. In the general anisotropic case, S has 21 independent degrees of freedom [5].

The goal of our study is to obtain homogenized material properties of the trabecular bone tissue. In other words – to find the stiffness tensor of a homogeneous material which would have the same macro-level properties as our RVE. Our approach follows the numerical upscaling method from [1] (see also [6,7]).

The homogenization scheme requires finding Ω-periodic functions $\xi^{kl} = (\xi_1^{kl}, \xi_2^{kl}, \xi_3^{kl})$, $k, l = 1,2,3$, satisfying the following equation in a week formulation:

$$\int_\Omega \left(s_{ijpq}(x) \frac{\partial \xi_p^{kl}}{\partial x_q} \right) \frac{\partial \phi_i}{\partial x_j} d\Omega = \int_\Omega s_{ijkl}(x) \frac{\partial \phi_i}{\partial x_j} d\Omega, \tag{5}$$

for an arbitrary Ω-periodic variational function $\phi \in H^1(\Omega)$. After computing the characteristic displacements ξ^{kl}, from (5) we can compute the homogenized elasticity tensor s^H using the following formula:

$$s_{ijkl}^H = \frac{1}{|\Omega|} \int_\Omega \left(s_{ijkl}(x) - s_{ijpq}(x) \frac{\partial \xi_p^{kl}}{\partial x_q} \right) d\Omega. \tag{6}$$

From (5) and due to the symmetry of the stiffness tensor (3), we have the relation $\xi^{kl} = \xi^{lk}$. Therefore the solution of only six problems (5) is required to obtain the homogenized stiffness tensor.

The periodicity of the solution implies the use of periodic boundary conditions. Rotated trilinear (Rannacher-Turek) finite elements [8] are used for the numerical solution of (5). This choice is motivated by the additional stability of the nonconforming finite element discretization in the case of strongly heterogeneous materials [9]. Construction of a robust non-conforming finite element method is generally based on application of mixed formulation leading to a saddle-point system. By the choice of non continuous finite elements for the dual (pressure) variable, it can be eliminated at the (macro)element level. As a result we obtain a symmetric positive semi-definite finite element system in primal (displacements) variables. We utilize this approach, which is referred as the *reduced and selective integration* (RSI) [10].

For the solution of the arising linear system, the preconditioned conjugate gradient is used. For the construction of the preconditioner the isotropic variant of the displacement decomposition (DD)[11] was used. We write the DD auxiliary matrix in the form

$$C_{DD} = \begin{bmatrix} A & & \\ & A & \\ & & A \end{bmatrix} \tag{7}$$

where A is the stiffness matrix corresponding to the bilinear form

$$a(u^h, v^h) = \sum_{e \in \Omega^h} \int_e E \left(\sum_{i=1}^{3} \frac{\partial u^h}{\partial x_i} \frac{\partial v^h}{\partial x_i} \right) de. \tag{8}$$

Such approach is motivated by the second Korn's inequality, which holds for the RSI FEM discretization under consideration. More precisely, in the case of isotropic materials, the estimate

$$\kappa(C_{DD}^{-1} K) = O((1 - 2\nu)^{-1})$$

holds uniformly with respect to the mesh size parameter in the FEM discretization, where ν is the Poisson ratio.

As the arising linear systems are large, the problems are solved in parallel. Parallel MIC(0) preconditioner for scalar elliptic systems [12] is used to approximate (7). Its basic idea is to apply MIC(0) factorization of an approximation B of the stiffness matrix A. Matrix B has a special block structure. Its diagonal blocks are diagonal matrices. This allows the solution of the preconditioning system to be performed in parallel. The condition number estimate $\kappa(B^{-1}A) \leq 3$ holds uniformly with respect to mesh parameter and possible coefficient jumps (see for the related analysis in [12]). This technique is applied three times – once for each diagonal block of (7). Thus we obtain the parallel MIC(0) preconditioner in the form:

$$C_{DDMIC(0)} = \begin{bmatrix} C_{MIC(0)}(B) & & \\ & C_{MIC(0)}(B) & \\ & & C_{MIC(0)}(B) \end{bmatrix}.$$

More details on applying this preconditioner for the proposed homogenization technique can be found in [1].

3 Principal Directions of Anisotropy

We follow the procedure for determining the PDA described in [13]. A coordinate system is said to coincide with the PDA of a material, when the material, subjected to "all-around uniform pure extension state," forms a "pure tension state."

Let us introduce the bulk modulus tensor

$$K = \begin{bmatrix} K_{11} & K_{12} & K_{13} \\ K_{21} & K_{22} & K_{23} \\ K_{31} & K_{32} & K_{33} \end{bmatrix}. \tag{9}$$

The elements of K are defined as

$$K_{ij} = \sum_{k=1}^{3} s_{ijkk} \tag{10}$$

We write the "all-round uniform extension" as $\varepsilon_{ij} = \tilde{\varepsilon}\delta_{ij}$, where $\tilde{\varepsilon}$ is a constant reference strain and δ_{ij} is the Kronecker delta. Then, the stress components are

$$\sigma_{ij} = K_{ij}\tilde{\varepsilon}. \tag{11}$$

Hence the principal directions of the tensor K coincide with the stress principal directions. The stress values in these principal directions are

$$\sigma_{ij} = \lambda_i\tilde{\varepsilon}\delta_{ij}, \tag{12}$$

where λ_i are the eigenvalues of the tensor K. To ensure uniqueness of the transformation, we order the eigenvalues $\lambda_3 \geq \lambda_2 \geq \lambda_1$, i.e. the biggest eigenvalue is the third and the smallest is the first. With this order, we enforce the material to orient its strongest direction in z axis and its weakest in x. The case of equal eigenvalues, leads to equivalence of the material in two or more directions. The transformation matrix T, which rotates the coordinate system to the one which coincides with the PDA, is given by the corresponding normalized eigenvectors v^i of K:

$$T = \begin{bmatrix} v_1^1 & v_2^1 & v_3^1 \\ v_1^2 & v_2^2 & v_3^2 \\ v_1^3 & v_2^3 & v_3^3 \end{bmatrix}. \qquad (13)$$

Now we are able to rotate the stress tensor using formula

$$\bar{s}_{klst} = s_{mnpr} T_{km} T_{ln} T_{sp} T_{tr}. \qquad (14)$$

Here summation over repeating indexes is assumed.

4 Numerical Experiments

To solve the above described upscaling problem, a portable parallel FEM code is designed and implemented in C++. The parallelization has been facilitated using the MPI library [14].

The analyzed test specimens are parts of trabecular bone tissue extracted from a high resolution computer tomography image [15]. The voxel size is 37 μm. The trabecular bone has a strongly expressed heterogeneous microstructure composed of solid and fluid phases.

Homogenized properties of two different RVEs with sizes of $128 \times 128 \times 128$ are shown, see Fig. 1. The RVEs are different, but part of the same vertebra. The Young modulus and the Poisson ratio of the solid phase, taken from [16], are $E^s = 14.7$GPa and $\nu^s = 0.325$. Our intention is to obtain the homogenized elasticity tensor of the RVE, taking into account the elastic response of the solid phase only. We interpret the fluid phase as a fictitious domain. Thus, we set Young modulus $E^f = \zeta E^S$ for the voxels corresponding to the fluid phase. The parameter ζ is set to 10^{-5}. The choice of ζ is studied in [1]. We also set $\nu^f = \nu^s$. The chosen values of E^f and ν^f practically do not influence the homogenization result.

The iteration stopping criterion is $||\mathbf{r}^j||_{C^{-1}}/||\mathbf{r}^0||_{C^{-1}} < 10^{-6}$, where $\mathbf{r}^j$ is the residual at the j-th iteration step of the preconditioned conjugate gradient method and C stands for the used preconditioner.

Numerical experiments are performed on a Blue Gene/P machine. It is a massively parallel computer consisting of quad-core computing nodes. The PowerPC based low power processors run at 850 MHz. Each node has 2 GB of RAM. The nodes are interconnected with several specialized high speed networks—3D mesh network for peer to peer communications and tree network for collective communications, among others.

The computations were performed on 256 processors. The computations took between 4 and 5 h for each of the auxiliary problems. This has notable increase

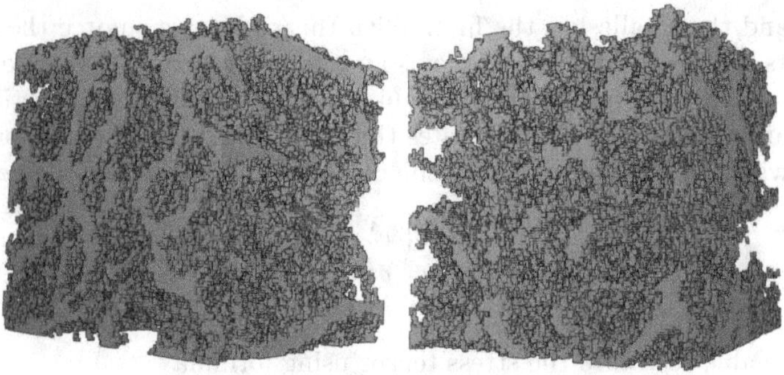

Fig. 1. Structure of the two RVEs

from the case where a truly periodic media is considered [1]. In that case the number of iterations (and thus the compute time) for similar problems was around six times less.

The computed homogenized stiffness matrix for the first specimen is

$$S_1^H = \begin{bmatrix} 802 & 218 & 212 & -11.7 & -1.31 & 72.8 \\ 218 & 566 & 167 & -16.2 & 0.25 & 48.5 \\ 212 & 167 & 133 & -71.4 & 31.8 & 22.8 \\ -11.7 & -16.2 & -71.4 & 206 & 31.7 & 2.91 \\ -1.31 & 0.25 & 31.8 & 31.7 & 313 & -9.11 \\ 72.8 & 48.5 & 22.8 & 2.91 & -9.11 & 197 \end{bmatrix}, \qquad (15)$$

and for the second one —

$$S_2^H = \begin{bmatrix} 372 & 127 & 74.6 & -4.75 & 4.01 & -20.0 \\ 127 & 436 & 81.3 & -9.03 & 3.60 & -16.2 \\ 74.6 & 81.3 & 606 & -44.5 & 27.8 & -11.8 \\ -4.75 & -9.03 & -44.5 & 98.2 & -21.0 & 1.68 \\ 4.01 & 3.60 & 27.8 & -21.0 & 100 & -6.17 \\ -20.0 & -16.2 & -11.8 & 1.68 & -6.17 & 120 \end{bmatrix}. \qquad (16)$$

All values are measured in megapascals (MPa). The transformation procedure, described in Sect. 3, is applied to the stiffness matrices S_1^H and S_2^H. As a result, the stiffness matrices $\bar{S}_1^H$ and $\bar{S}_2^H$, characterizing the properties of the considered RVEs in the coordinate systems aligned with their PDA are obtained:

$$\bar{S}_1^H = \begin{bmatrix} 501 & 221 & 154 & 8.36 & -14.0 & -17.8 \\ 221 & 847 & 224 & 8.41 & -6.89 & 18.1 \\ 154 & 224 & 1340 & -16.7 & 20.9 & -0.28 \\ 8.36 & 8.41 & -16.7 & 320 & 14.2 & 9.20 \\ -14.0 & -6.89 & 20.9 & 14.2 & 196 & 1.64 \\ -17.8 & 18.1 & -0.28 & 9.20 & 1.64 & 204 \end{bmatrix}, \qquad (17)$$

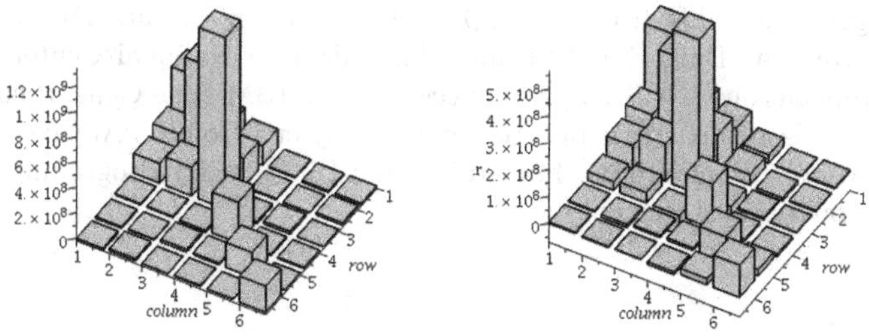

First specimen   Second specimen

Fig. 2. Structure of the two transformed stiffness matrices

$$\bar{S}_2^H = \begin{bmatrix} 343 & 121 & 848 & 31.7 & 3.75 & 5.45 \\ 121 & 372 & 139 & 38.8 & -10.7 & -3.90 \\ 84.8 & 139 & 573 & -70.6 & 7.02 & -1.54 \\ 31.7 & 38.8 & -70.6 & 165 & 8.17 & -9.42 \\ 3.75 & -10.7 & 7.02 & 8.17 & 97.3 & 22.5 \\ 5.45 & -3.90 & -1.54 & -9.42 & 22.5 & 119 \end{bmatrix}. \tag{18}$$

The matrices $\bar{S}_1^H$ and $\bar{S}_2^H$ are visualized on Fig. 2. The degree of anisotropy η can be defined as the ratio

$$\eta = \bar{s}_{3333}/\bar{s}_{1111}. \tag{19}$$

The degrees of anisotropy for the two RVEs η_1 and η_2 are 2.67 and 1.67. One can see that although part of the same vertebra, the two specimens have different degrees of anisotropy and different magnitudes of the elastic moduli. This demonstrates the importance of the material microstructure for the elastic response.

It is well known, that the trabecular bone tissue adapts to the stresses it experiences (a fact referred to as a Wolffs law) [17]. In agreement with this, the presented homogenized stiffness tensors show considerable level of anisotropy. Our results evidently confirm that the anisotropy cant be neglected in the simulations. As a next step in this study, the analysis of a representative set of CT images is needed to provide data for correlation analysis of the homogenized stiffness tensors. Then, the map of principle directions of the experienced loads for a particular bone at the organ level will provide new opportunities for more realistic patient specific simulations using the clinically available information for the bone density.

In this context, let us remind that the presented results use very high resolution X-ray CT scans. Due to the level of radiation intensity, such a full length organ-level scanning is not applicable in-vivo. In this sense, the more standard multiscale approach is not applicable due to the lack of data.

In addition, the fluid phase of the bone plays an important part in its elastic response. One possible approximation of this two-phase system is to interpret the fluid as an almost incompressible elastic material (see, e.g., [18]). One important future goal is to verify the related results in a comparison with some more general poroelastic (say Biot) models.

Acknowledgments. This work is supported in part by Grants DFNI I01/5 and DCVP-02/1 from the Bulgarian NSF and the Bulgarian National Center for Supercomputing Applications (NCSA), giving access to the IBM Blue Gene/P computer.

The research is also partly supported by the project AComIn "Advanced Computing for Innovation", grant 316087, funded by the FP7 Capacity Programme (Research Potential of Convergence Regions)

References

1. Margenov, S., Vutov, Y.: Parallel MIC(0) preconditioning for numerical upscaling of anisotropic linear elastic materials. In: Lirkov, I., Margenov, S., Waśniewski, J. (eds.) LSSC 2009. LNCS, vol. 5910, pp. 805–812. Springer, Heidelberg (2010)
2. Wirth, A.J., Mueller, T.L., Vereecken, W., Flaig, C., Arbenz, P., Mller, R., van Lenthe, G.H.: Mechanical competence of bone-implant systems can accurately be determined by image-based micro-finite element analyses. Arch. Appl. Mech. **80**(5), 513–525 (2010)
3. Fung, Y.C.: Foundations of Solid Mechanics. Prentice-Hall, Englewood Cliffs (1965)
4. Nayfeh, A., Pai, P.: Linear and Nonlinear Structural Mechanics. Wiley, New York (2004)
5. Sokolonikoff, I.: Mathematical Theory of Elasticity. Mc-Graw-Hill, New York (1956)
6. Hoppe, R.H.W., Petrova, S.I.: Optimal shape design in biomimetics based on homogenization and adaptivity. Math. Comput. Simul. **65**(3), 257–272 (2004)
7. Bensoussan, A., Lions, J.L., Papanicolaou, G.: Asymptotic Analysis for Periodic Structures. Elsevier, Amsterdam (1978)
8. Rannacher, R., Turek, S.: Simple nonconforming quadrilateral Stokes element. Numer. Meth. Partial Differ. Equ. **8**(2), 97–112 (1992)
9. Arnold, D.N., Brezzi, F.: Mixed and nonconforming finite element methods: Implementation, postprocessing and error estimates. RAIRO. Model. Math. Anal. Numer. **19**, 7–32 (1985)
10. Malkus, D., Hughes, T.: Mixed finite element methods – reduced and selective integration techniques: an uniform concepts. CMAME **15**, 63–81 (1978)
11. Blaheta, R.: Displacement decomposition-incomplete factorization preconditioning techniques for linear elasticity problems. NLAA **1**(2), 107–128 (1994)
12. Arbenz, P., Margenov, S., Vutov, Y.: Parallel MIC(0) preconditioning of 3D elliptic problems discretized by Rannacher-Turek finite elements. Comput. Math. Appl. **55**(10), 2197–2211 (2008)
13. Rand, O., Rovenski, V.: Analytical Methods in Anisotropic Elasticity: With Symbolic Computational Tools. Birkhauser, Boston (2004)
14. Walker, D., Dongarra, J.: MPI: a standard Message Passing Interface. Supercomputer **63**, 56–68 (1996)
15. Beller, G., Burkhart, M., Felsenberg, D., Gowin, W., Hege, H.-C., Koller, B., Prohaska, S., Saparin, P.I., Thomsen, J.S.: Vertebral body data set esa29-99-l3. http://bone3d.zib.de/data/2005/ESA29-99-L3/
16. Cowin, S.: Bone poroelasticity. J Biomech. **32**, 217–238 (1999)
17. Wolff, J.: The Law of Bone Remodeling. Springer, Heidelberg (1986)
18. Kosturski, N., Margenov, S.: Numerical homogenization of bone microstructure. In: Lirkov, I., Margenov, S., Waśniewski, J. (eds.) LSSC 2009. LNCS, vol. 5910, pp. 140–147. Springer, Heidelberg (2010)

How to Make a Domain Decomposition Method More Robust

Nicole Spillane(✉)

Laboratoire Jacques Louis Lions,
UPMC, 4 Place Jussieu, 75005 Paris, France
spillane@ann.jussieu.fr
http://www.ljll.math.upmc.fr/~spillane/

Abstract. The simulation of processes in highly heterogeneous media comes with many challenges. In particular many domain decomposition methods do not perform well in this case, specially if the decomposition into subdomains does not accommodate the coefficient variations. For three popular domain decomposition methods (two level additive Schwarz, BDD and FETI) we have proposed a remedy to this problem in previous work with coauthors. Here we present the strategy which was used by applying it to the Hybrid Schwarz preconditioner. It is based on identifying a bottleneck estimate in the proof of convergence which cannot be satisfied for the entire solution space. Then the part of the solution which is problematic is isolated *via* a generalized eigenvalue problem and solved separately.

Keywords: Domain decomposition · Robustness · Heterogeneous problems · Hybrid Schwarz

1 Introduction

The method presented here is a different application of a strategy devised in [7,8], in collaboration with Victorita Dolean, Patrice Hauret, Frédéric Nataf, Clemens Pechstein and Robert Scheichl, and generalized in [3] with Daniel J. Rixen. It is also closely related to the work of [6] and many references therein.

Our objective is to develop black box domain decomposition methods which are scalable and robust even for hard problems. The hybrid Schwarz method is a preconditioner for the conjugate gradient (CG) iterative solver. Within each application of the preconditioner smaller problems are solved using direct solvers on each subdomain and on an additional space, V_0, which is shared between all subdomains. Our strategy is to take full advantage of the hybrid (iterative/direct) framework: the solution space is divided into a space where the preconditioner does a good job and a space where it does not give such a good approximation. On the first we apply the CG (iterative) solver and theoretical results guarantee fast convergence. On the second, CG will not perform well, and so we set V_0 to be this space. Since a direct solve is applied in V_0 robustness

I. Lirkov et al. (Eds.): LSSC 2013, LNCS 8353, pp. 355–362, 2014.
DOI: 10.1007/978-3-662-43880-0_40,

will not be an issue. Of course it remains to define the meaning of the word 'good'. We do this by using a generalized eigenvalue problem [5] which is built just from the element matrices and connectivity graph. For this reason the range of problems that we can solve is very wide. The requirements are stated in the next subsection. In the final section we give an illustration for linear elasticity.

2 Two Level Schwarz Method with Projection (also Known as Hybrid Schwarz)

2.1 One Level Schwarz Method

Maybe the most straightforward of the domain decomposition method is Additive Schwarz (see [4] and references therein). The information needed to build the additive Schwarz preconditioner is the following:

- A set $\omega = \{1, \dots, n\}$ of degrees of freedom,
- A set of symmetric positive semi-definite element matrices $\{A_\tau \in \mathbb{R}^{n\times n}; \tau \in \mathcal{T}_h\}$, which give the weights of the connections between degrees of freedom,
- The connectivity graph for each connection $\tau \in \mathcal{T}_h$ which is the list $dof(\tau) \subset \omega$ of degrees of freedom which are connected to others through τ.

If the problem stems from the finite element approximation of a partial differential equation, these have geometrical interpretations: $\mathcal{T}_h$ is the mesh of the global domain, τ is an element of this mesh and $dof(\tau)$ is the set of degrees of freedom attached to τ, *e.g.* its vertices in the case of $\mathbb{P}_1$ Lagrange finite elements.

The global problem matrix is assembled as: $A := \sum_{\tau\in\mathcal{T}_h} A_\tau$. We assume that A is symmetric positive definite (spd). Then, given a right hand side $f \in \mathbb{R}^n$ the objective is to solve:

$$\text{Find } x^* \in \mathbb{R}^n \text{ such that } Ax^* = f. \tag{1}$$

The idea behind the Additive Schwarz preconditioner is to approximate the global inverse of A by a sum of local inverse A_j^{-1}. The local inverses are based on an overlapping partition of the set of degrees of freedom ω:

$$\omega = \omega_1 \cup \dots \cup \omega_N,$$

and on the corresponding interpolation matrices between global unknowns and local unknowns: for any $j = 1, \dots, N$ let n_j be the cardinality of ω_j, then the restriction matrix $R_j \in \mathbb{R}^{n_j\times n}$ is the Boolean matrix with one 1 entry on each line which corresponds to a degree of freedom in ω_j. With this the one level Schwarz preconditioner writes:

$$M^{-1} := \sum_{j=1}^{N} R_j^\top A_j^{-1} R_j, \quad A_j := R_j A R_j^\top. \tag{2}$$

The matrices A_j are built by extracting the coefficients in the global matrix A which correspond to degrees of freedom in ω_j so they are also symmetric positive definite. Unfortunately, the conjugate gradient algorithm applied to (1) and preconditioned with M^{-1} is usually not scalable and lacks robustness when simulating many real life problems such as phenomena in heterogeneous media. One way to improve this is to use a projected operator.

2.2 Adding Projection Steps

In [1] the idea to use a projection as a preconditioner for a Krylov method is introduced. Having chosen a set V_0 of vectors in $\mathbb{R}^n$ which are spanned by the n_0 (linearly independent) columns of $R_0^\top \in \mathbb{R}^{n \times \# V_0}$, the A-orthogonal projection operator onto V_0 is

$$P_0 := R_0^\top (R_0 A R_0^\top)^{-1} R_0 A. \tag{3}$$

With this the original problem (1) rewrites as two independent problems: Find x_1^* $(= (I - P_0)x^*)$ and x_2^* $(= P_0 x^*)$ such that

$$Ax_1^* = (I - P_0^\top)f, \text{and } Ax_2^* = P_0^\top f. \tag{4}$$

The number of vectors in V_0 is supposed to be sufficiently small so that the projected part of the solution, x_2^* can be obtained by computing the inverse of $R_0 A R_0^\top$ with a direct solver. Conjugate gradient iterations are then used only for the first equation in (4). The rationale behind this splitting of the solution is that even if A is ill-conditioned, in many cases the ill-conditioning is caused only by a small number of vectors. As long as these vectors can be identified the projection framework allows to deal with them with a guaranteed success (thanks to the direct solver) and reserve the iterative solver for the *'easier'* part of the solution on which it will perform efficiently.

In our case we use both preconditioning and projection. With M^{-1} as in (2), the projected and preconditioned problem is: Find $x_1^* \in \text{range}(I - P_0)$ such that

$$M^{-1} A x_1^* = M^{-1}(I - P_0^\top) f. \tag{5}$$

The problem is now to find an estimate for the convergence rate of the projected preconditioned conjugate gradient algorithm applied to (5). A well known result [2, Theorem 6.29] is that the convergence rate of a conjugate gradient method depends only on the condition number of the operator at stake ($M^{-1}A$ restricted to range $(I - P_0)$ in this case). For this reason our objective is the following.

> Identify a space V_0 which is sufficiently small for $R_0 A R_0^\top$ to be inverted using a direct solver and such that the condition number of $M^{-1}A$ on range $(I - P_0)$ does not depend on the number of subdomains (scalability) or on any of the parameters in the original set of equations (robustness).

Intuitively, the 'troublesome' vectors which we are looking for are the parts of the solution space where the preconditioner does not do a good job ($M^{-1}Ax$ is very different from x).

3 Choosing the Projection Space

3.1 Abstract Schwarz Theory

Fortunately the additive Schwarz preconditioner has already been thoroughly analyzed and an abstract presentation of this theory can be found in [4] (see Theorem 2.13 which relies on Assumption 2.12). We rely on this analysis.

Preconditioner M^{-1} is invertible (with its inverse denoted by M) and $M^{-1}A$ is self adjoint with respect to the inner product induced by M and thus has real-valued eigenvalues whose extrema can be characterised by Rayleigh quotients: if there exist constants C_1 and C_2 such that

$$C_1\langle Mx, x\rangle \leq \langle Ax, x\rangle \leq C_2\langle Mx, x\rangle,\ \forall x \in \text{range}(I - P_0) \tag{6}$$

then $\lambda_{max} \leq C_2$, $\lambda_{min} \geq C_1$ and the condition number of $M^{-1}A$ restricted to range $(I - P_0)$ is bounded by C_2/C_1. The constants measure, on range $(I - P_0)$, the difference between the energy norm with respect to the original operator A and the energy norm with respect to its approximation M.

In practice we never need to compute the inverse M of the preconditioner. In our analysis we will use the expression given in Lemma 2.5 of [4][1]:

$$\langle Mx, x\rangle = \min_{\{x_j \in \mathbb{R}^{n_j};\, x=\sum_{j=1}^N R_j^\top x_j\}} \sum_{j=1}^{N} \langle A_j x_j, x_j\rangle. \tag{7}$$

The energy norm of x with respect to M minimizes the sum, over all possible decompositions of x onto the subdomains, of the local energies.

Next we recall the proof of an upper bound for λ_{max} [4, Lemma 2.10 and Theorem 2.13]. It depends on the maximal number $\mathcal{N}^c$ of colors that are needed to color each of the sets ω_j in such a way that two subsets with the same color are A-orthogonal. More precisely let $color(j) \in \{1, \dots, \mathcal{N}^c\}$ denote the color of a subdomain j then

$$\langle AR_k^\top u_k, R_l^\top u_l\rangle = 0,\ \forall u_k \in \omega_k \text{ and } u_l \in \omega_l \text{ if } color(k) = color(l).$$

Given the decomposition $x = \sum_{j=1}^N R_j^\top x_j$ which realizes the minimum in (7) the proof from [4] reads

$$\begin{aligned}\langle Mx, x\rangle &= \textstyle\sum_{j=1}^N \langle AR_j^\top x_j, R_j^\top x_j\rangle = \sum_{c=1}^{\mathcal{N}^c} \langle A \sum\limits_{\{i; color(i)=c\}} R_i^\top x_i, \sum\limits_{\{i; color(i)=c\}} R_i^\top x_i\rangle \\ &\geq \textstyle\frac{1}{\mathcal{N}^c} \langle A \sum_{j=1}^N R_j^\top x_j, \sum_{j=1}^N R_j^\top x_j\rangle = \frac{1}{\mathcal{N}^c}\langle Ax, x\rangle.\end{aligned} \tag{8}$$

The argument for the second last inequality is Cauchy's inequality in $\mathbb{R}^{\mathcal{N}_c}$. This proves that $\lambda_{max} \leq \mathcal{N}_c$. We don't need to work to improve this estimate because it already does not depend on the number of subdomains and it holds independently of the choice of the projection space.

[1] In the book $P_{ad} = M^{-1}A$ so $AP_{ad}^{-1} = M$.

3.2 Identifying the Bottleneck

Deriving a bound for λ_{min} is a trickier job. We proceed by beginning to write the proof of the convergence theorem from [4, Sect. 2.5.2] for a constant C to be specified later on. We do not go to the end of the proof. Instead, we exhibit a sufficient condition for the bound on λ_{min} to be true which has the nice feature of being local. Finally the projection space is built by identifying (*via* generalized eigenvalue problems) the parts of the solution space that do not satisfy the aforementioned condition.

Let $x \in \text{range}(I - P_0)$, lets reformulate the lower bound in (6) using (7):

$$\begin{aligned}
&\langle Ax, x\rangle \geq C\langle Mx, x\rangle \\
&\Leftrightarrow \langle Ax, x\rangle \geq C \min_{\{x_j \in \mathbb{R}^{n_j}; x=\sum_{j=1}^N R_j^\top x_j\}} \textstyle\sum_{j=1}^N \langle A_j x_j, x_j\rangle \\
&\Leftrightarrow \langle Ax, x\rangle \geq C \min_{\{x_j \in \mathbb{R}^{n_j}; x=\sum_{j=1}^N R_j^\top x_j\}} \textstyle\sum_{j=1}^N \langle A(I-P_0)R_j^\top x_j, (I-P_0)R_j^\top x_j\rangle.
\end{aligned} \tag{9}$$

The last equivalence is too long to prove here. The idea is to look at the projected Additive Schwarz preconditioner as the textbook additive Schwarz preconditioner for the projected operator $(I - P_0)^\top A(I - P_0)$ with prolongation operators $R_j^\top$ replaced by $(I - P_0)R_j^\top$.

A sufficient condition for (9) to be true is that this inequality hold for one particular choice of the decomposition of x so we choose one. Let $D_j \in \mathbb{R}^{n_j \times n_j}$ be diagonal weighting matrices which form a partition of unity: $\sum_{j=1}^N R_j^\top D_j R_j = I$ (I is the identity in $\mathbb{R}^{n\times n}$) and let $x_j := D_j R_j x$ then $\sum_{j=1}^N R_j^\top x_j = x$ and

$$\langle Ax, x\rangle \geq C \sum_{j=1}^N \langle A(I-P_0)R_j^\top D_j R_j x, (I-P_0)R_j^\top D_j R_j x\rangle \Rightarrow (9).$$

The final step to make the condition local is to make the left hand side local. We recall that A is assembled as a sum of element matrices $A = \sum_{\tau\in\mathcal{T}_h} A_\tau$. Each of these element matrices was supposed to be symmetric positive semi definite. This means that if we assemble the element matrices over a subset $\mathcal{T}_h^j$ of $\mathcal{T}_h$ the resulting energy norm will be bounded with respect to $\langle A\cdot, \cdot\rangle^{1/2}$. In particular, let $\mathcal{T}_h^j = \{\tau; dof(\tau) \subset \omega_j\}$ be the set of connections which are completely in subdomain j and define the corresponding local matrix as $\tilde{A}_j = \sum_{\tau\in\mathcal{T}_h^j} R_j^\top A_\tau R_j$ then $\sum_{j=1}^N \langle \tilde{A}_j R_j x, R_j x\rangle \leq \mathcal{N}^c\langle Ax, x\rangle$, and the sufficient condition becomes local:

$$\langle \tilde{A}_j R_j x, R_j x\rangle \geq \frac{C}{\mathcal{N}^c}\langle A(I-P_0)R_j^\top D_j R_j x, (I-P_0)R_j^\top D_j R_j x\rangle \Rightarrow (9). \tag{10}$$

We do not know how to simplify this condition further without using information on the underlying set of partial differential equations and on the partition of unity so this is the bottleneck estimate which discriminates between the 'good' vectors and the ones that need to be in the projection space V_0. The bottleneck estimate is not exactly the same as in [7,8] where the decomposition of u also requires a component in the projection space. The generalized eigenvalue which we propose is also slightly different.

3.3 Building the Projection Space to Satisfy the Bottleneck

Definition 1. *For any subdomain $j = 1, \dots, N$, find the eigenpairs $(p_j^k, \lambda_j^k) \in \mathbb{R}^{n_j} \times \mathbb{R}^+$ of*

$$\tilde{A}_j p_j^k = \lambda_j^k D_j A_j D_j p_j^k. \tag{11}$$

Then for a given threshold $\mathcal{K}$ let the coarse space be defined by

$$V_0 = \sum_{j=1,\dots,N} R_j^\top D_j (V_0^j);\ V_0^j = span\{p_j^k; \lambda_j^k < \mathcal{K}\}.$$

Let $R_0^\top$ be the interpolation operator whose columns are the vectors $R_j^\top D_j p_j^k$ such that $\lambda_j^k < \mathcal{K}$ and let P_0 be the A-orthogonal projection operator onto V_0 introduced in (3).

This way we have completed the definition of our domain decomposition method. All that is left to do is to make sure that the projection space introduced in Definition 1 does indeed do its job in the sense that the bottleneck estimate (10) holds for any $x \in \text{range}(I - P_0)$.

Suppose that the eigenvectors have been normalized so that $\langle D_j A_j D_j p_j^k, p_j^k \rangle = 1$. If P_0^j is the A-orthogonal projection onto the span of $\{R_j^\top D_j p_j^k; \lambda_j^k < \mathcal{K}_j\}$ then

$$\langle A(I - P_0)u, (I - P_0)u \rangle \leq \langle A(I - P_0^j)u, (I - P_0^j)u \rangle,\ \forall u \in \mathbb{R}^n. \tag{12}$$

This is true in particular for $u = R_j^\top D_j R_j x$.

Lets assume that the eigenvectors have been normalized so that $\langle D_j A_j D_j p_j^k, p_j^k \rangle = 1$. The $(D_j a_j D_j)$-orthogonal projection onto V_0^j is $\Pi_j : \mathbb{R}^{n_j} \to \mathbb{R}^{n_j}$ such that $\Pi_j x_j = \sum_{\{k; \lambda_j^k < \mathcal{K}\}} \langle D_j A_j D_j x_j, p_j^k \rangle p_j^k$. This projection operator was already introduced in [8, Lemma 2.11]. We notice that $P_0^j R_j^\top D_j = R_j^\top D_j \Pi_j$ so

$$\langle A(I - P_0^j) R_j^\top D_j R_j x, (I - P_0) R_j^\top D_j R_j x \rangle = \langle D_j A_j D_j (I - \Pi_j) R_j x, (I - \Pi_j) R_j x \rangle. \tag{13}$$

Finally, we apply the abstract Lemma 2.11 from [8] and use the fact that $\langle \tilde{A}_j p_j^k, p_j^l \rangle = 0$, $k \neq l$, to get

$$\begin{aligned} \langle D_j A_j D_j (I - \Pi_j) R_j x, (I - \Pi_j) R_j x \rangle &\leq \tfrac{1}{\mathcal{K}} \langle \tilde{A}_j (I - \Pi_j) R_j x, (I - \Pi_j) R_j x \rangle \\ &\leq \tfrac{1}{\mathcal{K}} \langle \tilde{A}_j R_j x, R_j x \rangle. \end{aligned} \tag{14}$$

Putting (14) together with (12) and (13) proves the condition in (10) for $C/\mathcal{N}^c = \mathcal{K}$ and thus the following theorem. This is the main contribution in this work. Thanks to the projection steps, the condition number estimate is improved compared to the fully Additive Schwarz algorithm considered in [7,8].

Theorem 1. *The lowest eigenvalue of $M^{-1}A$ on range $(I - P_0)$ satisfies $\lambda_{min} \geq \mathcal{K}/\mathcal{N}^c$ so the condition number of the preconditioned operator is bounded by $(\mathcal{N}^c)^2/\mathcal{K}$. Hence (according to [2, Theorem 6.29]) if x^* is the exact solution of the original problem (1), x^0 is the initial guess, and x^m is the approximate* *solution given by the m-th step of the preconditioned conjugate gradient algorithm with the projected Additive Schwarz preconditioner, the error decreases at least as*

$$\frac{\|x^* - x^m\|_A}{\|x^* - x^0\|_A} \leq 2 \left[\frac{\mathcal{N}^c - \sqrt{\mathcal{K}}}{\mathcal{N}^c + \sqrt{\mathcal{K}}} \right]^m , \tag{15}$$

where $\|\cdot\|_A = \langle A\cdot, \cdot \rangle^{1/2}$, $\mathcal{K}$ *is the* ***chosen*** *threshold used to select eigenvectors for the projection space in Definition 1, and* $\mathcal{N}^c$ *is the number of colors that are needed to color the subdomains in such a way that two subdomains with the same color are orthogonal.*

4 Numerical Illustration

In this section for lack of space we have chosen to illustrate the way that the method works rather than a set of performance tests. The implementation uses matlab and Freefem++. We solve the two dimensional linear elasticity equations discretized with $\mathbb{P}_1$ (piecewise linear) finite elements on a 121×16 regular mesh with simplicial elements. More precisely, the problem is to find $\mathbf{u} = (u_1, u_2)^{\mathrm{T}}$, such that $-\mathrm{div}(\sigma(\mathbf{u})) = \mathbf{f}$ where the stress tensor $\sigma(\mathbf{u})$, the Lamé coefficients λ and μ and the right hand side are

$$\begin{cases} \sigma_{ij}(\mathbf{u}) = 2\mu\varepsilon_{ij}(\mathbf{u}) + \lambda\delta_{ij}\mathrm{div}(\mathbf{u}),\ \varepsilon_{ij}(\mathbf{u}) = \frac{1}{2}\left(\frac{\partial u_i}{\partial x_j} + \frac{\partial u_j}{\partial x_i}\right) \\ \mu = \frac{E}{2(1+\nu)},\ \lambda = \frac{E\nu}{(1+\nu)(1-2\nu)}, \mathbf{f} = (0, g)^T = (0, 10)^T. \end{cases} \tag{16}$$

The domain is an 8×1 rectangle which we decompose into 8 side by side unit squares. Then we add one layer of overlap to each. The medium is a soft material ($E = 10^7$, $\nu = 0.4$) with two layers of a harder material ($E = 10^{12}$, $\nu = 0.4$). In Fig. 1 we have plotted the original configuration for one subdomain as well as the first eigenmodes for eigenproblem (11). In Fig. 2 we show that the

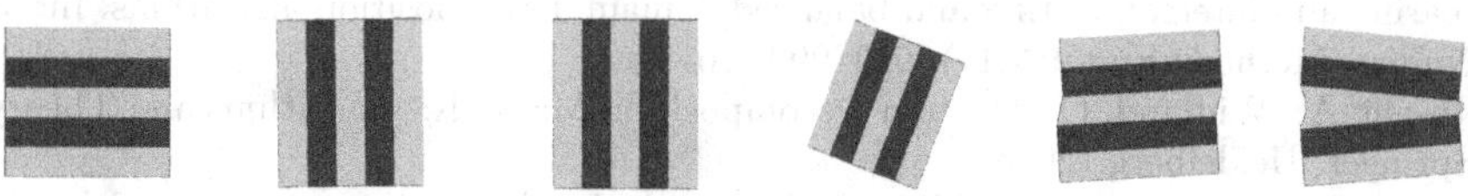

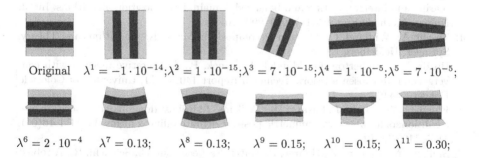

Original $\lambda^1 = -1 \cdot 10^{-14}$;$\lambda^2 = 1 \cdot 10^{-15}$;$\lambda^3 = 7 \cdot 10^{-15}$;$\lambda^4 = 1 \cdot 10^{-5}$;$\lambda^5 = 7 \cdot 10^{-5}$;

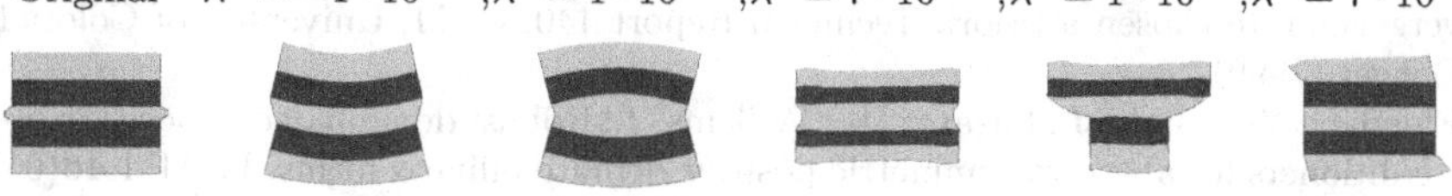

$\lambda^6 = 2 \cdot 10^{-4}$ $\lambda^7 = 0.13$; $\lambda^8 = 0.13$; $\lambda^9 = 0.15$; $\lambda^{10} = 0.15$; $\lambda^{11} = 0.30$;

Fig. 1. Original configuration and first eleven eigenvectors for a floating subdomain (dark or red: hard material). With $\mathcal{K} = 0.1$ we select six eigenvectors for the projection space. Among these, the first three correspond to the rigid body modes ($\tilde{A}_j p_j^{1,2,3} = 0$). In total the size of the projection space is 46 (Color figure online).

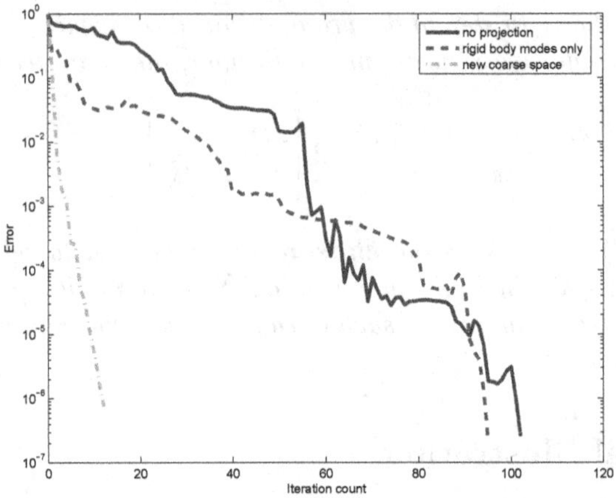

Fig. 2. Error versus the iteration count for three methods: no projection (blue full line), projection onto the rigid body modes (red hashes), projection onto the space from definition 1 (green hashes and dots). The new projections space does its job. The condition number is reduced from 3576 to 13. With just the rigid body modes it is 1808 (Color figure online).

new method converges very fast and that the projection step does its job since it reduces the condition number from 3576 to 13 using only 46 projection vectors. open question.

References

1. Dostàl, Z.: Conjugate gradient method with preconditioning by projector. Int. J. Comput. Math. **23**(3–4), 315–323 (1988)
2. Saad, Y.: Iterative Methods for Sparse Linear Systems. SIAM, Philadelphia (2003)
3. Spillane, N., Rixen, D.J.: Automatic spectral coarse spaces for robust finite element tearing and interconnecting and balanced domain decomposition algorithms. Int. J. Numer. Methods Eng. **95**(11), 953–990 (2013)
4. Toselli, A., Widlund, O.: Domain Decomposition Methods: Algorithms and Theory. Springer, Heidelberg (2005)
5. Brezina, M., Heberton, C., Mandel, J., Vaněk, P.: An iterative method with convergence rate chosen a priori. Technical Report 140, CCM, University of Colorado Denver (1999)
6. Efendiev, Y., Galvis, J., Lazarov, R., Willems, J.: Robust domain decomposition preconditioners for abstract symmetric positive definite bilinear forms. ESAIM **46**(05), 1175–1199 (2012)
7. Spillane, N., Dolean, V., Hauret, P., Nataf, F., Pechstein, C., Scheichl, R.: A robust two level domain decomposition preconditioner for systems of PDEs. C. R. Math. **349**(23–24), 1255–1259 (2011)
8. Spillane, N., Dolean, V., Hauret, P., Nataf, F., Pechstein, C., Scheichl, R.: Abstract robust coarse spaces for systems of PDEs via generalized eigenproblems in the overlaps. Numer. Math. **126**(4), 741–770 (2014)

Large-Scale Models: Numerical Methods, Parallel Computations and Applications

Assessment of the Air Quality in Bulgaria - Short Summary Based on Recent Modelling Results

Hristo Chervenkov[1](✉), Dimiter Syrakov[2], Maria Prodanova[2], and Kiril Slavov[2]

[1] National Institute of Meteorology and Hydrology, Branch Plovdiv - Bulgarian Academy of Sciences, "Russki" 139, 4000 Plovdiv, Bulgaria
hristo.tchervenkov@meteo.bg

[2] National Institute of Meteorology and Hydrology, Sofia - Bulgarian Academy of Sciences, "Tsarigradsko Shose" 66, 1784 Sofia, Bulgaria

Abstract. The European Union has developed an extensive body of legislation which establishes health based standards and objectives for a number of pollutants in air. An appropriate method for evaluating the air quality of a certain EU-country is to contrast the actual air pollution levels to the critical ones, prescribed in the legislative standards. This study, which is part of greater one, is focused of the pollutants of most concern - the ground level ozone, particulate matter, and the sulphur dioxide, which is still a problem in Southeastern Europe. High-resolution data from the Bulgarian chemical weather forecasting and information system for the last three years are used to compute certain pollutant levels, which are further compared with the critical ones, prescribed in the EU-legislation directives. The obtained results can be treated as objective pattern of the situation over Bulgaria in the end of the first decade of the present century.

Keywords: Air pollution simulation · High resolution · EC-directives · Air pollution levels · Pollutant exceedances

1 Introduction

The adverse health effects from short and long-term exposure to air pollution range from premature deaths caused by heart and lung disease to worsening of asthmatic conditions, which often leads to a reduced quality of life and increased costs of hospital admissions. Poor outdoor air quality can be a contributing factor also to reduced agricultural crop yields, changes in ecosystem species composition, damage to physical infrastructure and cultural heritage due to material deterioration [3] and etc.. Despite of the emissions abatement, many European countries still do not comply with one or more emission ceilings set under European Union (EU) and United Nations (UN) agreements. Furthermore, due to

I. Lirkov et al. (Eds.): LSSC 2013, LNCS 8353, pp. 365–372, 2014.
DOI: 10.1007/978-3-662-43880-0_41,

the complex links between emissions and air quality, emission reductions do not always produce a corresponding drop in atmospheric concentrations, especially for particulate matter (PM) and ozone (O_3). The 2008 ambient air quality directive [7] sets legally binding limits for concentrations in outdoor air of major air pollutants that impact public health and environment. An appropriate method for evaluating the air quality of a certain area is to contrast the actual air pollution levels to the critical ones, prescribed in the legislative standards. The application of numerical simulation models for assessing the real air quality status is allowed by the legislation of the European Community (EC). In the presented work we use most recent high-resolution modelling data for region centered over Bulgaria to assess the compliance with the air quality limit and target values set out in the current EU legislation. The study is focused on the PM and O_3 which, at present, are Europe's most problematic pollutants in terms of harm to human health [14,15]. The sulphur dioxide (SO_2) is also studied, because, in contrast to the bigger part of Europe, still causes concern in the Southeastern region of the continent [2].

2 Short Summary of the Air Quality Legislation in the European Union

The principal driver of Bulgarian air quality legislation derives from EU Directives. Although various pieces of air quality legislation were produced within the EU, historically the 1996 Air Quality Framework Directive (96/62/EC) [7] was the first overarching strategy-level document. It is an important legal milestone in the EU's fight against air pollution. The Directive itself did not create any precise air quality objectives, but rather it set out a framework and basic principles for ambient air quality monitoring and management. These were to go into effect once daughter directives for specific pollutants had been adopted. The first daughter directive (1999/30/EC) [8] targets sulphur dioxide, nitrogen dioxide, particulate matter and lead, and was adopted in 1999. The second directive (2000/69/EC) [10] targets benzene and carbon monoxide, and was adopted in 2000. The third directive (2002/3/EC) [11] targets ground-level ozone and was adopted in late 2001. Once agreed upon, EU directives must be transposed into the legislation of Member States. It should be noted that individual Member States may enact legislation, which is more stringent than required by an EU Directive but cannot weaken any of the numerical limits laid down in any Directive. In 2003, the European Commission proposed a policy for updating and streamlining EU legislation. This arose as a result of the Better Regulation initiative. In 2005, the European Commission, in collaboration with the Clean Air For Europe (CAFE) team, presented a proposal for a Directive of The European Parliament and of The Council on Ambient Air Quality and Cleaner Air for Europe. This is part of the wider implementation of the Thematic Strategy on air pollution, itself one of the seven key pillars of the Sixth Environmental

Action Plan. The final Directive (2008/50/EC) [13] was adopted on 21 May 2008 and includes the following key elements:

- that most of existing legislation (Framework Directive 96/62/EC, 1–3 daughter Directives 1999/30/EC, 2000/69/EC, 2002/3/EC, and Decision on Exchange of Information 97/101/EC [9]) be merged into a single directive (except for the fourth daughter directive [12]) with no change to existing air quality objectives
- new air quality objectives for $PM_{2.5}$ (fine particles) including the limit value and exposure related objectives - exposure concentration obligation and exposure reduction target
- the possibility to discount natural sources of pollution when assessing compliance against limit values
- possibility for time extensions of three years PM_{10} (coarse particles) or up to five years (NO_2, benzene) for complying with limit values, based on conditions and the assessment by the European Commission

Table 1. Synthesis table of the prescribed AQLV and AQTV (threshold concentration in $\mu g/m^3$/number of allowed exceedances) for the considerated pollutants

Av. period/parameter pollutant	1 h	Max. daily 8 h mean	1 day	Winter (1.10–31.03)	Calendar year
PM_{10}			50/35 50/7		40/0 20/0
$PM_{2.5}$			25/0		25/0 20/0
O_3		120/25			
SO_2	350/24		125/3	20/0	20/0

Quantitative expression of the air quality objectives are the air quality limit (AQLV) and target values (AQTV) (Table 1). EU Limit values are legally binding EU parameters that must not be exceeded. They are set for individual pollutants and are made up of a concentration value, an averaging time over which it is to be measured, the number of exceedances allowed per year, if any, and a date by which it must be achieved. Some pollutants have more than one limit value covering different endpoints or averaging times. For example the stricter 1 day and calendar year AQLV (50 $\mu g/m^3$/7 occurrences and 20 $\mu g/m^3$/0 occurrences respectively) for PM_{10} are, according [8], "Indicative limit values to be reviewed in the light of further information on health and environmental effects, technical feasibility and experience in the application of Stage 1 limit values in the Member States". Actually this intention is not reconfirmed in the newer ambient air quality directive [13]. Similar, the second criterion for the $PM_{2.5}$ is indicative limit value, which have to be reviewed by the Commission in 2013.

The World Health Organization (WHO) has developed own air quality standardization, namely air quality guidelines (AQG), which for many pollutants are different, often stricter than the EU requirements. So the 1 day (24 h mean) for $PM_{2.5}$, shown in the table is such WHO AQLV. Nevertheless this criterion is taken into account, because, from one site, there is not EU-alternative for this averaging period, and from other, it is important milestone, similar to the above mentioned indicative values, for the performed comparisons in the next chapter. Target values are used in some EU Directives and are set out in the same way as limit values. They are to be attained where possible by taking all necessary measures not entailing disproportionate costs.

3 Methodology and Performed Computations

The harmful effects of the air pollution, mentioned in the previous section, invokes a project financially supported by the National Science Fund with the Ministry of Education and Science aiming at creating the Bulgarian Chemical Weather Forecast and Information System (BgCWFS). It is intended to provide timely, informative and reliable forecast products tailored to the needs of various users and decision-makers. The system has nesting structure, starting from the region of Bulgaria and nearest territories as a whole (background pollution) and zooming to smaller and/or bigger areas of interest. The model grid consists of 54×40 grid cells with size 10×10 km and covers Bulgaria entirely, together with the border regions of the neighboring countries and the most western part of the Black sea. The country part of BgCWFS is designed in a way to fit the real-time constraints and to deliver forecasts twice a day (00 and 12 UTC) for the next 48 hours. BgCWFS has a modular structure and all building blocks, operational design and data flow are described in detail in [16,17]. To calculate the corresponding statistical quantities is essential to know the value of the concentration for each pollutant in every hour for at least one year, i.e. it is necessary to reconstruct it's time series. This is done in the following manner: the 12 h subset of concentration data for the first day of the year, between the first and second run is extracted. Then the second subset, between the second and third (i.e. the first for the second day) run is extracted and so on until the end of the year. At the end all subsets are pieced together to continuous series. Due to the validated reliability and consistency of the BgCWFS from one site and to the nature of the data, obtained in prescribed way (forecast data for very short range) from other, is reasonable to expect that such sets do not differ significantly from the reality. They are completely suitable for assessing the (background) air pollution continuously over the whole model domain. On the next step we have to be able to answer if the AQLV or AQTV for each pollutant under interest is breached or not. This is done in the following manner: If, say, for a given pollutant the directive allows m exceedances of the corresponding statistical average, we find the *(m+1)-th* maximum from all possible values. If this value is greater than the prescribed AQLV or AQTV, the legislation is breached. This task, namely to find the *(m+1)-th* largest among n, where n is the number of all possible values, has to be repeated many times in each grid cell and that's why it is essential to optimize the corresponding numerical routine. This is done by setting especially effective procedure for $m \ll n$, because this condition is very well expressed. In all cases we find the maximal value also due to its relevance as indicative of the air pollution. According the requirements of the second daughter directive and to exclude short time variations, input data for the last three years, 2010, 2011 and 2012, are used and the obtained annual results are finally averaged.

4 Results

The key manmade contributions to ambient SO_2 derive from sulphur containing fossil fuels and biofuels used for domestic heating, stationary power generation and transport. Different studies suggest that the sulphur dioxide is the core reason of variety of processes with deep environmental consequences [1]. These processes span in the time scale from minutes to years and this is the main reason for the fact, that in the contemporary directives this pollutant is with the greatest number of defined AQLV. The calculated statistics are shown on Figs. 1 and 2.

Particulate matter is partly directly emitted into the atmosphere and partly formed in the atmosphere. The formation depends on a variety of chemical and

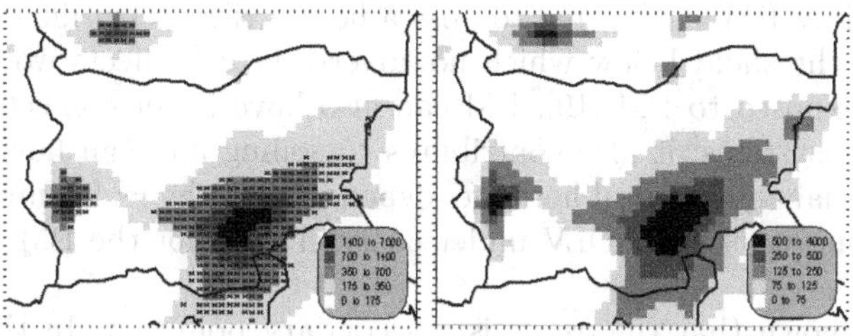

Fig. 1. 99.71 percentile (25th hourly maximum) of the hourly concentration of the sulphur dioxide (left panel, unit: $\mu g/m^3$). The gridcells, where the hourly alert threshold is bridged at least once for the whole period, are marked with crosses. 98.90 percentile (4th daily maximum) of the daily mean concentration of the sulphur dioxide (right panel, same unit)

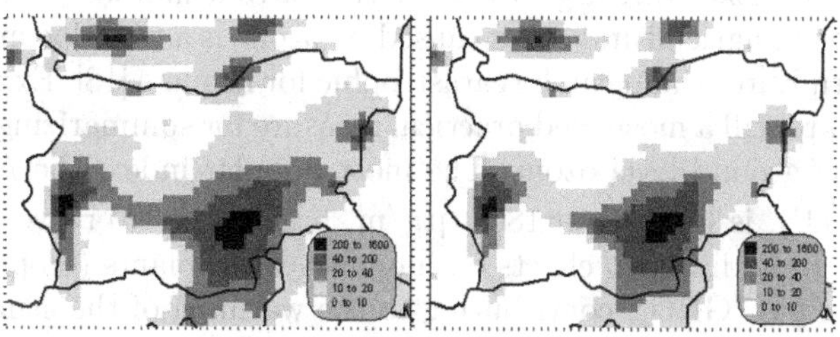

Fig. 2. Winter (left panel) and annual mean concentration (units: $\mu g/m^3$) of the sulphur dioxide

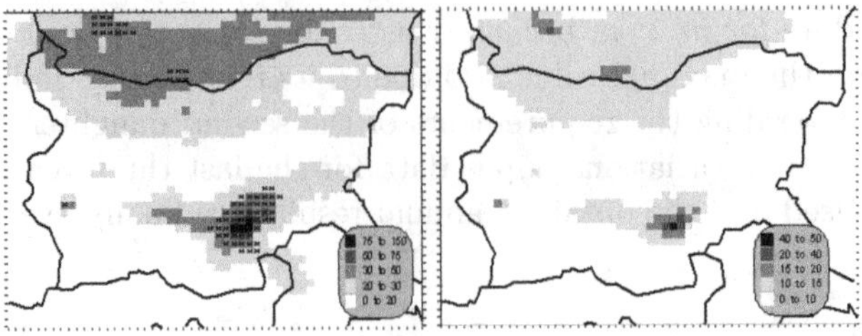

Fig. 3. 97.81 percentile (8th daily maximum) of the daily mean concentration of the PM_{10} (left panel, unit: $\mu g/m^3$). The gridcells, where the daily mean AQLV is bridged at least once (1th daily maximum), are marked with crosses. Annual mean concentration of the PM_{10} (right panel, same unit)

physical factors, such as the concentrations of the main precursors, the reactivity of the atmosphere and meteorological conditions. Due to the interplay and variability of the above factors, it is difficult to relate ambient concentrations of formed substances, present in ambient PM, to the emissions of precursor gases. Epidemiological studies attribute the most severe health effects from air pollution to PM and, to a lesser extent, O_3. Even at concentrations below current air quality guidelines PM is expected to pose a health risk. Scientific evidence does not suggest a threshold below which no adverse health effects would be anticipated when exposed to PM [19]. PM can also have adverse effects on climate change and ecosystems; it also contributes to soiling and can have a corrosive effect on material and cultural heritage, depending on the PM composition. The spatial distribution of the AQLV under consideration for the PM_{10} are shown on Fig. 3.

It is well-known that the ozone is a secondary pollutant. In the lower troposphere it is mainly formed through chemical reactions between nitrogen oxides and volatile organic compounds (VOC) in the presence of short-wavelength radiation from the sun during a timescale from hours to days. Second, it's concentration depends to a greater extent, then those of the others, on meteorological conditions, especially the sunlight intensity and the temperature. Due to these reasons the link between (precursor) emissions and concentration is much more complicated [4,6]. Excessive O_3 in the air can have a marked effect on human health. Damage to agricultural crops caused by ozone is a well-documented problem in Southern Europe and can be catastrophic for farmers [18]. Exposure-based ozone metrics are still a most used practical measure for summarizing ambient air quality relating ground-level ozone. The most popular index, the AOT40 crops (AOTc - see [11]), is limited on $18000\,\mu g/m^3\,h$ averaged over five (in this case - three) years. In addition to effects on human health, plants (crops), the ozone is a greenhouse gas (GHG), contributing to the warming of the atmosphere and also increases the rate of degradation of buildings and physical cultural heritage. The obtained results for this pollutant are shown on Fig. 4.

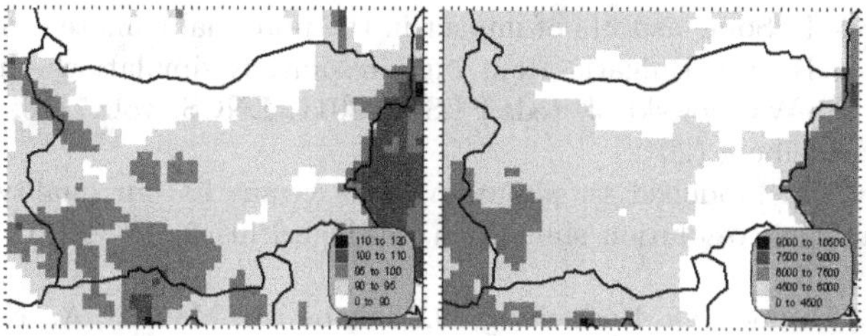

Fig. 4. 92.88 percentile (26th maximum of this statistics) of the running 8 h mean concentration of the ozone (left panel, unit: $\mu g/m^3$). AOTc on right panel (unit: $\mu g/m^3$ h)

5 Comments and Conclusion

The presented study confirms the significant (in most drastic cases up to factor of 10) overrun of all AQLV for the sulphur dioxide over a large territory, known from other studies [2,5]. The most polluted areas are in the vicinity of the TPP "Maritsa-Istok" - a set of three coal burning thermal power plants, which are the most relevant source in whole SE Europe at all. With broad maxima (below, but close to the thresholds) around the strongest sources, the distribution of the PM-air pollution is also typical. Keeping in mind that the $PM_{2.5}$ is fraction of PM_{10} and analyzing Fig. 3, indirectly can be stated that the $PM_{2.5}$ annual AQLV can be breached only in small vicinity of the TPP "Maritsa-Istok". Similar to the sulphur dioxide, the calculated results for the PM-air pollution are generally in agreement with recent AirBase measurements [1]. The ozone pollution is pronounced in regions with strong photochemical activity, such as the Mediterranean basin and Balkan Peninsula. In the presence of volatile organic compounds, the equilibrium favours the formation of higher levels of ozone, confirmed in previous works [5]. In this sense the calculated relatively low pollution levels are surprising, formal reason for which are the unusual small concentrations in the first and second year. Obviously the problem has to be investigated further in detail.

Despite of the methodological conventionality of the applied approach, the presented results can be treated as evidence of the need for further mitigation efforts - keeping always in mind that any level of air pollution is a matter of concern, and the existence of guideline values never means a license to pollute.

References

1. Air quality in Europe — 2012 report, EEA Report No 4/2012, ISBN 978-92-9213-328-3. doi:10.2800/55823
2. Chervenkov, H., Syrakov, D., Prodanova, M.: On the sulphur pollution over the Balkan region. In: Lirkov, I., Margenov, S., Waśniewski, J. (eds.) LSSC 2005. LNCS, vol. 3743, pp. 481–489. Springer, Heidelberg (2006)
3. Chervenkov, H.: Assessment of material deterioration in Bulgaria owing to air pollution. Int. J. Environ. Pollut. **31**(3/4), 385–393 (2007)

4. Chervenkov, H.: Some aspects of impact in the potential climate change on ozone pollution levels over Bulgaria from high resolution simulations. In: Lirkov, I., Margenov, S., Waśniewski, J. (eds.) LSSC 2011. LNCS, vol. 7116, pp. 275–282. Springer, Heidelberg (2012)
5. Chervenkov, H.: Modelled air pollution levels versus EC air quality legislation - results from high resolution simulation. SpringerPlus **2**, 78 (2013). doi:10.1186/2193-1801-2-78
6. Colette, A., Granier, C., Hodnebrog, Ø., Jakobs, H., Maurizi, A., Nyiri, A., Rao, S., Amann, M., Bessagnet, B., D'Angiola, A., Gauss, M., Heyes, C., Klimont, Z., Meleux, F., Memmesheimer, M., Mieville, A., Rouïl, L., Russo, F., Schucht, S., Simpson, D., Stordal, F., Tampieri, F., Vrac, M.: Future air quality in Europe: a multi-model assessment of projected exposure to ozone. Atmos. Chem. Phys. **12**, 10613–10630 (2012)
7. Council Directive 96/62/EC of 27 September 1996 on ambient air quality assessment and management Official Journal L 296, 21/11/1996, pp. 0055–0063
8. Council Directive 1999/30/EC of 22 April 1999 relating to limit values for sulphur dioxide, nitrogen dioxide and oxides of nitrogen, particulate matter and lead in ambient air Official Journal L 163, 29/06/1999, pp. 0041–0060
9. Decision 97/101/EC - Official Journal L 35, 5.2.1997, p. 14
10. Directive 2000/69/EC of the European Parliament and of the Council of 16 November 2000 relating to limit values for benzene and carbon monoxide in ambient air, Official Journal L 313, 13/12/2000, pp. 0012–0021
11. Directive 2002/3/EC of the European Parliament and of the Council of 12 February 2002 relating to ozone in ambient air, Official Journal L 67, 9/03/2002, pp. 0014–0030
12. Directive 2004/107/EC of the European Parliament and of the Council of 15 December 2004 relating to arsenic, cadmium, mercury, nickel and polycyclic hydrocarbons in ambient air, Official Journal L 23, 26/01/2005, pp. 003–0016
13. Directive 2008/50/EC of the European Parliament and of the Council of 21 May 2008 on ambient air quality and cleaner air for Europe, Official Journal L 152, 11/6/2008, pp. 001–0044
14. Emberson, L.D., Kitwiroon, N., Beevers, S., Buker, P., Cinderby, S.: Scorched earth: how will changes in ozone deposition caused by drought affect human health and ecosystems? Atmos. Chem. Phys. Discuss. **12**, 27847–27889 (2012). doi:10.5194/acpd-12-27847-2012
15. Fang, Y., Naik, V., Horowitz, L.W., Mauzerall, D.L.: Air pollution and associated human mortality: the role of air pollutant emissions, climate change and methane concentration increases during the industrial period. Atmos. Chem. Phys. Discuss. **12**, 22713–22756 (2012). doi:10.5194/acpd-12-22713-2012
16. Syrakov, D., Prodanova, M., Etropolska, I., Ganev, K., Miloshev, N., Slavov, K., Jordanov, G.: Automated system for chemical weather forecast in Bulgaria. Bulgarian J. Meteorol. Hydrol. (BJMH) **16**(1), 30–40 (2011)
17. Syrakov, D., Etropolska, I., Prodanova, M., Ganev, K., Miloshev, N., Slavov, K.: Operational pollution forecast for the region of Bulgaria. AIP Conf. Proc. **1487**, 88 (2012). doi:10.1063/1.4758945
18. Vlachokostas, C., Nastis, S.A., Achillas, C., et al.: Economic damages of ozone air pollution to crops using combined air quality and GIS modelling. Atmos. Environ. **44**, 3352–3361 (2010)
19. WHO.: Air quality guidelines — global update 2005, World Health Organization Regional Office for Europe, Copenhagen (2006)

Application of POD-DEIM Approach for Dimension Reduction of a Diffusive Predator-Prey System with Allee Effect

Gabriel Dimitriu[1(✉)], Ionel M. Navon[2], and Răzvan Ştefănescu[2]

[1] Department of Mathematics and Informatics, "Grigore T. Popa" University of Medicine and Pharmacy, 700115 Iaşi, Romania
dimitriu.gabriel@gmail.com

[2] Department of Scientific Computing, The Florida State University, Tallahassee, FL 32306, USA
{inavon,rstefanescu}@fsu.edu

Abstract. In this work we carry out an application of DEIM combined with POD to provide dimension reduction of a system of two nonlinear partial differential equations describing the spatio-temporal dynamics of a predator-prey community, where the prey *per capita* growth rate is damped by the Allee effect. DEIM improves the efficiency of the POD approximation reducing the computational complexity of the nonlinear term and regains the full model reduction expected from the POD model. Numerical results show that the dynamics of the predator-prey model in the full-order system of dimension 2048 can be captured accurately by the POD-DEIM reduced system with the computational time reduced by a factor of $\mathcal{O}(10^4)$.

1 Introduction

Proper Orthogonal Decomposition (POD) – see [2,4,7,8,10] and the references therein – is probably the mostly used and most successful model reduction technique, where the basis functions contain information from the solutions of the dynamical system at pre-specified time-instances, so-called *snapshots*. Due to a possible linear dependence or almost linear dependence, the snapshots themselves are not appropriate as a basis. Hence a singular value decomposition is carried out and the leading generalized eigenfunctions are chosen as a basis, referred to as the POD basis.

Unfortunately, for nonlinear PDEs, the efficiency in solving the reduced-order systems constructed from standard Galerkin projection with any reduced globally supported basis set, including the one from POD, is limited to the linear or bilinear part, both for finite volume and finite difference schemes. In the case of quadratic nonlinearities a so-called precomputed POD technique achieves the same level of reduction as in the case of linear terms.

A considerable reduction in complexity is achieved by DEIM – a discrete variation of Empirical Interpolation Method (EIM), proposed by Barrault et al. in [3].

I. Lirkov et al. (Eds.): LSSC 2013, LNCS 8353, pp. 373–381, 2014.
DOI: 10.1007/978-3-662-43880-0_42,

According to this method, the evaluation of the approximate nonlinear term does not require a prolongation of the reduced state variables back to the original high dimensional state approximation required to evaluate the nonlinearity in the POD approximation.

In this study we carry out an application of DEIM combined with POD to provide dimension reduction of a system of two nonlinear partial differential equations describing the spatio-temporal dynamics of a predator-prey community, where the prey *per capita* growth rate is damped by the Allee effect. This model was introduced and analyzed in an infinite space by Petrovskii et al. [14], together with properties of the solution and biologically significant dependence on the parameter values.

2 The Predator-Prey Model with Allee Effect

The spatio-temporal dynamics of a predator-prey system can be described by the equations [13]:

$$\frac{\partial U(X,T)}{\partial T} = D\frac{\partial^2 U}{\partial X^2} + f(U)U - r(U)V, \quad (1)$$

$$\frac{\partial V(X,T)}{\partial T} = D\frac{\partial^2 V}{\partial X^2} + \kappa r(U)V - g(V)V, \quad (2)$$

where U and V are the densities of prey and predator, respectively, at position X and time T. The function $f(U)$ is the *per capita* growth rate of the prey and the term $r(U)V$ stands for predation. κ is the coefficient of food utilization, and $g(V)$ is the *per capita* mortality rate of predator. Here, the first term on the right-hand side of Eqs. (1) and (2) describes the spatial mixing caused either by self-motion of individuals [15] or by properties of the environment, for example, for plankton communities the mixing is attributed to turbulent diffusion [9]. D is the diffusion coefficient, which we assume to be the same for both prey and predator.

For different species, functions f, r, and g can represent different functional responses (logistic, Gompertz, Holling, etc.). We assume that the prey dynamics is subjected to the Allee effect [1,6,12], so that its *per capita* growth rate is not a monotonically decreasing function of the prey density, but possesses a local maximum. In this model, the standard parametrization [11] is defined by

$$f(U) = \alpha(U - U_0)(K - U),$$

where K denotes the prey carrying capacity and U_0 is a certain measure of the Allee effect. Regarding the *per capita* predator mortality, one assumes that it is described by the following function:

$$g(V) = M + d_0V^2$$

where M and d_0 are positive parameters. Function $g(V)$ gives the so-called *closure term* because it is supposed not only to describe the process taking place inside the predator population (such as natural mortality, competition, possibly cannibalism, etc.) but also, virtually to take into account the impact of higher predators that are not included into the model explicitly. We assume that the predator shows a linear response to prey according to the classical Lotka-Volterra model, that is, $r(U) = \mu U$. Then, Eqs. (1)–(2) take the form

$$\frac{\partial U(X,T)}{\partial T} = D\frac{\partial^2 U}{\partial X^2} + \alpha U(U - U_0)(K - U) - \mu UV, \tag{3}$$

$$\frac{\partial V(X,T)}{\partial T} = D\frac{\partial^2 V}{\partial X^2} + \kappa\mu UV - MV - d_0 V^3. \tag{4}$$

A common procedure for solving the system of Eqs. (3)–(4) is to first nondimensionalize the system, and then obtain the numerical solution by employing a discretization scheme. We define the nondimensional variables and parameters to be:

$$u = \frac{U}{K}, \quad v = \frac{\eta V}{\alpha K^2}, \quad x = X\sqrt{\frac{\alpha K^2}{D}}, \quad t = T\alpha K^2.$$

From Eqs. (3) and (4) one obtains

$$u_t = u_{xx} - \beta u + (\beta + 1)u^2 - u^3 - uv, \tag{5}$$

$$v_t = v_{xx} + kuv - mv - \delta v^3, \tag{6}$$

where $\beta = U_0 K^{-1}$, $k = \kappa\eta(\alpha K)^{-1}$, $m = M(\alpha K^2)^{-1}$ and $\delta = d_0 \alpha K^2 \eta^{-2}$ are positive dimensionless parameters, subscripts x and t stand for the partial derivatives with respect to dimensionless space and time, respectively. Here we consider Eqs. (5) and (6) in a bounded domain Ω with homogeneous Dirichlet boundary conditions. The initial conditions given by $u(x, 0) = u_0(x)$ and $v(x, 0) = v_0(x)$ will be specified in Sect. 4.

3 The POD and POD-DEIM Reduced Order System

In this section we provide some details for constructing the reduced-order system of the full-order system (5)–(6) applying Proper Orthogonal Decomposition (POD) and Discrete Empirical Interpolation Method (DEIM).

POD is an efficient method for extracting orthonormal basis elements that contain characteristics of the space of expected solutions which is defined as the span of the snapshots [7,8]. In this framework, *snapshots* are the sampled (numerical) solutions at particular time steps or at particular parameter values. POD gives an optimal set of basis vectors minimizing the mean square error from approximating these snapshots. In this finite dimensional setting, POD is closely related to the singular value decomposition (SVD).

The projected nonlinearities in Eqs. (5)–(6) are approximated by DEIM in the form that enables precomputation, so that evaluating the approximate nonlinear terms using DEIM does not require a prolongation of the reduced state variables back to the original high dimensional state approximation, as it is required for nonlinearity evaluation in the original POD approximation. Only a few entries of the original nonlinear term, corresponding to the specially selected interpolation indices from DEIM must be evaluated at each time step [3,5,16]. We give formally the DEIM approximation in Definition 1, and the procedure for selecting DEIM indices is shown in Algorithm DEIM. Each DEIM index is selected to limit growth of a global error bound using a greedy technique relating the DEIM approximation to the full optimal POD approximation [5].

Definition 1. [5] Let $\mathbf{f} : \mathcal{D} \mapsto I\!R^n$ be a nonlinear vector-valued function with $\mathcal{D} \subset I\!R^d$, for some positive integer d. Let $\{\mathbf{u}\}_{\ell=1}^m \subset I\!R^n$ be a set of linearly independent vectors, for $m = 1, \ldots, n$. For $\tau \in \mathcal{D}$, the DEIM approximation of order m for $\mathbf{f}(\tau)$ in the space spanned by $\{\mathbf{u}\}_{\ell=1}^m$ is given by

$$\widehat{\mathbf{f}}(\tau) := \mathbf{U}(\mathbf{P}^T\mathbf{U})^{-1}\mathbf{P}^T\mathbf{f}(\tau), \tag{7}$$

where the basis $\mathbf{U} = [\mathbf{u}_1, \ldots, \mathbf{u}_m] \in I\!R^{n\times m}$ can be constructed effectively by applying the POD method on the nonlinear snapshots $f(\tau_i)$, $\tau_i \in \mathcal{D}$ and $\mathbf{P} = [\mathbf{e}_{\varrho_1}, \ldots, \mathbf{e}_{\varrho_m}] \in I\!R^{n\times m}$ with $\{\varrho_1, \ldots, \varrho_m\}$ being the output from Algorithm DEIM with the input basis $\{\mathbf{u}_i\}_{i=1}^m$.

Algorithm DEIM:
INPUT: $\{\mathbf{u}\}_{\ell=1}^m \subset I\!R^n$ linearly independent
OUTPUT: $\boldsymbol{\varrho} = [\varrho_1, \ldots, \varrho_m]^T \in I\!R^m$

1. $[|\rho|\ \varrho_1] = \max\{|\mathbf{u}_1|\}$
2. $\mathbf{U} = [\mathbf{u}_1]$, $\mathbf{P} = [\mathbf{e}_{\varrho_1}]$, $\boldsymbol{\varrho} = [\varrho_1]$
3. **for** $\ell = 2$ to m **do**
4. Solve $(\mathbf{P}^T\mathbf{U})\mathbf{c} = \mathbf{P}^T\mathbf{u}_\ell$ for $\mathbf{c}$
5. $\mathbf{r} = \mathbf{u}_\ell - \mathbf{U}\mathbf{c}$
6. $[|\rho|\ \varrho_\ell] = \max\{|\mathbf{r}|\}$
7. $\mathbf{U} \leftarrow [\mathbf{U}\ \mathbf{u}_\ell]$, $\mathbf{P} \leftarrow [\mathbf{P}\ \mathbf{e}_{\varrho_\ell}]$, $\boldsymbol{\varrho} \leftarrow \begin{bmatrix} \boldsymbol{\varrho} \\ \varrho_\ell \end{bmatrix}$
8. **end for**

The notation *max* in Algorithm DEIM is the same as the function *max* in Matlab. Thus, $[|\rho|\ \varrho_\ell] = \max\{|\mathbf{r}|\}$ implies $|\rho| = |r_{\varrho_\ell}| = \max_{i=1,\ldots,n}\{|r_i|\}$, with the smallest index taken in case of a tie. According to this algorithm, the DEIM procedure generates a set of indices inductively on the input basis in such a way that, at each iteration, the current selected index captures the maximum variation of the input basis vectors. The vector $\mathbf{r}$ can be viewed as the error between the input basis $\{\mathbf{u}\}_{\ell=1}^m$ and its approximation $\mathbf{U}\mathbf{c}$ from interpolating the basis $\{\mathbf{u}\}_{\ell=1}^{m-1}$ at the indices $\varrho_1, \ldots, \varrho_{m-1}$. The linear independence of the input basis $\{\mathbf{u}\}_{\ell=1}^m$ guarantees that, at each iteration, $\mathbf{r}$ is a nonzero vector and the output indices $\varrho_1, \ldots, \varrho_m$ are not repeating [5].

4 Numerical Results

We shall present three numerical experiments. The system (5)–(6) was solved numerically using a finite difference discretization. Let $0 = x_0 < x_1 < \cdots < x_n < x_{n+1} = 1$ be equally spaced points on the x-axis for generating the grid points on the dimensionless domain $\Omega = [0, 1]$, and take time domain $[0, T] = [0, 1]$. The corresponding spatial finite difference discretized system of (5)–(6) becomes a system of nonlinear ODEs. The semi-implicit Euler scheme was used to solve the discretized system of full dimension, as well as the POD and POD-DEIM reduced order systems.

Case 1. The parameters used here are $m = \beta = 4$, $\kappa = 15$, and $\delta = 0.25$. The initial conditions were set to $u_0(x) = \sin x \sin(\pi x) \exp(x)$, and $v_0(x) = x(1-x)^3$. The number of spatial inner grid points on the x-axis – which defines the dimension of the full-order system – was successively taken as 16, 32, 64, 128, ..., 2048. It shows that POD-DEIM reduces more than 400 times in dimension and reduces the computational time by a factor of $\mathcal{O}(10^4)$ as shown in Table 1. From Table 1, the CPU time used in computing POD reduced system clearly reflects the dependency on the dimension of the original full-order system. Table 1 also shows a significant improvement in computational time of the POD-DEIM reduced system as compared to both the POD reduced and the full-order systems.

Table 1. CPU time of full-order system, POD and POD-DEIM reduced systems with the corresponding average relative errors for u and v – Case 1.

Internal Nodes N	CPU Time Full Dim	CPU Time POD	CPU Time POD-DEIM	$Error^{rel}$ POD – u	$Error^{rel}$ POD – DEIM – u	$Error^{rel}$ POD – v	$Error^{rel}$ POD – DEIM – v
16	3.632462e-001	7.122489e-001	1.811141e-002	1.170196e-005	1.857876e-004	2.103251e-004	1.222462e-002
32	4.169362e-001	7.154559e-001	1.860040e-002	1.169715e-005	1.463095e-004	2.099999e-004	1.230968e-002
64	6.164471e-001	7.516708e-001	2.826825e-002	1.169587e-005	9.926064e-005	2.099146e-004	1.128253e-002
128	6.529374e-001	8.020902e-001	1.812896e-002	1.169554e-005	1.560804e-004	2.098927e-004	1.165304e-002
256	1.631008e+000	8.673314e-001	1.819947e-002	1.169545e-005	1.481253e-004	2.098871e-004	1.168835e-002
512	6.377997e+000	1.012015e+000	1.823390e-002	1.169543e-005	1.323507e-004	2.098857e-004	1.166019e-002
1024	2.924355e+001	1.291486e+000	1.827065e-002	1.169542e-005	1.330641e-004	2.098853e-004	1.172391e-002
2048	1.675980e+002	2.788567e+000	1.825973e-002	1.169542e-005	1.340120e-004	2.098852e-004	1.171443e-002

Table 2. CPU time of full-order system, POD and POD-DEIM reduced systems with the corresponding average relative errors for u and v – Case 2.

Internal Nodes N	CPU Time Full Dim	CPU Time POD	CPU Time POD-DEIM	$Error^{rel}$ POD – u	$Error^{rel}$ POD – DEIM – u	$Error^{rel}$ POD – v	$Error^{rel}$ POD – DEIM – v
16	3.553413e-001	6.865113e-001	1.793760e-002	2.356905e-005	1.580829e-005	3.047587e-004	4.552821e-004
32	4.284408e-001	6.980146e-001	1.802235e-002	2.360341e-005	2.358619e-005	3.067127e-004	2.008601e-004
64	4.867406e-001	7.455845e-001	1.855140e-002	2.360549e-005	2.207394e-005	3.070870e-004	7.456335e-004
128	6.408845e-001	7.897221e-001	1.834978e-002	2.360580e-005	3.493150e-005	3.071769e-004	4.430065e-004
256	1.098676e+000	8.647706e-001	1.822993e-002	2.360586e-005	3.485313e-005	3.071994e-004	5.934588e-004
512	3.885919e+000	1.006938e+000	1.858914e-002	2.360588e-005	3.497038e-005	3.072050e-004	6.145148e-004
1024	1.917511e+001	1.256279e+000	1.878149e-002	2.360588e-005	3.495507e-005	3.072064e-004	6.035787e-004
2048	1.148451e+002	1.883425e+000	1.870667e-002	2.360588e-005	3.507343e-005	3.072068e-004	5.992118e-004

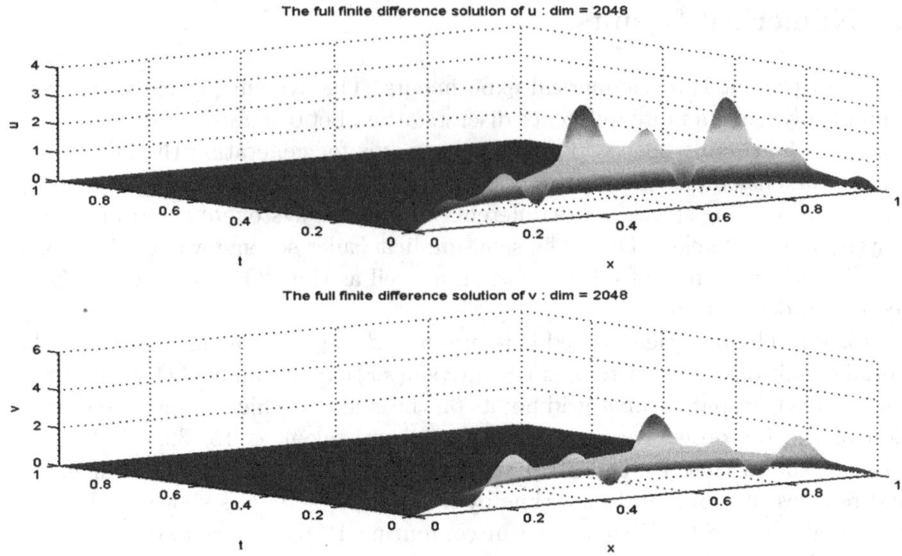

Fig. 1. Solution plots of the model from the full-order system of dimension 2048.

Case 2. The numerical results obtained in this case (see Table 2 and Fig. 1) were generated with parameters: $m = \beta = 1.1$, $\kappa = 5$, and $\delta = 1$. In Figs. 2 and 3, the solutions for state variables (u and v) from POD and POD-DEIM reduced systems, with dimPOD = 10 and dimDEIM = 5, are depicted with the corresponding ones from the full-order system, as well as the corresponding average relative errors at the grid points. We used the following initial conditions: $u_0(x) = 10x(1-x)(1+0.8\sin(30x)\cos(10x))$, $v_0(x) = 10x(1-x)(1+0.8\sin(10x)\cos(30x))$. In comparison with Case 1, here the densities of the species present initially large fluctuations along the whole space domain, damped very fast by the Allee effect.

Case 3. In this experiment we use the same initial conditions and values of the parameters as those indicated in Case 1. Here we performed the computations with dimPOD = 45 and dimPOD-DEIM = 90. The numerical results are contained in Table 3. We note that the POD-DEIM relative errors for both state variables, u and v, are 10 times smaller than those obtained in Case 1.

Table 3. CPU time of full-order system, POD and POD-DEIM reduced systems with the corresponding average relative errors for u and v – Case 3.

Internal Nodes N	CPU Time Full Dim	CPU Time POD	CPU Time POD–DEIM	$Error^{rel}$ POD – u	$Error^{rel}$ POD – DEIM – u	$Error^{rel}$ POD – v	$Error^{rel}$ POD – DEIM – v
128	5.741809e-001	8.289299e-001	5.605313e-002	1.169554e-005	1.106775e-005	2.098927e-004	5.454479e-003
256	1.144514e+000	9.969025e-001	5.958563e-002	1.169545e-005	1.055126e-005	2.098871e-004	2.347417e-003
512	3.807256e+000	1.154277e+000	6.668969e-002	1.169543e-005	1.486580e-005	2.098857e-004	5.295462e-003
1024	1.886075e+001	1.411974e+000	6.400274e-002	1.169542e-005	1.264476e-005	2.098853e-004	5.813729e-003
2048	1.098164e+002	2.144956e+000	6.639338e-002	1.169542e-005	1.548229e-005	2.098852e-004	7.346586e-003

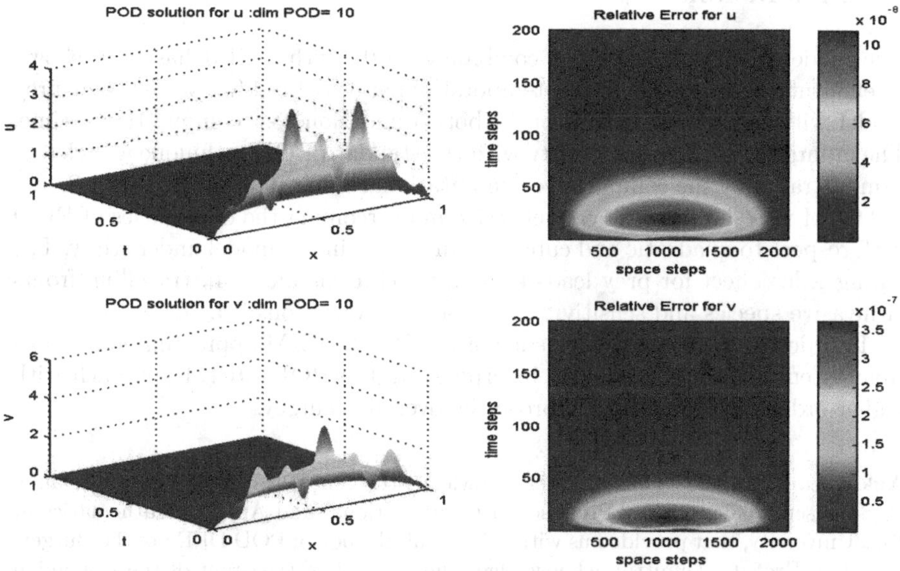

Fig. 2. Solution plots of the model from POD reduced system (dimPOD = 10) with the corresponding average relative errors at the inner grid points – Case 2.

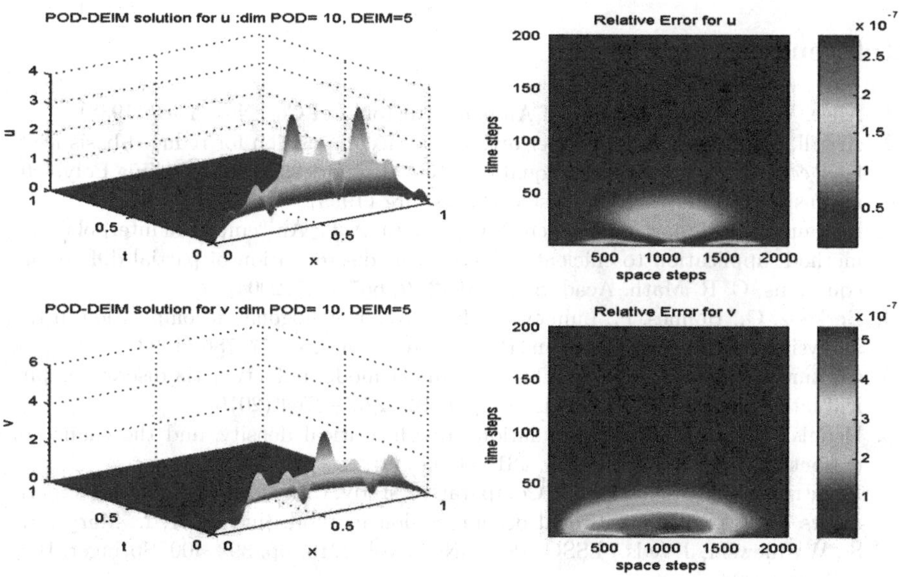

Fig. 3. Solution plots of the model from POD-DEIM reduced system (dimPOD = 10, dimDEIM = 5), with the corresponding average relative errors at the inner grid points – Case 2.

5 Conclusions

The model reduction technique combining POD with DEIM has been shown to efficiently capture the spatio-temporal dynamics of a diffusive predator-prey model with substantial reduction in both dimension and computational time. The failure to decrease complexity with the standard POD technique was clearly demonstrated by the comparative computational times shown in Tables 1, 2, and 3. DEIM was shown to be very effective in overcoming the deficiencies of POD with respect to quadratic and cubic nonlinearities in the model under study. The strong Allee effect for prey leads to a very rich dynamics [14], travelling fronts of invasive species and sensitivity to parameter variations [14,17].

In order to increase the efficiency of the POD-DEIM approximation, a possible extension would consist in incorporating the POD-DEIM approach with higher-order FD schemes to improve the overall accuracy.

Acknowledgement. The authors acknowledge the generous help of Professor Danny C. Sorensen Noah Harding Professor of Computational and Applied Mathematics at Rice University, that provided us with a 1-D Matlab code of POD DEIM of the Burgers equation. Prof. G. Dimitriu acknowledges the support of the grant of the Romanian National Authority for Scientific Research, CNCS - UEFISCDI, project number PN-II-ID-PCE-2011-3-0563, contract no. 343/5.10.2011 "Models from medicine and biology: mathematical and numerical insights". Prof. I.M. Navon and Dr. R. Ştefănescu acknowledge the partial support of NOAA grant NA10NES4400008.

References

1. Allee, W.C.: The Social Life of Animals. Norton and Co., New York (1938)
2. Atwell, J.A., King, B.B.: Proper orthogonal decomposition for reduced basis feedback controllers for parabolic equations. ICAM report 99–01-01, Virginia Polytechnic Institute and State University, Blacksburg (1999)
3. Barrault, M., Maday, Y., Nguyen, N.C., Patera, A.T.: An "empirical interpolation" method: application to efficient reduced-basis discretization of partial differential equations. C. R. Math. Acad. Sci. Paris **339**, 667–672 (2004)
4. Berkooz, G., Holmes, P., Lumley, J.: The proper orthogonal decomposition in the analysis of turbulent flows. Ann. Rev. Fluid Mech. **25**, 777–786 (1993)
5. Chaturantabut, S., Sorensen, D.C.: Nonlinear model reduction via discrete empirical interpolation. SIAM J. Sci. Comput. **32**, 2737–2764 (2010)
6. Dennis, B.: Allee effects: population growth, critical density, and the chance of extinction. Nat. Res. Model. **3**, 481–538 (1989)
7. Dimitriu, G., Apreutesei, N.: Comparative study with data assimilation experiments using proper orthogonal decomposition method. In: Lirkov, I., Margenov, S., Waśniewski, J. (eds.) LSSC 2007. LNCS, vol. 4818, pp. 393–400. Springer, Heidelberg (2008)
8. Dimitriu, G., Apreutesei, N., Ştefănescu, R.: Numerical simulations with data assimilation using an adaptive POD procedure. In: Lirkov, I., Margenov, S., Waśniewski, J. (eds.) LSSC 2009. LNCS, vol. 5910, pp. 165–172. Springer, Heidelberg (2010)

9. Dubois, D.: A model of patchiness for prey-predator plankton populations. Ecol. Model. **1**, 67–80 (1975)
10. Kunisch, K., Volkwein, S.: Control of the Burgers' equation by a reduced order approach using proper orthogonal decomposition. J. Optim. Theor. Appl. **102**(2), 345–371 (1999)
11. Lewis, M.A., Kareiva, P.: Allee dynamics and the spread of invading organisms. Theor. Popul. Biol. **43**, 141–158 (1993)
12. Morozov, A.Y., Petrovskii, S.V., Li, B.-L.: Bifurcations and chaos in a predator-prey system with the Allee effect. Proc. R. Soc. B **271**, 1407–1414 (2004)
13. Murray, J.D.: Mathematical Biology, 2nd edn. Springer, Berlin (1993)
14. Petrovskii, S.V., Malchow, H., Li, B.-L.: An exact solution of a diffusive predator-prey system. Proc. R. Soc. A **461**, 1029–1053 (2005)
15. Skellam, J.G.: Random dispersal in theoretical populations. Biometrika **38**, 196–218 (1951)
16. Ştefănescu, R., Navon, I.M.: POD/DEIM nonlinear model order reduction of an ADI implicit shallow water equations model. J. Comput. Phys. **237**, 95–114 (2013)
17. Volpert, V., Petrovskii, S.: Reaction-diffusion waves in biology. Phys. Live Rev. **6**, 267–310 (2009)

FARSITE and WRF-Fire Models, Pros and Cons for Bulgarian Cases

Nina Dobrinkova[1(✉)] and Georgi Dobrinkov[2]

[1] Institute of Information and Communication Technologies, Bulgarian Academy of Sciences, Sofia, Bulgaria
nido@math.bas.bg

[2] Institute of Mathematics and Informatics, Bulgarian Academy of Sciences, Sofia, Bulgaria
g.dobrinkov@gmail.com

Abstract. Statistics for the southern memeber states in EU from 2007 show increasement of occurence of wildland fires for the last 25 years. This is true also for Bulgaria. That is the reason a team from Bulgarian Academy of Sciences to start experimenting with different type models for fire spread prediction. The models which have been applied in Bulgaria in the last years for fire behaviour modelling range from typical mathematical approaches such as cell automata ones to the semi-empirical models like WRF-Fire (Weather Research Forecast - Fire) and FARSITE (Fire Area Simulator Model). Our article will focus on the last two models where GIS tools are applied for territories in south Bulgaria and we will show comparison between the time needed for calculation of burned area. We will focus on places for which the needed input data for both models is available and we will conclude with comparison between the models in case either one is used in operational mode from fire fighter brigades.

1 Introduction

Forest and field fires occur in nature mostly after thunder storms or single lightning hit on the ground. Such fires are considered as part of the life cycle in the forests and they are good for the biological species renewal. In the fires of this type there is a reduce of the accumulated organic matter in the autumn-winter period. However most of the Bulgarian forest or field fires occur mainly because of human mistake when working with fire in open space or as it is in the last decade in Bulgaria the fires are artificially made for easier timber collection afterwards. From official statistics published on the web-site of the ministry of forests, food and agriculture written in Bulgarian [1] fires between 1970 and 2005 have huge incensement in the year 2000 (Fig. 1).

New official statistics has not been published on the same site for the years from and after 2006 until present. That is why we have downloaded satellite information from NASA official web site http://www.earthdata.nasa.gov/data/nrt-data/firms/active-fire-data in GIS-point format and created statistics on

I. Lirkov et al. (Eds.): LSSC 2013, LNCS 8353, pp. 382–389, 2014.
DOI: 10.1007/978-3-662-43880-0_43,

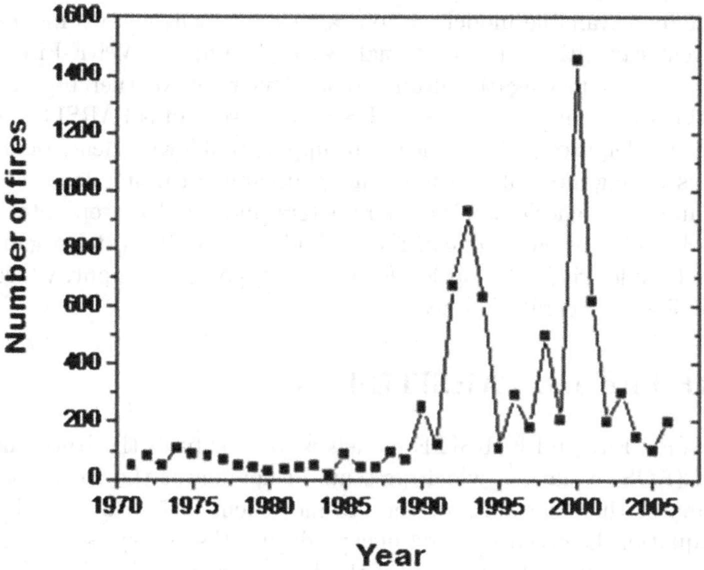

Fig. 1. Forest fires in Bulgaria between 1970–2005.

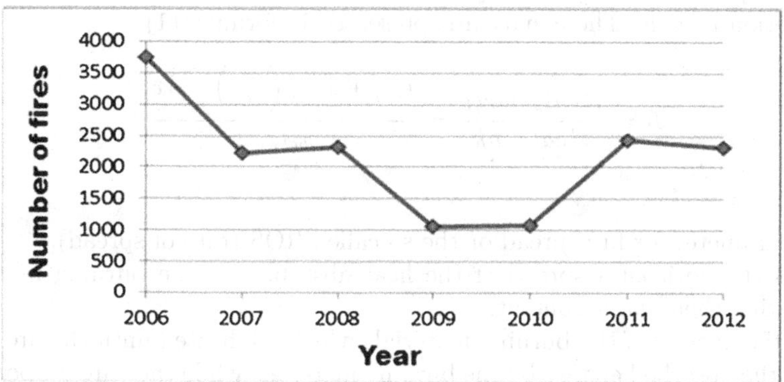

Fig. 2. NASA satellite images information about fires on the Bulgarian territory.

what the satellites has registered as fires on the Bulgarian territory from the period 2006 until 2012. The results are summarized on the diagram from Fig. 2.

From the two diagrams we can see that the wildland fire occurence on the territory of Bulgaria has been increased after the year of 1990. The first figure show the official statistics published by the Ministry of Agriculture, Food and Forests in Bulgaria, while the second figure is overview information prepared by the authors from sources of the NASA hotspots database for the territory of Bulgaria. As outcome from the diagrams we think that there is need for better development of the scientific research on the fire behavior modeling in Bulgaria for the different available models. In our article we focus on two models which have different capacities for operational usage. The first is WRF-Fire which is presenting the fire propagation from point of view of weather inputs, paying less attention on burning materials and relief. The second is FARSITE, which is stand alone model, used in US as decision support tool for incident commanders. That tool is having more attention to the landscape of the simulated area, have weather input for general weather parameters and wind as separate one and include fuel models as either one of the 13 Anderson or 40 Scott-Burgan classes. This second model gives also option for custom type of fuel input, which makes it more flexible for operational needs.

2 WRF-Fire and FARSITE Basis

For both WRF-Fire and FARSITE models is used as basis the Rothermel Rate Of Spread (ROS) equation, which presents in nominator the Heat Source and in denominator the Heat sink for the fire movement [3,7]. Rothermel describe with its equation how the released energy during the pyrolysis process is distributed in the nature. He shows how the burning materials release its energy while burning and how the surrounding materials absorbed it by evaporating its water quantities reaching the moment of setting their organic matter into fire. His equation is revolutionary and gives base for most of the nowadays fire propagation models. The equation is presented in formula (1):

$$R = \frac{HeatSource}{HeatSink} = \frac{I_{xig} + \int_{-\infty}^{0} \left(\frac{\partial I_z}{\partial_z}\right)_{Z_c} \mathrm{d}x}{\rho_{be} Q_{ig}}. \tag{1}$$

where

R - is parameter for fire spread or the so called ROS (rate of spread),
I_{xig} - is the horizontal spread of the heat absorbed by the burning materials evaporating their water content,
ρ_{be} - is the density of the burning materials which are heated until the fire start,
Q_{ig} - is the absorbed energy by the burning materials while they are evaporating their water content,
$\frac{\partial I_z}{\partial_z}$ - is the gradient of vertical intensity in the plane, where the energy is released.

Horizontal and vertical coordinates are x and z [2]. All parameters are in English system, where the spread rate is in ch/m, the fuels are in lb/ft^3 and lb.

This equation is modified in WRF-Fire model in a way to present ROS as domain where the fire is burning, there are also included level-set functions giving sub-domains presenting the active burning and already burned areas in a fire. In WRF-Fire ROS and the domains from the level-set functions are connected to the weather part of the model WRF (Weather Research Forecasting) there is constant feedback from the fire part to the meteorological one and that's why the fire part can not be externalized in order to be considered the surrounding area weather conditions. That is how the fire propagation is presented in WRF-Fire.

On the other hand FARSITE (Fire Area Simulator Model) incorporates existing models of surface fire, crown fire, point-source fire acceleration, spotting, and fuel moisture, where the ROS idea is taken into consideration. It is a standalone decision support tool, which is used by operational teams as fire growth simulator. The models integrated in FARSITE are using vector propagation for fire perimeter spread that is controlling both space and time resolution of fire propagation over the landscape. FARSITE results are vector fire perimeters the so called in GIS polygons at specified time intervals. The vertices of these polygons contain information on the fire's spread rate and intensity, from which can be produced raster maps of fire behavior useful in wild land fire management.

3 WRF-Fire Input for Real Data Calibration

In order to apply WRF-Fire model for real data calibration purposes there is a need for preprocessing of the input. The preprocessor compilation for real data cases need to process the available data with setting up place on the earth where the fire occurs. The location coordinates and all needed inputs are set by using file called *namelist.wps*, where is written section *geogrid* defining the weather domain from where the weather conditions will be taken during the burning processes. In order to define also the place where the fire is running there is a set up of smaller domain inside the weather one and that is done by including section *share* in the input file *namelist.wps*. From this two sections our simulation has weather and fire growth domains where the coordinates and the time step of getting new weather input for fire spread prognosis is done. The weather data is not available automatically for Bulgaria that is why is preferable to be used the 1 degree resolution from U.S. National Center for Environmental Protection (NCEP) as free source. Additional information which is included in the input file is also the topology. It is available through the Shuttle Radar Topography Mission (SRTM) for Bulgarian territory and is having 90 m resolution. Having this basis runs of simulations for the Bulgarian case near by the village of Leshnikovo, Harmanly region has been done on US and Bulgarian super computers. The US architecture used was the Janus cluster at the University of Colorado. The computer consists of nodes with dual Intel X5660 processors (total 12 cores per node), connected by QDR InfiniBand. The results there showed that runs using 360 and above as number computational cores can give simulation results faster than real time, but the fire burned shape will not match very well the real fire growth. This was reason of the data accuracy [3].

In Bulgaria runs of the same case simulated fire near by the village of Leshnikovo were computed on the IBM Blue Gene/P wich configuration is consisted of two racks with 2048 computational nodes connected by PowerPC 450 processors (32 bits, 850 MHz), 8192 cores and total 4 TB operational memory. The simulation could run successfully on 100 processors and it took 7 h and 43 min to run the case of Leshnikovo village. This result gives as real-time coefficient the number of - 0,0054, which is much below 1, which is not good for operational needs [4].

Having these results we could say that WRF-Fire model has the abilities for scientific fire investigation purposes of surface fires only, but as tool for operational fire growth predictions is very sophisticated and not every hardware can support it easily.

4 FARSITE Input for Real Data Calibration

Fire behavior models used in wildland fire management are distinguished as four main types - surface fires, crown fires, spotting and point-source fire acceleration. FARSITE combines in its structure all of them. This approach makes FARSITE more flexible than WRF-Fire because it does not focus only on surface fire type behavior.

The inputs needed in FARSITE for Bulgarian fire behavior calibration simulation has not been enough for the case nearby Leshnikovo village that is why we will describe how the model works with real data using as example the Ashley Lakes Fire in US. FARSITE is a fire growth simulator and as such requires spatial data that comprise the fuels, weather, and topographic information of the fire behavior. A simulation with FARSITE weather and winds are input as streams of data, whereas fuels and topography are GIS raster files for spatial data incorporation. The raster data resolution should not be higher than 25–50 m, because above that level the details in heterogeneous landscapes are not good for simulations.

The raster inputs can be divided in seven main categories, which we will briefly describe. The first category is the elevation of the terrain used in both meters and feet. It is used in a simulation for adjustment of temperature and humidity according to elevation input of the weather stream. The slope and aspect are presented in percents in the input data and are used directly for computing the angle of incident solar radiation (along with latitude, date and time of the day), which gives the spread rates on the surface. The third input is the fuel model, which in FARSITE is possible to be inserted as fuel type according to the Anderson 13 or Scott-Burgan 40 classes [5,6]. Also the fuels can be represented as custom type if the needed measurements are done on the field and the landscape coverage is determined correctly.

In Bulgaria fuel types division for the wild lands in the forests has never been developed. That is why our team started such classification by adopting the structures and ideas of the two main classes in US. That work is still ongoing and this is the reason that a Bulgarian simulation case with FARSITE will not be presented in this article. Canopy cover is the fifth element needed in the raster input which is inserted as percent and determine the average shading of the surface fuels that affects the moisture calculations. Also the canopy cover can give wind decrease in cases of 6.1 m and above high vegetation. Crown height is presented in meters or feet and it affects the wind profile that is extended above terrain. Canopy cover and crown height are factors for wind reduction. Crown base height and Crown bulk density are parameters in the raster input which are related to crown fires and both are giving to the simulations the option of a surface fire can go to a transition of active crown fire.

The third component for a simulation in FARSITE after we presented the fuel types and landscape characteristics is weather and wind inputs as two separate files. The input for the weather provides daily observations on temperature and humidity as well as precipitation that depicts a temporal weather stream. The input counts the month and the day for which we need information for daily amount of rain in hundredths of inch or millimeters as integer value. For the same day and month we insert information also about minimum and maximum temperatures of the day rounded to integers in degree Fahrenheit or Celsius, along with minimum and maximum humidity in percentage. The last information inserted is elevation in feet or meters above the sea level along with minimum and maximum rain amount if any for the observed day. This format of input for weather information allow limitation of the amount required data for a simulation.

The wind as part of the weather is inserted in a simulation as a separate file with its own characteristics, because of its importance for ROS components. Winds are usually variable in space and time. FARSITE however, assumes winds to be constant in space for a given wind stream but variable in time. That is because, there is no topographic effects on winds. The input format of winds is similar to that for a weather file. In wind file is needed information about month and the observed day and the nearest hour to the fire start for which as integer is presented the windspeed in 20 ft miles per hour or 10 m windspeed in kilometers per hour. The other component is the wind direction clockwise from north which is given as integer from 0 to 360. The last information in the wind input is the cloud cover specified in percentage integer number from 0 to 100 according to [7].

The FARSITE simulator model is created to run on a various conditions, but what is the best of it is that it could run also on personal computer. We will test it with Ashley Lakes Fire input data on computer with configuration Intel core Duo 1.40 GHz, 32-bit machine with 3 GB RAM with Operational system Windows 7.

The first thing which has to be included in a FARSITE simulation is the spatial data in landscape file. The landscape file can be generated if there are available five raster files referring to — elevation, slope, aspect, fuel model and canopy cover of the area for simulation. This file contains the main input for the simulation and can be changed every time when we change the five initial files as recreating it. The next step is to generate a project file which has all the initial conditions for a simulation. It consist of landscape file, adjustment file for the terrain, moisture file for the canopy cover, custom models for the fuel types, weather and wind files along with the road and water streams shape files which will give additional information when the simulation is running. We set the simulation duration by having the month and day information, because this is the required information from the FARSITE inputs. The weather always is having for first day the day before the actual fire appearance. In our case we set the simulation for August and the dates are 10th, 11th and 12th and the hours while we are interested to run our Ashley Lakes Fire is from 12.00 h noon

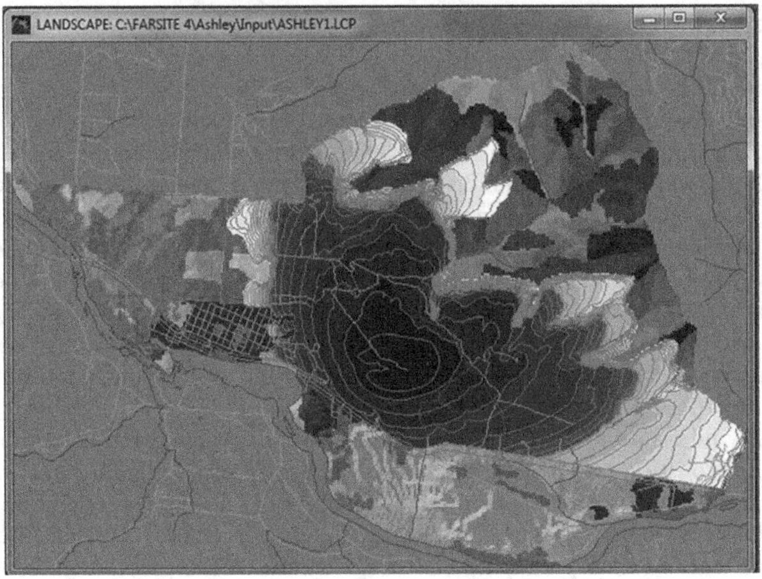

Fig. 3. Ashley Lake Fire - Burned area.

time on August 11th until 18.00 h on August 12th. The first day we simulate the weather conditions before the fire occurrence, so we don't set time for it. After we set the duration we have to set the ignition type. In our Ashley case we set it to ignition point and we get as fuel type 1, which is corresponding to short grass and gives surface fire spread. For the outputs we set shape and raster GIS files which will give the final picture of the simulation result. The simulation runs for 1 min and 6 s and provide information about the intensity of fire spread with GIS shape file, which gives the burned area in black and the non or less burned area with lighter or white color on Fig. 3.

5 Conclusion

From the presented results we can say that for research needs WRF-Fire and FARSITE models are very good options, but when it comes to operational tool for real fire simulations the Fire Area Simulator (FARSITE model) has better characteristics. Our work is still ongoing for meeting the hardware limits of Blue Gene/P, the Bulgarian supercomputer for the additional runs of WRF-Fire. For the FARSITE model we need to finish the fuel type categories according to the two US classifications available, but we believe that in near future in Bulgaria will be used more and different capacity models for operational need for the firebrigades and volunteer groups helping them in their everyday job.

Acknowledgments. This work was supported by the National Science Fund of the Bulgarian Ministry of Education, Youth and Science under Grants DID02/29 and I01/0006. Also the paper is supported by the project OUTLAND, funded under second call of Greece-Bulgaria Territorial Cooperation Program 2007–2013.

References

1. Ecopolis, bulletin 48 (2001), Forest fires reach catastrophic scales (In Bulgarian). http://www.bluelink.net/bg/bulletins/ecopolis12/1_os_1.html
2. Rothermel, R.C.: A mathematical model for predicting fire spread in wildland fuels. Research Paper INT-115. Ogden, UT: US Department of Agriculture, Forest Service, Intermountain Forest and Range Experiment Station, pp. 1–40 (1972)
3. Jordanov, G., Beezley, J.D., Dobrinkova, N., Kochanski, A.K., Mandel, J., Sousedík, B.: Simulation of the 2009 Harmanli fire (Bulgaria). In: Lirkov, I., Margenov, S., Waśniewski, J. (eds.) LSSC 2011. LNCS, vol. 7116, pp. 291–298. Springer, Heidelberg (2012)
4. Dobrinkova, N., Jordanov, G., Vassilev, P.: Generalized net model of decision support system of wildland fire estimation. The Case of Harmanli Fire (Bulgaria) 2009. In: Twelfth International Workshop on Intuitionistic Fuzzy Sets and Generalized Nets, WIFSGN'2013, Warsaw, 11 October 2013 (accepted)
5. Anderson, H.E.: Aids to determining fuel models for estimating fire behavior. USDA For. Serv. Gen. Techn. Rep. INT-122 (1982)
6. Scott, J.H., Burgan, R.E.: Standard fire behavior fuel models: a comprehensive set for use with Rothermel's surface fire spread model. Gen. Tech. Rep. RMRSGTR-153. FortCollins, CO: U.S. Department of Agriculture, Forest Service, Rocky Mountain Research Station, 72 p. (2005)
7. Finney, M.A.: FARSITE: Fire Area Simulator - model development and evaluation. Res. Pap. RMRS-RP-4. Ft. Collins, CO: U.S. Department of Agriculture, Forest Service, Rocky Mountain Research Station, 47 p. (1998)

Analysis of the Processes Which Form the Air Pollution Pattern over Bulgaria

Georgi Gadzhev[1], Kostadin Ganev[1(✉)], Nikolay Miloshev[1], Dimiter Syrakov[2], and Maria Prodanova[2]

[1] National Institute of Geophysics Geodesy and Geography, Bulgarian Academy of Sciences, Acad. G. Bonchev Str., Bl.3, 1113 Sofia, Bulgaria
{ggadjev,kganev}@geophys.bas.bg

[2] National Institute of Meteorology and Hydrology, Bulgarian Academy of Sciences, "Tsarigradsko Shose" 66, 1784 Sofia, Bulgaria

Abstract. The air pollution transport is subject to different scale phenomena, each characterized by specific atmospheric dynamics mechanisms, chemical transformations, typical time scales etc. The air pollution pattern is formed as a result of interaction of different processes. The present study attempts to make some evaluations of the contribution of different processes to the local to regional pollution over Bulgaria. The US EPA Model-3 system is chosen as a modelling tool. As the NCEP Global Analysis Data with one degree resolution is used as meteorological background, the MM5 and CMAQ nesting capabilities are applied for downscaling the simulations to a 3 km resolution over Bulgaria.

The TNO emission inventory is used as emission input. Special preprocessing procedures are created for introducing temporal profiles and speciation of the emissions. The biogenic emissions are estimated by the model SMOKE.

The Models-3 "Integrated Process Rate Analysis" option is applied to discriminate the role of different dynamic and chemical processes for the pollution formation. The processes that are considered are: advection, diffusion, mass adjustment, emissions, dry deposition, chemistry, aerosol processes and cloud processes/aqueous chemistry.

The simulations are carried out for several years. The obtained results make it possible to evaluate the impact of the above listed processes in many different terms - spatial pattern, averaged over the country or for selected points, seasonal behaviour.

Keywords: Atmospheric composition · Regional scale modelling · US EPA Models-3 system · Integrated process rate analysis

1 Introduction

Recently extensive studies for long enough simulation periods and good resolution of the atmospheric composition status in Bulgaria have been carried out using up-to-date modeling tools and detailed and reliable input data [5,7,8].

I. Lirkov et al. (Eds.): LSSC 2013, LNCS 8353, pp. 390–396, 2014.
DOI: 10.1007/978-3-662-43880-0_44,

The air pollution transport is subject to different scale phenomena, each characterized by specific atmospheric dynamics mechanisms, chemical transformations, typical time scales etc. The air pollution pattern is formed as a result of interaction of different processes, so knowing the contribution of each for different meteorological conditions and given emission spatial configuration and temporal behavior is by all means important. That is why the one of the overall study goals is to make some evaluations of the contribution of different processes to the local to regional pollution over the Balkans and/or Bulgaria.

2 Approaches, Tools, Data, Domains and Nesting

All the simulations are based on the US EPA Model-3 system. The system consists of three components: MM5 [4,8], used as meteorological pre-processor, CMAQ [1,2], the Chemical Transport Model of the system and SMOKE [3] - the emission pre-processor of Models-3 system.

The large scale (background) meteorological data used by the study is the NCEP Global Analysis Data with $1° \times 1°$ resolution. At the moment the created database contains all the necessary information since year 2000.

The TNO high resolution emission inventory [9] is exploited. A more detailed description of the emission modeling is given in [5].

As far as the background meteorological data is the NCEP Global Analysis Data with $1° \times 1°$ resolution, it is necessary to use MM5 and CMAQ nesting capabilities as to downscale to 3 km step for the innermost domain (Bulgaria).

The Models-3 "Integrated Process Rate Analysis" option is applied to discriminate the role of different dynamic and chemical processes for the air pollution pattern formation. The procedure allows the concentration change for each compound for an hour ΔC to be presented as a sum of the contribution of the processes, which determine the concentration. The processes that are considered are: advection, diffusion, mass adjustment, emissions, dry deposition, chemistry, aerosol processes and cloud processes/aqueous chemistry.

3 Some Examples of Process Analysys Simulations

Due to the limited volume of the present paper only few examples will be given here, just to demonstrate the kind of information and knowledge that can be gained from the process analysis. All the results are retrieved from a pretty extensive (8 years) simulation ensemble, so they can be considered representative for the atmospheric composition "climate" behaviour.

An example of the annually averaged special distribution of the processes contribution to the surface ozone is given in Fig. 1. It can be seen that the chemical processes have mostly negative impact. In particular the big cities and the road network (powerful nitrogen oxide sources) can be clearly followed as ozone sinks.

The vertical diffusion impact is mostly positive (turbulent transport of ozone from the upper layers). The effect is very prominent in the big cities, where the

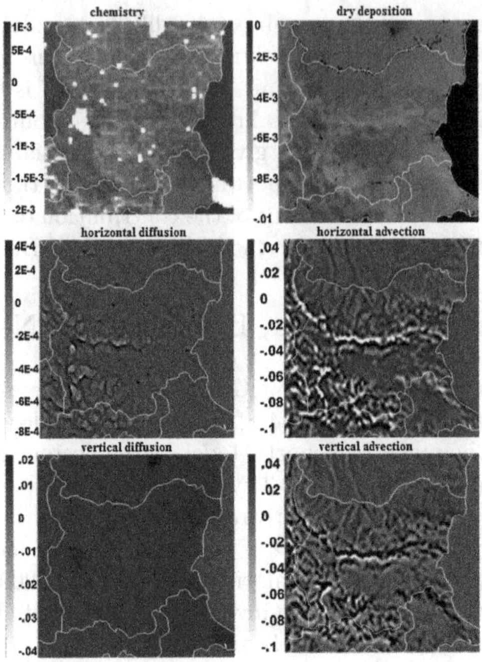

Fig. 1. Horizontal distribution of the contributions [µg/hour] of different processes to the hourly surface ozone changes at 06.00 GMT (08.00 local time)

very large nitrogen oxide surface sources cause big ozone deficiency (big negative vertical gradients) and so the turbulent transport is more intensive. Some small spots of vertical diffusion negative impact can be seen at the location of big power plants. This is probably due to the fact that these are high sources of nitrogen oxide, which cause ozone deficiency aloft, so the ozone vertical gradients near surface are positive.

The horizontal and vertical advection contributions pattern is very complex and clearly reflects landscape induced local circulation systems. The horizontal and vertical advection contributions have mostly opposite signs, which is a direct and apparent consequence of the atmosphere continuity equation.

The horizontal diffusion, as it should, acts for compensating the ozone deficiency and so is generally in counter-phase with the chemical processes.

The same processes contribution to the surface SO_2 changes are shown in Fig. 2. What can be immediately seen from the plots is that the most prominent SO_2 sources - the thermal power plants (TPP) can be detected in the fields of practically all the processes. The explanation of these effects is rather straightforward: the large SO_2 sources form small areas of very high concentrations around the TTP's, thus zones with large positive vertical diffusion (high SO_2 sources) and negative chemical processes (formation of sulfate radicals), horizontal advection and diffusion contributions. To some extend these effects can

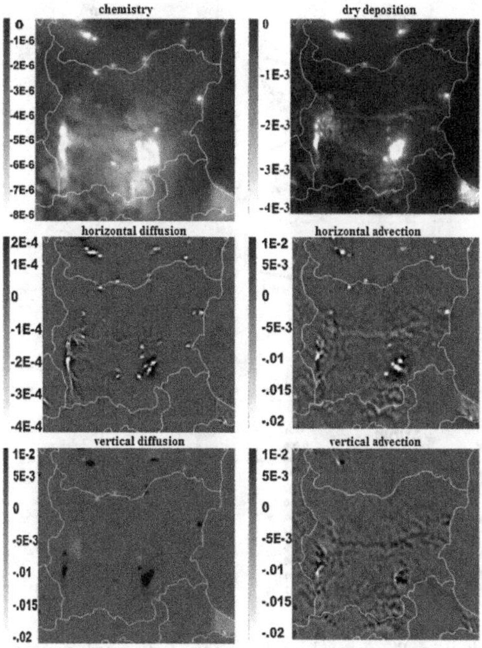

Fig. 2. Horizontal distribution of the contributions [µg/hour] of different processes to the hourly surface SO_2 changes at 17.00 GMT (19.00 local time).

be observed also for the large cities with the difference that for the cities the vertical diffusion is negative (low SO_2 sources in the cities).

The landscape effects in the horizontal and vertical advection contributions are again well manifested.

The averaged over the territory of Bulgaria contributions of some of the processes to the surface ozone concentrations will be also demonstrated Fig. 3. Very briefly the main characteristics, which can be seen from the plots, are the following: (1) There are well manifested seasonal differences and diurnal variations; (2) The ozone concentration change is formed as a rather small sum of processes with larger values and different signs; (3) Averaged for the territory of Bulgaria the impacts of horizontal diffusion and cloud processes/aqueous chemistry are negligible; (4) For all the seasons, except winter, and annually the vertical diffusion has a large positive impact, especially during the day (more intensive turbulence) - ozone transport from higher atmosphere to ground level; (5) The dry deposition has negative impact, but it is almost negligible during winter and significant for spring, summer (in particular) and autumn during daytime. This is easy to explain - the dry deposition is proportional to surface concentration, and so is large when the surface concentrations are large; (6) For all the seasons, except summer and especially in winter, and annually the horizontal advection has large positive impact. In summer around noon there is a

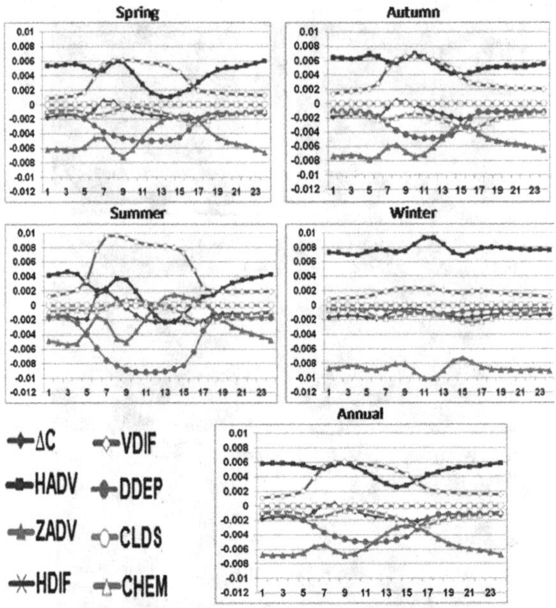

Fig. 3. Plots of the "typical" annual and seasonal diurnal course of the averaged for the territory of Bulgaria contributions of vertical advection (ZADV), vertical diffusion (VDIF), emissions (EMIS), dry deposition (DDEP), chemistry (CHEM), horizontal advection (HADV), vertical advection (ZADV), cloud processes/aqueous chemistry (CLDS) to the hourly changes (ΔC) of surface O_3.

period of horizontal advection negative impact. All this means that for most of the time there is ozone inflow trough the country boundary; (7) The impact of chemical processes is always negative, except during daytime in the summer; (8) The resulting hourly ozone surface concentration change ΔC is always negative, except for some hours during the day in summer (most prominent), spring and autumn.

The last four characteristic features of the processes behavior are sound evidence that the ozone/ozone precursors in Bulgaria are mostly of foreign origin.

In order the local heterogeneities of the different processes behavior to be demonstrated the annually averaged process contributions to SO_2 surface concentration changes for 4 different points in Bulgaria, together with the averaged for the country are shown in Fig. 4. It can be seen that the processes temporal behavior and interaction is different for the different points. It is remarkable how fast and chaotic the changes of the horizontal and vertical advection are for Sofia and Burgas. The horizontal and vertical advection contributions have mostly opposite signs, which effect had already been mentioned above. The diurnal course of horizontal and vertical advection contributions for Rojen is a very typical and good example of the role of mountain circulation.

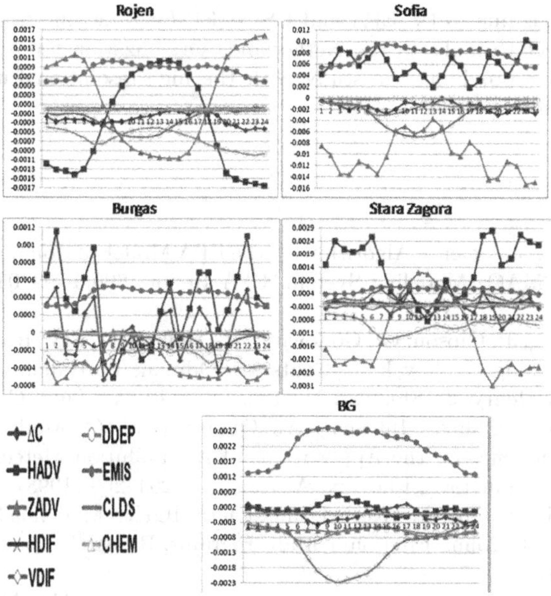

Fig. 4. Plots of the "typical" annual diurnal course of the in some points and averaged for the territory of Bulgaria contributions of vertical advection (ZADV), vertical diffusion (VDIF), emissions (EMIS), dry deposition (DDEP), chemistry (CHEM), horizontal advection (HADV), vertical advection (ZADV), cloud processes/aqueous chemistry (CLDS) to the hourly changes (ΔC) of surface SO_2

4 Conclusions

The numerical experiments performed produced a huge volume of information, which can not be presented here. Some more general features of the processes behavior could be mentioned, however:

- The spatial/temporal behavior of the processes is very complex;
- For some processes the contribution sign is obvious (like emissions or dry deposition), but some can have different signs for different species, depending also on the emission configuration and the meteorological conditions;
- For most of the compounds some of the advection/diffusion processes have a significant role.

The analysis of the behavior of different processes does not give simple answers to the question how the air pollution in a given point or region is formed. The "Integrated Process Rate Analysis" is a fruitful approach, however and it would be worthwhile to provide a more general discussion of the simulated processes.

Acknowledgements. The present work is supported by the Bulgarian National Science Fund (grant no DCVP-02/1/29.12.2009) and by the EU FP7 project PASODOBLE, grant no 241557.

Deep gratitude is due to US EPA and US NCEP for providing free-of-charge data and software. Special thanks to the Netherlands Organization for Applied Scientific research (TNO) for providing the study with the high-resolution European anthropogenic emission inventory.

References

1. Byun, D., Ching, J.: Science Algorithms of the EPA Models-3 Community Multiscale Air Quality (CMAQ) Modeling System. EPA Report 600/R-99/030, Washington, DC (1999)
2. Byun, D., Young, J., Gipson, G., Godowitch, J., Binkowski, F.S., Roselle, S., Benjey, B., Pleim, J., Ching, J., Novak, J., Coats, C., Odman, T., Hanna, A., Alapaty, K., Mathur, R., McHenry, J., Shankar, U., Fine, S., Xiu, A., Jang, C.: Description of the Models-3 Community Multiscale Air Quality (CMAQ) Modeling System. In: 10th Joint Conference on the Applications of Air Pollution Meteorology with the A&WMA, 11–16 Jan 1998, Phoenix, Arizona, pp. 264–268 (1998)
3. CEP: Sparse Matrix Operator Kernel Emission (SMOKE) Modeling System. University of Carolina, Carolina Environmental Programs, Research Triangle Park, North Carolina (2003)
4. Dudhia, J.: A non-hydrostatic version of the Penn State/NCAR Mesoscale Model: validation tests and simulation of an Atlantic cyclone and cold front. Mon. Weather Rev. **121**, 1493–1513 (1993)
5. Gadzhev, G., Jordanov, G., Ganev, K., Prodanova, M., Syrakov, D., Miloshev, N.: Atmospheric composition studies for the Balkan region. In: Dimov, I., Dimova, S., Kolkovska, N. (eds.) NMA 2010. LNCS, vol. 6046, pp. 150–157. Springer, Heidelberg (2011)
6. Gadzhev, G., Ganev, K., Syrakov, D., Miloshev, N., Prodanova, M.: Contribution of biogenic emissions to the atmospheric composition of the Balkan region and Bulgaria. Int. J. Environ. Pollut. **50**(1/2/3/4), 130–139 (2012)
7. Gadzhev, G., Ganev, K., Miloshev, N., Syrakov, D., Prodanova, M.: Numerical study of the atmospheric composition in Bulgaria. Comput. Math. Appl. **65**, 402–422 (2013)
8. Grell, G.A., Dudhia J., Stauffer D.R.: A description of the Fifth Generation Penn State/NCAR Mesoscale Model (MM5). NCAR Technical Note, NCAR TN-398-STR, 138 pp. (1994)
9. Visschedijk, A., Zandveld P., van der Gon, H.: A high resolution gridded European emission database for the EU integrated project GEMS, TNO report 2007-A-R0233/B, The Netherlands (2007)

On the Adaptive Time-Stepping in Radio-Frequency Liver Ablation Simulation: Some Preliminary Results

K. Georgiev(✉), N. Kosturski, and Y. Vutov

Institute of Information and Communication Technologies,
Bulgarian Academy of Sciences, Sofia, Bulgaria
georgiev@parallel.bas.bg

Abstract. Radio-frequency ablation is a low invasive technique for treatment of liver tumors. This work concerns the mathematical modeling and computer simulation of the heat transfer process. The core is solving the time-dependent partial differential equation of parabolic type. Instead of a uniform discretization of the considered time interval, an adaptive time-stepping procedure is applied in an effort to decrease the simulation time. The procedure is based on the local comparison of the Crank Nicholson and backward Euler approximations. Results of some preliminary numerical experiments performed on a selected test problems are presented and discussed.

1 Introduction

The minimally invasive treatment called radio-frequency ablation (RFA), one of several types of ablation therapy, may be the alternative when open surgery of certain cancer types is not a good option. Guided by imaging techniques, the doctor inserts a thin needle through the skin and into the tumor. High-frequency electrical energy delivered through this needle heats and destroys the tumor. The circuit is closed with a ground pad applied to the patient's skin.

An important advantage of RF current (over previously used low frequency AC or pulses of DC) is that it does not interfere with the muscles and can be used without the need for general anaesthesia.

There is an ongoing research in RF probe design. The right procedure parameters are very important for the successful killing of all of the tumor cells with minimal damage on the non-tumor cells.

Computer simulation on geometry obtained from a magnetic resonance imaging (MRI) scan of the patient is performed. The influence of the position of the ground pad to the ablated volume is of special interest, both from the medical and simulation point of view. Often, in computer simulations reported in the literature e.g. [1,4–6,10,11], the position of the ground pad is neglected and a simple computational domain with a cubic shape is considered. In [12] the

I. Lirkov et al. (Eds.): LSSC 2013, LNCS 8353, pp. 397–404, 2014.
DOI: 10.1007/978-3-662-43880-0_45,

authors check the correctness of the assumption that when the pad is *far* from the probe then zero potential condition can be applied on the whole boundary of the domain and compare the resulting ablated volumes, when ground pads are put in different positions.

In this work, an adaptive time stepping algorithm is applied to the simulation in order to reduce the computational time.

The rest of the paper is organized as follows. In Sect. 2, the mathematical model is presented along with the space and time discretization schemes. Section 3 describes the adaptive time-stepping algorithm. Section 4 is devoted to the computer simulations and analysis of the results obtained on an IBM Blue Gene/P supercomputer. Finally, some concluding remarks can be found in Sect. 5.

2 The Model, Space and Time Discretization

As mentioned above, the RFA procedure destroys the unwanted tissue by heating, arising when the energy dissipated by the electric current flowing through a conductor is converted to heat. The considered RF probe consists of a stainless steel needle, insulated with polyurethane. The RFA procedure starts by placing the probe inside the tumor. The surgeon performs this under computed tomography (CT) or ultrasound guidance. The human liver has a complex structure, composed of materials with unique thermal and electrical properties. There are three types of blood vessels with different sizes and flow velocities. Here, a simplified test problem, where the liver consists of homogeneous hepatic tissue and only the large portal vein vessels is considered.

The bio-heat time-dependent partial differential equation [5,6] is the governing equation describing this process. It can be presented as follows:

$$\rho c_{heat} \frac{\partial T}{\partial t} = \nabla \cdot k \nabla T + J \cdot E - \alpha \, h_{\mathrm{B}} \, (T - T_{\mathrm{B}}), \tag{1}$$

where the thermal energy arising from the current flow is described by $J \cdot E$ in (1) and $\alpha \, h_{\mathrm{B}} \, (T - T_{\mathrm{B}})$ accounts for the heat loss due to blood perfusion in the capillaries. The heat produced from metabolic functions of the liver is neglected. The initial and boundary conditions which are used in this approach are as follows:

$$T = 37°C \qquad \text{when } t = 0 \text{ at } \Omega, \tag{2a}$$

$$T = 37°C \qquad \text{when } t \geq 0 \text{ at } \partial\Omega \backslash \Gamma_{\mathrm{R}}, \tag{2b}$$

$$-k \frac{\partial T}{\partial n} = \alpha (T - T_{\mathrm{B}}) \qquad \text{when } t \geq 0 \text{ at } \Gamma_{\mathrm{R}} \tag{2c}$$

The notations which are used in (1) and (2) are given below:

- Ω – the entire domain of the model;
- $\partial\Omega$ – the boundary of the domain;
- Γ_R – the boundary of the blood vessel;
- ρ – density [kg/m^3];
- c_{heat} – specific heat [J/kg K];
- k – thermal conductivity [W/m K];
- J – current density [A/m];
- E – electric field intensity [V/m];
- t – time [s];
- T – temperature [K];
- T_{B} – blood temperature (37°C);
- w_{B} – blood perfusion coefficient [s^{-1}];
- $h_{\mathrm{B}} = \rho_{\mathrm{B}} c_{\mathrm{B}} w_{\mathrm{B}}$ – convective heat transfer coefficient accounting for the blood perfusion in the model;
- α – tissue state coefficient;
- n – the outward-pointing normal vector of the boundary.

The cumulative damage integral $\Psi(t)$ is used as a measure of ablated region [1,12]:

$$\Psi(t) = \ln\left(\frac{c(0)}{c(t)}\right) = A \int e^{-\frac{\Delta E}{RT(t)}}\, dt, \tag{3}$$

where $c(t)$ is the concentration of living cells, R is the universal gas constant, A is the "frequency" factor for the kinetic expression [s^{-1}], and ΔE is the activation energy for the irreversible damage reaction [J mol^{-1}]. The values used $A = 7.39 \times 10^{39}$ s^{-1} and $\Delta E = 2.577 \times 10^5$ J mol^{-1} are taken from [1]. Tissue damage $\Psi(t) = 4.6$ corresponds to 99 % probability of cell death. The value of $\Psi(t) = 1$, corresponding to 63 % probability of cell death is significant, because at this point the tissue coagulation first occurs and blood perfusion stops.

The tissue state coefficient α is expressed as

$$\alpha(t) = \begin{cases} e^{-\Psi(t)} & \text{if } \Psi(t) < 1, \\ 0 & \text{if } \Psi(t) \geq 1. \end{cases}$$

In the presented algorithm the bio-heat problem (1) is solved in two steps (see [12] for more details):

1. Finding the heat source $J \cdot E$ using that: (a) $E = -\nabla V$ (V is the electric potential in the computational domain Ω), and (b) $J = \sigma E$, where σ is the electric conductivity [S/m];
2. Finding the temperature T by solving the heat transfer Eq. (1) using the heat source $J \cdot E$ obtained in the first step.

For the numerical solution of (1) the finite element method in space is used [7]. *Linear conforming tetrahedral elements* are used in this study. They are directly defined on the elements of the used unstructured mesh. An *algebraic multigrid* (AMG) preconditioner is used [3]. The time derivative is discretized via finite differences and the both the *backward Euler* [8] and the *Crank-Nicholson* schemes are used [9].

Let the matrices K and M be the stiffness and mass matrices from the finite element discretization of (1):

$$K = \left[\int_\Omega k \nabla\Phi_i \cdot \nabla\Phi_j d\mathbf{x}\right]_{i,j=1}^N, \qquad M = \left[\int_\Omega \rho c_{heat} \Phi_i \Phi_j d\mathbf{x}\right]_{i,j=1}^N.$$

Let us also denote with $\Omega_{\rm B}$ the subdomain of Ω where we account for the blood perfusion (the liver tissue) and with $M_{\rm B}$ the matrix

$$M_{\rm B} = \left[\int_{\Omega} \delta_{\rm B} h_{\rm B} \Phi_i \Phi_j d\mathbf{x}\right]_{i,j=1}^{N}, \quad \text{where} \quad \delta_{\rm B}(x) = \begin{cases} \alpha & \text{for } x \in \Omega_{\rm B}, \\ 0 & \text{for } x \in \Omega \backslash \Omega_{\rm B}. \end{cases}$$

The influence of the Robin boundary conditions given in (2c) and the electric field intensity is presented by:

$$M_{\rm R} = \left[\int_{\Gamma_{\rm R}} \alpha \Phi_i \Phi_j d\mathbf{x}\right]_{i,j=1}^{N}, \quad \text{and} \quad F = \left[\int_{\Omega} JE \Phi_i \Phi_j d\mathbf{x}\right]_{i,j=1}^{N},$$

Than, the spatially discretized parabolic Eq. (1) can be written in matrix form as:

$$M\frac{\partial T}{\partial t} + (K + M_{\rm B} + M_{\rm R})T = F + M_{\rm B}T_{\rm B} + M_{\rm R}T_{\rm B}. \tag{4}$$

3 Adaptive Time-Stepping Algorithm

To ensure accuracy and not waste computational effort, it is important to adapt the time steps to the behavior of the solution.

The time discretization for both backward Euler method and the Crank-Nicolson one can be written in the form

$$\begin{aligned}(M + \tau^n\theta(K + M_{\rm B} + M_{\rm R}))\, T^{n+1} &= (M - \tau^n(1-\theta)(K + M_{\rm B} + M_{\rm R}))\, T^n \\ &\quad + (\tau^n\theta + \tau^n(1-\theta))(F + M_{\rm B}T_{\rm B} + M_{\rm R}T_{\rm B}),\end{aligned} \tag{5}$$

where the current (n-th) time-step is denoted with τ^n, the unknown solution at the next time step – with T^{n+1}, and the solution at the current time step – with T^n. If we set the parameter $\theta = 1$, (5) gives a system for the backward Euler discretization. When $\theta = 0.5$ (5) becomes Crank-Nicolson one. The solution of the linear system (5) with $\theta = 1$ and $\theta = 0.5$ gives us $T_{\rm BE}$ and $T_{\rm CN}$ respectively.

A suitable adaptive time-stepping procedure is based on a local comparison of the backward Euler ($T_{\rm BE}$) and Crank-Nicolson ($T_{\rm CN}$) approximations for the current timestep, and is controlled by the ratio

$$\eta = \frac{||T_{\rm CN} - T_{\rm BE}||}{||T_{\rm BE}||}. \tag{6}$$

This approach has a down side, that solving two linear systems is required to obtain $T_{\rm BE}$ and $T_{\rm CN}$. This is, from the computational point of view, expensive. Nevertheless overall decrease in computational time is expected.

The algorithm below, describing our adaptive time-stepping procedure, is based on the one for adaptive time stepping for processes in spent nuclear fuel repositories [2]. It has several parameters:

1. τ^1 – initial timestep;
2. N_{Adapt} – a parameter showing how often the adaptive time stepping strategy is applied, e.g. $N_{\text{Adapt}} = 1$ shows that the adaptive time stepping is used on each step while $N_{\text{Adapt}} = 3$ – that the adaptive time stepping is performed at every third time step, $N_{\text{Adapt}} = 0$ indicates that all time steps are non-adaptive.
3. $\lambda_{\text{NonAdapt}}$ – a parameter showing whether and by how much the time step is multiplied, in non-adaptive time steps, e.g. $\lambda_{\text{NonAdapt}} = 1$ means that the time step is not changed, while $\lambda_{\text{NonAdapt}} = 1.2$ means that the time step on the current level is multiplied by 1.2 for the next time level.
4. ε_{min} and ε_{max} are minimal and maximal thresholds for the error estimate η.

Algorithm 1 (Adaptive Time-Stepping Procedure).

for $k = 1, 2, \ldots$ **until** *the end of time* **do**
    if *CurrentStepIsAdaptive*(N_{Adapt}, k)
    then
        do
            compute T_{BE}, T_{CN} *with* τ^k
            compute η
            if $\eta < \varepsilon_{min}$ **then** $\tau^{k+1} = 2\tau^k$
            if $\eta > \varepsilon_{max}$ **then** $\tau^k = 0.5\tau^k$
        while $\eta > \varepsilon_{max}$ // *if too big error, stay on the same timestep*
        $T^{k+1} = T_{BE}$
    else
        compute T_{BE} with τ^k
        $T^{k+1} = T_{BE}$
        $\tau^{k+1} = \tau^k \lambda_{NonAdapt}$
    end if
end for

The last timestep is always truncated to the time of simulation.

Inner PCG iteration with the BoomerAMG [3] preconditioner, part of the software package HYPRE, is used for the solution of (5). The preconditioner is reconstructed if the number of inner iterations goes above 12. The reconstruction takes place before the solution of the next timestep.

4 Computer Simulations and Analysis of the Output Results

The IBM Blue Gene/P computer, located at the Bulgarian Supercomputing Center, is used for the simulations and numerical experiments with the new adaptive time stepping algorithm. This machine consists of two racks, 2048 Power PC 450 based compute nodes, 8192 processor cores and a total of 4 TB random

Table 1. Vol_1 and $Vol_{4.6}$ as functions of the thresholds in the adaptive time-stepping algorithm

ε_{min}	ε_{min}	Vol_1 [cm^3]	Variation in %	$Vol_{4.6}$ [cm^3]	Variation in %
Without adaptive time stepping		22.15	-	15.60	-
5.0×10^{-3}	5.0×10^{-2}	23.72	7.08	16.37	4.90
5.0×10^{-3}	1.0×10^{-2}	23.69	6.99	16.36	4.83
1.0×10^{-3}	5.0×10^{-3}	23.01	3.89	16.06	2.91
5.0×10^{-4}	2.5×10^{-3}	22.94	3.57	16.05	2.84
2.5×10^{-4}	1.25×10^{-3}	22.72	2.57	15.93	2.08

access memory. Each processor core has a double-precision, dual pipe floating-point core accelerator. Sixteen I/O nodes are connected via fiber optics to a 10 Gbps Ethernet switch.

The material properties which are used in the simulations are taken from [5]. The blood perfusion coefficient is $w_B = 6.4 \times 10^{-3}$ s^{-1}. The applied electrical power is 15 W, and the simulation is done for 7 min.

We run several test to choose a suitable set of values for the threshold parameters ε_{min} and ε_{max}. As a quantitative criterion of quality of the solution we used two volumes – the volume Vol_1, which is the volume of the tissue, where the cumulative damage integral Ψ is greater than 1, and $Vol_{4.6}$ – the volume of the tissue, where $\Psi > 4.6$. The results of the nonadaptive algorithm with step $\tau = 1$ s were compared with the ones from adaptive runs. Some of the output results obtained on 128 processors on the IBM Blue Gene/P machine are presented in Table 1. Looking at the last four columns in this table one can see that an acceptable variation in the two important volumes less than 3 % occurs when the threshold interval is $[2.5 \times 10^{-4}, 1.25 \times 10^{-3}]$ and this interval is used in the computer simulations. Based on these preliminary tests, a number of runs were done both using 128 and 1024 processors. Uniformly refined mesh was used for the runs on 1024 processors. Some of the output results obtained during the simulations are presented in Tables 2 and 3. Comparing the total CPU times for 128 and 1024 processors (see the fifth column in both tables) and taking into account that we solve eight times bigger problems on eight times more processors we may conclude that the adaptive time stepping algorithm has excellent scalability. One can see in both tables that the best results with regards to CPU time and number of the inner iterations are obtained when the adaptive strategy is applied at each second time step and meanwhile, at the intermediate time steps τ is multiplied by 1.2. In this case, comparing the total CPU times of the algorithm without the adaptive time-stepping and using this strategy, it is seen that the time of the new algorithm is almost three times shorter.

Table 2. Number of iterations and the CPU time in the adaptive time-stepping algorithm in the Case of 128 processors.

N_{Adapt}	$\lambda_{\text{NonAdapt}}$	No. of inner iterations	No. of outer iterations	CPU time [s]	Vol_1 cm^3	$Vol_{4.6}$ cm^3
0	1.0	2233	420	7608	22.14	15.60
1	1.0	917	102	3968	22.72	15.93
	1.0	731	104	3137	22.63	15.87
2	1.2	535	71	2321	22.87	16.00
	1.3	587	77	2624	22.87	16.02
	1.0	700	113	3053	22.58	15.83
3	1.2	539	76	2329	22.88	16.03
	1.3	592	77	2559	22.81	15.97

Table 3. Number of iterations and the CPU time in the adaptive time-stepping algorithm in the Case of 1024 processors.

N_{Adapt}	$\lambda_{\text{NonAdapt}}$	No. of inner iterations	No. of outer iterations	CPU time [s]	Vol_1 cm^3	$Vol_{4.6}$ cm^3
0	1.0	604	420	7259	22.21	15.65
1	1.0	777	101	4234	22.70	15.92
	1.0	594	101	3488	22.70	15.92
2	1.2	478	71	2619	23.01	16.10
	1.3	539	77	2982	22.94	16.07
	1.0	549	104	3121	22.70	15.93
3	1.2	455	76	2530	22.85	16.01
	1.3	514	75	2740	22.94	16.06

5 Conclusions

An adaptive time stepping algorithm for simulating the radio-frequency ablation for treatment of liver tumors is presented. The procedure is based on the local comparison of the Crank-Nicholson and the backward Euler approximations. Results of some preliminary numerical experiments performed are presented and discussed. The first experimental results show that the new algorithm is scalable. The tests allowed us to find some suitable parameters and showed the practical usefulness of the developed solver for such kind of computer simulations. One can observe that the computing time is decreased more than three times, the number of outer iterations is decreased from 420 to 71, and the number of inner iteration decreases form 2233 to 535. This preliminary results are a good motivation for further improving the algorithm and doing more simulations.

Acknowledgments. This research is supported in part by Grants DFNI I01/5 and DCVP-02/1 from the Bulgarian NSF and the Bulgarian National Center for Supercomputing Applications (NCSA) giving access to the IBM Blue Gene/P computer.

References

1. Chang, I.A., Nguyen, U.D.: Thermal modeling of lesion growth with radiofrequency ablation devices. Biomed. Eng. OnLine **3**, 27 (2004)
2. Blaheta, R., Byczanski, P., Kohut, R., Stary, J.: Algorithms for parallel FEM modelling of thermo-mechanical phenomena arising from the disposal of spent nuclear fuel. In: Stephansson, O., Hudson, J.B., Jing, L. (eds.) Coupled Thermo-Hydro-Mechanical-Chemical Processes in Geosystems. Elsevier, Amsterdam (2004)
3. Henson, V.E., Yang, U.M.: BoomerAMG: a parallel algebraic multigrid solver and preconditioner. Appl. Numer. Math. **41**(1), 155–177 (2002). Elsevier
4. Scotch and PT-Scotch: Software package and libraries for sequential and parallel graph partitioning, static mapping, and sparse matrix block ordering, and sequential mesh and hypergraph partitioning. http://www.labri.fr/perso/pelegrin/scotch/
5. Tungjitkusolmun, S., Staelin, S.T., Haemmerich, D., Tsai, J.Z., Cao, H., Webster, J.G., Lee, F.T., Mahvi, D.M., Vorperian, V.R.: Three-dimensional finite-element analyses for radio-frequency hepatic tumor ablation. IEEE Trans. Biomed. Eng. **49**(1), 3–9 (2002)
6. Tungjitkusolmun, S., Woo, E.J., Cao, H., Tsai, J.Z., Vorperian, V.R., Webster, J.G.: Thermal-electrical finite element modelling for radio frequency cardiac ablation: effects of changes in myocardial properties. Med. Biol. Eng. Comput. **38**(5), 562–568 (2000)
7. Axelsson, O.: Iterative Solution Methods. Cambridge University Press, Cambridge (1996)
8. Brenner, S., Scott, L.: The mathematical theory of finite element methods. In: Antman, S.S., Holmes, P., Sreenivasan, K. (eds.) Texts in Applied Mathematics, vol. 15. Springer, New York (1994)
9. Hairer, E., Norsett, S.P., Wanner, G.: Solving Ordinary Differential Equations I, II. Springer Series in Comp. Math. Springer, Heidelberg (2002)
10. Kosturski, N., Margenov, S.: Supercomputer simulation of radio-frequency hepatic tumor ablation. In: AMiTaNS'10 Proceedings, AIP CP, vol. 1301, pp. 486-493(2010)
11. Kosturski, N., Margenov, S., Vutov, Y.: Comparison of two techniques for radio-frequency hepatic tumor ablation through numerical simulation. In: AIP Conference Proceedings, vol. 1404, p. 431 (2011)
12. Kosturski, N., Margenov, S., Vutov, Y.: Supercomputer simulation of radio-frequency hepatic tumor ablation. In: AMiTaNS'12 Proceedings, AIP CP, vol. 1487, pp. 120–126 (2012)

Nonlinear Forced Vibration Analysis of Elastic Structures by Using Parallel Solvers for Large-Scale Systems

Stanislav Stoykov(✉) and Svetozar Margenov

Institute of Information and Communication Technologies,
Bulgarian Academy of Sciences, Acad. G. Bonchev Str., Bl. 25A, 1113 Sofia, Bulgaria
{stoykov,margenov}@parallel.bas.bg

Abstract. Geometrically nonlinear forced vibrations of three dimensional structures, due to harmonic excitations, are investigated in the frequency domain. Structures of elastic materials are considered and the discretized equation of motion is derived by the finite element method, using Elmer software. The shooting and the continuation methods are applied to the resulting large scale FEM system by using scalable parallel solvers.

Periodic steady-state solutions are of interest and their computation is achieved by two techniques: shooting and continuation methods. The periodic solutions are obtained by shooting method, i.e. by solving a two-point boundary value problem defined by the periodicity condition. For that purpose, a time integration scheme, such as Newmark's method is used and the correction of the initial guess is accomplished through a Newton-Raphson method. The next solution of the bifurcation diagram is obtained by the arc-length continuation method. A prediction for the new point from the bifurcation diagram is defined by using the previous solution and the new solution is obtained by correcting the prediction, i.e. by shooting method.

The main objective of the current work is to investigate the potential of the proposed methods for the efficient computation of the bifurcation diagrams of large-scale dynamical systems, which result from the discretization in space of real-life structures, achieved by appropriate numerical techniques and parallel algorithms.

1 Introduction

The modern engineering structures are of complex geometry and made from composite materials. Many researchers have developed reduced order models, such as beams, plates or shells, in order to model the engineering structures with fewer degrees of freedom. Nevertheless, the efficient modeling of real structures requires a fine mesh of three dimensional finite elements, which results into a system of ordinary differential equations with large number of unknowns and its solution becomes computationally expensive. The necessity of using parallel algorithms and time effective solvers, for such large-scale problems is evident.

I. Lirkov et al. (Eds.): LSSC 2013, LNCS 8353, pp. 405–412, 2014.
DOI: 10.1007/978-3-662-43880-0_46,

Because nonlinear effects occur frequently in structural dynamics [1] and the linear models have limited validity, the investigations of the dynamical behavior of elastic structures, taking into account nonlinearity due to large displacements or due to plasticity [2], is of great importance for their optimal and successful modeling and performance. In nonlinear structural dynamics, the frequency of vibration and the amplitude are dependent, and small changes of the excitation frequency may lead to big changes of the amplitude of vibration or even to chaotic, quasi-periodic or multimodal motions [3]. These phenomenon appears due to bifurcations which arise when the excitation frequency pass a certain value.

The parametric study of large-scale nonlinear dynamical systems is presented at this work. Two techniques are implemented: shooting method [4], which finds the periodic solution for a fixed excitation frequency due to initial guess of the initial conditions, and arc-length continuation method [5], which defines a predictor for the initial conditions of the next point from the bifurcation diagram. The equation of motion is derived by the Elmer software [6] and Newmark's method is used for the time integration. The shooting and continuation methods are implemented within Elmer. Most of the research related with shooting method, which exists in the literature, is done for systems of first order ordinary differential equations. However, in the current work, it was preferred to develop the shooting method for systems of second order ordinary differential equations, because, as will be seen in the next section, for second order ODE, the shooting method requires to solve $2N$ equations with N degrees of freedom (DOF) each, while for first order ODE, the shooting method requires to solve $2N$ equations with $2N$ DOF each, which will significantly increase the computational time. Furthermore, the solution of the equation of motion, which Elmer solves for each time step, is obtained from the second order ODE and the same time discretization, which is performed by Elmer, is used to find the variation of the solution when the initial conditions are perturbed, for the same time step.

2 Numerical Computation of Periodic Steady-state Responses

The equation of motion of any elastic structure [2], considering geometrical type of nonlinearity, can be written in the following way:

$$\mathcal{F} \equiv \mathbf{M}\ddot{\mathbf{q}}(t) + \mathbf{C}\dot{\mathbf{q}}(t) + \mathbf{K}(\mathbf{q}(t))\mathbf{q}(t) - \mathbf{F}(t) = \mathbf{0} \tag{1}$$

with the initial conditions

$$\mathbf{q}(0) = \mathbf{q}_0 \tag{2a}$$

$$\dot{\mathbf{q}}(0) = \dot{\mathbf{q}}_0 \tag{2b}$$

where $\mathbf{M}$ is the mass matrix, $\mathbf{C}$ is the damping matrix, $\mathbf{K}$ is the stiffness matrix which depends on the vector of generalized coordinates $\mathbf{q}$, also called displacement vector, and $\mathbf{F}$ is the vector of external forces. The dimension of these matrices and vectors is N, where N is the total number of DOF, which depends on the type and the number of elements used for the discretization. The vectors of generalized coordinates and the external forces depend on time t which will be omitted.

In order to outline the dependence of the initial conditions on the displacement vector and velocity, these vectors will be written as:

$$\mathbf{q} = \mathbf{q}(t, \mathbf{q}_0, \dot{\mathbf{q}}_0) \tag{3a}$$

$$\dot{\mathbf{q}} = \dot{\mathbf{q}}(t, \mathbf{q}_0, \dot{\mathbf{q}}_0) \tag{3b}$$

The shooting method consists of finding, on iterative way, the initial conditions that perform a periodic motion. The initial value problem is converged into a two point boundary value problem [3]. One seeks initial conditions $\mathbf{q}(0) = \mathbf{q}_0$, $\dot{\mathbf{q}}(0) = \dot{\mathbf{q}}_0$ and solution $\mathbf{q}(t, \mathbf{q}_0, \dot{\mathbf{q}}_0)$, $\dot{\mathbf{q}}(t, \mathbf{q}_0, \dot{\mathbf{q}}_0)$ with period T such that:

$$\mathbf{q}(T, \mathbf{q}_0, \dot{\mathbf{q}}_0) = \mathbf{q}_0 \tag{4a}$$

$$\dot{\mathbf{q}}(T, \mathbf{q}_0, \dot{\mathbf{q}}_0) = \dot{\mathbf{q}}_0 \tag{4b}$$

It is assumed that the period of vibration T is known, and the initial conditions $\mathbf{q}_0$ and $\dot{\mathbf{q}}_0$, which define the steady-state response with period T, will be obtained. Let $\mathbf{s}_0$ and $\dot{\mathbf{s}}_0$ are initial guess of the initial conditions. By application of the shooting method, the initial guess is corrected by $\delta\mathbf{s}_0$ and $\delta\dot{\mathbf{s}}_0$. The corrections are obtained by minimizing the difference between the initial conditions and the response of the system at time T:

$$\mathbf{q}(T, \mathbf{s}_0 + \delta\mathbf{s}_0, \dot{\mathbf{s}}_0 + \delta\dot{\mathbf{s}}_0) - (\mathbf{s}_0 + \delta\mathbf{s}_0) = 0 \tag{5a}$$

$$\dot{\mathbf{q}}(T, \mathbf{s}_0 + \delta\mathbf{s}_0, \dot{\mathbf{s}}_0 + \delta\dot{\mathbf{s}}_0) - (\dot{\mathbf{s}}_0 + \delta\dot{\mathbf{s}}_0) = 0 \tag{5b}$$

Applying Taylor's formula to the above equations and neglecting quadratic and higher order terms, the following equations are obtained:

$$\mathbf{q}(T, \mathbf{s}_0 + \delta\mathbf{s}_0, \dot{\mathbf{s}}_0 + \delta\dot{\mathbf{s}}_0) = \mathbf{q}(T, \mathbf{s}_0, \dot{\mathbf{s}}_0) + \frac{\partial\mathbf{q}(T, \mathbf{s}_0, \dot{\mathbf{s}}_0)}{\partial\mathbf{s}}\delta\mathbf{s}_0 + \frac{\partial\mathbf{q}(T, \mathbf{s}_0, \dot{\mathbf{s}}_0)}{\partial\dot{\mathbf{s}}}\delta\dot{\mathbf{s}}_0 \tag{6a}$$

$$\dot{\mathbf{q}}(T, \mathbf{s}_0 + \delta\mathbf{s}_0, \dot{\mathbf{s}}_0 + \delta\dot{\mathbf{s}}_0) = \dot{\mathbf{q}}(T, \mathbf{s}_0, \dot{\mathbf{s}}_0) + \frac{\partial\dot{\mathbf{q}}(T, \mathbf{s}_0, \dot{\mathbf{s}}_0)}{\partial\mathbf{s}}\delta\mathbf{s}_0 + \frac{\partial\dot{\mathbf{q}}(T, \mathbf{s}_0, \dot{\mathbf{s}}_0)}{\partial\dot{\mathbf{s}}}\delta\dot{\mathbf{s}}_0 \tag{6b}$$

Replacing Eqs. (6a, 6b) into Eqs. (5a, 5b), the following system for the corrections $\delta\mathbf{s}_0$ and $\delta\dot{\mathbf{s}}_0$ is obtained:

$$\left(\frac{\partial\mathbf{q}(T, \mathbf{s}_0, \dot{\mathbf{s}}_0)}{\partial\mathbf{s}} - \mathbf{I}\right)\delta\mathbf{s}_0 + \frac{\partial\mathbf{q}(T, \mathbf{s}_0, \dot{\mathbf{s}}_0)}{\partial\dot{\mathbf{s}}}\delta\dot{\mathbf{s}}_0 = \mathbf{s}_0 - \mathbf{q}(T, \mathbf{s}_0, \dot{\mathbf{s}}_0) \tag{7a}$$

$$\frac{\partial\dot{\mathbf{q}}(T, \mathbf{s}_0, \dot{\mathbf{s}}_0)}{\partial\mathbf{s}}\delta\mathbf{s}_0 + \left(\frac{\partial\dot{\mathbf{q}}(T, \mathbf{s}_0, \dot{\mathbf{s}}_0)}{\partial\dot{\mathbf{s}}} - \mathbf{I}\right)\delta\dot{\mathbf{s}}_0 = \dot{\mathbf{s}}_0 - \dot{\mathbf{q}}(T, \mathbf{s}_0, \dot{\mathbf{s}}_0) \tag{7b}$$

It is necessary to evaluate the coefficients of the matrices $\frac{\partial\mathbf{q}(T,\mathbf{s}_0,\dot{\mathbf{s}}_0)}{\partial\mathbf{s}}$ and $\frac{\partial\mathbf{q}(T,\mathbf{s}_0,\dot{\mathbf{s}}_0)}{\partial\dot{\mathbf{s}}}$, at time T, in order to proceed further. Equation (1) is differentiated with respect to the initial conditions $\mathbf{s}_0$ and $\dot{\mathbf{s}}_0$:

$$\frac{\partial \mathcal{F}}{\partial \mathbf{s}_0} \equiv \mathbf{M}\frac{\partial \ddot{\mathbf{q}}}{\partial \mathbf{s}_0} + \mathbf{C}\frac{\partial \dot{\mathbf{q}}}{\partial \mathbf{s}_0} + \mathbf{J}\frac{\partial \mathbf{q}}{\partial \mathbf{s}_0} = \mathbf{0} \tag{8a}$$

$$\frac{\partial \mathcal{F}}{\partial \dot{\mathbf{s}}_0} \equiv \mathbf{M}\frac{\partial \ddot{\mathbf{q}}}{\partial \dot{\mathbf{s}}_0} + \mathbf{C}\frac{\partial \dot{\mathbf{q}}}{\partial \dot{\mathbf{s}}_0} + \mathbf{J}\frac{\partial \mathbf{q}}{\partial \dot{\mathbf{s}}_0} = \mathbf{0} \tag{8b}$$

Differentiation of Eqs. (2a) and (2b), leads to the following initial conditions for Eqs. (8a, 8b):

$$\frac{\partial \mathbf{q}}{\partial \mathbf{s}_0}(0) = \mathbf{I}, \frac{\partial \dot{\mathbf{q}}}{\partial \mathbf{s}_0}(0) = \mathbf{0} \tag{9a}$$

$$\frac{\partial \mathbf{q}}{\partial \dot{\mathbf{s}}_0}(0) = \mathbf{0}, \frac{\partial \dot{\mathbf{q}}}{\partial \dot{\mathbf{s}}_0}(0) = \mathbf{I} \tag{9b}$$

In Eqs. (8a, 8b), $\mathbf{J}$ is the Jacobian of the system defined by:

$$\mathbf{J} = \frac{\partial \mathbf{K}(\mathbf{q})\mathbf{q}}{\partial \mathbf{q}} \tag{10}$$

Once $\mathbf{q}(t)$ is obtained by time integration method, $\frac{\partial \mathbf{q}(T)}{\mathbf{s}_0}$ and $\frac{\partial \mathbf{q}(T)}{\dot{\mathbf{s}}_0}$ might also be determined by applying the same time integration method for the systems defined in Eqs. (8a, 8b) with initial conditions (9a, 9b). It should be noted that the system of second order ordinary differential equations (8a, 8b) is linear, while the system (1) is nonlinear. Equations (7a, 7b) are solved and the corrections of the initial conditions $\delta \mathbf{s}_0$ and $\delta \dot{\mathbf{s}}_0$ are obtained. After the corrections are determined, it is verified if Eqs. (5a, 5b) are satisfied within desired accuracy. If not, the initial guess is updated and the procedure is repeated.

Once the initial conditions, which lead to a periodic solution are obtained, initial guess, for the next point from the bifurcation diagram, is defined by applying the continuation method [5] and the shooting method is repeated.

The systems defined by Eqs. (8a) and (8b), have to be solved for N different initial conditions defined in (9a) and (9b), for each time step, where N is the total number of degrees of freedom of the equation of motion (1) and each system consists of N equations. Often the discretization of the elastic structure leads to enormous degrees of freedom, especially in the cases when three dimensional finite elements are used. Thus, the process of correcting the initial conditions becomes computationally expensive and the necessity of using parallel algorithms is significantly important and may reduce the CPU time essentially.

3 Numerical Examples

The shooting method, for systems of second order ordinary differential equations, presented in the previous section is implemented within Elmer [6]. Elmer is an open source finite element software, it has modules for different physical problems, including finite elasticity, and direct and iterative solvers of linear systems suitable for sequential and parallel runs.

First, the proposed method is validated with a beam model, where the computation of the bifurcation diagram was achieved by the harmonic balance and

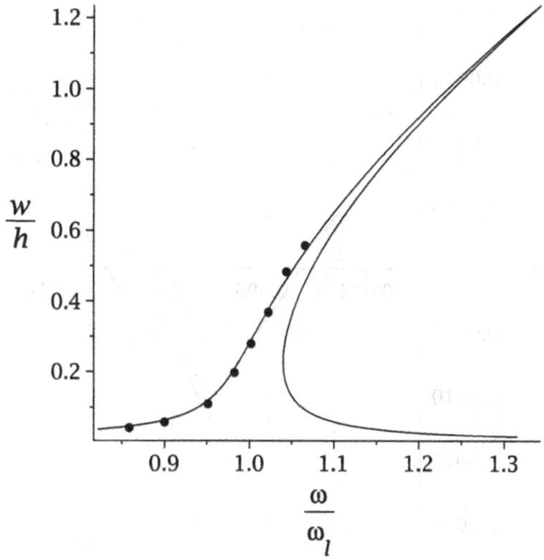

Fig. 1. Bifurcation diagram of beam due to external harmonic force, — results from beam model and HBM method, • results from three-dimensional discretization and shooting method, h - thickness, w - amplitude at the middle of the beam for $t = 0$, ω - excitation frequency, ω_l - fundamental linear frequency.

continuation methods [7]. At this stage, the shooting method is run on one processor, but its implementation on parallel processors is considered. Thus, the efficiency and the acceleration of Elmer is also investigated. The numerical experiments are performed on GRID-cluster.

The beam structure from reference [7] is implements here, i.e. the dimensions are $l = 0.406\,\mathrm{m}$, $b = 0.02\,\mathrm{m}$, $h = 0.002\,\mathrm{m}$ and the material is isotropic and homogeneous with the following properties (aluminium): $E = 7e10\,\mathrm{N/m^2}$, $\rho = 2800\,\mathrm{kg/m^3}$, $\nu = 0.33$, where l is the length of the beam, b is width, h is height, E is Young modulus, ρ is density and ν is Poisson's ratio. The beam is with clamped-clamped boundary conditions. An external harmonic and uniformly distributed force is applied in transverse direction with amplitude of $0.134\,\mathrm{N/m^2}$. The results from shooting method, obtained from the discretization of the beam structure by three-dimensional finite elements are in very good agreement with results from the beam model [7], based on Timoshenko's theory for bending and obtained by the harmonic balance method (HBM), where harmonics up to third order were used. The comparison of the results is presented on Fig. 1. It can be seen that the hardening effect, which arise due to the geometrical nonlinearity, can be successfully obtained by the shooting method implemented with Elmer and considering three-dimensional finite elements. On Fig. 2 are presented the time responses of the beam due to the initial conditions obtained from the shooting method on each shooting iteration.

The usage of three-dimensional finite elements gives opportunities to model complex structures, for which the reduced models, such as beams or plates, are

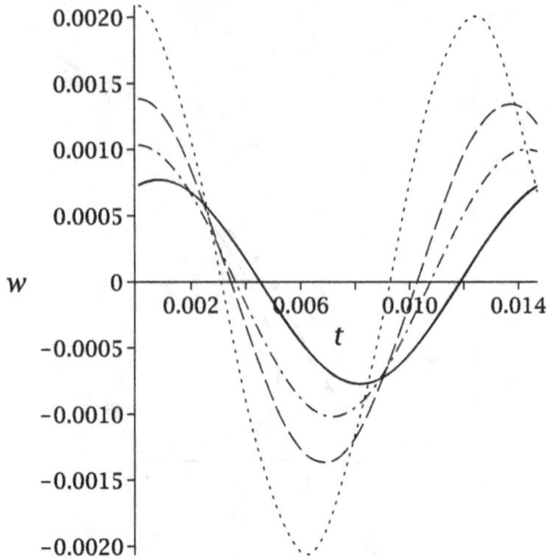

Fig. 2. Time responses for one period due to excitation frequency of $\omega/\omega_l = 1.02$, $\cdots$ response due to initial conditions obtained after the first iteration, - - - response after the second iteration, - · - · - response after the third iteration, — periodic steady-state solution, t - time, w - amplitude at the middle of the beam.

not appropriate. The future implementation of the shooting method will focus on parallel computing, which will reduce the CPU time, when complex structures are investigated and systems of enormous degrees of freedom are obtained. Because Elmer is used for solving the nonlinear equation of motion (1) in time domain, as well for solving the linear time dependent equations (8a, 8b) and for solving the linear system (7a, 7b), for correcting the initial conditions, it is essential to investigate how the acceleration and efficiency of Elmer change with increasing the number of parallel processors. The efficiency and acceleration of the solvers implemented in Elmer are investigated in the next paragraphs.

The same beam structure is modelled with fine mesh of quadratic tetrahedrons. The mesh is generated by Gmsh [8], it has 384 735 elements, which have 591 358 nodes, each node has 3 DOF, i.e. the resulting system has 1 774 074 DOF.

Three cases are studied for the generated mesh of the beam structure: solving a linear static problem, solving a geometrically nonlinear static problem and solving the eigenvalue problem to obtain the natural frequencies of the beam. MUMPS library (MUltifrontal Massively Parallel Solver) [9] which is a parallel direct sparse solver is used. The CPU time obtained with 16 parallel processors is used as a reference time and it is compared with the CPU time for the equivalent problems obtained with 32, 64 and 128 parallel processors.

The results of the linear static case, due to uniformly distributed external load of 5 N applied in the transverse direction, are given in Table 1. By increasing the number of processors, the CPU time decreases with very good acceleration. For the case of 128 parallel processors, an efficiency of 90 % is achieved.

Table 1. Strong scalability results of linear static problem.

P	CPU (s)	Speed up	Efficiency %
16	1311.92	1	-
32	666.85	1.97	98.37
64	301.08	4.36	108.93
128	180.9	7.25	90.65

Table 2. Strong scalability results of nonlinear static problem.

P	CPU (s)	Speed up	Efficiency %
16	4375.24	1	-
32	2213.03	1.98	98.85
64	1038.75	4.21	105.30
128	612.24	7.15	89.33

Table 3. Strong scalability results of eigenvalue problem.

P	CPU (s)	Speed up	Efficiency %
16	766.53	1	-
32	395.10	1.94	97.00
64	270.92	2.83	70.73
128	243.96	3.14	39.27

Similar results, but for the nonlinear static case, due to uniformly distributed external load of 30 N, are obtained and presented in Table 2. The results for the nonlinear problem were obtained by solving 7 linear systems, which result from application of Newton's method. Again, an efficiency of about 90 % is achieved when 128 parallel processors are used.

Finally, the acceleration and the efficiency is investigated for the natural frequencies of the beam, i.e. for solving the eigenvalue problem. In that case, the CPU time decreases with the number of parallel processors, but the solver is not as efficient as in the linear and nonlinear static cases. The efficiency is about 40 % for the case of using 128 parallel processors. The results are presented in Table 3.

It should be pointed out that the periodic solution is achieved mostly by iterative solutions of linear systems, i.e. Newmark's method, which is used for time integration in the shooting method, is applied to the equation of motion (1) and to the systems of Eqs. (8a, 8b) and finds the solution for each time step by solving an algebraic system of nonlinear equations, which results from Eq. (1), i.e. by solving several linear systems which result due to Newton's linearisation, and by solving an algebraic linear systems which result from Eqs. (8a, 8b). The eigenvalue problem is used once, for the case of free vibration analysis only, i.e. the eigenvector of the linear mode, which is scaled with appropriate amplitude is used as initial guess of the first periodic solution, and each next solution from the bifurcation diagram is obtained by defining an initial guess from the previous solution. Thus, the efficiency achieved by Elmer and MUMPS is appropriate for such large-scale problems.

4 Conclusion

A numerical procedure for investigating the nonlinear forced vibrations of elastic structures was developed and presented. The procedure which finds the periodic steady-state solutions in frequency domain, is based on shooting and continuation methods. The methods were implemented in Elmer software, which allows to investigate the dynamics of structures with complex geometry and allows to use parallel solvers.

The efficiency and the acceleration of the solvers within the Elmer's environment was studied, it was shown that linear and nonlinear solvers are efficient when the number of parallel processors is increased, while the solver for the eigenvalue problems is less efficient. The efficiency of the solvers guarantees successful implementation of the proposed methods in parallel computations, which paves the way for future investigation of the dynamics of complex real life structures.

Acknowledgement. The support of this work through the project AComIn "Advanced Computing for Innovation", grant 316087, funded by the FP7 Capacity Programme (Research Potential of Convergence Regions) and through the Bulgarian NSF Grant DCVP 02/1 is gratefully acknowledged.

References

1. Kerschen, G., Worden, K., Vakakis, A., Golinval, J.-C.: Past, present and future of nonlinear system identification in structural dynamics. Mech. Syst. Signal Process. **20**, 505–592 (2006)
2. Nayfeh, A., Pai, P.: Linear and Nonlinear Structural Mechanics. Wiley, Weinheim (2004)
3. Nayfeh, A., Balachandran, B.: Applied Nonlinear Dynamics. Wiley, Weinheim (1995)
4. Ribeiro, P.: Non-linear forced vibrations of thin/thick beams and plates by the finite element and shooting methods. Comput. Struct. **82**, 1413–1423 (2004)
5. Borst, R., Crisfield, M., Remmers, J., Verhoosel, C.: Non-linear Finite Element Analysis of Solids and Structures. Wiley, New York (2012)
6. Elmer web site: www.csc.fi/elmer. Accessed 10 March 2013
7. Stoykov, S., Ribeiro, P.: Stability of nonlinear periodic vibrations of 3D beams. Nonlinear Dyn. **66**, 335–353 (2011)
8. Geuzaine, C., Remacle, J.-F.: Gmsh: a three-dimensional finite element mesh generator with built-in pre- and post-processing facilities. Int. J. Numer. Meth. Eng. **79**, 1309–1331 (2009)
9. Amestoy, P., Guermouche, A., L'Excellent, J., Pralet, S.: Hybrid scheduling for the parallel solution of linear systems. Parallel Comput. **32**, 136–156 (2006)

A Multy-Domain Operational Chemical Weather Forecast System

Dimiter Syrakov[1](✉), Maria Prodanova[1], Iglika Etropolska[1], Kiril Slavov[1], Kostadin Ganev[2], Nikolay Miloshev[2], and Todor Ljubenov[3]

[1] National Institute of Meteorology and Hydrology, Bulgarian Academy of Sciences, Tsarigradsko Shose 66, 1784 Sofia, Bulgaria
dimiter.syrakov@meteo.bg

[2] National Institute of Geophysics Geodesy and Geography, Bulgarian Academy of Sciences, Acad. G. Bonchev Str., Bl.3, 1113 Sofia, Bulgaria

[3] Space Research and Technology Institute, Bulgarian Academy of Sciences, Acad. G. Bonchev Str., Bl.1, 1113 Sofia, Bulgaria

Abstract. Lately, together with the numerical weather forecast, in many European countries Systems for Chemical Weather Forecast operate, Chemical Weather being understood as concentration distribution of key pollutants in a particular area and its changes during some forecast period. In Bulgaria, a prototype of such a system was built in the frame of a project with the National Science Fund. It covers a relatively small domain including Bulgaria that requires using chemical boundary conditions from similar foreign systems. As far as this data is prepared abroad and transferred by Internet, many failures took place during the operation of the system. To avoid this problem, a new version of the system was built on the base of the nesting approach. This version is realized on five domains: Europe, Balkan Peninsula, Bulgaria, Sofia-Region and Sofia-City with increasing space resolution - from 81 km (Europe) to 1 km (Sofia-City). For the Mother domain (Europe) climatic boundary conditions are applied. All other domains take there boundary conditions from the senior one. Computations start automatically at 00 UTC every day and the forecast period is 3 days. The System is based on the well known models WRF (Meso-meteorological Model) and US EPA dispersion model CMAQ (Chemical Transport Model). As emission input the TNO data is used for the two biggest domains. For the 3 Bulgarian domains the current emission inventory prepared by Bulgarian environmental authorities is exploited.

Keywords: Air pollution modeling · Chemical transport model · Nesting approach

1 Introduction

The Air Quality (AQ) is a key element for the well-being and quality of life of European citizens. There is increasing evidence for adverse effects of air pollution on both the respiratory and the cardiovascular system as a result of both

I. Lirkov et al. (Eds.): LSSC 2013, LNCS 8353, pp. 413–420, 2014.
DOI: 10.1007/978-3-662-43880-0_47,

acute and chronic exposure. There is considerable concern about air quality conditions over many areas in Europe, especially in urbanized areas, in spite of about 30 years of legislation and emission reduction. Current legislation, e.g. the Ozone daughter directive 2002/3/EC [3], requires informing the public on AQ by assessing air pollutant concentrations throughout the whole territory of Member States. For the purpose, modeling tools must be used in parallel with air pollution measurements. In last years the concept of "chemical weather" arises and in many countries respective forecast systems are being developed along with the usual meteorological weather forecasts (see, for instance [9] and the COST Action ES0602 web-portal http://www.chemicalweather.eu/Domains.

In Bulgaria, a prototype of such a System was created by the support of the National Science Fund (BgCWFIS, ver.1, see [10]). It covers a relatively small domain around Bulgaria that requires using chemical boundary conditions from similar foreign systems. Such data has been prepared abroad and transferred by Internet causing many failures during the System operation. Later on, partly in the frame of EU FP7 project PASODOBLE, new versions (BgCWFIS, ver.2) was elaborated on the base of the nesting approach that allowed downscaling the service from resolution of 81 Km over Europe to resolution of 1 Km over Sofia city. Here, the current level to which the BgCWFIS, ver.2, has been developed will be described and its end-user products will be presented.

2 Models Used, Domains, Information Flow

BgCWFIS is designed on the base of Models-3 air quality modeling system (US EPA):

CMAQ v.4.6 - Community Multi-scale Air Quality model,
http://www.cmaq-model.org/ [2], the Chemical Transport Model (CTM);
WRF v.3.2.1 - Weather Research and Forecasting Model,
http://www.wrf-model.org/, [7] the meteorological pre-processor to CMAQ;
SMOKE v.2.4 - Sparse Matrix Operator Kernel Emissions Modelling System,
http://www.smoke-model.org/, [6] the emission pre-processor to CMAQ.

In its mother domain, WRF is driven by the NCEP GFS (Global Forecast System) data that can be accessed freely from http://www.ftp.ncep.noaa.gov/data/nccf/com/gfs/prod/. This data is global weather forecast in GRIB-2 format with space resolution $1^\circ \times 1^\circ$ and 6 h time resolution. Its downloading is invoked every day at 00Z. 84 h forecast starting at 12Z of the previous day is exploited. The first 12 h of this period are used for WRF spinning-up followed by 3-day forecast. The chemical weather forecast duration is from 00Z of the current day to 00Z of the forth day after (3-day forecast).

The nesting capabilities of WRF and CMAQ are used to downscale the forecasts from European region to Sofia-city area. The resolution of the mother domain (Europe) is 81 km, big enough as to correspond to the GFS met-data space resolution. Four other domains are nested in it and in each other - Balkan

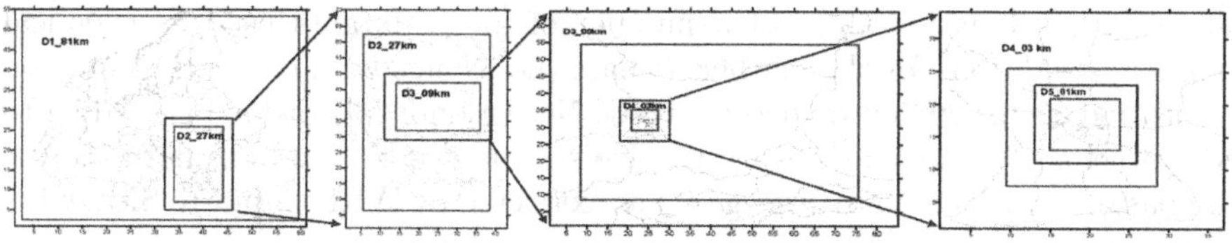

Fig. 1. Five computational domains of BgCWFIS, ver.2 (CMAQ domain nested in WRF one)

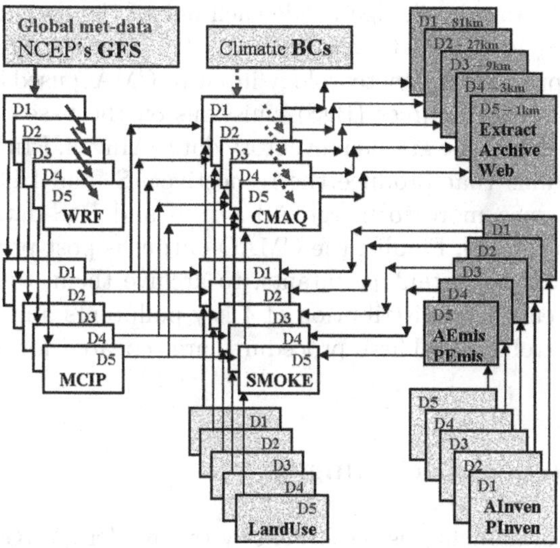

Fig. 2. BgCWFIS, ver.2, information flow diagram

Peninsula (27 km resolution), Bulgaria (9 km), Sofia district (3 km) and Sofia city (1 km) as shown in Fig. 1.

In version 2 of BgCWFIS, climatic data is used for chemical boundary conditions following the presumption that the errors introduced by this assumption will decrease quickly to the center of the domain due to the continuous acting of the pollution sources. All other domains receive their boundary conditions from the previous domain in the hierarchy.

The models indicated above are linked with a number of Linux scripts and FORTRAN interface programs in a way to be able to calculate the future levels of many air pollutants for each of system domains as indicated in Fig. 2.

The Models-3 elements are denoted with white boxes on the diagram. The dark grey boxes present FORTRAN programs aimed at emission input modeling of Area Sources (AS) and Large Point Sources (LPS) as well as data post-processing (archiving, images). The light grey boxes present the different kinds of input information. Those are: the NCEP GFS data drives the meteorological pre-processor of the System (WRF) downloaded in real-time from NCEP's GFS web-site; the climatic values of a number of air pollutants used as chemical boundary conditions for the mother domain (no change with time); the emission inventory data for 2005 provided by TNO, Netherlands [5], gridded according to the System's domains, and respective five sets of gridded land-use data (USGS data base, http://landcover.usgs.gov/) as extracted by WRF (input to SMOKE's biogenic processor).

The data exchange in the System is denoted with arrows. The thick arrows show the data exchange between computational domains - transfer of boundary conditions for WRF (compact arrows) and for CMAQ (dotted arrows). The thin arrows indicate the data exchange inside each one of the domains. Shortly, the WRF outputs feed MCIP (Meteorology-Chemistry Interface Processor) module of CMAQ which prepares the meteorology input to CMAQ used also by SMOKE for calculating Biogenic Sources (BgS) emissions on the base of the respective gridded LandUse data. The gridded inventory data (AInven, PInven) feed AEmis and PEmis programs that produce the respective AS- and LPS-emission files. SMOKE is used once more to merge AS-, LPS- and BgS-data in a common emission input to CMAQ. Finally, the CMAQ output is post-processed in a way to extract the most important pollutants, to archive them, to produce hourly images with concentration distribution of 4 key pollutants and to upload them to the respective web-sites. These procedures are repeated for all 5 domains of the system.

3 Meteorological Modeling

The meteorological modeling is performed by two models: WRF and MCIP.

The Weather Research and Forecasting (WRF) Model is a next-generation meso-scale **numerical weather prediction** system, an evolutionary successor to the MM5 model. The WRF is a fully compressible and non-hydrostatic model with terrain-following hydrostatic pressure coordinate. In BgCWFIS, ver.2, WRF-ARW (Advanced Research WRF), version 3.2.1, is exploited. The vertical structure consists of 27 levels. The Analysis Nudging option (Four-Dimensional Data Assimilation) is switched on for the first computational domain (Europe), only, nudging the WRF forecast to the meteorological driving data (GFS forecast). WRF offers multiple physics options that can be combined in any way. Here, well-tried schemes are used.

The coupling of meteorological and chemical transport models is not a trivial issue because all meteorological models are not built for air quality modeling purposes. Interface processing is needed and such an element in CMAQ system is MCIP (Meteorology-Chemistry Interface Processor). Here, ver.3.6., is exploited. MCIP deals with issues related to data format translation, diagnostic estimations of parameters not provided by WRF (like dry deposition velocities), extraction of data for appropriate window domains, and reconstruction of meteorological data on different grid and layer structures. Here, MCIP interpolates the vertical structure of WRF to those of CMAQ.

4 Emission Modeling

CMAQ demands its emission input in specific format reflecting the time evolution of all pollutants accounted for by the chemical mechanism used (CB-IV in this case). Emission inventories are the row data for anthropogenic emissions processing. The inventories are made on annual basis for big territories and many pollutants are estimated as groups (Volatile Organic Compounds - VOC, for instance). Three operations must be applied to this data, preliminary: ***gridding, temporal allocation and speciation***. Obviously, emission processors are needed. Such component in Models-3 system is SMOKE but it is partly used, here, because its strong relation to US emission sources specifics. In BgCWFIS, SMOKE is used only for calculating BgS emissions and for merging AS-, LPS- and BgS-files into a common emission input for CMAQ. The AS- and the LPS-emission files are prepared by the interface programs AEmis and PEmis (Fig. 2).

For the moment, TNO inventory for 2005 is exploited for the two senior domains (Europe and Balkans). The TNO has been produced several sets of inventories for different years. The anthropogenic sources in this inventories are distributed over 10 SNAPs (Selected Nomenclature for Air Pollution) classifying them according to the processes leading to harmful material release to the atmosphere [4]. This inventory has resolution of $0.125° \times 0.0625°$ (about 7×8 km) distributed as a comma-delimited text-file. Each line of the file contains data for a single source, namely the mesh coordinates, the country abbreviation, the type of source (A/P), the SNAP, and the yearly emissions of 8 pollutants. The SNAP 7 (road transport) is presented as 5 sub-SNAPs.

For Bulgarian domains the Bulgarian inventory for 2010, as provided by Bulgarian Executive Environmental Agency, is used.

The first emission modeling procedure, ***the gridding***, is recalculation of the inventory data to the grids used (5 domains with different resolution, here). A web-based GIS system is created for the purpose. Number of so-called custom grids can be defined on the base of the standard grid description (projection, window parameters, resolution). After the TNO data is introduced in the system, it recalculates inventory values for each cell and for each type of source of the current custom grid. Additional functionalities of the system are linked to the Bulgarian inventory. A specific feature of this inventory is that the data for SNAP 1-6 is attributed to particular sources with there coordinates and the system aggregates them to the custom grid's cells. For SNAP 7-10 total country amounts of the released pollutants are available. They are disaggregated to the same grid cells using different GIS elements (surrogates) like road distribution, airports, agriculture areas etc. The gridded inventories are introduced in the emission processing programs AEmis and PEmis, where the remaining two procedures are applied.

The ***temporal allocation*** is made on the base of temporal profiles, provided by TNO [1]. According to the anthropogenic activity the profiles are divided in three groups - Monthly, Weekly, and Hourly profiles. In addition TNO provides a vertical profile for large point sources.

The ***speciation*** is splitting of group pollutants (NO_X, SO_X, VOC, $PM_{2.5}$) to several simpler or "lump" pollutants required by the chemical mechanism. The speciation profiles, used here, are elaborated following the US EPA ones on the base of coincidence between US and European source categorization.

Both programs produce respective emission files. The AEmis output is 2-dimensional, the PEmis one - 3-dimensional. They contain hourly data for the whole forecast period.

The biogenic emissions are prepared by SMOKE by the BEIS-3.13 mechanism [8] on the base of the gridded LandUse data. SMOKE merges the 3 emission files in a common CMAQ emission input in IO/API NetCDF format.

5 Operational Performance of BgCWFIS, Ver.2

Fourteen σ-levels with varying thickness determine the vertical structure of CMAQ. The Planetary Boundary Layer (PBL) is presented by the lowest 8 of these levels.

The CMAQ v.4.6 input consists of various files containing concentration, deposition, visibility and other variables. The concentration output is a NetCDF file with 3-D hourly data for 78 pollutants - gases and aerosols.

The post-processing program XtrCON extracts part of the pollutants for archiving and further handling. Only surface values of the most important pollutants are saved - 8 gases and 11 aerosols (including PM_{10} and $PM_{2.5}$). Part of these pollutants is more or less monitored and they are referred in the European legislation with the respective thresholds.

As to make the results of BgCWIS operation public, specialized web-site was created on the NIMH server (http://www.meteo.bg/en/cw/). For the moment it presents 4 main pollutants - Ozone, NO_2, SO_2 and PM_{10}. It is fed by images

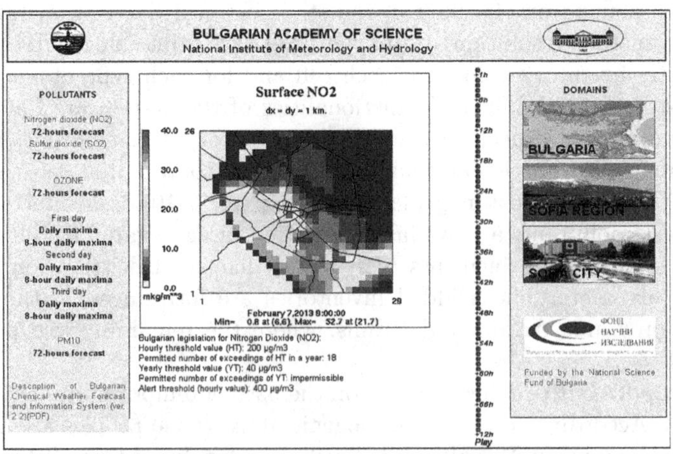

Fig. 3. View of BgCWFIS, ver.2.2. web-site

created using the PAVE package [11]. PAVE supports its own meta-language that allows drawing plots in an automatic way invoking the respective scripts.

In Fig. 3, an example of BgCWFIS web site is displayed. A particular pollutant is invoked by clicking in the list at the left side of the page. Note, that together with ozone, two types of daily maxima can be visualized. The region of forecast can be chosen by clicking one of the small images in the right. In the center of the page, hourly concentration field of the chosen pollutant is situated. Putting the mouse cursor on one of the points from the point column in the right side of the image invokes the forecasted field for the respective hour. Putting the cursor over "Play" invokes animation of the forecast. Under each pollutant's view, respective thresholds according to Bulgarian legislation (harmonized with European one) are shown. At the bottom of pollutants a link to a pdf-file with description of Bulgarian Chemical Weather Forecast and Information System, version 2, is placed.

6 Conclusion

The Bulgarian Chemical Weather Forecast and Information System is designed on the base of US EPA Models-3 System: WRF, SMOKE and CMAQ. The meteorological input to the system is the NCEP Global Forecast data. At this stage, the emission input exploits the high resolution inventory for year 2005 produced by TNO, The Netherlands. The Bulgarian national emission inventory for 2010 is used as well. The system is realized on 5 nested domains with increasing resolutions.

At the moment, the system is running automatically once a day (00Z). The forecast period is 3 days (72 h). The results of each System's run are post-processed in a way to archive the most important pollutants. Part of these pollutants is visualized as sequences of maps giving the evolution of the air quality over Europe and Bulgaria and can be seen on the system's web-site http://www.meteo.bg/en/cw/.

Acknowledgements. This study is made under the financial support of Bulgarian National Science Fund (Grants No. 002-161/16.12.2008 and 02/1/29.12.2009). The presented results were not possible without the experience obtained during the participation in the FP5 project BULAIR, the FP6 Network of Excellence ACCENT, the FP6 Integrated Projects QUANTIFY and CECILIA, the FP7 projects PASODOBLE and EGI-InSpire.

Deep gratitude is due to all organizations providing free of charge data and software used in this study, namely US EPA, US NCEP and European institutions like EMEP, EEA, TNO, UBA and many others.

References

1. Builtjes, P.J.H., van Loon, M., Schaap, M., Teeuwisse, S., Visschedijk, A.J.H., Bloos, J.P.: Project on the modelling and verification of ozone reduction strategies: contribution of TNO-MEP, TNO-report, MEP-R2003/166. Apeldoorn, The Netherlands (2003)

2. Byun, D., Schere, K.L.: Review of the governing equations, computational algorithms, and other components of the models-3 community multiscale air quality (CMAQ) modeling system. Appl. Mech. Rev. **59**, 51–77 (2006)
3. European Parliament (2002) DIRECTIVE 2002/3/EC of 12 February 2002 relating to ozone in ambient air, Official Journal of the European Communities (9.3.2002) L67: 14–30
4. EMEP/CORINAIR (2002) Atmospheric emission inventory guidebook, third edition, European Environmental Agency. http://reports.eea.europa.eu/EMEPCORINAIR3/en/page002.html
5. Denier van der Gon, H., Visschedijk, A., van de Brugh, H., Dröge, R.: A high resolution European emission data base for the year 2005, TNO-report TNO-034-UT-2010-01895_RPT-ML, Apeldoorn, The Netherlands (2010)
6. Houyoux, M.R., Vukovich, J.M.: Updates to the Sparse Matrix Operator Kernel Emission (SMOKE) Modeling System and Integration with Models-3, The Emission Inventory: Regional Strategies for the Future. Raleigh, NC, Air and Waste Management Association (1999)
7. Michalakes, J., Dudhia, J., Gill, D., Henderson, T., Klemp, J., Skamarock, W., Wang, W.: The weather research and forecast model: software architecture and performance. In: Proceeding of the Eleventh ECMWF Workshop on the Use of High Performance Computing in Meteorology, Reading, UK, 25–29 October 2004
8. Schwede, D., Pouliot, G., Pierce, T.: Changes to the Biogenic Emissions Invenory System Version 3 (BEIS3) (2005)
9. Sofiev, M., Siljamo, P., Valkama, I.: A dispersion modeling system SILAM and its evaluation against ETEX data. Atmos. Environ. **40**, 674–685 (2006)
10. Syrakov, D., Etropolska, I., Prodanova, M., Ganev, K., Miloshev, N., Slavov, K.: Operational pollution forecast for the region of Bulgaria. AIP Conf. Proc. **1487**, 88 (2012). doi:10.1063/1.4758945
11. PAVE package. http://www.ie.unc.edu/cempd/EDSS/pave_doc/index.shtml

Automatic Data Quality Control for Environmental Measurements

A. Tchorbadjieff(✉)

Institute for Nuclear Research and Nuclear Energy–BAS,
72 Tzarigradsko Chaussee, Blvd., 1784 Sofia, Bulgaria
assen@inrne.bas.bg

Abstract. The modern physics requires a large uninterrupted data for advanced study. The research data must be reliable, precise, adequate and available on time. Therefore advanced information systems should be developed and used. These systems must implement the most advanced technologies, algorithms and knowledge of informatics, programming and mathematics. This article describes a neural network model of automatic data quality control for large amount of real time uninterrupted data, implemented in the Institute for Nuclear Research and Nuclear Energy (INRNE) at the Basic Environmental Observatory (BEO) at Moussala.

1 Introduction

In recent years the amount of data acquired for research expanded to unobservant quantity. However, the large part of it is not useful due to wrong measuring process, mainly caused from technical glitches in equipment. This creates a lot of data to useless and uninformative records. Moreover, the most of measurement parameters are part of multivariate models and thus wrongness in only single parameter has enormous influence on further results from the rest of parameters usage. Likewise, the time reliance is critical for permanent daily operations and requirements for maximizing effectiveness and implementing real-time monitoring of data. But, the implementation and installation of the required data and network systems is not enough for fulfill these requirements. And additional upgrade with appropriate data quality mechanism and process scheduler for its in-time operation and synchronization is crucial.

The neural networks are a possibility for modelling efficient solutions in which these requirements are implemented. The most important advantage of this new approach is it similarities to non-linear regression models for large number parameters, approximated to any continuous function. For this case, the parameters are assumed to measure different data records, resent afterwards through several intermediate computational or storage places, defined as layers, until reach final destination of requirements. This data traffic is measured with cost function and probability for possibility for failure of data acquisition or transfer through any hidden layer. The decision for effectiveness of data quality network design is based on cost minimizing. But because of the probabilistic nature of cost function, the selected model is particle-like modelling the network and resulting Entropy as measure for optimal cost function.

I. Lirkov et al. (Eds.): LSSC 2013, LNCS 8353, pp. 421–427, 2014.
DOI: 10.1007/978-3-662-43880-0_48,

2 Network Design

A neural network for real-time measurements and analysis is assumed an infrastructure for data measurements, data storage, data quality classification and if possible an event predicting and notify system. For model input parameters the raw measurement records are used. They are serial data records different in size, precision, time intervals between consecutive time intervals and meta data which holds additional information about process of measurement. However, every record has probability for erroneous due to technical glitches or wrong interrupting process contamination. Likewise, there is a possibility for work interruptions, causing data serial line gaps.

The verification for error is process dedicated to be performed on hidden layer. This is required intermediate step before adequate data analysis. However, the process of data filtering is unique, but dependent on every distinct parameter property. Moreover, for the most of them it is multivariate process, dependent on another parameter that require own filtering. Thus, data quality process may take several separated consecutive computational operations, separated in different hidden layers. After filtering, the remained data is ready for usage. How it is done depends on the purpose and parameter itself, but in the most part of the cases are required additional multivariate computations before final output. The simplified graphics of neural network is shown in Fig. 1. With implementation of multivariate data processing, every parameter is allowed to take part in a different number of computation operations. This rises a cost of reallocated and used resources in every hidden layer of neural network.

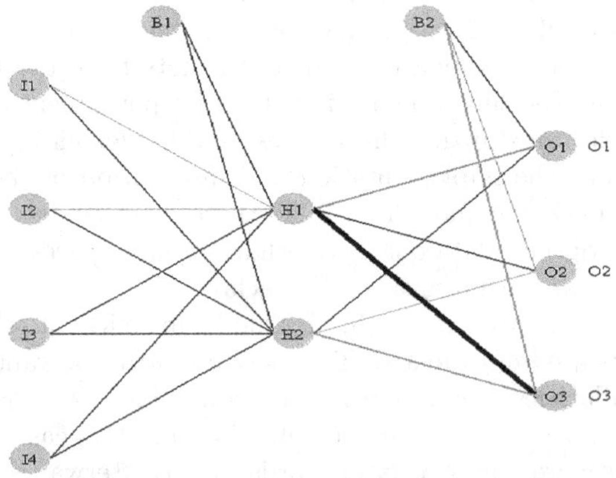

Fig. 1. The graphics shows schema of neural model from random generated connections for 4 input parameters and 3 output ones. They are numbered and indexed respectively with I and O. The hidden layers are named with prefix H. The constant values are added to input and output for normalization. They are labeled with B. For graphic generation free libraries from R are used, as shown in http://www.r-bloggers.com/visualizing-neural-networks-from-the-nnet-package/

The measure for computational cost is a utility function U_i - counting required price of allocated memory, used CPU time or data as required traffic bandwidth for specific data quantity proceeding. However, the resource costs are produced from computations, based on repeating executions based on the same algorithm with the same data size. Moreover, with permanent possibility for hardware upgrade, the resource allocation is less problematic.

However, the reliability of such type of network system depends on its time for result producing and minimizing the periods of standby caused by failure. Synchronization between different parameters acquisition is with main importance for improvement and optimization of system performance, especially when time resolution of them is different. With expanding number of hidden layers and parameters, the complication of synchronization algorithm grows. Moreover, the utility of every layer operation is a probabilistic function with decomposition on deterministic part V_i and stochastic ϵ_i:

$$U_i = V_i + \epsilon_i \tag{1}$$

The probabilistic part is measure for uncertainties of measurement process, such as probabilities for measurement interruption, erroneous measurement, missing parameter in multivariate analysis, etc. Due to these uncertainties every activity j of alternative i is estimated with probability function $p_{i,j}(x)$. These probabilities are ratio of used and assigned resource R' units for activity j. However, available connection between two activities is a binomial discrete value - $P(x_i = 1|j) = p_{i,j}$ and $P(x_i = 0|j) = 1 - p_{i,j}$. Thus, for optimization criteria is assumed the Entropy [1]

$$H_{i,j} = -p_{i,j}log_2 p_{i,j} - (1 - p_{i,j})log_2(1 - p_{i,j}) \tag{2}$$

3 Probability Distribution and Optimization

The most prominent statistics for this class of models is the Generalized Logistic model with independently and identically distributed random components. Thus, the probability of linear combination $V_{i,j} = y_{i,j} = x'_{i,j}\beta_j$ derived from (1) of explanatory variable $x_{i;j}$ is equal to [2]:

$$P(y_i = j) = \pi_{i,j} = \frac{e^{V_{i,j}}}{\sum_k e^{V_{i,k}}} \tag{3}$$

where the parameter β_j is the corresponding vector for j-th alternative and not dependent on alternative. Moreover, the explanatory vector $x_{i,j}$ consists of either zero or x_i value vectors and represents availability of the j-th alternative.

The probabilities in (3) are known functions for estimation that data from subject i is valid because the j-th category. Thus, after summarizing probabilities over all possible categories, the general probability, $\sum_j \pi_{i,j}$, is equal to 1. Moreover, the odds rates are derived directly from (1), (3) and equal to [2]:

$$ln\frac{P(y_i = j)}{P(y_i = n)} = V_{i,j} - V_{i,n} = x'_i \tag{4}$$

where n is a total number of all available alternatives. Thus, the log-likelihood, similar to cross-Entropy for binary, is equal to [2]:

$$H(\beta) = \sum_i \sum_j y_{i,j} ln \pi_{i,j} = \sum_i \left[\sum_j^n y_{i,j} x'_{i,j} \beta - ln(\sum_l^n e^{x'_{i,l}\beta}) \right] \tag{5}$$

Therefore, the yielded regression coefficients β, denoted as weights or odds, from derivatives of $\partial H(\beta)/\partial\beta$ are the maximum log-likelihood estimator. They are usually initialized with random values derived from a standard normal distribution. Next, they are iterated recursively until weights are smaller than a given threshold [3].

4 Real Test Scenario

For real test of a simplified model of data correction procedure for measurements from two devices - atmosphere pressure and muon telescope, is selected. The pressure is measured by meteorology equipment and is explanatory by itself - it is real physical parameter. Conversely, the muon telescope measure secondary muon particle flux in different coincidences. But their real meaning is dependent on atmosphere in anti correlation relation. Thus, from 12 raw parameters measured with telescope, the output consists of 5 real meaning values of fluxes in different directions, corrected with atmosphere pressure. Therefore, the data quality procedure requires two steps - first atmosphere pressure and muon data filtering. Then, the results are ready for usage, one of which is correction of muon parameters.

Thus, the neural model consists of two import parameters - atmosphere station and telescope. The hidden layers are assumed 2, as many as filters are, and output parameters are 6 - atmosphere pressure and 5 muon parameters. The input parameters are ratios of real operational time to complete assigned time. The outputs are merged ratios between input ratio and ratios of correctness after data filtering. However, because muon output parameters are multivariate for the input ratio the minimal one is taken. Likewise, the ratio of correctness for every muon output parameter is different and depends on validity of pressure measurements. Therefore, the output ratios are remained completely valid data ratios normalized to all available measurements.

The test scenario compares two real cases - available network and proposed improved model. For substitute model availability of additional atmosphere sensor integrated in muon telescope is proposed. Thus, the pressure measurements for muons corrections run in parallel with the rest 12 raw data series. Therefore, muon data is independent from meteorological station failures, and dependent only on telescope failures. As a result neural network adds an additional output for added pressure sensor. The annual data for both cases for last 6 years are shown in Tables 1 and 2.

The neural network odds are computed with R-statistical software and dedicated neuralnet library [3]. At first, both cases are run with precise threshold of

Table 1. Case 1 input and output parameters

Year	Input ratios		Output ratios					
	Pressure	Muon	Press	V.	WE	EW	SN	NS
2006	0.94	0.95	0.931	0.876	0.93	0.93	0.917	0.912
2007	0.95	0.94	0.941	0.876	0.93	0.93	0.917	0.921
2008	0.855	0.955	0.847	0.797	0.846	0.845	0.834	0.83
2009	0.895	0.778	0.886	0.722	0.768	0.768	0.757	0.753
2010	0.987	0.952	0.978	0.889	0.942	0.942	0.929	0.924
2011	0.98	0.946	0.971	0.883	0.936	0.936	0.923	0.918

Table 2. Case 2 input and output parameters

	Input ratios		Output ratios						
Year	Pressure	Muon	PressI	PressII	V	WE	EW	SN	NS
2006	0.94	0.95	0.931	0.94	0.885	0.939	0.939	0.926	0.912
2007	0.95	0.94	0.941	0.94	0.876	0.939	0.93	0.917	0.921
2008	0.855	0.955	0.847	0.945	0.89	0.944	0.944	0.929	0.924
2009	0.895	0.778	0.886	0.768	0.723	0.767	0.767	0.757	0.753
2010	0.987	0.952	0.978	0.942	0.889	0.941	0.941	0.929	0.924
2011	0.98	0.946	0.971	0.936	0.882	0.935	0.935	0.923	0.918

0.01 for cross-entropy (5). The required recursive iterations for optimization are respectively 261 and 30 (Fig. 2) and (Fig. 3). The smaller number of interactions suggests that Case 2 is better optimized than currently working one. However, due to large number of outputs the complete output error of case 2 (0.0757) is larger than this of case 1 (0.0341). The main difference between two cases is independence of computation of muon flux from meteorological station operation ratio. This minimizes disorder in synchronization between data and support suggestions that the waiting time and reliance of muon telescope is improved. This assumption is supported in numerical data, derived from generalized weights analysis [4]. The generalized weights (GW) estimates effect of each regression coefficient in general model, but depending on all other covariates. It is equal to [4]

$$w_i = \frac{\partial log(\frac{\pi_i}{1-\pi_i})}{\partial_{x_i}} \tag{6}$$

As suggestions shows, the operation ratio of muon telescope and meteorological station has almost linear effect for muons during Case 1 tests, confirmed with generalized wights close to 1. But, this degrade the reliance due to double reliance on two different data sources, with high level of divergence in periods of interruptions. Therefore, installing additional sensor in telescope, such as assumed in Case 2, may improve reliance due to better time synchronization between input parameters. This is derived from significantly higher level of outputs dependence on added pressure sensor, than on independent one. The assumption follows from

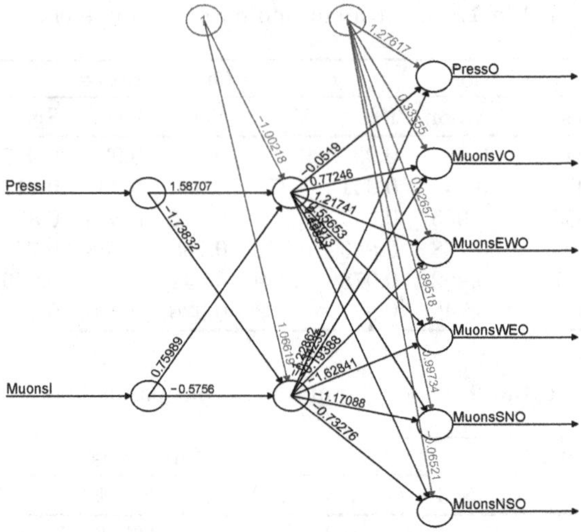

Fig. 2. The resulting neural network computed for Case 1 test for error threshold 0.01. The regression intercepts are enumerated with 1. The every odd value is associated with connection line between different layers.

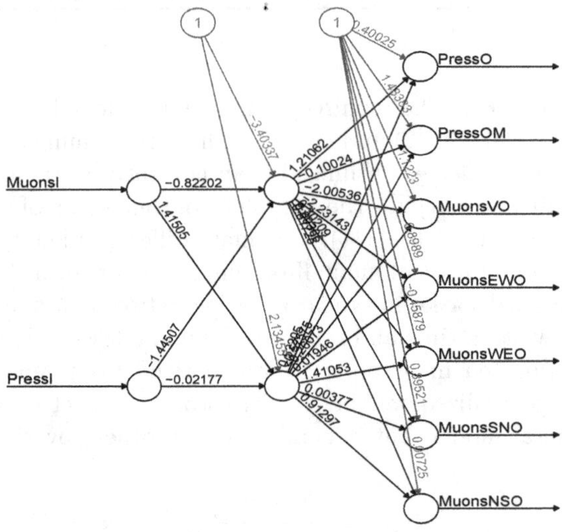

Fig. 3. The resulting neural network computed for Case 2 test for error threshold 0.01. The regression intercepts are enumerated with 1. The every odd value is associated with connection line between different layers.

very large values (>4) of GW for muons with exponential effect, while pressure from independent meteo sensor almost preserve its close to linear GW due to separate meaning as physical parameter (Fig. 4).

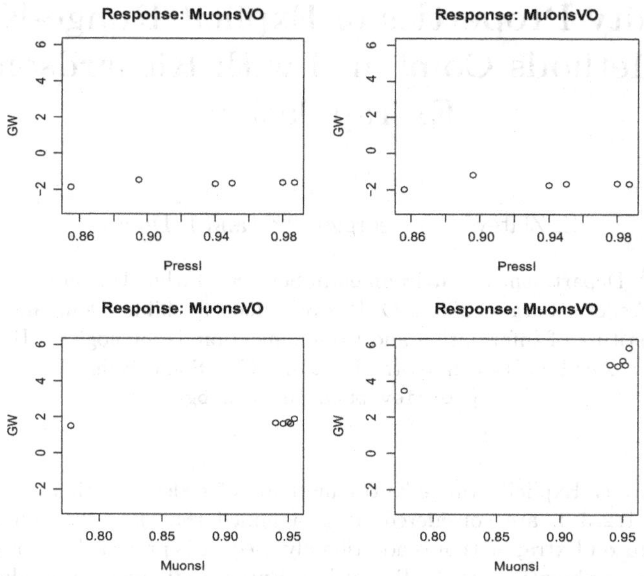

Fig. 4. Graphics shows generalized wights for pressure station and muon telescope for vertical muon flux responses. The upper row graphics shows weights of pressure and telescope responses are below. The results for Case 1 test are on left side, whereas the Case 2 tests are right sided.

5 Conclusion

The neural network models are reliable analytic tool for estimation of effectiveness of complicated computational systems. Their successful implementation is demonstrated in this paper as implementation of special case scenario - partial analysis of real operational data quality system for operational data acquisition system. Moreover, it is shown that the neural network modelling is convenient, effective and easy to use tool for system effectiveness and reliability improvements.

References

1. Christodoulou, S., Ellinas, G., Aslani, P.: Entropy-based scheduling of resource-constrained construction projects. Autom. Constr. **18**, 919–928 (2009)
2. Brockwell, P., Davis, R.: Introduction to Time Series and Forecasting. Springer, New York (2002)
3. Gunther, F., Fritsch, S.: neuralnet: Training of neural networks. R J. **2**(1), 30 (2010)
4. Intrator, O., Intrator, N.: Interpreting neural network results: a simulation study. Comput. Stat. Data Anal. **37**, 373–393 (2001)

Stability Properties of Explicit Runge-Kutta Methods Combined with Richardson Extrapolation

Z. Zlatev[1], K. Georgiev[2(✉)], and I. Dimov[2]

[1] Department of Environmental Science, Aarhus University, Frederiksborgvej 399, P.O. Box 358, 4000 Roskilde, Denmark
[2] Institute of Information and Communication Technologies – BAS, Acad. G. Bonchev Str., Bl. 25-A, 1113 Sofia, Bulgaria
georgiev@parallel.bas.bg

Abstract. Explicit Runge-Kutta methods of order p with m stages, $m = 1, 2, 3, 4$, are considered. It is assumed that $p = m$ and that Richardson Extrapolation is additionally used. It is proved that not only are the combinations of the Richardson Extrapolation with the selected explicit Runge-Kutta methods more accurate than the underlying numerical methods, but also their absolute stability regions are considerably larger. Sometimes this fact allows us to apply larger time-stepsizes during the numerical solution when Richardson Extrapolation is used. The possibility to achieve such a positive effect is verified by numerical experiments carried out with a carefully chosen example. It is pointed out that the application of Richardson Extrapolation together with explicit Runge-Kutta methods might be useful when some large-scale mathematical models, including models that are arising in air pollution studies, are handled numerically.

1 Selection of Numerical Methods

Consider the classical initial value problem for non-linear systems of ordinary differential equations (ODEs):

$$y\prime = f(t, y), \quad t \in [a, b], \quad b > a, \quad y \in D \subset R^s, \quad s \geq 1, \quad y(a) = \eta \in D, \quad (1)$$

and assume that one of the following four explicit Runge-Kutta methods is used in the numerical treatment of (1):

$$y_n = y_{n-1} + hf(t_{n-1}, y_{n-1}), \quad (2)$$

$$y_n = y_{n-1} + \frac{1}{2}h\left(k_1 + k_2\right), \; k_1 = f\left(t_{n-1}, y_{n-1}\right), \; k_2 = f\left(t_{n-1} + h, y_{n-1} + h\right). \quad (3)$$

I. Lirkov et al. (Eds.): LSSC 2013, LNCS 8353, pp. 428–435, 2014.
DOI: 10.1007/978-3-662-43880-0_49, © Springer-Verlag Berlin Heidelberg 2014

$$y_n = y_{n-1} + \frac{1}{4}h\,(k_1 + 3k_3)\,, \quad k_1 = f\,(t_{n-1}, y_{n-1})\,, \tag{4}$$
$$k_2 = f\left(t_{n-1} + \frac{1}{3}h, y_{n-1} + \frac{1}{3}hk_1\right), k_3 = f\left(t_{n-1} + \frac{2}{3}h, y_{n-1} + \frac{2}{3}hk_2\right).$$

$$y_n = y_{n-1} + \frac{1}{6}h\,(k_1 + 2k_2 + 2k_3 + k_4)\,, \quad k_1 = f\,(t_{n-1}, y_{n-1})\,, \tag{5}$$
$$k_2 = f\left(t_{n-1} + \frac{1}{2}h, y_{n-1} + \frac{1}{2}hk_1\right),$$
$$k_3 = f\left(t_{n-1} + \frac{2}{2}h, y_{n-1} + \frac{1}{2}hk_2\right), \quad k_4 = f\,(t_{n-1} + h, y_{n-1} + hk_3)\,.$$

In the above formulae, y_{n-1} and y_n are approximations of the values $y(t_{n-1})$ and $y(t_n)$ of the exact solution of (1) at the points t_{n-1} and t_n which belong to the grid:

$$t_0 = a, \quad t_n = t_{n-1}+h = t_0+nh \quad (n = 1, 2, \ldots, N), \quad t_N = b, \quad h = \frac{b-a}{N}. \tag{6}$$

The numerical methods defined by (2)–(5) are one-, two-, three- and four-stage explicit Runge–Kutta methods (see [3,6–9,12]). The order p of each of these methods is equal to the number m of stages used.

2 Application of Richardson Extrapolation

Assume that some $y_{n-1} \approx y(t_{n-1})$ has been calculated by any of the listed above four numerical methods. Perform one large step and two small steps with stepsizes h and $h/2$ respectively. Let z_n and w_n be the computed approximations and form

$$y_n = \frac{2^p w_n - z_n}{2^p - 1}, \tag{7}$$

where p is the order of the method applied to calculate z_n and w_n. The process of obtaining y_n by using (7) is called Richardson Extrapolation [5,10,18,19]. Its order of accuracy, being $p+1$ is higher than the orders of both z_n and w_n. Thus, the accuracy can be improved when (7) is used. In this paper we shall show that also the stability of computations can be improved by applying the Richardson Extrapolation.

3 Stability Considerations

The linear stability theory developed by Dahlquist [4] is commonly used (see also [3,6–9,12,18,19]). It is based on the application of the scalar test-problem $y' = \lambda y$ with $\lambda = \alpha + \beta i$ where it is assumed the real part α is non-negative. Under this assumption the exact solution of the test problem $y' = \lambda y$ is bounded and, therefore, it is desirable that the approximate solution produced by the selected numerical method also remains bounded. The application of explicit

Runge-Kutta methods in the solution of the test-problem leads to the recurrent relationship $y_n = R(h\lambda)y_{n-1}$, where $R(h\lambda)$ is called a stability polynomial. This polynomial is of order m, where m is the number of stages used in the selected method. Moreover, if $m = p$ (i.e. when the number of stages is equal to the order of the explicit Runge-Kutta method, which can only be achieved when $m = 1, 2, 3, 4$), then as shown in [9] we have:

$$R(h\lambda) = 1 + \sum_{i=1}^{p} \frac{(h\lambda)^i}{i!}. \tag{8}$$

Assume now that the selected explicit Runge-Kutta method with $m = p$ is combined with Richardson Extrapolation and the resulting numerical method is applied in the solution of the Dahquist's test-problem. Then a recurrent relationship $y_n = \bar{R}(h\lambda)y_{n-1}$ can be obtained with

$$\bar{R}(h\lambda) = \frac{2^p R(0.5h\lambda) - R(h\lambda)}{2^p - 1} \tag{9}$$

where $R(h\lambda)$ is defined in [8].

The computations carried out in the solution of the scalar equation $y\prime = \lambda y$ with any of the methods (2)–(5) be stable for a given value of $h\lambda$ if the condition $|R(h\lambda)| \leq 1$ is satisfied. The set of all points for which this inequality is satisfied forms **the absolute stability region** of the method.

It is clear that similar statements hold for the numerical procedure, which is a combination of an explicit Runge-Kutta method and Richardson Extrapolation. It will only be necessary to replace $|R(h\lambda)| \leq 1$ with $|\bar{R}(h\lambda)| \leq 1$ in this case.

4 Drawing the Absolute Stability Regions

Consider the case where explicit Runge-Kutta methods are used directly. The boundaries of the absolute stability regions are obtained in the following way. Let $\tilde{\lambda} = h\lambda$ be equal to $\alpha + betai$ and ε be some small increment. Start with $\alpha = 0$ and test the values of the stability polynomial $R(\tilde{\lambda}$ for $\beta = 0, \varepsilon, 2\varepsilon, 3\varepsilon, \ldots$. Continue this process as long as $R(\tilde{\lambda}) \leq 1$ and denote by β_0 the last value for which the inequality)$R(\tilde{\lambda}) \leq 1$ was satisfied. Set $\alpha = -\varepsilon$ and repeat the same computations with $\beta = 0, \varepsilon, 2\varepsilon, 3\varepsilon, \ldots$, to obtain the largest value β_ε for which $R(\tilde{\lambda}) \leq 1$ is satisfied. Continuing in this way it will be possible to calculate the coordinates of a set of the points $(0, \beta_0), (-\varepsilon, \beta_\varepsilon), (-2\varepsilon, \beta_{2\varepsilon}), \ldots$ in the negative part of the complex plane. More precisely, all of these points are located close to the boundary of the part of the absolute stability region which is located over the real axis and to the left of the imaginary axis. Moreover, all these points lie inside the absolute stability region. Therefore, the curve connecting these points will be a close approximation of the boundary of the part of the stability region which is located over the real axis and to the left of the imaginary axis. It should be mentioned here that $\varepsilon = 0.001$ was actually used in the preparation of all plots that are presented in this section. It can easily be shown that the absolute stability region is symmetric with regard to the real axis. Therefore there is no need to repeat the process that was sketched above for negative values of β. Some people are drawing parts of the stability regions which are located to the right of the imaginary axis (see, for example, [9]). In our opinion this is not necessary and in the most of the cases it will not be desirable either. This can be explained as follows. Consider $y\prime = \lambda y$ and let again $\tilde{\lambda}$ be equal to $\alpha + \beta i$ but assume this time that α is positive. Then the exact solution of $y\prime = \lambda y$ is not bounded and it is clearly not desirable to search for numerical methods which will produce bounded approximate solutions (the concept of relative stability, see [9], p. 75, is more appropriate in this situation, but this topic is beyond the scope of the present paper).

The same principle can be applied when the explicit Runge-Kutta methods are combined with Richardson Extrapolation. In this case, the polynomial $R(\tilde{\lambda})$ should be replaced with $\bar{R}(\tilde{\lambda})$.

The stability regions of the methods studied in this paper are given in Fig. 1. Note that there exists only one first-order one-stage explicit Runge-Kutta method (the Forward Euler Formula) which is given by (2). When m-stage explicit Runge-Kutta methods of order p with $p = m$ and $p = 2, 3, 4$ are used for each $p = m$ there exists a big class of explicit Runge-Kutta methods. Single representatives of these classes are given by formulae (3), (4) and (5). However, all numerical methods from any of these three classes have the same absolute stability region, the stability given in Fig. 1.

The above remark and the plots drawn in Fig. 1 show clearly that the following theorem holds:

Theorems 1. The use of Richardson Extrapolation together with the explicit m-stage Runge-Kutta methods of order p leads always to larger absolute stability regions than the absolute stability regions of the underlying method when $p = m$.

5 Preparation of an Appropriate Numerical Example

Consider the problem defined by

$$y\prime = Ay,\ A \in R^{3\times 3},\ y = (y_1, y_2, y_3)^T \in R^3,\ t \in [0, 13.1072],\quad y(0) = (1, 0, 3)^T, \tag{10}$$

$$a_{11} = 741.4,\quad a_{12} = 749.7,\quad a_{13} = 741.7, \tag{11}$$

$$a_{21} = 765.7,\quad a_{22} = 758.0,\quad a_{23} = 757.7, \tag{12}$$

$$a_{31} = 725.7,\quad a_{32} = 741.7,\quad a_{33} = 734.0. \tag{13}$$

The eigenvalues of the matrix A are $(750, 0.3 + 8i, 0.3 - 8i)$, which means that the problem defined by (10)–(13) is moderately stiff and the requirements for achieving stable computations are much more restrictive than the requirements for achieving sufficiently high accuracy. The solution of the example is given in Fig. 2.

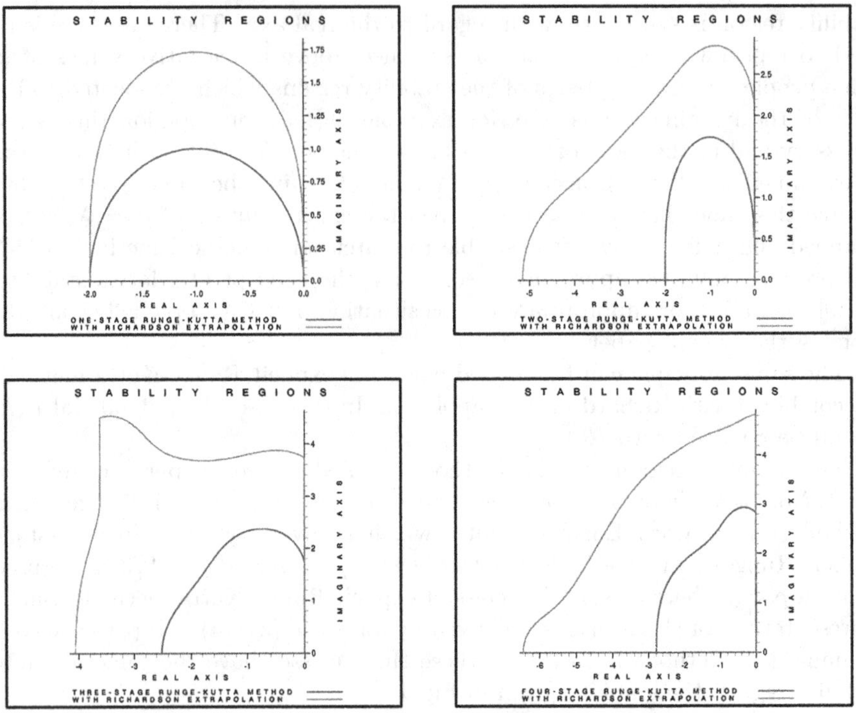

Fig. 1. Absolute stability regions of explicit Runge-Kutta methods with m = p when these are applied directly and together with Richardson Extrapolation.

6 Organization of the Computations

The integration interval [0, 13.1072] was divided into 128 equal sub-intervals and the accuracy of the results obtained by any of the selected numerical methods was evaluated at the end of each sub-interval. Let $\hat{t}_j, j = 1, 2, \ldots, 128$, be the end of any sub-interval. Then the following formula is used to evaluate the accuracy achieved at this point:

$$ERROR_I = \frac{\sqrt{\sum_{i=1}^{k} \left(y_i(\hat{t}_j) - \hat{y}_{ij}\right)^2}}{max\left[\sqrt{\sum_{i=1}^{k} \left(y_i(\hat{t}_j)\right)^2}, 1.0\right]}. \tag{14}$$

The global error is computed as

$$ERROR = \max_{j=1,2,\ldots,128} \left(ERROR_j\right). \tag{15}$$

Ten runs were performed with every numerical method. The first run is carried out by using $h = 0.00512$. In each of the next nine runs the stepsize is halved (which leads automatically to performing twice more time-steps). All calculations

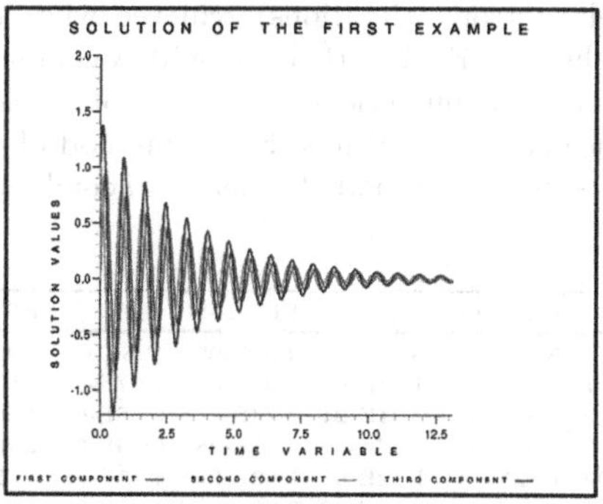

Fig. 2. Plots of the three components of the solution of the problem defined by (10)–(13).

were carried out at the computers of the Centre of Scientific Computing at Technical University of Denmark [11].

7 Numerical Results

Accuracy results, which are obtained when eight numerical methods for the solution of systems of ODEs are used, are given in Table 1. Convergence rates observed for the eight tested numerical methods are shown in Table 2.

Two important conclusions can immediately be drawn by investigating the results presented in the two tables: (a) for the largest stepsize the methods are producing unstable results (but the Richardson Extrapolation may succeed where the original Runge-Kutta methods fail) and (b) the results show that the calculated (as ratios of two consecutive error estimations) convergence rates of the Runge-Kutta method of order p are about 2^p while for the combinations of the Runge-Kutta methods and the Richardson Extrapolation the corresponding convergence rates are 2^{p+1} which means that the order of accuracy is increased by one. Some more conclusions are given in [18].

8 Major Concluding Remarks and Plans for Future Research

It is well known that the application of the Richardson Extrapolation leads to an improvement of the accuracy of the underlying numerical method, not only the explicit Runge-Kutta methods with $p = m, m = 1, 2, 3, 4$, see [5,10,16]. In the present paper it was shown that the combined methods (any of the explicit Runge-Kutta methods with $p = m, m = 1, 2, 3, 4$, plus the Richardson Extrapolation) have additionally larger regions of absolute stability.

It should also be emphasized here that non-stiff systems of ODEs appear after some kind of discretization and/or splitting of mathematical models appearing in

Table 1. Accuracy results (error estimations) achieved when the first example from Sect. 5 is run by using eight numerical methods on a SUN computer by using quadruple precision. "N.S." means that the numerical method is not stable for the stepsize used. "ERKi", $i = 1, 2, 3, 4$, means explicit Runge-Kutta method of order p=i. "ERKi+R" refers to the explicit Runge-Kutta method of order $p = i$ combined with the Richardson Extrapolation.

	Stepsize	Steps	ERK1	ERK1+R	ERK2	ERK2+R	ERK3	ERK3+R	ERK4	ERK4+R
1	0.00512	2560	N.S.	N.S.	N.S.	2.39E-05	N.S.	6.43E-03	N.S.	4.49E-10
2	0.00256	5120	2.01E-01	4.22E-02	4.22E-02	2.99E-06	5.97E-06	7.03E-09	2.46E-08	1.41E-11
3	0.00128	10240	9.21E-02	2.91E-04	2.91E-04	3.73E-07	7.46E-07	4.40E-10	1.54E-09	4.39E-13
4	0.00064	20480	4.41E-02	7.27E-05	7.27E-05	4.67E-08	9.33E-08	2.75E-11	9.62E-11	1.37E-14
5	0.00032	40960	2.16E-02	1.82E-05	1.82E-05	5.83E-09	1.17E-08	1.72E-12	6.01E-12	4.29E-16
6	0.00016	81920	1.07E-02	4.54E-06	4.54E-06	7.29E-10	1.46E-09	1.07E-13	3.76E-13	1.34E-17
7	0.00008	163840	5.32E-03	1.14E-06	1.14E-06	9.11E-11	1.82E-10	6.71E-15	2.35E-14	4.19E-19
8	0.00004	327680	2.65E-03	2.84E-07	2.84E-07	1.14E-11	2.28E-11	4.20E-16	1.47E-15	1.31E-20
9	0.00002	655360	1.33E-03	7.10E-08	7.10E-08	1.42E-12	2.85E-12	2.62E-17	9.18E-17	4.09E-22
10	0.00001	1310720	6.66E-04	1.78E-08	1.78E-08	1.78E-13	3.56E-13	1.64E-18	5.74E-18	1.28E-23

Table 2. As Table 1 but convergence rates are given instead of accuracy results

	Stepsize	Steps	ERK1	ERK1+R	ERK2	ERK2+R	ERK3	ERK3+R	ERK4	ERK4+R
1	0.00512	2560	N.A.	N.A.	N.A.	N.A.	N.A.	N.A.	N.A.	N.A.
2	0.00256	5120	N.A.	N.A.	N.A.	7.99	N.A.	Very big	N.A.	31.84
3	0.00128	10240	2.18	145.02	145.02	8.02	8.00	15.98	15.97	32.12
4	0.00064	20480	2.09	4.00	4.00	7.99	8.00	16.00	16.01	32.04
5	0.00032	40960	2.04	3.99	3.99	8.01	7.97	15.99	16.01	31.93
6	0.00016	81920	2.02	4.01	4.01	8.00	8.01	16.07	15.98	32.01
7	0.00008	163840	2.01	3.98	3.98	8.00	8.02	15.95	16.00	31.98
8	0.00004	327680	2.01	4.01	4.01	7.99	7.98	15.97	15.99	31.98
9	0.00002	655360	1.99	4.00	4.00	8.03	8.00	16.03	16.01	32.03
10	0.00001	1310720	2.00	3.99	3.99	7.98	8.01	15.98	15.99	31.95

different areas of science and engineering. As an example large-scale air pollution models (see [1,2,13,15]) should be mentioned. Such models can be used in many studies. The most important is perhaps the investigation of impact of climate change on air pollution levels(for example, [14,17,19]). The advection terms of the air pollution models can be treated with explicit methods (see again [1,2,13,15]). An attempt to use the explicit Runge-Kutta methods discussed in this paper will be carried out in the near future.

Acknowledgments. The research of K. Georgiev and I. Dimov is supported in part by Grants DCVP-02/1 and I01/5 from the Bulgarian National Science Found. The authors thanks Centre of Scientific Computing at Technical University of Denmark for giving access to their computers for making computations.

References

1. Alexandrov, V., Owczarz, W., Thomsen, P.G., Zlatev, Z.: Parallel runs of large air pollution models on a grid of SUN computers. Math. Comput. Simul. **65**, 557–577 (2004)

2. Alexandrov, V., Sameh, A., Siddique, Y., Zlatev, Z.: Numerical integration of chemical ODE problems arising in air pollution models. Environ. Model. Assess. **2**, 365–377 (1997)
3. Butcher, J.C.: The Numerical Analysis of Ordinary Differential Equations. Runge-Kutta Methods and General Linear Methods. Wiley, Chichester (1987)
4. Dahlquist, G.: A special stability problem for linear multistep methods. BIT **3**, 27–43 (1963)
5. Farago, I., Havasi, Á., Zlatev, Z.: Efficient implementation of stable Richardson extrapolation algorithms. Comput. Math. Appl. **60**(8), 2309–2325 (2010)
6. Hairer, E., Norsett, S.P., Wanner, G.: Solving Ordinary Differential Equations I. Nonstiff Problems. Springer, Heidelberg (1987)
7. Hairer, E., Wanner, G.: Solving Ordinary Differential Equations II. Stiff and Differential-Algebraic Problems. Springer, Heidelberg (1991)
8. Hundsdorfer, W., Verwer, J.G.: Numerical Solution of Time-Dependent Advection-Diffusion-Reaction Equations. Springer, Heidelberg (2003)
9. Lambert, J.D.: Numerical Methods for Ordinary Differential Equations. Wiley, New York (1991)
10. Richardson, L.F.: The deferred approach to the limit I-single lattice. Philos. Trans. Roy. Soc. Lond. A **226**, 299–349 (1927)
11. WEB-site of the Centre for Scientific Computing at the Technical University of Denmark: Sun High Performance Computing Systems (2002). http://www.hpc.dtu.dk
12. Stetter, H.J.: Analysis of Discretization Methods for Ordinary Differential Equations. Springer, Heidelberg (1973)
13. Zlatev, Z.: Computer Treatment of Large Air Pollution Models. Kluwer (now Springer), Dordrecht (1995)
14. Zlatev, Z.: Impact of future climate changes on high ozone levels in European suburban areas. Clim. Change **101**, 447–483 (2010)
15. Zlatev, Z., Dimov, I.: Computational and Numerical Challenges in Environmental Modelling. Elsevier, Amsterdam (2006)
16. Zlatev, Z., Farago, I., Havasi, Á.: Stability of the Richardson extrapolation applied together with the θ - method. J. Comput. Appl. Math. **235**(2), 507–520 (2010)
17. Zlatev, Z., Georgiev, K., Dimov, I.: Influence of climatic changes on air pollution levels in the Balkan Peninsula. Comput. Math. Appl. **65**(3), 544–562 (2013)
18. Zlatev, Z., Georgiev, K., Dimov, I.: Absolute stability properties of the Richardson extrapolation combined with explicit Runge-Kutta methods (2013). http://parallel.bas.bg/dpa/BG/dimov/index.html, http://parallel.bas.bg/dpa/EN/publications_2012.htm, http://parallel.bas.bg/dpa/BG/publications_2012.htm
19. Zlatev, Z., Havasi, Á., Farago, I.: Influence of climatic changes on pollution levels in Hungary and its surrounding countries. Atmosphere **2**, 201–221 (2011)

Numerical Solvers on Many-Core Systems

Peta-Scale Hierarchical Hybrid Multigrid Using Hybrid Parallelization

Björn Gmeiner(✉) and Ulrich Rüde

Department of Computer Science 10,
University of Erlangen-Nürnberg, Erlangen, Germany
{bjoern.gmeiner,ulrich.ruede}@fau.de

Abstract. In this article we present a performance study of our finite element package Hierarchical Hybrid Grids (HHG) on current European supercomputers. HHG is designed to close the gap between the flexibility of finite elements and the efficiency of geometric multigrid by using a compromise between structured and unstructured grids. A coarse input finite element mesh is refined in a structured way, resulting in semi-structured meshes. Within this article we compare and analyze the efficiencies of the stencil-based code on those clusters.

Keywords: Parallel multigrid · Performance analysis · HHG

1 Introduction

In electro-chemistry, *density functional theory* (DFT) plays an important role as a class of models to calculate the electrical potential imposed by the charges of an ensemble of atom nuclei and electrons [7,11]. One essential step in the DFT is the solution of the potential equation that reduces to Poisson's equation in the case of a homogeneous dielectricity coefficient [12]. However, often effects of an ionic solvent with varying dielectricity cannot be neglected. The governing equation in this case is given as

$$-\nabla \cdot k(x,y,z)\nabla u(x,y,z) = f(x,y,z), \tag{1}$$

where k denotes the dielectricity constant, u the potential field and f the right-hand side. For the sake of simplicity, we assume Dirichlet conditions at the boundaries of the simulation domain Ω. The paper is structured as follows:

The remaining part of this section introduces the software package HHG (Hierarchical Hybrid Grids), and three peta-scale class HPC systems. The second section describes a novel hybrid parallelization strategy implemented within HHG to allow extreme scale simulations on the clusters JUQUEEN and SuperMUC. Scalability experiments and performance analysis on different clusters are presented.

I. Lirkov et al. (Eds.): LSSC 2013, LNCS 8353, pp. 439–447, 2014.
DOI: 10.1007/978-3-662-43880-0_50,

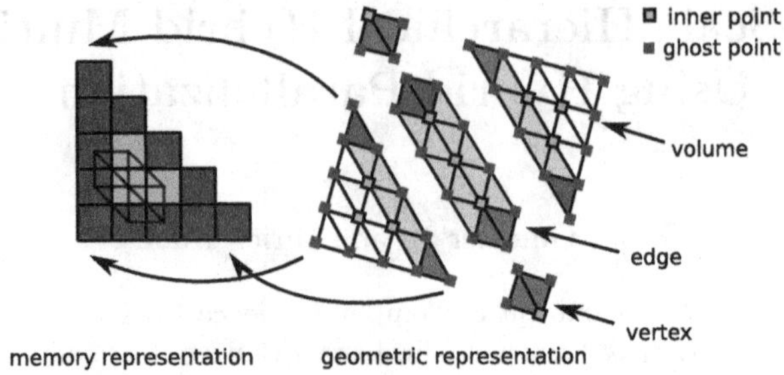

Fig. 1. Splitting of two triangle input elements into HHG grid primitives after two steps of refinement. Additionally, the memory representation of a refined triangle with a 7-point stencil for the lower left inner point is sketched [9].

1.1 Parallel Multigrid with Hierarchical Hybrid Grids

For solving partial differential equations (PDEs), finite elements (FE) methods are a popular discretization scheme, since they allow flexible, unstructured meshes. The framework HHG [2,8] is designed to combine the flexibility of the FE method and the superb performance of geometric multigrid [3,10] by using a compromise between structured and unstructured grids. A coarse input FE mesh is organized into the grid primitives vertices, edges, faces, and volumes. The primitives are then refined in a structured way (see Fig. 1), resulting in semi-structured meshes. The regularity of the resulting grids may be exploited in such a way that it is no longer necessary to explicitly assemble the global discretization matrix. In particular, given an appropriate input grid, the discretization matrix may be defined implicitly using stencils for each structured patch. Here a stencil represents a row of the global stiffness matrix. Within HHG, we have implemented an MPI[1]-parallel geometric multigrid method that operates on the resulting block-structured grid hierarchy. The settings of the multigrid components and parameters used in this paper are three Gauss-Seidel iterations for pre- and post-smoothing steps, linear interpolation between six multigrid levels, parallel Conjugated Gradient algorithm to solve the coarsest grid problem, and direct coarse grid approximation with coefficient averaging.

The stencils can be stored in registers when the dielectricity is piecewise constant, or it can be assembled on-the-fly for a variable dielectricity. In both cases this results in a so-called *matrix-free* implementation. This can have significant performance benefits since it reduces memory traffic, possibly at the expense of redundant computations.

1.2 Architectures

Within this article we compare the performance of HHG on three European supercomputers: JUGENE, JUQUEEN are both located at FZ Jülich, and

[1] www.mcs.anl.gov/mpi

Table 1. System overview of the IBM clusters JUGENE, JUQUEEN, and SuperMUC.

	JUGENE	JUQUEEN	SuperMUC
System	BlueGene/P	BlueGene/Q	System x iDataPlex
Processor	IBM PowerPC 450	IBM PowerPC A2	Intel Xeon E5-2680
Clock frequency	0.85 GHz	1.6 GHz	2.8 GHz
Number of nodes	73 728	24 576	9 216
Cores per node	4	16	16
HW threads per Core	1	4	2
Memory per HW thread	0.5 GB	0.25 GB	1 GB
Network topology	3D torus	5D torus	Tree
Gflop/s per Watt	0.44	2.54	0.94

SuperMUC is located in the LRZ supercomputing center in Garching. Table 1 presents a system overview of these clusters.

JUGENE, was the largest BlueGene/P installation with 294 912 compute cores and 1 petaflop/s peak performance. Each node was equipped with a PowerPC 450 quadcore processor running at a low clock frequency. The architecture provided a very high main memory performance. A three-dimensional torus network in combination with a tree-based collective network was available for communication. We were able to solve FE systems with in excess of 10^{12} degrees of freedom on JUGENE with the HHG package. We use these three year old performance results obtained on JUGENE as reference for our new results.

The BlueGene/Q system JUQUEEN is the successor of the JUGENE with a peak performance of 5.9 petaflop/s. Although the clock-frequency still remains relatively low, it is nearly doubled. Each of the 16 cores available for user applications has four hardware (HW) threads. The memory bandwidth has not scaled up accordingly, but in order to compensate this disadvantage in part, e.g. the prefetching and speculative execution facilities have been improved. The torus network is extended to five dimensions for shorter paths, and the collective network was fused into the torus network. The ratio of peak network bandwidth node performance and peak floating point performance is only 50 % of that of BlueGene/P. On the other hand, the cores within each node and consequently the intra-node communication performance has drastically increased.

SuperMUC is a 3.2 petaflop/s IBM x iDataPlex cluster. This machine consists of 18 thin islands, carrying 97.5 % of total performance, and one fat island for moderately parallel, memory intensive applications. Each thin island is equipped with 512 compute nodes. Two sockets with Sandy Bridge-EP Intel Xeon E5-2680 8C provide 16 physical cores. The Xeon processors deliver a significantly higher core and node performance than the PowerPCs in the IBM architectures, at the price of higher power consumption. The nodes within an island are linked by an Infiniband non-blocking tree, whereas a pruned 4:1 tree connects all islands.

2 Porting Hierarchical Hybrid Grids to BlueGene/Q

For the substantial changes that were necessary to use HHG on more than 30 000 parallel threads, we refer to [9]. This includes the design of data structures for generating tetrahedral input grids efficiently in parallel. However, for the current and upcoming systems, this alone proved not to be sufficient and thus this paper will present a new hybrid parallelization strategy.

The new system architectures with more powerful and complex compute nodes make a hybrid parallelization approach especially attractive and potentially profitable, since they provide better opportunities for a shared memory parallelization via OpenMP[2]. Thus a hybrid parallelization strategy, including message passing for coarse grain parallelism, and shared memory parallelism within a node for finer scale parallel execution, has been found essential for exploiting the full potential of architectures like JUQUEEN or SuperMUC.

In a pure MPI parallel setting, the available main memory per process is only 256 MB per process on JUQUEEN. This is too small for the three largest runs described in the next sections. In contrast, a hybrid parallelization increases the available main memory for each process. On SuperMUC, the scaling breaks down when too many MPI processes are being used. A hybrid parallelization helps to limit the total number of MPI processes and this helps to maintain scalability for extreme size simulations.

The current OpenMP implementation in HHG supports parallelism inside kernel executions and copy of ghost layers on several primitives. However, the MPI instructions are executed asynchronously, but not explicitly OpenMP-parallel. Further, OpenMP introduces an additional overhead for spawning threads, which is especially critical on the coarsest grids, where the workloads per thread is small. The quality of the MPI/OpenMP parallel execution is reflected in Table 2. All runs up to the last two are executed by four threads per compute core. From the timings we conclude that serial fraction of the code is still between $1-2\,\%$. We will use a hybrid parallelization with up to eight OpenMP-threads for the largest parallel run on JUQUEEN in the following scaling experiment as the performance loss is still not too high.

Table 2. Efficiency of the hybrid parallelization compared to a pure MPI parallel approach on JUQUEEN for moderate problem sizes.

MPI processes	OpenMP threads	Runtime	Efficiency (%)	MPI processes	OpenMP threads	Runtime	Efficiency (%)
4 096	1	3.09		64	64	5.22	59
2 048	2	3.10	99				
1 024	4	3.21	96	2 Threads/Core:			
512	8	3.45	90	64	32	5.77	54
256	16	3.95	78	1 Threads/Core:			
128	32	4.33	71	64	16	8.36	37

[2] www.openmp.org

Table 3. Weak scaling experiment on JUQUEEN solving a problem on the full machine

Number of threads	Number of unknowns	Time per V-cycle (s)	Number of threads	Number of unknowns	Time per V-cycle (s)
64	$1.33 \cdot 10^8$	2.34	16 384	$3.43 \cdot 10^{10}$	3.15
128	$2.67 \cdot 10^8$	2.41	32 768	$6.87 \cdot 10^{10}$	3.28
256	$5.35 \cdot 10^8$	2.80	65 536	$1.37 \cdot 10^{11}$	3.39
512	$1.07 \cdot 10^9$	2.82	131 072	$2.75 \cdot 10^{11}$	3.56
1 024	$2.14 \cdot 10^9$	2.82	262 144	$5.50 \cdot 10^{11}$	3.68
2 048	$4.29 \cdot 10^9$	2.84	524 288	$1.10 \cdot 10^{12}$	3.76
4 096	$8.58 \cdot 10^9$	2.96	1 048 576	$2.20 \cdot 10^{12}$	4.07
8 192	$1.72 \cdot 10^{10}$	3.09	1 572 864	$3.29 \cdot 10^{12}$	4.03

2.1 Weak Scaling on JUQUEEN

This section shows the scalability of the HHG approach on a current cluster. The program is compiled with the IBM XL compiler suite on both BlueGene clusters. As a test case we use a piecewise constant dielectricity, and thus can use constant stencils within each HHG block and each geometric primitive. Consequently, the numerical efficiency is extremely high and in a relative sense, the communication is very intensive. Therefore, this is quite a challenging setup for maintaining the parallel scalability as we will show in the performance study in the next section. Table 3 shows the run-time results of a scaling experiment. The smallest test run already solves a system of slightly more than 10^8 unknowns and one V-cycle takes approximately 2.3 s. Note that this is performed on a single compute node on JUQUEEN, demonstrating the high efficiency of the HHG approach. In each further row of the table, the problem size is doubled as well as the number of nodes. This is a classical weak scalability test. The full machine could eventually solve a linear system with $3.3 \cdot 10^{12}$ unknowns, corresponding to more than 10^{13} tetrahedral finite elements. In total, this computation uses 300, out of the almost 400 TB of main memory during the solution process.

Four hardware threads are necessary to saturate the performance of one processor core, leading to a parallel execution of more than *one million threads*. Although, the computational time increases only moderately, we note that the coarse grid solver is only a straightforward Conjugated Gradient (CG) iteration. Therefore in large runs, more than half a second of the V-cycle execution time is spent in the increasing number of CG iterations on the coarsest grid, that is caused by larger and larger coarse grids. This shows clearly, that for perfect asymptotic scalability a better coarse grid solver would be necessary. Nevertheless, we believe that our results with the CG solver indicate clearly that the coarse grid solver performance is not as critical for scalability, as has been discussed in the older literature on parallel multigrid methods.

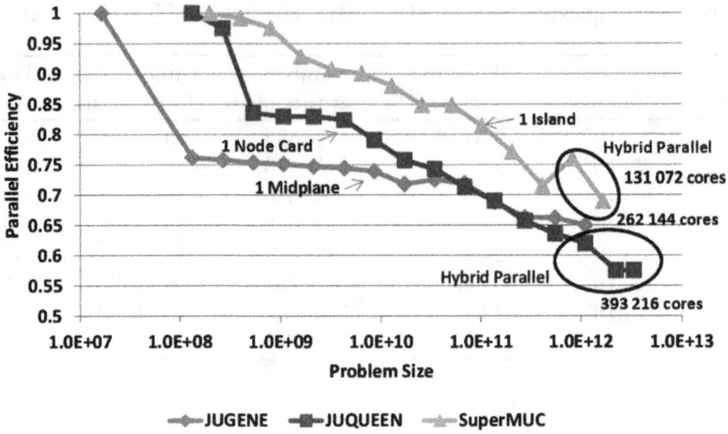

Fig. 2. Parallel efficiencies on different supercomputers in a weak scaling on three different supercomputers. The largest runs required a hybrid parallelization strategy. Some hardware structures (node card, midplane, island) of the clusters can be identified by the gradient of the parallel efficiency.

2.2 Comparison of Scalability Results on Other Peta-scale Clusters

In this section, we compare the parallel efficiency of our code on different HPC clusters. In contrast to the BlueGene systems, the program is compiled with the Intel compiler suite and IBM MPI for SuperMUC with *-O3 -xavx* compiler flags. As reference, one V-cycle takes 4.25 s for JUGENE, and 1.18 s on SuperMUC on one compute node. Figure 2 shows strong efficiency drops when advancing from one node to several nodes. This is especially prominent on both BlueGene systems. However, from there onwards to larger parallel runs, the parallel efficiency stays nearly constant. Only the transition from a single Midplane on BlueGene/P, or one Node Card on BlueGene/Q to larger sub-portions of the architecture, induce again more significant performance drops. On SuperMUC the efficiencies up to a quarter island ($2.6 \cdot 10^{10}$ unknowns) differ between the multigrid cycles. We believe that this is caused by perturbations due to other applications running simultaneously on the same island. From quarter of an island to half of an island ($5.2 \cdot 10^{10}$ unknowns) the performance even improves. However, when leaving a single island of the architecture, the parallel efficiency drops significantly. This is likely caused by the reduced communication performance beyond each island in the pruned 4:1 tree. For more than two islands we also disable hyperthreading to obtain substantially more reproducible run-times. In contrast to this observation, the run-times of both BlueGene machines remain more stable for all problem sizes. SuperMUC presents the best parallel efficiency. However, we could not map our mesh onto the torus networks, since the coarsest mesh is basically unstructured.

First scaling experiments on SuperMUC showed a breakdown at 65 536 MPI processes, resulting in roughly four times longer run-times, as well as fluctuations

Table 4. Single node and parallel efficiencies (scaling), as well as power consumptions of used parts of the clusters while running HHG.

	JUGENE	JUQUEEN	SuperMUC
Single node			
Peak flop/s (constant dielectricity)	6 %	7 %	12 %
Peak flop/s (variable dielectricity)	9 %	10 %	13 %
Peak bandwidth (constant dielectricity)	11 %	53 %	60 %
Parallel efficiencies (at ≈ 0.8 Pflop peak)			
Scaling (constant dielectricity)	65 %	64 %	72 %
Scaling (variable dielectricity)	94 %	93 %	96 %
Scaling – without CG (constant dielectricity)	75 %	70 %	79 %
Number of processes	262 144	262 144	32 768
Energy improvement compared to JUGENE (const.)	1	6.6	4.7
Energy improvement compared to JUGENE (var.)	1	6.4	3.2

in the timings of up to 15 s between the single V-cycles compared to runs with 32 768 MPI processes. Figure 2 displays that there have already been problems on 32 768 MPI processes (corresponding to $4.1 \cdot 10^{11}$ unknowns or 4 islands).

The results on larger machine sizes use the hybrid parallelization that allows us to execute the largest two runs with only 16 384 MPI processes, leading to a significantly improved parallel efficiency. The largest run was carried out on 16 islands of the cluster. Different from the behavior of SuperMUC, the hybrid parallelization on JUQUEEN, as used for the largest runs, clearly decreases the parallel efficiency. However a hybrid parallelization is still necessary as explained above in order to have enough main memory available.

Table 4 shows the single node performance, parallel efficiencies, and energy consumptions relatively to JUGENE. The runs were carried out for a node allocation providing ≈ 0.8 Pflop/s nominal peak. Even though a major design goal of the BlueGene/P was to have a low energy consumption, the next generation could improve the energy consumption by a factor between six to seven for our application. The SuperMUC turns out not to be as energy efficient as JUQUEEN, however it does not require such a high degree of parallelism from the application.

2.3 Single Node Performance Analysis

This section will analyze the single node performance as given in Table 4. This is for the case of constant dielectricity.

On JUGENE one MPI process is assigned to each compute core. Since the processors provide high memory bandwidth, codes tend to be more limited by instruction throughput than by memory bandwidth. However, the kernel that applies the stencil is affected on JUGENE from a serialization within the PowerPC multiply-add instructions. Additionally, a correct memory alignment for vectorized loads (for the SIMD units) is not assured due to the varying loop sizes that are caused by the tetrahedral macro elements. The limited-issue width and the in-order architecture of the processor leads to further performance limitations and eventually result in a node performance of only 6.1 % of the peak performance (see Table 4). This is for a complete multigrid cycle.

For a comparison, we refer to results of Datta et al. who present autotuning results for an averaging 29-point stencil on this architecture [5,6]. Their baseline implementation achieves about 0.035 GStencil/s updates which corresponds to 7.7 % of the peak performance for a reference (in-cache) implementation. Basically two of eleven optimization techniques (e.g. padding, core blocking, software pre-fetching) techniques can achieve a significant speedup: common subexpression elimination and register blocking. While the first inherently cannot be applied for our stencil, since we do not have redundant calculations, the register blocking results in our case roughly in a speedup of two. In principle we could use this code optimization, but it leads to very small sub-blocks that will suffer from non-constant loop sizes. Moreover, the issue with serialization remains a bottleneck, which is not the case for the averaging stencil.

On JUQUEEN, we assign one MPI process to each HW thread. The stream benchmark shows that it is possible to run at a high fraction of ≈ 85 % of the effective maximal memory bandwidth of 27.8 GB/s by using one process per node. Two or four threads per node saturate the effective bandwidth completely. Going from one to two threads per core, HHG gives a factor of two improvement in performance. In these cases, the code is still instruction bound like on BlueGene/P. Going from two to four threads per core, the additional speedup is only a factor 1.3. Overall, in this case, a multigrid cycle utilizes in average about 18.1 GB/s of the main memory bandwidth. Only by reducing the main memory footprint and possibly improving the core performance itself, we see a chance for further reductions of the execution time.

Similarly to the situation on JUQUEEN, the code is mainly memory bandwidth limited on the SuperMUC node architecture. However, the nodes can saturate the bandwidth better and its machine balance suits better the characteristics of our code. Thus we achieve with a better flop/s performance than on JUQUEEN. However, hyperthreading improves the performance only insignificantly by at most a few percent.

3 Conclusion and Future Work

We presented a weak-scaling comparison of HHG on three different HPC petaflop clusters. To reach the full potential of the recent architectures, a hybrid parallelization approach turned out to be necessary for the growing node-level parallelism to compensate memory limitations and maintain scalability. Recently, we designed a Stokes solver for Earth mantle convection simulations [1,4] within HHG utilizing the presented multigrid performance.

Acknowledgements. The work was supported by the International Doctorate Program (IDK) within the Elite Network of Bavaria. The authors gratefully acknowledge the Gauss Centre for Supercomputing (GCS) for providing computing time through the John von Neumann Institute for Computing (NIC) on the GCS share of the supercomputer JUQUEEN at Jülich Supercomputing Centre (JSC). GCS is the alliance of the three national supercomputing centres HLRS (Universität Stuttgart), JSC (Forschungszentrum Jülich), and LRZ (Bayerische Akademie der Wissenschaften), funded by the German Federal Ministry of Education and Research (BMBF) and the German State Ministries for Research of Baden-Württemberg (MWK), Bayern (StMWFK) and Nordrhein-Westfalen (MIWF).

References

1. Baumgardner, J.R.: Three-dimensional treatment of convective flow in the earth's mantle. J. Stat. Phys. **39**(5/6), 501–511 (1985)
2. Bergen, B., Gradl, T., Hülsemann, F., Rüde, U.: A massively parallel multigrid method for finite elements. Comput. Sci. Eng. **8**(6), 56–62 (2006)
3. Brandt, A.: Multi-level adaptive solutions to boundary-value problems. Math. Comput. **31**(138), 333–390 (1977)
4. Burstedde, C., Ghattas, O., Gurnis, M., Stadler, G., Tan, E., Tu, T., Wilcox, L.C., Zhong, S.: Scalable adaptive mantle convection simulation on petascale supercomputers. In: Proceedings of the 2008 ACM/IEEE Conference on Supercomputing, p. 62. IEEE Press (2008)
5. Datta, K., Williams, S., Volkov, V., Carter, J., Oliker, L., Shalf, J., Yelick, K.: Auto-tuning the 27-point stencil for multicore. In: Proceedings of the Fourth International Workshop on Automatic Performance Tuning (iWAPT2009) (2009)
6. Datta, K., Yelick, K.: Auto-tuning stencil codes for cache-based multicore platforms. Ph.D. thesis, University of California, Berkeley (2009)
7. Fattebert, J.L., Gygi, F.: Density functional theory for efficient ab initio molecular dynamics simulations in solution. J. Comput. Chem. **223**, 662–666 (2002)
8. Gmeiner, B., Gradl, T., Köstler, H., Rüde, U.: Highly parallel geometric multigrid algorithm for hierarchical hybrid grids. In: Binder, K., Münster, G., Kremer, M. (eds.) NIC Symposium 2012. Publication series of the John von Neumann Institute for Computing, vol. 45, pp. 323–330. Jülich (2012)
9. Gmeiner, B., Köstler, H., Stürmer, M., Rüde, U.: Parallel multigrid on hierarchical hybrid grids: a performance study on current high performance computing clusters. Concurr. Comput. Pract. Exp. (2012). http://dx.doi.org/10.1002/cpe.2968
10. Hackbusch, W.: Multi-Grid Methods and Applications. Springer, Heidelberg (1985)
11. Sanchez, V., Sued, M., Scherlis, D.: First-principles molecular dynamics simulations at solid-liquid interfaces with a continuum solvent. J. Chem. Phys. **131**(17), 174108 (2009)
12. Schmid, R., Tafipolsky, M., König, P.H., Köstler, H.: Car-Parrinello molecular dynamics using real space wavefunctions. Phys. Status Solidi B **243**(5), 1001–1015 (2006)

Many–Core Sustainability by Pragma Directives

Andreas Kucher(✉) and Gundolf Haase

Institute for Mathematics and Scientific Computing,
University of Graz, Graz, Austria
andreas.kucher@edu.uni-graz.at

Abstract. Many–core hardware is well adopted in scientific computing for a number of applications in an academic setting. Uncertainty about upcoming architectures and large development times for this hardware result in a modest acceptance in industry for commercial use. An upcoming turn from language–based many–core programming towards directive–based frameworks, similar to OpenMP, is an attempt to tackle these issues.

We present a case study for a many–core acceleration of a large–scale commercial CFD solver by means of such frameworks. We achieved a local acceleration of up to 45 for hot spots with recent hardware but the global speedup remains below 2. The main obstacle for an efficient instrumentation is the design and the complexity of the original software. Further, restrictions given by the hardware and the frameworks exist. Based on the results we sketch a long term plan for a further acceleration.

Keywords: Computational fluid dynamics · General purpose GPU · Many–core · Parallelization · OpenACC · OpenHMPP · Accelerator frameworks

1 Introduction

Many–core processors such as GPUs can outperform recent CPUs with respect to their compute power in a fine–grained parallel setting at the cost of a higher total power consumption. However, for a successful application of many–core hardware in numerical software, algorithms have to be modified and then implemented in a distinct programming language like CUDA or OpenCL. This has been done for a number of algorithms and the results are available as libraries giving good or excellent performance (e.g. [6]). Without such libraries, a many–core migration of algorithms and their implementation can be challenging.

This results in a good acceptance of many–core hardware in small and research codes for certain applications, but in industry the acceptance is modest. The migration of code to many–core hardware has to be profitable, i.e., the performance gain must be high and the software has to be sustainable at a long term.

I. Lirkov et al. (Eds.): LSSC 2013, LNCS 8353, pp. 448–456, 2014.
DOI: 10.1007/978-3-662-43880-0_51,

The impact on CPU performance due to many–core support should be negligible and a many–core migration must not affect regular code enhancements, particularly for large groups of developers. Independence of the hardware used and bit-compatibility to CPU results are of advantage.

There is a trend to move from language based many–core programming (CUDA, OpenCL) towards directive–based frameworks (OpenACC, OpenHMPP) similar to OpenMP. This paradigm shift is supposed to reduce development time and allows programmers to invoke many–core hardware without having deep knowledge on either hardware or language. The frameworks are expected to make many–core hardware interesting for commercial use, as they meet the requirements listed above in parts.

In a case study we investigate the applicability and sustainability of directive–based frameworks for a many–core acceleration of a commercial large–scale structured flow solver for CFD simulations. Flux computations on GPUs have been investigated in [5]. The potential of GPUs for CFD with respect to performance has been proven in [3] if the code can be fully matched with the GPU. The rigidity of the solver design leaves little margin for modifications. Code complexity is high and the period of vocational adjustment for modifications can be up to several months. The level of complexity increases by integrating many–core hardware support, mainly because CPU and many–core accelerator do not share the same memory space.

The remaining paper is organized as follows: Sect. 2 gives a brief introduction to directive–based hardware accelerator frameworks. Section 3 is an outline of the case study. Section 4 and Sect. 5 are dedicated to results and remarks on a further acceleration. The paper finishes with a conclusion in Sect. 6.

2 Directive–Based Hardware Accelerator Frameworks

Directive–based hardware accelerator frameworks allow many–core programming by adding meta–information in form of directives to CPU code blocks. The frameworks generate accelerator code based on this information. Figure 1 shows the basic concept and indicates that large parts of accelerator programming are transparent to the programmer. An API and data–directives allow to control the accelerator and data transfers between system memory and accelerator. Membarth et al. performed an exhaustive evaluation of such frameworks for a small scale code in image registration [4] obtaining good results.

As of today, two major frameworks are established. OpenACC is an open standard by a consortium of numerous vendors [2]. It can be considered a consequence of PGI Accelerator by the Portland Group. The PGI implementation recently added support for Intel Xeon Phi processors. OpenHMPP is mainly directed by CAPS Enterprise and one implementation is HMPP Workbench [1]. OpenACC is more intuitive and strongly relies on a sophisticated code generator but lacks of flexibility compared to HMPP. The frameworks are accompanied by a porting methodology for existing codes. Due to restrictions given by hardware and frameworks, this methodology is applicable to a rather narrow class of code without major changes in software architecture and coding style.

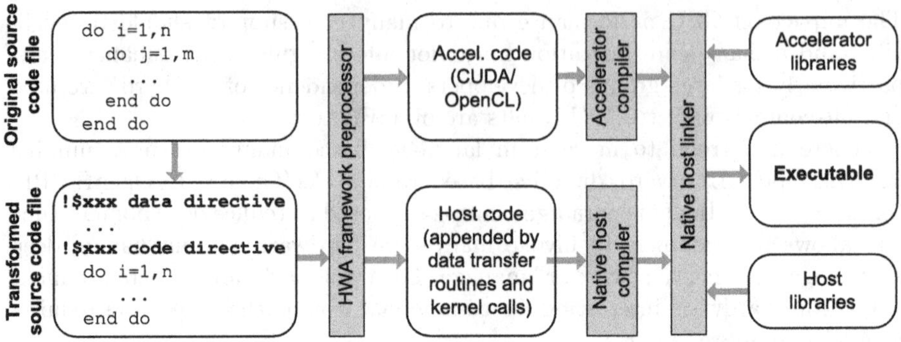

Fig. 1. Basic operation mode of directive–based frameworks.

There are efforts to eventually adopt OpenACC in the OpenMP standard. A recent technical report specifies accelerator directives proposed for a future OpenMP standard [7].

3 The Case Study

Subject to our case study is a many–core acceleration of an industrial MPI–based CFD solver for structured meshes and various CFD models. Restrictions as presented in Sect. 1 make directive–based frameworks the best option for an acceleration for now. The time frame for the case study was 10 months. It was split in 3 months for the evaluation of the best suited framework and for code analysis followed by 7 months for porting and testing.

3.1 Code Foundation

The solver has reached an advanced state of maturity and is in development for 20+ years. It is based on finite volume schemes, multilevel methods and distributed memory parallelization. It is implemented in the Fortran 77 language and the full code size can be estimated to be 400 K lines. Each CFD model has a separate, possibly redundant implementation.

Large initialization routines at program launch read in the configuration such as CFD model, geometry and load distribution amongst multiple processes for the simulation from the hard disk. A Fortran 77 work array gets allocated based on this configuration. It contains all data required for the particular CFD model. It further contains control flags for large mediator routines and space for temporary data, which is managed on a stack in this array. This results in spaghetti code. However, it can be considered representative for a number of industrial codes.

The work array is argument to all mediator routines in the solver and data within the array may be accessed by means of hash functions. The mediator routines control the program flow, e.g., make decisions about the right compute routines for a CFD model and allocate/deallocated temporary data on up to seven levels. Multi grid cycles within global iteration and time marching loops are performed until convergence after initialization. The dependent variables are advanced one time step per iteration in which fluxes and residuals are computed and boundary elements are communicated to other processes in parallel execution.

3.2 Approach and Work Done

An evaluation of HMPP Workbench and PGI Accelerator (similar to OpenACC) based on a coupled 7-point-stencil computation, as it may arise in finite volume schemes, showed that HMPP Workbench gives better performance results (Table 2). The compelling argument for excluding PGI Accelerator is the lack of applicability to multi grid as data transfers cannot be avoided between levels.[1] The solver prove to be sensitive to the many–core integration, resulting in side effects hard to resolve on high level routines. Thus, a minimal invasive bottom–up approach was chosen. We started instrumenting flux routines and established data transfers before and after they are called. Constant data is transferred only once. The flux computations are implemented in a face–centered scheme. In a first step we compute the fluxes of the faces in two/three iterations. Each iteration computes all faces of one spatial direction with a mapping of one thread per face. In a second step we compute the residuals by summing up the fluxes. The porting approach for these low–level routines is similar to the porting methodology indicated by framework vendors (e.g. [1]).

All code related to the accelerator was then offloaded to an external Fortran 90 module. This allows to use pointers for data management and establishes a separation from CPU code.

3.3 Confinements During the Porting Process

During the porting process we observed major difficulties. They can be characterized in three categories: Software design, framework–related issues and limitations given by the hardware. Crucial points are discussed below. Issues marked with † are fundamental and related to technological limitations. Issues marked with ⋆ are related to software design and may be resolved by manpower.

⋆*The Solver Architecture* does not implement a separation between logic/ program flow, data and algorithms. This can be resolved at the end of the call tree, but not a priori for other routines. Hence, the percentage of code that can be accelerated is limited. A reorganization is expensive. Adding additional program logic to resolve this is error–prone and hard to maintain.

†*Hardware Accelerator Frameworks* like HMPP Workbench are a rather new piece of technology. They have reached some maturity but cannot be compared to well established compiler suites in terms of stability and functionality.

[1] The same statement holds for OpenACC 1.0. A recent proposal for OpenACC 2.0 resolves this issue.

This also holds for debugging. In our case, all accelerator related code is part of a Fortran 90 module. A modification of one code line requires a recompilation of the full module and results in large compilation times. Certain e.g. GPU specific features are not supported. This has an impact on the performance achievable.

†*Bit–compatibility* as binary equivalence of arithmetic results to previous versions of the solver on the same hardware architecture is of big importance for the vendor and demanded by some customers. For internal testing purpose it is considered to be valuable. By invoking many–core hardware it cannot be preserved. Even minor code modifications, which are mandatory for a good many–core performance, are likely to cause different results at bit–level.

†*Compute Performance* of one code targeting two different hardware architectures is a tricky matter. A directive–based framework creates accelerator code by means of CPU code. A compromise between CPU and many–core performance is evident. Code redundancies for CPU and many–core hardware for large parts of the solver reduce the application of directive–based frameworks to absurdity. In general, CFD algorithms for CPU execution are well studied. This is not true for many–core hardware, except for standard algorithms.

⋆*Data Management and Data Transfers* to accelerator memory have to be performed explicitly and depend on the CFD model launched. An increasing number of routines accelerated requires complicated if/else constructs to decide when and which data has to be allocated/transferred.

Consequences are that the issues above may not be resolved by so called workarounds. Doing so is error-prone, reduces the reliability and maintainability of the solver and would result in fragile code.

4 Results

4.1 Framework Evaluation and Compute Routines

Table 1 shows the hardware used for benchmarking. We always use 1 core for CPU execution. Table 2 shows a performance evaluation of hardware accelerator frameworks for a coupled 7–point stencil arising from a CFD flux computation on system (A). After familiarization with the frameworks, approximately one man week was invested for each implementation. The better performance of HMPP compared to CUDA for large grids can be explained by this time frame. In CUDA the memory wall can be shifted for small grids by texture caches. HMPP gives better results than PGI Accelerator. The table shows that memory transfers are the main bottleneck reducing the best speedup from 21.7 to 1.6.

Table 3 shows timings for routines accelerated. An acceleration pays off only if data transfers can be reduced between calls for several routines. This is not true for routines with high arithmetic density. We could limit the CPU performance loss to an acceptable level. The CPU run time of Flux 2 could be reduced significantly.

Table 1. Description of the hardware platforms used.

	(A)	(B)
CPU	Intel Core i7 2600 K	Intel Xeon X5650
# CPUs	1	2
Cores used	1	1
Clock [GHz]	3.40	2.67
Memory [GB]	16	96
GPU	GTX 580	Tesla C2070
# GPUs	1	4
Compiler	gcc 4.5	gcc 4.1
HMPP Workb.	3.1	3.1
PGI Accel.	12.04	12.04

Table 2. CPU and GPU timings and speedups for a coupled 7–point stencil (single precision).

Grid points	CPU	PGI	HMPP	CUDA	PGI	HMPP	CUDA
		[ms]				speedup	
Excluding data transfers							
61 712	6.7	0.7	0.7	0.2	9.6	9.6	**27.9**
474 525	29.2	2.4	1.8	1.5	6.2	16.2	**19.5**
1 868 907	115.2	6.7	5.3	6.1	17.2	**21.7**	18.9
Including data transfers							
61 712	6.7	38.8	4.6	n/a	0.2	**1.5**	n/a
474 525	29.2	53.6	22.3	n/a	0.5	**1.3**	n/a
1 868 907	115.2	96.4	72.9	n/a	1.2	**1.6**	n/a

Table 3. CPU and GPU timings and speedups for the accelerated routines excluding and including data transfers with a block size of 149 × 113 × 113.

Routine	CPU			GPU		Speedup	
	orig. [ms]	mod. [ms]	loss/gain [%]	excl. transf. [ms]	incl. transf. [ms]	CPU orig./GPU excl. transf.	incl. transf.
Flux 1	115.2	118.0	−2.4	5.3	72.9	21.74	1.58
Flux 2[a]	133.0	49.9	+265.3	2.4	n/a	20.79	n/a
Flux 3	503.5	515.0	−2.3	13.7	43.4	36.75	11.60
Flux 4	1019.5	1034.5	−1.45	15.9	46.2	64.12	22.06
Flux 6[b]	2149.5	2195.6	−2.1	17.0	47.5	126.44	45.25
Flux 7	n/a	817.3	n/a	27.1	76.5	30.16	10.46
Precon. 1	n/a	273	n/a	10.0	51.0	27.30	5.35
Tridiag.	27.0	31.1	−15.1	2.3	24.0	11.74	1.13

[a] Here we calculate the GPU speedup based on the modified CPU implementation because we could accelerate the CPU code by a factor of approx. 2.5.
[b] Dense arithmetics with few memory accesses and few data transfers is performed in this routine. Still, there may be room for improvements of the CPU code.

4.2 Test Case

Table 4 shows timings for a CFD computation for an annular tube based on a steady–state laminar Navier–Stokes model. The mesh consists of seven blocks, each having three levels of discretization (a total of 500 MB memory). The finest discretization is computed on the GPU, the two remaining discretizations are computed on the CPU because coarse discretizations penalize GPU performance strongly. One can observe that the routines instrumented take approx. 65 % (35 %) of the total run time of the solver for the original (modified) CPU code. The maximal achievable speedup is less than 3 (1.5) compared to the original (modified) CPU code. We could achieve an overall speedup of 1.66 for the

Table 4. CPU and GPU timings for a sequential computation of an annular turbine (laminar Navier-Stokes).

	CPU orig. [s]	CPU mod. [s]	CPU Perf. Loss/Gain [%]	GPU [s]	Speedup CPU orig./GPU
Total run time	1303	955	+36	784	1.66
Incl. data transfer					
Flux 1	91.91	90.71	+1.32	74.76	1.23
Flux 3	134.89	132.80	+1.58	21.78	6.19
Precond. 1	415.08	140.91	+294.57	67.98	6.11
Excl. data transfer					
Flux 1	91.91	90.71	+1.32	17.97	5.11
Flux 3	134.89	132.80	+1.58	9.45	14.27
Precond. 1	415.08	140.91	+294.57	11.43	36.31

computation compared to the original CPU code. The modified CPU code runs faster than the original CPU code on (A).

A parallel execution of the test case on test system (B) with 1 master process and three compute processes (each having 1 GPU attached) gave a total speedup of 1.10. The MPI load–balancing strategy used for CPU execution did not prove to be suitable for GPUs, as their performance has a highly nonlinear relation to the size of domains processed.

5 Long Term Approach

The results show, that the acceleration of hot spots only results in a modest performance gain. Data transfers between CPU and accelerator are the main bottleneck. They can be reduced if all compute related routines are accelerated and share data without passing it via CPU memory first. For this, these routines must be prepared for many–core execution and reimplemented for a latter application of HMPP directives. Then it suffices to only transfer boundary elements within a multi grid cycle for MPI–based parallelism. We now outline an approach for a full acceleration. It may be applicable for a many–core acceleration of similar Fortran codes that use a work array for dynamic data management.

1. Implement automated accelerator data allocation and management based on the compute configuration instead of a manual data allocation in mediator routines. We consider this to be essential for the code integrity of the solver.
2. Perform a full many–core acceleration of a simple test case. I.e.:
   - For each compute related routine invoked in the test case:
     - Separate computations from logic and logging/error handling, i.e., define code parts for accelerator execution.
     - Instrument the routine by means of HMPP Workbench.
     - If the routine shares data with other routines already instrumented, omit redundant CPU–accelerator data transfers.

3. Enable asynchronous execution for independent compute routines to fully occupy the accelerator.
4. Refine the support for MPI–based parallelism. (CPU–accelerator transfers of boundary elements).
5. The repeated porting of code invoked for other test cases.

One consequence is the reimplementation of the solver in large parts for many–core execution. Some of the fundamental aspects discussed in Sect. 3.3 still can not be tackled. We estimate labor costs of approximately 24 man months for an acceleration of a rather simple CFD model within the solver and a speedup for this model of 20 at best.

6 Conclusion

We performed a case study for a many–core acceleration of a large–scale numerical CFD code by means of directive–based frameworks. Hot spots could be accelerated significantly, whereas the overall performance gain is modest.

The frameworks are convenient, as they reduce implementation time for accelerator code dramatically. But the major challenge in software engineering is not necessarily implementation work, but also algorithm and software design. This is particularly true for many–core hardware, as it is not fully compatible with well–established software engineering techniques: Constraints due to solver architecture, complexity and technological limitations prevent from a further acceleration of the solver without intricate work. In our case, many–core hardware and large code complexity can be hardly combined.

The usability of the frameworks is high for new or small scale codes that are primarily dedicated to many–core hardware. In such cases the limitations of hardware and frameworks can be met, without having to consider aspects related to CPU execution or existing code designs. For other cases, open questions have to be answered by hardware and framework vendors.

Acknowledgments. This work is supported by the CleanSky Joint Undertaking trough grant JTI–CS-2010–1–GRA–02–008 within the Seventh Framework Programme of the European Union.

References

1. CAPS Enterprise: HMPP 3.2 Workbench Directives (2012)
2. Cray Inc., CAPS Enterprise, NVidia, Portland Group: OpenACC 1.0 Specification (2011)
3. Emans, M., Liebmann, M.: Velocitypressure coupling on GPUs. Computing **94**, 1–21 (2012). http://dx.doi.org/10.1007/s00607-012-0228-6
4. Membarth, R., Hannig, F., Teich, J., Korner, M., Eckert, W.: Frameworks for GPU accelerators: a comprehensive evaluation using 2D/3D image registration. In: Proceedings of the 2011 IEEE 9th Symposium on Application Specific Processors, SASP '11, pp. 78–81. IEEE Computer Society, Washington, DC (2011)

5. Micikevicius, P.: 3D finite difference computation on GPUs using CUDA. In: Kaeli, D.R., Leeser, M. (eds.) Proceedings of 2nd Workshop on General Purpose Processing on Graphics Processing Units, pp. 79–84. ACM International Conference Proceeding Series. ACM (2009)
6. nVidia Corporation: CUDA CUBLAS Library, Aug 2010. http://developer.download.nvidia.com/compute/cuda/3_2_prod/toolkit/docs/CUBLAS_Library.pdf
7. OpenMP Consortium: OpenMP technical report 1 on directives for attached accelerators. Technical report, OpenMP architecture review board, November 2012. http://www.openmp.org/mp-documents/TR1_167.pdf

Towards Efficient Decomposition and Parallelization of MPDATA on Hybrid CPU-GPU Cluster

Roman Wyrzykowski, Lukasz Szustak(✉), Krzysztof Rojek, and Adam Tomas

Czestochowa University of Technology, Dabrowskiego 73, 42-201 Czestochowa, Poland
{roman,lszustak,krojek,atomas}@icis.pcz.pl

Abstract. EULAG (Eulerian/semi-Lagrangian fluid solver) is an established computational model for simulating thermo-fluid flows across a wide range of scales and physical scenarios. The multidimensional positive definite advection transport algorithm (MPDATA) is among the most time-consuming components of EULAG.

New supercomputing architectures based on multi- and many-core processors, such as hybrid CPU-GPU platforms, offer notable advantages over traditional supercomputers. In our previous works we considered adaptation of 2-dimensional (2D) MPDATA computations to a single CPU-GPU node. The main goal of this paper is to study tenets of optimal parallel formulation of 3D MPDATA on heterogeneous CPU-GPU cluster. Such supercomputer architecture requires not only a different philosophy of memory management than traditional massively parallel supercomputers, but also a comprehensive look at load balancing in the heterogeneous co-processing computing model.

In this paper we propose an approach to implementation of 3D MPDATA algorithm on hybrid CPU-GPU cluster, using a mixture of MPI, OpenMP, and CUDA programming standards. This approach focuses on the donor-cell numerical scheme, and is based on a hierarchical decomposition including level of cluster, as well as distribution of computations between CPU and GPU components of each node, and within CPU and GPU devices. We discuss preliminary performance results for the proposed approach running on a single cluster node consisting of two AMD Opteron Interlagos CPUs and one or two NVIDIA Fermi GPUs.

1 Introduction

The Multidimensional Positive Definite Advection Transport Algorithm (MPDATA) is among the most time-consuming calculations of the EULAG model [8]. In our previous works [7,11] we proposed two decompositions of 2D MPDATA computations, which provide adaptation to CPU and GPU architectures separately. The achieved performance results showed the possibility of achieving high performance both on CPU and GPU platforms.

In this paper, we develop a hybrid CPU-GPU version of 2D MPDATA, to fully utilize all the available computing resources by spreading computations

I. Lirkov et al. (Eds.): LSSC 2013, LNCS 8353, pp. 457–464, 2014.
DOI: 10.1007/978-3-662-43880-0_52,

across the entire machine. When adapting MPDATA to modern hybrid architectures, consisting of GPU and CPU components, the main challenge is to provide high performance for each component, taking into account their properties, as well as efficient cooperation.

The proposed approach to parallelization of the 2D MPDATA algorithm is the starting point for the implementation of 3D MPDATA on hybrid CPU-GPU clusters. We propose a hierarchical decomposition including the level of cluster, as well as distribution of computations between CPU and GPU components of each node, and within CPU and GPU devices. Hybrid clusters offer a fast solution, but understanding the parallel trade-offs is crucial for providing efficiency of computations. These architectures allow for creating many thousands of threads, which has a significant influence on performance of parallel codes [3].

2 Architecture Overview

In our research we use the Cane cluster located at the Poznan Supercomputing and Networking Center, Poland [1]. This machine includes 227 nodes, connected with each other by the InfiniBand QDR network. Each node consists of two AMD Opteron 6234 CPUs (codenamed Interlagos) and one or two NVIDIA Tesla M2050 GPUs, as well as 64 GB of the main memory. The architecture of a single CPU-GPU node is shown in Fig. 1.

Each of AMD Opteron 6234 CPUs [1] includes two dies, each containing 6 cores and 8 MB of L3 cache. All dies are connected by AMD HyperTransport links. For the clock frequency of 2.4 GHz, the peak performance of these two CPUs is respectively 460.8 GFlop/s and 230.4 GFlop/s in a single and double precision.

The NVIDIA Tesla M2050 GPU [5] is based on the Fermi architecture, and includes 14 streaming multiprocessors, each consisting of 32 CUDA cores with 48 KB of shared memory and 16 KB of L1 cache. It gives a total number of 448 available CUDA cores with the clock rate of 1147 MHz. It provides the peak performance of 1.03 TFlop/s and 512 GFlop/s in a single and double precision, respectively. This graphics accelerator card includes 3 GB of global memory with the peak bandwidth of 148.4 GB/s. All the accesses to the global memory go through the L2 cache of size 512 KB.

3 Introduction to MPDATA Algorithm

The MPDATA algorithm belongs to the group of nonoscillatory forward in time algorithms [8]. The 2D MPDATA is based on the first-order-accurate advection equation:

$$\frac{\partial \Psi}{\partial t} = -\frac{\partial}{\partial x}(u\Psi) - \frac{\partial}{\partial y}(v\Psi), \tag{1}$$

where x and y are space coordinates, t is time, $u, v = const$ are flow velocities, and Ψ is a nonnegative scalar field. Equation (1) is approximated according to

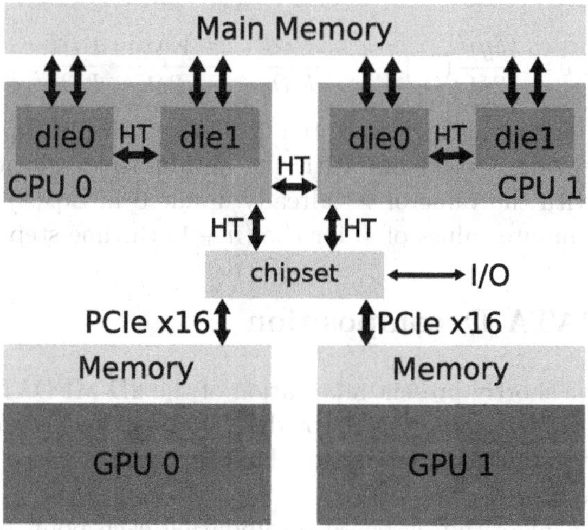

Fig. 1. Architecture of hybrid CPU-GPU node

the donor-cell scheme, which for the $(n+1)$-th time step $(n = 0, 1, 2, \ldots)$ gives the following equation:

$$\begin{aligned}\Psi^*_{i,j} = \Psi^n_{i,j} &- [F(\Psi^n_{i,j}, \Psi^n_{i+1,j}, U_{i+1/2,j}) - F(\Psi^n_{i-1,j}, \Psi^n_{i,j}, U_{i-1/2,j})]\\ &- [F(\Psi^n_{i,j}, \Psi^n_{i,j+1}, V_{i,j+1/2}) - F(\Psi^n_{i,j-1}, \Psi^n_{i,j}, V_{i,j-1/2})].\end{aligned} \quad (2)$$

Here the function F is defined in terms of the local Courant number U:

$$F(\Psi_L, \Psi_R, U) \equiv [U]^+\Psi_L + [U]^-\Psi_R, \quad (3)$$

$$U \equiv \frac{u\delta t}{\delta x};\ [U]^+ \equiv 0,5(U + |U|);\ [U]^- \equiv 0,5(U - |U|). \quad (4)$$

The same definition is true for the local Courant number V.

The first-order-accurate advection equation can be approximated to the second-order in δx, δy and δt, defining the advection-diffusion equation:

$$\begin{aligned}\frac{\partial\Psi}{\partial t} = &-\frac{\partial}{\partial x}(u\Psi) + \frac{(\delta x)^2}{2\delta t}(|U| - U^2)\frac{\partial^2\Psi}{\partial x^2}\\ &-\frac{\partial}{\partial y}(v\Psi) + \frac{(\delta y)^2}{2\delta t}(|V| - V^2)\frac{\partial^2\Psi}{\partial y^2}\\ &-\frac{UV\delta x\delta y}{\delta t}\frac{\partial^2\Psi}{\partial x\partial y}.\end{aligned} \quad (5)$$

The antidiffusive pseudo velocities $\tilde{u}$ and $\tilde{v}$ in respectively x and y directions are defined according to the following equations:

$$\tilde{u} = \frac{(\delta x)^2}{2\delta t}(|U| - U^2)\frac{1}{\Psi}\frac{\partial\Psi}{\partial x} - \frac{UV\delta x\delta y}{2\delta t}\frac{1}{\Psi}\frac{\partial\Psi}{\partial y}, \quad (6)$$

$$\tilde{v} = \frac{(\delta y)^2}{2\delta t}(|V| - V^2)\frac{1}{\Psi}\frac{\partial \Psi}{\partial y} - \frac{UV\delta x \delta y}{2\delta t}\frac{1}{\Psi}\frac{\partial \Psi}{\partial x}. \tag{7}$$

Therefore, in order to compensate the first-order error of Eq. (1), once again the donor-cell scheme is used but with the antidiffusive velocity $\tilde{u} = -u_d$ in place of u, and with the value of Ψ^* already updated in Eq. (2) in place of Ψ^n. It allows us to compute values of Ψ for the $(n+1)$-th time step.

4 2D MPDATA Decomposition

In this section, we shortly present adaptation of the 2D MPDATA algorithm to the hybrid CPU-GPU architecture, providing trade-off between communication and computation within its components. This approach is based on the efficient use of a single node.

The MPDATA algorithm is based on updating each point of the grid with values from neighboring grid points. Typically the neighborhood structure is fixed, in which case it is called a stencil [2,9]. Our previous research show that MPDATA is a memory-bound algorithm [7,12].

The main task here is the decomposition of MPDATA grid into CPU and GPU domains. We propose the basic strategy of grid partitioning, that assigns two stripes of grid rows to CPU and GPU. Data transfers between CPU and GPU domains are minimized by providing extra computations within both domains. Therefore, the CPU has to compute more rows, because some rows, which originally were assigned to the GPU domain only, are now duplicated in the CPU domain, and vice versa. This approach allows us to avoid communication between CPU and GPU domains within each time step of the MPTADA algorithm, since CPU and GPU components compute their domains separately. As is shown in Fig. 2, the CPU-GPU cooperation, including communication and synchronization, is required only after each time step.

When adapting MPDATA to the hybrid CPU-GPU architecture, the next task is to provide efficient performance for each component. Hence, two different adaptations of the MPDATA algorithm to CPU and GPU processors are required. Each of these adaptations takes into account constraints for the memory bandwidth.

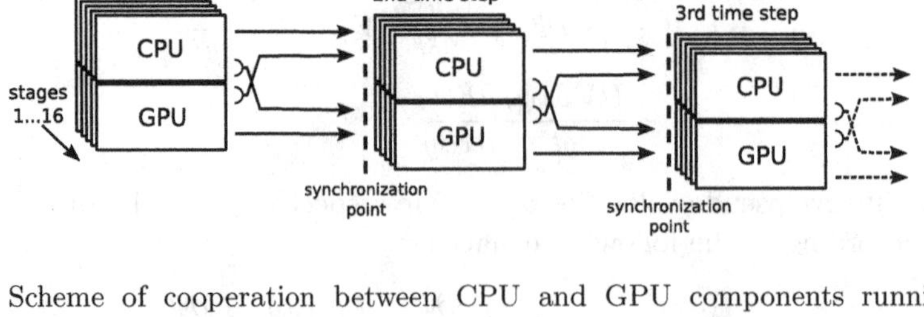

Fig. 2. Scheme of cooperation between CPU and GPU components running the MPDATA algorithm

For CPU, this goal can be achieved by taking advantage of cache memory reusing, as high as possible. This requires to apply an appropriate block decomposition strategy, when the intermediate results of computations for a single block are placed in the cache memory. Only the final results are returned to the main memory. Such an approach is commonly called the temporal blocking [4,10]. Computations within each block are distributed across available CPU cores, and the SIMD processing is applied inside each core. Each AMD Interlagos CPU contains groups of cores (or dies) connected each other by AMD HyperTransport links [1]. Dies have direct access to their own cache memory, and indirect access to caches of other dies. To eliminate inter-cache communications among dies, at the cost of extra computations, we use exactly the same grid decomposition as in the case of adaptation of MPDATA to CPU-GPU architecture. Another advantage of this approach is possibility to apply the NUMA "first-touch" policy.

For GPUs, their global memory allows us to decrease the intensity of access to the main memory, since results of GPU computations performed within a single time step can be stored in GPU only. As a result, performance restrictions due to the memory bandwidth saturation can be alleviated, and the high density of computing resources is better utilized.

The GPU parallelization of MPDATA is based on three levels of GPU parallel hierarchy: (i) overlapping data transfers between the host memory and GPU global memory with GPU computations; (ii) parallel computations across threads running on GPU cores; (iii) vectorization within a GPU thread. The first level requires to apply an appropriate decomposition of data domain into streams, in order to use the streams processing mechanism. It allows us to alleviate bandwidth constraints of PCIe connection between CPU and GPU. The second level concerns parallel processing of GPU threads, which are assembled into CUDA blocks. The last level allows for increasing the amount of computations within a single GPU thread, and reducing overheads of access to GPU global and shared memories.

5 2D MPDATA Performance Results

Table 1 presents execution times of the 2D MPDATA algorithm for 500 time steps and different sizes of grid, using a single node of the target cluster. The achieved performance results correspond to different configuration including the basic serial version running on a single CPU core without using block decomposition and SIMD vectorization, and parallel versions using configurations with 1CPU, 2CPUs, 1GPU, and 2GPUs, as well as hybrid configuration with 2CPUs and 2GPUs. In all the parallel versions, the block decomposition and SIMD vectorization techniques are applied to speedup MPDATA computations over the basic serial version, which does not use these techniques. The speedup of parallel versions over the basic serial version is shown in Fig. 3. For all the grid sizes, the hybrid version allows us to achieve the highest performance. In particular, for the grid of size 4096×4096, it gives speedup of 93.46 over the serial version.

Table 1. Execution times of 2D MPDATA for 500 time steps

size	serial	1CPU	2CPUs	1GPU	2GPUs	2CPUs+2GPUs
1024 × 1024	99.24	5.21	2.55	2.81	1.47	1.38
2048 × 2048	384.68	19.14	9.66	10.83	5.48	4.74
3072 × 3072	869.91	40.91	20.78	26.12	13.07	9.45
4096 × 4096	1568.22	74.50	37.81	53.99	22.43	16.78

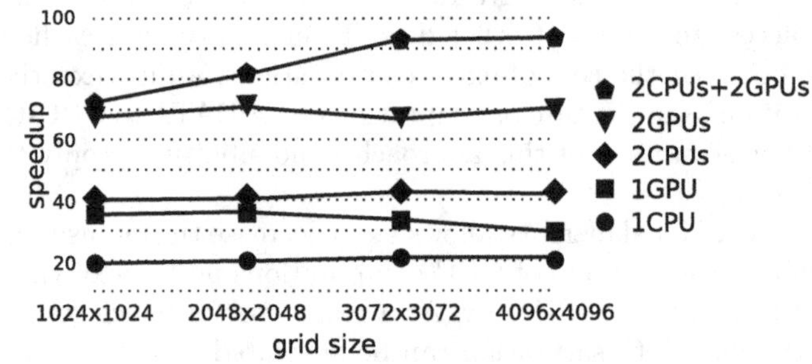

Fig. 3. Speedup of parallel versions over basic serial version

For the largest grid size, the hybrid version is about 2.25 times faster in comparison with using 2CPUs, and about 1.33 times faster than the 2GPUs version.

6 3D MPDATA Decomposition

The achieved performance results show a high perspectives of using the hybrid architecture to the MPDATA algorithm in the 3D case, as well. Following these results, in this section we propose an approach to adaptation of 3D MPDATA to the CPU-GPU cluster, employing both CPU and GPU computing resources. Our approach is based on a hierarchical decomposition including level of cluster, as well as distribution of computations between CPU and GPU components of each node, and within CPU and GPU devices. To take advantage of CPU-GPU cluster, the MPI standard is used across nodes, while OpenMP and CUDA are applied within each node.

This adaptation consists of two basic steps. The first step (Fig. 4a) takes into account the decomposition on the cluster level, and provides data distributions across a 2D mesh of nodes. Each node consists of a group of components, which include CPU and GPU resources. The second step takes into account the data decomposition within a single CPU-GPU node (Fig. 4b). This step is based on the approach previously developed for 2D MPDATA.

The 3D MPDATA algorithm performs simulations determined by size $n \times m \times l$ of the grid. In case of simulations in the EULAG numerical weather prediction [6], the size of grid is usually specified by the following constraints: $n = 2 * m$,

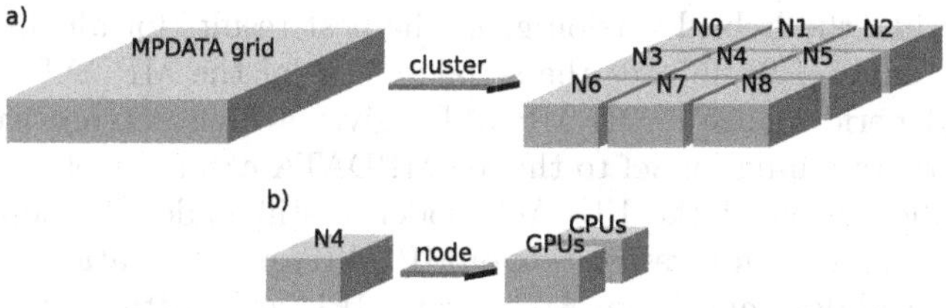

Fig. 4. Grid decomposition of 3D MPDATA onto CPU-GPU cluster

$l \leq 128$, and $n, m >> l$. Such a 3D grid is mapped on a 2D mesh of CPU-GPU nodes of size $r \times c$. As a result, the MPDATA grid is partitioned into subdomains of size $n_p \times m_p \times l$, where each node is responsible for computing within a single subdomain, and:

$$n_p = \frac{n}{r}; \quad m_p = \frac{m}{c}. \tag{8}$$

Every subdomain is further partitioned into two parts, assigned to CPU and GPU resources separately. This partitioning is given by the following equations:

$$S_{GPU} = (G * n_p) \times m_p \times l, \tag{9}$$

$$S_{CPU} = (C * n_p) \times m_p \times l, \tag{10}$$

where parameters G and C characterize GPU and CPU parts, respectively, satisfying the following constraints:

$$G + C = 1; \; G, C \in [0; 1]. \tag{11}$$

Currently, to provide the load balancing between CPU and GPU components, values of G and C parameters are evaluated in an empirical way. However, a dynamic load balancing model will be developed in future work, allowing us to increase the portability of MPDATA code across a variety of hybrid clusters.

7 Conclusions and Further Work

New strategies for memory and computing resources management allow us to ease memory bounds, and better exploit the theoretical floating point efficiency of hybrid architectures. The hybrid computing is a promising approach for increasing performance of numerical simulations of geophysical flows using the EULAG model.

We propose the basic strategy of partitioning the MPDATA grid, that assigns two stripes of grid rows to CPU and GPU components. Thus, data transfers between CPU and GPU domains are minimized by providing extra computations within both domains. Moreover, two separate adaptations of MPDATA algorithm to CPU-GPU hybrid architecture are required, to better utilize features of hybrid architectures.

For 2D grids, the hybrid version gives the best results for all the grid sizes, providing speedup of 93.46 over the serial version of the MPDATA algorithm. The achieved performance of 2D MPDATA gives a high perspectives of using the hybrid programming model to the 3D MPDATA case, as well.

Our parallelization of the EULAG model is still under development. The future work will focus on investigation of MPDATA parallelization based on the proposed 3D grid decomposition. Apart from GPU architectures, the particular attention will be paid to other accelerators such as Intel Xeon Phi.

Acknowledgments. This work was supported by the Polish National Science Centre under grant no. UMO-2011/03/B/ST6/03500.

References

1. AMD and GPGPU cluster, https://hpc.man.poznan.pl/modules/resourcesection/item.php?itemid=61
2. Datta, K., Kamil, S., Williams, S., Oliker, L., Shalf, J., Yelick, K.: Optimization and performance modeling of stencil computations on modern microprocessors. SIAM Rev. **51**(1), 129–159 (2009)
3. Kurzak, J., Bader, D., Dongarra, J.: Scientific Computing with Multicore and Accelerators. Chapman & Hall/CRC , Boca Raton (2010). (Chapman & Hall/CRC Computer and Information Science Series)
4. Nguyen, A., Satish, N., Chhugani, J., Changkyu, K., Dubey, P.: 3.5-D blocking optimization for stencil computations on modern CPUs and GPUs. In: Proceedings of the 2010 ACM/IEEE International Conference for High Performance Computing, Networking, Storage and Analysis, pp. 1–13 (2010)
5. NVIDIA Best Practices Guide, http://developer.nvidia.com/nvidia-gpu-computing-documentation
6. Piotrowski, Z., Wyszogrodzki, A., Smolarkiewicz, P.: Towards petascale simulation of atmospheric circulations with soundproof equations. Acta Geophys. **59**, 1294–1311 (2011)
7. Rojek, K., Szustak, L.: Parallelization of EULAG model on multicore architectures with GPU accelerators. In: Wyrzykowski, R., Dongarra, J., Karczewski, K., Waśniewski, J. (eds.) PPAM 2011, Part II. LNCS, vol. 7204, pp. 391–400. Springer, Heidelberg (2012)
8. Smolarkiewicz, P.: Multidimensional positive definite advection transport algorithm: an overview. Int. J. Numer. Meth. Fluids **50**, 1123–1144 (2006)
9. Venkatasubramanian, S., Vuduc, R.: Tuned and wildly asynchronous stencil kernels for hybrid CPU/GPU systems. In: ICS, pp. 244–255 (2009)
10. Wittmann, M., Hager, G., Treibig, J., Wellein, G.: Leveraging shared caches for parallel temporal blocking of stencil codes on multicore processors and clusters. Parallel Process. Lett. **20**(4), 359–376 (2010)
11. Wyrzykowski, R., Rojek, K., Szustak, L.: Model-driven adaptation of double-precision matrix multiplication to the cell processor architecture. Parallel Comput. **38**, 260–276 (2012)
12. Wyrzykowski, R., Rojek, K., Szustak, Ł.: Using blue gene/P and GPUs to accelerate computations in the EULAG model. In: Lirkov, I., Margenov, S., Waśniewski, J. (eds.) LSSC 2011. LNCS, vol. 7116, pp. 670–677. Springer, Heidelberg (2012)

Cloud and Grid Computing for Resource-Intensive Scientific Applications

Distributed System for Query Processing with Grid Authentication

E.I. Atanassov, D. Georgiev, T. Gurov, A. Karaivanova(✉), and Y. Nikolova

IICT-BAS, Acad. G. Bonchev str., bl. 25A, 1113 Sofia, Bulgaria
anet@parallel.bas.bg

Abstract. When using Grid resources, usually a specific type of middleware is deployed. For example the European infrastructure is based on EMI-GLite and GLOBUS. The deployment of this middleware enforces a number of limitations concerning the OS, libraries and software packages. On the other hand the typical Grid middleware does not provide elasticity in the way Cloud services do. In our system we propose an entirely Java-based solution designed with elasticity in mind, providing distributed processing of queries from a single endpoint. The system is lightweight and easy to deploy on Cloud resourced and depends only on a few EMI packages. In this paper we describe the architecture and workflow of the system, providing benchmarks which show the effectiveness of our solution during processing of queries related to monitoring computational jobs in the Grid, using a distributed Cassandra database.

1 Introduction

1.1 Grids and Clouds

The Grid is a computational infrastructure which ensures transparent access to geographically and institutionally distributed computational resources and data. The Grid has been studied extensively during the last two decades, here we refer to some works of Foster and Kesselman, and also to the description of the largest grid (the European Grid Infrastructure) [2–5]. Ian Foster and coauthors gave a three point checklist [5] to help determine what the Grid is, and what is not:

1. The Grid coordinates resources that are not subject to centralized control,
2. The Grid uses standard, open, general-purpose protocols and interfaces, and
3. The Grid delivers non-trivial qualities of service.

Although Cloud Computing is a relatively newer technology, it has an intrinsic connection to the Grid Computing paradigm. There is little consensus on how to define the Cloud and here we accept the definition of Foster in [6]: "*A large-scale distributed computing paradigm that is driven by economies of scale, in which a pool of abstracted, virtualized, dynamically-scalable, managed computing power, storage, platforms, and services are delivered on demand to external customers over the Internet.*"

I. Lirkov et al. (Eds.): LSSC 2013, LNCS 8353, pp. 467–475, 2014.
DOI: 10.1007/978-3-662-43880-0_53,

In the same paper [6], Foster et al. compare Grid and Cloud and conclude that only point 3 from the above check-list holds true for Cloud computing, but neither point 1 nor point 2 are valid for Clouds. Common characteristics of Grid and Cloud include: utility computing, aggregation of heterogeneous resources, access transparency for the end user, reconfigurability, service negotiation based on SLA, capacity provisioned on demand, continuous availability, single sign on.

In Europe there is a strong tradition in using Grids for scientific applications, especially in the domain of High Energy Physics, while Clouds are an attractive option for development of new applications or porting the existing ones due to expected economies of scale and certain usability advantages.

That is why the combined use of Grid and Cloud resources is an interesting topic from both theoretical and practical viewpoint [9]. In our previous work [7] we studied the security issues which arise when using both of them as part of a scientific application. Some of the use cases and conclusions from there motivated our desire to develop a lightweight framework, that can facilitate easy development and deployment of Grid services using distributed resources that are local or Cloud-based.

The authentication and authorization mechanism of the framework is based on the usage of X.509 certificates and proxy certificates, thus being easily accessible from both Grid jobs and users from their local machine. In this paper we describe our framework and an efficient approach of deploying it on multiple, possibly not publicly accessible machines, with the usage of a load balancer. A natural use case for such a distributed service would be to provide access to a distributed database organized around the NoSQL paradigm with a single point of entry. We provide benchmarks of an example application based on our framework, which grants access to a distributed Cassandra database with Grid authentication enabling storing and collection of statistics of the usage of the infrastructure.

1.2 Authentication and Authorization in Grids

The main authentication mechanism in Grids is based on the use of X.509 certificates. Certain Certification Authorities (CA) are recognized throughout the international research community. The EUGRIDPMA organization [8] is responsible for vetting the certification authorities that support researchers from European countries. As an example, the Bulgarian certification authority is called BG.ACAD CA and provides certificates for Bulgarian scientists and students, that can be used to access the resources of the European Grid Infrastructure.

Traditionally, as a way to introduce a level of indirection and to mitigate the dangers that a stolen certificate may pose to the whole infrastructure, proxy certificates are first generated and then used instead of the full user certificates. Proxy certificates have shorter validity and may also be limited in what can be done with them - the so-called "limited certificates".

The cornerstone of the authorization part of security management in Grid is the notion of the so-called Virtual Organization (VO), which enable their members to share resources through the Grid. Multiple institutions may provide the resources of the virtual organization, which become available to all members of the VO. It is important to note that only the biggest VOs, comprising of thousands of scientists, define groups inside the virtual organization and enforce different priorities and access rights for members of these groups and the respective job priorities. It is important to note that in most practical cases a user of the VO has access to all the data files of the VO and can not only retrieve data, but also delete data, even if the file has been created by another user. Although X.509 certificates are the main authentication mechanism, other authentication mechanisms are also used. For example, it may be possible to use an identity provider to obtain an X.509 certificate on-the-fly, thus allowing more decentralized management of the authentication process.

The protocols that are used in the Grid are mostly relaying on transport-level security provided by SSL. The proxy certificates may also carry information about the VO membership of the user through a mechanism called "VOMS extensions". The VOMS extensions in a X.509 proxy certificate also provide the Grid services with the finer-grained authorization information, related to groups and roles within the VO. This authorization information is available if one has access to the full stack of X509 certificates sent by the client when opening the SSL connection.

On the other hand, the Grid services authenticate themselves with X.509 certificates, corresponding to their names in DNS. Traditionally, the clients check not only forward, but also reverse DNS resolution, which is a hindrance if the provider of a service can not ensure that the IP address maps back to the same server name that is presented to the client. In practice this means that there will be a problem if the provider uses Cloud service like Amazon EC2 [1], because the reverse DNS will not work as required.

In practical terms we should either relax this requirement on the client side, or provide an entry point that is under control of the service provider, for example from the campus network, using Cloud resources only as a back-end.

2 Description of the Framework

2.1 Main Considerations

In our system, the security is based on X.509 certificates and proxy certificates, with support for VOMS extensions. The certification authorities that we support are defined by the EGI International Grid Trust Federation (IGTF) release, which contains all the Certification Authorities (CAs) accredited by the IGTF. Naturally, specific deployments may contain only a subset of these CAs or organization-specific CAs as per the users' and provider's requirements. We assume that one entry point for the Grid service is necessary and it will be a server with proper X.509 certificate, issued by some of these CAs. The Grid service operations are described via the Web Services Description Language (WSDL). In this way we can leverage the established mechanisms for development of web services, adding the necessary bits for Grid authentication and authorization. Behind this entry point we can have either single server or a distributed load-balanced server farm. The client software is based on the WSDL template of the service. It can be developed using any language and web service framework that can incorporate the Grid-specific authentication mechanism. We should note that it is allowed for the certificate of the Grid server to be a proxy certificate, something that general frameworks may reject.

2.2 Basic Implementation of the Framework

If a grid service running on a single server is to be implemented, we can follow the natural approach of enhancing the Apache Tomcat servlet container with support for the Grid Security Infrastructure (GSI) by using GSI implementations supported by the Europeam Middleware Initiative (EMI – [14]). EMI provides two main packages for processing secured web-requests based on GSI proxies - EMI Trustmanager and EMI GridSite. The EMI Trustmanager is a Java-based authentication solution, which contains client libraries, along with Tomcat and Axis integration classes. On the other hand the GridSite package provides an efficient Apache HTTP module [10] written in *C*, which can be used for both authentication and authorization. As we mentioned above, our client is entirely based on the Trustmanager, while for the server we compared two approaches - secured Apache Tomcat [11] using the EMI Trustmanager, and Apache Tomcat with security offloaded to a GridSite-enabled Apache HTTP server.

While the first approach is easier to setup, after extensive stress-testing, the second approach showed superior performance, providing a substantial difference in the achieved throughput of requests. However, if performance is not critical, the use of EMI Trustmanager Tomcat library may be a better approach as it is faster to set-up and provides simpler control over the client credentials.

Integrating Apache HTTP server with Apache Tomcat can also be done in several ways. The most popular approaches include the Apache modules mod_jk, mod_proxy_http and mod_proxy_ajp, which all make possible the passing of SSL attributes, but differ greatly in their configuration and installation. Using mod_proxy_http, necessitates a lot of effort for configuration, as it requires SSL attributes to be inserted in the HTTP header and then extracted in Tomcat via a specially configured "valve". Its main advantage over the other approaches is that it provides an easy mechanism for encrypting traffic between HTTP and Tomcat. However, since in our system the HTTP server and Tomcat run on the same machine and Tomcat is running on a port which is not accessible from outside the machine, this is unnecessary. On the other hand, mod_proxy_ajp and mod_jk both provide proxying to Tomcat by using the Apache JServ Protocol (AJP). While both provide almost the same functionality, mod_proxy_ajp is compliant with the mod_proxy API and thus is easier to configure than mod_jk. However, one major feature of mod_jk is not yet implemented in the mod_proxy_ajp official releases - forwarding of the complete SSL chain. Additionally, a slight improvement is noticeable when using mod_jk, which is often credited to its lack of compliance to the mod_proxy API. Based on these differences between the three approaches we chose to use mod_jk because it is relatively easy to install

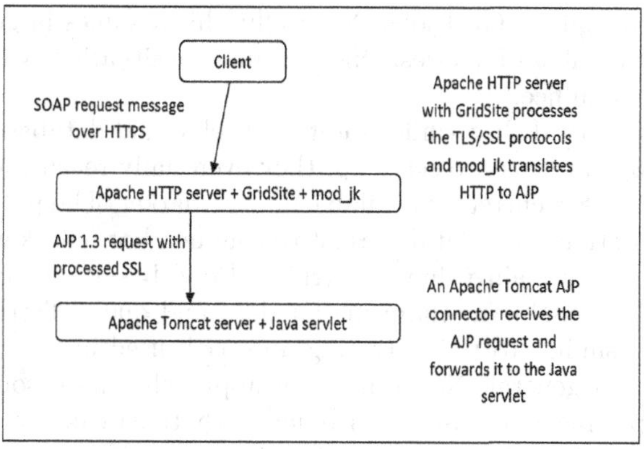

Fig. 1. Schematic representation of a single-server deployment.

from source, can pass SSL attributes to an AJP connector and forwarding of the SSL chain is implemented in the official release. GridSite [13] may be installed from the EMI repository either as a binary package or as a source package to be built on the target system. Configuration of the GridSite Apache module is straight-forward - since mod_gridsite internally uses mod_ssl it also inherits most of its configurations. Along with authentication, GridSite offers also authorization mechanisms, based on Grid Access Control List (GACL), but we used trivial GACL and relayed on the Java servlet to implement its own authorization logic based on the forwarded SSL attributes.

Our Java servlet was developed in a WSDL-first manner. Using Apache CXF [12] the created WSDL is translated to abstract Java classes which are inherited in the web service in order to implement the defined Remote Procedure Calls (RPC). This approach, often called contract-driven, is widely preferred as it allows easier requirements management. On Fig. 1 one can see a graphical representation of the software pieces and their interactions.

2.3 Implementation of the Framework in Distributed Setting

Setting up multiple instances of our system on different machines is organized along the idea of load-balancing at the level of TCP. In our system we used Balance [15], a lightweight TCP balancing proxy, written in C, which utilizes heavily the available Linux system calls. On the client side, instead of directly addressing an instance, the client sends requests to the machine running Balance, which forwards traffic in a round-robin manner. This approach enables the execution servers to be deployed on non-public resources and requires only the balancer to be accessible via Internet.

The underlying Java servlet can implement any RPC that requires Grid authentication. In our example system, we implemented a distributed query processing engine that can be easily accessed from both users and Grid jobs running on a given site. The engine can be used to store and access statistics, input and even output of Grid jobs. Naturally, this assumes large quantities of data and a large number of queries. One can imagine situations when traditional databases will not suffice.

It has been observed that while traditional relational database management systems (RDBMS) are easy to scale up, they eventually reach the limits set by the high-end hardware currently available on the market. The problems of effective query distribution turns databases into a common bottleneck when the usage of a system reaches certain values. Distributed databases aim to mitigate this issue by providing reliable mechanisms for data and query distribution among multiple servers, similar to the mirroring approach used in web servers and in scalable services in general. Naturally such approaches have some drawbacks, for example replacing database consistency with the weaker eventual consistency. In our example system we chose Apache Cassandra [16] for our database management system as it provides several important features. Unlike most of the available solutions, Cassandra is completely decentralized and every instance in a Cassandra cluster has the same role. This allows Tomcat servers to query the instance running on the same machine, thus needing only Cassandra inter-cluster communications. Additionally, by setting a replication factor, each entry will be written on multiple Cassandra nodes, mitigating the single point of failure problem. Additionally, the Cassandra Thrift API grants the ability to request different levels of consistency before a write or a read, thus allowing fine-tuning

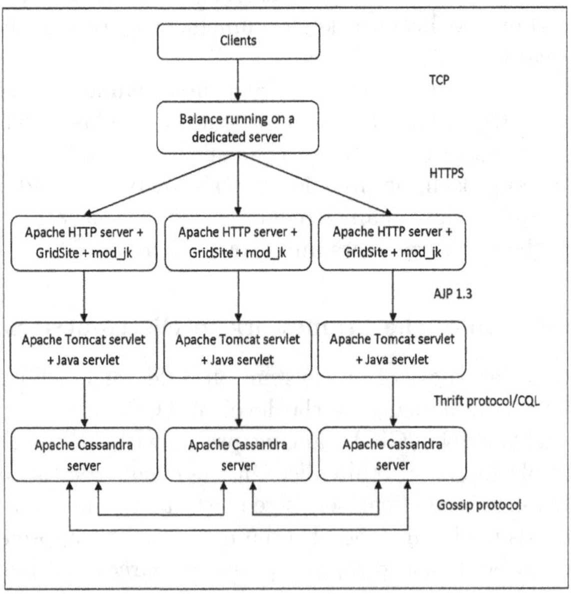

Fig. 2. Representation of a multiple-server deployment scheme along with the associated protocols at the different layers.

Table 1. Benchmark results for simple authenticated request-reply, measured in number of messages per second

Single node	Two nodes	Three nodes
7211 messages	14054 messages	20840 messages

queries even at runtime. On Fig. 2 one can see a graphical representation of the interactions of the components, described above.

Using persistent HTTP connections, i.e., reusing a TCP connection for multiple HTTP requests and replies, provides a substantial improve in the throughput, when clients initiate a number of sequential requests with no or relatively small time in-between. Apart from the obvious speedup due to the reduced latency, network congestion and usage of CPU and memory, a substantial speedup is achieved by reusing the initial SSL/TLS handshake. Reusing TCP connections also causes the balancer to redirect sequential requests originating from the same client to the same server, thus reducing the overall load on Balance. Since each write and read request spawns a new connection to the Cassandra database, naturally the same approach is not applicable here. Instead Cassandra is explicitly configured to destroy all TCP connection after the execution of the query. This results in a massive speedup, due to using a small pool of connections.

3 Benchmarking

Our benchmarks were performed on several HP ProLiant BL280c G6 blades, each equipped with two Intel Xeon 5560 processors and 24 GB RAM. Each processor has 4 physical (8 logical) CPUs for a total of 16 logical CPUs per node. All nodes are interconnected via non-blocking DDR InfiniBand. In our experiments we benchmarked several aspects of our application. The first tests, consisting of a simple authenticated request-reply, allowed us to evaluate the throughput of our distributed system. The tests ran on one to three nodes show excellent scaling in terms of throughput, see Table 1.

Due to the low-level approach adopted in Balance, it can handle a large amount of connections, while also allowing a massive throughput of packets. The second benchmark tested the writing capabilities of our system, with replication factor of 1 and 2. In the presented results Table 2, a write consists of 5 insertions in different column families (the rough equivalent of an SQL table) in Cassandra.

Table 2. Benchmark results for the writing capabilities of the system, measured in number of writes per second

Replication factor	Single node	Two nodes	Four nodes
1	393 writes	774 writes	1480 writes
2	N/A	472 writes	964 writes

Table 3. Benchmark results for peak performance, measured in number of writes per second

Replication factor	Single node peak performance	Two nodes peak performance	Four nodes peak performance
1	3786 writes	7462 writes	15233 writes
2	N/A	3913 writes	7790 writes

Table 4. Benchmark results for processing reads with a fixed consistency requirement, measured in number of messages per second

Single node	Two nodes	Four nodes
896 messages	1656 messages	3312 messages

It is important to note, however, that these benchmarks were obtained during 12 h period of heavy load. A similar benchmark, which simulates peaks in the client activity, was used to measure the peak performance of our system and its ability to handle rapid changes in load, see Table 3.

The ability of our system to process reads was tested in a similar manner, with a fixed consistency requirement i.e. data confirmation from different nodes in Cassandra of 1, see Table 4.

In a way similar to the tests of the writing capabilities, these numbers were obtained after several hours of heavy load. Using the same technique to simulate peaks in the load, we noticed an increase in the throughput of the reads, although with a smaller factor of 4.2.

A more sophisticated load-balancing strategy may be applied to our system, as Balance often requires live resetting of the maximal connections per channel to ensure a proper balancing of long TCP sessions. An approach using the balancer included in mod_jk might be considered, although it relies on a single HTTP server as a front-end. Fine-tuning Cassandra and the way Tomcat connects to Cassandra, may also result in a significant speed up in the query processing. A more sophisticated approach for placing Tomcat servers and Cassandra instances may be considered. Separating them to different machines and increasing the number of Cassandra instances compared to Tomcat servers seems like a promising approach.

4 Conclusions and Future Work

In this work we presented an easy to install and lightweight framework for aggregating resources, possibly hidden behind firewalls or coming from outside cloud providers and offering access to them via Grid interface. Our approach is rather generic, follows established best practices and achieves acceptable scalability. The use of a Cassandra database as a back-end was provided rather as an example of using the framework to provide access to data, since replacing it with other databases or simply providing load-balanced computational capabilities is another option. We noticed some problems and limitations both in our use of Balance and in the particular way we have deployed Cassandra, so in our future efforts we will attempt to mitigate these problems or to employ alternative approaches.

Acknowledgment. The research work reported in the paper is partly supported by the project AComIn "Advanced Computing for Innovation", grant 316087, funded by the FP7 Capacity Programme (Research Potential of Convergence Regions), and by the Bulgarian NSF grant DVCP02/1 CoE SuperCA++.

References

1. Amedro, B., Baude, F., Huet, F., Mathias, E.: Combining grid and cloud resources by use of middleware for SPMD applications. In: Proceedings of 2nd International Conference on Cloud Computing Technology and Science, Indianapolis, IN, USA, pp. 177–184 (2010)
2. Bird, I., Jones, B., Kee, K.: The organization and management of grid infrastructures. Computer **36–46**, 42 (2009). (IEEE Computer Society, ISSN 0018–9162/09)
3. Foster, J., Kesselmann, C.: The Grid: Blueprint for a New Computing Infrastructure. Morgan Kaufmann, San Francisco (1999)
4. Foster, J., Kesselmann, C., Tuecke, S.: The anatomy of the grid. Int. J. Supercomput. Appl. **15**(3), 200–222 (2001)
5. Foster, I.: What is the grid? A three Point Checklist, Grid Today, July 2002. http://www.mcs.anl.gov/~itf/Articles/WhatIsTheGrid.pdf
6. Foster, I., Zhao, Y., Raicu, I., Lu, S.: Cloud Computing and Grid Computing 360-Degree Compared. IEEE (2008)
7. E. Atanassov, T. Gurov, A. Karaivanova, Cloud and Grid Computing: Security Aspects. In: BGSIAM'11 Proceedings, pp. 21–25 (2012). ISSN: 1313–3357
8. EUGridPMA - Bulding Trust for Authentication in e-Science. http://www.eugridpma.org/
9. Jha, S., Merzky, A., Fox, G.: Using clouds to provide grids with higher levels of abstraction and explicit support for usage models. Concurrency and Comput. Pract. Experience **21**(8), 1087–1108 (2009)
10. Apache HTTP Server Project, Apache Software Foundation. http://httpd.apache.org/
11. Apache Tomcat, Apache Software Foundation. http://tomcat.apache.org/
12. Apache CXF, Apache Software Foundation. http://cxf.apache.org/
13. GridSite. https://www.gridsite.org/
14. European Middleware Initiative. http://eu-emi.eu/
15. Balance, Inlab Software GmbH. http://www.inlab.de/balance.html
16. Apache Cassandra Project, Apache Software Foundation. http://cassandra.apach.org/

Performance Analysis of Cloud-Based Application

Peter Budai(✉) and Balazs Goldschmidt

Department of Control Engineering and Informatics,
Budapest University of Technology and Economics, Budapest, Hungary
{budai,balage}@iit.bme.hu

Abstract. Cloud computing, and IaaS cloud services in particular suit well to resource-intensive applications by offering on-demand allocation of computing power, storage space and network bandwidth together with pay-as-you-go billing system.

Typical cloud applications consist of several interdependent components, all residing on one or more dedicated virtual computers. In order to be able to accurately estimate the resource requirements of a specific component, one must carry out detailed performance analysis.

In this paper, we present the general concepts and pitfalls of performance analysis in the cloud environment. Then we present a lightweight distributed framework that is capable of generating load to and collecting performance metrics from the component instances. The capabilities of our framework will be demonstrated on a case study of the scalability analysis of a distributed MySQL relational database management system.

Keywords: Cloud computing · Performance analysis · Load generation · Test framework

1 Introduction

From an emerging buzzword that it was a few years ago, cloud computing has became a more widely accepted solution for cost-effective provisioning of computational resources. The technology behind cloud computing is not completely new, it links and wraps existing technologies such as hardware virtualization, grid computing and service-oriented architectures. There is no universally acceptable definition of cloud computing, in general we could say that it covers computing power, storage capacity and software appearing as services. There is some distinguishing features that differentiates it from the aforementioned existing technologies like self-service operation, pay-per-use billing, elasticity, and options for customization [6].

There are researches that investigate the possibility of using infrastructure (IaaS) cloud computing services for resource-intensive scientific applications that demonstrated practical examples of use [12], involved financial [7], and performance [10] analysis of the field.

I. Lirkov et al. (Eds.): LSSC 2013, LNCS 8353, pp. 476–483, 2014.
DOI: 10.1007/978-3-662-43880-0_54,

1.1 Performance Analysis in the Cloud

Being it a scientific or standard multi-tier web application, we believe that performance testing and analysis plays an important role in resource and operational cost planning. Popular public IaaS cloud providers, like Amazon [2] offer several types of virtual resources each with a different pricing. To be able to make a well established choice regarding the types and the number of virtual machines that the application requires, resource consumption must be correctly measured for its designed capacity and even beyond.

However, there are many difficulties that have to be addressed when performing performance tests for applications that are hosted on the cloud environment compared to the traditional deployment model on physical servers. In the virtualized environments the properties of the host system are usually completely hidden from the end users, and it is not possible to obtain exact performance metrics for virtualized resources like CPU and memory utilization or I/O usage, or even deduce the status of the host system from the performance metrics regarding the virtual machine.

Applications deployed into the cloud are often a subject to performance degradation caused by the lack of resources which itself is caused by the behavior of other virtual environments hosted on the same physical server, which is called the noisy neighbor problem. Typical benchmarks are performed in sterile environments and do not take this kind of background load into consideration.

The remainder of the paper is organized as follows. In Sect. 2 we introduce our test framework for performance analysis of distributed systems. In Sect. 3 we present a case study to demonstrate the applicability of our framework on a practical example. Section 4 contains an overview of related commercial and academic solutions. The paper is closed with Sect. 5 presenting our conclusions and future plans regarding the test framework.

2 Framework for Performance Testing

In this section we present a distributed test framework that aims to overcome the difficulties discussed above by allowing the test engineer to generate various types of artificial workloads on large number of remote computers simultaneously.

2.1 Architecture and Operation

The framework consists of highly autonomous software components called *agents* that are deployed on all the virtual machines involved in the test. They form a homogenous distributed framework, every target machine runs exactly the same piece of agent software. To prevent unnecessary interference with the system under test, the agents are designed to have low resource footprint when they are idle. Basically, during their lifetime the agents are performing two kind of operations, messaging and job execution, which are discussed in details below.

Messaging. The agents are exchanging information regarding their knowledge about the statuses of the active agents and the currently executing jobs in the system by sending and receiving *messages* to and from other agents. Messages can be directed at a specific agent or a job within the agent, but can be broadcasted to every agents and/or every jobs within an agent.

Job execution. The main task of the agents is to execute various types of *jobs* determined by the user. Jobs are not scheduled, they are run parallel on their own thread of execution. Jobs can be started or stopped dynamically by sending an appropriate message to the agent. A special job called *watchdog* is automatically started on all agents by default. The watchdog is responsible for periodically connecting all known other agents and keeping their status information up to date.

The agent continuously maintains a list of known peer agents based on status information obtained from other agents. Upon startup, the agent connects to an arbitrary agent on the network, signals its presence and requests peer information. By propagating their peer list to each other, eventually all agents will become aware of the complete network status, forming a fully connected network of agents. The maximum number of peer requests needed for this process is $O(n^2)$.

Suppose we have a fully connected network of N agents labelled $a_1, a_2, \ldots, a_n$ and a new agent a_0 joins the network by connecting to agent a_0 and requesting the peer information about all other agents. At this point (after one peer request) only agents a_0 and a_1 are aware of the correct network status. Now the watchdog job on an arbitrary agent a_2 starts scanning all peers known to it $(a_1, a_3, \ldots, a_n)$ one by one in unspecified order. In the worst case it takes $N-2$ requests to reach agent a_1 and obtain information about the new agent a_0. Now perform this procedure on agent a_3. This time there are 3 agents (a_0, a_1, a_2) with up to date information, thus it takes a maximum of $N-3$ peer request for agent a_3 to became up to date itself. If we repeat the procedure for all remaining agents, the number of peer request in the worst case is

$$1 + (N-2) + (N-3) + \ldots + 2 + 1 = 1 + \sum_{i=1}^{N-2} i = 1 + \frac{(N-2)(N-1)}{2} = O(N^2)$$

This number could be greatly reduced using a more sophisticated algorithm for peer status propagation – our ongoing work aims to employ an $O(\log N)$ algorithm –, but it was not in focus upon design. The test framework only requires a vertex-connectivity (κ) of 1 for correct operation, fully connected agent network only serves fault tolerance purposes in case of one or more virtual machines hosting the agents become unavailable.

After the initialization phase described above, any agent can serve as a gateway for sending control (e.g. starting and stopping jobs) or informational (e.g. querying job status) messages to other agents in the network. The agents and jobs are controlled via a command-line client interface which is built into the agent software or can be run standalone.

2.2 Implementation

The framework was implemented in Java programming language. The main reason we chose Java was the platform independence it provides, the agents running on different operating systems are able cooperate seamlessly. Java also provides an easy-to-use, lightweight remote method invocation (RMI) library, which serves as a basis of the messaging subsystem used in the test framework. Request and response messages are transferred as arguments and return values of standard RMI remote procedure calls.

Since it would be impossible to prepare for all existing test scenarios and load generators, extensibility was key concern upon designing the test framework. RMI ensures extensibility via dynamic classloading and class serialization. Test authors are able to define new jobs and message types as standard Java classes and utilize them in the test framework even without recompiling the agents.

2.3 Features and Benefits

When designing the test framework, our primary goal was to make it possible to apply artificial load on several virtual machines simultaneously and have fine-grained control over each load generator component. We distinguish two types of artificial workloads:

In-band. In-band load is caused by the normal functionality of the application being under test. Request numbers approximating or exceeding the designed capacity may cause heavy load on the application. A practical example for this is issuing a large number of HTTP requests towards a web server from many clients.

Out-of-band. This type of load is independent of the application under test, but it still affects its performance by using up shared resources on the virtual host machine. For the hypothetic web server from above, a high I/O demanding backup process run by the operating system is an example of out-of-band load.

Our test framework supports generation of both load types. It features built-in job types with either CPU-intensive or memory-consuming operations and jobs that induce high disk or network I/O traffic. Built-in job types can be parameterized which allows fine-grained control over the generated load volume. Because of their dependence on the actual application under test, there are no built-in jobs for generating in-band load, but the test framework is designed to be easily supplemented with new job and message types that can be adapted for a specific application domain.

These properties of the test framework allows us to effectively perform load and stress tests on various cloud applications as it is demonstrated in the next section.

3 Case Study — Performance Test of MySQL Cluster

In this section we present an example case study on the performance analysis of a distributed application which employs the test framework for generating in-band requests and collecting performance metrics.

3.1 MySQL Cluster

The MySQL Cluster is an open source distributed database management system, which is a development branch of the popular MySQL database. The main difference is the database engine, in the MySQL Cluster the NDB engine is used instead of the usual InnoDB or MyISAM engines. MySQL Cluster offers high availability, redundancy and increased performance for large number of parallel transactions. One of the main distinguishing features of NDB is that it stores all data in memory by default and only the transaction logs are written to the disk.

A MySQL Cluster deployment consists of three different software components which are usually deployed on separate physical or virtual servers that are interconnected via high-speed TCP/IP network links. The three software components are as follows.

Management node. This component is responsible for the entire system administration. Its task is to register and manage all the other components in the system and administer any changes in the architecture or the configuration parameters. It is required to run at configuration time but it does not have any jobs during normal operation of the cluster besides monitoring other nodes and receiving node logs.

Data node. This components stores all the actual data. A maximum of 48 data nodes can be present in a cluster. The number of servers is defined by the degree of redundancy and the amount of data that needs to be stored. For an R-times redundancy a number of $P \times R$ servers are needed, where P is a positive integer. The MySQL cluster divides the data to $2 \times P$ partitions, where every two partition is served by a *node group* containing R instances as shown on the following figure. This structure provides the scalability and high availability property of the database system, as it is able to serve SQL request when one or more data nodes are down, provided that one data node in each partition is still operational.

SQL node. This component acts as a traditional SQL server, it provides an interface for the clients. This node interprets the incoming queries, computes SQL execution plans and retrieves data from the data nodes. For load balancing, higher throughput and fault tolerance purposes this component could also be multiplied.

3.2 Configuration of the Test Framework

The performance test discussed in this section was intended to provide performance and scalability metrics for the MySQL Cluster database management system by measuring query execution times at various different system

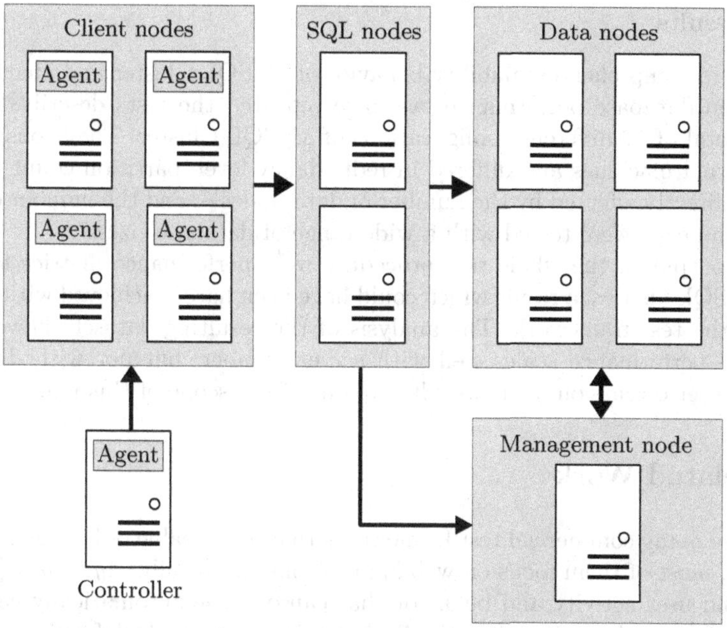

Fig. 1. Schematic configuration of the MySQL Cluster performance test showing the virtual servers hosting each software component

configurations. Figure 1 shows an overview of the test configuration, which consisted of several virtual machines hosted on the Amazon Elastic Compute Cloud (EC2) [2]. Besides the virtual machines that hosted software components of the MySQL Cluster, we have allocated multiple instances to act as database clients. We have also implemented new specialized job types for performing different types of SQL queries using the Java Database Connectivity (JDBC) API. These jobs and related message classes were included in the test framework agents that were deployed on the client nodes.

Upon initialization a sample database was created on the MySQL Cluster, then all client nodes began to execute the query jobs simultaneously. The client jobs selected a random SQL node for each query and measured the execution time of it. A total of six query types were implemented, and they were performed repeatedly by the corresponding jobs to minimize the effects of transient errors: *(a)* inserting rows; selecting rows from a single table on *(b)* indexed; or on *(c)* unindexed columns; *(d)* selecting rows from joined tables; performing aggregation functions on *(e)* a single; or on *(f)* multiple joined tables.

After the test run finished, the query execution time metrics were gathered from the agents by the controller machine for further analysis.

3.3 Results

In order to comprehend scalability behavior of MySQL Cluster and identify key factors on database performance, we have repeated the tests described above with a total of 17 different configurations of MySQL Cluster. They consisted of 8–21 virtual machines and differed in redundancy level, partition count – these two are directly affected by the number of data nodes –, and the number of SQL nodes, and each were tested with a wide range of database sizes.

The output of the whole test procedure were performance metrics for over 350 000 SQL queries, a result which could have been hardly achieved without the help of the test framework. The analysis of the resulting dataset showed that database performance scales well with server number, but not with database size. Further discussion of the results are out of the scope of this paper.

4 Related Work

There are many commercial test frameworks that are capable of load generation, however, most of them focus on web applications specifically. *AgileLoad* [1] simulates end user activity and behavior that can even be automatically captured instead of manual specification. Similarly to our solution, AgileLoad load injectors are separate software components deployed on physical or virtual machines. *LoadImpact* [4] is a SaaS solution for load testing websites, it runs entirely from the cloud and does not employ any deployable client-side software. *Keynote* [3] offers a similar service with an addition of geographically distributed load generation network.

This area is also the target of many academic research. Dumitrescu et al. developed and used *DiPerF* [9,11], a distributed performance evaluation framework with great success for performance analysis of client/server applications running on various grid computing environments. DiPerF uses an approach very similar to our framework, allowing users to submit workload generator jobs to a pool of client agents and it also provides clock synchronization between the clients, a feature which is not supported by our solution yet. However, unlike our fully distributed and homogenous test framework, DiPerF relies on a single central component to manage and control client components.

The CLIF [8] project aims to provide a generic, scalable, and user-friendly platform for performance testing. Besides load injector components similar to the jobs in our test framework, it features so called probes that can be deployed either on the workload generator or the test target system and are constantly collecting performance metrics.

5 Conclusions and Future Work

In this paper we have discussed that obtaining accurate performance metrics for virtual machines and applications hosted on an IaaS cloud service has many difficulties. Then we have presented a lightweight, distributed framework that is capable of generating artificial load to simulate the cases that may cause performance degradation of an application, caused by either normal operation or an environmental factor unrelated to the application. As we have demonstrated on a case study, our test framework suits well to perform load testing and performance testing of practical applications.

Our ongoing work involves improving the messaging subsystem using peer-to-peer (P2P) technologies, to avoid unnecessary network communications and adding performance monitoring capabilities similarly to the idea presented by Bizenhöfer et al. [5] to offer a more complete solution for performance analysis.

Acknowledgements. The work reported in the paper has been developed in the framework of the project "Talent care and cultivation in the scientific workshops of BME". This project is supported by the grant TÁMOP-4.2.2.B-10/1–2010-0009.

References

1. AgileLoad website. http://www.agileload.com/ (2013). Accessed 15 Mar 2013
2. Amazon Elastic Compute Cloud website. http://aws.amazon.com/ec2/ (2013). Accessed 15 Mar 2013
3. Keynote Internet Testing Environment website. http://kite.keynote.com/ (2013). Accessed 15 Mar 2013
4. Load Impact website. http://loadimpact.com/ (2013). Accessed 15 Mar 2013
5. Binzenhöfer, A., Tutschku, K., Graben, B., Fiedler, M., Arlos, P.: A P2P-based framework for distributed network management. In: Cesana, M., Fratta, L. (eds.) Euro-NGI 2005. LNCS, vol. 3883, pp. 198–210. Springer, Heidelberg (2006)
6. Buyya, R., Broberg, J., Goscinski, A.: Cloud Computing: Principles and Paradigms. Wiley, New York (2010). (Wiley Series on Parallel and Distributed Computing)
7. Deelman, E., Singh, G., Livny, M., Berriman, B., Good, J.: The cost of doing science on the cloud: the montage example. In: Proceedings of the 2008 ACM/IEEE Conference on Supercomputing, SC '08, pp. 50:1–50:12. IEEE Press, Piscataway, NJ, USA (2008)
8. Dillenseger, B.: CLIF, a framework based on fractal for flexible, distributed load testing. Ann. Telecommun. **64**(1), 101–120 (2009)
9. Dumitrescu, C., Raicu, I., Ripeanu, M., Foster, I.: DiPerF: an automated distributed performance testing framework. In: Proceedings of 5th IEEE/ACM International Workshop on Grid Computing, pp. 289–296 (2004)
10. Ostermann, S., Iosup, A., Yigitbasi, N., Prodan, R., Fahringer, T., Epema, D.: A performance analysis of EC2 cloud computing services for scientific computing. In: Avresky, D.R., Diaz, M., Bode, A., Ciciani, B., Dekel, E. (eds.) Cloudcomp 2009. LNICST, vol. 34, pp. 115–131. Springer, Heidelberg (2010)
11. Raicu, I., Dumitrescu, C., Ripeanu, M., Foster, I.: The design, performance, and use of DiPerF: an automated distributed performance evaluation framework. J. Grid Comput. **4**(3), 287–309 (2006)
12. Vecchiola, C., Pandey, S., Buyya, R.: High-performance cloud computing: a view of scientific applications. In: 2009 10th International Symposium on Pervasive Systems, Algorithms, and Networks (ISPAN), pp. 4–16. IEEE (2009)

Some Basic Facts About the Atmospheric Composition in Bulgaria – Grid Computing Simulations

Georgi Gadzhev[1], Kostadin Ganev[1(✉)], Nikolay Miloshev[1], Dimiter Syrakov[2], and Maria Prodanova[2]

[1] National Institute of Geophysics Geodesy and Geography, Bulgarian Academy of Sciences, Acad. G. Bonchev Street, Bl.3, 1113 Sofia, Bulgaria
{ggadjev,kganev}@geophys.bas.bg

[2] National Institute of Meteorology and Hydrology, Bulgarian Academy of Sciences, Tsarigradsko Shose 66, 1784 Sofia, Bulgaria

Abstract. The present work aims at studying the local to regional atmospheric pollution transport and transformation processes over Bulgaria and at tracking and characterizing the main pathways and processes that lead to atmospheric composition formation in the region.

The US EPA Models-3 system is chosen as a modelling tool. As the NCEP Global Analysis Data with 1 degree resolution is used as meteorological background, the MM5 and CMAQ nesting capabilities are applied for downscaling the simulations to a 9 Km resolution over Balkans and 3 Km over Bulgaria. The TNO emission inventory is used as emission input. Special pre-processing procedures are created for introducing temporal profiles and speciation of the emissions.

The study is based on a large number of numerical simulations carried out day by day for years 2000–2007 and five emission scenarios - with all the emissions and with biogenic emissions, emissions from energetics, road transport and none industrial combustion reduced. Results from the numerical simulations concerning the main features of the atmospheric composition in Bulgaria and the contribution of the different emission categories are demonstrated in the paper.

Keywords: Atmospheric composition · Regional scale modelling · US EPA Models-3 system · Grid computing

1 Introduction

Recently extensive studies for long enough simulation periods and good resolution of the atmospheric composition status in Bulgaria have been carried out using up-to-date modeling tools and detailed and reliable input data [7,9–11].

The simulations aimed at constructing of ensemble, comprehensive enough as to provide statistically reliable assessment of the atmospheric composition climate of Bulgaria - typical and extreme features of the special/temporal behavior, annual means and seasonal variations, etc.

I. Lirkov et al. (Eds.): LSSC 2013, LNCS 8353, pp. 484–490, 2014.
DOI: 10.1007/978-3-662-43880-0_55,

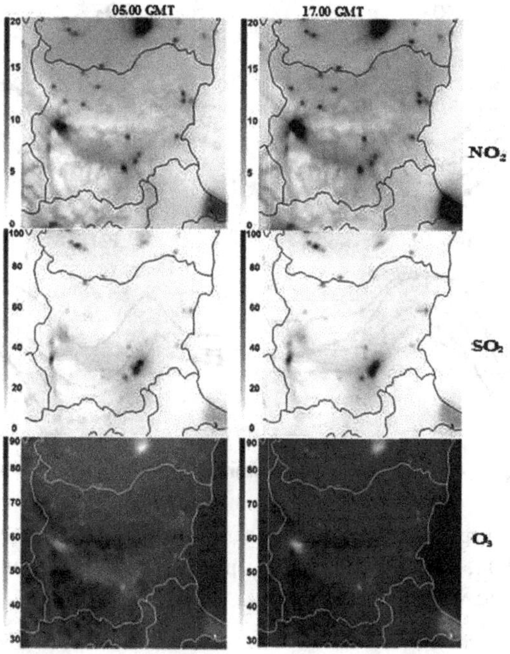

Fig. 1. Surface concentrations of NO_2, SO_2 and O_3 [µg/m^3] averaged annually in 05.00 and 17.00 GMT

The present paper will focus on two important characteristics of the atmospheric composition climate of Bulgaria - the concentrations of different compounds and the evaluation of the contribution of different emission categories to the overall air pollution in the country. The problem of the role of different processes for the atmospheric composition formation will not be considered here, because it is a subject of a separate paper [9,10].

2 Approaches, Tools, Data, Domains, and Nesting

All the simulations are based on the US EPA Model-3 system. The system consists of three components: MM5 [5,11] used as meteorological pre-processor, CMAQ [2,3], the Chemical Transport Model of the system and SMOKE [4] - the emission pre-processor of Models-3 system.

The large scale (background) meteorological data used by the study is the NCEP Global Analysis Data with $1° \times 1°$ resolution. The MM5 and CMAQ nesting capabilities are used to downscale the problem to a 3 Km horizontal resolution for the innermost domain (Bulgaria).

The TNO high resolution emission inventory [12] is exploited. A more detailed description of the emission modeling is given in [9].

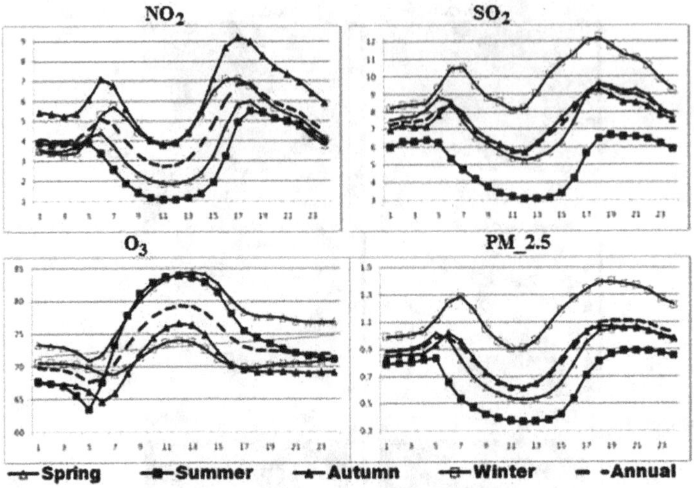

Fig. 2. Plots of the “typical” diurnal course of the averaged for the territory of Bulgaria concentrations of NO_2, SO_2, O_3 and $PM_{2.5}$ [µg/m^3] averaged annually, for the spring, summer, autumn, and winter.

The study is based on a large number of numerical simulations carried out day by day for years 2000–2007 and five emission scenarios - with all the emissions and with biogenic emissions, emissions from energetics, road transport and none industrial combustion reduced. This makes it possible to evaluate the contribution of different emission categories to the formation of the overall atmospheric composition pattern. Performing extensive simulations of this kind with up to date highly sophisticated numerical models obviously requires large computer resource. That is why grid computing [1,6] was applied for the present simulations. Details about the performance of this grid application can be seen in [9].

3 Some Examples of the Numerical Simulation Results

The most simple atmospheric composition evaluations are, of course, the surface concentrations. By averaging over the 8-year simulated fields ensemble the mean annual and seasonal surface concentrations can be obtained and treated as respective “typical” daily concentration patterns. Plots of some of these “typical” annual surface concentrations are shown in Fig. 1 for some of the most popular compounds - NO_2, SO_2, ozone. What can be seen from the plots is not surprising: the big cities and the road network are clearly outlined in the NO_2 surface concentrations, the big power plants in the SO_2 surface concentrations.

The ozone fields are much more complex. What should be mentioned is the expected effect of ozone minimums over big cities. The road network can also be followed in the plots as lines with lower ozone concentrations. This is in a good

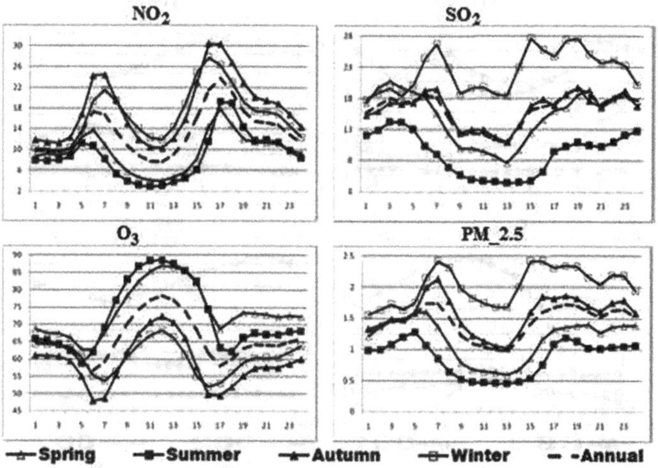

Fig. 3. Plots of the "typical" diurnal course of the concentrations of NO_2, SO_2, O_3 and $PM_{2.5}$ [µg/m^3] in Sofia, averaged annually, for the spring, summer, autumn, and winter.

agreement with the ozone chemistry scheme. The averaged over the ensemble 2D concentration fields can be themselves averaged over the territory of Bulgaria and thus some more easy to comprehend plots of the "typical" diurnal course of the concentrations for the year and the four seasons to be obtained. Such plots are given in Fig. 2 for several compounds.

The behavior of the same compounds for a chosen point - the city of Sofia (Fig. 3) is not qualitatively much different. Of course, as it should be expected, the values of NO_2, SO_2, and $PM_{2.5}$ in the city are significantly larger than the averaged over the country.

Five emission scenarios will be considered in the present paper: Simulations with all the emissions, simulations with biogenic emissions and the emissions of categories 1 (energetics), 2 (none industrial combustion) and 7 (road transport) for Bulgaria reduced by a factor of 0.8. This makes it possible, according to (1), to evaluate the contribution of emission categories to the atmospheric composition in Bulgaria. These relative contributions were calculated day by day and then, by averaging over the 8-year ensemble the "typical" contributions for the four seasons and annually were obtained.

2D plots of the diurnal evolution of the "typical" relative emission contributions are given in [9]. Plots of this kind are rather spectacular and can give a good qualitative impression of the spatial complexity of the emission contribution. In order to demonstrate the emission contribution behavior in a more simple and easy to comprehend way, the respective fields can be averaged over some domain (in this case the territory of Bulgaria), which makes it possible to jointly follow and compare the diurnal behavior of the respective contributions for different species. Such plots for some of the compounds are given in Fig. 4.

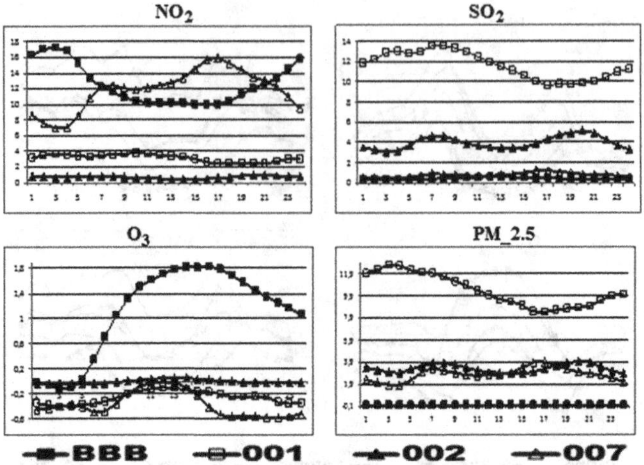

Fig. 4. Plots of the "typical" annual diurnal course of the averaged for the territory of Bulgaria relative contributions [%] of emissions from categories 1 (001), 2 (002) and 7 (007) and of the biogenic emissions (BBB) to the concentrations of NO_2, SO_2, O_3 and $PM_{2.5}$.

There is no need to describe the plots in details, but some comments on them could be made. First of all it could be seen that the different emissions relative contribution to the concentration of different species could be rather different. The contributions of different emission categories to different species surface concentrations have different diurnal course and different importance. The energetics is the major contributor to SO_2 and $PM_{2.5}$ concentrations, while the biogenic emissions have near zero or even negative contributions. The major contributors to the NO_2 concentrations are the road transport and biogenic emissions. Their diurnal courses are in counter- phase, which can be easily explained by the ozone photochemistry cycle.

One can not help but notice the small contribution of biogenic emissions to surface ozone. This fact was extensively discussed in [9] and was explained by the fact that for Bulgaria the local O_3 production rate is limited by the availability of NO_X concentration, a regime which is called NO_X-limitiation. The contribution of the emission from categories 1 and 7, which are the major sources of the other ozone precursor - nitrogen oxides, is also small. This, once again is an indirect indicator, that the surface ozone in Bulgaria is to a small extend due to domestic sources, but is mostly imported.

The picture is completely different for the city of Sofia. The NO_2 concentrations are totally dominated by road transport emissions. The none industrial combustion has big contribution in SO_2 formation (probably mostly from the city heating plants and domestic heating). The NO_2 also has dominating (negative) contribution to the surface ozone. It is particularly large in morning and late afternoon, when the city traffic is most intensive. In the afternoon the

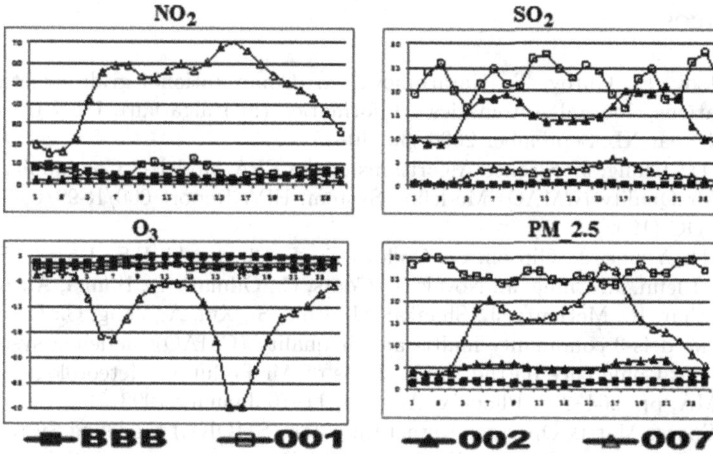

Fig. 5. Plots of the "typical" annual diurnal course of the relative contributions [%] of emissions from categories 1 (001), 2 (002) and 7 (007) and of the biogenic emissions (BBB) to the concentrations of NO_2, SO_2, O_3 and $PM_{2.5}$ for the city of Sofia.

contribution of road transport to the $PM_{2.5}$ levels becomes even bigger than the contribution of the energetics.

4 Conclusions

The numerical experiments performed produced a huge volume of information, which have to be carefully analyzed and generalized so that some final conclusions could be made. Simulations for emission scenarios concerning the contribution of the other emission categories have to be performed.

The obtained ensemble of numerical simulation results is extensive enough to allow statistical treatment - calculating not only the mean concentrations and different emission categories contribution mean fields, but also standard deviations, skewness, etc., with their dominant temporal modes (seasonal and/or diurnal variations). Some advanced and sophisticated methods for statistical treatment of the results should also be appropriately applied.

Acknowledgements. The present work is supported by the Bulgarian National Science Fund (grant DCVP-02/1/29.12.2009) and by the EU FP7 project PASODOBLE, grant №241557.

Deep gratitude is due to US EPA and US NCEP for providing free-of-charge data and software. Special thanks to the Netherlands Organization for Applied Scientific research (TNO) for providing the study with the high-resolution European anthropogenic emission inventory.

References

1. Atanassov, E., Gurov, T., Karaivanova, A.: Computational grid: structure and applications, Journal Avtomatica i Informatica (in Bulgarian), ISSN 0861–7562, 3/2006, year XL, September 2006, pp. 40–43
2. Byun, D., Ching, J.: Science Algorithms of the EPA Models-3 Community Multiscale Air Quality (CMAQ) Modeling System. EPA Report 600/R-99/030, Washington DC (1999)
3. Byun, D., Young, J., Gipson, G., Godowitch, J., Binkowski, F.S., Roselle, S., Benjey, B., Pleim, J., Ching, J., Novak, J., Coats, C., Odman, T., Hanna, A., Alapaty, K., Mathur, R., McHenry, J., Shankar, U., Fine, S., Xiu, A., Jang, C.: Description of the Models-3 community multiscale air quality (CMAQ) modeling system. In: 10th Joint Conference on the Applications of Air Pollution Meteorology with the A&WMA, pp. 264–268. Phoenix, Arizona, 11–16 January 1998
4. CEP: Sparse Matrix Operator Kernel Emission (SMOKE) Modeling System. University of Carolina, Carolina Environmental Programs, Research Triangle Park, North Carolina (2003)
5. Dudhia, J.: A non-hydrostatic version of the Penn State/NCAR mesoscale model: validation tests and simulation of an Atlantic cyclone and cold front. Mon. Wea. Rev. **121**, 1493–1513 (1993)
6. Foster, J., Kesselmann, C.: The Grid: Blueprint for a New Computing Infrastructure. Morgan Kaufmann, San Francisco (1998)
7. Gadzhev, G., Jordanov, G., Ganev, K., Prodanova, M., Syrakov, D., Miloshev, N.: Atmospheric composition studies for the Balkan Region. In: Dimov, I., Dimova, S., Kolkovska, N. (eds.) NMA 2010. LNCS, vol. 6046, pp. 150–157. Springer, Heidelberg (2011)
8. Gadzhev, G., Ganev, K., Syrakov, D., Miloshev, N., Prodanova, M.: Contribution of biogenic emissions to the atmospheric composition of the Balkan Region and Bulgaria. Int. J. Environ. Pollut. **50**(1/2/3/4), 130–139 (2012)
9. Gadzhev, G., Ganev, K., Miloshev, N., Syrakov, D., Prodanova, M.: Numerical study of the atmospheric composition in Bulgaria. Comput. Math. Appl. **65**, 402–422 (2013)
10. Gadzhev, G., Ganev, K., Miloshev, N., Syrakov, D., Prodanova, M.: Analysis of the processes which form the air pollution pattern over Bulgaria. In: The Proceedings of the 9th International Conference on "Large-Scale Scientific Computations". Sozopol, Bulgaria, 3–7 June 2013 (accepted for publishing)
11. Grell, G.A., Dudhia, J., Stauffer, D.R.: A description of the Fifth Generation Penn State/NCAR Mesoscale Model (MM5). NCAR Technical Note, NCAR TN-398-STR, 138 pp. (1994)
12. Visschedijk, A., Zandveld P., van der Gon, H.: A high resolution gridded European emission database for the EU integrated project GEMS, TNO report 2007-A-R0233/B, The Netherlands (2007)

Harnessing Wasted Computing Power for Scientific Computing

Sándor Guba, Máté Őry, and Imre Szeberényi(✉)

Budapest University of Technology and Economics, Budapest, Hungary
{guba.sandor,orymate,szebi}@iit.bme.hu

Abstract. Nowadays more and more general purpose workstations installed in a student laboratory have a built in multi-core CPU and graphics card providing significant computing power. In most cases the utilization of these resources is low, and limited to lecture hours. The concept of utility computing plays an important role in technological development. In this paper, we introduce a cloud management system which enables the simultaneous use of both dedicated resources and opportunistic environment. All the free workstations are allocated to a resource pool, and can be used like ordinary cloud resources. Our solution leverages the advantages of *HTCondor* and *OpenNebula* systems. Modern graphics processing units (GPUs) with many-core architectures have emerged as general-purpose parallel computing platforms that can dramatically accelerate scientific applications used for various simulations. Our business model harnesses the computing power of GPUs as well, using the needed amount of unused machines.

Keywords: Cloud · GPGPU · Grid · HTC · Utility computing

1 Introduction

In universities there is a huge demand for high performance computing, but the smaller research groups can not afford buying a supercomputer or a large compute cluster. However significant unused computing capacity is concentrated in the student laboratories, as most of our student labs have quite new PCs with modern multi-core CPUs and high performance graphics cards. The total computing performance of the laboratory resources could be significant. The open questions are: (a) how can we collect and use these resources; (b) what is the time limit of the usage; (c) what happens if one or more jobs are not finished during the given time slot; (d) what management software and management rules are needed to support the various software environments which must be flexible and on demand.

In this paper we are investigating these problems and we introduce a solution based on a new approach. We show that the cloud technology, based on hardware accelerated virtualization, can be the right answer to these questions. First of all the management of the cloud based systems is easier and they are more flexible.

I. Lirkov et al. (Eds.): LSSC 2013, LNCS 8353, pp. 491–498, 2014.
DOI: 10.1007/978-3-662-43880-0_56,

According to the literature [12] and our experiences the modern virtualization has minimal overhead compared to the native systems and has more advantages than disadvantages.

Our basic idea is to run only a minimal host operating system on the bare metal and virtualize everything else. In this manner we can easily solve the questions raised up. We do not need a time consuming cloning process for configuration management. We can save an ongoing scientific computing process at any time, and we can restore and continue it even on another host machine. One can say, yes, these goals are solved already by various cloud management systems in corporate environment. So what is the novum on this?

The main difference between the corporate cloud infrastructure running 24/7 and our laboratory environment is that the corporate infrastructure is used only for serving the virtual machines. However, functions of our student laboratory are twofold: (1) During the scheduled lab exercises, the workstations act as a cloud host which serves only the virtual machines owned by the student sitting in front of the workstation or act as a simple cloud client. (2) While the lab is not used for teaching, the workstations are acting as a normal cloud host running computing intensive jobs like a normal HTCondor executing machine.

Our solution, *CIRCLE* (Cloud Infrastructure for Research and Computer Labs in Education) is not only harnessing the idle CPU cycles for scientific computing, but it provides an easy and flexible web-portal for the usage and the management as well. The user can easily manage their virtual machines and access the files stored on the online storage. Nevertheless the lecturers can easily customise a new virtual machine image and share this image with the students. In this way all the students have the same and clean learning environment which enables to concentrate on the real task.

In the following sections we present the applied technologies and components used in our pilot system.

2 Virtualization

Most IaaS (infrastructure as a service) cloud systems are based on virtual machines. Although the technique is available since the end of 1960's [3], widespread adoption of x86 based systems in the server segment made it almost entirely disappear. Later, some vendors started implementing different software based solutions for virtualizing operating systems or even emulating CPUs. The renaissance of virtualization began with manufacturers extending the x86 instruction set to support low-overhead virtualization. The currently available such extensions are *Intel VT-x* and *AMD-V*.

Current popular techniques are operating system virtualization and full hardware accelerated virtualization. The former typically takes shape in *chroot* environments and in namespacing of some kernel resources. This does not even allow running different kernels, nor different kinds of operating systems. The latest technique is full hardware accelerated virtualization, which is based on the CPU support for isolating the concurrently running instances. This approach is normally extended with paravirtualized device drivers, which eliminate the need for emulating real world storage and network controllers.

Hardware accelerated virtualization requires CPU support, however it is not available currently on the low-end product line of the main x86 CPU manufacturers: some models of Intel Atom, Celeron, and Pentium. This hardware acceleration provides a near-native performance for HPC applications [12].

Currently there are many competing full virtualization solutions. The most notable free ones are *KVM* and *XEN*. At the time of our decision, the installation of a *XEN* hypervisor required modifications to the *Linux* kernel, and it was unacceptable for us. This is no longer the case, but we are satisfied with KVM.

Additionally, we use all KVM functions through the *libvirt* library, which provides an abstract interface for managing virtual machines [9]. This has the benefit of theoretically flawless migration to other hypervisors like XEN, *ESXi*, or *Hyper-V* [1].

Physically accessible computers are normally used with directly attached devices like display and keyboard. These devices are emulated by KVM, and you can access virtual machines' consoles via the VNC protocol. This is useful for installing the operating system or troubleshooting, but Windows and Linux both provide better alternatives for remote access.

We use the *remote desktop protocol* (RDP) for accessing Windows hosts, and *secure shell* (SSH) for text-based Linux machines. Remote graphical login to *X11* servers has always been available, but this is not reliable even on local network connections because it is stateless. We use *NoMachine NX* [8] instead.

3 Networking

Most virtual machines in a cloud have a network connection. On the physical layer, the KVM hypervisor provides a virtual network interface controller, which is an emulated or paravirtualized NIC on the side of the guest operating system, and a virtual NIC on the host side. The communication between the two endpoints is KVM emulating the PCI signals to the virtual machine (emulated case), or the *virtio* protocol, which is optimized for software implementation (paravirtualized case).

Virtual machines are connected to virtual networks provided by manageable virtual switches (Fig. 1). The *Open vSwitch* [7], what we are using, is a high performance multi-layer virtual switch with *VLAN*, *QoS* and *OpenFlow* support, merged into the mainline Linux kernel.

Virtual networks do not necessarily differ from physical ones in the upper layers. The most important different condition is the frequency of changes. Our system from the point of traditional physical networks' view is like someone changes the cabling hundred times in the middle of the day. The developed CIRCLE networking module consists of an *iptables* gateway, a *tinydns* name server and an *ISC* DHCP server. All of these are configured through remote procedure calls, and managed by a relational database backed object model.

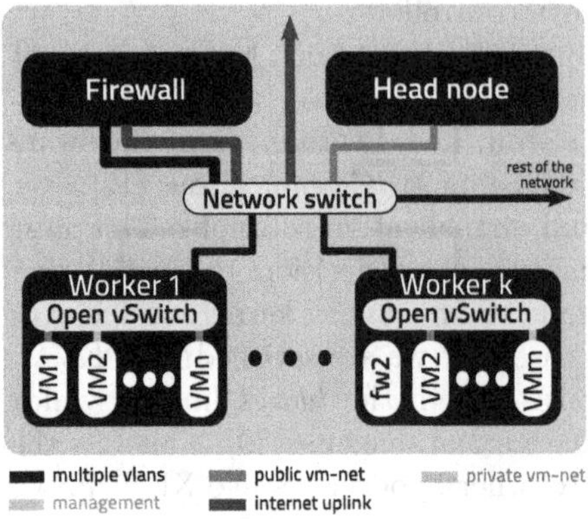

Fig. 1. The structure of the network

Fig. 2. Technologies used for CIRCLE

Our solution groups the VMs to two main groups. The public vm-net is for machines which provide public services to more people, the private vm-net is for those which are used only by one or two persons. Public vm-net machines have public IPv4 and IPv6 addresses, and are protected with a simple *ipset*-based input filter. On the private vm-net, machines have private IPv4 and public IPv6 addresses. The primary remote connection is reached by automatically configured IPv4 port forward, or directly on the IPv6 address. As connecting to the standard port is a more comfortable solution, users who load our web portal from an IPv6 connection, get a hostname with public AAAA and private A records. If the user has no IPv6 connection, we display a common hostname with a single A record, and a custom port number. As IPv6 is widely available in the central infrastructure of our university, IPv6-capable clients are in majority. Users can open more ports, which means enabling incoming connections, and setting up IPv4 port forwarding in the background.

4 Storage

Virtual machines' hard drives are provided for the hypervisors as read-write NFS shares managed by OpenNebula. Our cluster has a legacy InfiniBand SDR network, which is despite its age much faster than the gigabit Ethernet network. InfiniBand has its own data-link protocol, and Linux has mainline support for remote direct memory access (RDMA) over it, which provides near-local access times and no CPU load [2]. Unfortunately this kernel module causes random cluster-wide kernel panics, which is unacceptable in a production system. We decided to use *NFS4* over *IP* over *InfiniBand*, which also provided near-local timing. One problem remained: intensive random writes made the local file access on the NFS server slow (both with *RDMA* and IP over IB). Switching to the *deadline* scheduler solved this.

Disk images are stored in *qcow2* (QEMU copy on write) format, which allows images with large free space to be stored in a smaller file, and also supports copy-on-write differential images. The latter feature is used for virtual machines, which eliminates the need of copying the whole base image file before launching a new instance. Saving a template consists of merging the base and differential images to a single one.

Since our usual virtual machines have temporary disks, there is a common need for a permanent online storage that can be easily accessed. It allows the user to use the same resources from different virtual computers or even from home, and it helps sharing data between virtual machines and local computers on a simple interface.

Our solution—CIRCLE File Server—is a multi-protocol file server, that runs on a virtual machine. Each user gets an amount of disk space, which is automatically mounted on our prepared appliances.

Windows VMs access the storage over *SMB/CIFS*. The authentication is handled by CIRCLE with automatically generated passwords. For security reasons we do not allow SMB access outside vm-net. Linux guests mount the remote files with *SSHFS* [6], a userspace SSH/SFTP virtual file system. For virtual machines the manager automatically generates key-pairs. SFTP service is also accessible over the internet. Users can set public keys on the web portal and immediately access their folder.

It is also possible to manage files on the cloud portal with an *AJAX* based web interface. Its backend consists of a *Celery* worker and an *Nginx* httpd.

5 Putting It Together

The main goal was to give a self-service interface to our researchers, lecturers, and students. Cloud management frameworks like *OpenNebula* and *OpenStack* promise this, but after learning and deploying OpenNebula, we found even its Self-Service portal's abstraction level too low.

Our solution is a new cloud management system, called CIRCLE, built up from various open source software components (Fig. 2). It provides an attractive web interface where users can do independently all the common tasks including launching and managing/controlling virtual machines, creating templates based on other ones, and sharing templates with groups of users.

This cloud management system is based on *Django* [5]. This popular *Python* framework gives us among other things a flexible object-relational mapping system. Although the Django framework is originally designed for web applications, the business logic is not at all web specific. That's why it is easy to provide command line or remote procedure call interfaces to the model.

As the primary interface is web, which is in some aspect a soft real-time system, the model can not use synchronous calls to external resources, nor execute system commands. This is the reason why all remote procedure calls are done asynchronously through a standard task queue. Our choice is the Celery distributed task queue. This is the most popular among such systems, which are integrated with Django. Celery is configured to use an implementation of *AMQP* [10] protocol—called *RabbitMQ*—as its message broker.

Celery workers set up the netfilter firewall, the domain name and DHCP services, the IP blacklist, execute file server operations, and also communicate with OpenNebula. This distributed solution enables to dynamically alter the subsystems.

In the opposite direction, some subsystems notify others of their state transitions through Celery. Based on this information further Celery tasks are submitted, and the models are updated.

CIRCLE manages the full state space of the resources. Some of it is also stored by the underlying OpenNebula, but most of this redundant information is bound to its initial value as OpenNebula does not handle changes in meta information. This behavior arises of design decisions, and not expected to be improved. The thin slice of OpenNebula used by our system is continuously shrinking, and we intend dropping OpenNebula in favor of direct bindings to libvirt and the also considerably customized storage and network hooks.

6 Execution on Workstations

The cloud system at our institute takes a big role in education and in general R&D infrastructure, but there is a significant demand for high-throughput scientific computing. This requirement usually appears in form of many long-running, independent jobs. On most parts of the world there is no fund to build dedicated HPC clusters with enough resources for these jobs.

The highest load on the cloud takes place during office hours and the evenings, in more than half of the time we have many free resources, so it is possible to run these jobs on low priority virtual machines in the cloud. If interactive load is increasing, we can even suspend these machines, and resume them later.

Running scientific batch jobs on student laboratory computers also has a long history. Our idea is to run some of these jobs on virtual machines in the computer laboratories overnight and on weekends. We can suspend in the morning all virtual machines to a memory image, and resume on the same or some other hypervisor next evening. This solution makes possible to run individual jobs virtually continuously through months or a year, without any specific efforts. This is important because of our observation that the runtime of similar jobs have a high standard deviation, and it also protects against losing the partial result of months long computations in case of hardware or power failure. HTCondor has a similar result with its checkpoint support, but it needs to modify the software, which is often impossible or sometimes the users are not able to do this modification by themselves.

To be able to resume suspended machines, we have to copy back the differential image and the memory dump. Our choice for this is rsync.

The lab network is exposed to unauthorized access, so we have to limit access to confidential material. As a physically accessible general purpose workstation does not currently provide a way to reliably authenticate itself to a server, nor to protect the stored data, we can not employ any solution against these attacks other than security through obscurity and not using these systems for any confidential executions.

Another important aspect is energy efficiency. We have successfully used HTCondor to automatically turn on and off the compute nodes of a HPC cluster. This is also working with Wake on LAN and SSH on the workstations.

7 GPUs in the Cloud

The most significant HPC performance in our student laboratories is provided by the mid-level GPUs in all the workstations used for teaching computer graphics. There is a technology we applied successfully to use GPGPUs from the dedicated cluster's virtual machines: *PCI passthrough* [11]. However, this technology requires both CPU and motherboard support of *IOMMU*, which is a high-end feature nowadays. The implementations are called *Intel VT-d* and *AMD-Vi* technologies, and they appear in the server- and high-end workstation segments.

As none of our laboratory computers support IOMMU, we have to find a different solution. The first one is using *rCUDA*, which is a small framework makes possible to run the host and device side of a *CUDA* program on different hosts, communicating over TCP/IP or InfiniBand network [4]. With this, we can launch user-prepared virtual machines on each host, and run the device code via local (*virtio*-based) network on the hypervisor. rCUDA is also capable to serve more clients with a single device. This is useful if the host code uses the GPU only part time.

The other option is using directly the host machine to execute GPGPU jobs. This is a simpler approach, but necessarily involves a more complicated scheduler. Our choice for this type of problems is HTCondor, which can manage this scenario without much customization. The disadvantage is that the user can not customize the host-side operating system.

8 Conclusions and Future Plans

Our cloud system is built up in a modular manner. We have implemented all the main modules which enabled us to set up a production system. The system is now used as an integral part of our teaching activity, and also hosts several server functions for our department to use. At the time of writing this paper, there are 70 running and 54 suspended machines, using 109 GiB of memory and producing not more than 3 % cumulated host cpu load on the cluster. In the first two months' production run, more than 1500 virtual machines have been launched by 125 users.

The students found the system useful and lecturers use it with pleasure because they can really set up a new lab exercise in minutes. The feedbacks from the users are absolutely positive, which encourages us to proceed and extend the system with the GPGPU module. We are working on making it fully functional, and releasing the whole system in an easily deployable and highly modular open source package. We are planning to finish the current development phase until the end of August.

Acknowledgements. We thank to the other members of the developer team: Dániel Bach, Bence Dányi, and Ádám Dudás.

References

1. Bolte, M., Sievers, M., Birkenheuer, G., Niehörster, O., Brinkmann, A.: Non-intrusive virtualization management using libvirt. In: Proceedings of the Conference on Design, Automation and Test in Europe, pp. 574–579. European Design and Automation Association (2010)
2. Callaghan, B., Lingutla-Raj, T., Chiu, A., Staubach, P., Asad, O.: NFS over RDMA. In: Proceedings of ACM SIGCOMM Summer 2003 NICELI Workshop (2002)
3. Creasy, R.J.: The origin of the vm/370 time-sharing system. IBM J. Res. Dev. **25**(5), 483–490 (1981)
4. Duato, J., Pena, A.J., Silla, F., Fernández, J.C., Mayo, R., Quintana-Orti, E.: Enabling CUDA acceleration within virtualmachines using rCUDA. In: 18th International Conference on High Performance Computing (HiPC) 2011, pp. 1–10. IEEE (2011)
5. Holovaty, A., Kaplan-Moss, J.: The Definitive Guide to Django: Web Development Done Right. Apress, Berkeley (2009)
6. Hoskins, M.E.: SSHFS: super easy file access over SSH. Linux J. **146**, 4 (2006)
7. Pfaff, B., Pettit, J., Koponen, T., Amidon, K., Casado, M., Shenker, S.: Extending networking into the virtualization layer. In: Proceedings of HotNets, October 2009 (2009)
8. Pinzari, G.F.: Introduction to NX technology. Technical report, NoMachine Technical, Report 309 (2003)
9. Victoria, B.: Creating and controlling kvm guests using libvirt. University of Victoria (2009)
10. Vinoski, S.: Advanced message queuing protocol. IEEE Internet Comput. **10**(6), 87–89 (2006)
11. Yang, C.T., Wang, H.Y., Ou, W.S., Liu, Y.T., Hsu, C.H.: On implementation of GPU virtualization using PCI pass-through. In: 2012 IEEE 4th International Conference on Cloud Computing Technology and Science (CloudCom), pp. 711–716. IEEE (2012)
12. Younge, A.J., Henschel, R., Brown, J.T., von Laszewski, G., Qiu, J., Fox, G.C.: Analysis of virtualization technologies for high performance computing environments. In: 2011 IEEE International Conference on Cloud Computing (CLOUD), pp. 9–16. IEEE (2011)

Performance Analysis of Windows Azure Data Storage Options

Istvan Hartung(✉) and Balazs Goldschmidt

Department of Control Engineering and Information Technology,
Budapest University of Technology and Economics,
Magyar tudósok körútja 2, Budapest 1094, Hungary
{hartung,balage}@iit.bme.hu

Abstract. Windows Azure provides an IaaS cloud service with virtual machines, web and worker roles and practically unlimited, pay-as-you-go storage options which can be used for applications requiring big data or parallel computing which is important in many fields including biology, astronomy, nuclear physics and economics.

When moving an application or computation task to the cloud it is very important to perform proof of concept performance testing and to carefully choose the proper building blocks for the given tasks. Windows Azure provides multiple data management options with a relational SQL database for transactional data access, Azure Tables for auto scalable storage of unstructured data, and a blob storage for storing large amounts of binary data which is easily mountable to a given virtual machine.

In this paper we present a general performance analysis of the Windows Azure cloud with focus on cloud storage options. We present an environment to perform automated testing of the major features of Azure storage and we also present the preliminary results and suggestions regarding the usage of the different services.

1 Introduction

Recently the cloud computing paradigm [2] has led to novel solutions for storing and processing data both in the industry and in the academic world. Any resource-intensive task may be moved to the cloud which provides scalable and practically unlimited resources where the provisioning of 1000 virtual machines for one hour costs as much as provisioning one instance for 1000 hours. Many companies including Amazon, Google, HP, IBM and Microsoft offer public cloud infrastructures available to anyone providing virtual machine instances, novel storage services and traditional database solutions among other enterprise solutions.

The use of public cloud for scientific calculations in many fields is a logical consequence of the requirement of both large scale data storage options and parallel computing. Clouds provide a cheap alternative to specialized clusters and supercomputers. Amazon Web Services provides public data sets in a form of

I. Lirkov et al. (Eds.): LSSC 2013, LNCS 8353, pp. 499–506, 2014.
DOI: 10.1007/978-3-662-43880-0_57,

a centralized repository which can be accessed in a few minutes and are completely free [1]. These datasets contain terabytes of data from a wide spectrum of domains including biology, astronomy, economics, chemistry, mathematics and geography and are freely available with pre-configured Amazon EC2 instances.

The Windows Azure cloud provides Infrastructure as a Service (IaaS) and Platform as a Service (PaaS) components that support parallel computing and storage of large data. The platform provides novel storage options for large binary data in blob storage, Tables storage with NoSQL capabilities for unstructured data, and a queue service for asynchronous communication between elements of a distributed system. Besides the large and scalable solutions Azure also provides a full featured traditional database-as-a-service, which is built on technologies of SQL Server and provides a traditional interface. The Azure platform is already used in various scientific projects, [8] but the proper storage solution must be carefully chosen for any application.

In this paper we introduce the storage options for the Azure cloud and provide a solution to benchmark the PaaS components of the Azure platform and present our first results and recommendations for the use of storage options.

The rest of the paper is organized as follows. In the second section, related work is examined. In the third section, the Azure platform is introduced, the measurement architecture is shown and finally the evaluation of the results is provided. The fourth section summarizes the results.

2 Related Work

In the past few years cloud computing has drawn attention from many researchers. Back in 2008 Vaquero et al. analyzed 22 different definitions for cloud computing and compared the paradigm with Grid technology [11].

Performance of cloud computing has been studied by many research groups. Many investigated the low level performance of virtualization in general and the performance of Xen, which is used for virtualization by many public and private cloud providers [3,7].

Others compared the performance of major public IaaS cloud providers in terms of network capabilities, memory, disk and CPU utilization, binary object storage and queue access [6]. Li et al. focused on comparing the common services provided by the public providers with an application that could be deployed on a virtual machine of the cloud. We present a distributed testing application that is freely scalable and provides testing of big parallel workloads on Azure.

Jackson et al. provide a comparison of the performance of Amazon EC2 and the standard supercomputing centers and showed that there is a correlation between the amount of time a given application spends communicating through network and its performance on Amazon EC2 [5]. The use of Amazon EC2 for scientific calculations was analyzed in aspects of virtual machine performance, resource acquisition and virtualization by Ostermann et al. [9].

There are papers about the performance of Windows Azure, [4] but storage services over went a major performance change in December of 2012 so none of them provide up to date information about the capabilities of the PaaS storage service. Additionally our goal was to compare the performance of SQL Database with Azure Tables storage.

Windows Azure has been used in scientific computation in many fields including biology [10], where it has been shown that moving computation to data, supporting large data sets by the cloud simplify the implementation of some problems over traditional systems.

3 Performance Measurements in Windows Azure

This section introduces the Azure platform, the measurement architecture and the results of the measurements.

3.1 Windows Azure

The Windows Azure platform is the public cloud provider of Microsoft. It offers both Linux and Windows based virtual machines as well as scalable compute instances, known as roles. The role instances can be categorized into Web Roles, which have a public interface, run a preconfigured IIS and are accessible via HTTP or HTTPS and Worker Roles that typically run background computation tasks. Azure also provides components that aid communication between roles and solutions to store terabytes of data as well as an SQL database as a service component for relational data.

The Azure Blob storage provides storage for binary data and metadata in containers. The container provides a logical grouping and defines the level of sharing. There first of the two types of supported blobs, block blobs are targeted for streaming workloads, have a maximum size limit of 200 GB and are updated with commit-based semantics. Each blob consists of a list of blocks, which can be modified by first uploading the new uncommitted blocks for the blob file, then a single call with the list of the new blocks will commit all changes on the blob. Page blobs may be files up to 1 TB of size and support random write workloads. Each page blob consists of an array of pages, which can be immediately updated and support change in only a portion of the file.

The Azure Tables service allows storage of enormous amounts of data with efficient querying and insertion. Each created table contains a set of entities which can hold up to 255 properties where the size of the entities must be under 1 MB. A single table may contain different types of entities the only restriction is that each entity within a table must have a unique PartitionKey and RowKey. These are the only two columns in a table that are indexed, further on the partition key is used by the Tables service to distribute data across the storage instances automatically. Entities with the same PartitionKey within a table will always be stored on the same instance.

The Windows Azure SQL Database is a relational database with size up to 150 GB with almost full Transact-SQL support, including creating and executing stored procedures, functions, transactions and triggers, supporting security via multiple logins and users, and a configurable firewall. The database provides federation via data partitioning to support larger data set sizes, but lacks transaction support over multiple databases.

3.2 Measurement Architecture

When choosing the proper environment for testing the Azure service it was very important to keep certain requirements in mind. For testing the capabilities of Azure storage it is very important for the network connection to be sufficient, because it could easily become a bottleneck when testing. It is also important to have multiple instances to run tests to provide multiple endpoints. Otherwise load balancers, firewalls or other network components may throttle connections or filter traffic in order to defend against denial of service based attacks. Another important aspect is for the clients to store all measurement data persistently and in a way that does not limit the measurement itself. The measurement data should be collected independently and should be converted to in an easy to process form.

Our constructed measurement system is built from three layers of components (see Fig. 1). The input layer consists of a scheduler role, which is an Azure Web Role which provides a web service interface and accepts a set of queries and tasks that are to be run on the system under test. The virtual machine places these received tasks in an Azure Queue and are later processed and executed by one or all of the tester roles with given amount of execution counts.

The number of tester roles can be scaled up from the Azure Management Portal dynamically during the execution of the test. The results of the given test (elapsed time, query identifier, number of results) are logged using Windows Azure Diagnostics. The data this way will be automatically stored in memory or on the local computer and will be periodically transferred to a given storage account by a different process. Data here is stored in a standard form in Azure Tables.

The third layer (containing the report role) reads this table and parses the log entries. It also puts the entries in chronological order and converts the data to a standard csv form. The test results are stored in merged and consolidated

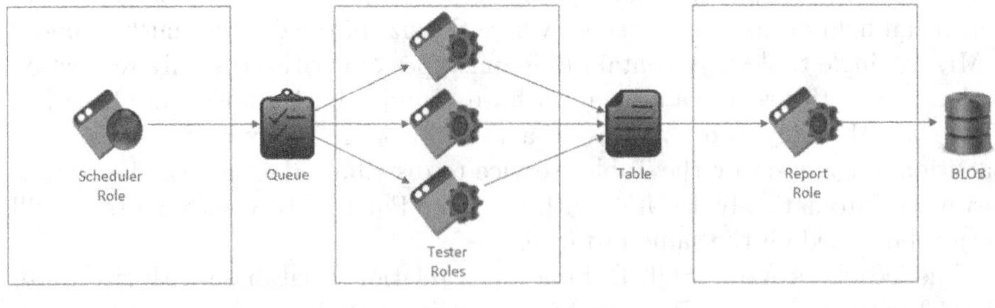

Fig. 1. Measurement architecture

format on a blob storage and are downloadable for later processing. It is not required to run the input and the output layer on the same cloud environment as the tester roles.

3.3 Azure Storage Evaluation

The first goal of our research was to test the capabilities of Azure Tables storage compared to Azure SQL Database. Azure Tables promises to provide a scalable NoSQL service which is simultaneously accessible by many clients. The known advantage of SQL Database is the full potential of a relational database and the ability to add additional indexes and constraints on the dataset. The other advantage is the power of SQL query language, with aggregate functions, wildcards and joins. The disadvantages are that in SQL only less than 100 GB data, has limitations in parallel executable queries and costs more in Azure. For the Tables storage one must choose the PartitionKey and RowKey carefully, because it determines the indexes, the placement of data and implicitly the performance of queries.

In our first experiment we converted a free flight database containing 10 million records of world-wide flight data to a proper Azure Tables and an SQL table format. We filled the database continuously with batch insertions and queried the database with four different types of queries simultaneously. The first type of query selected a single row from the database which was identified by the primary key in SQL and by the PartitionKey in Azure Tables. The second type of query selected rows via a portion of the primary key (PartitionKey) and via a portion of an other indexed row (the RowKey in Azure Tables). The third query selects multiple rows based only on an indexed row (in the case of Azure Tables only the RowKey). The fourth query selects rows based on a portion of the primary key and a non-indexed column.

The test was executed multiple times with different storage accounts across regions and we got similar results with each run. One instance was continuously inserting entities with multiple threads into the given data provider while other 10 instances where executing queries parallel. The experiments showed that in case of Azure tables the insert operation is slightly faster than in case of the SQL database (see Table 1). The first, second and the last query ran with no significant difference in case of a few rows and millions of rows. Both Azure SQL and tables performed well with few million rows.

Based on only these numbers only there is no real gain in using Azure Tables over SQL Database. But there is one more aspect that must be examined.

Table 1. SQL and Tables performance (Milliseconds/Returned row)

	Insert	1. Query	2. Query	3. Query	4. Query
Azure Tables	4	102	620	274399	56
SQL database	17	995	1015	14767	7.16

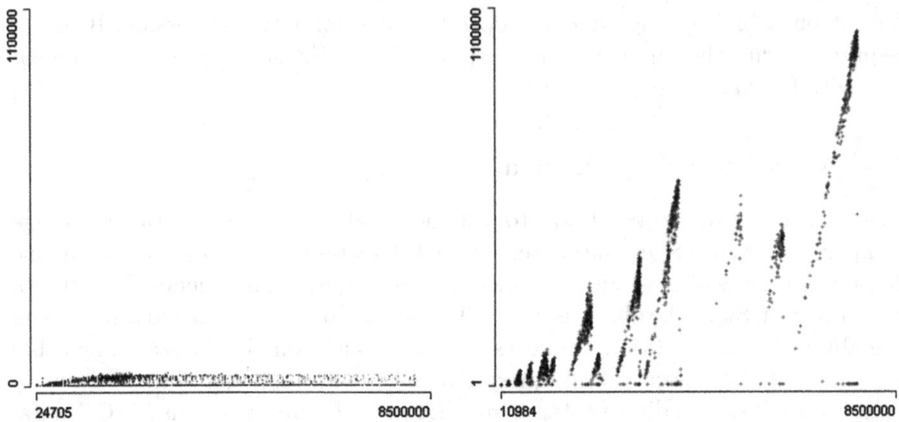

Fig. 2. *Left*: The 3. Query in SQL, horizontal: number of entities in database, vertical: retrieval time for a single row (in milliseconds), *Right*: The 3. Query in Azure Tables, horizontal: number of entities in database, vertical: retrieval time for a single row (in milliseconds)

The percent of failed requests in case of the SQL database was close to 10 % while the number of failed requests in case of the Tables test was under 0.1 %. One SQL Database can only execute 180 worker threads simultaneously, so all other connection attempts will be rejected. Based on the documentation Azure Tables detects denial of service attacks, so it may filter incoming requests, but in case of a slowly growing load it is able to server 20,000 entities per second, though we where not able to get near this limit (our maximum value was 9,000 entities per second) due to the lack of instances in our tests.

The other significant difference was the round trip times of the third query. The execution of this query required a full table scan and the size of the returned data is also significant. Azure Tables stores data on multiple servers and can only return data in batches containing a maximum of 1000 entries. A single query is executed relatively fast (within a few seconds), but with more requests the runtime of the query linearly grows with the number of parallel requests. If we look at the results when calling the SQL Database there is a significant difference. After a certain amount of data is in the database the retrieval time per row is relatively constant (see Fig. 2). The exact cause of the phenomenon requires more research, but it is highly probable that it's due to the fact that only 180 parallel queries are served by the database simultaneously, while during the high response times Tables was serving more than 2,000 concurrent requests.

4 Conclusion

In this paper we presented a framework that may be generally used to execute any kind of performance tests in the Azure cloud. The application may be deployed in the same data center as the system under test thus providing low network latency. The architecture provides option to scale to virtually any number of instances and can execute various types of tasks queued by a scheduler. The results are stored persistently in Azure Tables and are later processed by a different virtual machine.

By executing tests we ascertained that storing data in the novel storage options of Azure can be very efficient, the throughput of 20,000 entities per seconds can be reached with enough parallel endpoints with an ingress bandwidth of 10 GB/s and egress bandwidth of 15 GB/s. The greatest challenge is to choose a proper PartitionKey and RowKey for stored data. The format of the stored data specifies the queries that will run efficiently on the table service due to the fact that additional indexes can not be added to the entities. It is also very important to consider the PartitionKey in particular, because entities with different values may be stored on different instances, so queries that affect multiple PartitionKeys may have to call multiple server instances. The other important factor when choosing Azure Tables over SQL may be the fact that it supports extremely high values of simultaneous requests. On the other hand the Azure SQL Database provides multiple indexes, stored procedures and database constraints. If our dataset needs these kinds of capabilities then we should consider using a federated database to achieve higher degree of parallelism.

Acknowledgments. The work reported in the paper has been developed in the framework of the project "Talent care and cultivation in the scientific workshops of BME" project. This project is supported by the grant TÁMOP-4.2.2.B-10/1–2010-0009.

References

1. Amazon Web Services: Public data sets catalog, March 2013. http://aws.amazon.com/datasets
2. Armbrust, M., Fox, A., Griffith, R., Joseph, A.D., Katz, R.H., Konwinski, A., Lee, G., Patterson, D.A., Rabkin, A., Stoica, I., Zaharia, M.: Above the clouds: a berkeley view of cloud computing. Technical report UCB/EECS-2009-28, EECS Department, University of California, Berkeley, February 2009. http://www.eecs.berkeley.edu/Pubs/TechRpts/2009/EECS-2009-28.html
3. Barham, P., Dragovic, B., Fraser, K., Hand, S., Harris, T., Ho, A., Neugebauer, R., Pratt, I., Warfield, A.: Xen and the art of virtualization. In: Proceedings of the Nineteenth ACM Symposium on Operating Systems Principles, SOSP '03, pp. 164–177. ACM, New York (2003). http://doi.acm.org/10.1145/945445.945462
4. Hill, Z., Li, J., Mao, M., Ruiz-Alvarez, A., Humphrey, M.: Early observations on the performance of windows azure. In: Proceedings of the 19th ACM International Symposium on High Performance Distributed Computing, HPDC '10, pp. 367–376. ACM, New York (2010). http://doi.acm.org/10.1145/1851476.1851532
5. Jackson, K.R., Ramakrishnan, L., Muriki, K., Canon, S., Cholia, S., Shalf, J., Wasserman, H.J., Wright, N.J.: Performance analysis of high performance computingapplications on the amazon web services cloud. In: Proceedings of the 2010 IEEE Second International Conference on Cloud Computing Technology and Science, CLOUDCOM '10, pp. 159–168. IEEE Computer Society, Washington, DC (2010). http://dx.doi.org/10.1109/CloudCom.2010.69

6. Li, A., Yang, X., Kandula, S., Zhang, M.: Cloudcmp: comparing public cloud providers. In: Proceedings of the 10th ACM SIGCOMM conference on Internet measurement, IMC '10, pp. 1–14. ACM, New York (2010). http://doi.acm.org/10.1145/1879141.1879143
7. Menon, A., Santos, J.R., Turner, Y., Janakiraman, G.J., Zwaenepoel, W.: Diagnosing performance overheads in the xen virtual machine environment. In: Proceedings of the 1st ACM/USENIX International Conference on Virtual Execution Environments, VEE '05, pp. 13–23. ACM, New York (2005). http://doi.acm.org/10.1145/1064979.1064984
8. Microsoft research: cloud research projects, March 2013. http://research.microsoft.com/en-us/projects/azure/projects.aspx
9. Ostermann, S., Iosup, A., Yigitbasi, N., Prodan, R., Fahringer, T., Epema, D.: A performance analysis of EC2 cloud computing services for scientific computing. In: Avresky, D.R., Diaz, M., Bode, A., Ciciani, B., Dekel, E. (eds.) Cloudcomp 2009. LNICST, vol. 34, pp. 115–131. Springer, Heidelberg (2010)
10. Qiu, X., Ekanayake, J., Beason, S., Gunarathne, T., Fox, G., Barga, R., Gannon, D.: Cloud technologies for bioinformatics applications. In: Proceedings of the 2nd Workshop on Many-Task Computing on Grids and Supercomputers. MTAGS '09, pp. 6:1–6:10. ACM, New York (2009). http://doi.acm.org/10.1145/1646468.1646474
11. Vaquero, L.M., Rodero-Merino, L., Caceres, J., Lindner, M.: A break in the clouds: towards a cloud definition. SIGCOMM Comput. Commun. Rev. **39**(1), 50–55 (2008). http://doi.acm.org/10.1145/1496091.1496100

Performance Analysis of the Regional Grid Resources for an Environmental Modeling Application

Radoslava Hristova[1,2](✉), Sofiya Ivanovska[1], and Mariya Durchova[1]

[1] Institute of Information and Communication Technologies, Bulgarian Academy of Sciences, Acad. G. Bonchev str., bl. 25A, 1113 Sofia, Bulgaria
[2] Faculty of Mathematics and Informatics, Sofia University "St. Kliment Ohridski", 5 James Bourchier, 1164 Sofia, Bulgaria
radoslava@fmi.uni-sofia.bg

Abstract. The speed of execution of resource intensive application depends mostly on the performance of the underlying hardware and network infrastructure. The overall performance of complex Grid applications that include different types of processing in the same Grid job is difficult to predict reliably. In this paper we define several key performance indicators and collect data from the execution of a resource intensive environmental modeling application on the regional resources of the European Grid Infrastructure. The application is based on the Models-3 system, consisting of three components: meteorological preprocessor MM5, chemical transport model CMAQ and emission preprocessor SMOKE. The computations are resource intensive with respect to the input and output data which stress both the computational and data capabilities of the resource centers. In the paper we analyze the relative importance of these indicators and draw conclusions, regarding the optimal use of available resource centers.

1 Introduction

The speed of execution of resource intensive application in Grid environment depends mostly on the performance of the underlying hardware and network capacities. Even more, the overall performance of the complex Grid applications that include different types of processing (MPI processing, I/O data transfer processing, etc.) in the same Grid job is difficult to predict reliably. The computations of the resource intensive application can stress both the computational and data capabilities of the underling infrastructure. Example of resource intensive application is presented in [1]. The application requires "extensive simulations with up to date highly sophisticated numerical models" and therefore strongly rely on the computer resources of the Grid infrastructure. But is the Grid infrastructure well-balanced for resource intensive application? The answer of this question motivates us to analyze the performance of the regional Grid resources, especially for resource intensive applications. For the study we use an

I. Lirkov et al. (Eds.): LSSC 2013, LNCS 8353, pp. 507–514, 2014.
DOI: 10.1007/978-3-662-43880-0_58,

Table 1. Computer requirements for 3-day real time simulations [6]

Domain	Computational time	HDD(input/output) (GB)
CMAQ on 4 CPUS	8 h 00 min	28.84
CMAQ on 8 CPUS	6 h 00 min	28.84
CMAQ on 16 CPUS	4 h 00 min	28.84
CMAQ on 32 CPUS	3 h 00 min	28.84

environmental modeling application and the regional SEE-Grid infrastructure which is a part of the global European Grid Infrastructure. The environmental modeling application [2] is based on the Models-3 system, which consists of three components:

- MM5 [3] a fifth generation PSU/NCAR Meso-meteorological Model, used as meteorological pre-processor;
- CMAQ (Community Multiscale Air Quality System) [4] the Chemical Transport Model (CTM) of the system;
- SMOKE (Sparse Matrix Operator Kernel Emissions Modelling System) [5] the emission pre-processor of the Models-3 system.

The CMAQ component processes the input meteo data (MM5) and the input emission data (SMOKE) to produce output data. The CMAQ is the resource intensive part of the application with respect to the computational resources; the MM5 and the SMOKE are resource intensive with respect to the storage. In [6] the computer requirements are presented for 3-day real time simulations for the application (Table 1). This experimental data are achieved from running the application locally on the HPC cluster of IICT-BAS. The capabilities of the HPC cluster are described in [7]. The application was tested on 4, 8, 16, and 32 CPUs in parallel mode. The total input/output data of the application are 28.84 GB. As it can be seen, the computer resource requirements are quite big. In [6] the authors also conclude that "the successful execution of the jobs on the Grid is quite probable". However, such tests were not done. In order to test the performance of the regional Grid infrastructure we define several key performance indicators for the environmental modeling application and collect data from its execution into the Grid. The key performance indicators (KPI) [8] are key indicators that are used to monitor the overall performance of a system. In the paper we analyze the relative importance of the defined indicators and draw conclusions, regarding the optimal use of available resource centers.

2 Running Application on the Regional Grid

The environmental modeling application is compiled and implemented with MPI and requires Intel compiler and MPI libraries in order to be run. The MPI implementation is used to achieve better performance for the application. The application requires:

- mpiexec-0.83-intel a replacement program for the mpirun script. It is used to initialize a parallel job from within a PBS batch [9];
- NetCDF (network Common Data Form) a set of software libraries and machine-independent data formats that support the creation, access, and sharing of array-oriented scientific data;
- Intel Compiler (Version 11.1) a Intel compiler for Linux that includes C compiler, C++ compiler, and Fortran compiler;
- Intel MPI Library for Linux (Version 4.0.0.028).

Taking into account these requirements an appropriate JDL file and bash scripts were developed. Because of the resource intensity of the application, especially for the input/output data operations, the traditional job submission cannot be applied for the application. The maximum size of the InputSandbox and the OutputSandbox files for a job usually is limited to 100 MB. The size can vary according to the WMS settings, but nevertheless it is far away from 8 GB (which is the approximate size of the archive with all necessary data and libraries for the application). In order to submit the environmental modeling application into the Grid we use the following approach:

1. Finding appropriate Grid sites, satisfying the requirements of the application
2. Creating archive with all necessary data and libraries (approximate size of the archived file is 8 GB)
3. Copy the archived file to the site storage (SE)
4. Arrange the KPIs into a bash script
5. Submit the script into the Grid
6. Copy the archived file from site storage to the worker node (WN)
7. Execute the application
8. Collect data for the KPIs
9. Archive the output data and copy them back to the site storage (SE)
10. Get the output data

As it can be seen the essential part of the successful job execution is the successful file transfer.

The Grid sites in the regional Grid infrastructure that respond to the above JDL requirements are:

Table 2. Sites' grid resources

Site name	Computing element (CE)	Storage element (SE)
AEGIS01-IPB-SCL	ce64.ipb.ac.rs:8443/cream-pbs-seeGrid	dpm.ipb.ac.rs
BG01-IPP	cr1.ipp.acad.bg:8443/cream-pbs-envir	se001.ipp.acad.bg
GR-10-UOI	cream01.grid.uoi.gr:8443/cream-pbs-envir	se01.grid.uoi.gr
HG-06-EKT	cream02.athena.hellasgrid.gr:8443/cream-pbs-envir	se01.athena.hellasgrid.gr
MK-03-FINKI	ce.hpgcc.finki.ukim.mk:8443/cream-pbs-env	se.hpgcc.finki.ukim.mk

- AEGIS01-IPB-SCL - Located in Belgrade, Serbia;
- BG01-IPP - Located in Sofia, Bulgaria;
- GR-10-UOI - Located in Ioannina, Greece;
- HG-06-EKT - Located in Athens, Greece;
- MK-03-FINKI - Located in Skopje, Macedonia;

The overall information for the Grid resources of the sites is shown in Table 2. For the job submission we are using the env.see-grid-sci.eu virtual organization.

3 KPIs Definition

The key performance indicators (KPI) are key indicators that are used to monitor the overall performance of a system. The KPIs are extremely useful for detecting and distinguishing the performance issues in a system or in an infrastructure. Usually, KPIs measure the performance of the system according to previously defined metrics. In the context of the Grid infrastructure, example KPIs could be the average time for job submission, the average time for job execution, the number of failed jobs on a given site, etc. Taking into account the approach mentioned above we define KPI measurement for each subtask of the approach. The subtask description and according KPI measurement are presented in Table 3.

For each subtask we measure the average time necessary for task finalization. Different scenarios are possible. For the purposes of the current investigation we

Table 3. KPIs definitions

Subtask description	KPI measurement
Copy the archived file to the site storage	AVG time to transfer
Copy the archived file from site storage to the worker node	AVG time to transfer
Unzip the archived file	AVG time to unzip the file
Execute the CMAQ application	AVG time for execution
Archive the output data	AVG time for archived
Copy the archived data back to the site storage	AVG time to transfer
Get the output data	AVG time to transfer

Table 4. KPIs scenarios

KPI name	KPI measurement	Scenarios
UI_FT_SE	AVG time for file transfer	From UI to the site SE
UI_NT_SE	AVG speed for file transfer	From UI to the site SE
SE_FT_WN	AVG time for file transfer	From site SE to the WN
WN_UTAR	AVG time to unzip the file	Unzip the file on WN
WN_EX_JB	AVG time for execution	CPU 4, CPU 8, CPU 16
WN_TAR	AVG time for archived	Make archive on data output
WN_FT_SE	AVG time for file transfer	From WN to the site SE
SE_FT_UI	AVG time for file transfer	From site SE to the UI

Table 5. Average time for each KPI in minutes

N	Site name	AEGIS01-IPB-SCL	BG01-IPP	GR-10-UOI	HG-06-EKT	MK-03-FINKI
1	UI_FT_SE	4	7	5	14	7
2	SE_FT_WN	5	5	4	3	13
3	WN_UTAR	13	7	15	14	4
4	WN_TAR	13	29	30	15	22
5	WN_FT_SE	11	14	10	12	13
6	SE_FT_UI	5	6	4	3	15

use the KPIs and the scenarios presented in Table 4. All of the measurements in each scenario, except UI_NT_SE are in minutes, rounded to integer. Also other KPI's can be defined, for example measuring the time for application execution on 16 CPUs and compare it with the time for execution on 2 WN with 8 CPUs. This however is a matter of further investigation.

4 Performance Analysis of the Regional Grid

In Table 5 the results from measurement of all KPIs are shown: for file transfer from the UI - gw.ipp.acad.bg to the site storage of each sites (UI_FT_ST), for file transfer from the site storage to the site WN of each site (SE_FT_WN), for the time which is needed to unzip the archived file of the application on the WN of each site (WN_UTAR), for the time which is needed to create archive of the output file on the WN (WN_TAR), for file transfer from the WN to the site storage of each sites (WN_FT_ST) and for the file transfer from the site storage to the UI - gw.ipp.acad.bg (SE_FT_UI).

The general observations are that the average time for file transfer from the UI to the site storage of each site is around 7 min. The file transfer is an operation that depends on the network and the site storage state. Very slow file transfer can be caused by some temporary network issues or storage problems. An example for detected temporary network issues is the site HG-06-EKT. The average time for file transfer from the UI to the SE of the site is around 14 min (UI_FT_ST). From the second indicator (SE_FT_WN) however, we can see that the time for file transfer from the SE to the WN for the site is around 3 min. These values show that SE issues seem unlikely. In order to prove network issues we do further tests and measure the speed for file transfer from the UI to the site storage of each site. The results (Table 6) shows us that at the time the test were done the network transfer from UI to the SE of the site HG-06-EKT indeed was slower than the network transfer to the other sites.

Furthermore, the average time for file transfer of each SE to the according WN of the site (SE_FT_WN) is around 6 min (Table 5). However, this is not the case for the site MK-03-FINKI. This is another indication for existing temporary network or HDD problems. The difference from the previous case is that these issues concerns the local sites' infrastructure. They are three possible reasons

Table 6. Network performance (MB/sec)

Site Name	AEGIS01-IPB-SCL	BG01-IPP	GR-10-UOI	HG-06-EKT	MK-03-FINKI
UI_NT_SE	60	22	25	12	20

Table 7. Overall time - I/O operations

Time	AEGIS01-IPB-SCL	BG01-IPP	GR-10-UOI	HG-06-EKT	MK-03-FINKI
Overall from UI to WN	22	19	24	31	24
Overall to WN from UI	29	49	44	30	50
Overall Time	51	68	63	61	74

for that: SE issue, WN issue or network connectivity problem between the SE and the WN of the site. The values for the third KPI (WN_UTAR) for the site show that temporary WN problem of the site seems unlikely. From the first KPI (UI_FT_SE) value for the site we can conclude that SE issues also is not the reason. Therefore, temporary network connectivity problems between the SE and the WN of the site are very likely.

The measured values for indicators (3) and (4) show us that the process of archiving and unzipping the archive file to the site WNs is time consuming. We cans say that on the same WN, the time for creating archive is approximately twice more than the time which is needed to unzip the archive. From the results - values for the indicators (1), (2) and (6) for all the sites, except the two ones with detected issues, we can conclude that the SE file transfer is well-balanced.

If we analyze the overall times in Table 7 we can say that the file transfer from the WN of each site to the UI is around 40 min (twice more than the overall time for file transfer from the UI to the WN). The whole overall time for the transfer from the UI to the WN and back is around 63 min. If we reconsider the data from Table 1 and the time of 63 min, we can see that the I/O operations are essential part of the application processing into the Grid. For the case of 32 CPUs we can say that it is 25 % from the whole time, which is needed for job execution. This is an indication that the current grid infrastructure is not well balanced for resource intensive applications with respect to the I/O data.

5 Detected Issues

All the issues that were detected during the test can be due to the temporarily problems in the Grid infrastructure. Nevertheless, they are mentioned in this section in order to show the relative importance of the KPIs defined in previous sections. During the tests we come across on two problems. The first problem the time which have to wait in order to submit a job in to the Grid. The results are generalized for each site in Table 8. From the data we can say that the sites GR-10-UOI and HG-06-EKT are most overloaded. We have to mention that all

Table 8. Waiting time for job submission issue

Site name	AEGIS01-IPB-SCL	BG01-IPP	GR-10-UOI	HG-06-EKT	MK-03-FINKI
Waiting time	50	2	73	100	74

Table 9. Detected HDD issues

WN (HG-06-EKT)	SE_FT_WN	WN_UTAR
wn126	12	39
wn126	6	38
wn118	6	14
wn126	4	43
wn138	13	23

of the tests were done with env.see-grid-sci.eu credentials, which can affect the result. The second problem we detect is that two or more jobs were scheduled on the same WN. This problem is more essential and can reflect mostly on the HDD performance, but also and to the network performance. The results are shown in Table 9. We can see that three of the same application jobs are scheduled on the same WN. Respectively, this reflects on the KPIs (WN_UTAR) and (WN_TAR). Compared with the KPIs for the other WNs (wn118, wn138) the time for (WN_UTAR) on the WN (wn126) is two times more - around 40 min.

6 Conclusions

The defined indicators have their meaning in the terms of Grid computing. They have been selected for the specific application, taking into account its specific structure, but could be applied with minor modifications to other similar applications. The measurement of the KPIs gives the idea of the current state of the considered Grid sites, and thus provides guidelines to avoid problems, associated with the instability of the infrastructure. The current regional Grid infrastructure is not well balanced for resource intensive applications with respect to the I/O data. Nevertheless, we can conclude that the overall performance of the regional Grid for resource intensive applications is satisfactory.

Acknowledgments. The research work reported in the paper is partly supported by the project AComIn - Advanced Computing for Innovation, grant 316087, funded by the FP7 Capacity Programme (Research Potential of Convergence Regions), and by the National Science Fund of Bulgaria under Grants DCVP02/1 (SuperCA++).

References

1. Ganev, K., Syrakov, D., Prodanova, M., Hristov, H., Atanasov, E., Gurov, T., Karaivanova, A., Miloshev, N.: Grid computing for air quality and environmental studies in Bulgaria. In: Proceedings of 23rd EnviroInfo 2009 Conference, Berlin, 9–11 Sept 2009, pp. 147–155 (2009)
2. Ganev, K., Syrakov, D., Gadzhev, G., Prodanova, M., Jordanov, G., Miloshev, N., Todorova, A.: Joint analysis of regional scale transport and transformation of air pollution from road and ship transport. In: Lirkov, I., Margenov, S., Waśniewski, J. (eds.) LSSC 2009. LNCS, vol. 5910, pp. 180–187. Springer, Heidelberg (2010)
3. Grell, G.A., Dudhia, J., Stauffer, D.R.: A description of the Fifth Generation Penn State/NCAR Mesoscale Model (MM5), NCARTechnical Note, NCAR TN-398-STR, 138 pp. (1994)
4. Byun, D., Ching, J.: Science Algorithms of the EPA Models-3 Community Multiscale Air Quality (CMAQ) Modeling System, EPA Report 600/R-99/030, Washington, DC (1999)
5. CEP, Sparse Matrix Operator Kernel Emission (SMOKE) Modeling System, University of Carolina, Carolina Environmental Programs, Research Triangle Park, North Carolina (2003)
6. Gadzhev, G.K., Ganev, K.G., Miloshev, N.G., Syrakov, D.E., Prodanova, M.: Numerical study of the atmospheric composition in Bulgaria. Comput. Math. Appl. **65**(3), 402–422 (2013)
7. Atanassov, E., Gurov, T., Karaivanova, A.: Capabilities of the HPC cluster at IICT-BAS. J. Automatika and Informatika **2**, 7–11 (2011). ISSN: 0861–7562 (in Bulgarian)
8. Hristova, R.: Monitoring of business processes in the EGI. In: Conference Proceedings of the 6th International Conference ISGT, 01–03 June 2012, Sofia, Bulgaria, pp. 294–300 (2012)
9. MPIEXEC Tutorial. https://www.osc.edu/djohnson/mpiexec/

Framework for Genetic Algorithms Using Pilot Jobs in Adaptive Grid Workflows

Boro Jakimovski(✉), Bojan Ilijoski, Goran Velinov, and Dragan Sahpaski

Faculty of Computer Science and Engineering,
Ss. Cyril and Methodius University in Skopje, Skopje, Macedonia
boro.jakimovski@finki.ukim.mk

Abstract. The performance of Grid applications may be very unstable, especially when using workflows for job distribution. This is mainly due to the Grid overheads, like scheduling and queuing, introduced before the job is executed on a worker node. Optimization problems using Genetic Algorithms (GAs) can be easily and efficiently implemented on Grids using Grid workflows. Due to the file dependencies introduced in the Grid workflows for GAs, mainly for genetic material interchange, these overheads are cumulative and thus very noticeable. This problem is also very evident when the jobs are short compared to the Grid overheads, i.e. the job spends more time waiting in a queue to be executed than the execution itself.

In this paper we introduce a framework that enables users to easily utilize the Grid infrastructure for their optimization using GAs. It allows a user to preallocate certain number of pilot jobs, and also to dynamically manage their number for optimal availability of resources during the optimization process. In this way, once an application starts to execute the workloads, it will have at least one available pilot for execution of pooled tasks. This introduces better utilization of the Grid resources, as well boost the confidence in the infrastructure from users point of view.

Keywords: Parallel genetic algorithms · Grid infrastructure · Pilot jobs

1 Introduction

The Genetic Algorithms (GAs) are widely used for problems which can be solved by simulation of the process of natural evolution. One feature of these GAs is that they can run in parallel. This makes them suitable for implementation on a Grid infrastructure [2]. A Grid computing allows us to use resources from multiple administrative domains in order to get better performances for executing complex tasks. But naturally, the resources are shared by many users and applications and always there are large number of tasks waiting to use these resources. For this purpose, there are several layers of schedulers at different levels: Cluster, Grid and Workflow. Because of this, one task might spend more time waiting to be executed in queues, than its execution time. In cases where applications

I. Lirkov et al. (Eds.): LSSC 2013, LNCS 8353, pp. 515–522, 2014.
DOI: 10.1007/978-3-662-43880-0_59,

consist of many tasks (dependent or independent), and most of the tasks are spending their time waiting instead of executing, the overall application is very inefficient.

Our previous research was focused on overcoming this problem of efficient execution of Grid Genetic Algorithms (GGAs) by using adaptive workflows [5,10]. In this paper we further optimize the performance of the Adaptive GGAs (AGGAs) by introducing a framework that enables an easy implementation of pilot jobs [3] for its execution. Pilot jobs are increasingly used to improve scheduling and reliability on production grids [11,12]. For performance and reliability reasons, pilot jobs install the user's own job management system on the resources provisioned from the grid, and then execute user tasks through this system. Some of the common pilot job tools are DIANE [13], glideinWMS [14], GridBot [15] and DIRAC [12]. Pilot jobs are very useful for achieving better execution times in several cases of Grid usage, one being workflow applications that have large number of consecutive small jobs. These applications are mainly parallel algorithms having very coarse level of parallelism, implemented as Workflow grid applications with large number of small jobs. Due to the need for inter job communication, these jobs need to finish, exchange files and then start all over again. The pilot job execution of such scenario helps to avoid the subsequent Grid job overheads when inter job communication is needed.

AGGAs are such applications where pilot jobs can be successfully utilized. When we submit an optimization problem, we know the number of tasks that will be executed in parallel at one time. This helps us to plan to occupy a sufficient number of worker nodes, so before the optimization tasks of the GGAs are executed, we submit a certain number of pilot jobs. This makes the resource allocation fairness and efficiency better and easier to manage.

The rest of this paper is organized as follows. Section 2 describes how GGAs can run using adaptive workflow. In Sect. 3 we present an architecture of the initial framework. Section 4 presents the results produced by using this framework and comparison between time required for execution of the program with and without pilot jobs. The last section, Sect. 5, concludes this paper and gives future development issues.

2 Background Work

The Parallel GAs (PGAs) are extensions of the single population GAs. The well-known advantage of PGAs is their ability to perform speciation, a process by which different subpopulations evolve in diverse directions simultaneously. This kind of PGA is suitable for message passing parallel environments. Another implementation of PGAs is to implement them as GGAs using Grid workflows for distribution and file transfer for chromosome migration. Grid workflows represent a network of interconnected Grid jobs. Interconnected jobs are data dependent, i.e. data output from one job is fed into the dependent jobs for further processing. This model of parallel execution allows easy and efficient use of the Grid for data parallelization problems that can be easily divided into parallel independent jobs [5].

As shown in [10] the workflow execution performance is very influenced by the Grid non-deterministic behaviour. Grid represents an environment that behaves very chaotically, and neither users nor services can predict its behaviour regarding the job queuing times. In [6] they identify waiting times at batch schedulers as "the most relevant and prevalent source of performance variability". This means that big Grid job overheads due to queuing times and Grid heterogeneity influences greatly on workflow execution by forcing dependent jobs to wait for the slowest job (job that due to Grid overheads finishes the latest on an overloaded cluster). This greatly impacts the level of concurrency, and thus the speed-up factor of the parallelization.

In our previous research regarding GGAs [5] we have elaborated the different approaches for overcoming the problem of Grid overheads. Our solutions has evolved in several phases, every time obtaining better performance and new knowledge regarding the Grid workflow execution. The first GGA implementation was by using DAG workflows, very soon followed by the second implementation that improved the DAG by optimization of the job description in the GGA resulting in twice as less dependant jobs. The third implementation used the High-Level Petri nets (HLPN) for the description of Grid Workflows [1], that enabled more flexibility in the definition of the workflow [1,8]. The HLPN implementation of GGA was named Adaptive GGA (AGGA). This paper focuses on better execution model for AGGA by using Pilot jobs.

The main advantage that the HLPN model gives is the ability to define a non-deterministic connection between jobs, as opposed to the rigid DAG workflows. This means that in the AGGA after the submission of the initial wave of *Breeder* jobs, the workflow does not need to specify which particular jobs need to finish, for the migration to take place. Hence the new submission of *Migration/Breeder* jobs will not be influenced by the stalling jobs which was the case in DAG workflows. This kind of non-determinism in the execution strategy can be used due to the specific nature of the PGAs to be able to produce output and continue to work with partial or out of order results.

Another ability of the PN workflow model is the availability of conditional loops, which can enable optimizations to be executed until certain threshold is met, as opposed to DAGs that have very static nature. In this paper we used a simpler finishing condition, counting number of finished islands, in order to be able to compare the performance of the pilot and regular job submission. The execution of the algorithm is described in detail in [5].

3 The Framework

As mentioned in the introduction, there are several projects that enable users to use Pilot jobs. We have decided to implement a lightweight, per user Pilot system in order to be able to better integrate with the Adaptive workflow. Existing systems are VO oriented, and give small and limited workflow execution possibilities.

The system is made from three internal parts: *Pilot job*, *Task*, and *Manager*, and one external module (not part of the framework), an *Application*.

A *Pilot job* [3] is a job which is submitted to the Grid, that is intended to occupy a computational resource (worker node) and does not execute any computational task. It is a job that serves as an already allocated worker node that waits for a task for execution. When a task is assigned to a pilot job, it is immediately executed, bypassing queuing and workload management services.

An *Task* is a computational task that is part of a bigger problem that needs to be executed on the Grid. It is equivalent to a Grid job, for the computational part, but is executed on the Grid using a *Pilot job*. It can be defined as a single "Job" type JDL [4] Grid job and has the following attributes: InputSandbox, OutputSandbox and ApplicationID.

An *Application* is the real problem that needs to be solved on the Grid. It consists of many *Tasks*, that need to be submitted and computed on the Grid. Users can submit multiple *Applications* at the same time. The Application workflow is independent from the framework and is defined by the problem.

The last part of the framework is the *Manager*. The *Manager* is the core of the system and it is responsible for creation of pilot jobs, monitoring pilot jobs, matchmaking of tasks and pilot jobs, and the applications. It consists of three parts: web services, database service, and matchmaker. The web services are the interface of the *Manager* with the other components. The database service is used for application, pilot and task persistence into a database. The third part, the matchmaker, is used as a gateway between applications and pilots. It is currently implemented to work as per Application matchmaking using the policy last come - first served.

The framework is implemented in Java as a Java web application and all communications are implemented as Web Services (WS). There are 3 main user oriented scenarios in the framework: *Pilot job creation*, *Task submission*, and *Results retrieval*. All scenarios are depicted in Fig. 1. As previously mentioned, and also from the picture, the *Application* acts as a client in the framework and initiates all three web services.

The first web service, *Pilot job creation*, is depicted as the scenario *createPilots(number)*. This web service contacts the Manager as an asynchronous call, and instructs the *Manager* to create *number* pilot jobs for the purpose of the running application. As a consequence, the *Manager* starts the creation of the pilot jobs by contacting the Grid infrastructure using the gLite WMS service. As a result it retrieves a Grid job identifier that is later used for monitoring purposes. The Pilot jobs that are submitted by the WMS service, run the *Pilotjob* part of the Framework. When it starts on a Grid Worker Node (WN), it immediately reports to the Manager using a WS. At this moment, the Manager marks this pilot job as "available", and the Pilot can be mapped with a pending *Task*.

The second web service *Task submission* is depicted as the scenario *submitTask*. This is also an asynchronous call to the Manager and carries a single task that needs to be executed by the "available" Pilot jobs. The Manager collects all submitted tasks and pools them for execution to the specified Application. The matching between the pilots and the tasks are done only by application ID that is specified both at Pilot job creation and at Task submission. The *Pilot jobs*

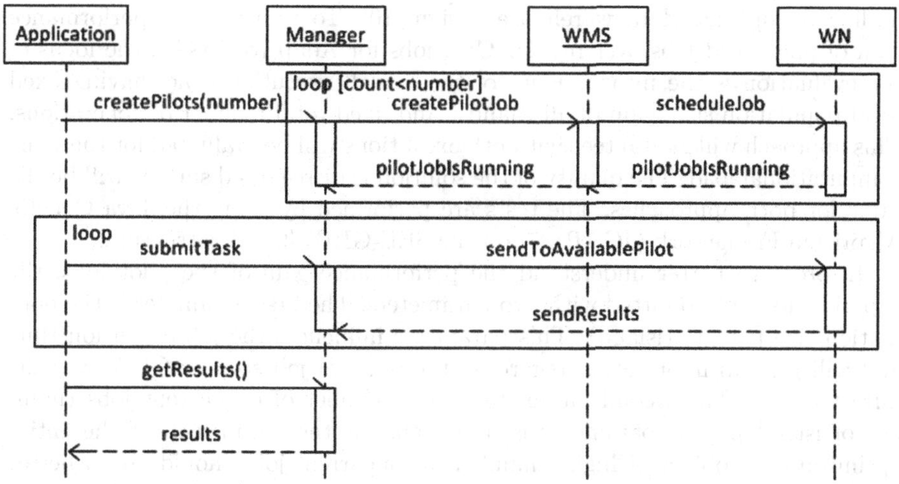

Fig. 1. The Sequence diagram of the operations that are performed in the framework

pull *Tasks* from the *Manager*. This operation is initiated by the "available" pilot jobs, and they contact the *Manager* on a fixed time interval. If there are no tasks to be executed, the manager instructs the pilot job to try again later. Otherwise it returns the web service call by sending a selected *Task*.

The *Tasks* are represented by a single .zip file which must include .jdl file. The *Manager* does not manipulate the task. It only resubmits it to the *Pilot job* for execution. Once the *Pilot job* receives the *Task*, it unpacks the content and parses the .jdl file in order to know which files to execute and which files it should send back as a result from that execution process. Then the *Pilot job* starts the execution and reports its new status to the *Manager* as "running". When the execution finishes with success, the results are packed together, and the achieve is sent back to the *Manager*, the *Task* is marked as "done", and the *Pilot* reports again to the *Manager* as "available".

The third web service *Results retrieval* is depicted as the scenario *getResults*. This scenario can be used for both for *Task* status retrieval, and for results retrieval. The Application submits to the WS the task ID and application ID and retrieves the results.

The *Application* implements all workflow logics. It can be a Grid Workflow Management System that executes the Application workflow and instead of submitting jobs directly to the Grid WMS, it can utilize the Pilot job framework for its execution.

4 Experimental Results

In this section we present the results obtained by execution of the framework on a real problem. The evaluation is based on real optimization problem for finding an optimal data warehouse design [10]. To measure the performance gain of pilot Grid jobs over regular Grid jobs for Adaptive GGAs, the focus of the evaluation is the measurement of the overall execution time, having fixed total population size (sum of all islands) and fixed total number of generations. This approach will guarantee that both executions will be evaluated for the same computational load. The quality of the solution is disregarded since it will be the same for both approaches. The tests are performed by using the Java Genetic Algorithm Framework (JGAP) [7] on the SEE-GRID [9] infrastructure.

In order to better understand the performance gain of the pilot jobs, the experiments are submitted with two parameters. The first parameter is the population size per job (island). This parameter influences the job execution time and will give an information regarding the optimal job size for GGA over the infrastructure. The second parameter is the number of concurrent jobs (number of islands). This parameter is important for the parallelism of the entire optimization problem. A higher number of concurrent jobs should give a better speed-up, but also more jobs introduces more overhead from the Grid. Finding an optimal level of concurrency is crucial for achieving best performance.

For the performance comparison, we have chosen to solve the optimization problem having a total population of 1000 chromosomes iterated for 3000 generations over the population. In order to find the best performance of the optimization process, we keep the product of the two parameters, population size per island and number of islands, to be always 1000. We have evaluated the problem for the following cases: 5 jobs with 200 chromosomes per job, 10 jobs with 100 chromosomes per job, 20 jobs with 50 chromosomes per job and 40 jobs with 25 chromosomes per job. The experiments for all cases are executed in parallel at the same time, for both pilot and regular jobs. Using this approach for experiment execution, we will evaluate the performance gain between both approaches having the same Grid resources, and also the jobs in all cases will compete for the same resources at the same time.

In Fig. 2 we can see the mean time for execution, and its standard deviation, of the optimization of 1000 chromosomes in all cases, for both pilot and regular jobs. The results obtained, depict an evident performance gain when using pilot jobs. As expected, the overall time of the executions for regular jobs decreases until it reaches a point where waiting time is longer than the execution time. At this point the overall execution time start to increase as the concurrency increases. What is even more important, is that using pilot jobs, the deviation of execution time is dramatically much smaller than the regular job cases. For the pilot jobs, the deviation is approximately 4 min as opposed to 3 hours for the regular jobs.

In order to see the pilot jobs influence of the execution time, in Fig. 3 we represent the cumulative time in respect to the number of finished jobs. The depicted experiments presents the cases with 10 concurrent jobs each having 100 chromosomes and 20 concurrent jobs each having 50 chromosomes. The total number of jobs is 30 for the first case and 60 for the second case. As it can be concluded, the execution time for the pilot jobs is very close to a straight line,

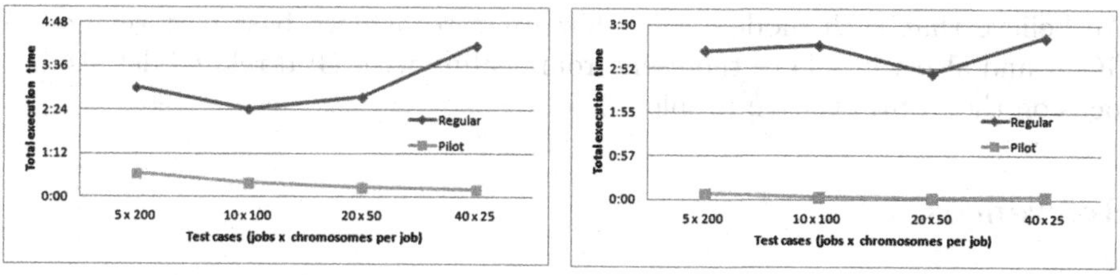

Fig. 2. Execution time with and without pilot jobs: (a) mean time (b) stdev

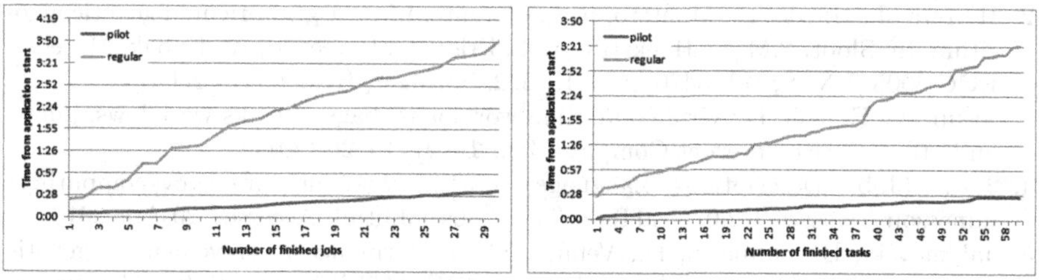

Fig. 3. (a)10 parallel jobs, 100 gen. per job (b)20 parallel jobs, 50 gen. per job

without big jump, that results in more stable and predictable total optimization time. On the other hand, the regular jobs line experiences large jumps from time to time, that finally result in less predictable overall time and the final results have large standard deviation.

The nature of the pilot jobs is to occupy resources on the infrastructure and wait for the tasks, this leads to inefficient utilization of resources. For this reason, we have measured the average idle time per pilot job. The results show that the average idle time for each of the four testing cases (5, 10, 20 and 40 concurrent) pilot jobs are 7 min, 1 min, 20 s and 1 min respectively. This is very low compared to the overall execution time, or on average 5 %.

5 Conclusions and Future work

In this paper we have shown an implementation of GGA using Pilot jobs in Adaptive Grid Workflows. We have defined a Framework for easy implementation of pilot jobs, and used it to implement the GGAs. The experimental results are very promising and show that the new approach using pilot jobs gives both better results in relation to the overall performance of the optimizations and better stability and predictability of the execution time. Also we have analysed the dependence of the execution time on the number of the parallel tasks and the job size (single task execution time). Even though these parameters are very influenced by the current load on the Grid infrastructure, one can easily predict the state of the infrastructure and dynamically adjust the parameters.

Our future work will be focused on defining a model for adaptive parameter adjustment depending on the past execution statistics of a single workflow.

We believe that such models will allow users to achieve best performance of GGAs and also will relieve the users from dealing with Grid related details and focus on their optimization problems.

References

1. Alt, M., Hoheisel, A., Pohl, H.W., Gorlatch, S.: Using high level Petri-nets for describing and analysing hierarchical grid workflows. In: Proceedings of the CoreGRID Integration Workshop, pp. 267–276, Pisa, Italy, Nov 2005
2. Herrera, J., Huedo, E., Montero, R.S., Llorente, I.M.: A grid-oriented genetic algorithm. In: Sloot, P.M.A., Hoekstra, A.G., Priol, T., Reinefeld, A., Bubak, M. (eds.) EGC 2005. LNCS, vol. 3470, pp. 315–322. Springer, Heidelberg (2005)
3. Iosup, A., Epema, D.: Grid computing workloads: bags of tasks workflows, pilots, and others. IEEE Internet Comput. Mag. **15**(2), 19–26 (2011)
4. JDL (Job Description Language). http://www-numi.fnal.gov/computing/minossoft/releases/R2.0/GridTools/docs/jobs_jdl.html. Accessed 20 Apr 2013
5. Jakimovski, B., Sahpaski, D., Velinov, G.: Performance improvement of genetic algorithms by adaptive grid workflows. In: SYNACS '09, pp. 221–228, Timisoara, Romania, Sept 2009
6. Khalili, O., He, J., Olschanowsky, C., Snavely, A., Casanova, H.: Measuring the performance and reliability of production computational grids. In: Proceedings of the 7th IEEE/ACM International Conference on Grid Computing, pp. 293–300 (2006)
7. Meffert, K.: JGAP - Java Genetic Algorithms and Genetic Programming Package. http://jgap.sf.net
8. Pellegrini, S., Giacomini, F.: Design of a Petri net-based workflow engine. In: Proceedings of the 3rd International Conference on Grid and Pervasive Computing Workshops 2008, pp. 81–86, Kunming, May 2008
9. South Eastern European GRid-enabled eInfrastructure Development (SEE-GRID). http://www.see-grid.org/. Accessed 20 Apr 2013
10. Velinov, G., Jakimovski, B., Cerepnalkoski, D., Kon-Popovska, M.: Improvement of data warehouse optimization process by workflow gridification. In: Atzeni, P., Caplinskas, A., Jaakkola, H. (eds.) ADBIS 2008. LNCS, vol. 5207, pp. 295–304. Springer, Heidelberg (2008)
11. Luckow, A., et al.: Towards a common model for pilot-jobs. In: Proceedings of the 21st International Symposium on High-Performance Parallel and Distributed Computing, HPDC12, pp. 123–124, Delft, The Netherlands, June 2012
12. Tsaregorodtsev, A., et al.: DIRAC3 - the new generation of the LHCb grid software. J. Phys. Conf. Ser. **219**(6), 062029 (2009)
13. Sarrut, D., Guigues, L.: Region-oriented CT image representation for reducing computing time of Monte Carlo simulations. Med. Phys. **35**(4), 1452–1463 (2008)
14. Sfiligoi, I.: glideInWMS: a generic pilot-based workload management system. J. Phys. Conf. Ser. **119**(6), 062044 (2008)
15. Silberstein, M., et al.: Gridbot: execution of bags of tasks in multiple grids. In: Proceedings of the Conference on High Performance Computing Networking, Storage and Analysis, pp. 1–12 (2009)

Solvation of Fluoroform in Liquid Krypton: A Theoretical Cryospectroscopy Approach on a HPC Environment

Emilija Kohls[1], Dragan Sahpaski[2(✉)], Anastas Mishev[2], and Ljupco Pejov[1]

[1] Institute of Chemistry, Faculty of Natural Sciences and Mathematics, University "Ss. Cyril and Methodius", Skopje, Macedonia
[2] Faculty of Computer Science and Engineering, University "Ss. Cyril and Methodius", Skopje, Macedonia
dragan.sahpaski@finki.ukim.mk

Abstract. The anharmonic CH stretching vibrational frequency was calculated for a dilute solution of fluoroform (CF_3H) in liquid krypton at 131 K from classical Monte Carlo (MC) simulations followed by electronic structure calculations (at various levels of theory, including B3LYP, MP2 etc.) for small clusters including fluoroform and few solvent atoms residing in its neighborhood. Nuclear dynamics calculations were also quantum mechanical, i.e. the vibrational Schrodinger equation was solved at grid of points representing an intersection through the vibrational potential energy surface of fluoroform, corresponding to the CH stretching motion. The calculated Raman bands are compared with experimental results, and an in-depth physical insight is gained into the factors influencing the CH stretching frequency shifts upon solvation. On the basis of Kitaura-Morokuma and RVS SCF analysis of the vibrational potentials, it was concluded that the solvent electrostatics influence (both classical and non-classical) would induce frequency red shifts, while the exchange Pauli repulsion induces frequency blue shifts. This robust and complex computational methodology was implemented in a HPC environment.

Keywords: Monte Carlo · Condensed phases · Liquids · Fluctuating environments · Fluoroform · Liquid krypton · Intermolecular interaction potentials · Hybrid statistical physics - quantum mechanical approach

1 Introduction

Properties of molecules in condensed phases have always attracted interdisciplinary scientific attention due to numerous reasons. First, nearly all of the relevant processes to both chemistry and chemical engineering, as well as in biomedical

This paper is based on the work done in the framework of the HP-SEE FP7 EC funded project.

I. Lirkov et al. (Eds.): LSSC 2013, LNCS 8353, pp. 523–531, 2014.
DOI: 10.1007/978-3-662-43880-0_60,

sciences occur in condensed phases (liquids in particular). Second, understanding of the solute-solvent interactions on a profound quantum-theoretical level is of certain fundamental significance. The simplest, but perhaps still most widely used approach to study such systems has been based on exploration of potential energy surfaces of clusters containing a solute and several solvent molecules, which mimic the condensed phase. However, it has certain drawbacks and inherent limitations. One of the most relevant gross errors inherent to this "supermolecular" approach lies in the fact that considering only the PES of such complex system actually implies a complete neglect of its dynamical properties. If one is primarily focused on technological applications of condensed-phase systems, then it is important to be able to understand their properties at finite temperatures. This is so since virtually all devices operate at finite temperatures. As mentioned before, however, virtually all studies based on PES explorations actually refer to 0 K, and the models do not explicitly include the molecular motions in the description of the system of interest. The present paper aims to implement a different, robust methodology that accounts for the molecular motions within the mentioned systems. Our approach is based on a general hybrid statistical physics quantum mechanical methodology which explicitly accounts for the dynamical phenomena in the description of the condensed phases. In the present paper we study the solvation of fluoroform in liquid Kr at cryogenic conditions, through the intramolecular C-H stretching vibrational frequency shift of the fluoroform moiety upon solvation.

2 Computational Details and Algorithms

The methodology that we implement for the purpose of the present study is a hybrid statistical physics - quantum mechanical one. Statistical physics phase in the present paper was based on the Monte Carlo (MC) approach [1]. Rigid body MC simulations of fluoroform in liquid Kr at cryogenic conditions were carried out. Intermolecular interactions were described by a sum of Lennard-Jones 12-6 site-site interaction energies plus Coulomb terms:

$$U_{ab} = \sum_{i}^{a} \sum_{j}^{b} 4\varepsilon_{ij} \left(\left(\frac{\sigma_{ij}}{r_{ij}} \right)^{12} - \left(\frac{\sigma_{ij}}{r_{ij}} \right)^{6} \right) + \frac{q_i q_j e^2}{4\pi\varepsilon_0 r_{ij}} \tag{1}$$

where i and j are sites in interacting molecular systems a and b, r_{ij} is the interatomic distance between sites i and j, while e is the elementary charge. The "geometric mean" combination rules were used to generate two-site Lennard-Jones parameters ε_{ij} and σ_{ij} from the single-site ones. Model potential parameters $\sigma = 3.895\,\text{Å}$ and $\varepsilon = 0.308\,\text{kcal mol}^{-1}$ were used for Kr, while the charge distribution in the case of fluoroform was computed as follows. The calculated MP2/6-31++G(*d*,*p*) electronic density (corresponding to the minimum on the MP2/6-31++G(*d*,*p*) PES) was fitted to a set of point charges placed at the nuclear positions with the CHelpG point-selection algorithm [2]. Fitting to the molecular electrostatic potential was carried out imposing a constrain that the derived point-charge distribution reproduces the "correct" dipole moment of fluoroform, as computed from the MP2/6-31++G(*d*,*p*) electronic density. The mentioned geometry corresponding to the minimum on MP2/6-31++G(*d*,*p*) PESs of fluoroform was used throughout the rigid-body Monte Carlo simulations. Non-bonded LJ parameters for fluoroform were taken from the OPLS-AA force field [3]. MC simulations were performed in the canonical (NVT) ensemble, with the Metropolis sampling algorithm, at T = 131 K. Experimental density of liquid Kr of 2.3406 g cm^{-3} at these conditions was used throughout the current study. The simulation conditions have actually been chosen to correspond to the experimental ones under which the Raman spectra of CF_3H in liquid Kr cryosolutions were recorded [4]. In particular, MC simulations were done for one fluoroform molecule surrounded by 1249 Kr atoms placed in a cubic box with side length of 42.04 Å, imposing periodic boundary conditions. Thermalization phase of simulations consisted of $6.25 \cdot 10^7$ MC steps. It was subsequently followed by averaging (simulation) phase of $2.50 \cdot 10^8$ MC steps. All MC simulations in the present study were performed with the DICE statistical mechanics Monte Carlo code [5].

Subsequent quantum mechanical phase of the computational methodology consisted of several steps, each of which was required in order to finally compute the anharmonic C-H Stretching frequencies of fluoroform in liquid Kr and to generate the corresponding probability distribution functions. For that purpose, first, series of snapshots from the equilibrated MC runs were chosen as "representative configurations", aimed to be used to calculate frozen-field "in-liquid" C-H stretching potential of the fluoroform molecule. Due to the high computational cost of the sequential QM phase, a method of choice is to use a relatively small number of structures generated by MC simulations. The statistical correlation between MC-generated configurations which are sufficiently close to each other is high. Therefore, performing QM calculations on such configurations would be a waste of time, as new statistically relevant contribution wont be added to the results. A much better approach is to choose configurations with low mutual statistical correlation and perform the QM computational part only on these configurations [6]. In the present study, we have chosen statistically uncorrelated configurations on the basis of computation and subsequent analysis of the energy autocorrelation function C(n):

$$C_n = \frac{\langle \delta E_i E_{i+n} \rangle}{\langle \delta E^2 \rangle} = \frac{\sum_i (E_i - \langle E \rangle)(E_{i+n} - \langle E \rangle)}{\sum_i (E_i - \langle E \rangle)^2} \tag{2}$$

This parameter may serve as a clear indicator for mutual statistical correlation between subsequent MC-generated configurations [6]. The computed C(n) from the MC simulation phase of CF_3H and 1249 Kr atoms, together with the least-squares fit to exponential decay function of the form:

$$C_n = \sum_{i=1}^{k} A_i exp(-n/\tau_i) \tag{3}$$

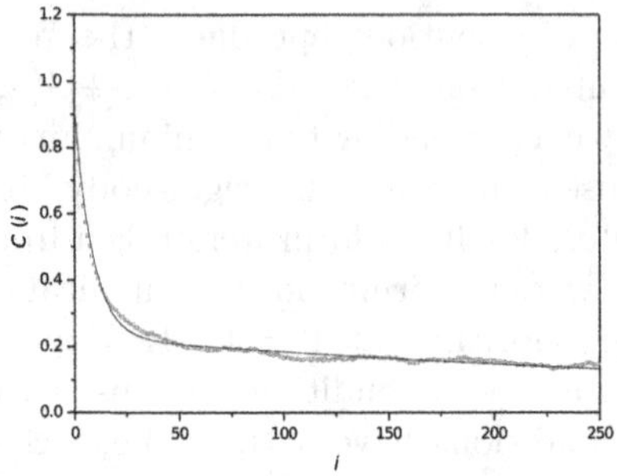

Fig. 1. The computed C(n) from the MC simulation phase of CF_3H and 1249 Kr atoms, together with the exponential decay function fit

is shown in Fig. 1. The presented fit in Fig. 1 is actually based on only 2 exponential functions in (3). Test calculations were, however, done with various numbers of exponentials in (3). Integrating the energy autocorrelation function from 0 to ∞, correlation step of 111 was obtained. Therefore, MC configurations mutually separated by 500 steps ($> 4\tau$) are correlated less than 10 %, and for sequential QM calculations, we have chosen 50 uncorrelated configurations from the equilibrated MC runs, separated by as much as 3500 MC steps. The mutual correlation between these configurations is negligible (less than 0.01 %).

Subsequent computations were performed by a quantum-mechanical (QM) approach, denoted as $\text{QM}_{electronic} + QM_{nuclei}$. Such notation has been adopted since both the electronic and nuclear subsystems are treated quantum mechanically [7]. Series of $\text{QM}_{electronic}$ + QM_{nuclei} calculations were carried out for 50 supermolecular clusters containing one fluoroform molecule and a representative part of the first solvation shell around it. A given Kr atom was included in the relevant part of the first solvation shell if the distance between the Kr atom and the fluoroform H atom was smaller than 4.0 Å. In the $\text{QM}_{electronic}$ phase of calculations, vibrational potential energy function ($V = f(r_{CH})$) for each chosen CH oscillator was computed performing a series of 20 pointwise HF, DFT (B3LYP) or MP2 energy calculations. To generate the $V(r_{CH})$ function, C-H distances were varied from 0.900 to 1.375 Å, keeping the center-of-mass of the vibrating C-H fragment fixed. The obtained energies were least-squares fitted to a fifth-order polynomial in Δr_{OH}, the resulting potential energy functions were subsequently cut after fourth order and transformed into Simons-Parr-Finlan (SPF) type coordinates [8]. 1D vibrational Schrödinger equation was subsequently solved variationally and the fundamental *anharmonic* C-H stretching frequency was computed from the energy difference between the ground and first excited vibrational states. All $\text{QM}_{electronic}$ calculations were carried out with the 6-31++G(d, p) basis set for orbital expansion, with the Gaussian03 series of codes [9].

3 Results and Discussion

Analysis of the C-Kr radial distribution function for CF_3H in liquid Kr, computed from the equilibrated MC run revealed existence of clearly pronounced five solvation shells. The first one, which is most relevant to our study, starts at about 3.5 Å and ends at about 6 Å, containing 14 Kr atoms on average. Theoretical Raman C-H stretching bands of fluoroform dissolved in liquid Kr under cryogenic conditions (i.e. the vibrational density of states histograms generated from the computed anharmonic in-liquid C-H stretching frequencies) at the employed theoretical levels are shown in Fig. 2. Solving the vibrational Schrödinger equation for a free fluoroform molecule, a value of $3227.5\,\mathrm{cm}^{-1}$ is obtained for the fundamental anharmonic C-H stretching frequency at the HF level of theory, $3002.9\,\mathrm{cm}^{-1}$ at B3LYP and $3110.7\,\mathrm{cm}^{-1}$ at MP2 level. The corresponding experimental value is $3035.2\,\mathrm{cm}^{-1}$ [10]. One can therefore conclude that the advanced computational method applied in the present study predicts a blue shift of the fluoroform CH stretching frequency (Table 1) upon its solvation in liquid Kr. This is in contrast with the experimental data, according to which the CH stretching frequency experiences a small red shift ($\sim -2\,\mathrm{cm}^{-1}$). Our further focus in the study will therefore be to explain such apparent discrepancy. As intuitively expected, the fluoroform CH stretching frequency shift is mostly dependent on the presence of Kr atoms residing on the vibrating H-atom side.

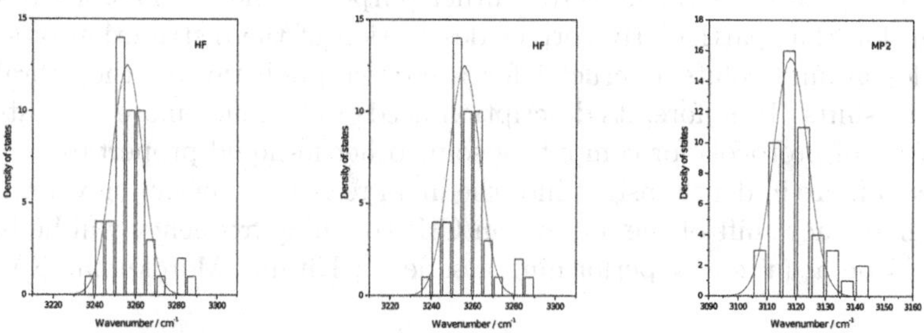

Fig. 2. Theoretical Raman CH stretching bands of fluoroform dissolved in liquid Kr under cryogenic conditions computed at different employed levels of theory

The computed C-H stretching frequency shifts with respect to the gas phase value as a function of the radial coordinate (i.e. C…Kr distance R) at B3LYP level of theory for a CF_3H…Kr dimer upon a collinear approach of Kr atom with the C-H bond are schematically presented in Fig. 3. As can be seen, B3LYP level of theory predicts a very small red shift of the CH stretching frequency as the Kr atom approaches CF_3H along the C-H axis, up to a distance of about 6 Å. After this point, the frequency shift experiences certain fluctuations and afterwards changes its sign (i.e. becomes a blue shift). Upon further diminishing of the R coordinate, the shift becomes larger (in absolute value) as the distance R decreases. At much smaller values of R it starts to rise abruptly. The maximum

Table 1. Anharmonic C-H stretching frequencies of free fluoroform, as well as of fluoroform solvated in liquid Kr under cryogenic conditions, together with the corresponding frequency shifts with respect to the free molecule.

Level of theory	Free CF_3H	CF_3H in liquid Kr	
	ν/cm^{-1}	$\langle\nu\rangle/\text{cm}^{-1}$	$\langle\Delta\nu\rangle/\text{cm}^{-1}$
MP2	3110.7	3120.2	+9.5
B3LYP	3003.5	3017.6	+14.1
HF	3234.6	3257.0	+22.4
Exp.	3035.2	3033.1	-2.1

of the C...Kr radial distribution function computed from MC simulation falls at about 4.15 Å. It is therefore obvious that most of the Kr atoms within the first shell are located at such positions at which the CH frequency shifts has either small negative or small positive values according to Fig. 3. So, even the actual *sign* of the frequency shift is strongly dependent on the actual value of R. This parameter, on the other hand, depends critically on the correct description of the atomic positions of solvent molecules within the first solvation shell around CF_3H generated by the statistical physics approach. This depends crucially on the type and quality of the intermolecular interaction potentials. Though the interaction potentials on which we relied in the present study could be considered as sufficiently accurate for numerous other purposes, one is undoubtedly led to a conclusion that particularly correct description of the distribution of solvent molecules around solute is crucial for a correct prediction of the vibrational frequency shifts. Therefore, its description needs to be much more accurate that in the case of protocols for computing some other in-liquid properties.

Even a more in-depth insight into the main reasons determining whether the overall frequency shift of the fluoroform CH stretching frequency will be blue or red could be acquired by performing a series of Kitaura-Morokuma (KM [11])

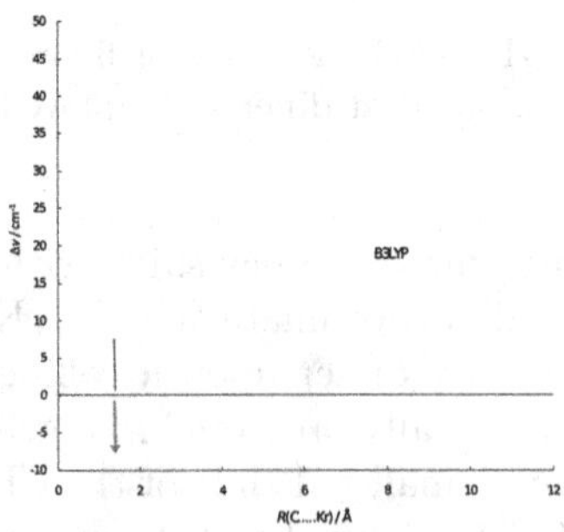

Fig. 3. Anharmonic C-H stretching frequency shift as a function of the CKr distance (R) for the CF_3H ...Kr dimer upon a collinear approach of Kr atom with the C-H bond, computed at B3LYP level of theory.

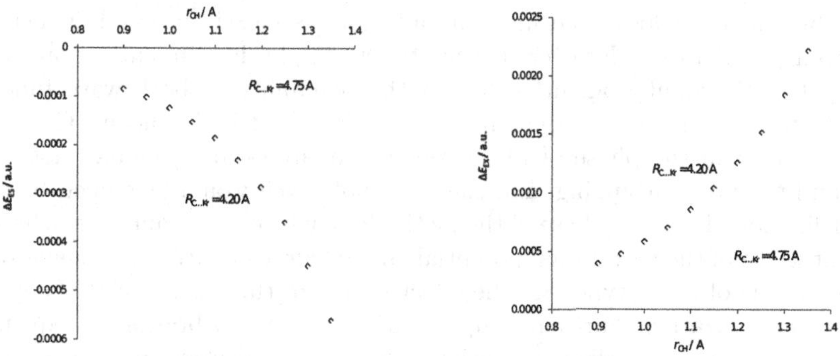

Fig. 4. Dependence of selected interaction energy components on rCH as computed by Kitaura-Morokuma (KM) energy partitioning analysis at two values of R(C... Kr).

and RVS-SCF [12] energy partitioning analyses for each of the sampling points at which the C-H stretching potentials were computer at two values of R(C... Kr): 4.20 Å and 4.75 Å. At R = 4.20 Å, the frequency shift is $+7.3\,\text{cm}^{-1}$, while at R = 4.75 Å the frequency shift exhibits a red shift of $-2.2\,\text{cm}^{-1}$. Selected results are summarized in Fig. 4. While the electrostatic (ES), polarization (POL) and charge-transfer (CT) energy components all favor a frequency red shift (stabilization of the system by C-H bond stretching), the exchange (EX) and mixed (MIX) terms favor a blue shift. These conclusions are valid for both of the considered R(C... Kr) values (4.20 Å and 4.75 Å). The derivatives $\partial E_i/\partial r_{CH}$ have been found to be significantly larger at smaller R, along with the absolute values of each ΔE_i. The ratio $(\partial E_i/\partial r_{CH})^{R1}/(\partial E_i/\partial r_{CH})^{R2}$ is largest for the polarization, electrostatic and exchange terms. As the absolute value of the exchange interaction energy component is significantly higher than the other ones and the derivative ratio $(\partial E_i/\partial r_{CH})^{R1}/(\partial E_i/\partial r_{CH})^{R2}$ for the exchange is close to that for the electrostatic term, it is the enhanced exchange interaction energy component that determines the sign of frequency shift (blue-shift) at lower values of R.

At higher values of R, it is the electrostatic (+ polarization) terms that induce a small frequency red shift. The relative smallness of the electrostatic contribution to frequency redshift, which is due to the negative permanent dipole moment derivative of fluoroform with respect to the C-H stretching vibrational coordinate ($\partial\mu^0/\partial r_{CH} < 0$), enables the electronic exchange effects to come into play and cause a more significant frequency blue shift at smaller R values.

The computational methodology that we use is a hybrid one. It consists of several steps, each of which demands computational resources to a various extent, and scales differently with the number of processors/cores. As the procedure has not been fully automated yet, and due to the need to check certain results manually, it is best to judge on its overall scalability on the basis of scalability of its component phases. Such analysis, as explained below, provides a good overview of the overall scalability of the method. The first phase of the hybrid methodology (the MC computation) is based on a code which has not been parallelized. However, if the statistical physics simulations are done e.g. by CPMD simulation, in such case the scalability in both wave function optimization and molecular dynamics simulation itself is significant. Computations involved in the phase of MC trajectory analyses are generally not much time-and resource-consuming, and therefore not much would be gained by their parallelization. The next phase of the methodology involves quantum mechanical computations of the vibrational potential energy curve or surface, or single-point computations of other type. So, the effectiveness of this phase heavily depends on the parallelization. Further computation of e.g. the vibrational frequencies involves either standard diagonalization procedures or Fourier-transform - based techniques. In the case of one-dimensional problems, diagonalization and FFT computations are quick, and do not benefit much from parallelization. However, in the case of multi-dimensional vibrational problems the scalability could be significant.

4 Conclusions and Directions for Future Work

Implementing a hybrid statistical physics - quantum mechanical methodology, in the present work we have gained certain physical insights into the reasons behind the magnitude and direction of the fluoroform CH stretching frequency shift upon its solvation in liquid krypton under cryogenic conditions. Following the implications concerning the quality of the intermolecular interaction potentials acquired in the present study, our further efforts will be directed towards development of a new class of intermolecular potentials for statistical physics simulations, allowing more accurate description of the structure of the first solvation shell around the solute molecule.

References

1. Allen, M.P., Tildesley, D.J.: Computer Simulation of Liquids. Oxford University Press, New York (1997)
2. Breneman, C.M., Wiberg, K.B.: Determining atom-centered monopoles from molecular electrostatic potentials. The need for high sampling density in formamide conformational analysis. J. Comp. Chem. **11**, 361 (1990)
3. Kaminski, G., Duffy, E.M., Matsui, T., Jorgensen, W.L.: Free energies of hydration and pure liquid properties of hydrocarbons from the OPLS all-atom model. J. Phys. Chem. **98**, 13077 (1994)
4. Van den Kerkhof, T., Bouwen, A., Goovaerts, E., Herrebout, W.A., Van der Veken, B.J.: Raman spectroscopy of cryosolutions: the van der Waals complex of dimethyl ether with fluoroform. Phys. Chem. Chem. Phys. **6**, 358 (2004)
5. Coutinho, K., Canuto, S.: DICE: A Monte Carlo Program for Molecular Liquid Simulation. University of São Paulo, São Paulo (2003)
6. Coutinho, K., Canuto, S.: Solvent effects from a sequential Monte Carlo-quantum mechanical approach. Adv. Quantum Chem. **28**, 89 (1997)

7. Pejov, L., Spångberg, D., Hermansson, K.: Al3+, Ca2+, Mg2+, and Li+ in aqueous solution: calculated first-shell anharmonic OH vibrations at 300K. J. Chem. Phys. **133**, 174513 (2010)
8. Simons, G., Parr, R.G., Finlan, J.M.: New alternative to the Dunham potential for diatomic molecules. J. Chem. Phys. **59**, 3229 (1973)
9. Frisch, M.J., et al.: Gaussian 03 (Revision C.01). Gaussian Inc., Pittsburgh (2003)
10. Dubal, H.R., Ha, T.K., Lewerenz, M., Quack, M.: Vibrational spectrum, dipole moment function, and potential energy surface of the CH chromophore in CHX3 molecules. J. Chem. Phys. **91**, 6698 (1989)
11. Kitaura, K., Morokuma, K.: A new energy decomposition scheme for molecular interactions within the Hartree-Fock approximation. Int. J. Quantum Chem. **10**, 325 (1976)
12. Stevens, W.J., Fink, W.H.: Frozen fragment reduced variational space analysis of hydrogen bonding interactions. Application to the water dimer. Chem. Phys. Lett. **139**, 15 (1987)

Image Classification Optimization of High Resolution Tissue Images

M. Kozlovszky(✉), K. Hegedűs, G. Windisch, L. Kovács, and G. Pintér

John von Neumann Faculty of Informatics, Óbuda University, Budapest, Hungary
kozlovszky.miklos@nik.uni-obuda.hu
http://nik.uni-obuda.hu

Abstract. Generic image classification methods are not performing well on tissue images. Such software solutions are producing high number of false negative and positive results, which prevents their clinical usage. We have created the MorphCeck high resolution tissue image processing framework, which enables us to collect morphological and morphometrical parameter values of the examined tissues. Size of such tissue images can easily reach the order of 100 MB–1 GB. Therefore, the image processing speed and effectiveness is an important factor. Our main goal is to accurately evaluate high resolution H-E (hematoxilin-eozin) stained colon tissue sample images, and based on the parameters classify the images into differentiated sets according to the structure and the surface manifestation of the tissues. We have interfaced our MorphCheck tissue image measurement software framework with the WND-CHARM general purpose image classifier and tried to classify high resolution tissue images with this combined software solution. The classification is by default initiated with a large training set and three main classes (healthy, adenoma, carcinoma), however the new image classification process' wall-clock time was intolerably high on single core PC. The processing time is depending on the size/resolution of the image and the size of the training set. Due to the tissue specific image parameters the classification effectiveness was promising. So we have started a development process to decrease the processing time and further increase the accuracy of the classification. We have developed a workflow based parallel version of the MorphCheck and WND-CHARM classifier software. In collaboration with the MTA SZTAKI Application Porting Centre the WND-CHARM has been ported to some distributed computing infrastructure (DCI). The paper introduces the steps that were taken to optimize WND-CHARM applications running faster using DCIs and some performance results of the tissue image classification process.

Keywords: Application porting · Medical image processing workflow · HP-SEE · gUSE · Scalability · MorphCheck · WND-CHARM

M. Kozlovszky would like to thank Semmelweis University and Major & Co. to provide us annotated tissue samples for processing and classification. Authors would like to thank for the technical support of the WND-CHARM developer team as well.

I. Lirkov et al. (Eds.): LSSC 2013, LNCS 8353, pp. 532–539, 2014.
DOI: 10.1007/978-3-662-43880-0_61,

1 Introduction

High resolution tissue image analysis and classification is a hot research topic nowadays. Main examples of the existing, widely used state-of-the-art high-resolution tissue image analyzer software solutions are: 3DHistech's Histo-, Nuclear- and MembraneQuant [1] (which are capable to do colorization based segmentation, analyze cell nucleus and analyze membrane structures between the cellular matrix) Aperio's [2] ePathology, Definiens's Tissue Studio [3] and Visiopharm's Visio/TissuemorphDP™ [4] (which are providing tissue image analysis and cell population analysis). High-resolution image processing using large image databases/training sets is a highly data-, and compute-intensive challenge. Parallelization of such applications can decrease the computational time significantly [7]. This paper shows how we have created an image classification service using the grid and cloud User Support Environment (gUSE [5]) to evaluate images processed by our Morphcheck (MC) software and evaluated it by "Weighted Neighbor Distance Using Compound Hierarchy of Algorithms Representing Morphology" (WND-CHARM [6]) software.

1.1 MorphCheck

MorphCheck (MC) is a high resolution tissue image analyzer framework, which processes high resolution digital tissue images. MorphCheck software framework is capable to effectively recognize -with its extendable algorithm repository- large number of differentiated tissue structures (such as surface epithelium, gland structures, lamina muscularis, submucosa etc.), and measure their morphological and morphometrical properties. The software supports both some vendor specific tissue scanner image formats and regular image standards (such as Tagged Image File Format /tiff/, Joint Photographic Experts Group /jpg/). It supports various colorization schemes like: HE (Hematoxilin-Eozin), DAB (3,3'-Diaminobenzidin), multi-color FISH. It contains various texture-based algorithms, furthermore intensity and structure-based algorithms (such as K-means, region growth, etc.) [8] (Fig. 1).

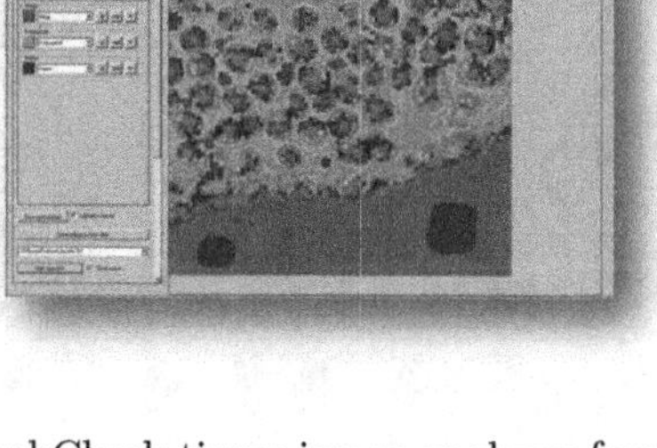

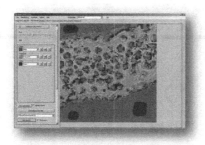

Fig. 1. MorphCheck tissue image analyzer framework GUI

1.2 WND-CHARM

WND-CHARM is an acclaimed open source image classifier application developed at National Institute of Aging (NIA, NIH) [9] that supports generic image analysis methods. WND-CHARM extracts a large set of (normal: 1025, or extended: 2873) image features including polynomial decompositions, high contrast features, pixel statistics, and textures. These features are computed on: the raw image; transforms of the image; and second transforms of the image. The feature values are then used to classify test images into a set of pre-defined image classes. WND-CHARM is using WND-5 as image classifier. We are using recently the v1.30.227 as the baseline application for our MC-WND image classification service and for performance measurements as well. We have modified the original WND-CHARM software to include our MC tissue parameters. We have modified both the training and the classification part of the software.

1.3 gUSE

gUSE is basically a virtualization environment providing large set of high-level DCI (Distributed Computing Infrastructure, such as supercomputers, grids or clouds) services by which interoperation among classical service and desktop grids, clouds and clusters, unique web services and user communities can be achieved in a scalable way. gUSE has a graphical user interface, which is called WS-PGRADE. All part of gUSE is implemented as a set of Web services. gUSE supports various DCIs, and its execution concept is heavily based on workflows. The definitions (graph, etc.) of workflows and their jobs are stored in a local storage. Job executions on DCIs requires user level authentication, and this can be managed transparently via the WS-PGRADE.

HP-SEE's Life Science Portal (Bioinformatics eScience Gateway). The Bioinformatics eScience Gateway is based on gUSE and operates within the Life Science VO of the HP-SEE [10] infrastructure. We have used this facility to implement our workflow and create the MC-WND tissue image classification service (Fig. 2).

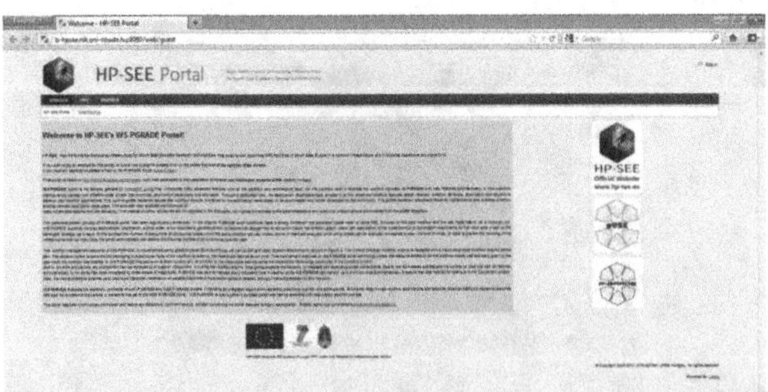

Fig. 2. HP-SEE Bioinformatics eScience Gateway

1.4 MC-WND Tissue Image Classification Service

The close collaboration between John von Neumann Faculty of Informatics, Obuda University and 2nd Department of Internal Medicine, Semmelweis University enabled us to define 75 tissue parameters of the human colonic region. Identified main tissue parameter groups:

- morphological and morphometrical properties of surface epithelium,
- morphological and morphometrical properties of gland structures,
- morphological and morphometrical properties of lamina muscularis,
- morphological and morphometrical properties submucosa,
- morphometrical properties of the cells.

We have extended our generic MorphCheck medical (tissue) image analysis framework to accurately measure these tissue image parameters. The objective numerical values of the pre-defined tissue parameters calculated by MorphCheck enables us to integrate and adapt a generic image classifier software solution, which can do effective tissue image classification automatically based on our parameter set. The two software solutions (MorphCheck and WND-CHARM) have been loosely coupled together to realize a single tissue image classification service (MC-WND). Data exchange between the two software solutions is realized with simple file exchange mechanisms. The MC-WND (Fig. 3) tissue image classification service allows researchers to process and categorize medical high resolutions tissue images using HPC infrastructure in a fast and easy way.

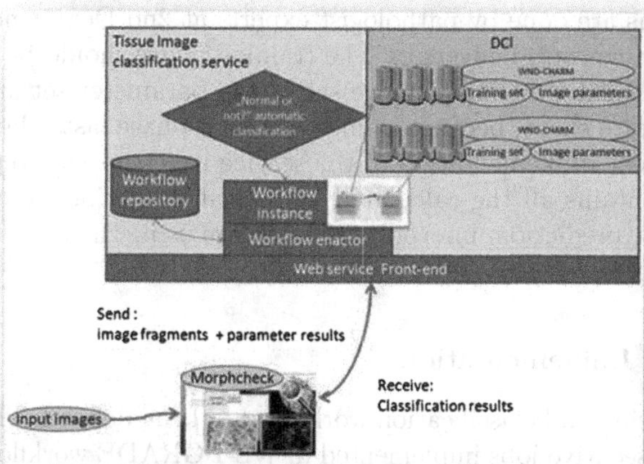

Fig. 3. MC-WND tissue image classification service schematic overview

1.5 MC-WND Tissue Image Classification Workflow

We have defined the image classification tasks within a single workflow.

- Inputs of the MC-WND workflow are:
  - Fragment of the high resolution tissue image:
    * size: (512×512), in TIFF format,
    * resolution/zoom level is the same which was defined during (automatic/manual) ROI definition/annotation,
  - MC calculated parameter results exported into a *csv* file.

- Output of the MC-WND workflow are:
  - WND-CHARM calculated image parameter results stored in a *csv* file classification process results, , stored in a *html* file (contains all the calculated statistical results)

1.6 MC-WND Training Set

The modified WND-CHARM application is using our tissue image specific parameter as classifier parameters both during the training phase and during the image classification phase. The training phase is using the 2873 internal (WND-CHARM) parameters plus 75 tissue parameters measured by MorphCheck (MC). The results of the training phase are dumped into a single file (with WND-CHARM's internal coding format), and can be reused as an offline file for all the classification processes. Our tissue image training set contains more than 90 annotated HE (Hematoxilin-Eozin) colon tissue image samples with the following main categories: healthy, malignant (adenoma and carcinoma). All tissue image annotations are done by pathologist experts at 2nd Department of Internal Medicine, Semmelweis University. The training phase should be re-launched each time the annotated training image set or the parameter set are extended. Luckily this is a rare event, because a single training phase lasts about 10 h normally. In our recent implementation of the service it generates a report in *html* format, which contains all the calculated statistical results of the classification process (accuracy, prediction, interpolation). We are using the *stdout* to monitor the process and receive status information.

1.7 Workflow Implementation

An image processing and classification workflow has been defined. This workflow contains two consecutive jobs implemented in WS-PGRADE workflow language. The first job is a pre-processor, the second job utilizes WND-CHARM in a parametric manner. The second job contains the WND-CHARM execution and it is launched in parallel as many times the service receives tissue images from outside. WND-CHARM is installed and launched in the so called user space, which was a hard task to realize. We are using LibTIFF [11] and FFTW [12] as external software packages inside our service. We are collecting the results

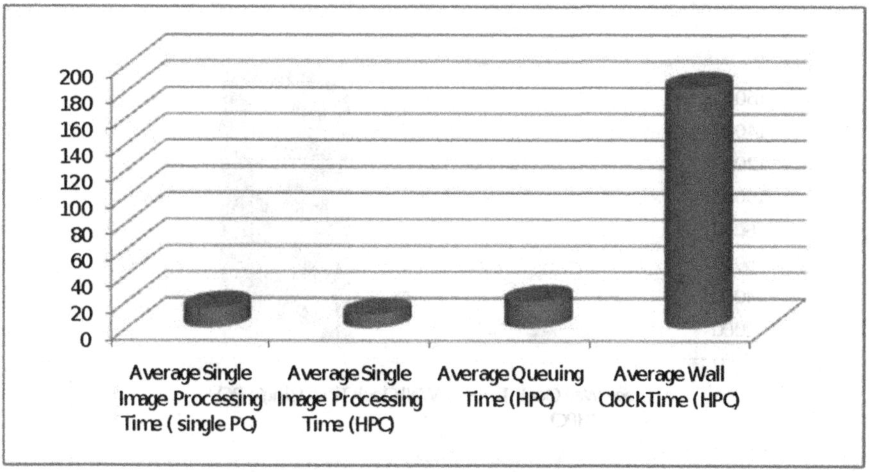

Fig. 4. Average MC-WND service execution time using HPC infrastructure (in min)

from all the WND-CHARM instances both from the *stdout* (as a file) and the generated *html* files. We have done performance evaluation to see how the WND-CHARM can run on HPC infrastructure. The workflow has been executed on one of the HPC centers operated by NIIF called "Budapest", which is an HP fat-node cluster using CP4000BL blades, consisting of 32 nodes with 24 Magny Cours CPU cores each (i.e. total number of CPU cores: 768). It has a mesh like topology with an Infiniband internal network. It has 1.96 TB memory and the total performance of the system is about 5,48 TFlops. Each measurement was executed 10 times, the average of the 10 executions was taken as the final result.

2 Performance Measurement Results

A single run of the image classification process is about 10 min for a 512×512 px tissue image size.

Nowadays a normal high resolution tissue image size (whole size) is about 4096×4096. This is about 64 times larger than our 512×512 unit size. We have launched 991 tissue image units against the HPC infrastructure. The following graphs show the result of the multiple executions of WND-CHARM on the HPC infrastructure. Figure 4 shows the processing time result. Average queuing overhead was 20 min. Average execution time was 10 min per image. The average total Wall Clock time was 3.02 h for the whole image set (991 images), which means 11,503 s was the average processing time for the whole image set. Figure 5 shows the WND-CHARM wall clock (execution) time compared to a single CPU. We gain significant speedup with parallel execution of the WND-CHARM, even if the average queuing time and result collection took us some negligible time.

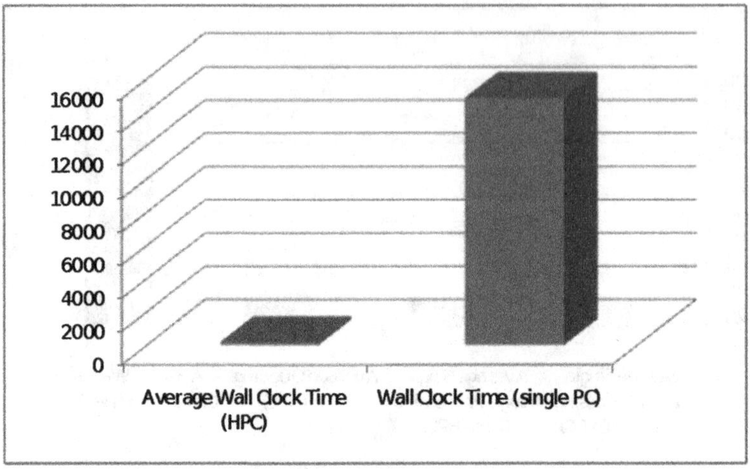

Fig. 5. MC-WND service execution time (HPC vs. single PC) (for 991 images in min)

2.1 Classification Accuracy

To create a usable classification procedure for colonic tissue images based on automatic pre-filtering solution we have introduced a cut-off number which defines a tolerance level of the classification certainty. Below this certain numerical value the image marked automatically as malignant and forwarded to manual evaluation. In our large-scale tissue image classification tests with more than 200 tissue images and with the cut-off value below 60 % we are able to distinguish between healthy and malignant colonic tissue images with 100 % accuracy at healthy category (that means healthy category strictly contains only healthy patient's images).

3 Conclusion

In this paper we have described how we have ported the modified WND-CHARM image classification software to work on distributed computing infrastructure (on HP-SEE supercomputing infrastructure) as a service. The parameter field of the WND-CHARM application is extended with our high resolution tissue image analysis parameter field (+75 tissue image parameters). We have created the workflow structure for the MC-WND image classification service. The service is tested with HE stained tissue images and capable to separate healthy and malignant tissue images automatically with a high accuracy. The service and the internal workflow was developed at Obuda University and hosted on the HP-SEE Life Science/Bioinformatics eScience Gateway. The service can be used to do tissue image classification of the colonic region against our large tissue training databases in a short time using the HP-SEE supercomputing infrastructure at NIIF, Hungary. We also describe the performance analysis has been done on the applications. As a future work we are trying to include other image classifier solutions into our MC-WND service portfolio in a plug-in like manner. We are planning to open up our image classification service for the wider research community. So far the tissue image classification service can only be executed from the MorphCheck software. We are planning to create a portlet-based web user interface to let pathologists manually upload and evaluate tissue images.

Acknowledgments. The authors would like to thank two projects for their financial support. This work was supported by the HP-SEE (High-Performance Computing Infrastructure for South East Europe's Research Communities, under contract no. RI-261499) project and by the 3DHistech08 project, the Hungarian National Technology Programme, A1, Life sciences, the "Development of integrated virtual microscopy technologies and reagents for diagnosing, therapeutical prediction and preventive screening of colon cancer". Authors would like to thank Semmelweis University and Major & Co. to provide us annotated tissue samples for processing and classification. Authors would like to thank for the technical support of the WND-CHARM developer team as well.

References

1. 3DHistech. http://www.3DHistech.com. Accessed 09 Feb 2013
2. Aperio. http://www.aperio.com. Accessed 09 Jan 2013
3. Definiens. http://www.definiens.com. Accessed 09 Jan 2013
4. Visiopharm. http://www.visiopharm.com. Accessed 15 Feb 2013
5. Kacsuk, P., Farkas, Z., Sipos, G., Hermann, G., Kiss, T.: Supporting workflow-level PS applications by the P-GRADE grid portal. In: Towards Next Generation Grids Proceedings of the CoreGRID Symposium (2007)
6. Orlov, N., Shamir, L., Macura, T., Johnston, J., Mark Eckley, D., Goldberg, I.G.: WND-CHARM: Multi-purpose image classification using compound image transforms. Pattern Recognit Lett. **29**(11), 1684–1693 (2008). (PMCID: PMC2573471, NIHMSID: NIHMS53719)
7. Szénási, S., Vámossy, Z., Kozlovszky, M.: Evaluation and comparison of cell nuclei detection algorithms. In: 16th International Conference on Intelligent Engineering Systems (INES), Lisbon, July 2012, pp. 469–475 (2012). ISBN: 978-1-4673-2694-0
8. MorphCheck. http://biotechweb.nik.uni-obuda.hu/web/en/research/projects/3dhist08/morphcheck2. Accessed 09 Feb 2013
9. http://www.grc.nia.nih.gov/branches/lg/iicbu/iicbu.htm. Accessed 09 Feb 2013
10. HP-SEE. http://www.hp-see.eu. Accessed 09 Feb 2013
11. Libtiff. http://www.libtiff.org. Accessed 09 Feb 2013
12. FFTW. http://www.fftw.org/download.html. Accessed 09 Feb 2013

On the Management of Cloud Services in Multi-Clouds for Scientific Applications

Dana Petcu(✉)

West University of Timişoara, Timişoara, Romania
petcu@info.uvt.ro

Abstract. Cloud computing is a new enabler for scientific discovery. The current barriers in adopting it on large scale are including low code portability and low availability of the proper support tools. This paper discuss the special needs of the scientific applications consuming Cloud services and how they are supported by the Multi-Cloud middlewares. A recent developed middleware for managing the consumption of Multi-Cloud services is considered as case study.

Keywords: Multi-Cloud · Service management · Scientific applications

1 Introduction

The availability of Cloud resources to everyone with a credit card is changing not only the society perspective on the e-infrastructure availability, but also how research and innovation is expected to perform. Small research groups can now initiate experiments using customizable services for e-infrastructure and software. The current preferences are clearly in favor for Infrastructure-as-a-Service (IaaS) and Software-as-a-Service (SaaS), and less for Platform-as-a-Service (PaaS). At a first look, this is surprising, since the PaaS offers the tools for programming applications on Clouds and releases the developer from the burden of managing the e-infrastructure resources. At a closer look, we see that most of current PaaS are mainly targeting the web applications, not compliant in structure with most of research codes. Small steps have been made in the last two years to attract the research community to use also PaaS, including Multi-Cloud services.

In this context, we discuss here the challenges and requirements at PaaS level to offer support for scientific communities. Moreover, we point towards a recent Multi-Cloud middleware, namely mOSAIC, that has prove its usefulness in the context of several research experiments. Therefore, the main contribution of this paper can be resumed in highlighting the expectations at PaaS level to support scientific applications and measurement of the degree in which mOSAIC's PaaS fulfills these expectations.

The paper is organized as follows. First we identify when the Cloud resources are proper to be used for research, highlighting also the specific needs of the

I. Lirkov et al. (Eds.): LSSC 2013, LNCS 8353, pp. 540–548, 2014.
DOI: 10.1007/978-3-662-43880-0_62,

research community and gaps in fulfilling them. Secondly, we point towards the need of the usage of multiple Clouds and current technical barriers. Thirdly, we resume mOSAIC's approach for multiple Cloud for scientific applications. Finally, we position this approach in the context of related work.

2 Consuming Cloud Resources by Scientific Applications

Peaks in the usage of scientific codes are more often encountered then in production codes. Moreover, most of the codes are not running continuously. Often e-infrastructures in research and academic institutions are under-utilized. Therefore, at a first glance, a pay per use model for consuming e-infrastructure or software services, as in the case of Cloud computing, is more adequate than an investment in e-infrastructure or software licenses with fast depreciation rate.

The take-up by the scientific community of the Cloud computing paradigm is very slow. The reasons are various: the specificity of the codes, the knowledge needed, the appropriateness of the tools, the administrative issues. The scientific codes have often special requirements related to parallel processing, particular libraries, or large data sets. The comparative studies of the performance on parallel computers versus the Cloud resources, are still disappointing (e.g. [1,2]). Fortunately, recent tests using the cluster-on-demand services are more positive.

Special services are usually offered by data centers (by humans) to their clients to set the environments for proper execution. The degree of customization of Cloud services is quite limited. The preference for IaaS is a consequence of this fact. Using an IaaS the scientist can set the environment and import the needed libraries (time consuming task, possible only if elementary knowledge about system administration is available).

The costs of local services or of a virtual organization (like in Grids) are not directly charged to a certain research project or initiative (indirect costs). When accessing the external services of a Cloud, the cost are measurable and directly chargeable. Compute resources, storage resources, as well as applications, are dynamically provisioned on a pay per use basis (moreover, these resources can be released when they are no more needed). Data transfers are also chargeable, as not done usually before. However most of the institutions and research agencies have not yet adopted the proper cost models to allow such acquisitions.

A particular issue is the vendor lock-in: the fact that the Cloud providers have not yet adopted standards in their service offers or agree upon similar access mechanisms and interfaces, is leading to the lock of the applications with the Cloud provider interfaces for which they have been written. This is not compliant with the expected free market of the services for public procurement.

Despite the above mentioned barriers, several positive reports in consuming Cloud resources appeared in the last three years (e.g. as reported in [3], medium sized parallel compute problems, typically present in industrial engineering applications, are successfully using CFD codes on Clouds). Most successful stories on using Cloud services in scientific applications are related to bio-informatics (like in [4]). In this field and others, the Cloud applications are small-memory-footprint embarrassingly parallel or loosely coupled, and are requiring little to no inter-processor communication (e.g. for Many-Task Computing: loosely coupled applications comprising many tasks). Moreover, according [5] Cloud offers better performance and value for processor- and memory-limited applications than for I/O-bound applications.

Data-intensive applications are also profiting from the availability of Cloud services. Simplifying the processing of large-scale distributed data volumes, the MapReduce programming paradigm has been embraced by a large number of researchers, from social sciences to high energy physics (e.g. the experiments from [6]). In the transition phase towards a data-intensive science, the Cloud can play an important role as enabler of research and innovation.

3 The Need for Multiple Clouds in Scientific Applications

As stated in the previous section, the vendor lock-in is hindering the freedom in a Cloud service market. This is not only a problem for the public procurement, but also for a long life of the scientific codes, as well as for the reproductivity of the experiments by other scientists. Moreover, it is not only a problem of the Cloud vendor, who can find an interest in the lock-in of its clients, but of the middleware tools that should cover the heterogeneity not only to the low level of the infrastructure or operating system, but to the level of services. Such middleware should ensure the seamless migration of the scientific codes consuming the services from one Cloud to another Cloud. Resource monitoring, analysis, and configuration tools are expected to be developed to offer the ability to dynamically provide and respond to information about the application state [7].

A particular use case is the Cloud bursting. The developer of an scientific application expects to be able to deploy and debug it on his desktop or a local cluster and then, to a certain moment in time, to port the application into the Cloud, or use extra resources from the Cloud when the local one are not sufficient. This scenarios is not matching the business model of most of the Cloud services, that are requiring from the start the consumption of remote resources. There are only few deployable tools that are assisting the developer to design and debug the application locally, and are allowing Cloud bursting.

The open-source code production, common for scientific applications, is affected also by the licenses of the interfaces to the Cloud services that are consumed. Only few Cloud providers are offering their application programming interfaces as open-source code. Moreover, the property protection rights are stopping the development of exactly same interfaces for the Cloud services hindering therefore the migration from one Cloud to another.

The variety of services offered by Cloud providers has also a positive aspect: a high probability to find the proper one. However, two new challenges are also generated. One is related to the complexity of the selection procedure and its semi-automatizing. The second is the architectural design of applications that are consuming simultaneously services from at least two different Clouds.

The use of services from multiple Clouds either sequentially (in the case of migration for example) or simultaneously (as described above) is the subject for Cloud Federation or Multi-Cloud. In the first case, the Cloud user expects to consume the resources of a certain Cloud, while the selected Cloud is subcontracting the services of another Cloud, based on an agreement between them. Interoperability between the Clouds involved in the agreement is the main barrier to be overcome [8]. In the second case, a third party is helping the Cloud client to find the proper service to be consumed (e.g. using some brokerage mechanisms), is offering service management for the selected services, and is watching for the fulfillment of the service level agreements. Portability between the Clouds that are registered to the third party is the main barrier to be overcome. While the interoperability is not a concern for the scientific application developer (but of the Cloud provider), the portability and service level agreements should be.

Another issue is the non-availability of a market for specific services for the scientific applications. Marketplaces for Cloud service are slowly developing, hindering also the development of the Multi-Cloud third-party middleware.

The Multi-Cloud is expected to overcome the issues related to the variety of Cloud services that are registered by providing extra services for search, brokerage, selection, customization, monitoring, registering, authentication, migration, scheduling, load balancing and so on. All of these extra services are associated to technical challenges. Note that currently virtual machines images are rarely replicable in other Clouds, the format of the data store varies, and the network setting are different from one Cloud to another.

Cloud management systems to support a Multi-Cloud are now in early stages of development. A such system, in its simplest form, is a library offering uniform APIs (like jclouds, libcloud, δ-cloud). In a more complex one, it can take the form of a service, hosted (like enStratus, Kaavo or Rightscale) or deployable (like Aoleus). What most the current management services for Multi-Clouds are offering is a uniform view of the infrastructure resources that are available or are already consumed. A basic functionality is to provision infrastructure resources.

From the point of view of the developer or user of an scientific applications the unique interface to many Clouds in a Multi-Cloud is essential (one-click far from the Cloud). Moreover, such interface is expected to be a web-based one as are the scientific portals used earlier. The management systems of service type are therefore preferable to the libraries. For reasons related to the openness and testing in own environment, the deployable ones are highly recommended. However their number is limited. We point in the next section to one of them.

4 mOSAIC's Approach for Science in Multi-Cloud

mOSAIC has started as a multi-national collaborative project partially funded by the European Commission in the frame of FP7 programme, and has run from Autumn 2010 until Spring 2013. One of the main results is a deployable platform, named mOSAIC PaaS, that offers services for Multi-Cloud. The open-source codes are available at https://bitbucket.org/mOSAIC. The main target is to ensure portability of codes that are consuming infrastructure services. The decision which Cloud to use is postponed until the deployment phase, in opposition to the usual request to take the decision at design phase.

Several scientific reports about various architectural components, as well as about their usage in the context of scientific applications are already available: a complete list of them is available at the project site (http://www.mosaic-cloud.eu, link to scientific publications). We point here to the support for scientific applications.

The hosting services of several IaaS providers are currently accessible through the deployable middleware: Amazon, Flexiant, GoGrid, CloudSigma and OnApp. Several deployable services for Private Clouds on own premises can be used in conjunction with the PaaS: OpenStack, CloudStack, Eucalyptus, OpenNebula, VMware and δ-cloud. A Cloud vendor template is available for further developments. The services of the same type, e.g. for data stores or communications, are represented in the programming libraries (for Java, Python, Erlang) by the same connector (e.g. one for key-value stores, one for distributed file systems).

A scientific application which can profit from full set of services is expected to be a component-based application. Component-based programming has been earlier applied, e.g. in [9], to address the requirements of large scale applications from sciences and engineering (in particular in [10] is described an experiment in bioinformatics involving a Hybrid Cloud). The main reason in mOSAIC to use components is to enable their elasticity (lowering the level of elasticity from the virtual machine to the application component) – details can be found in [11]. Moreover, components written in different languages can be composed into a loosely coupled applications. Furthermore, scientific work-flows can be easily seen as component-based applications – the mOSAIC application descriptor is allowing to specify the order of execution of the components as well as the dependences between them.

One of the particularity is the compliance with just-in-time Clouds: the resources are allocated only when demanded until there is use for them. This is possible due to the fact that the user is controlling the life-cycle of the application and the PaaS ensures that the resources are allocated at the deployment phase only for that application and are stopped when the application is stopped. Moreover, targeting long-time running applications, mOSAIC is suited to build and control Research-as-a-Service (exposing research services).

Subscribing to the PaaS idea to eliminate the burden of system administration encountered at IaaS level, mOSAIC is providing several application deployment tools. One such tool is the Personal Testbed Cluster (PTC) which can be installed on the own desktop or notebook to simulate the Cloud in the development phase of the application, and which allows together with a resource allocator to seamless deploy the final application on a selected Cloud if the credentials for that Cloud are available. Moreover, several Eclipse plug-ins are available for various editors (e.g. for the call of proposals for Cloud service offers).

The middleware is featuring the specific services for a Multi-Cloud. In particular, a broker based on service level agreements and multi-agent technologies has been implemented [12]. Semantic processors are assisting in discovering the features of the platform, the associated libraries, or the new services offered by the providers beyond the currently connected ones [13].

The first proof-of-the-concept applications developed in the frame of the project are related to various fields of research: information extraction, civil engineering, Earth observation, sensor data mining, and social simulations. Early reports about these applications are provided in [14–17]. Several codes are open-source and available in the mOSAIC repository, e.g. related to the information extraction service from batches of scientific papers for indexing purposes, and the Matlab-in-Cloud solution for the analysis of the building structure. Fortunately, the overhead introduced by the Multi-Cloud middleware compared with the one of the virtualization layer is considerable smaller. Most of the applications have been re-designed to be adapted to the component-based approach, and to be fully deployed on the Clouds, and by this, to benefit from the elasticity properties. On the opposite situation is the Earth observation application that had from a start a high degree of complexity, restrictions in moving part of the codes and data in the Cloud, and special requirements for specific libraries and tools; in this case, Hybrid Clouds were build based on in-premises data catalogs and data processing in Public Clouds selected using the broker; manual intervention for the special libraries and tools was necessary (one time, to prepare the virtual machine image; the mOSAIC platform is installed on top a such machine).

Scheduling, load balancing, monitoring, and model-driven code generation are Multi-Cloud services under development.

Drawbacks in conjunction with scientific applications are related to the availability of the libraries only for Java, Python, Erlang and Node.js, the event-driven programming style for interaction with the Cloud services (that has a higher complexity than the REST-based interaction for example), the communication system based on a message queuing system (AMQP compliant), the lack of single sign on, and hands-on when special libraries are requested.

5 Related Work

As earlier stated, the PaaS are less used for scientific research. Few reports are related to experiments made with a PaaS, like mOSAIC. In [18] Microsoft Azure is successfully used for a genomics application, while in [19], Google Application Engine is used for a Monte Carlo simulations with a large middleware overhead.

In what concerns the Multi-Cloud management tools, we already mentioned in Sect. 3 several libraries and hosted services. Few others are added here. FutureGrid [20] aims to offer access to a number of IaaSs, including Nimbus, Eucalyptus, OpenStack, and OpenNebula, with a catalog and repository to store virtual images. ViteraaS [21] is a PaaS, based on OpenNebula, that allows to dynamically create a cluster of virtual machines on idle resources or dedicated servers; it uses single sign-on and a quality of service monitoring module to monitor the performance and status of these virtual machines. In [22] an experimental Hybrid Cloud is reported capable of utilizing both local and remote computational services for single large embarrassingly parallel applications.

The Helix Nebula team (http://www.helix-nebula.eu) is currently building a strategic plan for a Scientific Cloud Computing Infrastructure in Europe, the long term goal being to create a multi-tenant open market place for science. EGI's architecture proposal for Federated Cloud for science (www.egi.eu/infrastructure/cloud/) is relying upon standards (OCCI, CDMI, GLUE) and is connecting OpenNebula, OpenStack, WNoDeS, StratusLab, Okeanos; monitoring is done with Nagios SAM, image management with Marketplace vmcatcher, accounting with APEL, AAI with VOMS, and message bus with ActiveMQ.

In executing scientific applications on a Cloud environment, it will clearly desirable to exploit its elasticity, by increasing or decreasing the number of instances or components during the execution, to meet time or cost constraints. Elasticity is a propriety that the scientific applications are not usually able to profit from. The paper [23] describes a preliminary work towards making existing applications elastic, while the paper [24] introduces the Elastic Cluster.

Another potential for enhancing scientific application consists in adding Cloud-based interactivity. The paper [25] proposes to build a user-interactive service: the adaptive method computes the completion times and prices of ongoing jobs in real-time, which enables users to start same jobs with different configurations simultaneously and select the best one at an early stage.

6 Conclusions

The support for developing, deploying and control the scientific applications on Clouds is currently incomplete. We have try in this paper to identify the needs and the state-of-the-art in fulfill them, in particular for the case of Multi-Cloud. This study is done also to identify the missing pieces for a particular open-source and deployable middleware for Multi-Cloud, namely mOSAIC, in order to improve it in the near future to better support the scientific community needs.

Acknowledgments. This research is partially supported by the Romanian grant PN-II-ID-PCE-2011-3-0260 (AMICAS), currently extending the platform developed in the frame of the grant FP7-ICT-2009-5-256910 (mOSAIC).

References

1. Jackson, K.R., Ramakrishnan, L., Muriki, K., Canon, S., Cholia, S., Shalf, J., Wasserman, H.J., Wright, N.J.: Performance analysis of high performance computing applications on the Amazon web services cloud. In: CloudCom 2010, pp. 159–168 (2010). doi:10.1109/CloudCom.69
2. Iosup, A., Ostermann, S., Yigitbasi, M.N., Prodan, R., Fahringer, T., Epema, D.H.J.: Performance analysis of cloud computing services for many-tasks scientific computing. IEEE Trans. Parallel Distrib. Syst. **22**(6), 931–945 (2011). doi:10.1109/TPDS.2011.66
3. Zaspel, P., Griebel, M.: Massively parallel fluid simulations on Amazon's HPC cloud. In: NCCA 2011, pp. 73–78 (2011). doi:10.1109/NCCA.2011.19

4. Ekanayake, J., Gunarathne, T., Qiu, J.: Cloud technologies for bioinformatics applications. IEEE Trans. Parallel Distrib. Syst. **22**(6), 998–1011 (2011). doi:10.1109/TPDS.2010.178
5. Berriman, G.B., Juve, G., Deelman, E., Regelson, M., Plavchan, P.: The application of cloud computing to astronomy: a study of cost and performance. In: e-Science Workshops 2010, pp. 1–7 (2010). doi:10.1109/eScienceW.2010.10
6. Ekanayake, J., Pallickara, S., Fox, G.: MapReduce for data intensive scientific analyses. In: eScience 2008, pp. 277–284 (2008). doi:10.1109/eScience.59
7. Brandt, J., Gentile, A., Mayo, J., Pebay, P., Roe, D., Thompson, D., Wong, M.: Resource monitoring and management with OVIS to enable HPC in cloud computing environments. In: IPDPS 2009, pp. 1–8 (2009). doi:10.1109/IPDPS.2009.5161234
8. Petcu, D.: Portability and interoperability between clouds: challenges and case study. In: Abramowicz, W., Llorente, I.M., Surridge, M., Zisman, A., Vayssière, J. (eds.) ServiceWave 2011. LNCS, vol. 6994, pp. 62–74. Springer, Heidelberg (2011). doi:10.1007/978-3-642-24755-2_6
9. Heron de Carvalho, J.F., Rezende, C.A.: Component-based refactoring of parallel numerical simulation programs: a case study on component-based parallel programming. In: SBAC-PAD 2011, pp. 199–206 (2011). doi:10.1109/SBAC-PAD.2011.28
10. Malawski, M., Meizner, J., Bubak, M., Gepner, P.: Component approach to computational applications on clouds. Procedia Comput. Sci. **4**, 432–441 (2011). doi:10.1016/j.procs.2011.04.045
11. Petcu, D., Macariu, G., Panica, S., Craciun, C.: Portable cloud applications - from theory to practice. Future Gener. Comput. Syst. **29**(6), 1417–1430 (2012). doi:10.1016/j.future.2012.01.009
12. Venticinque, S., Aversa, R., Di Martino, B., Petcu, D.: Agent based cloud provisioning and management. In: Design and Prototypal Implementation, CLOSER 2011, pp. 184–191 (2011). doi:10.5220/0003395901840191
13. Cretella, G., Di Martino, B.: Towards a semantic engine for cloud applications development. In: CISIS 2012, pp. 198–203 (2012). doi:10.1109/CISIS.2012.159
14. Cossu, R., Di Giulio, C., Brito, F., Petcu, D.: Cloud computing for earth observation. In: Data Intensive Storage Services for Cloud Environments. IGI Global (in print, 2013)
15. Panica, S., Neagul, M., Craciun, C., Petcu, D.: Serving legacy distributed applications by a self-configuring cloud processing platform. In: IDAACS 2011, vol. I, pp. 139–145 (2011), doi:10.1109/IDAACS.2011.6072727
16. Skoda, P., Sperka, S., Smrz, P.: Extracting information from scientific papers in the cloud. In: CISIS 2012, pp. 775–780 (2012). doi:10.1109/CISIS.2012.176
17. Stankovski, V., Konig, M.: A sustainable building application design based on the mOSAIC API and platform. In: SKG 2012, pp. 249–252 (2012). doi:10.1109/SKG.2012.13
18. Simmhan, Y., van Ingen, C., Subramanian, G., Li, J.: Bridging the gap between desktop and the cloud for eScience applications. In: CLOUD 2010, pp. 474–481 (2010). doi:10.1109/CLOUD.2010.72
19. Prodan, R., Sperk, M., Ostermann, S.: Evaluating high-performance computing on Google App Engine. IEEE Softw. **29**(2), 52–58 (2012). doi:10.1109/MS.2011.131
20. von Laszewski, G., Diaz, J., Wang, F., Fox, G.C.: Comparison of multiple cloud frameworks. In: CLOUD 2012, pp. 734–741 (2012), doi:10.1109/CLOUD.2012.104
21. Doelitzscher, F., Held, M., Reich, C., Sulistio, A.: ViteraaS: virtual cluster as a service. In: CloudCom 2011, pp. 652–657 (2011). doi:10.1109/CloudCom.101

22. Brock, M., Goscinski, A.: Execution of compute intensive applications on hybrid clouds. In: CISIS 2012, pp. 995–1000 (2012). doi:10.1109/CISIS.2012.109
23. Raveendran, A., Bicer, T., Agrawal, G.: A framework for elastic execution of existing MPI programs. In: IPDPSW 2011, pp. 940–947 (2011). doi:10.1109/IPDPS.2011.240
24. Mateescu, G., Gentzsch, W., Ribbens, C.J.: Hybrid computing - where HPC meets grid and cloud computing. Future Gener. Comput. Syst. **27**(5), 440–453 (2011). doi:10.1016/j.future.2010.11.003
25. Li, X., Palit, H., Foo, Y.S., Hung, T.: Building an HPC-as-a-Service toolkit for user-interactive HPC services in the cloud. In: WAINA 2011, pp. 369–374 (2011), doi:10.1109/WAINA.2011.116

GPU Calculations of Unsteady Viscous Compressible and Heat Conductive Gas Flow at Supersonic Speed

Kiril S. Shterev[1(✉)], Emanouil I. Atanassov[2], and Stefan K. Stefanov[1]

[1] Institute of Mechanics, Bulgarian Academy of Sciences, Acad. G. Bonchev Str., Block 4, 1113 Sofia, Bulgaria
{kshterev,stefanov}@imbm.bas.bg
http://www.imbm.bas.bg

[2] Institute of Information and Communication Technologies, Bulgarian Academy of Sciences, Acad. G. Bonchev Str., Block 25A, 1113 Sofia, Bulgaria
emanouil@parallel.bas.bg
http://www.iict.bas.bg

Abstract. The recent trend of using Graphics Processing Units (GPUs) for high performance computations is driven by the high ratio of price performance for these units, complemented by their cost effectiveness. Such kinds of units are increasingly being deployed not only as accelerators for supercomputer installations, but also in GPU-enabled nodes in Grid and Cloud installations. At first glance computational fluid dynamics (CFD) solvers match perfectly to GPU resources, because these solvers make intensive calculations and use relatively small memory. Nevertheless, there are scarce results about the practical use of this serious advantage of GPU over CPU, especially for calculations of viscous, compressible, heat conductive gas flows with double precision accuracy. In our work we present calculation of unsteady, viscous, compressible and heat conductive gas with double precision accuracy using GPU-enabled version of the algorithm SIMPLE-TS, written on standard OpenCL. As a test case we model the flow past a square in a microchannel at supersonic speed with Mach number M = 2.43 on AMD Radeon HD 7950 GPU and achieve 90 GFlops, which is 46 times faster than the CPU serial code run on Intel Xeon X5560.

Keywords: GPU · OpenCL · SIMPLE-TS · First order upwind scheme · Unsteady · Viscous · Compressible and heat conductive gas flow

1 Introduction

Computational analysis of fluid dynamics problems depends strongly on the computational resources [10]. The computational demands are related mainly to the floating point performance and the memory size.

I. Lirkov et al. (Eds.): LSSC 2013, LNCS 8353, pp. 549–556, 2014.
DOI: 10.1007/978-3-662-43880-0_63,

In the last few years the performance of Graphics Processing Units (GPUs) overcame significantly the performance of Central Processor Units (CPUs). At first glance computational fluid dynamics (CFD) solvers match perfectly to GPU resources, because these solvers make intensive calculations and use relatively little memory. Nevertheless, there are scarce results about the practical use of this serious advantage of GPU over CPU, especially for calculations of viscous, compressible, heat conductive gas flows with double precision accuracy. The reported speedups of GPU code to CPU code strongly depend on the mathematical model and the precision of floating point operations. The calculation of Euler flow with single precision in [3] demonstrates speedup of over 40x when comparing GPU NVIDIA 8800GTX and CPU Intel Core 2 Duo. The calculation of incompressible fluid demonstrates speedup of 4.0x when comparing GPU Nvidia C2050 and two Intel Xeon X5650 (six cores CPU) [11], which is equivalent to a speedup of 48x when the GPU code is compared to one CPU core (serial code). The calculation of compressible fluid demonstrates speedup of 11x when comparing GPU NVIDIA Tesla S2070 and serial code executed on Intel Xeon X5650, [6].

The main idea in presented approach is minimization of data transfers between memories. We copy all simulation data to the GPU once at the beginning of the application. Therefore, almost no GPU $\leftrightarrow$ CPU data transfers are necessary during the simulation, similar as [11]. Data transfers between global and local device memories is other possible bottleneck. GPU version of algorithm SIMPLE-TS is developed so that an entire iteration of the iterative process for the calculated time step is calculated by one run of a kernel (see Fig. 3). The calculation of entire loop in a single kernel is a new approach, according to authors knowledge. The algorithm SIMPLE-TS is developed to be easily parallel organized, which makes possible realization of this single kernel concept. As a result data transfers between memories of host $\leftrightarrow$ device and global $\leftrightarrow$ local/private memories of the device are minimized.

The portability of the code and lower price of hardware are important. OpenCL (Open Computing Language) is royalty-free standard for cross-platform, parallel programming of modern processors found in personal computers, servers and handheld/embedded devices (see [4]). OpenCL is supported from wide range of devices, because implementers of OpenCL are: Intel, QUALCOMM, ARM, AMD, Apple Inc., Vivante Corporation, STMicroelectronics International NV, IBM Corporation, Imagination Technologies, Creative Labs, NVIDIA and Samsung Electronics, while CUDA is supported only by NVIDIA. On the other hand performance of AMD Radeon HD 7900 series GPU (AMD Radeon HD GHz edition double precision performance is 1 Tflop) corresponds to NVIDIA GTX Titan (double precision performance is 1 Tflop) and Tesla series (Tesla K20X double precision performance is 1.31 Tflop), while AMD's GPUs are cheaper compared to NVIDIA's GPUs. The portability of application written on OpenCL and performance/cost of AMD's GPUs motivate us to use OpenCL standard and AMD GPU.

In this paper we consider the problem of calculation of a two-dimensional unsteady state gaseous flow past a particle moving with supersonic speed in

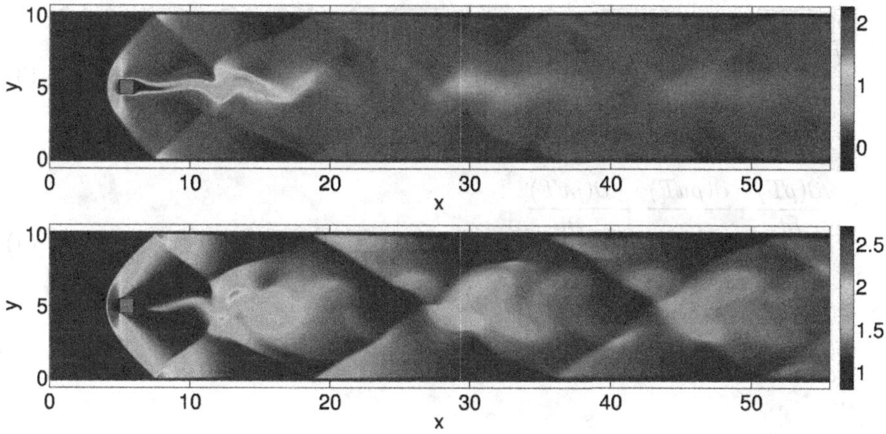

Fig. 1. Horizontal velocity (upper part) and temperature (lower part) fields calculated by GPU version of SIMPLE-TS.

a planar microchannel. We consider an unsteady supersonic flow with Mach number equal to 2.43. The shock wave formed in front of the particle reflects from the channel walls and interacts with the Karman vortex street (see Fig. 1) behind it. The shock wave has significant gradients of velocities, pressure and temperature. Thus, an accurate calculation of the flow requires the use of a very fine mesh, when first order upwind scheme is used for approximation of convective terms and density in middle points. The steady state calculations have been carried out for a set of gradually refined meshes. Finally, a mesh with 8000×1600 cells was found to give stable and accurate enough results [8]. Second order total variation diminishing (TVD) scheme SUPERBEE [5] reduces mesh nodes to 1000×200 cells [7], which is 64 times less. At the moment in GPU code we use 1st order upwing approximation scheme and present results and comparison using this scheme.

2 Continuum Model Equations

A two dimensional system of equations describing the unsteady flow of viscous, compressible, heat conductive fluid can be expressed in a general form as follows:

$$\frac{\partial \rho}{\partial t} + \frac{\partial(\rho u)}{\partial x} + \frac{\partial(\rho v)}{\partial y} = 0 \tag{1}$$

$$\begin{aligned} \frac{\partial(\rho u)}{\partial t} + \frac{\partial(\rho u u)}{\partial x} + \frac{\partial(\rho v u)}{\partial y} &= \rho g_x - A\frac{\partial p}{\partial x} + B\left[\frac{\partial}{\partial x}\left(\Gamma\frac{\partial u}{\partial x}\right) + \frac{\partial}{\partial y}\left(\Gamma\frac{\partial u}{\partial y}\right)\right] \\ &+ B\left\{\frac{\partial}{\partial x}\left(\Gamma\frac{\partial u}{\partial x}\right) + \frac{\partial}{\partial y}\left(\Gamma\frac{\partial v}{\partial x}\right) - \frac{2}{3}\frac{\partial}{\partial x}\left[\Gamma\left(\frac{\partial u}{\partial x} + \frac{\partial v}{\partial y}\right)\right]\right\} \end{aligned} \tag{2}$$

$$\frac{\partial(\rho v)}{\partial t}+\frac{\partial(\rho uv)}{\partial x}+\frac{\partial(\rho vv)}{\partial y}=\rho g_y-A\frac{\partial p}{\partial y}+B\left[\frac{\partial}{\partial x}\left(\Gamma\frac{\partial v}{\partial x}\right)+\frac{\partial}{\partial y}\left(\Gamma\frac{\partial v}{\partial y}\right)\right]$$
$$+B\left\{\frac{\partial}{\partial y}\left(\Gamma\frac{\partial v}{\partial y}\right)+\frac{\partial}{\partial x}\left(\Gamma\frac{\partial u}{\partial y}\right)-\frac{2}{3}\frac{\partial}{\partial y}\left[\Gamma\left(\frac{\partial u}{\partial x}+\frac{\partial v}{\partial y}\right)\right]\right\} \quad (3)$$

$$\frac{\partial(\rho T)}{\partial t}+\frac{\partial(\rho uT)}{\partial x}+\frac{\partial(\rho vT)}{\partial y}$$
$$=C^{T1}\left[\frac{\partial}{\partial x}\left(\Gamma^{\lambda}\frac{\partial T}{\partial x}\right)+\frac{\partial}{\partial y}\left(\Gamma^{\lambda}\frac{\partial T}{\partial y}\right)\right]+C^{T2}.\Gamma.\Phi+C^{T3}\frac{Dp}{Dt} \quad (4)$$

$$p=\rho T \quad (5)$$

where:

$$\Phi=2\left[\left(\frac{\partial u}{\partial x}\right)^2+\left(\frac{\partial v}{\partial y}\right)^2\right]+\left(\frac{\partial v}{\partial x}+\frac{\partial u}{\partial y}\right)^2-\frac{2}{3}\left(\frac{\partial u}{\partial x}+\frac{\partial v}{\partial y}\right)^2 \quad (6)$$

u is the horizontal component of velocity, v is the vertical component of velocity, p is pressure, T is temperature, ρ is density, t is time, x and y are coordinates of a Cartesian coordinate system. The parameters A, B, g_x, g_y, C^{T1}, C^{T2}, C^{T3} and diffusion coefficients Γ and Γ^{λ}, given in Eqs. (1)–(5), depend on the gas model and the equation in non-dimensional form. A first order upwind scheme is used for the approximation of the convective terms and a second order central difference scheme is employed for the approximation of the diffusion terms.

The Navier-Stokes-Fourier equations (1)–(5) are given in general form. For gaseous microflow description we use the model of a compressible, viscous hard sphere gas with diffusion coefficients determined by the first approximation of the Chapman-Enskog theory for low Knudsen numbers [9]. The Knudsen number (Kn), a nondimensional parameter, determines the degree of appropriateness of the continuum model. It is defined as the ratio of mean free path ℓ_0 to macroscopic length scale of the physical system L ($Kn=\ell_0/L$). For the calculated case the Knudsen number is equal to $Kn=0.01$. For a hard-sphere gas, the viscosity coefficient μ and the heat conduction coefficient λ (first approximations are sufficient for our considerations) read as follows:

$$\mu=\mu_h\sqrt{T},\ \mu_h=(5/16)\rho_0\ell_0V_{th}\sqrt{\pi} \quad (7)$$

$$\lambda=\lambda_h\sqrt{T},\ \lambda_h=(15/32)c_p\rho_0\ell_0V_{th}\sqrt{\pi} \quad (8)$$

The Prandtl number is given by $Pr=2/3$, $\gamma=c_p/c_v=5/3$. The dimensionless system of Eqs. (1)–(5) is scaled by the following reference quantities, as given in [9]: molecular thermal velocity $V_0=V_{th}=\sqrt{2RT_0}$ for velocity, for length - square size a (Fig. 1), for time - $t_0=a/V_0$, the reference pressure (p_0) is pressure at the inflow of the channel, the reference temperature (T_0) is equal to the channel walls, reference density (ρ_0) is calculated using equation of state

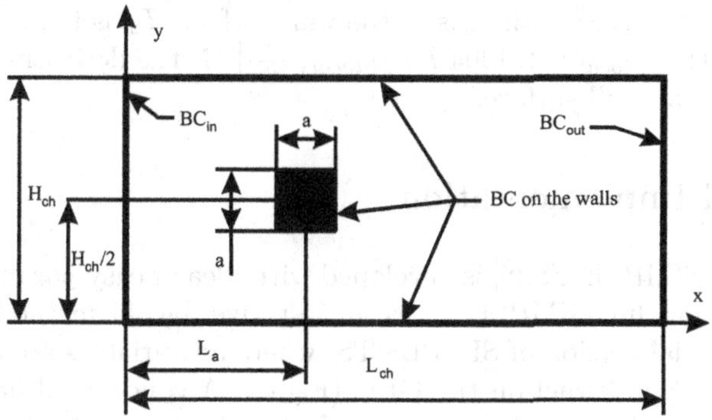

Fig. 2. Flow geometry for a square-shaped particle with size a confined in a channel with length L_{ch} and height H_{ch}.

(5). The corresponding non-dimensional parameters in the equation system (1) –(5) are computed by using the following formulas:

$$\begin{aligned} &A = 0.5,\ B = \frac{5\sqrt{\pi}}{16} Kn,\ \Gamma = \Gamma^{\lambda} = \sqrt{T} \\ &C^{T1} = Kn\sqrt{\pi\frac{225}{1024}},\ C^{T2} = \frac{\sqrt{\pi}}{4} Kn,\ C^{T3} = \frac{2}{5} \end{aligned} \tag{9}$$

3 Test Case Formulation

As a test case we use flow past a square particle in a microchannel at supersonic speed $M = 2.43$. Fig. 2 shows the geometry. The blockage ratio $B = a/H_{ch}$ is equal to $B = 10$, channel length is $L_{ch} = 176$. The problem is considered in a local Cartesian coordinate system, which is moving with the particle. Thus for an observer moving along with the particle the problem is transformed to a consideration of a gas flow past a stationary square confined in a microchannel with moving walls. Velocity-slip and temperature-jump boundary conditions [2] are imposed on the walls of the channel and the square. The velocity-slip BC is given as:

$$v_s - v_w = \zeta \left.\frac{\partial v}{\partial n}\right|_s , \tag{10}$$

where v_s is velocity of the gas at the solid wall surface, v_w is velocity of the wall, $\zeta = 1.1466.Kn_{local} = 1.1466.Kn/\rho_{local}$, Kn_{local} is the local Knudsen number, ρ_{local} is the local density, $\left.\frac{\partial v}{\partial n}\right|_s$ is the derivative of velocity normal to the wall surface. The temperature-jump boundary condition is:

$$T_s - T_w = \tau \left.\frac{\partial T}{\partial n}\right|_s , \tag{11}$$

where T_s is temperature of the gas at the wall surface, T_w is temperature of the wall, $\tau = 2.1904.Kn_{local} = 2.1904.Kn/\rho_{local}$, $\frac{\partial T}{\partial n}\big|_s$ is the derivative of temperature normal to the wall surface.

4 Parallel Implementation

The algorithm SIMPLE-TS [8] is developed with idea of easy parallel implementation. The algorithm SIMPLE-TS is an iterative Jacobi method. The entire loop 2 in the serial version of SIMPLS-TS, where all variables are calculated, is executed as a single kernel on the GPU (Fig. 3). A conditional of the form if-then-else generates branching (see [1]). To reduce branching (code serialisation) we used the Kronecker delta function Fig. 4. Thus the number of floating point operations in the GPU code becomes 917, while the CPU code has nearly 1.5

CPU (serial) [8]

Initialize variables.
Start loop 1:

Set the initial condition for the calculated time step.
Start loop 2 (calculating a state for a new time step):
  Calculate convective and diffusion fluxes.
  Calculate pseudo velocities (velocities, without pressure term), coefficients for pressure equation.
  Start loop 3:
    Calculate the coupled equations for energy and pressure.
  Stop loop 3. In most cases two iterations are sufficient.
  Calculate velocities using pseudo velocities and pressure (calculated within loop 3).
  Compute density, using pressure and temperature calculated within loop 3.
  Convergence of loop 2: Check for convergence of the iteration process for the current time step.
Convergence of loop 1: If the final time is not reached continue.

GPU code

Initialize variables.
Start loop 1:

Set the initial condition for the calculated time step.
Start loop 2 (calculating a state for a new time step):
  Run a kernel on GPU:
    Calculate equation for energy.
    Calculate pseudo velocities (velocities, without pressure term), coefficients for pressure equation and store data in local memory.
    Calculate equation for pressure.
    Calculate velocities using pseudo velocities and pressure.
  End kernel.
  Convergence of loop 2: Check for convergence of the iteration process for the current time step.
Convergence of loop 1: If the final time is not reached continue.

Fig. 3. Algorithm SIMPLE-TS for CPU (serial) and GPU.

```
                CPU code                                GPU code

if(0>u(i,j)) rho_u(i,j)=rho(i-1,j);  rho_u(i,j)=(0>u(i,j))*rho(i-1,j)
else         rho_u(i,j)=rho(i,j);                +(!(0>u(i,j)))*rho(i,j);
```

Fig. 4. Correspondence of if/else statement for CPU code and Kronecker delta expression for GPU code.

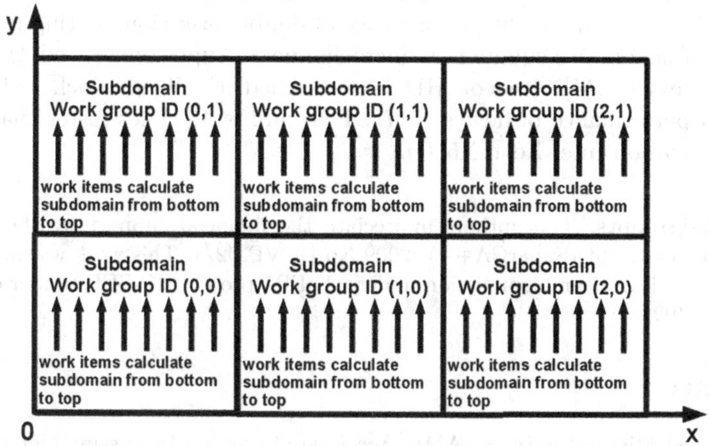

Fig. 5. Domain decomposition of computational domain.

times less floating point operations. At least two or three iterations of loop 2 are needed for the GPU algorithm SIMPLE-TS to become convergent.

A domain decomposition (data partitioning) approach is used. Each work group corresponds to a subdomain, Fig. 5. All threads of a work group calculate the same j. Threads calculate nodes in a subdomain from bottom to top. Some preliminary calculations of a subdomain have to be done.

5 Speedup Analysis

The mesh of the test case is 3528×200 points. A first order upwind scheme is used for the approximation of the convective terms, and a second order central difference scheme is employed for the approximation of the diffusion terms.

Computational domain is divided to 28 subdomains, which correspond to number of compute units of AMD Radeon HD 7950. Each work group has 256 threads, calculates 252 nodes on OX axis and 4 threads calculate halo region of subdomain.

We compare serial CPU code and GPU code running on AMD Radeon HD 7950 with 717 GF/s maximum performance, when using double precision floating point calculations. GPU code reached 90 GF/s, which is 1/8th of the maximum performance of the device and requires 50 % more floating point operations than the CPU serial code. Nevertheless the GPU code is 46 times faster than the CPU serial code run on Intel Xeon X5560.

6 Conclusions

SIMPLE-TS algorithm calculates Navier-Stokes-Fourier system of partial differential equations describing unsteady, viscous, compressible, heat-conductive gas flows. When we compare the performance of double precision floating point calculations of flow past a square in a microchannel at supersonic speed $M = 2.43$, the GPU device AMD Radeon HD 7950 reached 90 GF/s, which is 1/8th of maximum performance of device (717 GF/s) and is 46 times faster than serial CPU code run on Intel Xeon X5560.

Acknowledgments. The authors appreciate the financial support by the NSF of Bulgaria under Grant (SuperCA++)- 2009 No DCVP 02/1. This work was supported in part by the European Commission under EU FP7 project HP-SEE (under contract number 261499).

References

1. Advanced Micro Devices, I.: AMD Accelerated Parallel Processing OpenCL Programming Guide (v2.8) (2012). http://developer.amd.com/appsdk
2. Cercignani, C.: Theory and Application of the Boltzmann Equation. Scottish Academic Press, Edinburgh (1975)
3. Elsen, E., LeGresley, P., Darve, E.: Large calculation of the flow over a hypersonic vehicle using a GPU. J. Comput. Phys. **227**(24), 10148–10161 (2008). http://www.sciencedirect.com/science/article/pii/S0021999108004476
4. The open standard for parallel programming of heterogeneous systems - OpenCL (Open Computing Language). http://www.khronos.org/opencl/
5. Roe, P.L.: Some contributions to the modelling of discontinuous flows. Lect. Appl. Math. **22**, 163–193 (1985)
6. Salvadore, F., Bernardini, M., Botti, M.: GPU accelerated flow solver for direct numerical simulation of turbulent flows. J. Comput. Phys. **235**, 129–142 (2013). http://www.sciencedirect.com/science/article/pii/S0021999112006018
7. Shterev, K.S., Ivanovska, S.: Comparison of some approximation schemes for convective terms for solving gas flow past a square in a microchannel. AIP Conf. Proc. **1487**(1), 79–87 (2012). http://link.aip.org/link/?APC/1487/79/1
8. Shterev, K.S., Stefanov, S.K.: Pressure based finite volume method for calculation of compressible viscous gas flows. J. Comput. Phys. **229**(2), 461–480 (2010). http://dx.doi.org/10.1016/j.jcp.2009.09.042
9. Stefanov, S., Roussinov, V., Cercignani, C.: Rayleigh-Bénard flow of a rarefied gas and its attractors. I. Convection regime. Phys. Fluids **14**(7), 2255–2269 (2002). http://link.aip.org/link/?PHF/14/2255/1
10. Versteeg, H.K., Malalasekra, W.: An Introduction to Computational Fluid Dynamics: The Finite Volume Method, 2nd edn. Prentice Hall, Pearson (2007)
11. Zaspel, P., Griebel, M.: Solving incompressible two-phase flows on multi-GPU clusters. Comput. Fluids **80**, 356–364 (2013). http://www.sciencedirect.com/science/article/pii/S0045793012000308

Pseudorandom Bit Generator with Parallel Implementation

Borislav Stoyanov(✉) and Krasimir Kordov

Department of Computer Informatics, Faculty of Mathematics and Informatics, Konstantin Preslavski University of Shumen, Shumen, Bulgaria
borislav.stoyanov@shu-bg.net, krasimir.kordov@shu-bg.net

Abstract. In this work we present a new filtering variant of pseudorandom bit generation which combines a 2-adic Feedback with Carry Shift Register and the Editing bit-search generator, based on I. Erguler and E. Anarim research. The generated algorithm uses dynamic blocks for data encryption. Shrinking the block is necessary if low memory is available. The algorithm divides the data block among the cores (the processors) and every core encrypts part of the data file. We show the advantage of using parallel implementation. The experimental statistical results establish the time difference of performance on serial and parallel encryption.

The security of the generated bit streams are proven by using NIST, DIEHARD and ENT testing systems.

Keywords: Feedback with carry shift register · Editing generator · Parallel implementation

1 Introduction

Over the last few years, Feedback Shift Registers (FSR) in combination with different filtering rules have been used broadly in the pseudorandom bit generation. The most used FSRs are the Linear Feedback Shift Register (LFSR) [5] and the Feedback with Carry Shift Registers (FCSR) [9–11]. The registers offers an additional security by implementing more nonlinearity in the cipher schemes.

2 Previous Research

The Self-shrinking generator [15] is a well studied scheme. Its algorithm clocks two bits (a_{2i}, a_{2i+1}) from a single LFSR; if the second bit is 1 the output is the first bit. In [21] the Self-shrinking scheme is applied on FCSR.

In [1] another variant of the self-shrinking generator is introduced, where if a generated bit pair (a_{2i}, a_{2i+1}) equals the value (1, 0) or (0, 1) the scheme produces 0 or 1 respectively. It is performed a LFSR based simulation with this

I. Lirkov et al. (Eds.): LSSC 2013, LNCS 8353, pp. 557–564, 2014.
DOI: 10.1007/978-3-662-43880-0_64, © Springer-Verlag Berlin Heidelberg 2014

algorithm. This rule is combined with a FCSR in [19]. The Bit-Search Generator (BSG), based on a single input infinite sequence, is presented in [6]. Its decimation type construction is closely related to the Self-shrinking generator.

ABSG and MBSG are the improved versions of the BSG scheme which were proposed in [7]. MBSG stands for Modified Bit Search Generator, ABSG stands for Another Bit Search Generator.

The scheme of the Editing bit-search generator is investigated in [4]. The algorithm is a new bit inserting technique and increases balancing in the input sequence.

A novel hardware-oriented filtered FCSR stream cipher is presented in [2]. Two variants of the software oriented FCSR-based ciphers are proposed in [3].

In the past few years many researchers have proposed symmetric ciphers with parallel encryption/decryption parts. The schemes of two transformations of the FCSR into sub-sequences generators are investigated in [12]. In [20] new multi-threaded model for stream cipher algorithm is presented. Implementations of the Advanced Encryption Standard algorithm using parallel computing are introduced in [8] and [17]. Parallel encryption algorithms for dual-core processor based on chaotic maps are proposed in [13] and [23].

3 Basic Primitives

3.1 Feedback with Carry Shift Registers

FCSR is a FSR with a small amount of auxiliary memory. Let us fix an odd positive integer q, so named connection integer, $q \in \mathbf{Z}$, and let $r = \lfloor log_2(q+1) \rfloor$ (where $\lfloor \quad \rfloor$ denotes the integral part). Write $q = q_1 2 + q_2 2^2 + \cdots + q_r 2^r - 1$ for a binary representation of the integer $q+1$ (so $q_r = 1$). The feedback connections are given by the numbers from q_1 to q_r. The shift array uses $\lfloor log_2(r) \rfloor$ additional bits of memory, denoted initially m_{n-1}, and r elements, denoted by $a_{n-1}, a_{n-1}, \ldots, a_{n-r+1}, a_{n-r}$. On every clock the shift array forms the integer sum

$$\sigma_n = \sum_{k=1}^{r} q_k a_{n-k} + m_{n-1} \tag{1}$$

and shifts the contents one step to the right, outputting the rightmost binary value a_{n-r}. Then it assigns $a_n = \sigma_n \pmod 2$ into the leftmost cell of the shift register and replaces the memory integer m_{n-1} with $m_n = \lfloor \sigma_n / 2 \rfloor$.

The output sequence $\mathbf{a} = (a_0, a_1, a_2 \ldots)$ is strictly periodic if the following conditions are satisfied [2,11]:

- q is a prime number of $r+1$ bits.
- 2 is a primitive root modulo q.
- $q = 2d+1$ with d prime number.
- The Hamming weight $wt(q)$ of the binary representation of q is greater than $r/2$.

3.2 Editing Bit-Search Generator

The Editing bit-search generator is a new algorithm which is a modified version of ABSG. Assume that we have a binary input string **s** and the corresponding output string **z**. Lets denote the inserted memory bit as t with initial loading of 0. The generator searches the codeword $\bar{e}e^i\bar{e}$ for $i \geq 0$ and the inserted bit is t. If i is odd, then value of t is changed to its complementary as $t = \bar{t}$, otherwise nothing changes.

The EBSG produces bit strings with high period, perfect linear complexity and good randomness for the input sequence generated by a maximum length LFSR.

4 The Proposed Pseudorandom Bit Generator

Taking into account their positive features we introduce new scheme which combines a single feedback with carry shift register and the Editing bit-search rule.

The algorithm of the proposed scheme begins with choice of a big connection integer q, following the conditions from Subsect. 3.1, which defines the feedback taps. The initial filling of the FCSR is from the key K of $r + 1$ bitsize. The first r bits are directly used to initialize the shift register and the last one is stored in the auxiliary memory. The next step is clock the created FCSR $4 \times r$ times. The last step is regularity clock of automation and filter the output with the Editing bit-search rule.

Due to the software experiments we have used our previously generated 256 bit connection number [19]:

$$\begin{aligned} q = \; & 6914042789045916538839448685662679412843546653910203344 \\ & 574256414391833368993 9. \end{aligned} \tag{2}$$

The binary expansion form of the number is:

$$\begin{aligned} & 100110001101110000011000011010111110110010101100110111 01 \\ & 0111110001111101100100000111011101111101110110101110101 1 \\ & 1010010101111010110011001110101000101111110111101110101 0 \\ & 1111010100101001011010001001101010000111111001100010001 \\ & 1011011101110010000000000001010011. \end{aligned} \tag{3}$$

The Hamming weight of the number is 142. There are 141 feedback connections and 255 stages in the shift register.

5 Security Analysis

5.1 Resistance Against Attacks

The attacks against the Editing bit-search rule applied on the LFSR will be unrealistic in the new combined scheme. Because of the nonlinear output, it is impossible to apply directly the rational approximation attack [11].

Table 1. NIST statistical test results

NIST statistical tests	*P-value*	Pass rate
Frequency (monobit)	0.012128	990/1000
Block-frequency	0.818343	994/1000
Cumulative sums (Forward)	0.823725	991/1000
Cumulative sums (Reverse)	0.496351	993/1000
Runs	0.500279	993/1000
Longest run of ones	0.160805	988/1000
Rank	0.693142	994/1000
FFT	0.190654	989/1000
Non-overlapping templates	0.511966	990/1000
Overlapping templates	0.794391	987/1000
Universal	0.217857	989/1000
Approximate entropy	0.749884	994/1000
Random-excursions	0.401045	608/614
Random-excursions variant	0.637757	610/614
Serial 1	0.435430	986/1000
Serial 2	0.099513	989/1000
Linear complexity	0.881662	986/1000

5.2 Statistical Analysis

To determine the randomness of arbitrary long binary strings produced by a new proposed scheme the NIST [18], DIEHARD [14] and ENT [22] test packages are used.

The NIST statistical test suite includes 15 tests, which focus the attention on a variety of different types of non-randomness that could exist in a string. For the NIST tests, we generated 1000 keystreams of length 10^6 bits. We tested all outputs using the 15 tests with default parameters. The probability *P-value* should be $\geq$0.0001 and the Pass rate should be >980. The results are given in Table 1.

The minimum pass rate for each statistical test with the exception of the Random-excursion Variant test is larger than 980 for a sample size = 1000 binary sequences. The minimum pass rate for the Random-excursion Variant test is approximately = 600 for a sample size = 614 binary sequences. All tests are passed successfully. The result shows that the binary sequences generated by the new scheme have properties like a truly random sequence.

The DIEHARD package consists of 18 different statistical tests. For the DIEHARD tests, we generated a file with 80 million bits. The results are given in Table 2. All *P-values* are in acceptable range of $(0, 1)$. All tests passed.

The ENT package performs 6 tests to sequences of bytes stored in files and outputs the results of those tests. We tested output string of 125000000 bytes of the proposed scheme. The results are summarized in Table 3. It is obvious that the entropy value and the arithmetic mean value are very close to the theoretical ones of 8 and 127.5. The proposed generator passed all the tests of ENT.

Table 2. DIEHARD statistical test results

DIEHARD statistical tests	*P-value*
Birthday spacings	0.220956
Overlapping 5-permutation	0.588574
Binary rank (31 × 31)	0.625575
Binary rank (32 × 32)	0.552443
Binary rank (6 × 8)	0.630167
Bitstream	0.491649
OPSO	0.507526
OQSO	0.420654
DNA	0.456384
Stream count-the-ones	0.529472
Byte count-the-ones	0.426509
Parking lot	0.928680
Minimum distance	0.942912
3D spheres	0.021456
Squeeze	0.417780
Overlapping sums	0.837666
Runs up	0.522235
Runs down	0.589728
Craps	0.554436

Table 3. ENT statistical test results

ENT statistical tests	Results
Entropy	7.999998 bits per byte
Optimum compression	OC would reduce the size of this 125000000 byte file by 0 %.
χ^2 distribution	For 125000000 samples is 267.00, and randomly would exceed this value 29.02 % of the times.
Arithmetic mean value	127.5001
Monte Carlo π estimation	3.141471026 (error 0.00 %)
Serial correlation coefficient	−0.000028 (totally uncorrelated = 0.0)

Table 4. Parallel performance comparison with one or more processors with different block sizes in seconds

Number of processors	Bytes of a block		
	100	500	1000
1	62.02	64.31	61.54
2	64.04	54.24	52.27
4	90.30	57.07	55.20
8	104.40	54.19	51.94

6 Experimental Multicore Encryption

The parallel experiments were performed on a computer consisting of Pentium (R) Dual-Core T4400, with 4 GB of RAM. The operating system is Windows 7. Tests were run with Open MPI [16], release 1.6.1.

The source code provides solution for both sequential and parallel realization by checking available processors. The parallelization scheme is plaintext based which divides the input file into equal blocks first and then assigns the blocks to available processors. The sequential pseudorandom generation begins on a master processor. When enough pseudorandom bits are generated a block of data is being sent for encryption to one of the slave processors. The slave processors combine the pseudorandom bits with the plaintext using exclusive or operation (XOR) by keeping the offsets. When the encryption of the block is done the slave processor sends the cryptext back to the master processor. The master processor receives the data blocks in the same order as sent and combines the crypted data blocks into the encrypted file. These operations are repeated until the end of file is reached.

The experiments were made using 1, 2, 4, and 8 processors, testing file size 100 MB and block size 100, 500, and 1000 Bytes. The advantages of parallel implementation compared with sequential implementation is demonstrated in Table 4.

Table 4 indicates that using more processors reduces the time necessary for generating the bit streams. The use of larger size of the memory blocks also has positive effect on the performance. In the cases of 500 and 1000 bytes of block the performance of the pseudorandom generator is improved. Although the improvements are small under 20 %, we can summarize that the parallelizing has a positive result.

7 Conclusion

We have presented a novel pseudorandom cryptographic scheme constructed from Feedback with Carry Shift Register, filtered by the Editing Bit-Search Generator. The security analysis results show that the proposed pseudorandom derivative system can assure security in digital communications.

Acknowledgements. This paper is supported by the Project BG051PO001-3.3.06-0003 "Building and steady development of PhD students, post-PhD and young scientists in the areas of the natural, technical and mathematical sciences". The Project is realized by the financial support of the Operative Program "Development of the human resources" of the European social fund of the European Union.

The authors would like to thank Miroslav Kolev, Delian Sarmov, Nikolay Yankov, and Georgi Dimitrov for their comments and suggestion on earlier drafts of this paper.

References

1. Al Jabri, A.: Shrinking generators and statistical leakage. Comput. Math. Appl. **32**(4), 33–39 (1996)
2. Arnault, F., Berger, T.P.: F-FCSR: design of a new class of stream ciphers. In: Gilbert, H., Handschuh, H. (eds.) FSE 2005. LNCS, vol. 3557, pp. 83–97. Springer, Heidelberg (2005)
3. Arnault, F., Berger, T.P., Lauradoux, C., Minier, M.: X-FCSR – a new software oriented stream cipher based upon FCSRs. In: Srinathan, K., Rangan, C.P., Yung, M. (eds.) INDOCRYPT 2007. LNCS, vol. 4859, pp. 341–350. Springer, Heidelberg (2007)
4. Erguler, I., Anarim, E.: The editing bit-search generator. In: National Cryptology Symposium II, pp. 154–165, Ankara (2006)
5. Golomb, S.: Shift Register Sequences. Aegean Park Press, Laguna Hills (1982)
6. Gouget, A., Sibert, H.: The bit-search generator. In: The State of the Art of Stream Cipher: Workshop Record, pp. 60–68 (2004)
7. Gouget, A., Sibert, H., Berbain, C., Courtois, N.T., Debraize, B., Mitchell, C.: Analysis of the bit-search generator and sequence compression techniques. In: Gilbert, H., Handschuh, H. (eds.) FSE 2005. LNCS, vol. 3557, pp. 196–214. Springer, Heidelberg (2005)
8. Karthikeyan, S., Sairam, N., Manikandan, G., Sivaguru, J.: A parallel approach for improving data security. J. Theor. Appl. Inf. Technol. **39**(2), 119–125 (2012)
9. Klapper, A.: Feedback with carry shift register over finite fields. In: Preneel, B. (ed.) FSE 1994. LNCS, vol. 1008, pp. 170–178. Springer, Heidelberg (1995)
10. Klapper, A.: On the existence of secure feedback registers. In: Maurer, U.M. (ed.) EUROCRYPT 1996. LNCS, vol. 1070, pp. 256–267. Springer, Heidelberg (1996)
11. Klapper, A., Goresky, M.: Feedback Shift registers, 2-adic span, and combiners with memory. J. Cryptol. **10**(2), 111–147 (1997)
12. Lauradoux, C., Röck, A.: Parallel generation of ℓ-sequences. In: Golomb, S.W., Parker, M.G., Pott, A., Winterhof, A. (eds.) SETA 2008. LNCS, vol. 5203, pp. 299–312. Springer, Heidelberg (2008)
13. Liu, J., Song, D., Xu, Y.: A parallel encryption algorithm for dual-core processor based on chaotic map. In: Zeng, Z., Li, Y. (eds.) ICMV 2011. Proceedings of SPIE 8350, pp. 83500B-1–83500B-7 (2012)
14. Marsaglia, G.: DIEHARD: a Battery of Tests of Randomness. http://www.stat.fsu.edu/pub/diehard/
15. Meier, W., Staffelbach, O.: The self-shrinking generator. In: De Santis, A. (ed.) EUROCRYPT 1994. LNCS, vol. 950, pp. 205–214. Springer, Heidelberg (1995)
16. Message Passing Interface Forum: MPI: A Message-Passing Interface Standard, Version 3.0. High Performance Computing Center, Stuttgart (2012)
17. Pachori, V., Ansari, G., Chaudhary, N.: Improved performance of advance encryption standard using parallel computing. Int. J. Eng. Res. Appl. **2**(1), 967–971 (2012)
18. Rukhin, A., Soto, J., Nechvatal, J., Smid, M., Barker, E., Leigh, S., Levenson, M., Vangel, M., Banks, D., Heckert, A., Dray, J., Vo, S.: A statistical test suite for random and pseudorandom number generators for cryptographic application. Natl. Inst. Stand. Technol. Spec. Publ. 800–22rev1a (2010)
19. Stoyanov, B., Kolev, M., Nachev, A.: Design of a new self-shrinking 2-adic cryptographic system with application to image encryption. Eur. J. Sci. Res. **78**(3), 362–374 (2012)

20. Suwais, K., Samsudin, A.: High performance multithreaded model for stream cipher. Int. J. Comput. Sci. Netw. Secur. **8**(3), 228–233 (2008)
21. Tasheva, Z., Bedzhev, B., Stoyanov, B.: Self-shrinking p-adic cryptographic generator. In: Milovanović, B. (ed.) XL International Scientific Conference on Information, Communication and Energy Systems and Technologies, pp. 7–10. Niš (2005)
22. Walker, J.: ENT: A Pseudorandom Number Sequence Test Program. http://www.fourmilab.ch/random/
23. Wang, W., Wang, X., Song, D.: A parallel chaotic cryptosystem for dual-core processor. In: 2nd International Conference on Information and Engineering, pp. 920–923. IEEE Press, New York (2010)

Reengineering and Extending the Agents in Grid Ontology

Paweł Szmeja[1], Katarzyna Wasielewska[1], Maria Ganzha[1], Michał Drozdowicz[1], Marcin Paprzycki[1], Stefka Fidanova[2], and Ivan Lirkov[2](✉)

[1] Systems Research Institute, Polish Academy of Sciences, ul. Newelska 6, 01-447 Warsaw, Poland
{pawel.szmeja,katarzyna.wasielewska,drozdowicz}@gmail.com,
{maria.ganzha,marcin.paprzycki}@ibspan.waw.pl

[2] Institute of Information and Communication Technologies, Bulgarian Academy of Sciences, Acad. G. Bonchev, bl. 25A, 1113 Sofia, Bulgaria
{stefka,ivan}@parallel.bas.bg

Abstract. Ontology engineering, despite considerable progress, is still relatively new and dynamically evolving discipline. As a result, the *universal* standards for creating and/or editing an ontology, have not been established. This leads to problems with reusing and updating existing ontologies. It also makes writing an ontology from scratch seem like a good idea. The aim of this paper is two-fold. First, to discuss key issues encountered during re-engineering of an existing ontology. Second, to show how the good practices of ontology development were applied to model the area of computational linear algebra. Here, special attention is paid to the application of this ontology in the user support system.

1 Introduction

The context for this paper is provided by the *Agents in Grid* project (*AiG*; [2,5,8]), which aims at development of an agent-based infrastructure for intelligent resource management in the Grid. The *AiG* project combines software agents and semantic data processing. Specifically, all knowledge in the system is stored in/represented as an ontology, while communication protocols utilize messages with ontological content [4]. During the development of the system, three ontologies were designed to provide concepts necessary for: (i) resource and Grid structure description, (ii) contract and requirements specification, and (iii) content of messages exchanged in the system.

As the development of the system progressed, the ontological structures started to become complex (ontology consisting of 401 entities). Furthermore, when reasoning moved beyond simplistic examples (ontologies with a few concepts), we have been confronted with recurring errors generated by the reasoners. Therefore, the ontology reengineering became a necessity.

I. Lirkov et al. (Eds.): LSSC 2013, LNCS 8353, pp. 565–573, 2014.
DOI: 10.1007/978-3-662-43880-0_65,

2 *AiG* Ontology Reengineering

The original *AiG* grid ontology was created on the basis of the *Core Grid Ontology* (*CGO*; 217 entities). The *CGO* was extended (adding 88 entities) and modified to match the needs of the *AiG* project (see, [4,8]). During this process, features identified as most problematic, from the point of view of the *AiG* project, have been modified. Furthermore, constraints and messaging ontologies have been created (96 additional entities). However, no major "checking" of the *CGO* has been performed at that stage. Let us, therefore, summarize key issues encountered when such check was performed for the complete set of *AiG* ontologies.

2.1 Documentation Standards

It is crucial that the ontology is intuitive enough and that the intended use of its entities is clear. In OWL [10], this can be achieved through proper documentation, clear naming scheme, and overall consistency. The ontology should be uniform when representing the real world concepts and objects as OWL classes, properties and individuals. However, as we have found, the original *CGO* had problems in this area, and some of them carried over to the *AiG* ontology.

General ontology engineering standards state that names of OWL *classes* should be capitalized, whereas OWL *property* names should start with a lowercase letter, preferably in the format of *"has[Property]"* or *"is[Property]"*. This is particularly important for a hierarchy of ontologies, because the naming schemes carry over to all ontologies that import a given ontology. If ontologies in a hierarchy use different naming conventions, the overall naming scheme is broken.

Here, an example is the *operatingSystem* property that not only conflicted the naming scheme (it applied to properties such as *hasCPU* and *hasFileSystem*), but also could be easily confused with the *OperatingSystem* class. Recall that in OWL, IRIs should be unique in the scope of an ontology, *regardless* of the type of the entity. To solve the problem, the property has been renamed to *hasOperatingSystem*. The remaining (similar) problems have also been fixed.

Proper documentation should help reusability, e.g. by explaining how the ontology is intended to be used. While the OWL annotations can be used as the documentation, this was not the case with the *CGO* (only 7 classes had commenting annotations). The *AiG* ontologies are constantly updated with annotations that are to serve both as guidelines for the users and as reminders for the developers. In the future, we plan to use annotations in the dynamic user interface (see [3] for more details). Here, the GUI, in addition to adjusting to the ontology structure, would also display information (contained in annotations) to explain to the users (a) the entities in the ontology, and (b) their intended meaning.

2.2 Ontology Hierarchy

Recall, that the *AiG* grid ontology *extended* the *CGO* ontology to better fit the needs of the *AiG* system. When analyzing the interplay between these two

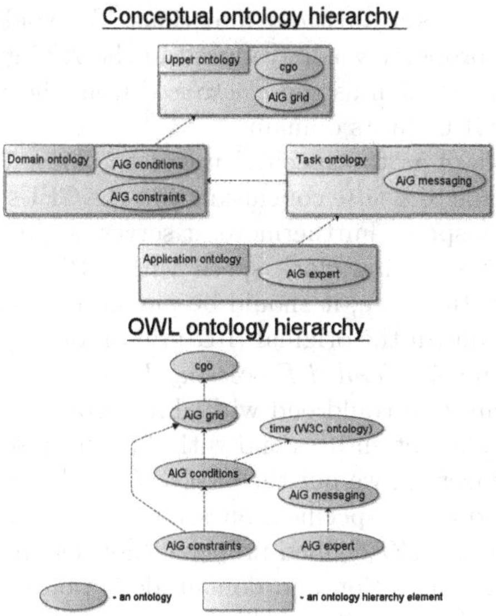

Fig. 1. Hierarchy of ontologies in *AiG*

ontologies it becomes clear that they are "conceptually" on the same level. Entities defined in the *CGO* could be transferred to the *AiG* grid ontology and vice-versa. This could be done without disturbing the main ideas underlying both ontologies and the *AiG* system. Furthermore, this could be done without impacting the *work* of the *AiG* system. However, this means that the *AiG* grid ontology must be used together with the *CGO*. This demonstrates a more general issue that is rarely discussed. The typical ontology hierarchy does not take into account the fact that, on each conceptual level, there may exist multiple *ontology files*. In the *AiG* system, the ontological base consists of the *CGO* and the *AiG* ontologies, with the messaging (*AigMessagingOntology*), the contract constraints (*AiGConstraintsOntology*), and the domain (expert, *AiGExpertOntology*) ontologies placed deeper in the hierarchy. Figure 1 presents the relation of ontology files within the actual ontology hierarchy.

The reengineering that started with the *CGO* involved changes that had to be immediately reflected in the *AiG* grid ontology, in order to preserve the connection between them, and to prevent introducing (new) errors. An example of how the original *CGO* was unsuitable for being extended was the *clockSpeed* property. It's original use, in the *CGO*, is summarized by two constructs: the restriction on the *CPU* class, and the domain specification on itself. The first states that every CPU needs a defined property *clockSpeed*. The latter restricts the *clockSpeed* to the CPUs *only*. The *AiG* grid ontology introduced the GPUs that had also to be described by the clock speed. Because of the domain restriction it was impossible to use the *clockSpeed* property from the *CGO*. Any GPU that used this property would be inferred to be a CPU. While technically correct, such inference was against our intentions. To avoid changing the *CGO* file, a *hasClockSpeed* property was introduced in the *AiG* grid ontology. There, it had the same interpretation as the *clockSpeed* from the *CGO*, only with the GPU, as well as the CPU, in its domain.

This is an example of a "too specific" upper ontology. When narrowing the domain, one might come to a *false* conclusion that the CPUs are the only objects characterized by clock speed. Furthermore, it serves no purpose in the scope of the ontology itself. Associating clock speed with CPUs is the *suggested* use, not the *only use* and, therefore, it should be put in annotations. Similarly, the extension of the domain (in the original *AiG* grid ontology) was incorrect. Note that, if we added some *Accelerated Processing Unit* to the ontology, we would face the same problem, and could end with three properties, each representing clock speed, but for different entities and with differently scoped domains.

During the reengineering, we put the *hasClockSpeed* property in place of the *clockSpeed*, with the domain specification set to $Thing$ or CPU or GPU. In other words, we use the *CGO* defined property (not defining our own) and add suggested use in both annotation and domain definition (without narrowing it down). In this way we fixed similar problems associated with other properties.

2.3 Cleaning Conceptual Inconsistencies

A number of entities with the same intended meaning were present (at the same time) in both in the *CGO* and the *AiG* ontologies. For example, both ontologies included a *CPU* class. They had a different IRI base and a different definition (e.g. one had the *clockSpeed* property in the definition). Individuals that should belong to a *single CPU* class were divided between them. As a result, the reasoning about individuals in the *CPU* class never gave a complete result (unless done in the scope of the IRI bases of both ontologies and then combined). As a consequence, multiple reasoners (tried in the system) had problems with creating an inferred hierarchy, or classifying the ontology. Note that, these errors became apparent only after reasoners started to be used in a working system on the full-blown ontology (400+ entities) rather than on mini-examples (10–20 entities) used in testing the agent infrastructure.

We have found that this problem resulted from a misconception (or a bug) in earlier versions of the Protegé platform that assumed that classes of newly created individuals belong to the active ontology, and asserted their existence (if they were not present). The newer versions of Protegé do not suffer from this problem. This shows that growth of knowledge about "ontologies in practice" leads to development of better tools, but leaves behind ontologies with limited usability. All such problems were fixed, which also solved the reasoning errors.

3 Lessons Learned

Here, let us note that literature considers mostly ontology creation, rather than long-term (re)use (see, for instance [1]). Furthermore, ongoing research concerns ontology merging, alignment, mapping, but almost nothing concerns "software engineering like" principles for ontology re-use. This being the case let us summarize the most important lessons learned from our work.

First, one should be mindful of the existing (or planned) ontology hierarchy, and how a new/modified ontology would fit into it. Hierarchies can vary but it's always good to remember that upper ontologies should contain "general concepts" and avoid introducing unnecessary conditions that would restrict usage of upper entities. Consequently, the hierarchy level should be reflected in the level of ontological specialization, when moving deeper into the imports chain.

Second, ontologies are meant to be reused. Thus, it is crucial to clearly communicate their intended use (e.g. by providing complete annotations and adhering to the naming standards). As seen in the examples above, this can help prevent misusing a concept or (re)defining it more times than intended.

Finally, the crucial lesson is that applying an ontology *in practice* is an indispensable for identifying the problems that exist in its design. It also helps to understand the importance of developed standards and best practices.

4 Adding a Domain Ontology

During the development of the *AiG* system, the need to add a new (created from scratch) ontology arose. Note that the *AiG* system is to provide support beyond the functionalities found in the existing Grid middlewares. Specifically, ontological representation of domain knowledge is to be a part of the decision support provided to the user. For instance, it should help the user to choose optimal algorithm and/or resource to solve her problem. Hence, this is another attempt (using modern tools) to achieve goals summarized in [6,7]. While work completed in 1990's did not gain traction, we believe that with help of ontologies and semantic data processing we may have more success. As a starting point, we have focused on computational linear algebra. The ontology under development is extending the existing *AiG* ontologies, and created taking into account the lessons learned from the reengineering of the *AiG* ontologies. The main goal of the *AiGExpertOntology* is to provide concepts necessary to capture three aspects of the domain: (i) problems to be solved, (ii) algorithms to solve them, (iii) objects that these algorithms operate on. Additionally, classes *DomainExpert* and *ExpertOpinion* where introduced to represent experts knowledge (recommendations) allowing matching of problems and algorithms. Therefore, the *ExpertOpinion* class has property *hasRecommendedResource*, which points to a resource that is most suitable for solving a specific problem (according to the expert). Obviously, resources originate from the *AiG* ontology. Let us now present the preliminary hierarchy of problems in computational linear algebra (Fig. 2). Here, we distinguished five types of problems represented with OWL classes: eigenproblem that can be further categorized into eigenvalue or eigenvector problem, least squares problem, solution of a system of linear equations, and calculation of a matrix norm.

The second part is the *Algorithm*; a superclass for classes (in Fig. 3, we present a fragment of this hierarchy) representing algorithms that can be used

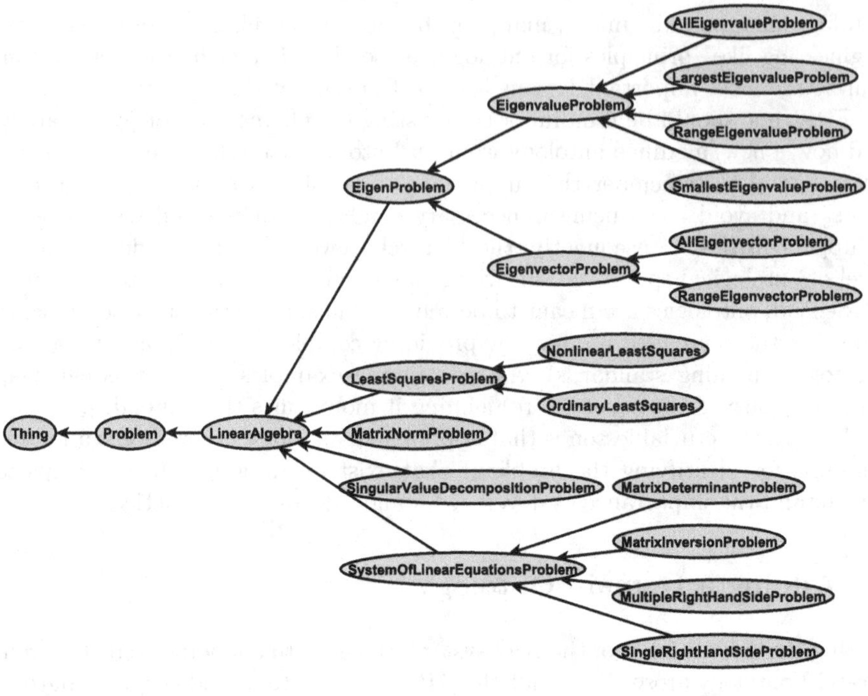

Fig. 2. Hierarchy of problems in *AiGExpertOntology*

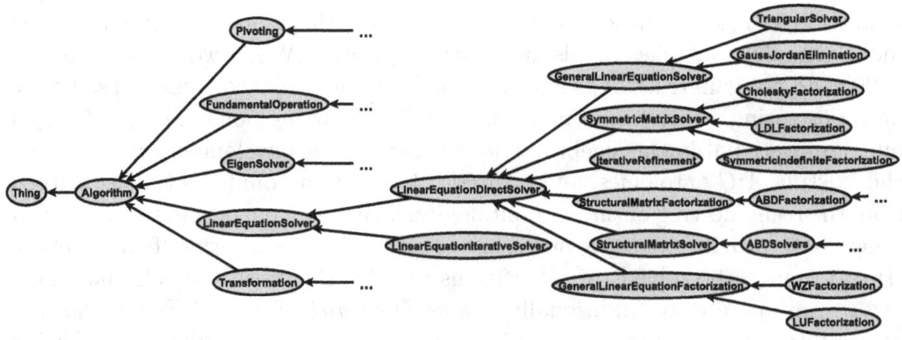

Fig. 3. Part of hierarchy of algorithms in *AiGExpertOntology*

to solve problems from Fig. 2, for a given input data (represented in the *Matrix* class). This part of the ontology is going to be most complex and is being developed based on domain expert knowledge.

Finally, we develop *Matrix* and *MatrixProperty* classes (Fig. 4) and the property *hasMatrixProperty* that defines their relationship. The *MatrixProperty* class is a superclass for a hierarchy of properties that describe the matrix (e.g.

symmetricity, density, structure, etc.). Obviously, in Fig. 4 we present only fragments of the ontology that is being extended on the basis of expert opinions.

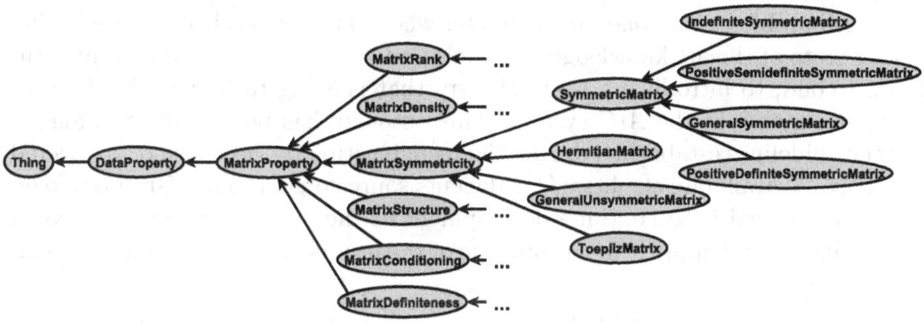

Fig. 4. Part of hierarchy of matrix properties in *AiGExpertOntology*

To illustrate how we plan to use the *AiGExpertOntology* ontology, let us consider a scenario, where the user is looking for a team to commission a job. Here, she could specify only requirements for resources needed to execute the job (assuming that she is certain about her needs). However, she could also indicate an individual of a subclass of the *Problem* class (Fig. 2), e.g. *SystemOfLinearEquationsProblem*. Such individual may have (optional) properties that specify (i) the input, e.g. individual of class *Matrix* with values of *hasProperty* being individuals of classes *PositiveDefiniteMatrix* and *SymmetricMatrix*, and (ii) algorithm, e.g. individual of class *CholeskyFactorization*. The first use case is to validate user request for a resource against resources recommended by experts for a combination of problem, input data, and algorithm. User's resource specification is evaluated against experts suggestions using Saaty's Analytical Hierarchy Process (AHP) for multicriterial assessment. The way to combine ontologies and the AHP method was introduced in [9].

Here, two use cases can be distinguished. First, when user requirements are significantly disjoint from the expert suggestions (e.g. request for *GeneralSolver* is made for a *SymmetricMatrix*), he will be provided with alternative suggestion(s) and thus may modify his request. Second, when user requirements are not very detailed, they will be made more specific by accommodating experts opinions. For instance, when user specified the problem, the matrix type (and size), and the algorithm, the system can additionally suggest the CPU / GPU type, and/or memory, and/or number of processors. Similarly, when the user specified only the problem and the matrix type, the expert knowledge and the AHP shall be utilized to suggest the algorithm and resources to be used.

In the *AiGExpertOntology* we follow, earlier specified, guidelines for ontology engineering, e.g. naming conventions for classes and properties, filling annotations for ontology elements. Moreover, we decided that new ontology has to become a new module (separated from previously designed ones).

5 Concluding Remarks

The aim of this paper was, first, to discuss important issues involved in ontology reengineering, based on our experiences with the *AiG* ontology. Here, we have discussed problems that one can encounter when ontology has been created using earlier state of the art knowledge and tools and has to be extended and modernized. Second, to introduce a new ontology that is going to be used in the user decision support in the *AiG* system. This ontology has been developed following the guidelines established during the reengineering process. The reengineered ontology is available at: http://gridagents.sourceforge.net/AiGGridOntology. Our current goal is to continue development of the ontology of computational linear algebra and apply it in a prototype of the user decision support subsystem.

Acknowledgments. Work of the authors was in part supported by bilateral grant between Polish Academy of Sciences and Bulgarian Academy of Sciences and grant DCVP 02/1 of the Bulgarian NSF.

References

1. Dieter, F.: Ontologies: A Silver Bullet for Knowledge Management and Electronic Commerce. Springer-Verlag, New York (2003)
2. Dominiak, M., Kuranowski, W., Gawinecki, M., Ganzha, M., Paprzycki, M.: Utilizing agent teams in Grid resource management-preliminary considerations. In: Proceedings of the IEEE John Vincent Atanasoff Conference, pp. 46–51. IEEE CS Press, Los Alamitos, CA (2006)
3. Drozdowicz, M., Ganzha, M., Wasielewska, K., Paprzycki, M., Szmeja, P.: Using ontologies to manage resources in grid computing: practical aspects. In: Ossowski, S. (ed.) Agreement Technologies, Law, Governance and Technology Series, vol. 8, pp. 149–168. Springer, Netherlands (2013)
4. Drozdowicz, M., Wasielewska, K., Ganzha, M., Paprzycki, M., Attaui, N., Lirkov, I., Olejnik, R., Petcu, D., Badica, C.: Ontology for contract negotiations in agent-based grid resource management system. In: Ivanyi, P., Topping, B. (eds.) Trends in Parallel, Distributed, Grid and Cloud Computing for Engineering, pp. 335–354. Saxe-Coburg Publications, Stirlingshire, UK (2011)
5. Kuranowski, W., Ganzha, M., Gawinecki, M., Paprzycki, M., Lirkov, I., Margenov, S.: Forming and managing agent teams acting as resource brokers in the grid-preliminary considerations. Int. J. Comput. Intell. Res. **4**(1), 9–16 (2008)
6. Lucks, M.: A knowledge-based framework for the selection of mathematical software. Ph.D. thesis, Southern Methodist University (1990)
7. Petcu, D., Negru, V.: Interactive system for stiff computations and distributed computing. In: Proceedings of IMACS'98: International Conference on Scientific Computing and Mathematical Modelling, pp. 126–129. IMACS (1998)
8. Wasielewska, K., Drozdowicz, M., Ganzha, M., Paprzycki, M., Attaui, N., Petcu, D., Badica, C., Olejnik, R., Lirkov, I.: Negotiations in an agent-based grid resource brokering systems. In: Ivanyi, P., Topping, B. (eds.) Trends in Parallel, Distributed, Grid and Cloud Computing for Engineering, pp. 355–374. Saxe-Coburg Publications, Stirlingshire, UK (2011)

9. Wasielewska, K., Ganzha, M.: Using analytic hierarchy process approach in ontological multicriterial decision making - preliminary considerations. In: Todorov, M. (ed.) Proceedings of 4th International Conference-AMiTaNS'12 Memorial Volume devoted to Prof. Christo I. Christov. AIP Conference Proceedings, vol. 1487, pp. 95–103 (2012)
10. OWL 2 Web Ontology Language. http://www.w3.org/TR/owl2-overview/

Contributed Papers

Fitting of Discrete Data with GERBS

Jostein Bratlie(✉), Rune Dalmo, and Peter Zanaty

R&D group in mathematical and geometrical modeling, numerical simulations, programming and visualization, Narvik University College, PO Box 385, 8505 Narvik, Norway
{jostein.bratlie,rune.dalmo,peter.zanaty}@hin.no
http://www.hin.no/Simulations

Abstract. In this paper, we present a study of fitting discrete data with Generalized Expo-rational B-splines. We investigate different ways to determine interpolation knots and generate GERBS local curves by partitioning the parametric space and solving a corresponding least-squares fitting problem. We apply our technique to discrete evaluations of continuous synthetic benchmark functions and compare the resulting GERBS to the original data with respect to errors and performance.

Keywords: Discrete data · Fitting · Interpolation · Significant point · Curvature · Inflection · GERBS

1 Introduction

In this work we investigate the properties of fitting Generalized Expo-Rational B-Spline (GERBS), introduced in [1], to regular discretized data. GERBS is a family of blending type spline constructions, where local functional coefficients are blended by GERBS basis functions. The choice of basis functions and the local enrichment functions determines the local and hence the global approximation properties of the resulting space.

One of the intrinsic properties of the GERBS bases are the minimal support of the basis functions, which allows for a simple approximation technique; instead of storing the individual data points, and then blending the corresponding local functions together, node by node, we can choose the interpolation knots and the accompanying local functions freely, depending on the data itself.

Using this, we investigate various techniques to partition the parametric space of the GERBS across the discrete data by changing the interpolation knots and simultaneously adjusting the corresponding coefficient functions. In addition, we look at the performance of the different constructions with respect to approximation.

Many papers have been published on the topic of data fitting, data reduction, compression and smoothing with B-splines using various methods. We mention here the knot removal technique presented by Lyche and Mørken in [4] and with a different approach by Eck and Hadenfeld in [2], and the shape-preserving knot

I. Lirkov et al. (Eds.): LSSC 2013, LNCS 8353, pp. 577–584, 2014.
DOI: 10.1007/978-3-662-43880-0_66,

removal method by Schumaker and Stanley in [7]. We also mention the work done by Saux and Daniel in [5,6] on estimating criteria for fitting and data reduction of polygonal curves using B-splines. We leave these topics for now and focus on a few simple methods for constructing GERBS local functions.

In Sect. 2 we start by giving a brief introduction to GERBS and its construction, as well as the partitioning and fitting setup we use throughout the article. Then in Sect. 3 we describe the different partitioning algorithms and then follows the description of the fitting method in Sect. 4. Finally in Sect. 5 we give some concluding remarks where we discuss our findings and future work.

2 Preliminaries

2.1 GERBS Basis Functions

Consider a strictly increasing knot vector $\boldsymbol{t} = \{t_k\}_{k=0}^{n+1}$, $t_0 < t_1 < \cdots < t_{n+1}$, $n \in \mathbb{N}$. The definition of the j-th GERBS is defined in [1] as follows.

$$B_j(t) = \begin{cases} F_j(t), & \text{if} \quad t \in (t_{j-1}, t_j], \\ 1 - F_{j+1}(t), & \text{if} \quad t \in (t_j, t_{j+1}), \\ 0, & \text{if} \quad t \in (-\infty, t_{j-1}] \cup [t_{j+1}, +\infty), \end{cases} \tag{1}$$

$j = 1, \ldots, n,$

where $\{F_i\}_{i=1}^{n+1}$ is a system of cumulative distribution functions such that for F_i, $i = 1, \ldots, n$,

1. the right-hand limit $F_i(t_{i-1}+) = F_i(t_{i-1}) = 0$,
2. the left-hand limit $F_i(t_i-) = F_i(t_i) = 1$,
3. $F_i(t) = 0$ for $t \in (-\infty, t_{i-1}]$,
4. $F_i(t) = 1$ for $t \in [t_i, +\infty)$, and $F(t)$ is monotonously increasing, possibly discontinuous, but left-continuous for $t \in [t_{i-1}, t_i]$.

2.2 GERBS Curves

Generalized expo-rational B-splines provide a blending type construction, where local functions at each knot are blended together by sufficiently smooth basis functions

$$s(t) = \sum_{i=1}^{n} \ell_i(t - t_i) B_i(t), \tag{2}$$

where $\boldsymbol{t} = \{t_k\}_{k=0}^{n+1}$ is a strictly increasing knot vector, and each basis function $B_j(t)$ is supported on (t_{j-1}, t_{j+1}) while possessing a Dirac property $B_j(t_i) = \delta_{ij}$.

The local functions ℓ_i throughout this paper shall be Taylor expanding polynomials up to a multiplicity μ_i

$$\ell_i(t - t_i) = \sum_{j=0}^{\mu_i} c_{i,j} \frac{(t - t_i)^j}{j!}, \tag{3}$$

and the corresponding GERBS base $B_i(t)$ is required to have vanishing derivatives of order up to, including, μ_i. For the rest of the paper we will use the Expo-rational B-spline basis described in [3] which is capable of transfinite Hermite interpolation, i.e. all of its derivatives vanish at all knots.

The knots ($\boldsymbol{t}$) and the multiplicities ($\boldsymbol{\mu}$) together define a spline space, where the coefficients ($\boldsymbol{c}$) have a natural meaning corresponding to a Hermite interpolation problem.

2.3 Partitioning and Fitting

In digital systems we often have to deal with continuous (analogue) input data. The way it is handled is that the analogue signal is converted to digital by sampling and quantizing (digitizing) and the resulting raw digital data is being used instead of the original data. Often it is a requirement to produce outputs which are either continuous or sampled at a higher rate than that of the input data, this problem translates into interpolation/extrapolation or approximation problems depending on the requirements.

In the current article, we are interested in comparing different strategies for representing uniformly sampled uni-variate functions with the use of GERBS based approximation, see Fig. 1. The partitioning algorithms work on a sampled data of two benchmark functions, given by

$$f_1(t) = \begin{pmatrix} \ln(t+1) \\ -t\sin(2t+1)) \end{pmatrix}, \qquad t \in [0,1], \tag{4}$$

$$f_2(t) = \begin{pmatrix} t \\ t\sin(\frac{1}{t}) \end{pmatrix}, \qquad t \in [0.01, 0.5], \tag{5}$$

and their task is to select the knot configuration and the corresponding local multiplicities of the spline space.

Then a fitting algorithm obtains the coefficients to the spline representation, finally we compare the resulting splines with both the original continuous benchmark functions $f_1(t)$, $f_2(t)$ and their sampled discrete versions $F_1[k]$, $F_2[k]$ and discuss some properties of the resulting transformations.

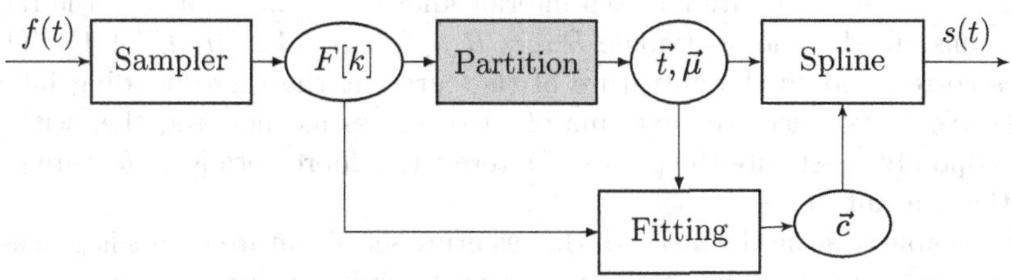

Fig. 1. Partitioning and fitting data

3 Partitioning Algorithms

In order to fit a GERBS construction (see Sect. 2.2) to discrete data, it is necessary to decide where to place the interpolation knots ($\boldsymbol{t}$) and to decide the corresponding multiplicities ($\boldsymbol{\mu}$) of the local functions. This can be done in a number of different ways. We describe three different algorithms for constructing local curves.

3.1 Uniform Partitioning

As a starting point with uniform sampling we define a knot for each discrete data point in $F[k]$. Next, the number of knots is reduced by selecting a subset of F in order to define the spline space. We add as a note here that the data is assumed to be appropriate for selection. (In some cases it is common to smooth the data before selecting to reduce errors or avoid problems related to oscillation.) It is possible to increase the degree by selecting derivatives for each knot. We illustrate uniform partitioning with three different examples:

1. Fixed sample rate
2. Specified number of knots
3. Parametric stride

In the first case, the sample rate simply states how many knots to skip between the selected knots. Hence, a sample rate of two selects every second knot, whereas a sample rate of 10 selects every 10th knot.

The number of knots in the second case defines the size of the resulting knot vector. This implies a computation of the sampling rate depending on the number of elements in F.

We consider parametric stride, where we select knots equidistant in parametric space, in the current article.

3.2 Curvature Based Partitioning

Moving away from uniform partitioning, we describe in brief a naive, curvature extrema based partitioning approach. From the discrete function $F[k]$, $k = 1, \ldots, M$, we compute for each interior knot t_i, $i = 2, \ldots, M-1$ the radius of circumscribed circle of triangle $R[i] = R_{circ}\triangle(F[i-1], F[i], F[i+1])$. These values correspond to the curvature of the curve at the corresponding interior points. Next, we select the extrema of these values as they, together with the two endpoints constitute the points of interest (for more details on feature point selection consult [6, 8]).

To be able to scale the method, the resulting set of feature points is processed further. Feature points that are too close are filtered out and new feature points are introduced uniformly between feature points that were too far away.

3.3 Partitioning Based on Inflexion

We look at two different approaches based on relative angular changes in the discrete data set. In both cases we consider the angle between the two vectors spanning a sample point. Where we in the first approach consider the change in the angle by tracing the curve, we start by sorting the angles into different buckets in the other.

Inline Traced Partitioning. In the first variation we look at an approach where we consider the linear interpolation between two neighboring data points to be a vector which provides a first derivative in one point. Given $\boldsymbol{a}$ and $\boldsymbol{b}$, two vectors, we use the dot product between vectors and the angular difference, γ, of $\boldsymbol{a}$ and $\boldsymbol{b}$

$$\cos(\gamma) = \frac{< \boldsymbol{a}, \boldsymbol{b} >}{|\boldsymbol{a}|\ |\boldsymbol{b}|} \tag{6}$$

Given the discrete data set $F[k]$, and an empty set $\mathbf{q}$ to store the detected feature points. Apply (6) to the vectors $\mathbf{a} = p_1 - p_0$ and let $\mathbf{b}$ "run" along the curve, starting with $\mathbf{b} = p_2 - p_1$, then $\mathbf{b} = p_3 - p_2$, and so on. By comparing the results of applying (6), whenever the sign of the gradient of the resulting "curve" changes, we find a point of inflection on the curve given by linear interpolation between points in $\mathbf{p}$.

Bucket Based Partitioning. The second approach is to do an angular difference based partitioning by segmenting knots into buckets, each bucket corresponding to a range subset of possible angular differences between the forward and the backward edge. From the discrete function $F[k]$, $k = 1, \ldots, M-1$, we compute for each interior knot the angle between the two adjacent vectors, $\boldsymbol{a}$ and $\boldsymbol{b}$, where $a = F[k+1] - F[k]$ and $b = F[k+2] - F[k+1]$. We sort the angles and divide the knots into equal sized buckets, this can be seen in Fig. 2. Next we run along the curve selecting feature knots, where if following knots is belonging to the same bucket, only the first is kept as a feature knot.

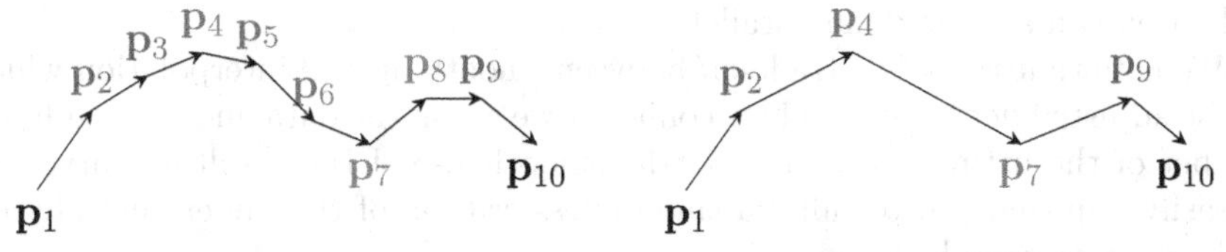

(a) The ten sample points. Interior sample points sorted into three different buckets.

(b) The interior feature points kept after partitioning.

Fig. 2. Ten points and the linear interpolation in-between drawn as vectors.

In addition to the found interior knots the end-point knots are also kept as feature knots. Finally, the resulting set of feature points is processed further, feature points that are too close are filtered out.

4 Fitting

Once the interpolation space is set up, by defining the knot vectors ($\boldsymbol{t}$) and the multiplicities ($\boldsymbol{\mu}$) the problem of finding the coefficients to best match the given discrete data set $F[k], k = \{k_1, k_2, \ldots, k_M\}$ of M points remains. For this purpose we will use the best L_2 approximant, given by the coefficient minimizing the mean squared errors

$$\|S - F\|^2 = \sum_{i=1}^{M} |s(k_i) - F[k_i]|^2, \tag{7}$$

the coefficients are obtained by the usual technique of solving least squares problems. We restrict our investigation to cases where the fitting problem is not ill-posed.

Figure 3 shows the performance of the considered methods introduced in Sect. 3. The x-axis shows the percentage of the original data that is being used while the y-axis displays the Signal to Noise Ratio (SNR) measured in dB, defined as

$$\text{SNR} = 10 \log_{10} \left(\frac{\|F\|^2}{\|F - S\|^2} \right), \tag{8}$$

where F stands for the original discrete data and S represent the reconstructed data and $\|\cdot\|^2$ stands for the square of the usual L_2 norm.

5 Concluding Remarks

Our technique is local, hence there is a small bandwidth in the resulting matrix in the least squares fitting, which in turn gives a computational advantage over global methods, i.e. classical polynomial B-splines.

The computational complexity of each feature detection method is linear and readily parallelizable, similarly since the splines are local the reconstruction and evaluation can also be done parallel.

We note that there is a trade-off between smoothing and interpolation which can be adjusted depending on how confident we are in the data and the condition number of the fitting. Furthermore, the smoothness of the resulting curve can be easily adjusted e.g. to fulfill a smoothness criteria of the underlying physics of the discrete sample points.

The primary utilization is to reduce an original data set and use GERBS type splines to represent the final data. From the two synthetic benchmarks we can conclude that the two types of feature extraction coupled with the coloring or the refining extension allowed for a construction of a series of tune-able spline spaces which performed at least as good as the least squares fitting for smooth inputs and proved to be much more stable for oscillating irregular input data.

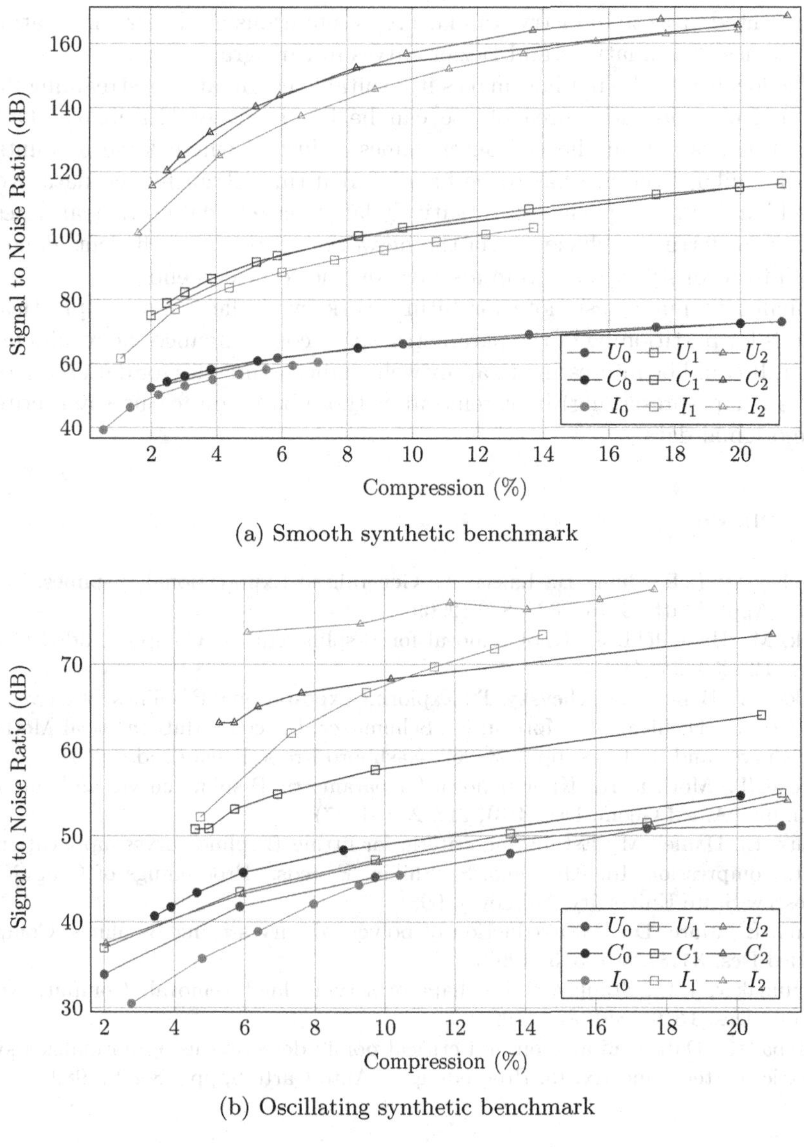

(a) Smooth synthetic benchmark

(b) Oscillating synthetic benchmark

Fig. 3. Error rates for the smooth (a) and oscillating (b) synthetic benchmarks. The teal lines U_0,U_1,U_2 represent the uniform algorithm, the purple lines C_0,C_1,C_2, stand for the curvature based refining method, while the orange marks I_0,I_1,I_2, show the performance of the bucketing algorithm based on angles. The lower indices correspond the applied uniform multiplicity, μ_i in (2).

5.1 Future Work

Future work related to applications includes the extension of the current study to applications in cartography and animation data, including the adaptation of the presented ideas to industry standard representations used there, i.e. Catmull-Rom splines in animations and Bézier curves in cartography.

The locality of the method makes it a suitable candidate for streaming data, one particular potential area of use can be in Massively Multiplayer Online (MMO) games within the computer games industry, where large amounts of data of similar structure has to be handled real time. Transferring data over a limited bandwidth, especially for relatively large discrete data sets, translates to simply transferring coefficients via the networks, since the coefficient alone are enough to reconstruct data from a sender on the receiver's end.

Finally, to put a last note for future work, we believe more sophisticated methods for partitioning of the parametric space could enhance the results much further. It could be interesting to apply well studied principles for data reduction, such as (shape-preserving) knot removal or those based on features and criteria of the original data.

References

1. Dechevsky, L.T., Bang, B., Lakså, A.: Generalized Expo-rational B-splines. Int. J. Pure Appl. Math. **57**(6), 833–872 (2009)
2. Eck, M., Hadenfeld, J.: Knot removal for B-spline curves. Comput. Aided Geom. Des. **12**, 259–282 (1994)
3. Lakså, A., Bang, B., Dechevsky, T.: Exploring expo-rational B-splines for curves and surfces. In: Dæhlen, M., Mørken, K., Schumaker, L. (eds.) Mathematical Methods for Curves and Surfaces, pp. 253–262. Nashboro Press, Brentwood (2005)
4. Lyche, T., Mørken, K.: Knot removal for parametric B-spline curves and surfaces. Comput. Aided Geom. Des. **4**(3), 217–230 (1987)
5. Saux, E., Daniel, M.: Estimating criteria for fitting B-spline curves: application to data compression. In: Klimenko, S., Shikin, E. (eds.) Proceedings of GraphiCon. Moscow State University, Moscow (1998)
6. Saux, E., Marc, D.: Data reduction of polygonal curves using B-splines. Comput. Aided Des. **31**(8), 507–515 (1999)
7. Schumaker, L.L., Stanley, S.S.: Shape-preserving knot removal. Comput. Aided Geom. Des. **13**(9), 851–872 (1996)
8. Thapa, K.: Data compression and critical points detection using normalized symmetric scattered matrix. In: Proceedings of Auto Carto 9, pp. 78–89 (1989)

Discrete Wavelet Compression of ERBS

Rune Dalmo[(✉)] and Jostein Bratlie

R&D Group in Mathematical and Geometrical Modeling,
Numerical Simulations, Programming and Visualization,
Narvik University College, P.O. Box 385, 8505 Narvik, Norway
{Rune.Dalmo,Jostein.Bratlie}@hin.no
http://www.hin.no/Simulations

Abstract. We study the application of Lorentz thresholding and composite Besov-Lorentz shrinkage to coefficients of wavelet-transformed discretized expo-rational B-spline (ERBS) data. ERBS constructions are well suited for interpolation in cases where a given continuity is required; including, but not limited to, applications within control theory, simulations and various interactive modeling. We provide different examples, which highlight the adaptivity features of the two methods. We show the performance with respect to compression rates and approximation errors for the two test cases.

Keywords: Compression · ERBS · Wavelets · Fitting · Shrinkage · Thresholding · Besov-Lorentz

1 Introduction

A comparison of Besov-Lorentz shrinkage to firm thresholding for use in wavelet compression of curves was investigated in [6]. One finding showed that it is possible to control the trade-off between error of approximation and rate of compression in the case of Besov-Lorentz shrinkage, whereas no such control was available with firm thresholding. In addition, Besov-Lorentz shrinkage was found to provide better fitting of singularities under the penalty of over-fitting smooth parts of the signal neighboring the isolated singularity. In the present paper we investigate the general application of the above mentioned methods to discretized expo-rational B-spline (ERBS) data, as well as principal characteristics between Lorentz type thresholding and Besov-Lorentz type shrinkage applied to ERBS.

In Sect. 2 we give a brief overview of the wavelet transform, coefficient shrinkage, compression and ERBS. Then in Sect. 3 we present an overview of the ERBS compression setup and the synthetic test surfaces. Finally in Sect. 4 we present findings, provide some concluding remarks and suggest topics for future work.

I. Lirkov et al. (Eds.): LSSC 2013, LNCS 8353, pp. 585–592, 2014.
DOI: 10.1007/978-3-662-43880-0_67,

2 Preliminaries

2.1 Wavelet Transform

The wavelet transform [3] is a technique to represent any arbitrary function f as wavelets, generated by dilations and translations from one single mother function ψ:

$$\psi_{a,b}(t) = |a|^{-1/2}\psi\left(\frac{t-b}{a}\right), \tag{1}$$

where a and b are constants defining dilatation and translation, respectively. It is required that the mother function has mean zero:

$$\int \psi(t)dt = 0, \tag{2}$$

which typically implies at least one oscillation of $\psi(t)$ across the t-axis. Following from the dilations of a single function, compared with the mother function, low frequency wavelets ($a > 1$) are wider in the t-direction, whereas high frequency wavelets ($a < 1$) are narrower.

For application within signal analysis, the parameters a and b in Eq. (1) are usually restricted through discretization. A dilation step $a_0 > 1$ and a translation step $b_0 \neq 0$ are fixed, leading to the wavelets for $j, k \in \mathbb{Z}$:

$$\psi_{jk}(t) = a_0^{-j/2}\psi(a_0^{-j}t - kb_0). \tag{3}$$

The discrete wavelet transform (DWT), T, associated with the discrete wavelets in Eq. (3), maps functions f to sequences indexed by $\mathbb{Z}^2$:

$$(Tf)_{jk} = \langle\psi_{jk}, f\rangle = a_0^{-j/2}\int \overline{\psi(a_o^{-j}t - kb_0)f(t)}dt. \tag{4}$$

Following the principle of decomposition, f can be reconstructed from its wavelet coefficients $\beta_{jk} = \langle\psi_{jk}, f\rangle$:

$$f = \sum_{j,k}\beta_{jk}\psi_{jk}(t). \tag{5}$$

An algorithm for indexing of discrete orthogonal wavelet spaces of different dimensions was introduced in [4]. We utilize this to perform discrete wavelet transforms efficiently on the graphical processing unit (GPU) by applying low-pass and high-pass filters iteratively.

2.2 Wavelet Shrinkage

Two types of wavelet shrinkage – thresholding and non-thresholding type of shrinkage rules – were discussed in [5]. With a thresholding rule, coefficients β whose absolute value is less than a given threshold, such that $|\beta| < \lambda$ where the threshold value $\lambda \geq 0$, are "eliminated" by setting them to zero, whereas in the case of a non-thresholding rule, all coefficients are shrunk *towards zero* without actually setting any of them to zero.

Soft and *hard* thresholding are two classic thresholding rules for smooth functions, proposed by Donoho and Johnstone in [9–11]. They consist of a continuous function ('shrink' or 'kill') or a discontinuous function ('keep' or 'kill'), respectively. An improved version combining the properties of both thresholding rules, *firm* shrinkage, was introduced by Gao and Bruce in [13] as a function taking two threshold parameters. It can be considered as a first example of a composite shrinking strategy [5].

According to [5], threshold type shrinkage methods are the preferred choice when estimating relatively regular functions, while non-threshold type methods are better suited for estimation of spatially in-homogeneous functions. However, there are down-sides with non-adaptive methods. Thresholding tends to over-smooth near singular points, whereas non-threshold methods tend to over-fit in regular points.

To improve on the above mentioned limitations, Dechevsky, Gundersen and Grip proposed an adaptive composite estimator in [5]: Lorentz-type thresholding based on decreasing re-arrangement of the wavelet coefficients, as outlined in [8], combined with the non-threshold shrinkage procedures for functions belonging to Besov spaces, also described in [8]. The regularity of a signal is there discussed in terms of the size of its semi-form in the homogeneous Besov spaces. Notably in [5], following from a criterion to use the real interpolation spaces, in which the Besov scale is closed, i.e., $(B^{\sigma}_{\pi\pi}, B^{s}_{pp})_{\theta,p(\theta)} = B^{s(\theta)}_{p(\theta)p(\theta)}$, was the announcement of a parameter θ determining to which degree the composite estimator is of Lorentz-type and Besov shrinkage-type, when $0 \leq \theta \leq 1$, where $\frac{1}{p(\theta)} = \frac{1-\theta}{\pi} + \frac{\theta}{p}$. Since general Lorentz shrinkage is a threshold method, while Besov shrinkage is of non-threshold type, θ can be used as a parameter to control the compression rate. (For details, see [5,8] and the references therein). Preliminary experiments on this new type of shrinkage for use in wavelet compression was explored in [6]. We will adapt the proposed, general 1D method to 2D surfaces by considering the whole signal as a general Besov space.

2.3 Wavelet Compression

Compressing a wavelet-transformed signal is essentially a two-step process:

1. Quantify the wavelet coefficients
2. Code-word assignment for the quantified coefficients

Errors are introduced in the quantification step, e.g. through shrinkage. We note that in order to achieve loss-less compression, that step has to be omitted. Thus, the compression we consider here is lossy.

The wavelet transformed and possibly quantified signal can be "packed" using error-free compression of the coefficients β. One such method is to encode the most frequent symbols with fewer bits rather than coding all symbols with an identical number of bits. A commonly used version of this technique is called

Huffman code [14]. Another method, which is effective on signals with less variations, is to apply run-length encoding (RLE). With RLE, whenever a symbol is repeated N times in a sequence, it is stored once together with the number N.

We investigate compression in this case with respect to quantification only by counting the relative number of coefficients which are set to zero.

2.4 Expo-rational B-splines

Expo-rational B-splines, as defined in [16], provide a blending type construction where local functions at each knot are blended together by C^∞-smooth basis functions:

$$f(t) = \sum_{k=1}^{n} \ell_k(t) B_k(t) \tag{6}$$

where $\mathbf{t} = \{t_k\}_{k=0}^{n+1}$ is an increasing knot vector, $\ell_k(t)$ are local (scalar or vector-valued) functions defined on (t_{j-1}, t_{j+1}), and each basis function $B_j(t)$ is supported on (t_{j-1}, t_{j+1}) with $B_j(t_k) = \delta_{jk}$ (Kronecker's delta). We consider the scalable subset of the ERBS basis (see [7]) with default set of intrinsic parameters, proposed by Lakså in [15]:

$$B_k(t) = \begin{cases} S \int_0^{w_{k-1}(t)} \psi(s) ds, & t_{k-1} < t \leq t_k \\ S \int_{w_k(t)}^{1} \psi(s) ds, & t_k < t < t_{k+1} \\ 0, & \text{otherwise}, \end{cases} \tag{7}$$

where $w_k(t) = \frac{t - t_k}{t_{k+1} - t_k}$, $\psi(s) = e^{-\frac{\left(s - \frac{1}{2}\right)^2}{s(1-s)}}$, and $S = \left(\int_0^1 \psi(s) ds \right)^{-1} \approx 1.6571$.

The motivation to explore Besov-Lorentz shrinkage applied to ERBS functions is due to their Hermite interpolation property, which facilitates representation of curves and surfaces with prescribed smoothness. Interestingly, this property encompass modeling of sharp edges, since an ERBS function completely interpolates values and all existing derivatives of its local functions in their respective knots (see the ERBS Hermite interpolation theorem described in [15] for details).

3 Wavelet Compression of ERBS Data

We consider parametric tensor product surfaces,

$$S(u, v) = \sum_{i=1}^{n_u} \sum_{j=1}^{n_v} s_{ij}(u, v) B_i(u) B_j(v), \tag{8}$$

where $s_{ij}(u, v)$, $i = 1, \ldots, n_u$, $j = 1, \ldots, n_v$ are $n_u \times n_v$ local Bézier patches, and $B_i(u), B_j(v)$ are the respective ERBS basis functions. We present two surfaces, one with less local variations (smooth) and another with sharp local variations (spikes).

The "smooth" data set is an ERBS approximation of a cubic Bézier surface representing a "hump". It is constructed from 3×3 local patches, where each local patch is a cubic Bézier surface.

The "spiked" data set is also an ERBS approximation of a cubic Bézier surface. However, in this case it is representing a "dip". Now, every second local patch, in u and v, is pulled through the surface and reduced to degree 0. This creates "spikes" and local irregularities. Plots of the two ERBS surface data sets can be seen in Fig. 1a and b, respectively.

It was noted in [6] that smooth parts of the signal neighboring isolated singularities were over-fitted due to the relatively large support of the orthonormal Daubechies 6 wavelet used. For this reason we apply the bi-orthogonal Cohen - Daubechies - Feauveau 5/3 wavelet, described in [2], also known as the LeGall 5/3 wavelet (see [12]). It forms the shortest symmetrical bi-orthogonal pair of order 2, hence, it combines sufficient smoothness with narrower support.

Using the notation from [5] to represent a signal as discrete orthonormal wavelets,

$$f(x) = \sum_{k \in \mathbb{Z}^d} \alpha_{0k} \phi_{0k}^{[0]}(x) + \sum_{j=0}^{\infty} \sum_{k \in \mathbb{Z}^d} \sum_{l=1}^{2^d-1} \beta_{jk}^{[l]} \psi_{jk}^{[l]}(x), \quad a.e. \;\; x \in \mathbb{R}^d, \tag{9}$$

where $\alpha_{0k} = \left\langle \phi_{0k}^{[0]}, f \right\rangle = \int_{\mathbb{R}^d} \phi_{0k}^{[0]}(x)\overline{f(x)}dx$ and $\beta_{jk}^{[l]} = \left\langle \psi_{jk}^{[l]}, f \right\rangle$ are the scaling and wavelet coefficients, respectively, surface fitting by wavelet shrinkage is performed here by shrinking the wavelet coefficients β_{jk} towards zero in the wavelet domain. We compare pure Lorentz thresholding (without invoking Besov-type non-threshold shrinking) with the composite Besov-Lorentz shrinkage outlined in [5].

The present wavelet implementation is based on the lifting scheme proposed by Sweldens in [17]. In the test case we apply two DWT levels on discretized 2D ERBS data of size 256×256.

Figure 2 shows the performance for the two cases of shrinkage applied to the considered ERBS surfaces. The horizontal axis displays the compression rate while the vertical axis shows the Signal to Noise Ratio (SNR) measured in dB, defined as

$$\text{SNR} = 20 \log_{10} \left(\frac{\| F \|^2}{\| F - S \|^2} \right), \tag{10}$$

where F is the original ERBS surface, S denotes the compressed (shrunk) surface and $\| \cdot \|^2$ stands for the square of the L_2 norm. Compression is measured in terms of how many of the wavelet coefficients which are set to zero relative to the size of the signal. The scaling coefficients α_{0k} in Eq. (9) remain un-touched. In the case of 2D data with $2^n \times 2^n$ resolution, the number of scaling coefficients is $\left(\frac{1}{2l}n\right)^2$. Thus, the highest possible compression rate with two DWT levels is 93.75 % when all wavelet coefficients β_{jk} in Eq. (9) are set to zero, since 6.25 % (or $\frac{1}{16}$) of all coefficients are α_{0k}.

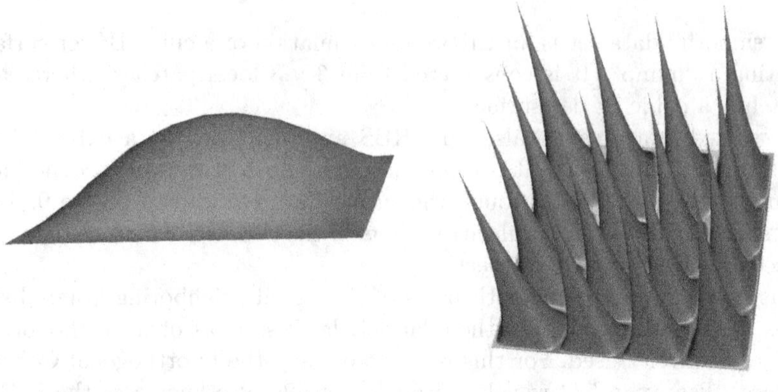

(a) Smooth surface with less local variations.

(b) Spiked surface, with *sharp* local variations.

Fig. 1. Visual representation of the synthetic ERBS surface test data.

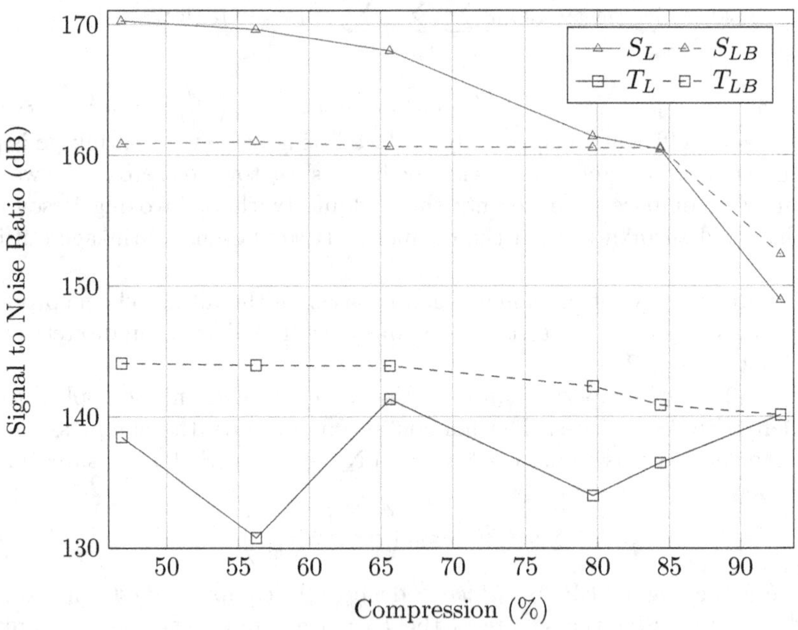

Fig. 2. Error rates for the synthetic benchmarks, as seen in Fig. 1. The locally smooth surface S is represented as teal lines with triangle markers (at the top), whereas the two purple lines with square markers further down show the performance for the spiked surface T with sharp local variations. Rates for Lorentz thresholding and composite Besov-Lorentz shrinkage are indicated with solid and dashed lines, respectively.

4 Concluding Remarks

Invoking Besov-type non-threshold shrinking reduces error when compressing ERBS surfaces with sharp local variations, despite the smoothness properties provided by the ERBS. In the case of "smoother" surfaces, in terms of less local variations, applying non-threshold shrinking has opposite effect. The compression rates are relatively high in both cases.

ERBS constructions are well suited for interpolation when a given continuity is required, which allows for modeling complex geometry as continuously evaluable functions. The flexibility provided by the composite shrinking seems to accommodate these possibly challenging properties well.

The primary utilization is to compress (and re-construct) ERBS data. There is obviously a potential use within data transferring and storage.

We note that there is a trade-off between compression rates and accuracy of data fitting which can be controlled efficiently by adjusting the control parameters. This may be useful in interactive applications, where user-guided compression is suitable, such as animation modeling.

The current study is performed on signals that belong to the general scale of Besov spaces. The wavelet transform is already a local method which is computed in parallel on the GPU. We note several advantages with extending the computation of composite shrinkage parameters to adapt to the local smoothness properties prescribed on large data sets, by analyzing the local sparseness or non-sparseness of the vector of wavelet coefficients. First of all, the compression performance would increase. In addition, the computation can be done in parallel. Moreover, a pure local method is suitable for data streaming, which is applicable to computer games, various controllers and simulations, among others.

On the topic of performance, it could be beneficial to apply well-known methods such as code-word assignment or packing for the quantified coefficients additional to shrinkage.

As a final remark, we note that according to [1], the LeGall 5/3 based on lifting is used also to obtain reversible or loss-less image compression in JPEG2000 since this particular wavelet, which has rational coefficients, can be implemented with integer operations only.

References

1. Boliek, M., Christopoulos, C., Majani, E.: JPEG 2000 part 1 final committee draft version 1.0, ISO/IEC JTC1/SC29 WG1 N1646R, Joint Bi-level Image Experts Group & Joint Photographic Experts, Group, March 2000
2. Cohen, A., Daubechies, I., Feauveau, J.-C.: Biorthogonal bases of compactly supported wavelets. Commun. Pure Appl. Math. **XLV**, 485–560 (1992)
3. Daubechies, I.: Orthonormal bases of compactly supported wavelets. Commun. Pure Appl. Math. **XLI**, 909–966 (1988)

4. Dechevsky, L.T., Bratlie, J., Gundersen, J.: Index mapping between tensor-product wavelet bases of different number of variables, and computing multivariate orthogonal discrete wavelet transforms on graphics processing units. In: Lirkov, I., Margenov, S., Waśniewski, J. (eds.) LSSC 2011. LNCS, vol. 7116, pp. 402–410. Springer, Heidelberg (2012)
5. Dechevsky, L.T., Grip, N., Gundersen, J.: A new generation of wavelet shrinkage: adaptive strategies based on composition of Lorentz-type thresholding and Besov-type non-threshold shrinkage. In: Truchetet, F., Laligant, O. (eds.) Proceedings of SPIE Wavelet Applications in Industrial Processing V, Boston, MA, USA, vol. 6763, pp. 1–14 (2007)
6. Dechevsky, L.T., Gundersen, J., Grip, N.: Wavelet compression, data fitting and approximation based on adaptive composition of Lorentz-type thresholding and Besov-type non-threshold shrinkage. In: Lirkov, I., Margenov, S., Waśniewski, J. (eds.) LSSC 2009. LNCS, vol. 5910, pp. 738–746. Springer, Heidelberg (2010)
7. Dechevsky, L.T., Lakså, A., Bang, B.: Expo-rational B-splines. Int. J. Pure Appl. Math. **27**, 319–362 (2006)
8. Dechevsky, L.T., Ramsay, J.O., Penev, S.I.: Penalized wavelet estimation with Besov regularity constraints. Mathematica Balkanica (N.S) **13**, 257–376 (1999)
9. Donoho, D.L., Johnstone, I.M.: Ideal spatial adaptation by wavelet shrinkage. Biometrika **8**, 425–455 (1994)
10. Donoho, D.L., Johnstone, I.M.: Minimax estimation via wavelet shrinkage. Ann. Stat. **26**, 879–921 (1998)
11. Donoho, D.L., Johnstone, I.M., Kerkyacharian, G., Picard, D.: Wavelet shrinkage: Asymptopia? J. Roy. Stat. Soc. **B57**, 301–369 (1995)
12. Gall, D.L., Tabatabai, A.: Subband coding of digital images using symmetric short kernel filters and arithmetic coding techniques. In: International Conference on Acoustic, Speech and Signal Processing, New York, NY, USA, vol. 2, pp. 761–765 (1988)
13. Gao, H.-Y., Bruce, A.G.: Waveshrink with firm shrinkage. Stat. Sinica **7**, 855–874 (1997)
14. Huffman, D.A.: A method for the construction of minimum-redundancy codes. Proc. I.R.E. **40**, 1098–1110 (1952)
15. Lakså, A.: Basic properties of Expo-rational B-splines and practical use in computer aided geometric design. Ph.D. thesis, University of Oslo (2007)
16. Lakså, A., Bang, B., Dechevsky, L.T.: Exploring expo-rational B-splines for curves and surfaces. In: Dæhlen, M., Mørken, K., Schumaker, L. (eds.) Mathematical Methods for Curves and Surfaces, pp. 253–262. Nashboro Press, Brentwood (2005)
17. Sweldens, W.: The lifting scheme: a custom-design construction of biorthogonal wavelets. Appl. Comput. Harmonic Anal. **3**, 186–200 (1996)

Numerical Method for Solving Free Boundary Problem Arising from Fixed Rate Mortgages

Juri D. Kandilarov(✉)

Department of Mathematics, University of Rousse, Rousse, Bulgaria
ukandilarov@uni-ruse.bg

Abstract. In this paper a mortgage contract with a given duration and a fixed mortgage interest rate is considered. The borrower is allowed to terminate the contract at any time at his choice by paying off the outstanding sum to the issuer. The mathematical model leads to a free boundary problem where the moving boundary is the optimal time of termination. A new numerical method, based on the immersed interface method (IIM) and integral representation of the solution is proposed. Using Thomas algorithm the nonlinear equation for the free boundary position is obtained and solved iteratively. Numerical analysis is presented and discussed.

1 Introduction

Mortgages are one of the most popular financial instruments in today's financial markets. A mortgage loan is a contract which allows the borrower to obtain funds from a financial instrument using a risky asset as collateral. The most common variety of mortgages, known as fixed rate mortgages, have a fixed contract rate and a fixed monthly payment. Residential mortgage contract typically grants the borrower several options to facilitate his reacting to the market movement, among which very important are the options of prepayment and refinancing. In this paper we are interested in the problem of whether it is better for the borrower to terminate the mortgage by prepaying it with a lump sum and the optimal time to do so.

Mortgages like many derivative securities can be valued using partial differential equations. As a variant of derivative securities of American style, there is no exact analytical solution to these free boundary problems. Explicit analysis from an option-theoretic viewpoint of such financial derivatives have been done by many researches in the last three decades [1,3,4,14,15].

The mortgage contract has a duration T and a fixed interest rate c. At any time t during the term of the mortgage, the outstanding balance owed $M(t)$, is reduced in the time period $[t, t+dt)$ by

$$dM(t) = cM(t)dt - mdt, \qquad \forall t \le T,$$

where $cM(t)dt$ is the interest accrued on the balance and mdt is a payment resulting from a constant continuous rate of payment of m. In order for the

I. Lirkov et al. (Eds.): LSSC 2013, LNCS 8353, pp. 593–601, 2014.
DOI: 10.1007/978-3-662-43880-0_68,

mortgage to be retired at $t = T$, the condition $M(T) = 0$ applies so that $M(t) = \frac{m}{c}\left(1 - \exp(c(t-T))\right)$. The borrower is allowed to terminate the contract at any time t ($t < T$) of his choice by paying off $M(t)$ or invest in the market with the amount $M(t)$ less the current obligatory payment of m per unit time, earning an instantaneous return rate r_t. We assume that this short term market return rate r_t follows Vasicek model [13], described by the stochastic differential equation

$$dr = \kappa(\theta - r)dt + \sigma dW_t\,,$$

where κ is the mean reverting speed, θ is the long term mean of r, σ is the volatility of r and W_t is the standard Wiener process. In this model the market price of risk has been incorporated into the drift $\kappa(\theta - r)$.

In [2], the function $V(r,t)$, being the expected value of the contract at time t and current market return rate r, is introduced and the following problem is obtained:

$$\begin{aligned} &\frac{\partial V}{\partial t} + \frac{\sigma^2}{2}\frac{\partial^2 V}{\partial r^2} + \kappa(\theta - r)\frac{\partial V}{\partial r} + m = rV \qquad \text{if} \quad V(r,t) < M(t), t < T, \qquad (1)\\ &0 \le V(r,t) \le M(t) := \frac{m}{c}\left(1 - e^{c(t-T)}\right) \qquad\qquad \forall t \le T, r \in \mathbb{R}. \end{aligned}$$

The value V is calculated according to the borrower's optimal decision to terminate the contract at the first time that the short term market return rate r is below $R(t)$. We call $r = R(t)$ the *optimal boundary* of mortgage contract termination.

The problem (1) has been studied analytically by Jiang et al. [4]. They proved that the problem is well-posed, i.e. there exist a unique solution which is smooth up to the free boundary $r = R(t)$. Also, the free boundary $R(t)$ is a smooth function, strictly increasing on $(-\infty, T)$ and has the asymptotic behavior

$$R(t) \sim c - \sigma\overline{\kappa}\sqrt{T-t} \quad \text{as} \quad t \to T, \quad \overline{\kappa} = 0.47386....$$

Many methods are used for solving free boundary problems, arising in financial mathematics. Some of them transform the free boundary to the strait line [6,10–12]. On the other hand, an algorithm using integral equations was proposed in [2,14]. In this paper a new numerical method, based on the IIM (see [5,7]) and integral representation of the solution is developed. Using Thomas algorithm a nonlinear equation for the free boundary position is obtained and solved iteratively. Numerical analysis is presented and discussed.

2 Mathematical Problem

Using a number of transformations (for more details see [2]) the Black-Scholes equation in problem (1) is reduced to a heat equation. First, a new variables *time to expiry* $\tau := T - t$ and *dimensionless quantity* $\psi(r,\tau) := c/m(M(t) - V(r,t))$ are introduced. Second, dependent variable change is made:

$$h(r,\tau) := \frac{\kappa}{\sigma^2}(r + \frac{\sigma^2}{\kappa^2} - \theta)^2 + (\kappa + \frac{\sigma^2}{\kappa^2} - \theta)\tau, \quad \phi(r,\tau) := e^{-h(r,\tau)}\psi(r,\tau).$$

Finally, independent variable change is done:

$$x = \frac{\sqrt{\kappa}e^{\kappa\tau}}{\sigma}(r + \frac{\sigma^2}{\kappa^2} - \theta), \quad s = e^{2\kappa\tau}, \quad u(x,s) = \frac{2\sqrt{\pi}\kappa^{3/2}}{\sigma}\phi(r,\tau).$$

After these transformations the following problem for the function u and the free boundary $X(s)$ is obtained:

$$\begin{aligned} &u_s - \frac{1}{4}u_{xx} = f(x,s)H(X(s)) \quad x \in \mathbb{R},\, s > 1 \\ &u(x,s) > 0 \quad \forall x > X(s),\, s > 1 \\ &u(x,s) = 0 \quad \forall x < X(s),\, s > 1 \\ &u(x,1) = 0 \quad \forall x \in \mathbb{R}, \end{aligned} \tag{2}$$

where $H(X(s))$ is right-continuous Heaviside function

$$H(X(s)) = \mathbf{1}_{[X(s),\infty)}(x) = \begin{cases} 1 \text{ if } x \geq X(s), \\ 0 \text{ if } x < X(s). \end{cases}$$

The right hand side function $f(x,s)$ is given by

$$f(x,s) = \sqrt{\pi}(s^\gamma - 1)s^{-\nu-1}(x - \beta\sqrt{s})e^{-\left(\frac{x}{\sqrt{s}}-\alpha\right)^2}, \tag{3}$$

where

$$\alpha = \frac{\sigma}{2\kappa^{3/2}}, \; \beta = \frac{\sqrt{\kappa}}{\sigma}\left(c - \theta + \frac{\sigma^2}{\kappa^2}\right), \; \gamma = \frac{c}{2\kappa}, \; \nu = 1 + \frac{\sigma^2}{4\kappa^3} + \frac{c-\theta}{2\kappa}.$$

Let $G(x,s)$ be the Green's function, associated with the heat operator $\partial_s - 1/4\partial_{xx}$:

$$G(x,s) = \frac{e^{-x^2/s}}{\sqrt{\pi s}}.$$

Then the solution $u(x,s)$ to the problem (2) can be expressed as

$$u(x,s) = \int_1^s d\xi \int_{X(\xi)}^\infty G(x-y, s-\xi)f(y,\xi)dy \quad \forall x \in \mathbb{R}, \;\; s \geq 1. \tag{4}$$

We call the free boundary *interface curve.* Also, let us denote by $[\cdot]_{X(s)}$ the jump of the corresponding function across the position of the free boundary $X(s)$. From the problem (2) we have the following *interface conditions*:

$$[u(x,s)]_{X(s)} = 0, \tag{5}$$

$$[u_x(x,s)]_{X(s)} = 0. \tag{6}$$

Differentiating (5) with respect to s we obtain

$$[u_x(x,s)]_{X(s)}\dot{X}(s) + [u_s(x,s)]_{X(s)} = 0, \tag{7}$$

and (6) leads to

$$[u_s(x,s)]_{X(s)} = 0. \tag{8}$$

From (2) and (8) we have

$$[u_{xx}(x,s)]_{X(s)} = -4f(X(s),s). \tag{9}$$

For the free boundary $X(s)$ in [2] the following estimates are proved:

$$X(s) < \beta\sqrt{s}, \quad \forall s, \tag{10}$$

$$X(s) = \beta - (0.334... + o(1))\sqrt{s-1}, \quad s \to 1. \tag{11}$$

The optimal boundary $r = R(t)$ for terminating the mortgage is given by

$$R(t) = c + \frac{\sigma}{\sqrt{\kappa}}\left(\frac{X(s)}{\sqrt{s}} - \beta\right). \tag{12}$$

3 Finite Difference Scheme

In order to solve the problem (2) numerically for given positive integers N and M we define the uniform meshes: $\overline{\omega}_h = \{0\} \cup \{L\} \cup \omega_h$, $\omega_h = \{x_i = ih,\ i = 1, \dots, (N-1),\ h = L/N\}$ and $\overline{\omega}_k = \{0\} \cup \{T\} \cup \omega_k$, $\omega_k = \{s^j = jk,\ j = 1, \dots, (M-1),\ k = T/M\}$. Our goal is to apply a finite difference method which is suitable for computing $u_i^j \approx u(x_i, s^j)$ for $(x_i, s^j) \in \omega_h \times \omega_k$ and associated front position $X^j \approx X(s^j)$ for $s^j \in \omega_k$. With I^j and $I^j + 1$ we denote the numbers of the mesh points on the time layer s^j, closely situated to the free boundary X^j: $x_{I^j} \le X^j < x_{I^j+1}$.

For discretization of the problem (4) we use Crank-Nicolson scheme with combination of the IIM [7]. The standard central finite difference approximation for the second derivative in space is corrected near the interface curve, using the interface jump conditions to improve the local truncation error. Then the difference scheme is:

$$\frac{u_i^j - u_i^{j-1}}{k} = \frac{1}{8}\left\{\frac{u_{i+1}^j - 2u_i^j + u_{i-1}^j}{h^2} + K_i^j + \frac{u_{i+1}^{j-1} - 2u_i^{j-1} + u_{i-1}^{j-1}}{h^2} + K_i^{j-1}\right\}$$
$$+ \frac{1}{2}\left\{f_i^j + f_i^{j-1}\right\}, \qquad i = 1, ..., N-1, \ \ j = 1, ..., M, \tag{13}$$

where the correction terms K_i^j depend on the free boundary position:

$$K_i^j = \begin{cases} 0 & i \ne I^j, I^j + 1 \\ \frac{2(x_{i+1} - X^j)^2}{h^2} f(X^j, s_j) & i = I^j \\ -\frac{2(x_i - X^j)^2}{h^2} f(X^j, s_j) & i = I^j + 1 \end{cases} \tag{14}$$

We set also initial and boundary conditions for the discrete system:

$$u_i^0 = 0 \quad \text{for} \quad i = 0, 1, ..., N; \quad u_0^j = 0, \quad u_N^j = P(L, s^j) \quad \text{for} \quad j = 1, ..., M. \tag{15}$$

The term $P(L, s^j)$ is the approximation of $u(L, s^j)$ and will be explained later. Next, we rewrite the scheme (9) in the following form [9]:

$$A_i u_{i-1}^j - C_i u_i^j + B_i u_{i+1}^j = -F_i^j, \quad i = 1, ..., N-1, \quad j = 1, ..., M \tag{16}$$

where

$$A_i = \frac{1}{8h^2}, \quad B_i = \frac{1}{8h^2}, \quad C_i = \frac{1}{k} + \frac{1}{4h^2},$$

and

$$F_i^j = \frac{1}{8}\left(\frac{u_{i+1}^{j-1} - 2u_i^{j-1} + u_{i-1}^{j-1}}{h^2} + K_i^{j-1} + K_i^j\right) + \frac{u_i^{j-1}}{k} + \frac{1}{2}\left(f_i^{j-1} + f_i^j\right). \tag{17}$$

For solving the obtained difference scheme we go from time layer $j-1$ to j and use on every stage a variant of Thomas algorithm [9]. We seek the solution u_i^j in the form

$$\begin{aligned}
u_{i+1}^j &= \zeta_{i+1}^j u_i^j + \eta_{i+1}^j, \qquad i = 0, 1, ..., N-1,\\
\zeta_N^j = 0, \quad \zeta_i^j &= \frac{A_i}{C_i - \zeta_{i+1}^j B_i}, \qquad i = N-1, ..., 1,\\
\eta_N^j = P(L, s^j), \quad \eta_i^j &= \frac{B_i \eta_{i+1}^j + F_i^j}{C_i - \zeta_{i+1}^j B_i}, \qquad i = N-1, ..., 1.
\end{aligned}$$

For a fixed j we require the solution u_i^j to be equal to zero for all $i \le I^j$. It follows that $\eta_{I^j}^j = 0$ and hence

$$\Phi(X^j) := B_{I^j}\eta_{I^j}^j + F_{I^j}^j = 0. \tag{18}$$

In this equation both $\eta_{I^j}^j$ and $F_{I^j}^j$ depend on the position of the free boundary X^j. This is a nonlinear equation for the unknown value X^j. To complete the difference scheme we set also $X^0 = \beta$, see (10).

Then the Newton method is applied to equation (18):

$$\begin{aligned}
\overset{0}{X^j} &= 2X^{j-1} - X^{j-2},\\
\overset{l+1}{X^j} &= \overset{l}{X^j} - \frac{\Phi(\overset{l}{X^j})}{\Phi'(\overset{l}{X^j})},\\
X^j &= \overset{l+1}{X^j} \quad \text{if } \left|\overset{l+1}{X^j} - \overset{l}{X^j}\right| < \varepsilon, \text{a given tolerance.}
\end{aligned} \tag{19}$$

Initial guess for $\overset{0}{X^1}$ is obtained by (11), putting $s = 1 + k$.

We focus on the term $P(L,s^j)$, the approximation of $u(L,s^j)$ which involves a double integral over for $y\in(X(\xi),\infty)$ and $\xi\in(1,s)$. To reduce the amount of calculations we make a simplification in the following way:

$$\begin{aligned} u(x,s) &= \int_1^s d\xi \int_{X(\xi)}^{\infty} G(x-y,s-\xi) f(y,\xi) dy \\ &= \int_1^s d\xi \int_{X(\xi)}^{\infty} \frac{e^{-\frac{(x-y)^2}{s-\xi}}}{\sqrt{s-\xi}} (\xi^{\gamma}-1)\xi^{-\nu-1}(y-\beta\sqrt{\xi}) e^{-\left(\frac{y}{\sqrt{\xi}}-\alpha\right)^2} dy \\ &= \int_1^s \frac{(\xi^{\gamma}-1)\xi^{-\nu-1}}{\sqrt{s-\xi}} e^{-\frac{(x-\alpha\sqrt{\xi})^2}{s}} d\xi \\ &\int_{X(\xi)}^{\infty} \left((y-\overline{\xi})+(\overline{\xi}-\beta\sqrt{\xi})\right) e^{-\frac{s(y-\overline{\xi})^2}{(s-\xi)\xi}} dy, \end{aligned}$$

where

$$\overline{\xi} = \frac{\xi x + (s-\xi)\alpha\sqrt{\xi}}{s}.$$

The inner integral is decoupled into two integrals. The first is solved exactly and the second one is solved by using co-error function:

$$\begin{aligned} J_1 &= \int_{X(\xi)}^{\infty} (y-\overline{\xi}) e^{-\frac{s(y-\overline{\xi})^2}{(s-\xi)\xi}} dy = \frac{1}{2}\frac{\xi(s-\xi)}{s} e^{-\frac{s(X(\xi)-\overline{\xi})^2}{\xi(s-\xi)}}, \\ J_2 &= \int_{X(\xi)}^{\infty} (\overline{\xi}-\beta\sqrt{\xi}) e^{-\frac{s(y-\overline{\xi})^2}{(s-\xi)\xi}} dy \\ &= \frac{\sqrt{\pi}}{2}(\overline{\xi}-\beta\sqrt{\xi})\sqrt{\frac{\xi(s-\xi)}{s}}\, Errc\left(\sqrt{\frac{\xi(s-\xi)}{s}}(X(\xi)-\overline{\xi})\right). \end{aligned}$$

Finally we obtain

$$u(L,s^j) = \frac{1}{2}\int_1^{s^j} \frac{(\xi^{\gamma}-1)\xi^{-\nu-1}\sqrt{s^j-\xi}}{s^j} e^{-\frac{(L-\alpha\sqrt{\xi})^2}{s^j}} e^{-\frac{s^j(X(\xi)-\overline{\xi})^2}{(s^j-\xi)\xi}} d\xi$$

$$+\frac{\sqrt{\pi}}{2}\int_1^{s^j} \frac{(\xi^{\gamma}-1)\xi^{-\nu-1/2}(\overline{\xi}-\beta\sqrt{\xi})}{\sqrt{s^j}} e^{-\frac{(L-\alpha\sqrt{\xi})^2}{s^j}} Errc\left(\sqrt{\frac{\xi(s^j-\xi)}{s^j}}(X(\xi)-\overline{\xi})\right)$$

This two integrals are computed by the trapezoidal rule.

Remark 1. The first integral is singular at $\xi=s^j$ but it is easy to see that the integral function goes to zero as $\xi\to s^j$, so that in the computation on the last stage we take $\xi\in[s^{j-1},s^j-eps]$, where *eps* is the precision number.

Also, in the Newton method we need of the following expression

$$\begin{aligned} \Phi'(X^j) &= B_{I^j}(\eta^j_{I^j})' + (F^j_{I^j})' \\ &= B_{I^j}\zeta^j_{I^j+1}\zeta^j_{I^j+2}\cdots\zeta^j_{N-1}(\eta^j_N)' + \zeta^j_{I^j+1}(F^j_{I^j+1})' + (F^j_{I^j})', \end{aligned}$$

Table 1. Mesh-refinement analysis and comparison with the upgraded method proposed by Xie et al. [2].

	Our method				Upgraded method of Xie et al. [2]			
N	M	$X(T)$	difference	*ratio*	M	$X(T)$	difference	*ratio*
10	8	0.245147766	-	-	8	0.2436451	-	-
20	16	0.244189707	0.000958058	-	16	0.2438225	1.8e-4	-
40	32	0.243992434	0.000197273	4.856	32	0.2439030	8.1e-5	2.2
80	64	0.243981988	1.04459e-05	18.885	64	0.2439357	3.3e-5	2.5
160	128	0.243961767	2.02213e-05	0.516	128	0.2439484	1.3e-5	2.6
320	256	0.243957111	4.65578e-06	4.343	256	0.2439531	4.7e-6	2.7
640	512	0.243955917	1.19425e-06	3.898	512	0.2439548	1.7e-6	2.7
1280	1024	0.243955797	1.19739e-07	9.973	1024	0.2439555	6.3e-7	2.8

where $(\cdot)'$ is differentiation with respect to $X(s^j)$ and then set $X(s^j) \approx X^j$. Differentiation under the integral sign gives (we follow again Remark 1):

$$(\eta_N^j)' = -\int_1^{s^j} \frac{(\xi^\gamma - 1)\xi^{-\nu-1}}{\sqrt{s^j - \xi}} e^{-\frac{(L-\alpha\sqrt{\xi})^2}{s^j}} \left((X(s^j) - \beta\sqrt{\xi}) \right) e^{-\frac{s^j (X(s^j) - \overline{\xi})^2}{(s^j - \xi)\xi}} \, d\xi \, .$$

4 Numerical Experiments

We consider problem (1) with parameter values $c = 0.055$, $\theta = 0.05$, $\sigma = 0.015$ and $k = 0.15$, $T = 1$, see [2]. Since there exists no analytical solution to the proposed free boundary problem, we use the mesh refinement analysis with doubling the mesh sizes h and k. In Table 1 we give the results for the free boundary position $X(s)$ at different number of grids N and M and final time T. Also, the difference between two consecutive values and the ratio are presented. The results show nearly second order of accuracy for the moving boundary at final time. The results obtained by the upgraded method of [2] are also presented, where the rate of convergence is near one and half. The number of iterations in our method is a little more, but only at first several time layers. Next, the free boundary becomes nearly straight line and the method needs of zero or 1 iteration on every layer. The oscillations in the Ratio are frequently phenomenon for the IIM, see [7], but in global the rate of convergence of our method is approximately two.

In Fig. 1 the numerical solutions of the curve $(t, R(t))$ obtained for $M = 32$ by the present method (solid line) and by the upgraded method of Xie et al. [2] (circles) are presented.

Notes and Comments. We have investigated a new numerical method for valuation of fixed rate mortgage contract which allows the mortgage holder to prepay the outstanding balance of the mortgage. The mathematical model is a free boundary problem, which is solved by the IIM, combined with Crank-Nicolson method and Newton method. The results show second order of convergence for

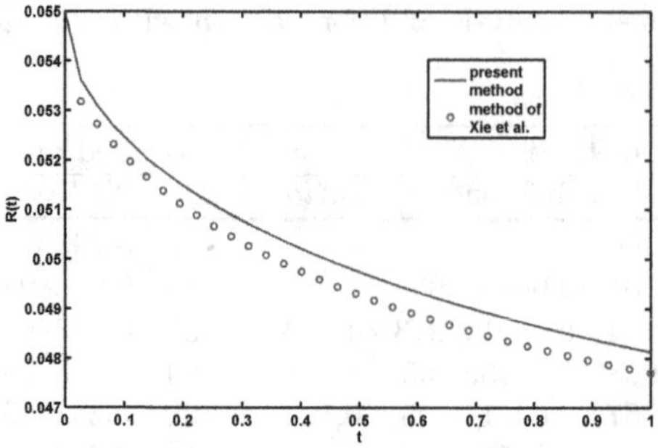

Fig. 1. The numerical solutions of the curve $(t, R(t))$ obtained for $M = 32$ by the present method (solid line) and by the upgraded method of Xie et al. (circles).

the unknown free boundary. However, the method is a little more computationally cost, but it gives a good results and it is appropriate for the problems in which not only the free boundary but also the value of some portfolio function is of interest for the researchers.

Acknowledgments. This research was supported by the European Union under Grant Agreement number 304617 (FP7 Marie Curie Action Project Multi-ITN STRIKE - Novel Methods in Computational Finance) and the Bulgarian National Fund of Science under Project DID 02/37-2009.

References

1. Buser, S.A., Hendershott, P.H.: Pricing default-free fixed rate mortgages. Hous. Financ. Rev. **3**, 405–429 (1984)
2. Chen, X., Chadam, J., Jiang, L., Zheng, W.: Convexity of the exercise boundary of the American put option on a zero dividend asset. Math. Fin. **18**, 185–197 (2008)
3. Epperson, J., Kau, J.B., Keenan, D.C., Muller, W.J.: Pricing default risk in mortgages. AREUEA J. **13**, 152–167 (1985)
4. Jiang, L., Bian, B., Yi, F.: A parabolic variational inequality arising from the valuation of fixed rate mortgages. Euro. J. Appl. Math. **16**, 361–383 (2005)
5. Kandilarov, J.: The immersed interface method for reaction-diffusion equation with a moving concentrated source. In: Dimov, I.T., Lirkov, I., Margenov, S., Zlatev, Z. (eds.) NMA 2002. LNCS, vol. 2542, pp. 506–513. Springer, Heidelberg (2003)
6. Kandilarov, J.D., Valkov, R.L.: A numerical approach for the American call option pricing model. In: Dimov, I., Dimova, S., Kolkovska, N. (eds.) NMA 2010. LNCS, vol. 6046, pp. 453–460. Springer, Heidelberg (2011)
7. Li, Z., Ito, K.: The Immersed Interface Method: Numerical Solutions of PDEs Involving Interfaces and Irregular Domains. SIAM, Philadelphia (2006)
8. Nielsen, B., Skavhaug, O., Tveito, A.: Penalty and front-fixing methods for the numerical solution of American option problems. J. Comp. Fin. **5**(4), 69–97 (2002)
9. Samarskii, A.: The Theory of Difference Schemes. Marcel Dekker, New York (2001)
10. Ševčovič, D.: Analysis of the free boundary for the pricing of an American call option. Eur. J. Appl. Math. **12**, 25–37 (2001)

11. Ševčovič, D.: Transformation methods for evaluating approximations to the optimal exercise boundary for linear and nonlinear Black-Sholes equations. In: Ehrhard, M. (ed.) Nonlinear Models in Mathematical Finance: New Research Trends in Optimal Pricing, pp. 153–198. Nova Sci. Publ., New York (2008)
12. Stamicar, R., Ševčovič, D., Chadam, J.: The early exercise boundary for the American put near expiry: numerical approximation. Canad. Appl. Math. Quart. **7**(4), 427–444 (1999)
13. Vasicek, O.: An equilibrium characterization of the term structure. J. Fin. Econ. **5**, 177–188 (1977)
14. Xie, D.: Fixed rate mortgages: valuation and closed form approximation. IAENG Int. J. Appl. Math. **39**(1), 9 (2003)
15. Xie, D., Chen, X., Chadam, J.: Optimal payment of mortgages. Euro. J. Appl. Math. **18**(3), 363–388 (2007)

A Splitting Numerical Scheme for Non-linear Models of Mathematical Finance

Miglena N. Koleva(✉) and Lubin G. Vulkov

Faculty of Natural Science and Education, University of Rousse,
8 Studentska Str., 7017 Rousse, Bulgaria
{mkoleva,lvalkov}@uni-ruse.bg

Abstract. We present and analyze a splitting numerical scheme for two non-linear models of mathematical finance. Each of the problems is split into two parts: a hyperbolic equation solved numerically by using a flux limiter technique and a parabolic equation computed by implicit-explicit finite difference scheme. We show that the presented splitting numerical schemes are convergent and positivity preserving. Numerical results are also discussed.

1 Introduction

Modelling financial derivative prices by PDE has been introduced in 1973, when a simple linear model was derived by F. Black and M. Scholes and independently by R. Merton, read for example [13]. Its simplicity is obtained by imposing a couple of limiting assumptions [13] which are too restrictive in practice. Therefore, further work has been done to relax one or more assumptions which lead to non-linear PDEs. A great part of the known non-linear modifications of the Black-Scholes equation can be summarized in the form [2,3]:

$$V_t = A(\cdot)V_{SS} + rSV_S - rV, A(\cdot) = \frac{1}{2}\tilde{\sigma}^2(S,t,V,V_{SS})S^2, S \in \Omega \equiv \mathbb{R}^+, 0 \le t \le T, \quad (1)$$

where $\tilde{\sigma}$ is the volatility function, r is the constant short rate and T is the maturity. The unknown function $V(S,t)$ is the option price, which depends on time t and spot price S of the underlying.

We will study (1) for European Call/Put option, i.e. the value $V(S,t)$ is the solution to (1), with the following initial and boundary conditions ($E > 0$ is the exercise price):

$$\begin{array}{lll} \textit{Call option} & \textit{Put option} & \\ V(S,0) = \max\{0, S-E\}, & V(S,0) = \max\{0, E-S\}, & 0 \le S < \infty, \\ V(0,t) = 0, & V(0,t) = Ee^{-rt}, & 0 \le t \le T, \\ V(S,t) = S - Ee^{-rt}, & V(S,t) = 0, & S \to \infty. \end{array} \quad (2)$$

I. Lirkov et al. (Eds.): LSSC 2013, LNCS 8353, pp. 602–610, 2014.
DOI: 10.1007/978-3-662-43880-0_69,

A quite different financial model was derived in [14]

$$\begin{aligned} &v_t v_{SS} + rSv_S v_{SS} - \lambda v_S^2 = 0, \quad S \in \Omega \equiv \mathbb{R}, \quad 0 \le t \le T \\ &v(S,T) = g(S), \quad S \in \Omega, \qquad v_{SS} < 0, \quad g'(S) > 0, \end{aligned} \tag{3}$$

where $v = v(S,t)$ is a *value function of the market model* presented in [14, Chap. 2]. The coefficient $\lambda = (c-r)/\varrho > 0$, where the positive constants r, c and ϱ are the *interest rate, the appreciation rate and the volatility* $(c-r>0)$, respectively. The model (3) describes the *simple market model in the case of one asset.* In a typical case, function $g(S)$ is given by $g(S) = 1 - e^{-\mu S}, \quad \mu > 0$.

Changing the "terminal condition" into the "initial condition" by setting $v(S,T-t) = \overline{v}(S,t)$ and then substituting [12]

$$V(S,t) = -\frac{\overline{v}_S(S,t)}{\overline{v}_{SS}(S,t)}, \quad S \in \Omega, \quad 0 \le t \le T,$$

we get the following initial-value (Cauchy) problem

$$\begin{aligned} &V_t = \lambda V^2 V_{SS} + rSV_S - rV, \quad S \in \Omega \equiv \mathbb{R}, \quad 0 \le t \le T \\ &V(S,0) = -\frac{g'}{g''} > 0, \quad S \in \Omega, \end{aligned} \tag{4}$$

In [10], a splitting numerical method is proposed for the quasi-linear heat equation

$$w_t - (w^m w_x)_x = w^p, \quad m > 0, \quad p \ge m+1, \quad t > 0, \quad w(x,0) = w_0(x), \quad x \in \mathbb{R}, \tag{5}$$

where w_0 is a function with compact support. Problem (5) is written as

$$u_t - \frac{1}{m}u_x^2 - uu_{xx} = mu^{q+1}, \quad t > 0, \quad u(x,0) = u_0(x) = w_0^m(x), \quad x \in \mathbb{R},$$

with $u = w^m$, $q = (p-1)/m$, $m > 0$, $q \ge 1$. This problem is splitted into two parts: a hyperbolic problem, the discrete version of which is solved explicitly on the new time level by the Hopf and Lax formula and a parabolic problem, which is solved by backward linearized Euler method. Different splitting method was applied in [11] to solve an optimal replication problem in incomplete markets, where the parabolic part is a linear Black-Scholes equation.

There exists many numerical methods and algorithms for different versions of the non-linear Black-Scholes equation [2,3,6,8]. In this work, having in mind maximum principle discussed in [1], we will present efficient, second order (in space), *positivity preserving* (i.e. the non-negativity of the numerical solution to be guaranteed) algorithms for solving the non-linear model problems (1)–(2) and (4). We develop the idea of [10] to split the problems (1)–(2) and (4) on two-parts: hyperbolic part and parabolic part. Then the hyperbolic problem is solved using van Leer flux limiter [5] and a parabolic part is solved by an implicit-explicit finite difference scheme.

The remaining part of this paper is organized as follows. In Sect. 2, we develop the numerical method. In the next section the properties of the presented algorithm are investigated. Finally, in Sect. 4 we discuss numerical examples.

2 Numerical Method

In order to solve the model problems, we separate them into two parts: a hyperbolic problem

$$0.5V_t = rSV_S, \quad S \in \Omega, \quad t > 0, \tag{6}$$

and a parabolic problem

$$0.5V_t = A(S, t, V, V_{SS})V_{SS} - rV, \quad S \in \Omega, \quad t > 0. \tag{7}$$

We denote by $\mathcal{L}_H$ and $\mathcal{L}_P$ the exact solution operator associated with the corresponding hyperbolic part (6) and parabolic subproblem (7), respectively. Then, introducing a time step τ_n, the solution of the original model problems is evolved in time in two substeps. First (6) is solved on the time interval $(t_n, t_n + \tau_n/2]$: $\hat{V}(S) = \mathcal{L}_H(\tau_n)V(S)$ and then the parabolic solution operator is applied to $\hat{V}$, which results in the following approximate solution at time $t_{n+1} = t_n + \tau_n$:

$$V(S, t_n + \tau) = \mathcal{L}_P\hat{V}(S) = \mathcal{L}_P(\tau_n)\mathcal{L}_H(\tau_n)V(S). \tag{8}$$

In general, if all solutions involved in the two-step splitting algorithm are smooth, the operator splitting method is second-order accurate at each time step and first-order accurate when it is applied for advancing the solution from $t = 0$ to the final time T, see [5].

In application, the exact solution operators $\mathcal{L}_H$ and $\mathcal{L}_P$ are replaced by their numerical approximations. The main advantage of the operator splitting technique is the fact that the hyperbolic, (6), and the parabolic, (7), subproblems, which are of different nature, can be solved numerically by different methods.

Instead of Ω, we consider a large enough computational interval $[L^-, L^+]$, where $L^+ > 0$, $L^- = 0$ for model (1)–(2) and $L^- < 0$ for (4) and define the mesh

$$\omega_h = \{S_{i+1} = S_i + h_{i+1},\ i = 0, \dots, N-1,\ S_0 = L^-,\ S_N = L^+\},\ \hbar_i = \frac{h_{i+1} + h_i}{2}.$$

Next, we denote by u_i^n the approximate solution at point (S_i, t_n). Similarly $u^n := [u_0^n, u_1^n, \dots, u_N^n]^T$. Further, for clarity of the exposition we set $u := u^n$, $\hat{u} := u^{n+1/2}$, $\hat{\hat{u}} := u^{n+1}$.

2.1 Hyperbolic Problem

In order to preserve the non-negativity of the numerical solution, we compute (6), implementing a flux limiter technique. In this paper we are going to use *van Leer limiter* [4,5]

$$\Phi(\theta) = (|\theta| + \theta)/(1 + |\theta|), \tag{9}$$

where $\Phi(\theta)$ is Lipschitz continuous, continuously differentiable for all $\theta \neq 0$, and

$$\Phi(\theta) = 0, \quad \text{if} \quad \theta \leq 0 \quad \text{and} \quad \Phi(\theta) \leq 2\min(1, \theta). \tag{10}$$

Following [4], the numerical flux $F_{i+1/2} = F(u_{i+1/2}) = u_{i+1/2}$ is constructed in a nonlinear way

$$F_{i+1/2} = u_i + \frac{1}{2}\Phi(\theta_{i+1/2})(u_i - u_{i-1}) \quad \text{with} \quad \theta_{i+1/2} = \frac{u_{i+1} - u_i}{u_i - u_{i-1}}. \tag{11}$$

Greek $\Delta = V_S$ in (6) is approximated by $u_{\mathring{x},i} = [u_{i+1/2} - u_{i-1/2}]/\hbar_i$. Next, reflecting the indices that appear in u_i (see (11)) about $i+1/2$ [4] and using the symmetry property $\Phi(\theta) = \theta\Phi(\theta^{-1})$, we modify $u_{\mathring{x},i}$, depending on the sign of S_i, such that

$$S_i u_{\mathring{x},i} = S_i^+ \Lambda_i^+ (u_{i+1} - u_i)/\hbar_i - S_i^- \Lambda_i^- (u_i - u_{i-1})/\hbar_i, \quad \text{where}$$
$$\Lambda_i^+ := 1 + \frac{1}{2}\Phi(\theta_{i+1/2}^{-1}) - \frac{1}{2}\Phi(\theta_{i+3/2}), \qquad \Lambda_i^- := 1 + \frac{1}{2}\Phi(\theta_{i+1/2}) - \frac{1}{2}\Phi(\theta_{i-1/2}^{-1}),$$
$$S^\pm = \max\{0, \pm S\}, \quad \text{and} \quad 0 \le \Lambda_i^+, \Lambda_i^- \le 2, \quad i = 1, \ldots, N \quad \text{in view of (9), (10).}$$

For values u_{-1} and u_{N+1} at outer grid nodes S_{-1} and S_{N+1} respectively, a second order extrapolation will be used: $u_{-1} = 3u_0 - 3u_1 + u_2$ and $u_{N+1} = 3u_N - 3u_{N-1} + u_{N-2}$ [4].

Thus, the numerical approximation of (6) is

$$\hat{u}_0 = V(L^-, t^{n+1}), \quad \hat{u}_i = u_i + \tau r S_i u_{\mathring{x},i}, \; i = 1, \ldots, N-1, \quad \hat{u}_N = V(L^+, t^{n+1}). \tag{12}$$

2.2 Parabolic Problem

Consider the case $A(\cdot) \ge 0$, independently of the approximation. For example, (4) and (1) with $\widetilde{\sigma}^2 = \sigma^2(1 + \sin\frac{\pi V}{E})$, see [9] and the lecture of C.H. Lai in [3], or with Leland volatility function $\widetilde{\sigma}^2 = \sigma^2(1 + Le\ \mathrm{sign}(V_{SS}))$, where $0 < Le \le 1$ is the Leland number, $\sigma > 0$ is the historical volatility, or Boyle-Vorst volatility $\widetilde{\sigma}^2 = \sigma^2(1 + Le\sqrt{\pi/2}\ \mathrm{sign}(V_{SS}))$, or Avellaneda-Parás volatility $\widetilde{\sigma}^2 = \begin{cases} \sigma^2_{\min}, & V_{SS} > 0; \\ \sigma^2_{\max}, & V_{SS} < 0 \end{cases}$, or stochastic volatility $\widetilde{\sigma}^2 = \sigma^2/(1 - \lambda V_{SS})^2$, where $\lambda(S) \ge 1$ depends on the pay-off function of the financial derivative, and s.o., see [3].

In this case, the finite difference approximation of (7) is

$$\frac{\hat{\hat{u}}_i - \hat{u}_i}{\tau} - \hat{A}_i\left(\frac{\hat{\hat{u}}_{i+1} - \hat{\hat{u}}_i}{\hbar_i h_{i+1}} - \frac{\hat{\hat{u}}_i - \hat{\hat{u}}_{i-1}}{\hbar_i h_i}\right) + r\hat{\hat{u}}_i = 0, \quad i = 1, \ldots, N-1.$$
$$\hat{A}_i = A(S_i, t^{n+1}, \hat{u}_i, \hat{u}_{\overline{x}x,i}), \qquad \hat{\hat{u}}_0 = V(L^-, t^{n+1}), \quad \hat{\hat{u}}_N = V(L^+, t^{n+1}). \tag{13a}$$

Next, let $A = A_1 + A_2 V_{SS}|V_{SS}|^p$, $A_1, A_2 > 0$. To this group belong Barles-Soner model ((1)–(2), $\widetilde{\sigma} = \sigma^2(1 + \Psi(e^{rt}a^2S^2V_{SS}))S^2$, $\Psi(s)$ solves an ODE [3], but often chosen $\Psi(s) = s$), i.e. $A_1 = \frac{1}{2}\sigma^2S^2$, $A_2 = \frac{1}{2}\sigma^2e^{rt}a^2S^4$, $p = 0$, where a is a parameter measure transaction cost and risk aversion, and Jandačka-Ševčovič model (1)–(2) [3], i.e. $A_1 = \frac{1}{2}\sigma^2S^2$, $A_2 = \frac{1}{2}\sigma^2\mu S^{8/3}$, $p = -2/3$, $\mu = 3(C^2M/(2\pi))^{1/3}$, where $M \ge 0$ is the transaction cost measure, $C \ge 0$ is the risk premium measure.

As before, we will use the notation $u^{\pm}_{\overline{x}x,i} = \max\{0, \pm u_{\overline{x}x,i}\}$, where $u_{\overline{x}x,i} = [(u_{i+1} - u_i)/h_{i+1} - (u_i - u_{i-1})/h_i]/\hbar_i$. Then, the discretization of (7) is

$$\frac{\hat{\hat{u}}_i - \hat{u}_i}{\tau} - [\hat{A}_{1_i} + \hat{A}_{2_i}(\hat{u}^{+}_{\overline{x}x,i})^{p+1}]\hat{\hat{u}}_{\overline{x}x,i} + r\hat{\hat{u}}_i = \hat{A}_{2_i}(\hat{u}^{-}_{\overline{x}x,i})^{p+2},$$
$$i = 1, \dots, N-1, \quad \hat{\hat{u}}_0 = V(L^-, t^{n+1}), \quad \hat{\hat{u}}_N = V(L^+, t^{n+1}). \tag{13b}$$

Finally, the two-stage algorithm (TSA) for solving problems (1)–(2) and (4) is:
stage 1. Knowing u, compute $\hat{u}$ from the difference scheme (12).
stage 2. Knowing $\hat{u}$, compute $\hat{\hat{u}}$ from the numerical scheme (13).
For the model problem (4) we impose $V(L^-, t^{n+1}) = V(L^-, 0)$, $V(L^+, t^{n+1}) = V(L^+, 0)$ just as in [7].

3 Properties of the Numerical Method

The aim of this section is to investigate some properties of the numerical solutions. We deal with classical solutions of the differential problems, cf. [1,12]. Comparison and maximum principle for problem (1)–(2) are presented in [1,12].

Proposition 1 (***Positivity preserving***). *If* $V(S, 0) \geq 0$ *and*

$$\tau \leq \min_{1 \leq i \leq N-1} \left\{ \frac{\hbar_i}{2r|S_i|} \right\}, \quad or \quad \Phi(\theta^{-1}_{i+1/2}) \leq \frac{2\hbar_i}{r\tau|S_i|} - 2, \; \tau \leq \min_{1 \leq i \leq N-1} \left\{ \frac{\hbar_i}{r|S_i|} \right\}, \tag{14}$$

then $u^n \geq 0$ *for all time levels* $n = 0, 1, 2, \dots$.

Proof. We apply the induction method. Let $V(S, 0) \geq 0$ and suppose that $u = u^n \geq 0$. We will proof that if the conditions (14) are fulfilled, then $u^{n+1} = \hat{\hat{u}} \geq 0$. Next, applying the same consideration at each time level, we will conclude that, starting with non-negative initial condition, the time integration procedure by TSA preserves the non-negativity of the numerical solution.

Hyperbolic part in TSA is approximated explicitly

$$\hat{u}_i = (1 - B^+_i - B^-_i)u_i + B^+_i u_{i+1} + B^-_i u_{i-1}, \quad \text{where} \quad B^{\pm}_i = \frac{\tau}{\hbar_i} r S^{\pm}_i \Lambda^{\pm}_i \geq 0,$$

and the non-negativity of the numerical solution $\hat{u}$ for $u \geq 0$ is guaranteed if $1 - B^+_i - B^-_i \geq 0$, which leads to inequalities (14).

Next, it is not difficult to verify that, if we rewrite the approximation (13) in equivalent matrix form $\mathcal{M}\hat{\hat{u}} = \hat{\mathcal{K}}\hat{u}$, the coefficient matrix $\mathcal{M}$ is an M-matrix (it guaranties that its inverse is non-negative) and $\hat{\mathcal{K}}\hat{u} = \hat{u}/\tau \geq 0$ (for (13a)) or $\hat{\mathcal{K}}\hat{u} = \hat{u}/\tau + \hat{A}_2(\hat{u}^{-}_{\overline{x}x,i})^{p+2} \geq 0$ (for (13b)), which is sufficient and necessary condition for the non-negativity of the numerical solution $\hat{\hat{u}}$. □

Note that in the second conditions in (14) the time step restriction is relaxed (two times) at the expense of the restriction of the flux limiter.

Proposition 2 (***First derivative sign-preserving***). *Let the conditions of Proposition 1 are fulfilled and u^0 is non-decreasing ($u^0_{\mathring{x},i} \geq 0$) or decreasing ($u^0_{\mathring{x},i} < 0$) function, then $u^n_{\mathring{x},i} \geq 0$ (< 0), respectively for all time levels t_n, $n = 0, 1, \ldots$.*

Proof (outline). Again we use induction method: suppose that $u_{\mathring{x},i} = u^n_{\mathring{x},i} \geq 0$ (<0), then subtracting from (12) and (13) at S_{i+1} the Eqs. (12) and (13), respectively, divide by $\hbar_i$ and presenting $\hat{u}_{\mathring{x},i}$ in (12) by $u_{\mathring{x},i}$, $u_{\mathring{x},i\pm1}$ and then $\hat{\hat{u}}_{\mathring{x},i}$, $\hat{u}_{\mathring{x},i\pm1}$ in (13a) by $\hat{u}_{\mathring{x},i}$ and $\hat{u}_{\mathring{x},i\pm1}$ in (13b), we conclude that $u^{n+1}_{\mathring{x},i} = \hat{\hat{u}}_{\mathring{x},i} \geq 0$ (< 0). Estimating (13b) we use Taylor expansion of power function $(\hat{u}^-_{\overline{x}x})^{p+2}$ around $\hat{u}^-_{\overline{x}x,i}$ and take into account that $\hat{u}^-_{\overline{x}x,i} > 0$, if $\hat{u}^-_{\overline{x}x,i} \neq 0$, $i = 0, \ldots, N$. □

On the base of (8), at each step the L^∞ - error of our method is a sum of three errors: the operator splitting error $E_S(\tau) = \mathcal{O}(\tau^2)$, the error of hyperbolic and parabolic substeps: $E_H(\tau) = \mathcal{O}(\tau + h^2)$ and $E_P(\tau) = C_3(\tau + h^2)$, $h = \max_i h_i$.

Let u^n_h denote the Lagrange interpolation of the numerical solution on the n-th time level. We define the function $u_{h\tau}$ by

$$u_{h\tau}(t) = u^n_h + \frac{t - t^n}{\tau}(u^{n+1}_h - u^n_h), \quad t^n \leq t \leq t^{n+1}.$$

Theorem 1 (***Convergence***). *Assume that the hypotheses of Propositions 1, 2 are satisfied. The sequence $u_{h\tau}$ (constructed on the base of scheme (12)–(13)) converges uniformly to the classical solution of (1)–(2) (respectively (4)).*

Proof (outline). The proof consists of two stages. On the first stage a number of estimates in L^∞ norm for the discrete solutions are obtained. Then, it is shown that one can extract from the sequence $\{u_{h\tau}\}$ a subsequence which converges uniformly on any compact subset of $(L^-, L^+) \times [0, T]$ to a function u that is a weak solution of the differential problem. Finally, conditions on which the weak solution can serves as a classical solution are found.

4 Numerical Experiments

We will test the efficiency of the presented TSA for model problems (1)–(2) and (4). The error $E_i = V(S_i, T) - u^T_i$, $i = 1, \ldots, N$ in maximal discrete norm is given by $\|E^N\|_\infty = \max_{1\leq i\leq N} |E_i|$ and the convergence rate is calculated using double mesh principle $CR_\infty = \log_2(\|E^N\|_\infty / \|E^{2N}\|_\infty)$.

The option and mesh parameters are: $a = 0.01$, $\lambda = 1$, $\sigma = 0.2$, $\sigma_{\max} = 0.25$, $\sigma_{\min} = 0.15$, $Le = 0.5$, $E = 5$, uniform mesh ω_h with step h, $\tau = h^2$, $T = 1$. We add a small positive number ($\sim 10^{-30}$) to both numerator and denominator of the gradient ratio in (11) in order to avoid division by zero in uniform flow regions.

Example 1 (Problem (1)–(2)). Consider the case of Call option with $A(\cdot) = 0.5\sigma^2(1 + \sin\frac{\pi V}{E})S^2$, $r = 0.1$, $L^- = 0$, $L^+ = 10$. In order to test the accuracy of

Table 1. Errors and convergence rates in maximal discrete norms, Example 1

	Solution		Greek delta		Greek gamma	
N	$\|E^N\|_\infty$	CR_∞	$\|E\Delta^N\|_\infty$	CR_∞	$\|E\Gamma^N\|_\infty$	CR_∞
40	1.28364e-2		1.04754e-2		2.13922e-2	
80	3.23238e-3	*1.9896*	2.65308e-3	*1.9813*	5.20567e-3	*2.0389*
160	8.13878e-4	*1.9897*	6.63108e-4	*2.0004*	1.29354e-3	*2.0088*
320	2.04151e-4	*1.9952*	1.65840e-4	*1.9995*	3.23613e-4	*1.9990*

Table 2. Errors and convergence rates in maximal discrete norms, Example 1

	Leland model		Avellaneda-Parás model		Barles-Soner model	
N	$\|E^N\|_\infty$	CR_∞	$\|E^N\|_\infty$	CR_∞	$\|E^N\|_\infty$	CR_∞
40	1.38147e-2		1.91471e-3		7.98768e-3	
80	3.45198e-3	*2.0007*	3.85852e-4	*2.3110*	1.96466e-3	*2.0235*
160	7.36847e-4	*2.2280*	1.42664e-4	*1.4354*	4.82159e-4	*2.0267*
320	1.81721e-4	*2.0196*	2.51395e-5	*2.5046*	1.16608e-4	*2.0478*

the method for non-smooth initial data (2), as exact solution we take the solution of linear Black-Scholes equation, adding appropriate function in the right hand-side of (1). We denote by $E\Delta_i = V_S(S_i,T) - u^T_{x,i}$, $u_{x,i} = (u_{i+1} - u_{i-1})/(2h)$ and $E\Gamma_i = V_{SS}(S_i,T) - u^T_{\overline{x}x,i}$ the errors of *Greeks* $\Delta = V_S$ and $\Gamma = V_{SS}$, computed by TSA (i.e. generating $u_{x,i}$ and $u_{\overline{x}x,i}$ from the computed u_i) and the one obtained by implemented in MATLAB `blsdelta` and `blsgamma` packages for solving Δ and Γ of linear Black-Scholes equation. Errors (in max. discrete norm) and convergence rates of TSA for solution and Greeks are listed in Table 1. Results show that the presented TSA is efficient even for uniform mesh in space and the method needed no special meshes. In Table 2 results from exact solution test for Leland, Avellaneda-Parás and Barles-Soner models are listed. The accuracy is better than those in [6] for the same mesh and model parameters.

Example 2 (Problem (4)*).* We compare the efficiency of TSA with $\mathcal{O}(\tau + h^2)$ scheme (12) ($\theta = 0$), based on Picard linearization ($A(V)$ is computed on the old time level) and Gauss-Seidel type iteration scheme (11), developed in our previous work [7] for model problem (4). Note that the above mentioned methods (scheme (11) and scheme (12) in [7]) preserve the positivity property of the numerical solution under additional restriction: $h \leq (\lambda/r) \min_{1\leq i\leq N-1} [(u_i)^2/S_i]$.

In Table 3 we list error, convergence rate and CPU time for TSA, $r = 0.5$, $L^- = -3$, $L^+ = 3$ and $V(S,t) = e^{-t}(-x^2 + x + 12)$, such that to compare the results with those in Table 1 in [7]. The accuracy for one and the same CPU of TSA for $\tau = h^2$ is more closer to the ones obtained by Gauss-Seidel scheme (12), than the ones obtained by scheme (11) in [7]. Regarding to positivity

Table 3. Errors, convergence rates and CPU time, Example 2

N	$\|E^N\|_\infty$	CR_∞	CPU
20	7.47049e-2		0.073
40	2.08905e-2	*1.8384*	0.347
80	5.33727e-3	*1.9687*	2.301
160	1.33766e-3	*1.9964*	17.024
320	3.35002e-4	*1.9975*	130.427

preserving, the schemes in [7] are much more time consuming than TSA, because the additional restriction for the space step size.

5 Conclusions

In this paper we have presented positivity preserving, second order (in space) algorithm for solving two nonlinear models in mathematical finance. To this aim a splitting method, combined with flux limiter technique is used. The resulting two-stage algorithm attains an optimal accuracy and preserves the non-negativity of the numerical solution if the time step restriction is fulfilled. The schemes proposed can be generalized to higher dimensional problem.

Acknowledgements. This research was supported by the European Union under Grant Agreement number 304617 (FP7 Marie Curie Action Project Multi-ITN STRIKE - Novel Methods in Computational Finance) and the Bulgarian National Fund of Science under Project DID 02/37-2009.

References

1. Agliardi, R., Popivanov, P., Slavova, A.: Nonhypoellipticity and comparison principle for partial differential equations of Black-Scholes type. Nonlinear Anal. Real Word Appl. **12**, 1429–1436 (2011)
2. Company, R., Navarro, E., Pintos, J., Ponsoda, E.: Numerical solution of linear and nonlinear Black-Scholes option pricing equations. Int. J. Comp. Math. Appl. **56**, 813–821 (2008)
3. Ehrhardt, M. (ed.): Nonlinear Models in Mathematical Finance: New Research Trends in Option Pricing. Nova Science Publishers, New York (2008)
4. Gerisch, A., Griffiths, D.F., Weiner, R., Chaplain, M.A.J.: A positive splitting method for mixed hyperbolic-parabolic systems. Numer. Meth. PDEs **17**(2), 152–168 (2001)
5. Hundsdorfer, W., Verwer, J.: Numerical Solution of Time-Dependent Advection-Diffusion-Reaction Equations. Springer, Heidelberg (2003)
6. Koleva, M.N.: Positivity preserving numerical method for non-linear Black-Scholes models. In: Dimov, I., Faragó, I., Vulkov, L. (eds.) NAA 2012. LNCS, vol. 8236, pp. 363–370. Springer, Heidelberg (2013)

7. Koleva, M.N., Vulkov, L.G.: A numerical study of a parabolic Monge-Ampère equation in mathematical finance. In: Dimov, I., Dimova, S., Kolkovska, N. (eds.) NMA 2010. LNCS, vol. 6046, pp. 461–468. Springer, Heidelberg (2011)
8. Koleva, M., Vulkov, L.: Quasilinearization numerical scheme for fully nonlinear parabolic problems with applications in models of mathematical finance. Math. Comput. Model. **57**(9–10), 2564–2575 (2013)
9. Lai, C.-H., Parrott, A.K., Rout, S.: A distributed algorithm for European options with nonlinear volatility. Comput. Math. Appl. **49**, 885–894 (2005)
10. Le Roux, A.-Y., Le Roux, M.-N.: Numerical solution of a Cauchy problem for nonlinear reaction process. J. Comput. Appl. Math. **214**, 90–110 (2008)
11. Le Roux, M.-N.: Numerical solution of a parabolic problem arising in finance. Appl. Numer. Math. **62**, 815–832 (2012)
12. Songzhe, L.: Existence of solution to initial value problem for a parabolic Monge-Ampère equation and application. Nonlinear Anal. **65**, 59–78 (2006)
13. Wilmott, P., Howison, S., Dewynne, J.: The Mathematics of Financial Derivatives: A Student Introduction. Cambridge University Press, Cambridge (1995)
14. Yong, J.: Introduction to mathematical finance. In: Yong, J., Cont, R. (eds.) Mathematical Finance-Theory and Applications. High Education Press, Beijing (2000)

Calibration of Parameters for Radio-Frequency Ablation Simulation

N. Kosturski(✉), S. Margenov, and Y. Vutov

Institute of Information and Communication Technologies,
Bulgarian Academy of Sciences, Sofia, Bulgaria
kosturski@parallel.bas.bg

Abstract. We consider the simulation of the thermal and electrical processes, involved in the radio-frequency (RF) ablation procedure. RF ablation is a low invasive technique for the treatment of hepatic tumors, utilizing AC current to destroy the tumor cells by heating. The procedure consists of inserting an RF probe in the patients liver and attaching a ground pad to the skin. After that the AC current is initiated and maintained for a prescribed duration.

We have conducted experiments with a pork liver and an RF ablation apparatus capable of measuring and recording some of the procedure parameters. Those include the applied power, the effective electrical impedance, and the temperature around the tip of the probe. A history of the values at each second of the test is obtained at the end.

Our aim is to adjust the material properties and other model parameters for the simulation to fit the experimentally obtained results. The electrical conductivity of the tissue can be deduced from the measured power and impedance. After that, we need to determine suitable heat conductivity and capacity coefficients. This is achieved via temperature curves comparison.

1 Introduction

RF ablation is an alternative, low invasive technique for the treatment of hepatic tumors, utilizing AC current to destroy the tumor cells by heating [8,9]. The destruction of the cells occurs at temperatures of 45 °C–50 °C. The procedure is relatively safe, as it does not require open surgery.

The considered RF probe consists of a stainless steel needle, insulated with polyurethane. The RF ablation procedure starts by placing the RF probe inside the tumor. The surgeon performs this under computed tomography (CT) or ultrasound guidance. Once the probe is in place, RF current is initiated. The surface area of the uninsulated part of the needle conducts RF current.

The human liver has a complex structure, with varying thermal and electrical properties – there are three types of blood vessels with different sizes and flow velocities. Here, we consider a simplified test problem, where the liver consists of one homogeneous hepatic tissue.

I. Lirkov et al. (Eds.): LSSC 2013, LNCS 8353, pp. 611–618, 2014.
DOI: 10.1007/978-3-662-43880-0_70,

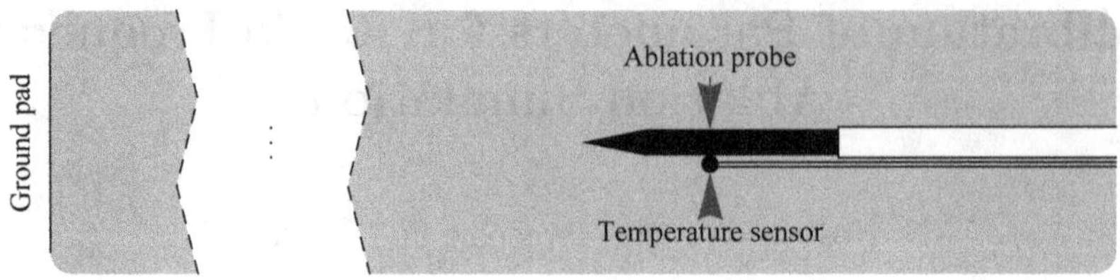

Fig. 1. Experimental procedure setup

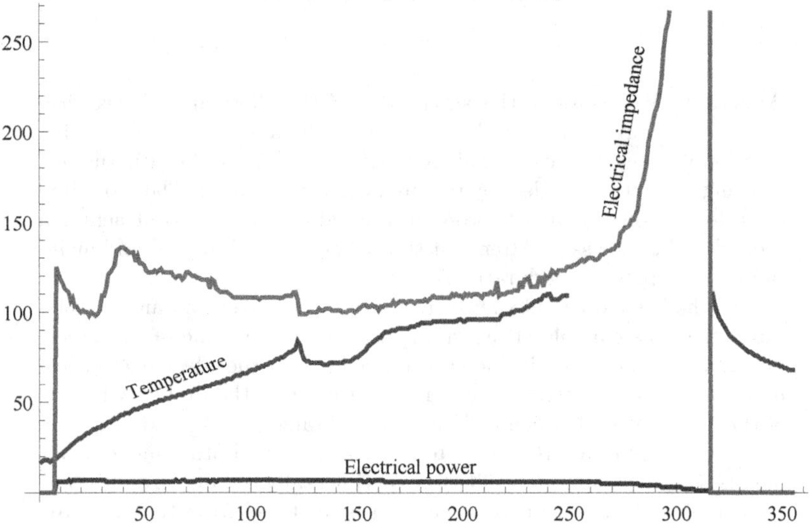

Fig. 2. Measurement data obtained during the experiment

2 Physical Experiment

We have conducted experiments with a pork liver and an RF ablation apparatus capable of measuring and recording some of the procedure parameters. The setup is illustrated on Fig. 1.

The pork liver was placed inside a metal plate, to which the ground pad was applied. The RF ablation probe was inserted in the liver. After that a temperature sensor was inserted in the hole, created by the probe, parallel to it, and the RF current was initiated. The apparatus measured the applied power, the effective electrical impedance, and the temperature. A history of the values at each second of the test can be seen on Fig. 2. Due to limitations of the used temperature sensor, only temperatures below 110 °C could be measured.

As can be seen, with the increase of the tissue temperature to a certain level, a steep increase of the electrical impedance is observed. This is attributed to tissue charring and vapor formation, which forms an isolating layer around the probe. As a result the electrical power quickly drops to zero and the ablation process stops. In this paper, we are concentrating on the ablation process before the impedance jump occurs, in order to check how well the simulation approximates the temperature field. Therefore we are considering the measurement data from the moment electrical current was initiated up to the moment the temperature at the sensor passed 110 °C.

3 Radio-Frequency Tumor Ablation Model

Let us turn our attention to the considered numerical simulation. The RF ablation procedure destroys the unwanted tissue by heating, arising when the energy dissipated by the electric current flowing through a conductor is converted to heat. A simplified bio-heat time-dependent partial differential equation [8,9]

$$s\frac{\partial T}{\partial t} = \nabla \cdot k\nabla T + J \cdot E \tag{1}$$

is used to model the heating process during the RF ablation. The simplification is due to the fact there is no blood perfusion and no metabolic heat production, as the experiment was performed on a dead pork liver. The term $J \cdot E$ in (1) represents the thermal energy arising from the current flow.

The following initial and boundary conditions are applied

$$\begin{aligned} T &= T_0 \text{ when } t = 0 \text{ at } \Omega, \\ T &= T_0 \text{ when } t \geq 0 \text{ at } \partial\Omega. \end{aligned} \tag{2}$$

The following notations are used in (1) and (2): Ω is the entire domain of the model, $\partial\Omega$ – the boundary of the domain, s – the volumetric heat capacity (J/m^3 K), k – the thermal conductivity (W/m K), J – the current density (A/m), E – the electric field intensity (V/m), and T_0 – body (or in this case room) temperature (°C).

The bio-heat problem is solved in two steps. The first step is finding the potential distribution V of the current flow. With the considered RF probe design, the current is flowing from the conducting electrodes to a dispersive electrode on the patient's body. The electrical flow is modeled by the Laplace equation

$$\nabla \cdot \sigma\nabla V = 0, \tag{3}$$

with boundary conditions

$$\begin{aligned} V &= 0 \text{ at } \partial\Omega_{gr}, \\ V &= V_0 \text{ at } \partial\Omega_{el}, \end{aligned}$$

where V is the potential distribution in Ω, σ – the electric conductivity (S/m), V_0 – the applied RF voltage, $\partial\Omega_{gr}$ – the part of the boundary connected to the ground pad, and $\partial\Omega_{el}$ – the surface of the conducting part of the RF probe.

After determining the potential distribution, the electric field intensity can be computed from

$$E = -\nabla V,$$

and the current density from

$$J = \sigma E.$$

The second step is to solve the heat transfer Eq. (1) using the heat source $J \cdot E$ obtained in the first step.

For the numerical solution of both of the above discussed steps of the simulation the Finite Element Method (FEM) in space is used [2]. Linear conforming elements are chosen in this study. The domain is represented by a voxel image with a resolution of 256 × 256 × 256. To apply the linear FEM discretization to the voxel domain, each voxel is split into six tetrahedra. To solve the bio-heat equation, after the space discretization, the time derivative is discretized via finite differences and the backward Euler scheme is used [4,5].

Let us denote with K^* the stiffness matrix coming from the FEM discretization of the Laplace Eq. (3). It can be written in the form

$$K^* = \left[\int_\Omega \sigma \nabla \Phi_i \cdot \nabla \Phi_j d\mathbf{x} \right]_{i,j=1}^N ,$$

where $\{\Phi_i\}_{i=1}^N$ are the FEM basis functions.

The system of linear algebraic equations

$$K^* X = 0 \tag{4}$$

is to be solved to find the nodal values X of the potential distribution.

The electric field intensity and the current density are then expressed by the partial derivatives of the potential distribution in each finite element. This way, the nodal values F for the thermal energy $E \cdot J$ arising from the current flow are obtained.

Let us now turn our attention to the discrete formulation of the bio-heat equation. Let us denote with K and M the stiffness and mass matrices from the finite element discretization of (1). They can be written as

$$K = \left[\int_\Omega k \nabla \Phi_i \cdot \nabla \Phi_j d\mathbf{x} \right]_{i,j=1}^N ,$$

$$M = \left[\int_\Omega s \Phi_i \Phi_j d\mathbf{x} \right]_{i,j=1}^N .$$

Then, the parabolic Eq. (1) can be written in matrix form as:

$$M \frac{\partial T}{\partial t} + KT = F. \tag{5}$$

If we denote with τ the time-step, with T^{n+1} the solution at the current time level, and with T^n the solution at the previous time level and approximate the time derivative in (5) we obtain the following system of linear algebraic equations for the nodal values of T^{n+1}

$$(M + \tau K) T^{n+1} = M T^n + \tau F. \tag{6}$$

The matrices of the linear systems (4) and (6) are ill-conditioned and large. Since they are symmetric and positive definite, we use the PCG [1] method, which is the most efficient solution method in this case.

A parallel AMG implementation – BoomerAMG [6,10] is used to precondition the linear systems. The matrix $A = M + \tau K$ from (6) is assembled only once on the first time step and not varied after that. The corresponding AMG preconditioner is also constructed only on the first time step. Additional details concerning the parallelization approach can be found in our paper [7].

4 Model Parameters Calibration

Our aim is to identify (tune) the liver tissue parameters, such that the simulation results better fit the results from the physical experiment. We start with the material properties in Table 1, taken from the literature [8].

Before we can calibrate the thermal properties s_l and k_l of the liver, we need to calibrate it's electrical conductivity σ_l in accordance with the experimental data. Since the electrical field in the model is static, we first computed the average values for the electrical power and impedance – $P = 6.653$ W and $R = 111.8\ \Omega$ respectively.

In order to match the electrical power P, we need to determine the potential V_0 for the second boundary condition of (3) that will yield the desired value. To do this, the Laplace equation is initially solved with a boundary condition $V = 1\,\text{V}$ at $\partial\Omega_{el}$. Then, E^* and J^* are obtained from the solution and the corresponding electrical power P^* can be computed as

$$P^* = \int_{\Omega} E^* \cdot J^* d\mathbf{x}.$$

Since the solution and all the components of E and J are proportional to the value of V_0 we can scale the obtained solution, instead of recomputing it, in the following way

$$V_0 = \lambda, \quad E = \lambda E^*, \quad J = \lambda J^*, \quad \text{where} \quad \lambda = \sqrt{P/P^*}.$$

Now we compute the effective electrical impedance $R^* = V_0^2/P$ and then by setting $\sigma_l = 0.333R^*/R \approx 0.2494$ and repeating the procedure we obtain a good match of both P and R with a potential $V_0 \approx 27.28\,\text{V}$.

Table 1. Thermal and electrical properties of the materials

Material	s (J/m^3 K)	k (W/m K)	σ (S/m)
Stainless steel	2.838×10^6	71	4×10^8
Liver	3.816×10^6	0.512	0.333
Polyurethane	73150	0.026	10^{-5}

After obtaining an electrical field matching the experimental data, we are now ready to calibrate the thermal properties of the tissue. In order to compare the numerical results to the measurement, we run the simulation with a time step of 1 s and consider the results in a single point that is selected on the boundary between the probe and the tissue around the middle of the uninsulated part of the needle. Let us denote the physically measured temperature on the i-th second of the procedure with T_i^m and the temperature from the simulation with $T_i^s(k_l, s_l)$ respectively. Note that we treat the simulated temperature as a function of the thermal properties of the tissue. Now we can formulate the calibration problem as a least-squares optimization problem

$$\min \sum_i \left(T_i^s(k_l, s_l) - T_i^m\right)^2 .$$

Since in the physical experiment the temperature sensor was not firmly attached to the ablation probe, we can attribute drops in the measured temperature to a displacement of the sensor with respect to the probe. Because of this, and also because the selected point normally has the highest temperature we add the following constraints

$$T_i^s(k_l, s_l) \geq T_i^m, \quad \forall i.$$

In order to enforce the constraints we use a penalty method, which consists of solving a series of unconstrained minimization problems

$$\min \Psi^j(k_l, s_l), \ \text{ for } j = 1, 2, \ldots$$

where

$$\Psi^j(k_l, s_l) = \sum_i \left(T_i^s(k_l, s_l) - T_i^m\right)^2 + \theta_j \sum_i \min\left(0, T_i^s(k_l, s_l) - T_i^m\right)^2 , \qquad (7)$$

until the minimum stops increasing. The penalty coefficient on the j-th iteration is selected as $\{\theta_j\}_{j=1}^{\infty} = \{0, 1, 10, 100, \ldots\}$. Each minimization result is used as an initial guess for the next minimization problem. We use the coefficients from the literature as an initial guess for the first unconstrained minimization. The iterations of the penalty method are illustrated on Fig. 3.

For the solution of each unconstrained minimization problem, we selected the principal axis method [3]. It is a derivative-free algorithm, where an approximate model is built up using only values from function evaluations. This algorithm consists of a series of linear searches, with directions, chosen in a way that ensures they are well aligned to the principal directions of a local quadratic model.

In our case each function evaluation meant running an RF ablation simulation with the corresponding material properties and evaluating $\Psi^j(k_l, s_l)$ from (7) with the resulting temperature samples. For the initial and boundary conditions (2) the value $T_0 = 18\,°\mathrm{C}$ from the measurement before starting the RF procedure is used.

Each unconstrained minimization required around 150 simulation runs. The constrained minimization had 10 steps. The implementation was run on 512 cores of the IBM Blue Gene/P computer and the maximum job duration of one

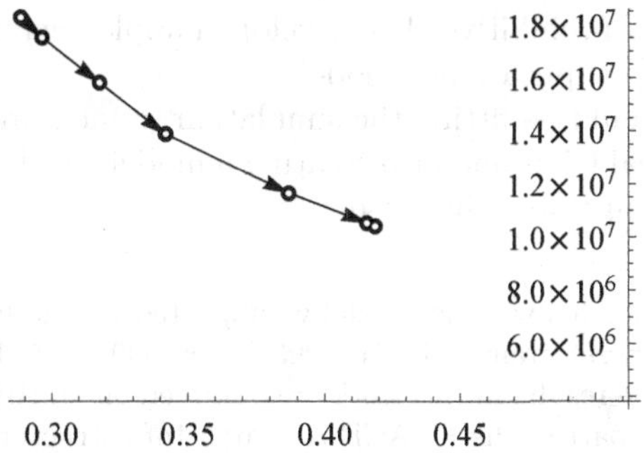

Fig. 3. Penalty method iterations

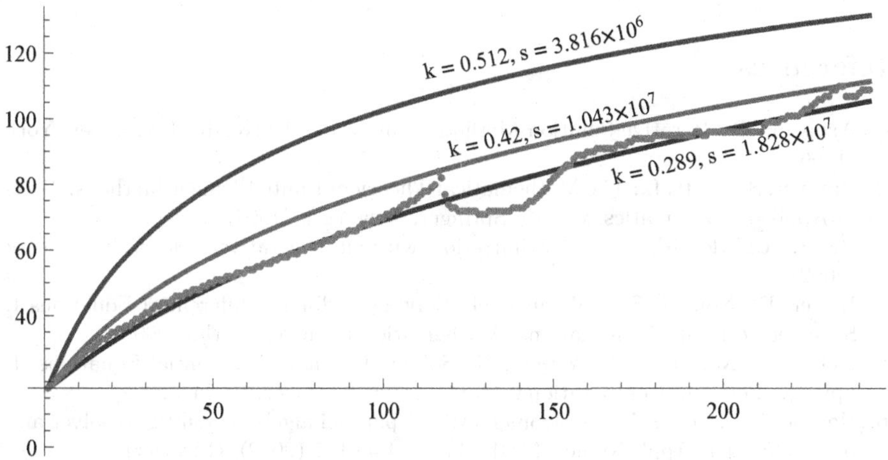

Fig. 4. Simulation results with different coefficients

week was not enough to complete the calibration. However, it was easy to restart the calibration procedure from where it finished and one restart was enough.

The simulation results with the literature coefficients, the unconstrained fit, and the constrained fit can all be seen on Fig. 4 along with the corresponding material properties.

5 Concluding Remarks

We have described a feasible, albeit time consuming, procedure for calibration of model parameters. No specific assumptions are made, therefore we think the procedure can be applied to any parameters fitting different measurements. In theory, we can calibrate more than two parameters at the same time with the developed implementation, although, this was not tested in practice and the performance might be prohibitive. The developed implementation should be very useful as we further complicate our model.

Our next steps would be fitting the simulation to measurements, taken from real patients in clinical trials and also creating a model which includes the experimentally observed impedance increase.

Acknowledgements. This work is partially supported by the EU Operational Programme "Competitiveness" grant BG161PO003-1.1.06-0004-C0001 and Bulgarian NSF Grant DCVP 02/1. The results in this work are obtained on an IBM BlueGene/P computer located in Sofia, part of the PRACE Research Infrastructure.

We would like to thank Theodor Popov and the engineering team at AMET Ltd. for providing the RF ablation equipment as well as conducting the experiments with us.

References

1. Axelsson, O.: Iterative Solution Methods. Cambridge University Press, New York (1996)
2. Brenner, S., Scott, L.: The Mathematical Theory of Finite Element Methods. Texts in Applied Mathematics, vol. 15. Springer, New York (1994)
3. Brent, R.: Algorithms for Minimization without Derivatives. Dover, New York (2002)
4. Hairer, E., Norsett, S.P., Wanner, G.: Solving Ordinary Differential Equations I. Springer Series in Computational Mathematics. Springer, Berlin (2000)
5. Hairer, E., Norsett, S.P., Wanner, G.: Solving Ordinary Differential Equations II. Springer Series in Computational Mathematics. Springer, Berlin (2002)
6. Henson, V.E., Yang, U.M.: BoomerAMG: a parallel algebraic multigrid solver and preconditioner. Appl. Numer. Math. **41**(1), 155–177 (2002). (Elsevier)
7. Kosturski, N., Margenov, S., Vutov, Y.: Improving the efficiency of parallel FEM simulations on voxel domains. In: Lirkov, I., Margenov, S., Waśniewski, J. (eds.) LSSC 2011. LNCS, vol. 7116, pp. 574–581. Springer, Heidelberg (2012)
8. Tungjitkusolmun, S., Staelin, S.T., Haemmerich, D., Tsai, J.Z., Cao, H., Webster, J.G., Lee, F.T., Mahvi, D.M., Vorperian, V.R.: Three-dimensional finite-element analyses for radio-frequency hepatic tumor ablation. IEEE Trans. Biomed. Eng. **49**(1), 3–9 (2002)
9. Tungjitkusolmun, S., Woo, E.J., Cao, H., Tsai, J.Z., Vorperian, V.R., Webster, J.G.: Thermal-electrical finite element modelling for radio frequency cardiac ablation: effects of changes in myocardial properties. Med. Biol. Eng. Comput. **38**(5), 562–568 (2000)
10. Lawrence Livermore National Laboratory, Scalable Linear Solvers Project. http://www.llnl.gov/CASC/linear_solvers/

Surfaces from Curves on Triangular Surfaces in Barycentric Coordinates

Arne Lakså(✉)

Narvik University College, P.O. Box 385, 8505 Narvik, Norway
Arne.Laksa@hin.no
http://www.hin.no/Simulations

Abstract. Barycentric coordinates are coordinates in which a position is provided by a blending of a weighted point set where the weights sum up to 1. Bezier-triangles and ERBS-triangles are typical examples of use of Barycentric coordinates.

We look at the framework for the description of curves on surfaces that are described in Barycentric coordinates and how we define surfaces in a Coons Patch like framework with the use of these curves on surfaces. The framework also includes pre-evaluation and other optimization technics for evaluation.

The background is to construct large complex surfaces. Given a surface constructed by a connected set of non-planar triangular surfaces. If the triangular surfaces are generalized expo-rational B-spline based, constructed by blending of triangular sub-surfaces from Bezier-patches, then the surface is smooth at the vertices but only continuous over the edges between the triangular surfaces. If we introduce a second set of vertices defined by the midpoint of each triangular surface, we can introduce a new set of edges constructed by straight lines from a vertex to the midpoint in the parameter plane of the respective triangular surface. In addition we also have information about the derivatives across these edges. This gives us the data to make a connected and smooth set of surfaces that are strongly connected to the set of triangular surfaces. The triangle based surface is easy to manipulate and reshape and then the smooth dual set of squared surfaces will automatically be updated.

Keywords: Curve · Surface · Barycentric coordinates · Blending

1 Introduction

A Bezier triangle is a surface parameterized with barycentric coordinates with the expression

$$s(u,v,w) = \sum_{i=0}^{n} c_i \; b_{d,i}(u,v,w)$$

where the coordinates c_i are points that describe a triangular control polygon. The basis functions $b_{d,i}(u,v,w)$ follow from expanding $(u \; v \; w)^d$ where d is the

I. Lirkov et al. (Eds.): LSSC 2013, LNCS 8353, pp. 619–627, 2014.
DOI: 10.1007/978-3-662-43880-0_71,

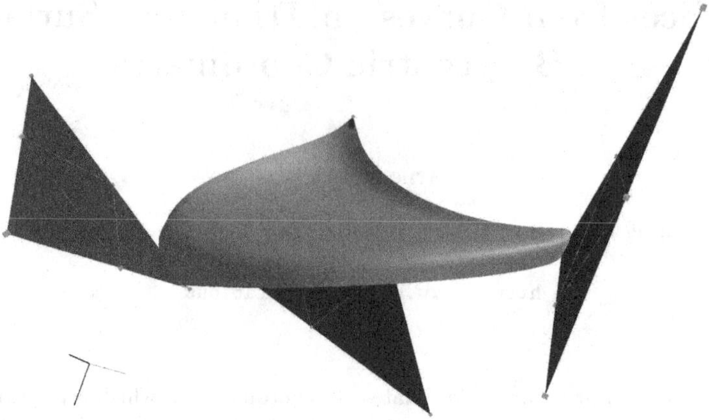

Fig. 1. A GERBS triangular surface, constructed by blending three Bezier triangles of degree 2 (but planar as we can see in the picture).

polynomial degree and $n = \sum_{i=1}^{d+1} i$. Bezier triangles and in general splines on triangulation are treated by many authors, see [4,6].

A GERBS triangle is a blending of triangular surfaces (see [2,5]), i.e.

$$s(u,v,w) = \sum_{i=1}^{3} c_i(u,v,w)\ b_i(u,v,w),$$

where $c_i(u,v,w), i = 1,2,3$ are three triangular surfaces and $b_i(u,v,w), i = 1,2,3$ are GERBS blending functions for triangular patches. In Fig. 1 we can see a GERBS triangle-surface constructed by blending three Bezier triangles. In the figure the control points that define the Bezier triangles are marked. We can see 6 control points on each Bezier triangle, which indicates that the degree of the Bezier triangles are 2, although the triangles seems to be planar.

A GERBS triangle interpolates the three local triangles in their respective vertices, not only the value but also with all derivatives up to a given order, depending on the choice of blending functions.

1.1 Surfaces over Triangular Structures

To make surfaces of all genius, it is convenient to use a set of connected triangular patches. The surface construction using GERBS-blending is as follows:

1. Given a point set where for each point is given surface normal and curvature.
2. The points must be ordered by a triangulation.
3. In each point we create a "local" Bezier-patch from the given surface normal and curvature. Each patch must cover the first neighborhood of points (points that are connected to the central point by an edge).

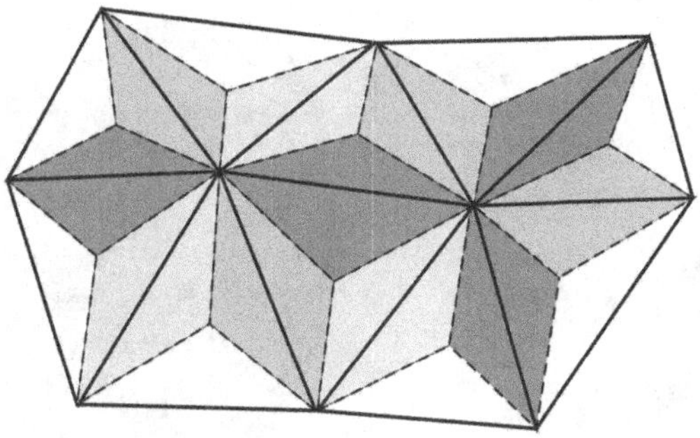

Fig. 2. The parameter plane of a surface constructed by a connected set of triangular surfaces (bold edges) and the dual set of squared patches over each internal edge.

4. Each of these Bezier surfaces are then divided into a series of contiguous triangular faces (sub-triangles). This is done by finding closest point from each of the neighboring points.
5. One sub-triangle from each of three neighboring points (the one that is connected to all three points) are then blended together to form one GERBS-triangular surface.

The resulting surface, that is a connected set of triangular surfaces, is smooth on the vertices but unfortunately only continuous over the edges.

To make a smooth surface, we would, related to the set of triangular patches, introducing a dual set of (this time) square patches. First, we introduce an additional set of vertices that are in the middle of each of the triangular patches. So, with the starting point of the two vertices defining an edge and the two new internal vertices in the inner of the corresponding triangular patches, we create a square surface over each internal edge. We use a Coons patch method where we get the needed boundary curves from curves in the triangular patches, i.e. from an original vertex to one new vertex. This can be seen in Fig. 2.

In Fig. 3 there is an example where we approximate a sphere at six vertices, using four point around equator and one at each poles. The Bezier patches at each vertex is then moved and rotated. We now have six Bezier patches, each connected to a specific vertex. This gives eight GERBS triangular patches that are smooth at the vertices, but only continues over the totally twelve edges that connect the triangular GERBS patches. On left hand side in Fig. 3 this is shown. On right hand side we can see the dual set of twelve squared patches that are covering and smoothing the twelve edges.

2 Barycentric Coordinates on a Bezier Patch

The domain of a Bezier patch $S(p)$ is $U = [0,1] \times [0,1] \subset \mathbb{R}^2$, commonly using Cartesian coordinates $p = (\mu, \nu)$.

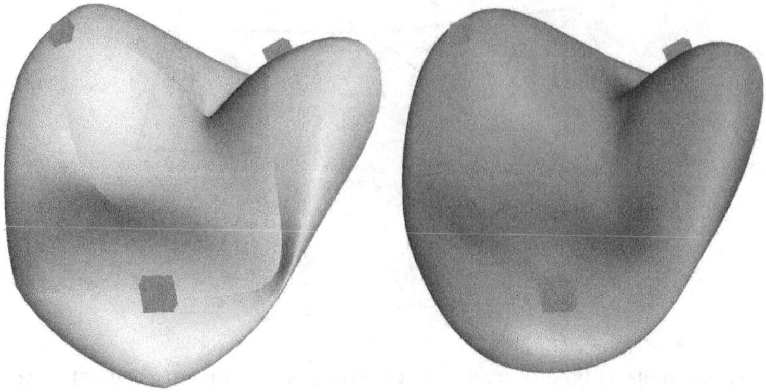

Fig. 3. On left hand side, we see a surface which consist of 8 GERBS triangular patches. The surface is made by approximating a sphere at 6 vertices. The local Bezier patches are then moved and rotated. On right hand side we see the dual set of 12 squared patches. The cubes in the figure are marking the position and orientation of the vertices.

We define a point set $p_i, i = 0, 1, ..., n$ on U describing a fan of n triangles. Each triangle $\triangle_i, i = 1, 2, ..., n$ is defined by the three points p_0, p_i, p_{i+1} (where $p_{n+1} = p_1$) which is the control polygon to a first degree Bezier triangle. The formula, including the partial derivatives, for $\triangle_i$ in barycentric coordinates is

$$\begin{aligned} \triangle_i(u,v,w) &= S(up_0 + vp_i + wp_{i+1}), \\ D_u\triangle_i(u,v,w) &= dS_{up_0+vp_i+wp_{i+1}}(p_0), \\ D_v\triangle_i(u,v,w) &= dS_{up_0+vp_i+wp_{i+1}}(p_1), \\ D_w\triangle_i(u,v,w) &= dS_{up_0+vp_i+wp_{i+1}}(p_2), \end{aligned} \qquad u+v+w=1.$$

The second order derivatives follow from that the columns of the matrix dS are the partial derivatives, i.e. $dS = [S_u \;\; S_v] : \mathbb{R}^2 \rightarrow \mathbb{R}^3$,

$$\begin{aligned} D_{uu}\triangle_i(u,v,w) &= [d(S_\mu)(p_0) \;\; d(S_\nu)(p_0)]\,(p_0), \\ D_{uv}\triangle_i(u,v,w) &= [d(S_\mu)(p_0) \;\; d(S_\nu)(p_0)]\,(p_i), \\ D_{uw}\triangle_i(u,v,w) &= [d(S_\mu)(p_0) \;\; d(S_v)(p_0)]\,(p_{i+1}), \\ D_{vv}\triangle_i(u,v,w) &= [d(S_\mu)(p_i) \;\; d(S_\nu)(p_i)]\,(p_i), \\ D_{vw}\triangle_i(u,v,w) &= [d(S_\mu)(p_i) \;\; d(S_\nu)(p_i)]\,(p_{i+1}), \\ D_{ww}\triangle_i(u,v,w) &= [d(S_\mu)(p_{i+1}) \;\; d(S_\nu)(p_{i+1})]\,(p_{i+1}), \end{aligned}$$

where the matrices $dS_\mu = [S_{\mu\mu} \;\; S_{\mu\nu}]$ and $dS_\nu = [S_{\mu\nu} \;\; S_{\nu\nu}]$.

If we want more freedom in the local parameterization, we can use second degree Bezier triangles in the domain of the Bezier patches associated with the vertices.

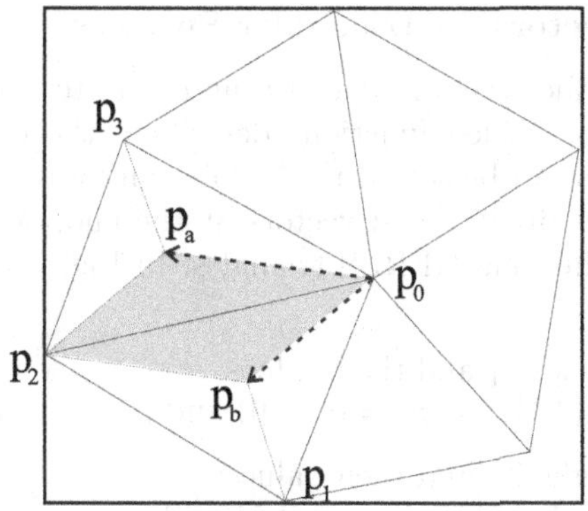

Fig. 4. The figure shows the parameter plane of a Bezier patch. In the parameter plane we can see 6 triangles located in a fan around point p_0. There is also a sketch of a squared patch (p_0, p_b, p_2, p_a), covering the edge (p_0, p_2).

2.1 Transferring Coordinates between Different Barycentric Coordinate System

Given two triangles, $\triangle_1(p_0, p_1, p_2)$ and $\triangle_2(p_0, p_2, p_3)$, in the domain of a Bezier patch and given a point $\bar{p}_a = (\bar{u}, \bar{v}, \bar{w})$ in barycentric coordinates with respect to triangle $\triangle_1$ (the situation is described in Fig. 4).

Lemma 1. *The change of coordinate system to a point p_a in triangle $\triangle_1$ to triangle $\triangle_2$ is: $(\bar{u}, \bar{v}, \bar{w}) \rightarrow (u, v, w)$, where*

$$\begin{aligned} u &= \tfrac{(p_a - p_3)\wedge(p_2 - p_3)}{(p_0 - p_3)\wedge(p_2 - p_3)}, \\ v &= \tfrac{(p_a - p_3)\wedge(p_0 - p_3)}{(p_2 - p_3)\wedge(p_0 - p_3)}, \\ w &= 1 - u - v, \end{aligned} \tag{1}$$

and where

$$p_a = \bar{u}\ p_0 + \bar{v}\ p_1 + \bar{w}\ p_2 \tag{2}$$

Proof. Changing coordinate system is as following using barycentric coordinates,

$$u\ p_0 + v\ p_2 + w\ p_3 = \bar{u}\ p_0 + \bar{v}\ p_1 + \bar{w}\ p_2.$$

Reorganizing, using expression (2) and $w = 1 - u - v$, we get,

$$u\ (p_0 - p_3) + v\ (p_2 - p_3) = p_a - p_3,$$

where we now have three vectors instead of points. Further, a wedge product of a vector with itself is zero, and therefore (1) follows.

2.2 Points and Vectors on Triangular Surfaces

To make a squared surface over an edge, it is necessary to have the four boundary curves and four vector valued functions describing the derivatives across the boundary curves. This can be solved in the following way.

We first find the endpoints and vectors at the endpoints in the parameter plane of the two neighboring GERBS triangles. In Fig. 4 we can see the points and vectors.

- The points in triangle $\triangle_1$ and their values:
  $p_0 = (1, 0, 0)$, $p_b = (\frac{1}{3}, \frac{1}{3}, \frac{1}{3})$, $p_2 = (0, 1, 0)$ and p_a.
- The points in triangle $\triangle_2$ and their values:
  $p_0 = (1, 0, 0)$, $p_a = (\frac{1}{3}, \frac{1}{3}, \frac{1}{3})$, $p_2 = (0, 0, 1)$ and p_b.

The points p_a in triangle $\triangle_1$ and p_b in triangle $\triangle_2$ has to be computed according coordinate change formula from expression (1). The computation (initially in the parameter space to the two GERBS triangles) has to be done in the parameter space of the Bezier patch connected to the central vertex. It is important to observe that there are two Bezier patches involved, one where the central vertex is p_0 and one where the central vertex is p_2. It follows that the points and thus vectors must be computed in the parameter plane of the Bezier patch where the vectors are connected to the central vertex.

We must find all four vectors in both triangular surfaces. For vectors we thus have:

- The vectors in triangle $\triangle_1$:
  $v_1 = p_a - p_2$, $v_2 = p_0 - p_b$, $v_3 = p_b - p_2$, $v_4 = p_0 - p_a$,
- The vectors in triangle $\triangle_2$:
  $v_5 = p_a - p_2$, $v_6 = p_0 - p_b$, $v_7 = p_b - p_2$, $v_8 = p_0 - p_a$,

where p_a in v_1 and p_b in v_3 has to be computed in the parameter plane to the surface where p_2 is the central vertex, while p_a in v_2 and p_b in v_4 has to be computed in the parameter plane to the surface where p_0 is the central vertex. The reason for this is that the surface over triangulation is smooth in the vertices and not over the edges (described in more detail in [5]).

Barycentric coordinates can be made more general so that it is possible to express a vertex, in a planar triangulation, as a convex combination of its neighboring vertices. This is called mean value coordinates and can be used to calculate the coordinates in the parameter plane of a Bezier patch for barycentric coordinates, see [3].

3 Curves and Vector Valued Functions on Triangular Surfaces

The general formula for a curve or a vector valued functions is,

$$\widehat{c}(t) = \sum_{i=0}^{d} c_i \; b_i(t),$$

where c_i, $i = 0, 1, ...d$ are points or vectors, and $b_i(t)$, $i = 0, 1, ..., d$ are basis functions spanning the function space.

If the basis functions are Bernstein polynomials, they sum up to 1, and the derivatives,

$$\sum_{i=0}^{j} b_i^{(j)}(t) = 0, \quad j = 1, 2, ..., d.$$

If the curve $\widehat{c}$ is in the parameter space of a surface S, then the formula for the space curve on the surface is,

$$c(t) = S \circ \widehat{c}(t)$$

and the derivative is,

$$c'(t) = dS(\widehat{c}'(t)).$$

If the curve is on a triangular surface using barycentric coordinates then,

$$dS = [S_u \; S_v \; S_w]$$

is a 3×3 matrix, and it follows that the second derivative is,

$$c''(t) = [dS_u(\widehat{c}'(t)) \; dS_v(\widehat{c}'(t)) \; dS_w(\widehat{c}'(t))] \, (\widehat{c}'(t)) + dS(\widehat{c}''(t)).$$

Now $\widehat{c}(t)$ is a point in barycentric coordinates where the sum of the coordinates is 1, and $\widehat{c}'(t)$ is a vector in barycentric coordinates where the sum of the coordinates is 0.

In the example shown in Fig. 4 the curve is linear in the parameter plane of the GERBS triangle. It follows that $\widehat{c}(t) = (1-t)p_i + tp_j$ where i, j are indices of two points defining a boundary curve on the squared patch covering an edge. The derivative $\widehat{c}'(t) = p_j - p_i$.

The vectors across an edge in the parameter plane are defined by the vectors describing the direction for the derivatives across the boundary curve at each end of a boundary curve v_i and v_j,

$$h(t) = v_i + b(t)(v_j - v_i),$$

where $b(t)$ is a GERBS basis function, i.e. $b(0) = 0, b'(0) = 0, b(1) = 1, b'(1) = 0$ (see [2] for the properties of a GERBS function).

A vector valued function describing the derivatives across the boundary curves will be on the form:

$$g(t) = dS(h(t))$$

and the derivative

$$g'(t) = [dS_u(\widehat{c}'(t)) \; dS_v(\widehat{c}'(t)) \; dS_w(\widehat{c}'(t))] \, (h(t)) + dS\,(h'(t))\,.$$

4 Surfaces over Edges in a Triangular Structure

One way to construct the surfaces over the edges is to use a Coons patch Bicubically Blending like procedure, [1]. In Fig. 4 we have a gray area covering an edge. If we name the boundary curves $c_i(t),\ i = 1,2,3,4$ we get,

$$\begin{aligned} c_1(t) &= \triangle_1(\widehat{c}_1(t)), \quad \text{where} \quad \widehat{c}_1(t) = p_2 + t(p_b - p_2) \\ c_2(t) &= \triangle_2(\widehat{c}_2(t)), \quad \text{where} \quad \widehat{c}_2(t) = p_a + t(p_0 - p_a) \\ c_3(t) &= \triangle_2(\widehat{c}_3(t)), \quad \text{where} \quad \widehat{c}_3(t) = p_2 + t(p_a - p_2) \\ c_4(t) &= \triangle_1(\widehat{c}_4(t)), \quad \text{where} \quad \widehat{c}_4(t) = p_b + t(p_0 - p_b), \end{aligned} \tag{3}$$

where $p_a = p_b = (\frac{1}{3}, \frac{1}{3}, \frac{1}{3})$ and the points else are defined in Sect. 2.2. The functions for the derivatives across the edges are

$$\begin{aligned} g_1(t) &= d\triangle_1(h_1(t)), \quad \text{where} \quad h_1(t) = v_1 + b(t)(v_2 - v_1) \\ g_2(t) &= d\triangle_2(h_2(t)), \quad \text{where} \quad h_2(t) = v_5 + b(t)(v_6 - v_5) \\ g_3(t) &= d\triangle_2(h_3(t)), \quad \text{where} \quad h_3(t) = v_7 + b(t)(v_8 - v_7) \\ g_4(t) &= d\triangle_1(h_4(t)), \quad \text{where} \quad h_4(t) = v_3 + b(t)(v_4 - v_3) \end{aligned} \tag{4}$$

and the vectors v_i are defined in Sect. 2.2.

The surface construction is, to make three surfaces:

$$S_1(u,v) = [c_1(u)\ c_2(u)\ g_1(u)\ g_2(u)] \begin{bmatrix} H_1(v) \\ H_2(v) \\ H_3(v) \\ H_4(v) \end{bmatrix},$$

$$S_2(u,v) = [H_1(u)\ H_2(u)\ H_3(u)\ H_4(u)] \begin{bmatrix} c_3(v) \\ c_4(v) \\ g_3(v) \\ g_4(v) \end{bmatrix},$$

$$S_3(u,v) = [H_1(u)\ H_2(u)\ H_3(u)\ H_4(u)] \begin{bmatrix} c_1(0) & c_1(1) & g_1(0) & g_1(1) \\ c_2(0) & c_2(1) & g_2(0) & g_2(1) \\ g_3(0) & g_3(1) & g_1'(0) & g_1'(1) \\ g_4(0) & g_4(1) & g_2'(0) & g_2'(1) \end{bmatrix} \begin{bmatrix} H_1(v) \\ H_2(v) \\ H_3(v) \\ H_4(v) \end{bmatrix},$$

where $H_i(t),\ i = 1,2,3,4$ are the third degree Hermite basis functions. The resulting surface is,

$$S(u,v) = S_1(u,v) + S_2(u,v) - S_3(u,v).$$

To create surfaces covering all internal edges, as described here, will result in a composite surface that is C^1-smooth all over, as we can see on right hand side in Fig. 3.

References

1. Coons, S.A.: Surfaces for computer-aided design of space forms. Project MAC report MAC-TR-41, MIT (1967)
2. Dechevsky, L.T., Bang, B., Lakså, A.: Generalized Expo-rational B-splines. Int. J. Pure Appl. Math. **57**(6), 833–872 (2009)
3. Floater, M.S.: Mean value coordinates. Comput. Aided Geom. Des. **20**, 19–27 (2003)
4. Lai, M.J., Schumaker, L.: Spline Functions on Triangulations. Cambridge University Press, Cambridge (2007)
5. Lakså, A.: Basic properties of Expo-rational B-splines and practical use in Computer Aided Geometric Design. Unipub, Oslo (2007)
6. Prautzsch, H., Boehm, W., Paluszny, M.: Bezier and B-Spline Techniques. Springer, Berlin (2002)

Robust Balanced Semi-coarsening Multilevel Preconditioning of Bicubic FEM Systems

M. Lymbery(✉)

Institute of Information and Communication Technologies, Bulgarian Academy of Sciences, Acad. G. Bonchev Str., bl. 25A, 1113 Sofia, Bulgaria
mariq@parallel.bas.bg

Abstract. This study presents the construction of a robust multilevel preconditioner for systems evolving from bicubic FEM discretizations of the second order elliptic boundary value problem. Robustness of the hierarchical two-level splitting of the FE space of continuous piecewise bicubic functions is achieved via the application of the balanced semi-coarsening technique. Behavior of the corresponding CBS constant which qualifies the hierarchical two-level splittings of the FEM stiffness matrices is analyzed and new uniform estimates are given. On the basis of the latter and the theory of the Algebraic Multilevel Iteration (AMLI) methods an optimal order multilevel algorithm is constructed whose total computation cost is proportional to the size of the discrete problem with a proportionality constant independent of the anisotropy ratio.

1 Introduction

Consider the linear system of algebraic equations

$$A_h \mathrm{u}_h = F_h, \tag{1}$$

which has been derived from the application of finite element discretization with conforming bicubic elements to the elliptic boundary value problem

$$-\nabla \cdot (a(x)\nabla u(x)) = f(x) \quad in \ \Omega, \tag{2a}$$

$$u = 0 \quad on \ \Gamma_D, \tag{2b}$$

$$(a(x)\nabla u(x)) \cdot \mathbf{n} = 0 \quad on \ \Gamma_N. \tag{2c}$$

In (2a) $\Omega \subset R^2$ is a domain composed of rectangles with a boundary $\Gamma = \Gamma_D \cup \Gamma_N$, $f(x) \in L_2(\Omega)$, $a(x) = (a_{ii}(x))$, $i = 1, 2$ is a diagonal positive definite (SPD) coefficient matrix, uniformly bounded in Ω and $\mathbf{n}$ is the outward unit vector normal to Γ. In (1) A_h denotes the global stiffness matrix, F_h is the given right hand side and h is the mesh parameter of the underlying partition $\mathcal{T}_h$ of Ω.

It is assumed that an initial mesh $\mathcal{T}_0$ has been introduced in Ω such that $a_{ii}, i = 1, 2$ are constants over each element of $\mathcal{T}_0$ and a recursive balanced semi-coarsening refinement procedure has been applied to it. As a result the nested

I. Lirkov et al. (Eds.): LSSC 2013, LNCS 8353, pp. 628–635, 2014.
DOI: 10.1007/978-3-662-43880-0_72,

meshes $\mathcal{T}_0 \subset \mathcal{T}_1 \subset ... \subset \mathcal{T}_\ell = \mathcal{T}_h$ are constructed where ℓ is an even number. The objective is to find a solution of (1) over the finest mesh $\mathcal{T}_\ell = \mathcal{T}_h$ best approximating (2a)–(2c).

Let A_e be the element stiffness matrix, then

$$A_h = \sum_{e \in \mathcal{T}_h} R_e^T A_e R_e, \tag{3}$$

where the operators R_e restrict a global vector to a given element $e \in \mathcal{T}_h$.

If the length and the height of an arbitrary element e from $\mathcal{T}_k$ are denoted by $h_x^{(k)}$ and $h_y^{(k)}$ respectively, the anisotropy ratio at the level k is given by

$$\varepsilon := (a_{22}/a_{11})\left(h_x^{(k)}/h_y^{(k)}\right)^2. \tag{4}$$

2 Balanced Semi-coarsening AMLI Algorithm

The AMLI preconditioner, e.g. [2,3,7], $M = M^{(\ell)}$ is defined recursively by

$$M^{(0)} = A^{(0)}, \qquad P^{(k)^{-1}} = [I - p_{\beta_k}(M^{(k)^{-1}}, A^{(k)})]A^{(k)^{-1}} \tag{5}$$

$$M^{(k+1)} = J^{(k+1)^{-T}} \begin{bmatrix} A_{11}^{(k+1)} & 0 \\ \tilde{A}_{21}^{(k+1)} & P^{(k)} \end{bmatrix} \begin{bmatrix} I & A_{11}^{(k+1)^{-1}} \tilde{A}_{12}^{(k+1)} \\ 0 & I \end{bmatrix} J^{(k+1)^{-1}}, \tag{6}$$

where β_k is the degree of the stabilization Chebyshev polynomial p_{β_k} which can be cyclicly varied resulting in the *hybrid V-cycle* AMLI algorithm, cf. [6], see also [3], p. 200, Theorem 9.1.

Theorem 1. *The PCG iteration method defined by the AMLI preconditioner is of optimal order if the properly scaled approximation $C_{11}^{(k+1)}$ satisfies the estimate*

$$\kappa(C_{11}^{(k+1)^{-1}} A_{11}^{(k+1)}) = O(1),$$

solving systems with $C_{11}^{(k+1)}$ requires $O(N_{k+1} - N_k)$ arithmetic operation and

$$\beta_k = 1 \quad \text{if}\,(k \bmod k_0) \neq 0, \qquad \frac{1}{\sqrt{1-\gamma^{(k_0)^2}}} < \beta_k < \rho_{k_0} \quad \text{if}\,(k \bmod k_0) = 0, \tag{7}$$

where N_k is the number of unknowns from $\mathcal{T}_k$, ρ_{k_0} is the mesh refinement ratio of k_0 consecutive mesh refinement steps while $\gamma^{(k_0)}$ is the constant in the strengthened CBS inequality related to the nested FE spaces $V_{(j+1)k_0}$ and V_{jk_0}.

3 FEM Matrices in Balanced Semi-coarsening Refinement

During the balanced semi-coarsening procedure lines parallel to the vertical edges at the odd steps ($k = 1, 3, 5, ...$) and lines parallel to the horizontal edges at the even steps ($k = 2, 4, 6, ...$) are added. Each element $e \in \mathcal{T}_k$ is therefore split into ρ congruent subelements, Fig. 1. If we set

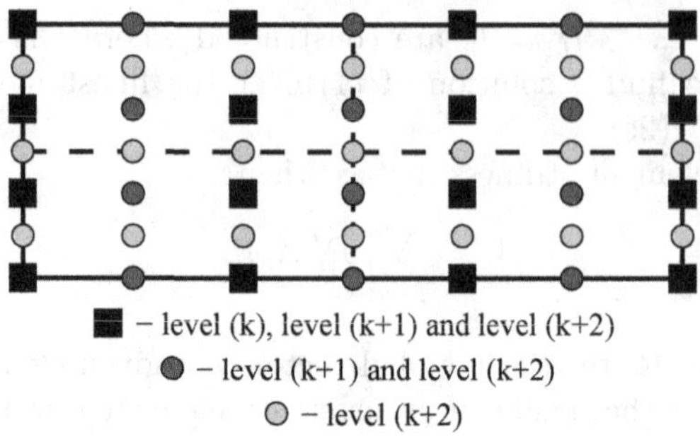

Fig. 1. Balanced semi-coarsening macroelement for $\rho = 2$

$$\delta_e^{(k)} := (a_{11}/45)(h_y^{(k)}/h_x^{(k)}), \tag{8}$$

the related element stiffness matrix then can be written in the block form

$$A_e^{(k)} = \delta_e^{(k)} \begin{pmatrix} A_{e:11}^{(k)} & A_{e:12}^{(k)} & A_{e:13}^{(k)} & A_{e:14}^{(k)} \\ A_{e:21}^{(k)} & A_{e:22}^{(k)} & A_{e:23}^{(k)} & A_{e:24}^{(k)} \\ A_{e:31}^{(k)} & A_{e:32}^{(k)} & A_{e:33}^{(k)} & A_{e:34}^{(k)} \\ A_{e:41}^{(k)} & A_{e:42}^{(k)} & A_{e:43}^{(k)} & A_{e:44}^{(k)} \end{pmatrix}, \tag{9}$$

where[1]

$$A_{e:11}^{(k)} = \frac{1}{7} \begin{pmatrix} \frac{148(1+\varepsilon)}{3} & \frac{1221}{32} - 63\varepsilon & -\frac{111}{8} + 18\varepsilon & \frac{703}{96} - \frac{13\varepsilon}{3} \\ \frac{1221}{32} - 63\varepsilon & \frac{999}{4} + 144\varepsilon & -\frac{999}{32} - 99\varepsilon & -\frac{111}{8} + 18\varepsilon \\ -\frac{111}{8} + 18\varepsilon & -\frac{999}{32} - 99\varepsilon & \frac{999}{4} + 144\varepsilon & \frac{1221}{32} - 63\varepsilon \\ \frac{703}{96} - \frac{13\varepsilon}{3} & -\frac{111}{8} + 18\varepsilon & \frac{1221}{32} - 63\varepsilon & \frac{148(1+\varepsilon)}{3} \end{pmatrix},$$

$$A_{e:12}^{(k)} = 3 \begin{pmatrix} -3 + \frac{407\varepsilon}{224} & -\frac{297}{128}(1+\varepsilon) & \frac{27}{448}(14+11\varepsilon) & -\frac{399+143\varepsilon}{896} \\ -\frac{297}{128}(1+\varepsilon) & \frac{27}{112}(-63+22\varepsilon) & \frac{27}{896}(63-121\varepsilon) & \frac{27}{448}(14+11\varepsilon) \\ \frac{27}{448}(14+11\varepsilon) & \frac{27}{896}(63-121\varepsilon) & \frac{27}{112}(-63+22\varepsilon) & -\frac{297}{128}(1+\varepsilon) \\ -\frac{399+143\varepsilon}{896} & \frac{27}{448}(14+11\varepsilon) & -\frac{297}{128}(1+\varepsilon) & -3 + \frac{407\varepsilon}{224} \end{pmatrix},$$

[1] Mathematica software tool is used in the derivation of the stiffness matrix.

$$A_{e:13}^{(k)} = \frac{1}{56}\begin{pmatrix} 144-111\varepsilon & \frac{81}{8}(11+14\varepsilon) & -\frac{81}{2}(1+\varepsilon) & \frac{3}{8}(57+26\varepsilon) \\ \frac{81}{8}(11+14\varepsilon) & 81(9-4\varepsilon) & \frac{81}{8}(-9+22\varepsilon) & -\frac{81}{2}(1+\varepsilon) \\ -\frac{81}{2}(1+\varepsilon) & \frac{81}{8}(-9+22\varepsilon) & 81(9-4\varepsilon) & \frac{81}{8}(11+14\varepsilon) \\ \frac{3}{8}(57+26\varepsilon) & -\frac{81}{2}(1+\varepsilon) & \frac{81}{8}(11+14\varepsilon) & 144-111\varepsilon \end{pmatrix}$$

$$A_{e:14}^{(k)} = \frac{1}{112}\begin{pmatrix} \frac{-416+703\varepsilon}{6} & -\frac{3(143+399\,\varepsilon)}{8} & \frac{3(26+57\varepsilon)}{4} & -\frac{247(1+\varepsilon)}{24} \\ -\frac{3(143+399\varepsilon)}{8} & 9(-39+38\varepsilon) & -\frac{9(-39+209\varepsilon)}{8} & \frac{3(26+57\varepsilon)}{4} \\ \frac{3(26+57\varepsilon)}{4} & -\frac{9(-39+209\varepsilon)}{8} & 9(-39+38\varepsilon) & -\frac{3(143+399\varepsilon)}{8} \\ -\frac{247(1+\varepsilon)}{24} & \frac{3(26+57\varepsilon)}{4} & -\frac{3(143+399\varepsilon)}{8} & \frac{-416+703\,\varepsilon}{6} \end{pmatrix}$$

$$A_{e:22}^{(k)} = \frac{1}{7}\begin{pmatrix} 144+\frac{999\varepsilon}{4} & \frac{81}{16}(22-63\varepsilon) & \frac{81}{8}(-4+9\,\varepsilon) & -\frac{9}{16}(-38+39\varepsilon) \\ \frac{81}{16}(22-63\varepsilon) & 729(1+\varepsilon) & -\frac{729}{16}(2+11\varepsilon) & \frac{81}{8}(-4+9\varepsilon) \\ \frac{81}{8}(-4+9\varepsilon) & -\frac{729}{16}(2+11\varepsilon) & 729(1+\varepsilon) & \frac{81}{16}(22-63\varepsilon) \\ -\frac{9}{16}(-38+39\varepsilon) & \frac{81}{8}(-4+9\varepsilon) & \frac{81}{16}(22-63\varepsilon) & 144+\frac{999\varepsilon}{4} \end{pmatrix}$$

$$A_{e:23}^{(k)} = \frac{1}{112}\begin{pmatrix} -\frac{9}{2}(352+111\varepsilon) & \frac{81}{8}(-121+63\varepsilon) & -\frac{81}{4}(-22+9\varepsilon) & \frac{9}{8}(-209+39\varepsilon) \\ \frac{81}{8}(-121+63\varepsilon) & -729(11+2\varepsilon) & \frac{8019(1+\varepsilon)}{8} & -\frac{81}{4}(-22+9\varepsilon) \\ -\frac{81}{4}(-22+9\varepsilon) & \frac{8019(1+\varepsilon)}{8} & -729(11+2\varepsilon) & \frac{81}{8}(-121+63\varepsilon) \\ \frac{9}{8}(-209+39\varepsilon) & -\frac{81}{4}(-22+9\varepsilon) & \frac{81}{8}(-121+63\varepsilon) & -\frac{9}{2}(352+111\varepsilon) \end{pmatrix}$$

and

$$\begin{aligned} A_{e:11}^{(k)} &= A_{e:44}^{(k)}, A_{e:22}^{(k)} = A_{e:33}^{(k)}, A_{e:12}^{(k)} = A_{e:21}^{(k)} = A_{e:34}^{(k)} = A_{e:43}^{(k)}, \\ A_{e:14}^{(k)} &= A_{e:41}^{(k)}, A_{e:23}^{(k)} = A_{e:32}^{(k)}, A_{e:13}^{(k)} = A_{e:31}^{(k)} = A_{e:24}^{(k)} = A_{e:42}^{(k)}. \end{aligned} \tag{10}$$

As a_{ii}, $i = 1, 2$ are piece-wise constant over $e_0 \in \mathcal{T}_0$, the matrices $A_e^{(k)}$ depend only on the level (k) and on the coarse element e_0 for which $e \subset e_0$. By performing ℓ refinement steps the element e is split into ρ^ℓ number of subelements which have equal heights $h_y^{(\ell)}$ and equal lengths $h_x^{(\ell)}$. Denote the coarsest mesh sizes along the y-axis and x-axis by $h_y^{(0)}$ and $h_x^{(0)}$ respectively. Then for the odd values of ℓ, $h_y^{(\ell)} = \rho^{-(\ell-1)/2}h_y^{(0)}$, $h_x^{(\ell)} = \rho^{-(\ell+1)/2}h_x^{(0)}$ while for even $h_y^{(\ell)} = \rho^{-\ell/2}h_y^{(0)}$, $h_x^{(\ell)} = \rho^{-\ell/2}h_x^{(0)}$. In the following two coupled consecutive refinement steps will be considered at once for the purposes of the analysis.

The macroelement stiffness matrix for $E \in \mathcal{T}^{(k+2)}$, Fig. 1, can be agglomerated from the matrices $A_e^{(k+2)}$ and further written in a two by two block form

$$A_E^{(k+2)} = \begin{pmatrix} A_{E:11}^{(k+2)} & A_{E:12}^{(k+2)} \\ A_{E:21}^{(k+2)} & A_{E:22}^{(k+2)} \end{pmatrix}. \tag{11}$$

The first diagonal block in (11) corresponds to the unknowns from $\mathcal{T}_{k+2}\backslash\mathcal{T}_k$ while the second is a (16×16) matrix, related to the nodes from $\mathcal{T}_k$.

Let $\tilde{A}_E^{(k+2)}$ be the hierarchical macroelement stiffness matrix corresponding to the two-level hierarchical nodal basis. It relates to the standard nodal macroelement matrix $A_E^{(k+2)}$ via the equation

$$\tilde{A}_E^{(k+2)} = J^T A_E^{(k+2)} J, \tag{12}$$

where the transformation matrix J between the standard and the hierarchical nodal bases has the form

$$J = \begin{pmatrix} I_k & \mathrm{Z} \\ 0 & I_{16} \end{pmatrix}, \qquad k = (3\varrho + 1)^2 - 16. \tag{13}$$

In accordance with the introduced splitting of the unknowns $\tilde{A}_E^{(k+2)}$ can also be written in a two by two block form

$$\tilde{A}_E^{(k+2)} = \begin{pmatrix} \tilde{A}_{E:11}^{(k+2)} & \tilde{A}_{E:12}^{(k+2)} \\ \tilde{A}_{E:21}^{(k+2)} & \tilde{A}_{E:22}^{(k+2)} \end{pmatrix}. \tag{14}$$

Then the assembled global standard and hierarchical nodal basis matrices after an appropriate ordering of the degrees of freedom are similarly expressed by

$$A^{(k+2)} = \begin{pmatrix} A_{11}^{(k+2)} & A_{12}^{(k+2)} \\ A_{21}^{(k+2)} & A_{22}^{(k+2)} \end{pmatrix}, \qquad \tilde{A}^{(k+2)} = \begin{pmatrix} \tilde{A}_{11}^{(k+2)} & \tilde{A}_{12}^{(k+2)} \\ \tilde{A}_{21}^{(k+2)} & \tilde{A}_{22}^{(k+2)} \end{pmatrix} \tag{15}$$

where the first diagonal blocks are related to the nodes from $\mathcal{T}_{k+2}\backslash\mathcal{T}_k$ while the second diagonal blocks are associated with the unknowns that belong within $\mathcal{T}_k$. $\tilde{A}^{(k+2)}$ is related to $A^{(k+2)}$ via the equation

$$\tilde{A}^{(k+2)} = J^{(k+2)^T} A^{(k+2)} J^{(k+2)} = \begin{pmatrix} A_{11}^{(k+2)} & \tilde{A}_{12}^{(k+2)} \\ \tilde{A}_{21}^{(k+2)} & A^{(k)} \end{pmatrix},$$

where $J^{(k+2)}$ is the global transformation matrix between the standard and the hierarchical nodal bases.

4 Uniform Estimates of the Constant in the Strengthened CBS Inequality

In the hybrid V-cycle AMLI algorithm $k_0 = 2$ in (7) for the balanced semi-coarsening mesh procedure. Set $\gamma^{(k_0)}$ as $\gamma^{(2)}$. Importantly, $\gamma^{(2)}$ can be estimated by the local CBS constants $\gamma_E^{(2)}$, $E \in \mathcal{T}_{k+2}$, e.g. [1,3], i.e.

$$\gamma^{(2)} \le \max_{E \in \mathcal{T}_{k+2}} \gamma_E^{(2)} \quad \text{where} \quad (\gamma_E^{(2)})^2 = 1 - \mu_1. \tag{16}$$

In (16) μ_1 is the minimal generalized eigenvalue of the macroelement Schur complement $S_E^{(k+2)}$ with respect to the element stiffness matrix $A_e^{(k)}$

$$S_E^{(k+2)} \mathbf{v}_{E:2} = \mu A_e^{(k)} \mathbf{v}_{E:2}, \quad \mathbf{v}_{E:2} \ne const. \tag{17}$$

Due to the size of the parameter dependent bicubic element matrices solving directly (17) becomes a very difficult task. Therefore, introduce the matrix

$$B := \begin{bmatrix} A_{E:11}^{(k+2)} & A_{E:12}^{(k+2)} \\ A_{E:21}^{(k+2)} & A_{E:22}^{(k+2)} - \underline{\mu} A_e^{(k)} \end{bmatrix}, \tag{18}$$

and use the fact that the constant $\underline{\mu} > 0$ for which B is symmetric positive semi-definite provides a lower bound for μ in (17) and consequently an estimate for $(\gamma^{(2)})^2$, e.g. [3]. Now the next lemma can be formulated.

Lemma 1. *Consider the balanced semi-coarsening AMLI algorithm with $\rho = 2$. The constant $\gamma^{(2)}$ in the strengthened CBS inequality corresponding to bicubic conforming finite elements is uniformly bounded with respect to the anisotropy ratio and the following estimate is valid*

$$(\gamma^{(2)})^2 \le \frac{203 + 5\sqrt{46}}{288} \approx 0.823, \qquad \varrho = 2. \tag{19}$$

Proof. First note the relations

$$h_x^{(k+2)} = \frac{h_x^{(k)}}{\ell}, \quad h_y^{(k+2)} = \frac{h_y^{(k)}}{\ell}, \quad \delta_e^{(k)} = \delta_e^{(k+2)}, \quad A_e^{(k+2)} \equiv A_e^{(k)}. \tag{20}$$

Therefore, the parameter $\delta_e^{(k)}$ appearing on both sides of (17) can be skipped from further consideration without loss of generality.

From (9), (18) and (20) it is evident that the matrix B can be written in the linear form

$$B = B_0 + \underline{\mu} B_{\underline{\mu}} + \varepsilon B_\varepsilon + \varepsilon \underline{\mu} B_{\varepsilon\underline{\mu}}, \tag{21}$$

where the matrices B_0, B_ε, $B_{\underline{\mu}}$ and $B_{\varepsilon\underline{\mu}}$ are symmetric and do not depend on any parameters. Therefore, B is a symmetric matrix independently of $\underline{\mu}$ and ε.

By performing an eigenvalue analysis on the right hand side matrices in (21) it is concluded that B_0 and B_ε are positive semi-definite while $B_{\underline{\mu}}$ and $B_{\varepsilon\underline{\mu}}$ are negative semi-definite matrices. Equation (21) is then rewritten in the form

$$B = (B_0 + \underline{\mu} B_{\underline{\mu}}) + \varepsilon(B_\varepsilon + \underline{\mu} B_{\varepsilon\underline{\mu}}). \tag{22}$$

As B has to remain semi-positive definite for any parameter $\varepsilon > 0$, then determine the values of $\underline{\mu}$ guaranteeing the semi-positivity of both $(B_0 + \underline{\mu} B_{\underline{\mu}})$ and $(B_\varepsilon + \underline{\mu} B_{\varepsilon\underline{\mu}})$ by first considering the generalized eigenvalue problem[2]

$$B_0 \mathrm{v} = \lambda(-B_{\underline{\mu}})\mathrm{v}.$$

Its minimal eigenvalue is therefore

$$\lambda'_{min} = 135/(32(17 + \sqrt{46})). \tag{23}$$

Then for any nonnegative $\underline{\mu} \leq 135/(32(17 + \sqrt{46}))$, $(B_0 + \underline{\mu} B_{\underline{\mu}})$ is semi-positive definite. Analogously, the generalized eigenvalue problem

$$B_\varepsilon \mathrm{v} = \lambda(-B_{\varepsilon\underline{\mu}})\mathrm{v}$$

is solved and its minimal eigenvalue is again found to be

$$\lambda''_{min} = 135/(32(17 + \sqrt{46})). \tag{24}$$

Then the necessary condition $(B_\varepsilon + \underline{\mu} B_{\varepsilon\underline{\mu}})$ to be semi-positive definite is also $\underline{\mu} \leq 135/(32(17 + \sqrt{46}))$. In conclusion, $135/(32(17 + \sqrt{46}))$ is the sharp uniform lower bound of μ_1 in (17). Combining with (16)

$$(\gamma^{(2)})^2 \leq \max_{E \in \mathcal{T}_{k+2}} \gamma_E^{(2)^2} \leq 1 - \mu_1 \leq 1 - 135/(32(17 + \sqrt{46})) = (203 + 5\sqrt{46})/288$$

and the proof is completed. ∎

5 Solving Systems with the Pivot Block $A_{11}^{(k+1)}$

By solving (1) via an AMLI algorithm the FE problem is reduced to a sequence of smaller subproblems with the pivot block matrices $A_{11}^{(k+1)}$. While $\kappa(A_{11}^{(k+1)})$ is uniformly bounded with respect to the related number of unknowns for isotropic problems, in the case of anisotropy the condition number deteriorates with the anisotropy ratio. Therefore, special robust preconditioning techniques are developed for the pivot blocks, e.g. [4], when uniform refinement is used in the AMLI methods for parameter dependent ill-conditioned elliptic problems.

However in a balanced semi-coarsening setting the unknowns of the bicubic FEM systems can be ordered such that the blocks $A_{11}^{(k+1)}$ are block diagonal with uniformly bounded semi-bandwidth [5]. Consequently, the computational complexity of any direct solver for banded matrices is of optimal order, i.e.

$$\mathcal{N}(A_{11}^{(k+1)^{-1}} \mathrm{v}) = O(N^{(k+1)} - N^{(k)}).$$

[2] Mathematica software tool is used in the presented computations.

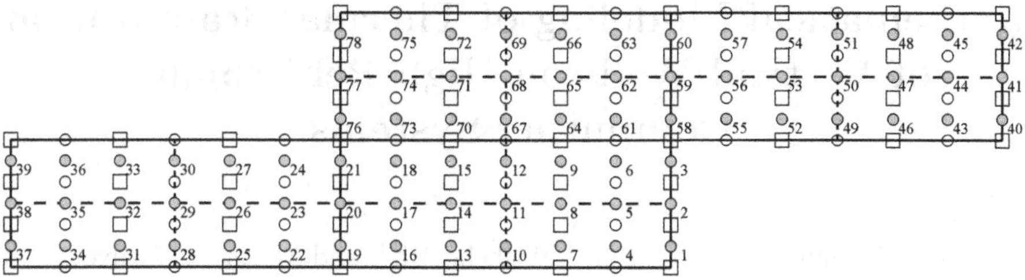

Fig. 2. Vertical numeration of the pivot block unknowns for $\rho = 2$

If the pivot block unknowns are numerated as depicted on Fig. 2, for the semi-bandwidth $d(\rho)$ of $A_{11}^{(k+1)}$ the next relations will hold true

$$d(\rho) = 9(\rho - 1) + 2 \quad \text{for} \quad \rho = 2, 3, \qquad d(\rho) = 9(\rho - 1) + 3 \quad \text{for} \quad \rho \geq 4.$$

Note that the ordering of the pivot block unknowns switches from vertical to horizontal direction and vice versa depending on whether k is odd or even.

From Theorem 1, the uniform estimates (19) and the last result, the main contribution of this paper is obtained.

Theorem 2. *The balanced semi-coarsening AMLI preconditioner (6) with parameters $\rho = 3$, $3 \leq \beta \leq 8$ and even k has an optimal order of computational complexity. The estimate is uniform with respect to mesh and coefficient anisotropy.*

Acknowledgment. The partial support of the Bulgarian NSF Grant DCVP 02/01 and project AComIn, grant 316087, funded by FP7 Capacity Programme, is highly appreciated.

References

1. Axelsson, O., Vassilevski, P.: Algebraic multilevel preconditioning methods I. Numer. Math. **56**, 157–177 (1989)
2. Kraus, J.: Additive Schur complement approximation and application to multilevel preconditioning. SIAM J. Sci. Comput. **34**, A2872–A2895 (2012)
3. Kraus, J., Margenov, S.: Robust Algebraic Multilevel Methods and Algorithms. De Gruyter, Berlin (2009)
4. Kraus, J., Margenov, S., Synka, J.: On the multilevel preconditioning of Crouzeix-Raviart elliptic problems. Num. Lin. Alg. Appl. **15**, 395–416 (2008)
5. Margenov, S.: Semicoarsening AMLI algorithms for elasticity problems. Num. Lin. Alg. Appl. **5**, 347–362 (1998)
6. Vassilevski, P.: Hybrid V-cycle algebraic multilevel preconditioners. Math. Comp. **58**, 489–512 (1992)
7. Vassilevski, P.: Multilevel Block Factorization Preconditioners: Matrix-based Analysis and Algorithms for Solving Finite Element Equations. Springer, New York (2008)

Mathematical Modeling of Thermal Stabilization of Vertical Wells on High Performance Computing Systems

Natalia V. Pavlova[1(✉)], Petr N. Vabishchevich[2], and Maria V. Vasilyeva[1]

[1] North-Eastern Federal University, Belinskogo Str 58, Yakutsk 677000, Russia
npav@rambler.ru

[2] Nuclear Safety Institute, B. Tulskaya Str 52, Moscow 115191, Russia

Abstract. Temperature stabilization of oil and gas wells is used to ensure stability and prevent deformation of a subgrade estuary zone. In this work, we consider the numerical simulation of thermal stabilization using vertical seasonal freezing columns.

A mathematical model of such problems is described by a time-dependent temperature equation with phase transitions from water to ice. The resulting equation is a standard nonlinear parabolic equation.

Numerical implementation is based on the finite element method using the package FEniCS. After standard purely implicit approximation in time and simple linearization, we obtain a system of linear algebraic equations. Because the size of freezing columns are substantially less than the size of the modeled area, we obtain mesh refinement near columns. Due to this, we get a large system of equations which are solved using high performance computing systems.

Keywords: Stefan problem · Finite element method · FEniCS · Thermal stabilization · High performance computing systems

1 Mathematical Model

We consider a mathematical model that describes the distribution of temperature with phase transitions at a given temperature T^* in a domain $\Omega = \Omega^- \cup \Omega^+$. Here $\Omega^+(t)$ is a domain with liquid phase, where the temperature is above the phase transition temperature

$$\Omega^+(t) = \{\boldsymbol{x} | \boldsymbol{x} \in \Omega, \quad T(\boldsymbol{x}, t) > T^*\}$$

and $\Omega^-(t)$ stands for a domain with solid phase,

$$\Omega^-(t) = \{\boldsymbol{x} | \boldsymbol{x} \in \Omega, \quad T(\boldsymbol{x}, t) < T^*\}.$$

The phase transition occurs at a phase change boundary $S = S(t)$.

I. Lirkov et al. (Eds.): LSSC 2013, LNCS 8353, pp. 636–643, 2014.
DOI: 10.1007/978-3-662-43880-0_73,

For simulation of heat transfer with phase transitions, the classical Stefan model is used [1,2]. This model describes the thermal processes accompanied by a phase change media, absorption and release of latent heat. We use

$$\left(\alpha(\phi) + \rho^{+} L \phi'\right) \frac{\partial T}{\partial t} - \operatorname{div}\left(\lambda(\phi) \operatorname{grad} T\right) = 0. \tag{1}$$

For the coefficients of the equation, we have the following relations

$$\alpha(\phi) = \rho^{-} c^{-} + \phi(\rho^{+} c^{+} - \rho^{-} c^{-}),$$
$$\lambda(\phi) = \lambda^{-} + \phi(\lambda^{+} - \lambda^{-}),$$

and

$$\phi = \begin{cases} 0, & T < T^{*}, \\ 1, & T > T^{*}, \end{cases}$$

where ρ^{+}, c^{+}, ρ^{-}, c^{-} are the density and specific heat capacity of the melt and frozen zone, respectively.

Since we consider the process of heat propagation in porous media, then for the coefficients we have:

$$c^{-}\rho^{-} = (1-m) c_{sc} \rho_{sc} + m c_{i} \rho_{i},$$
$$c^{+}\rho^{+} = (1-m) c_{sc} \rho_{sc} + m c_{w} \rho_{w},$$

where m is the porosity. Indexes sc, w, i denote the skeleton of the porous medium, water and ice. For the coefficients of thermal conductivity in the melt and frozen zone, we have similar relationships

$$\lambda^{-} = (1-m)\lambda_{sc} + m\lambda_{i},$$
$$\lambda^{+} = (1-m)\lambda_{sc} + m\lambda_{w}.$$

Note that, generally filtration processes must be considered in the soil. Furthermore, in some cases the effects of salinization also is important to take into account. In this paper, we consider a model without these effects.

In practice, phase transformations do not occur instantaneously and can occur in a small temperature range $[T^{*} - \Delta, T^{*} + \Delta]$ [1]. As the ϕ-function, we can take ϕ_{Δ}:

$$\phi_{\Delta} = \begin{cases} 0, & T \le T^{*} - \Delta, \\ \frac{T - T^{*} + \Delta}{2\Delta}, & T^{*} - \Delta < T < T^{*} + \Delta, \\ 1, & T \ge T^{*} + \Delta, \end{cases}$$

$$\phi'_{\Delta} = \begin{cases} 0, & T \le T^{*} - \Delta, \\ \frac{1}{2\Delta}, & T^{*} - \Delta < T < T^{*} + \Delta, \\ 0, & T \ge T^{*} + \Delta. \end{cases}$$

Then we obtain the following equation for the temperature in the domain Ω:

$$\left(\alpha(\phi_{\Delta}) + \rho_{l} L \phi'_{\Delta}\right) \frac{\partial T}{\partial t} - \operatorname{div}(\lambda(\phi_{\Delta}) \operatorname{grad} T) = 0. \tag{2}$$

The resulting equation (2) is a standard nonlinear parabolic equation.

The equation (2) is supplemented with the initial and boundary conditions

$$\begin{aligned} T(\boldsymbol{x},0) &= T_0, \quad \boldsymbol{x} \in \Omega, \\ T &= T_c, \quad \boldsymbol{x} \in \Gamma_D, \\ -k\frac{\partial T}{\partial n} &= 0, \quad \boldsymbol{x} \in \Gamma/\Gamma_D. \end{aligned} \tag{3}$$

Here Γ_D is a place of contact with the freezing columns.

2 Finite Element Realization

The equation (2) is approximated using a finite element method. We multiply the temperature equation by a test function v, and integrate it using the Green formula

$$\begin{aligned} &\int_\Omega (\alpha(\phi_\Delta) + \rho_l L\phi'_\Delta)\frac{\partial T}{\partial t}\, v\, dx \\ &\qquad + \int_\Omega (\lambda(\phi_\Delta) \operatorname{grad} T, \operatorname{grad} v)\, dx = 0, \quad \forall v \in H_0^1(\Omega). \end{aligned} \tag{4}$$

Here $H^1(\Omega)$ is a Sobolev space, which consists of the functions v such that v^2 and $|\nabla v|^2$ has a finite integral in the Ω and $H_0^1(\Omega) = \{v \in H^1(\Omega) : v|_{\Gamma_D} = 0\}$.

To approximate in time, we apply the standard fully implicit scheme. We use a simple linearization by setting the coefficients from the previous time layer. Let $t_n = n\,\tau$, $n = 0, 1, ...$, where τ – constant time step. The solution at time t_n is denoted by T^n. According to Eq. (4) we can write

$$\begin{aligned} &\int_\Omega (\alpha(\phi_\Delta^n) + \rho_l L\phi'^n_\Delta)\frac{T^{n+1} - T^n}{\tau}\, v\, dx \\ &\qquad + \int_\Omega \big(\lambda(\phi_\Delta^n) \operatorname{grad} T^{n+1}, \operatorname{grad} v\big)\, dx = 0. \end{aligned} \tag{5}$$

In such a way, we arrive at the following classical variational formulation of the problem: find $T \in H^1(\Omega)$, $[T(\boldsymbol{x},t) - T_c(\boldsymbol{x},t)]\,|_{\Gamma_D} \in H_0^1(\Omega)$ such that

$$\begin{aligned} &\frac{1}{\tau}\int_\Omega (\alpha(\phi_\Delta^n) + \rho_l L\phi'^n_\Delta) T^{n+1}\, v\, dx + \int_\Omega \big(\lambda(\phi_\Delta^n) \operatorname{grad} T^{n+1}, \operatorname{grad} v\big)\, dx \\ &\qquad = \frac{1}{\tau}\int_\Omega (\alpha(\phi_\Delta^n) + \rho_l L\phi'^n_\Delta)\, T^n\, v\, dx, \quad \forall v \in H^1(\Omega). \end{aligned} \tag{6}$$

To solve this equation numerically, we transform the continuous variational problem (6) to a discrete variational problem: find $T_h \in V_h$ such that

$$\begin{aligned} &\frac{1}{\tau}\int_\Omega (\alpha(\phi_\Delta^n) + \rho_l L\phi'^n_\Delta) T_h^{n+1}\, v\, dx + \int_\Omega \big(\lambda(\phi_\Delta^n) \operatorname{grad} T_h^{n+1}, \operatorname{grad} v\big)\, dx \\ &\qquad = \frac{1}{\tau}\int_\Omega (\alpha(\phi_\Delta^n) + \rho_l L\phi'^n_\Delta)\, T_h^n\, v\, dx, \quad \forall v \in \hat{V}_h. \end{aligned} \tag{7}$$

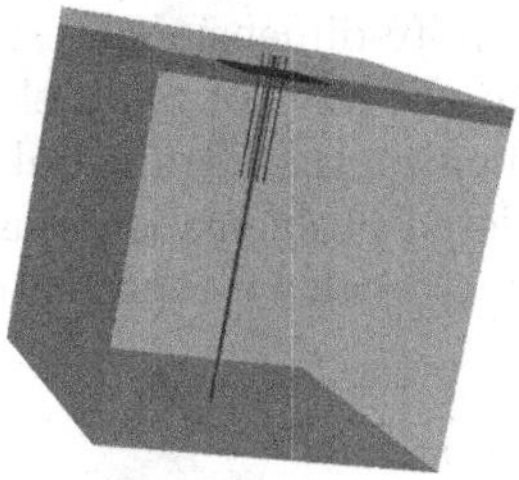

Fig. 1. The computational domain

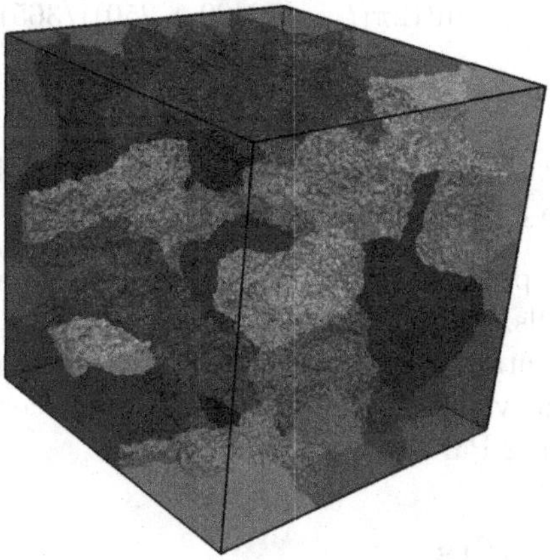

Fig. 2. Domain decomposition

Table 1. Problem parameters

Notation	Value	Metrics	Description
T_{cyl}	20.0	degree	Temperature of oil in the well
T_0	−5.0	degree	Initial temperature
T_*	0.0	degree	Phase change temperature
L	1.04e8	J/kg	Latent heat of the phase transition
$c\rho_{sc}$	2.17e6	J/m^3	Volumetric heat capacity of soil
$c\rho_l$	2.42e6	J/m^3	Volumetric heat capacity of water
$c\rho_{sa}$	1.34e6	J/m^3	Volumetric heat capacity of sand
$c\rho_{pe}$	0.20e6	J/m^3	Volumetric heat capacity of polystyrene
$c\rho_{ce}$	0.8e6	J/m^3	Volumetric heat capacity of cement
λ_{sc}	2.43	W/(m degree)	The thermal conductivity of soil
λ_l	2.22	w/(m degree)	The thermal conductivity of water
λ_{sa}	0.47	W/(m degree)	The thermal conductivity of sand
λ_{pe}	0.03	W/(m degree)	The thermal conductivity of polystyrene
λ_{ce}	0.21	W/(m degree)	The thermal conductivity of cement

Here $V_h \subset H_0^1$ and $\hat{V}_h \subset H^1$ – finite-dimensional test and trial spaces. The choice of V_h, $\hat{V}_h$ follows directly from the kind of finite elements. For our problem we use the first-order basis functions on the tetrahedral element. Using higher order approximations for the considered problems is unjustified.

The process of solving Eq. (7) can be represented as follows. While $t_{cur} < t_{max}$:

1. Calculate t_{cur}: $t_{cur} = t_{cur} + \tau$;
2. Save a previous time level values $T_{prev} = T$;
3. Recalculate the ambient temperature T_{air}:

$$T_{air} = 41 \sin((2\pi(t_{cur}/86400 + 250))/365) - 10.2;$$

4. If the temperature of the soil is less than T_{air}, then the freezing columns turn on, else turn off;
5. The temperature at the new time level are solved by a linear solver;
6. Write the results to a file.

Parameters of the problem are given in Table 1. The computational grid contains 10,903,946 cells.

Numerical implementation is performed using the FEniCS package [5]. For results visualization, the values at each time level were recorded in vtk file format that was visualized using ParaView program.

3 Numerical Results

As a model problem, we consider the process of thermal stabilization of the mouth of the oil or gas wells [3,4]. The geometric domain was built using the

Fig. 3. The temperature distribution after 5 years

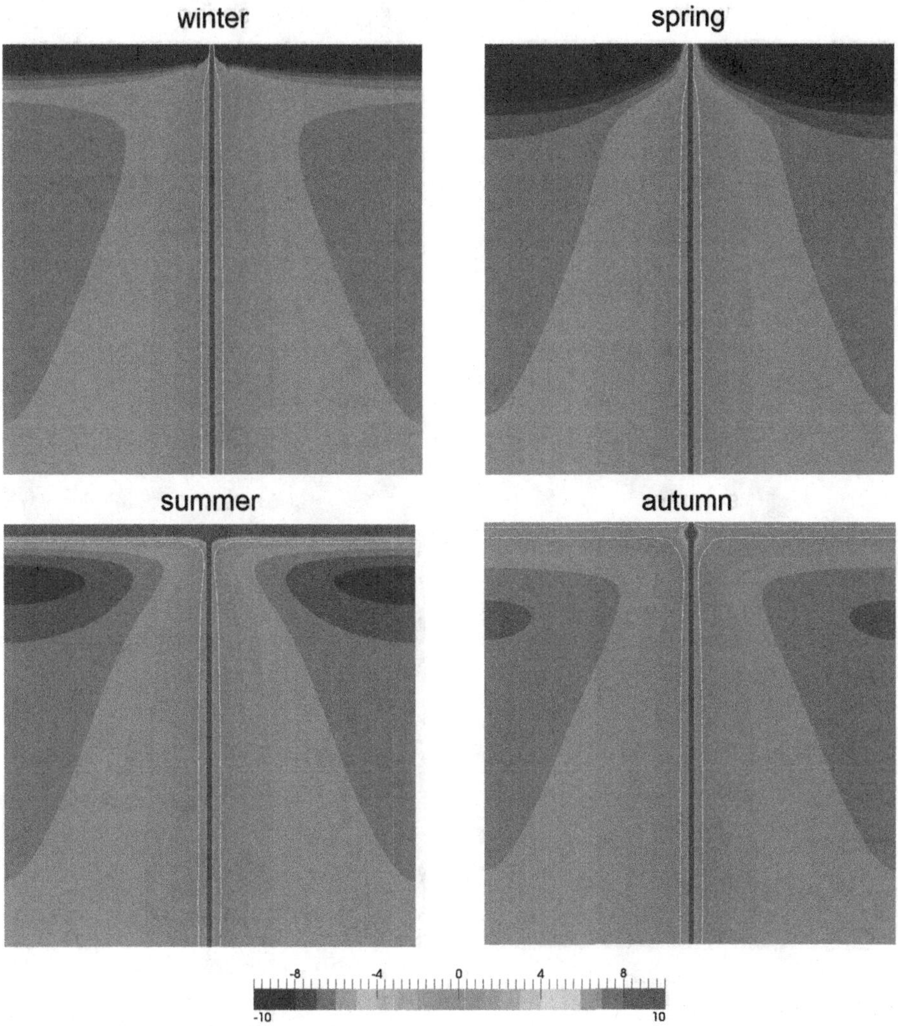

Fig. 4. Temperature after five years without freezing columns. Slice $x = 20$.

Netgen mesher. The computational domain is shown in Fig. 1 and has a length of 40 m in each direction, the well (radius 0.1 m) is located in the middle of a field, where oil flows with a given positive temperature. A cement layer with the thickness of 0.2 m is used for well heat insulation. 8 freezing columns with a radius of 0.05 m are deepened to 14 m around the well. The top sand layer has the thickness of 2 m. Near the well, there is laid penopleks (10 by 10 m, thickness of 200 mm).

The numerical results are presented in Fig. 3. Figure 4 illustrates the efficiency of the seasonal cooling devices, which accumulate the winter chill in the ground

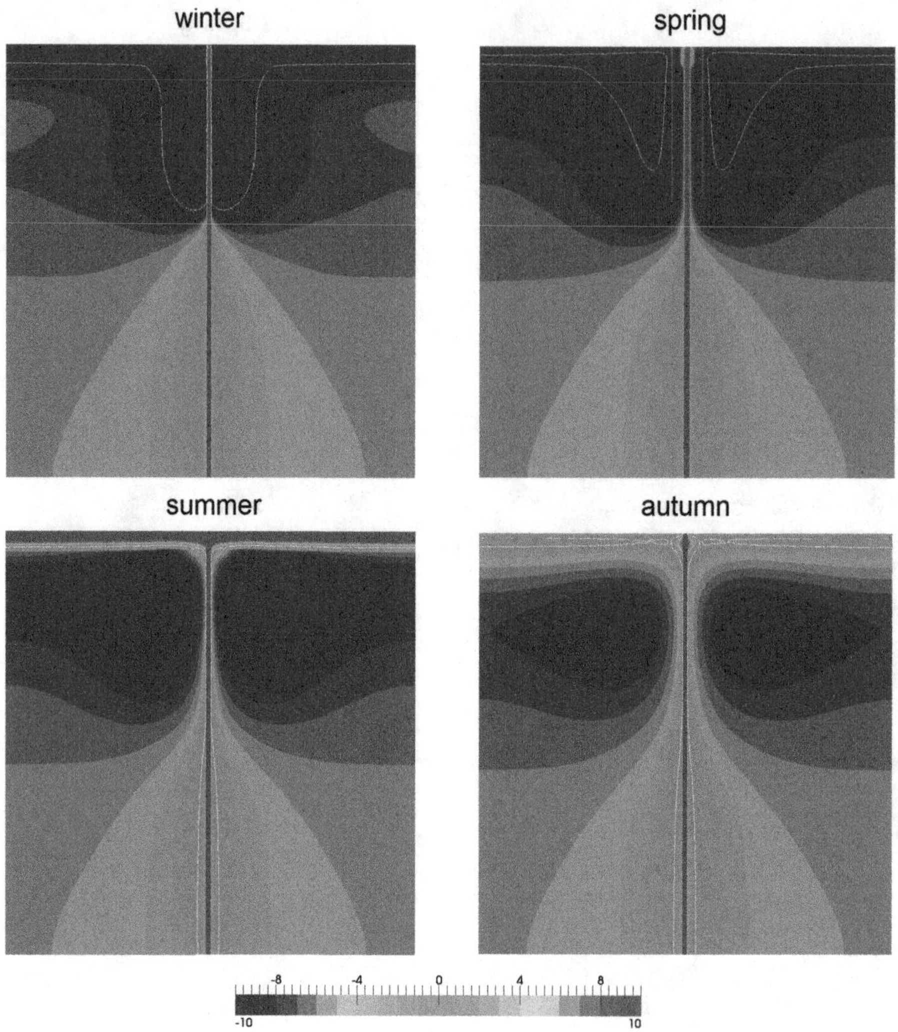

Fig. 5. Temperature after five years with freezing columns. Slice $x = 20$

and provide additional bearing capacity in the summer. By numerical simulation of the temperature stabilization of soils using freezing system, we can conclude that the presence of freezing columns can reduce soil thawing around wells. The calculations were performed using the NEFU computational cluster *Ariane Kuzmin*. The computation time using 64 processors (see Fig. 2) was about 14 min, with 32 processors - about 21 min, and with 16 processors - about 40 min, which shows good efficiency of parallelization (Fig. 4).

References

1. Samarskii, A.A., Vabishchevich, P.N.: Computational Heat Transfer. Wiley, Chichester (1995)
2. Vasilyev, V.I., Maksimov, A.M., Petrov, E.E., Cipkin, G.G.: Heat and Mass Transfer in Freezing and Thawing Soils. Nauka, Moscow (1996)
3. SNIP 2.02.04-88 Foundations on permafrost. State Construction Committee of Russia (2005)
4. Tishchenko, T.I., Gusev, A.Y.: Technical solutions for the thermal stabilization of soils mouths of oil and gas wells. In: Proceeding of the International Scientific-Practical Conference on Permafrost Engineering (2011)
5. Logg, A., Mardal, K.-A., Wells, G.: Automated solution of differential equations by the finite element method (2011). http://Fenicsproject.org

Large-Scale Simulation of Non-uniform Load Traffic in Studying the Throughput of a Crossbar Packed Switch

Tasho Tashev(✉) and Vladimir Monov

Institute of Information and Communication Technologies,
Bulgarian Academy of Sciences, Acad. G. Bonchev, bl. 2,
1113 Sofia, Bulgaria
{ttashev,vmonov}@iit.bas.bg
http://www.iict.bas.bg

Abstract. In the present paper we propose a family of patterns for non-uniform traffic simulating. The results from computer simulations of the throughput of a crossbar packet switch with these patterns are presented. The necessary computations have been performed on the grid-cluster of IICT-BAS. Our simulations utilize two algorithms for non-conflict schedule: the well known PIM-algorithm and an algorithm (MiMa-algorithm) proposed by the first author of the paper. Both algorithms are specified by the apparatus of Generalized Nets. It is shown that the throughput of the PIM-algorithm with the suggested family of patterns approaches 77.5 % while the throughput of the MiMa-algorithm tends to 100 %.

Keywords: Large-scale simulation · Generalized nets · Switch node

1 Introduction

A crossbar switch node routes traffic from the input to output where a message packet is transmitted from the source to the destination. The randomly incoming traffic must be controlled and scheduled to eliminate conflict at the crossbar switch. The goal of the traffic-scheduling for the crossbar switches is to maximize the throughput of packet through a switch and to minimize packet blocking probability and packet waiting time [1].

The problem of calculating of non-conflict schedule is NP-complete [2]. Algorithms are suggested which solve the problem partially such as PIM [3], iSLIP [4], etc. [1]. The latter is efficient enough for size of the matrix switch up to 32×32 lines. However, progress in optic fibers utilization as well as increasing the number of personal computers require larger size. One part of the investigations represent developing of modifications still using input buffering with VOQ (Virtual Output Queuing). The obtained results are effective when the size is 64×64 and larger, for example CTC(N) algorithm [5]. The approach of Birkhoff-von Neumann is very interesting, too [6]. Another group of researchers

I. Lirkov et al. (Eds.): LSSC 2013, LNCS 8353, pp. 644–651, 2014.
DOI: 10.1007/978-3-662-43880-0_74,

use input and intermediate buffering (CICQ), by applying a buffer associated with commutation field in different combinations, for example [7]. Of course, more investigations are directed to a completely optical commutation [8].

For description of switch algorithms many authors use different formal apparatus such as cellular automata, neural networks, queue theory, etc. [1]. We apply Generalized Nets (GN) apparatus [9,10] to specify our new algorithm: MiMa (Minimum from Maximal Matching). We have already used GN to describe the PIM-algorithm and the results were successful [11,12].

The efficiency check of the algorithms always begins with throughput modeling of the switch node with uniform load traffic. We have already studied the MiMa-algorithm with this traffic in [13]. The next step is the efficiency check for non-uniform traffic [5]. In the present paper, we propose a family of patterns for non-uniform load traffic simulation, based on the Rojas-Cessa unbalanced traffic model [14]. The aim is to use this family as a point of quick reference for variety of algorithms. For a correct comparison with results obtained with own MiMa-algorithm, we will use results for the well known PIM-algorithm. Therefore, we performed computer simulations for throughput of both algorithms by using the proposed patterns for non-uniform load traffic.

The paper is structured as follows. Section 2 briefly describes the MiMa-algorithm. Section 3 presents a GN-model of the MiMa-algorithm. Section 4 describes our patterns for non-uniform load traffic. Results from simulations using grid-resources are presented in Sect. 5, while Sect. 6 outlines the conclusions and some possible lines of future research.

2 MiMa-Algorithm of Non-conflicts Schedule

The requests for packet transmission through switching $n \times n$ line switch node is presented by an $n \times n$ matrix T, named traffic matrix (n is integer) [1]. Every element t_{ij}, ($t_{ij} \in \{0, 1, 2, \ldots\}$) of the traffic matrix represents a request for a packet from input i to output j. For example $t_{ij} = q$ means that q packets from the i-th input line have to be send to j-th output line of the switch node.

A conflict situation arises when in any row of the T matrix the number of requests is more than 1. This corresponds to the case when one source declares connection with more than one receiver. If any column of the T matrix hosts more than one elements different from zero, this also indicates a conflict situation. Avoiding conflicts is related to the switch node efficiency. In order to obtain a non-conflict schedule it is necessary to compute a sequence of non-conflict matrices $Q_1, Q_2, \ldots, Q_r$ such that their sum is equal to the traffic matrix T. Each row and column of every matrix Q_i, $i = 1, 2, \ldots, r$ has no more than one element equal to 1 and the rest of elements are equal to 0.

We will give a concise description of the MiMa-algorithm.

Initially, matrix T is introduced. A vector-column, which consists of the number of conflicts in each row (row conflict weights) is calculated. A vector-row, which consists of the number of conflicts in each column (column conflict weights), is calculated too. In the vector-row we choose the maximal element

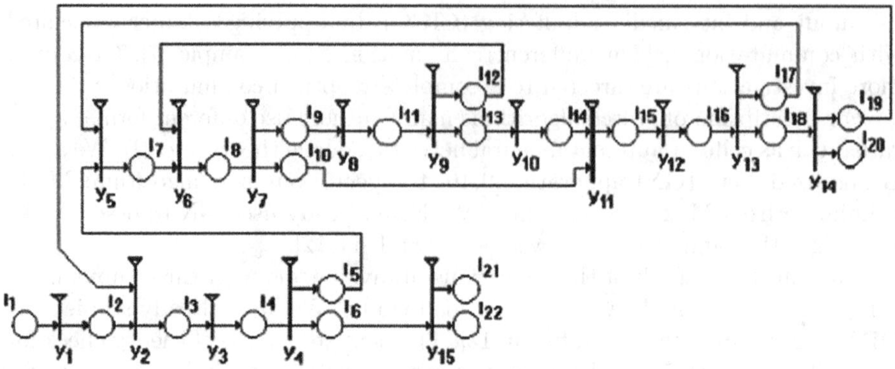

Fig. 1. Graphical form of GN model of the MiMa-algorithm

which determines the column with the most conflicts. In the vector-column we choose the maximal element which determines the row with the most conflicts. If there is a request in the place of intersection of the column and row with most conflicts, we take this request as an element of the non-conflict matrix Q_1. If there is no request, we choose the element in the vector-column which is closest in value to the maximal element. The element in the vector row remains the same. Again, we check if there is a request in the intersection, etc. As a result for the chosen column of T we will have a request selected for commutation (if such a request exists at all). The row and column containing the selected request are excluded from the computation of Q_1. The next elements of Q_1 are computed by repeating the above procedure.

As a result the first matrix Q_1 will consists of elements (requests) with maximal weight of conflicts in T. The next non-conflict matrices $Q_2, \ldots, Q_r$ are computed analogously. The last matrix Q_r will contain only the non-conflict requests in matrix T.

3 Generalized Net Model of MiMa-Algorithm

The algorithm MiMa can be described formally by the means of Generalized Nets. A model of the algorithm is developed for switch node with n inputs and n outputs. Its graphic form is shown on Fig. 1.

The descriptions of input and output places are: $l1$ - start; $l20$ - non-conflict matrices; $l21$ - stop; $l22$ - error;

The token α comes into place $l1$ with initial characteristic: $\langle \text{size} \rangle := n, \langle \text{trafficmatrix} \rangle := T, \langle \text{iteration} \rangle := 1$ $(i = 1)$. The parameter $\langle \text{size} \rangle$ has size $n \in N$ for a communication field $(n \times n)$ of matrix T and matrix Q. The parameter $\langle \text{traffic matrix} \rangle$ shows the traffic matrix T. The parameter $\langle \text{iteration} \rangle$ shows the number of iterations (decisions) i. The token α comes into place $l21$ with final characteristic $\langle \text{size} \rangle := n, \langle \text{traffic matrix} \rangle := T,$

$(T = 0), \langle\text{iteration}\rangle := r,\ (i = r)$. The tokens β_i come into place $l20$ with characteristic $\langle\text{size}\rangle := n, \langle\text{non-conflict matrix}\rangle := Q_i, \langle\text{iteration}\rangle := i,\ (i \in [1, r])$. The parameter $\langle$non-conflict matrix$\rangle$ shows the sequence of switching matrices: $Q_1, Q_2, \ldots, Q_r$.

Each of the transitions has one and the same priority. The same refers to the tokens. The model has possibilities to provide information about the number of switchings in the crossbar matrix. An analysis of the GN-model proves that we are obtaining a non-conflict schedule. Calculation complexity of the solution depends on the third power of the dimension n of the matrix T, $(O(n^3))$. Computer simulations should provide us with an answer to the question: do we have a better solution with this algorithm in comparison with existing ones?

4 Family of Patterns for Non-uniform Traffic

The matrix T defines a traffic demand matrix if the total number of packets in each row and each column are equal [1]. For our large-scale computer simulation we suggest several types of traffic matrices T, which will be called a family of patterns. They possess the following properties:

- easy generation for any size of the switch $(n \times n)$;
- generation does not depend on the type of hardware used, compiler and operation system;
- their exact, optimal, non-conflict schedule is known.

The proposed family of patterns is based on an unbalanced non-uniform traffic model [14] which we shall call Rojas model. This model is given by $\lambda_{ij} = \rho(\omega + (1-\omega)/n)$ for $i = j$ and $\lambda_{ij} = \rho(1-\omega)/n$ for $i \neq j$. Here, ρ is the load intensity of each input (i.i.d. Bermoulli), ω is the probability for unbalanced load ($\omega \in [0, 1]$). In our research, we work with $\omega = 0.5$ as the most informative value [5].

The first type matrix is called $Rojas_1$. It's optimal schedule requires $2n$ switchings of crossbar matrix for $n \times n$ switch. In general, this type matrix is denoted by R_i. It's optimal schedule requires $(2in)$ switchings of crossbar matrix for $n \times n$ switch. This type of matrices is shown in Fig. 2.

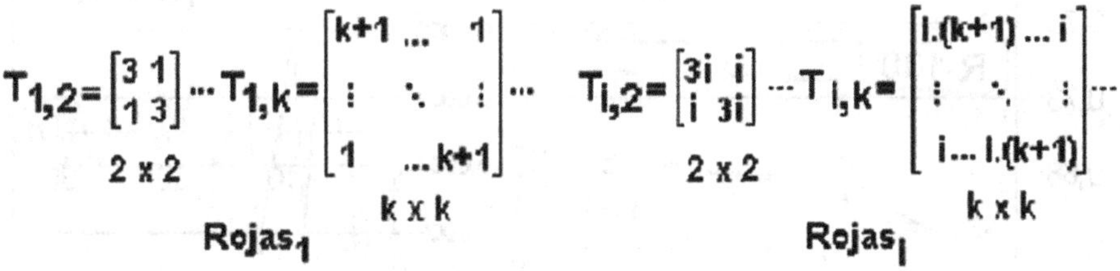

Fig. 2. Matrices of types $Rojas_1$ and $Rojas_i$

5 Result of Grid-Simulations

The transition from a GN-model to executive program is performed as in [15]. The program package Vfort of the Institute of mathematical modeling of Russian Academy of Sciences is used [16]. The source code has been tested on Vfort and then compiled by means of the grid-structure of the IICT-BAS. The resulting executive code is executed in the grid-structure. A main restriction is the time for execution.

In the figures, $Rojas_i$ is denoted as R-i for $i = 1, 2, \ldots$.

Figure 3 shows the results from computer simulation of the PIM-algorithm with input data $Rojas_{1,5,10,15,20}$. Sizes of the crossbar matrix from 2×2 to 130×130 are simulated. The resulting throughput and time of execution are the average for 10,000 simulations for each size. Figure 3 (left) shows that there is an upper bound of the throughput for this family of patterns. The right part of the figure indicates a linear dependence of the time of execution on the number of the pattern.

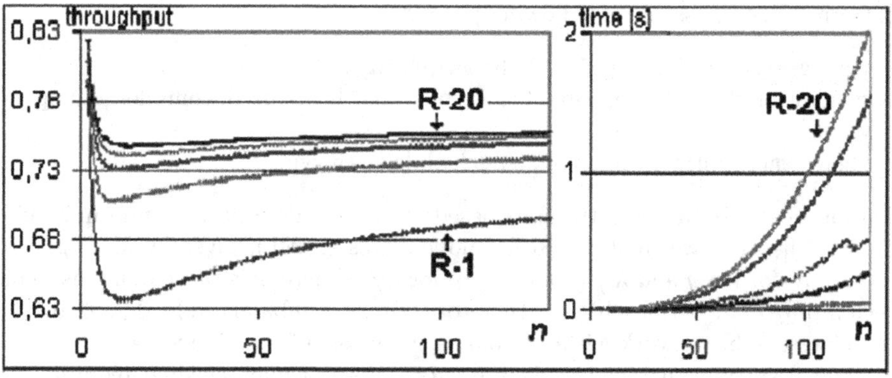

Fig. 3. Results for throughput and time with $Rojas_1$ to $Rojas_{20}$

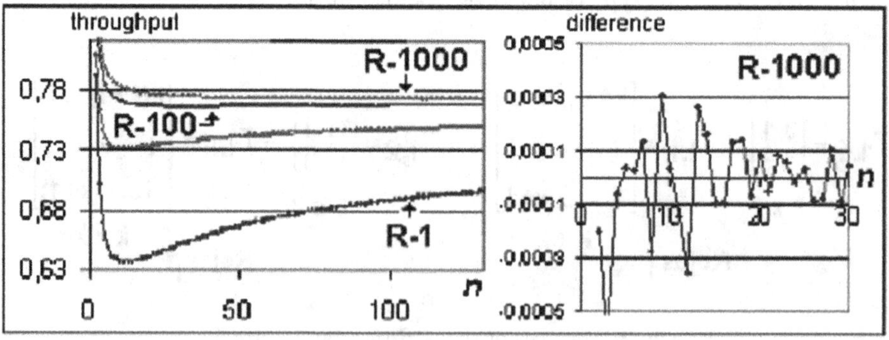

Fig. 4. Results for throughput with $Rojas_1$ to $Rojas_{1000}$

Figure 4 (left) presents the results with input data $Rojas_{1,10,100,1000}$. Sizes of the crossbar matrix from 2×2 to 130×130 are simulated. This simulation enables us to obtain a more precise value of the upper bound. Approximately, this bound is 0.775. Figure 4 (right) shows the difference in the throughput of $Rojas_{1000}$ for 1000 simulations and 100 simulations for each size of the commutation field n. This difference indicates that the upper bound can be estimated as 0.775±0.001.

Both Figs. 3 and 4 show the relevance of $Rojas_i$ with the conclusion that a larger input buffer produces a larger throughput. This simulations will be used as a basis for comparison of the simulation results obtained with the MiMa-algorithm. These results are shown in Figs. 5, 6, and 7.

Figures 5 and 6 show that the throughput of the MiMa-algorithm with the family of patterns $Rojas_i$ approaches 100 %. For values of the commutation field n equal to powers of 2, the throughput is exactly 100 %. The price of this result is approximately 2 times increased time for execution. This can be seen in Fig. 7 (left).

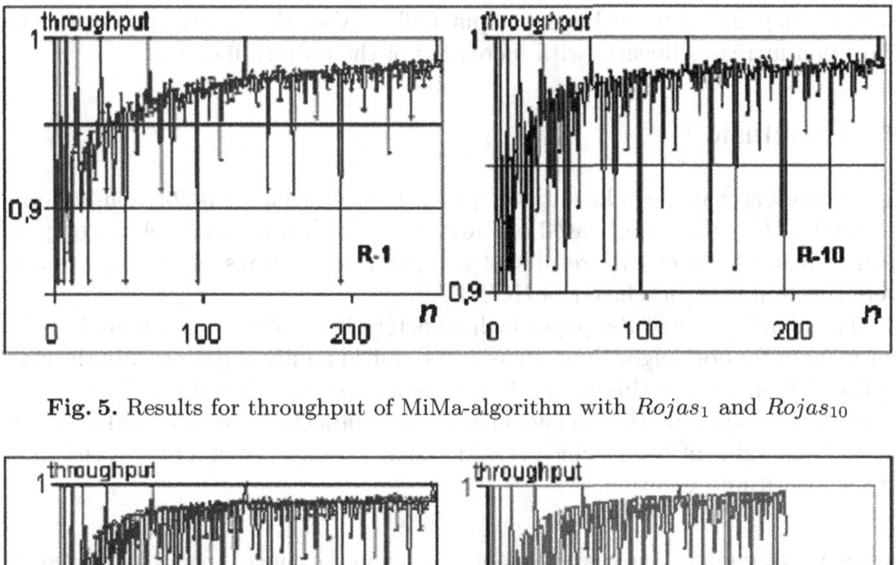

Fig. 5. Results for throughput of MiMa-algorithm with $Rojas_1$ and $Rojas_{10}$

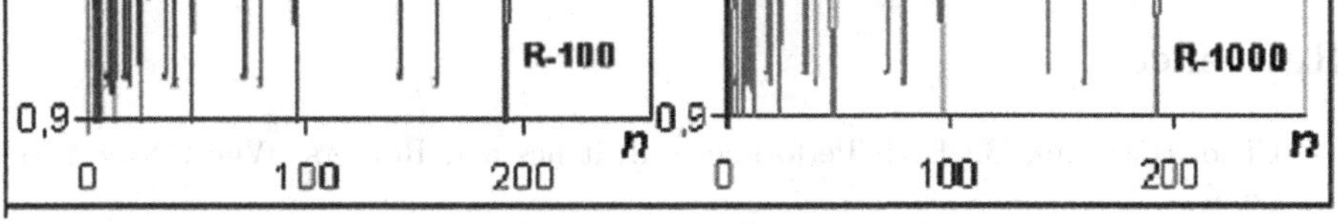

Fig. 6. Results for throughput of MiMa with $Rojas_{100}$ and $Rojas_{1000}$

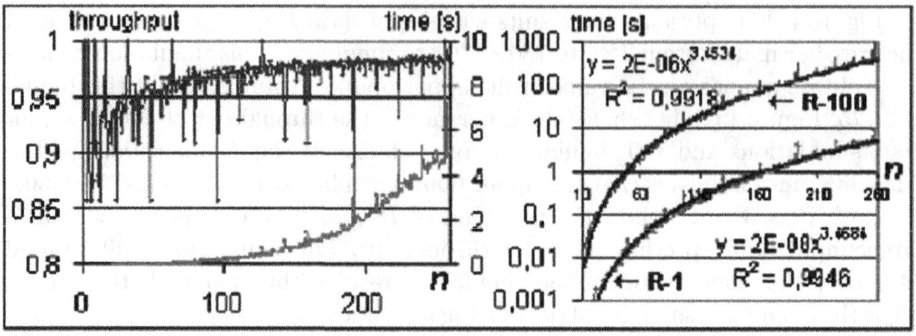

Fig. 7. Results for throughput and time with $Rojas_1$ and $Rojas_{100}$

Figure 7 (right) shows the execution time of the MiMa-algorithm with patterns $Rojas_1$ and $Rojas_{100}$. The picks of time coincide for the same values of the commutation field n. Thus, the increased time of execution for these values of n does not depend on the volume of input buffer. Also, the approximated time of execution increases linearly with increasing of the pattern index i.

6 Conclusion

In the present paper we have developed a family of patterns for non-uniform demand traffic simulating based on Rojas-Cessa unbalanced model. A comparison is made between the results of computer simulations of two algorithms performed on the grid-cluster of IICT-BAS.

The main results of the paper include determining of an upper bound of the throughput for both algorithms under the specified family of patterns. In the case of the PIM-algorithm this bound is determined to be $77.5 \pm 0.1\,\%$. The bound of the throughput of the second algorithm approaches the maximal possible throughput value of 100 % which is achieved at the expense of an increased time of the algorithm execution.

Acknowledgments. The research work reported in the paper is partly supported by the project AComIn "Advanced Computing for Innovation", grant 316087, funded by the FP 7 Capacity Programme (Research Potential of Convergence Regions).

References

1. Chao, H.J., Lui, B.: High Performance Switches and Routers. Wiley, New York (2007)
2. Chen, W., Mavor, J., Denyer, P., Renshaw, D.: Traffic routing algorithm for serial superchip system customisation. In: IEE Proceedings 137, [E]1 (1990)
3. Anderson, T., Owicki, S., Saxe, J., Thacker, C.: High speed switch scheduling for local area networks. ACM Trans. Comput. Syst. **11**(4), 319–352 (1993)

4. Gupta, P., McKeown, N.: Designing and implementing a fast crossbar scheduler. IEEE Micro **19**(1), 20–28 (1999)
5. Chang, H.J., Qu, G., Zheng, S.Q.: Performance of $CTC(N)$ switch under various traffic models. In: Qian, Z., Cao, L., Su, W., Wang, T., Yang, H. (eds.) Recent Advances in CSIE 2011. lnee, vol. 126, pp. 785–794. Springer, Heidelberg (2012)
6. Cheng, C., Chen, W., Huang, H.: Birkhoff-von Neumann input buffered crossbar switches. In: Proceedings of IEEE INFOCOM'00, vol. 3, pp. 1624–1633 (2000)
7. Dong, Z., Rojas-Cessa, R.: Throughput analysis of shared-memory crosspoint buffered packet switches. IET Commun. **6**(9), 1045–1053 (2012)
8. Yuang, M., Lin, Y.-M., Shih, J.-L., Tien, P.-L., Chen, J., Lee, S., Lin, S.-H.: A QoS optical packet switching system: architectural design and experimental demonstration. IEEE Commun. Mag. **48**, 66–75 (2010)
9. Atanassov, K.: Generalized Nets. World Scientific, Singapore (1991)
10. Atanassova, V.: Design of training tests on generalized nets. In: Proceedings of 5th IEEE International Conference on Intelligent Systems (IS), pp. 327–330, London, UK, 7–9 July 2010
11. Tashev, T., Monov, V.: Large-scale simulation of uniform load traffic for modeling of throughput on a crossbar switch node. In: Lirkov, I., Margenov, S., Waśniewski, J. (eds.) LSSC 2011. LNCS, vol. 7116, pp. 638–645. Springer, Heidelberg (2012)
12. Tashev, T., Monov, V.: Modeling of the hotspot load traffic for crossbar switch node by means of generalized nets. In: Proceedings of 6th IEEE International Conference on Intelligent Systems (IS), pp. 187–191, Sofia, Bulgaria, 6–8 Sept 2012
13. Tashev, T., Atanasova, T.: Computer simulation of MIMA algorithm for input buffered crossbar switch. Int. J. Inf. Technol. Knowl. **5**(2), 183–189 (2011)
14. Rojas-Cessa, R., Oki, E., Chao, H.J.: On the combined input-crosspoint buffered switch with round-robin arbitration. IEEE Trans. Commun. **53**(11), 1945–1951 (2005)
15. Tashev, T., Vorobiov, V.: Generalized net model for non-conflict switch in communication node. In: Proceedings of International Workshop on DCCN'2007, pp. 158–163. IPPI Publ., Moscow, 10–12 Sept 2007
16. http://www.imamod.ru/~vab/vfort/download.html

Author Index

Printed in the United States
By Bookmasters